MACROPODS

MACROPODS

THE BIOLOGY OF KANGAROOS, WALLABIES AND RAT-KANGAROOS

Editors: Graeme Coulson and Mark Eldridge

CSIRO
PUBLISHING

THE UNIVERSITY OF MELBOURNE
Department of Zoology
Australian Mammal Society
Australian museum
nature culture discover
The Koala and Kangaroo Contraception Program

National Library of Australia Cataloguing-in-Publication entry

Macropods: the biology of kangaroos, wallabies and rat-kangaroos/Graeme Coulson, Mark Eldridge.

9780643096622 (pbk)

Includes index.
Bibliography.

Macropus.
Kangaroos – Australia.
Wallabies – Australia.

Coulson, Graeme Maxwell.
Eldridge, Mark Derek Bruce.

599.2

Published by
CSIRO PUBLISHING
36 Gardiner Road, Clayton VIC 3168
Private Bag 10, Clayton South VIC 3169
Australia

Telephone: [+613] 9545 8555
Local call: 1300 788 000 (Australia only)
Fax: +61 3 9662 7555
Email: csiropublishing@csiro.au
Website: www.publishing.csiro.au

Front cover:
Original artwork by Julia Malcolm, Malcolm Studios

Back cover:
Top row (L to R): Male antilopine wallaroo, *Macropus antilopinus* (David Webb); Female tammar wallaby, *Macropus eugenii*, the model marsupial (Anna MacDonald); Brush-tailed bettong or woylie, *Bettongia penicillata* (Graeme Finlayson).

Middle row (L to R): Male brush-tailed rock-wallaby, *Petrogale penicillata* (Katherine Tuft); Transferring a brush-tailed rock-wallaby pouch young to a tammar wallaby surrogate mother (David Taggart); Female swamp wallaby, *Wallabia bicolor* (James Turner).

Bottom row (L to R): Radio-collared male western grey kangaroo, *Macropus fuliginosus* (Graeme Coulson); Radio-tracking kangaroos at Murray–Sunset National Park (Graeme Coulson); Radio-collared male bridled nailtail wallaby, *Onychogalea fraenata* (Dominique Sigg).

Set in 10/13 Adobe Minion and ITC Stone Sans
Edited by Adrienne de Kretser, Righting Writing
Cover and text design by James Kelly
Typeset by Desktop Concepts Pty Ltd, Melbourne
Printed by Ingram Lightning Source

CSIRO PUBLISHING publishes and distributes scientific, technical and health science books and journals from Australia to a worldwide audience and conducts these activities autonomously from the research activities of the Commonwealth Scientific and Industrial Research Organisation (CSIRO).

Mar26_RP_ILS

Contents

Contributors

Katherine Belov Faculty of Veterinary Science, University of Sydney, New South Wales 2006, Australia.

Roberta Bencini School of Animal Biology, University of Western Australia, Crawley, Western Australia 6009, Australia.

Ian Beveridge Department of Veterinary Science, University of Melbourne, Victoria 3010, Australia.

William G. Breed Discipline of Anatomical Sciences, School of Medical Sciences, Faculty of Health Sciences, University of Adelaide, South Australia 5005, Australia.

Brian K. Chambers School of Animal Biology, University of Western Australia, Crawley, Western Australia 6009, Australia.

Ronald Claassens Department of Biological Sciences, Macquarie University, New South Wales 2109, Australia.

Desmond W. Cooper School of Biological, Earth and Environmental Sciences, University of New South Wales, New South Wales 2052, Australia.

Graeme Coulson Department of Zoology, University of Melbourne, Victoria 3010, Australia.

Grey Coupland School of Plant Biology, University of Western Australia, Crawley, Western Australia 6009, Australia.

David B. Croft School of Biological, Earth and Environmental Sciences, University of New South Wales, New South Wales 2052, Australia.

Rebecca Dawson School of Earth and Geographical Sciences, University of Western Australia, Crawley, Western Australia 6009, Australia.

Terence J. Dawson School of Biological, Earth and Environmental Sciences, University of New South Wales, New South Wales 2052, Australia.

Janine E. Deakin ARC Centre for Kangaroo Genomics, Research School of Biological Sciences, Australian National University, Australian Capital Territory 0200, Australia.

Chris R. Dickman Institute of Wildlife Research, School of Biological Sciences, University of Sydney, New South Wales 2006, Australia.

Julian Di Stefano Department of Forest and Ecosystem Science, University of Melbourne, Creswick, Victoria 3363, Australia. Department of Zoology, University of Melbourne, Victoria 3010, Australia.

Mark D.B. Eldridge Australian Museum, Sydney, New South Wales 2010, Australia. Department of Biological Sciences, Macquarie University, New South Wales 2109, Australia.

Murray Evans Research and Planning Section, Parks, Conservation and Lands, ACT Government, Canberra, Australian Capital Territory 2601, Australia. Environmental Protection Agency, Toowoomba, Queensland 4350, Australia.

Lisa Farroway Science and Conservation Directorate, Department for Environment and Heritage, Adelaide, South Australia 5001, Australia.

Graeme R. Finlayson Institute of Wildlife Research, School of Biological Sciences, University of Sydney, New South Wales 2006, Australia.

Serena T. Finlayson School of Veterinary and Biomedical Sciences, Murdoch University, Western Australia 6150, Australia.

Diana O. Fisher School of Biological Sciences, University of Queensland, Queensland 4072, Australia.

Nancy N. FitzSimmons Institute for Applied Ecology, University of Canberra, Australian Capital Territory 2601, Australia.

Terry P. Fletcher Department of Zoology, University of Melbourne, Victoria 3010, Australia.

Anthony J. Friend Department of Environment and Conservation, Albany, Western Australia 6330, Australia.

Matthew S. Gibson Centre for Environmental Management, School of Science and Engineering, University of Ballarat, Victoria 3353, Australia.

Joshua Gilroy Department of Environment and Climate Change, Dubbo, New South Wales 2830, Australia.

Anne W. Goldizen School of Biological Sciences, University of Queensland, Queensland 4072, Australia.

Stacey A. Gowans Centre for Environmental Management, School of Science and Engineering, University of Ballarat, Victoria 3353, Australia.

Jennifer A.M. Graves ARC Centre for Kangaroo Genomics, Research School of Biological Sciences, Australian National University, ACT 0200, Australia.

Andrew Greenfield Department of Zoology, University of Melbourne, Victoria 3010, Australia.

Gordon C. Grigg School of Biological Sciences, University of Queensland, Queensland 4072, Australia.

Stephanie L. Hazlitt School of Biological Sciences, University of Queensland, Queensland 4072, Australia. Current address: Centre for Applied Conservation Research, Department of Forest Sciences, University of British Columbia, Vancouver, British Columbia V6T 1Z4, Canada.

Catherine A. Herbert Faculty of Veterinary Science, University of Sydney, New South Wales 2006, Australia. School of Biological, Earth and Environmental Sciences, University of New South Wales, New South Wales 2052, Australia.

Danielle Hickford Department of Zoology, University of Melbourne, Victoria 3010, Australia.

Karen Higginbottom Griffith School of Environment, Gold Coast Campus, Griffith University, Queensland 4222, Australia.

Ian D. Hume School of Biological Sciences, University of Sydney, New South Wales 2006, Australia.

Peter M. Johnson Queensland Parks and Wildlife Service, Pallarenda, Townsville, Queensland 4810, Australia.

Enhua Lee School of Biological, Earth and Environmental Sciences, University of New South Wales, New South Wales 2052, Australia. Current address: Eco Logical Australia Pty Ltd, Sydney, New South Wales 2000, Australia.

Roger G. Lentle Institute of Food Nutrition and Human Health, Riddet Institute, Massey University, New Zealand.

Geoff Lundie-Jenkins Environmental Protection Agency, Toowoomba, Queensland 4350, Australia.

Anna J. MacDonald Institute for Applied Ecology, University of Canberra, Australian Capital Territory 2601, Australia.

Clive McAlpine School of Biological Sciences, University of Queensland, Queensland 4072, Australia.

Andrew McMillan School of Plant Biology, University of Western Australia, Crawley, Western Australia 6009, Australia.

Norbert Menke School of Biological Sciences, University of Queensland, Queensland 4072, Australia. Biophysical Remote Sensing Group, University of Queensland, Queensland 4072, Australia.

Emily J. Miller Department of Biological Sciences, Macquarie University, New South Wales 2109, Australia. Current address: Faculty of Veterinary Science, University of Sydney, New South Wales 2006, Australia.

Harriet R. Mills School of Animal Biology, University of Western Australia, Crawley, Western Australia 6009, Australia.

David G. Morgan Department of Zoology, University of Melbourne, Victoria 3010, Australia.

Adam J. Munn Faculty of Veterinary Science, University of Sydney, New South Wales 2006, Australia.

Linda E. Neaves Department of Biological Sciences, Macquarie University, New South Wales 2109, Australia.

Graeme R. Newell Arthur Rylah Institute for Environmental Research, Department of Sustainability and Environment, Heidelberg, Victoria 3084, Australia.

Skye Page Griffith School of Environment, Gold Coast Campus, Griffith University, Queensland 4222, Australia.

Justyna Z. Paplinska Department of Zoology, University of Melbourne, Victoria 3010, Australia.

Nicole Payne Department of Environment and Climate Change, Dubbo, New South Wales 2830, Australia.

Philip Pegler Parks Victoria, Melbourne, Victoria 3000, Australia.

Stuart R. Phinn School of Biological Sciences, University of Queensland, Queensland 4072, Australia. Biophysical Remote Sensing Group, University of Queensland, Queensland 4072, Australia.

Maxine P. Piggott Fenner School of Environment and Society, Australian National University, Canberra, Australian Capital Territory 0200, Australia.

Anthony R. Pople Biosecurity Queensland, Queensland Primary Industries and Fisheries, Department of Employment, Economic Development and Innovation, Sherwood, Queensland 4075, Australia. School of Biological Sciences, University of Queensland, Queensland 4072, Australia.

Hugh P. Possingham School of Biological Sciences, University of Queensland, Queensland 4072, Australia.

Gavin J. Prideaux Department of Earth and Planetary Sciences, Western Australian Museum, Welshpool, Western Australia 6986, Australia. School of Biological Sciences, Flinders University, Bedford Park, South Australia 5042, Australia.

Daniel Ramp School of Biological, Earth and Environmental Sciences, University of New South Wales, New South Wales 2052, Australia.

Marilyn B. Renfree Department of Zoology, University of Melbourne, Victoria 3010, Australia.

Euan G. Ritchie School of Marine and Tropical Biology, James Cook University, Queensland 4811, Australia.

Michael W. Roberts Department of Biological Sciences, Macquarie University, New South Wales 2109, Australia. Current address: Biosis Research, Wollongong, New South Wales 2500, Australia.

Randy W. Rose School of Zoology, University of Tasmania, Hobart, Tasmania 7001, Australia.

Claire E. Sanderson Faculty of Veterinary Science, University of Sydney, New South Wales 2006, Australia.

Stephen D. Sarre Institute for Applied Ecology, University of Canberra, Australian Capital Territory 2601, Australia.

Bernadette Schmidt Department of Zoology, University of Melbourne, Victoria 3010, Australia. Ecology Australia, Fairfield, Victoria 3078, Australia.

David J. Schultz Conservation Ark, Royal Zoological Society of South Australia, Adelaide, South Australia 5001, Australia.

Geoff Shaw Department of Zoology, University of Melbourne, Victoria 3010, Australia.

Hannah V. Siddle Faculty of Veterinary Science, University of Sydney, New South Wales 2006, Australia.

Dominique P. Sigg School of Biological Sciences, University of Queensland, Queensland 4072, Australia. Current address: Ecosystems Branch, British Columbia Ministry of Environment, British Columbia V8W 9M1, Canada.

Ian G. Smith Conservation Ark, Royal Zoological Society of South Australia, Adelaide, South Australia 5001, Australia.

David M. Spratt CSIRO Sustainable Ecosystems, Canberra, Australian Capital Territory 2601, Australia.

Matthew Swan Department of Zoology, University of Melbourne, Victoria 3010, Australia. Current address: Biosis Research, Wollongong, New South Wales 2500, Australia.

David A. Taggart Conservation Ark, Royal Zoological Society of South Australia, Adelaide, South Australia 5001, Australia. School of Earth and Environmental Science, University of Adelaide, Adelaide, South Australia 5005, Australia. Department of Zoology, University of Melbourne, Victoria 3010, Australia. Discipline of Anatomical Sciences, School of Medical Sciences, Faculty of Health Sciences, University of Adelaide, South Australia 5005, Australia.

Peter D. Temple-Smith Department of Zoology, University of Melbourne, Victoria 3010, Australia. Centre for Reproduction and Development, Monash Institute of Medical Research, Monash University, Victoria 3800, Australia.

Timothy E. Trigg Peptech Animal Health Pty Ltd, Macquarie Park, New South Wales 2113, Australia.

Karl Vernes Ecosystem Management, University of New England, Armidale, New South Wales 2351, Australia.

Joanne Wann Corporate Services and Infrastructure, HMAS Stirling, Rockingham, Western Australia 6958, Australia.

Natalie M. Warburton Department of Earth and Planetary Sciences, Western Australian Museum, Welshpool, Western Australia 6986, Australia. School of Veterinary and Biomedical Sciences, Murdoch University, Western Australia 6150, Australia.

Koa N. Webster School of Biological, Earth and Environmental Sciences, University of New South Wales, New South Wales 2052, Australia. Current address: Department of Biological Sciences, Macquarie University, New South Wales 2109, Australia.

Martin E. Westbrooke Centre for Environmental Management, School of Science and Engineering, University of Ballarat, Victoria 3353, Australia.

Preface

This book arose from a symposium on the biology of kangaroos, wallabies and rat-kangaroos, collectively and colloquially known as macropods. The symposium was held on 6–7 July 2006 at the University of Melbourne, and was sponsored by the Australian Mammal Society. The symposium was attended by 117 macropod enthusiasts, and 55 papers were presented. This book contains 32 chapters derived from those papers, representing the work of 82 authors, refereed by at least two academic peers from Australia or overseas.

The Melbourne symposium was the second major scientific meeting on macropods. The first was held in 1988. It provided comprehensive reviews of our knowledge about this diverse group of Australasian marsupials at the time, and laid the foundations for work to follow (Grigg *et al.* 1989). The second was designed to build on the first, and bring together the many recent advances in the biology of macropods. However, the aim for the second symposium was not merely to update the earlier review papers, which remain valid and are still benchmarks in their areas. Instead, we aimed to highlight new developments in macropod biology and explore topics that have become prominent since the first symposium.

The subtitle for the Melbourne symposium was 'From genomics to GIS', and many authors took up the challenge. Deakin and Graves (Ch. 1) open the book with a summary of the release of the genome sequence of Australia's model marsupial, the tammar wallaby (*Macropus eugenii*); Chambers *et al.* (Ch. 25) describe the use of GIS technology to predict patterns of road-kill of the same species. Many other new techniques are covered: microsatellites (MacDonald *et al.*, Ch. 5), mitochondrial DNA (Hazlitt *et al.*, Ch. 8), stable isotopes (McMillan *et al.*, Ch. 14), assisted reproduction (Taggart *et al.*, Ch. 23), fertility control (Herbert *et al.*, Ch. 27) and population modelling (Pople *et al.*, Ch. 32).

Perhaps the most striking difference between the two macropod symposia is the balance of topics covered. Of the 61 chapters in Grigg *et al.* (1989), the majority were concerned with fundamental aspects of the biology of macropods – evolution, ecology, energetics, reproduction, development, behaviour and population biology. Only five (8%) covered management issues. In this book, 34% of the chapters are directly concerned with the management of macropods and others have some management elements. This change in focus undoubtedly arises from real concerns about the future of macropods, ranging from threatened taxa to overabundant populations. In part, however, the switch also reflects the changing funding landscape, with its greater emphasis on applied research.

Part I, 'Genetics, reproduction and development', is primarily concerned with fundamentals of macropod biology. Deakin and Graves (Ch. 1) describe the phylogenetic significance of the tammar wallaby genome, and its value in tracing the evolutionary history of mammalian chromosome evolution. Siddle *et al.* (Ch. 2) show how the availability of these genomic resources facilitated the isolation and characterisation of macropodid immune genes, which, contrary to historical speculation, are comparable to those of eutherians. Hickford (Ch. 3) explains how the genomic resources have contributed to the *in vitro* culture of marsupial embryos and fetuses. Eldridge *et al.* (Ch. 4) review the applications of hypervariable genetic markers to the study of wild macropod populations, and MacDonald *et al.* (Ch. 5) outline the potential of such markers on the sex chromosomes to shed light on sex-specific population and evolutionary processes. Paplinska *et al.* (Ch. 6) apply a comparative morphological analysis of relative testis size to the study of macropodid mating systems. Miller *et al.* (Ch. 7) use microsatellites to assign paternity and demonstrate that male body size is an important determinant of dominance status and reproductive success in the tammar wallaby. Hazlitt *et al.* (Ch. 8) use mitochondrial DNA to show striking matrilineal clusters within a single colony of brush-tailed rock-wallabies (*Petrogale penicillata*).

Part II, 'Morphology and physiology', addresses more fundamental questions. Dawson and Webster (Ch. 9) describe macropods as among the most athletic of mammals, with a combination of efficient locomotion and powerful skeletal muscles that enables them to sustain high speeds aerobically. Rose (Ch. 10) argues that skeletal muscle may allow the Tasmanian bettong (*Bettongia gaimardi*) to increase its metabolic rate by non-shivering thermogenesis. Munn and Dawson (Ch. 11) propose that juvenile red kangaroos (*M. rufus*) have high mortality rates during drought because they may be unable to meet their requirement for dietary fibre. Lentle and Hume (Ch. 12) argue that the occlusal forces involved in chewing fibre may not be as influential as nutritional factors in the progression of molars. Warburton and Prideaux (Ch. 13) analyse the pedal morphology of the extinct Pleistocene *Bohra*, confirming that tree-kangaroos occupied far broader geographic and climatic ranges than previously understood.

Vernes (Ch. 14) begins Part III, 'Ecology', with an analysis of the consumption of hypogeous ectomycorrhizal fungi in a community of four macropodid and two potoroid species; he concludes that the extent of mycophagy by macropods has been underestimated. McMillan *et al.* (Ch. 15) use stable isotope analysis of hair samples from a population of tammar wallabies to show their reliance on irrigated C4 grasses. Ritchie (Ch. 16) provides the only chapter on the ecology of a tropical macropod, the antilopine wallaroo (*M. antilopinus*), which has a broad but patchy distribution across northern Australia. Di Stefano *et al.* (Ch. 17) present a fine-scale study of space use by the swamp wallaby (*Wallabia bicolor*), showing that nocturnal use of harvested forest differs between males and females. At a large spatial scale, Pople *et al.* (Ch. 18) re-evaluate the population dynamics of red and grey (*M. fuliginosus* and *M. giganteus*) kangaroos from long-running aerial surveys, concluding that a number of intrinsic and extrinsic factors other than rainfall are influential. Chambers and Bencini (Ch. 19) highlight the influence of altered resources on tammar wallabies, which have higher population density, body weight and fecundity in modified habitats with abundant grass. Schmidt *et al.* (Ch. 20) demonstrate the role of fine-scale variation in habitat structure, rather than food resources, for coexistence of sympatric grey kangaroos and swamp wallabies. Finally, Beveridge *et al.* (Ch. 21) review the occurrence of helminth parasites (mostly nematodes) in macropods, showing that while helminth taxa are largely host-specific, some are able to parasitise multiple species where their macropod hosts occur in sympatry.

The theme of Management, Part IV, splits into two poles of an abundance scale. Rare and threatened species of macropods have been the focus of much conservation effort. Finlayson *et al.* (Ch. 22) review the successes and failures of translocations of potoroids, and the management lessons that have been learned. Taggart *et al.* (Ch. 23) review the promising techniques of pouch young isolation and cross-fostering to provide stock for translocation of threatened macropods. Ramp (Ch. 24) summarises the impact of roads as a threatening process for many macropod species, Chambers *et al.* (Ch. 25) analyse traffic, roadside and seasonal effects on road-kills of tammar wallabies, and Lee *et al.* (Ch. 26) link the flight behaviour of red and grey kangaroos as well as euros (*M. robustus erubescens*) to their differing susceptibility to road-kill. The management of high-density macropod populations is discussed by Herbert *et al.* (Ch. 27), who review developments in fertility control techniques. Roberts *et al.* (Ch. 28) present a darting protocol for the capture of eastern grey kangaroos, and Higginbottom and Page (Ch. 29) describe the fate of eastern grey kangaroos that were captured and translocated to a more secure site. Morgan and Pegler (Ch. 30) describe and evaluate a culling program designed to simulate natural predation on western grey kangaroos in a national park, and Gowans *et al.* (Ch. 31) report a corresponding improvement in vegetation condition in the park. Pople *et al.* (Ch. 32) evaluate the potential for using harvesting statistics at a regional level to reduce the frequency of costly aerial surveys and provide more intensive coverage of kangaroo populations.

The breadth of chapters show how much progress has been made since the first macropod symposium. Nonetheless, many aspects of the biology of the ~73 macropod species are yet to be revealed. Indeed, as noted in the macropod species list contained in the Appendix, even the actual number of species is still unclear and evolutionary relationships among many taxa are yet to be fully resolved. There is an enormous wealth of new data from fossil macropods to be integrated into our understanding of macropod biology and evolution.

For living macropod species, there remain many imbalances in our understanding even of basic biology. Tropical macropods remain poorly known: our ignorance of the biology of the black wallaroo

(*M. bernardus*), most of the tropical rock-wallabies (*Petrogale* spp.) and all 17 endemic macropods (*Dendrolagus*, *Dorcopsis*, *Dorcopsulus* and *Thylogale* spp.) from New Guinea is almost complete; the reproductive biology of the antilopine wallaroo is unresolved; the ecology of a dominant herbivore, the agile wallaby (*M. agilis*), has received little attention; the still widespread spectacled hare-wallaby (*Lagorchestes conspicillatus*) and northern nailtail wallaby (*Onychogalea unguifera*) have been little studied for decades. Macropods of the arid zone have been equally neglected, aside from specific studies of threatened species, mostly in predator-free facilities. Temperate species, which overlap the distribution of most Australian universities and research institutes, have received uneven attention. Until recently, most research on the swamp wallaby had been dietary studies. The biology of the western brush wallaby (*M. irma*) is essentially unknown, despite its being relatively common in the south-west of Western Australia. Despite decades of ecological, physiological, reproductive and behavioural research, the mating system of the widespread and abundant large kangaroos remains unconfirmed. Many aspects of the biology of the model marsupial – the tammar wallaby – are well understood, but its ecology has been studied only on Garden Island, Western Australia. Similarly, while we may have the complete genomic sequence of a Kangaroo Island tammar wallaby, our knowledge of the distribution of genetic diversity throughout the species' disjunct distribution, or even throughout Kangaroo Island, remains rudimentary. Indeed, the whole application of modern population genetics to macropods has been uneven, with most studies focusing on rare or threatened taxa. Very few widespread and common species have been examined, despite the enormous potential for this approach to reveal much about the species' ecology, reproduction, population biology and evolutionary history, especially when combined with detailed ecological, reproductive or behavioural data.

In light of these gaps and imbalances in our knowledge, the ongoing challenge for macropod biologists is to conduct high-quality research, in both pure and applied arenas, so our fundamental knowledge of this fascinating group of marsupials continues to advance, and macropods continue to prosper and intrigue us.

Publication of this work was assisted by publication grants from the University of Melbourne and the Australian Mammal Society.

G. Coulson and M.D.B. Eldridge

REFERENCES

Grigg G, Jarman P and Hume I (Eds) 1989. *Kangaroos, Wallabies and Rat-kangaroos*. Surrey Beatty and Sons, Sydney.

PART I

GENETICS, REPRODUCTION AND DEVELOPMENT

1 Mapping genes on tammar wallaby target chromosomes

J.E. Deakin and J.A.M. Graves

SUMMARY

This is an exciting time for marsupial genomics, with the release of the sequences of two marsupial genomes. One of these is the tammar wallaby (*Macropus eugenii*), Australia's model kangaroo, which has been the model macropodoid species for marsupial genetic studies for over 30 years. The tammar, and other marsupials, occupy a phylogenetic 'sweet spot' between birds and eutherian mammals that is very valuable for genome comparisons. To anchor the tammar genome sequence, we are generating a physical map of genes on tammar chromosomes. Of particular interest to our laboratory are genes on chromosomes X and 5. Both of these chromosomes harbour genes that are found on the human X chromosome, many of which are essential to mammalian sex differentiation, reproduction and development. Mapping these genes has assisted in tracing the evolutionary history of the mammalian X chromosome, identifying ancient and recently added regions. In addition, mapping genes to the tammar X is permitting an intense study of marsupial X chromosome inactivation, a process that involves silencing one X chromosome in females to compensate for the difference in the number of X chromosomes between males and females. Marsupial X inactivation is different at the phenotypic and molecular level from inactivation in human and mouse, so investigations of the tammar X will help to deconstruct this complex process. Tammar chromosome 5 contains, in addition to genes found on the human X, gene blocks from other human chromosomes; comparisons of gene organisation in this region between humans and other vertebrates have provided and will continue to provide insight into evolution of other regions of the mammalian genome.

VALUE OF MARSUPIALS IN COMPARATIVE GENOMICS

Comparative genomics is a powerful tool for identifying genes and regulatory sequences and for dissecting complex genetic pathways. The power of comparative genomic studies can be limited by the species included in the analysis. Genomes of closely related species are often too similar in sequence for meaningful comparisons, the high level of sequence similarity masking potentially important regulatory regions. Conversely, the comparison of very distantly related species is complicated by the difficulty in aligning their sequences, due to genome rearrangements, deletions or the level of sequence divergence. Marsupials fall at the phylogenetic 'sweet spot' in evolution, bridging the 310 million-year gap between birds and eutherians. Marsupial genomes are easily aligned to those of eutherians but have diverged enough for important conserved regions to be readily identified (Wakefield and Graves 2003).

Including marsupials in comparative genomic studies has made a major contribution to our understanding of the evolution of mammalian sex chromosomes (Graves 1995), X chromosome inactivation

(Davidow *et al.* 2007; Duret *et al.* 2006; Hore *et al.* 2007; Shevchenko *et al.* 2007), genomic imprinting (Edwards *et al.* 2007; Rapkins *et al.* 2006; Suzuki *et al.* 2005) and immune genes (Belov *et al.* 2006, 2007). Their genomes have also revealed previously unknown genes in human (Delbridge *et al.* 1999) and helped in the identification of regions that may be important for gene regulation (Deakin *et al.* 2006; Wakefield and Graves 2005).

One renowned example of the important role marsupials play in comparative genomics involves the discovery of the mammalian sex-determining gene. Many years of deletion mapping analysis on sex-reversed humans narrowed down the region responsible for testis determination to the short arm of the human Y chromosome. A gene cloned from this region, and thought to be the testis determining factor, was the zinc finger protein *ZFY*. The discovery that the marsupial orthologue of this gene mapped not to the Y chromosome but to a non-sex chromosome (autosome) provided the first evidence that *ZFY* was not the sex-determining gene (Sinclair *et al.* 1988). Subsequently, the *SRY* gene was discovered within the same region (Sinclair *et al.* 1990). Evidence for the role *SRY* plays in sex determination was demonstrated by the development of testes in XX mice transgenic for mouse *Sry* (Koopman *et al.* 1991). This gene was later mapped to the marsupial Y chromosome (Foster *et al.* 1992), confirming it as the mammalian sex-determining gene. Marsupials also provided information on the origin and evolution of *SRY* when a related gene (*SOX3*) was found on the X (Foster and Graves 1994).

Thus, the common ancestry of marsupials and eutherians provides a means to establish some of the ancestral mammalian genetic organisation and mechanisms. Similarly, comparisons between the two lineages can elucidate the evolution of unique marsupial features, such as the genes involved in their sophisticated lactation system and embryonic diapause.

MARSUPIAL GENOME PROJECTS

There are currently two marsupials whose genomes have been sequenced – the South American gray short-tailed opossum (*Monodelphis domestica*) and the tammar wallaby (*Macropus eugenii*). These two species diverged approximately 70 million years ago (Springer *et al.* 1994) so comparisons between the opossum and tammar will be similar in evolutionary terms to the human/mouse comparison that has been critical in understanding eutherian genomes. It is anticipated that there will be many important and informative differences between the two marsupial genomes.

The opossum genome has been sequenced by the Broad Institute (US) on average six times over (i.e. to a depth of six-fold) and the sequence has been assigned to chromosomes (Mikkelsen *et al.* 2007). Comparative analysis of this sequence has revealed a number of important findings in regards to mammalian genome evolution. One major finding is that approximately one-fifth of eutherian conserved non-coding elements (conserved sequences not involved in encoding protein-coding genes) have emerged after marsupial/eutherian divergence, with a substantial proportion of these eutherian-specific conserved non-coding elements (CNEs) having arisen from sequence inserted by transposable elements (mobile pieces of DNA). CNEs are thought to be involved in gene regulation. Hence, transposable elements may have played a major role in the evolution of mammalian gene regulation (Mikkelsen *et al.* 2007).

Although the opossum genome analysis has been very informative, comparison between the opossum sequence and that of an Australian marsupial would not only allow the detection of marsupial-specific genes and non-coding elements, but would also permit the reconstruction of the genome of their common ancestor (Mikkelsen *et al.* 2007).

Australia's model kangaroo, the tammar wallaby, has been sequenced jointly by the Australian Genome Research Facility (AGRF) and Baylor College of Medicine Human Genome Sequencing Center (US) to provide an approximately 2× coverage of the genome (http://www.genome.gov/12512299). To anchor this genome sequence to tammar chromosomes, the ARC Centre of Excellence for Kangaroo Genomics (http://kangaroo.genomics.org.au/) is generating a physical map by localising genes to chromosomes.

MAMMALIAN SEX CHROMOSOMES

As a start on the physical map, we have chosen to focus on chromosomes X and 5. These chromosomes are of special interest because they harbour genes that are found on the eutherian X chromosome.

Therian (marsupial and eutherian) sex chromosomes, the X and the Y, are very different in size and gene content. The X is a large gene-rich chromosome; the Y is small, gene-poor and highly enriched in repetitive sequences. The X and Y arose from a pair of ordinary autosomes some time between the divergence of monotremes and therian mammals (Veyrunes *et al.* 2008). Differentiation between the X and Y occurred due to a suppression of recombination between the X and the Y, which has ultimately resulted in the progressive degradation of the Y chromosome. Comparisons between the sex chromosomes of eutherians, marsupials and monotremes have been important for tracing this process and for identifying ancient and recently added regions of the human X and Y.

Therian X and Y chromosomes have many special properties arising from their different representation in XX females and XY males. The gene content of both chromosomes is biased toward those that are essential to mammalian sex differentiation, reproduction and development (Graves *et al.* 2006). Mapping these genes in marsupials will assist in tracing the evolutionary history of the eutherian X chromosome. In addition, mapping genes to the tammar X will permit an intense study of marsupial X chromosome inactivation, a process that involves silencing one X chromosome in females to compensate for the 2:1 difference in the number of X chromosomes between males and females.

MAPPING GENES IN THE TAMMAR

The tammar genome is similar in size to the human genome (about 3.5 billion base pairs), but is packaged into seven pairs of autosomes and one pair of sex chromosomes, with XX females and XY males. Genes have been assigned to tammar chromosomes by two different approaches – linkage or physical mapping.

Linkage mapping uses the frequency of recombination between two loci to provide the relative positions of markers, either genes or anonymous DNA markers (e.g. microsatellites) on a chromosome, determined on the basis of how frequently these markers are inherited together. Linkage mapping requires these markers to be polymorphic between individuals. The generation of a tammar linkage map has been facilitated by using crosses of Kangaroo Island and Garden Island subspecies, with polymorphic markers found between the two subspecies (Zenger *et al.* 2002). Linkage markers are currently being physically located onto chromosomes by FISH mapping to provide an integrated physical and linkage map.

The large and readily distinguishable chromosomes of the tammar are ideal for cytogenetic (physical) mapping. A skeleton physical map with 64 genes assigned to chromosomes was constructed with data collected over a 20-year period (Alsop *et al.* 2005). Many of the genes on this skeleton map were assigned to chromosomes by Radioactive *in situ* Hybridisation (RISH) using heterologous probes, typically human cDNA clones, or by Fluorescence *in situ* Hybridisation (FISH) using homologous probes.

Chromosome painting, a technique whereby individual chromosomes from one species can be hybridised to chromosomes from another species, has allowed comparison between distantly related marsupial genomes. For instance, chromosome-specific paints from the rufous rat-kangaroo (*Aepyprymnus rufescens*) (Rens *et al.* 2003) show that tammar chromosome 5 shares sequence with chromosomes 4 and 7 in the opossum (*M. domestica).* With the availability of the opossum genome sequence, it is possible to predict which genes will be on each tammar chromosome.

Although chromosome painting provides a global picture of the similarities between the tammar and opossum genomes, it is by physically mapping genes that a comparative map can be constructed between marsupials and other vertebrates. This will ultimately result in the determination of an in-depth evolutionary history of a region or chromosome. A new approach to mapping tammar genes is to use sequence from the tammar genome project to design small yet specific probes, called overgos, for screening a large insert genomic library (Baterical Artificial Chromosome library). These large insert clones are perfect for FISH mapping. This new efficient approach has allowed dense gene maps of the tammar X chromosome and the 'neo-X' region on chromosome 5 to be generated (Deakin *et al.* 2008). Genes marking the ends of evolutionary blocks of genes conserved between opossum and human have been mapped to tammar chromosome 5 (Fig. 1.1). This has permitted a virtual map of the chromosome to be constructed, by extrapolating from the opossum genome to infer the location of genes between conserved end markers to be present on tammar chromosome 5. In total, 73 and 141 genes have been

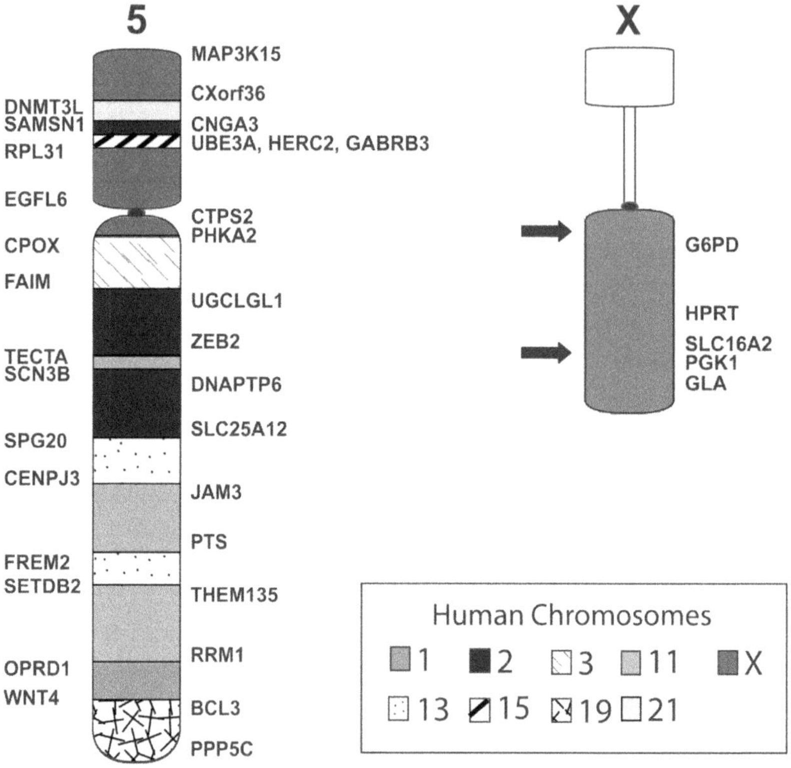

Figure 1.1: Tammar/human comparative map of tammar chromosomes 5 and X. Regions have been shaded based on the location of these genes in human. Genes located at the ends of evolutionary conserved blocks are indicated for chromosome 5. The five genes indicated on the X chromosome map are those for which XCI data is available in marsupials. Arrows indicate the position of genes flanking XIST in humans, demonstrating the disruption of this region in tammar.

assigned to the tammar chromosomes X and 5 respectively, with an additional 2320 genes virtually assigned to chromosome 5 (Deakin *et al.* 2008). This efficient approach is now being used to construct detailed physical and virtual maps of the remaining tammar autosomes.

By chromosome painting, tammar chromosome 5 was shown to consist of two segments conserved among marsupials, referred to as segments C11 and C12, with C11 spanning the entire short arm to just below the centromere and C12 covering the remainder of the chromosome (Rens *et al.* 2003). Conserved segment C11 corresponds to part of the short arm and the entire long arm of opossum chromosome 4. C12 spans the short arm and a small section of the long arm of opossum chromosome 7 (Fig. 1.2). Gene mapping on both tammar X and 5 has revealed internal rearrangements between tammar and opossum previously undetected by chromosome painting (Fig. 1.2) and provides a more accurate account of the extent of homology between tammar and opossum for these two chromosomes. Segment C12 does not actually span the entire short arm of tammar chromosome 5, as indicated by chromosome painting, but is restricted to the region surrounding the centromere. The remainder of the short arm consists of genes from the short arm of opossum chromosome 4 (Deakin *et al.* 2008). Cross-species chromosome painting did not provide information on the homology of the short arm of opossum chromosome 4 with other marsupials, including tammar (Rens *et al.* 2003). In addition, two rearrangements were detected in segment C11 between tammar and opossum.

Gene mapping and chromosome painting between the three major mammalian lineages (eutherians, marsupials and monotremes) has revealed that the human X chromosome can be divided into an ancient region and a recently added region (Deakin *et al.* 2008; Glas *et al.* 1999; Graves 1995; Wilcox *et al.* 1996). Genes on the human X that also map to the tammar X chromosome are part of the ancient region, having been part of the mammalian X chromosome since marsupials and eutherians last shared a common ancestor. Chromosome painting indicates that the long arm of the tammar X shares homology with the opossum X (conserved segment

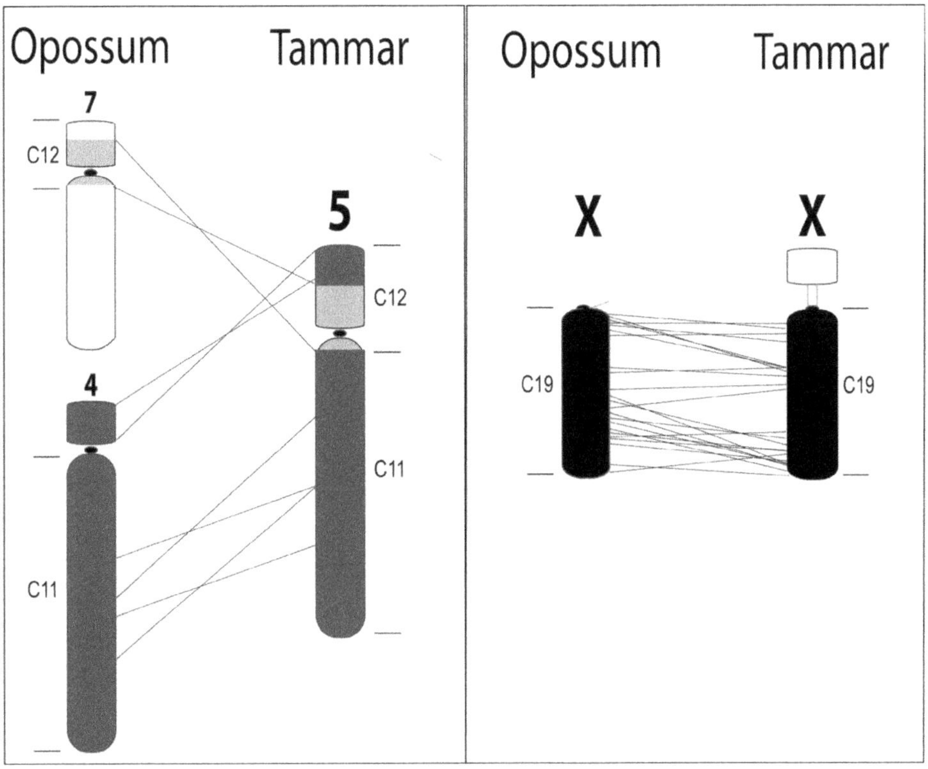

Figure 1.2: Rearrangements detected between tammar and opossum. Conserved segments C11, C12 and C19 as determined by chromosome painting are shown. Rearrangements are indicated by lines between chromosomes, revealing two large-scale rearrangements between these species within C11 and a highly rearranged gene order between the tammar and opossum X chromosomes.

C19) (Rens *et al.* 2003). Genes from this ancient region of the X are autosomal in birds and monotremes, being located on chromosome 4 in chicken and chromosome 6 in platypus (Veyrunes *et al.* 2008) (Fig. 1.3). Genes from the recently added region are located on chromosome 5 in tammar and were added to the eutherian X chromosome after the marsupial/eutherian divergence. Genes from this region are located on chromosome 1 in chicken and on chromosomes 15 and 18 in platypus (Edwards *et al.* 2007; Veyrunes *et al.* 2008). Gene mapping data have defined the limits of the ancient and recently added (neo-X) regions and narrowed down the fusion point of these regions to a 400kb region on the human X between genes *RP2* and *RBM10* (Deakin *et al.* 2008).

A comparison between the tammar and opossum gene order on the X chromosome has revealed many rearrangements between the opossum and tammar chromosomes (Fig. 1.2), rearrangements beyond the resolution of chromosome painting (Deakin *et al.* 2008). The highly rearranged order of genes on the X chromosome between the species was surprising, given the very conserved gene order for the ancient region of the X among eutherians, a feature thought to result from the eutherian mode of X inactivation (Mikkelsen *et al.* 2007).

Marsupial X chromosome inactivation

Most genes within the ancient region of the human X are subject to X chromosome inactivation (XCI), which equalises the level of their expression between females with two X chromosomes and males with one (Graves and Gartler 1986). However, many genes within the recently added region of the human X escape inactivation (Carrel and Willard 2005; Johnston *et al.* 2008). The best explanation for this phenomenon is provided by the evolutionary history of the human X chromosome. Genes within the recently added region were previously autosomal, so did not require dosage compensation prior to marsupial/eutherian divergence. Following addition of the region to the X and Y, degradation of genes on the Y left genes on the X with dosage differences, and there was selection for recruitment into the XCI system (Graves *et al.* 1998; Graves and Schmidt 1992).

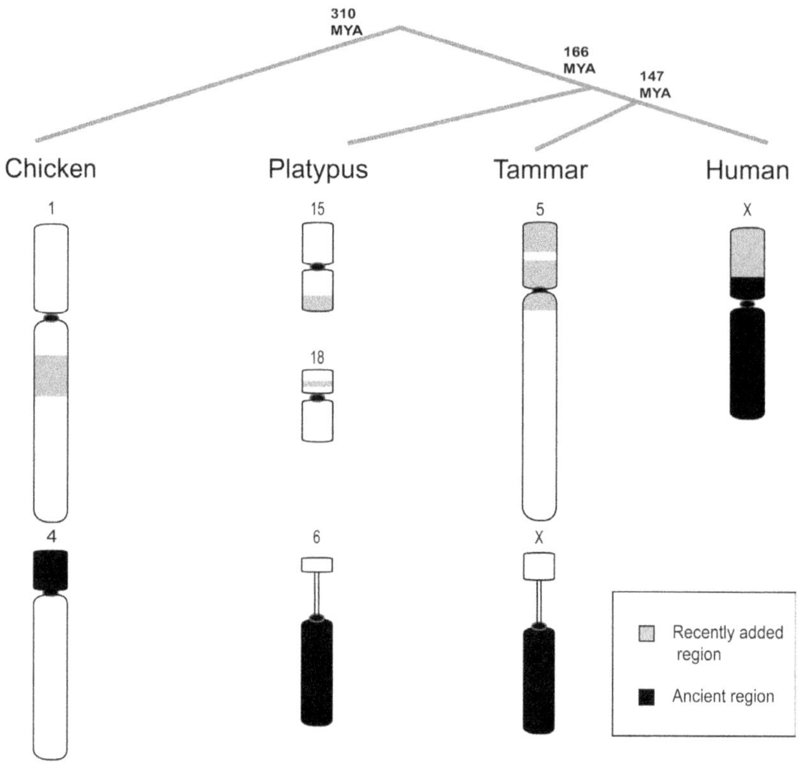

Figure 1.3: Evolution of the human X chromosome. The ancient region of the X (black) is located on the X chromosome in marsupials and eutherians but is autosomal in birds and monotremes. The recently added region (grey) is only found on the X chromosome in eutherians.

Relative to the human X, the mouse X has fewer genes that escape inactivation (Disteche *et al.* 2002), suggesting that recruitment to the inactivation system has been more rapid in the rodent lineage.

Eutherian XCI is a complex multi-stage process (Gartler *et al.* 1985), controlled by the *XIST* gene (X-Inactive Specific Transcript) which transcriptionally silences genes on one of the two X chromosomes in the somatic cells of females. Inactivation occurs in the early embryo and is random with regard to the parental origin of the X that is inactivated. It involves many variant histones and histone modifications such as deacetylation, and is stabilised by DNA methylation (Heard 2004). Marsupial XCI differs markedly from the random and fairly complete inactivation of eutherians. Marsupials have non-random X-inactivation, with the paternal X being preferentially silenced (Sharman 1971). It is tissue-specific, so that in some tissues both copies of an X-linked gene are active (Cooper *et al.* 1993). Variation in marsupial XCI exists between species; for example, the paternal *G6PD* allele is completely silenced in somatic tissues from two species of macropodids (*Macropus robustus* and *Macropus rufogriseus*) yet is partially active in tissues from the North American Virginia opossum (*Didelphis virginiana*). Variation in marsupial XCI also exists between genes within a species. Methylation does not appear to play a role (Hornecker *et al.* 2007; Kaslow and Migeon 1987; Loebel and Johnston 1996) but some histone modifications do appear to be involved in the XCI process (Koina *et al.* 2009; Wakefield *et al.* 1997). However, the most striking difference between marsupial and eutherian XCI is the apparent lack of an *XIST* gene in marsupials, with *XIST* flanking genes mapping to completely different regions of the X chromosome in both the tammar (Deakin *et al.* 2008) (Fig. 1.1) and opossum (Davidow *et al.* 2007; Hore *et al.* 2007; Shevchenko *et al.* 2007).

Unfortunately, our understanding of marsupial XCI is largely based on results obtained for just five genes and there has been no consistency in the species used for these studies. Older studies determined the inactivation status of *G6PD, PGK, GLA* and *HPRT* in the common brushtail possum (*Trichosurus vulpecula*), *Didelphis virginiana*, *Antechinus* species and a number of different macropodid species (Cooper *et al.* 1993). Inactivation status of the housekeeping genes *G6PD, PGK, GLA* and *HPRT* was determined mainly by isozyme analysis, although *G6PD* expression in two subspecies of wallaroo (*Macropus robustus*) (Watson *et al.* 2000) and *PGK1* and *G6PD* in *M. domestica* (Hornecker *et al.* 2007) have been examined using a molecular assay of transcription called SNuPE (Single Nucleotide Primer Extension). These techniques require differences in the two copies of the gene under study (polymorphism), which have been found haphazardly in different species. Differences in inactivation status between species, and even between genes within the one species, make it difficult to determine the true nature of marsupial X inactivation. Consistency in the species used for studies is required, as is a greater number of genes.

The construction of a dense gene map of the X allows an in-depth study of marsupial X inactivation in the tammar wallaby. Recently, a direct method for looking at RNA produced at a locus (FISH hybridising to the primary RNA transcript) was used to determine inactivation status of *SLC16A2* in the tammar (Koina *et al.* 2005). The advantage of this technique is that it can detect differences in expression of two gene copies without requiring polymorphisms. Hence, the inactivation status of genes mapped by BAC clones can be readily determined by RNA FISH. An activity map of the X is being generated.

TAMMAR CHROMOSOME 5 AND THE EVOLUTION OF GENOMIC IMPRINTING

Tammar chromosome 5 contains, in addition to genes from the human X, gene blocks from other human chromosomes (Fig. 1.1), including one that shows imprinted expression. Comparisons of gene organisation in this region between eutherians, marsupials and other vertebrates have provided insight into genome evolution and the origin of imprinting.

A block of genes from human chromosome 15, containing the Prader-Willi/Angelman syndrome region, is located on tammar chromosome 5. Genes in this region in human and mouse are imprinted, meaning that one copy of the gene is silenced in a parent-specific manner. Disruptions of the imprinting mechanism cause the Prader-Willi or Angelman syndromes (Horsthemke and Buiting 2006). *UBE3A*, the gene associated with Angelman syndrome, is paternally imprinted (silenced), meaning that only the maternal copy of the gene is expressed (Vu and Hoffman 1997). Also from this region in human is the paternally expressed and maternally silenced *SNRPN* gene (Reed and Leff 1994), deletions in which are responsible for Prader-Willi syndrome in humans.

Isolating and mapping the tammar orthologues of *UBE3A, SNRPN* and flanking genes has produced several surprises (Rapkins *et al.* 2006). Although *UBE3A* and non-imprinted flanking genes map to tammar chromosome 5 (Fig. 1.1), *SNRPN* lies on chromosome 1, close to a non-imprinted homologue. Other imprinted genes from the region seem not to exist at all in the tammar genome, and are also absent from the opossum database. These genes seem to have been derived from RNA copies of genes at other locations in the genome. This appears to be an ancestral arrangement, since it is the same in the platypus and chicken (Rapkins *et al.* 2006).

The discovery of a polymorphism allowed expression of the paternal and maternal copies of these genes to be distinguished, and allowed us to demonstrate that neither *UBE3A* nor *SNRPN* is imprinted in tammar. This led to the important conclusion that the Prader-Willi/Angelman imprinted region was recently derived in eutherians from a host of non-imprinted components gathered from around the genome (Rapkins *et al.* 2006).

CONCLUSIONS

This is an exciting time for marsupial genomics, and the tammar wallaby promises to be a vital inclusion in comparative genomic studies. Our intensive mapping of tammar chromosomes X and 5 will provide new information on the make-up of the eutherian X and Y chromosomes. Provision of many more X markers and their accurate locations is allowing the determination of not only the inactivation status of genes on the X, but whether genic activity is related to location on the X relative to an inactivation centre (if one exists). We can also see what genes escape inactivation in different tissues and explore the molecular mechanism of inactivation. Comparison with the genes and sequences on the X and regions of chromosome 5 that were autosomal in a common ancestor will allow answers to questions regarding the evolution of a biased gene content and mechanisms involved in X inactivation.

ACKNOWLEDGMENTS

We thank the Australian Research Council for funding the ARC Centre of Excellence for Kangaroo Genomics.

REFERENCES

Alsop AE, Miethke P, Rofe R, Koina E, Sankovic N, Deakin JE, Haines H, Rapkins RW and Graves JAM 2005. Characterizing the chromosomes of the Australian model marsupial *Macropus eugenii* (tammar wallaby). *Chromosome Res.* **13:** 627–636.

Belov K, Deakin JE, Papenfuss AT, Baker ML, Melman SD, Siddle HV, Gouin N, Goode DL, Sargeant TJ, Robinson MD, Wakefield MJ, Mahony S, Cross JG, Benos PV, Samollow PB, Speed TP, Graves JAM and Miller RD 2006. Reconstructing an ancestral mammalian immune supercomplex from a marsupial major histocompatibility complex. *PLoS Biol* **4:** e46.

Belov K, Sanderson CE, Deakin JE, Wong ES, Assange D, McColl KA, Gout A, de Bono B, Barrow AD, Speed TP, Trowsdale J and Papenfuss AT 2007. Characterization of the opossum immune genome provides insights into the evolution of the mammalian immune system. *Genome Res.* **17:** 982–991.

Carrel L and Willard HF 2005. X-inactivation profile reveals extensive variability in X-linked gene expression in females. *Nature* **434:** 400–404.

Cooper DW, Johnston PG and Graves JAM 1993. X-inactivation in marsupials and monotremes. *Sem. Dev. Biol.* **4:** 117–128.

Davidow LS, Breen M, Duke SE, Samollow PB, McCarrey JR and Lee JT 2007. The search for a marsupial XIC reveals a break with vertebrate synteny. *Chromosome Res.* **15:** 137–146.

Deakin JE, Papenfuss AT, Belov K, Cross JG, Coggill P, Palmer S, Sims S, Speed TP, Beck S and Graves JAM 2006. Evolution and comparative analysis of the MHC Class III inflammatory region. *BMC Genom.* **7:** 281.

Deakin JE, Koina E, Waters PD, Doherty R, Patel VS, Delbridge ML, Dobson B, Fong J, Hu Y, van den Hurk C, Pask AJ, Shaw G, Smith C, Thompson K, Wakefield MJ, Yu H, Renfree MB and Graves JA 2008. Physical map of two tammar wallaby chromosomes: a strategy for mapping in non-model mammals. *Chromosome Res.* **16:** 1159–1175.

Delbridge ML, Lingenfelter PA, Disteche CM and Graves JA 1999. The candidate spermatogenesis gene RBMY has a homologue on the human X chromosome. *Nat. Genet.* **22:** 223–224.

Disteche CM, Filippova GN and Tsuchiya KD 2002. Escape from X inactivation. *Cytogenet. Genome Res.* **99:** 36–43.

Duret L, Chureau C, Samain S, Weissenbach J and Avner P 2006. The Xist RNA gene evolved in eutherians by pseudogenization of a protein-coding gene. *Science* **312:** 1653–1655.

Edwards CA, Rens W, Clarke O, Mungall AJ, Hore T, Graves JAM, Dunham I, Ferguson-Smith AC and Ferguson-Smith MA 2007. The evolution of imprinting: chromosomal mapping of orthologues of mammalian imprinted domains in monotreme and marsupial mammals. *BMC Evol. Biol.* **7:** 157.

Foster JW and Graves JAM 1994. An SRY-related sequence on the marsupial X chromosome: implications for the evolution of the mammalian testis-determining gene. *Proc. Natl Acad. Sci. USA* **91:** 1927–1931.

Foster JW, Brennan FE, Hampikian GK, Goodfellow PN, Sinclair AH, Lovell-Badge R, Selwood L, Renfree MB, Cooper DW and Graves JAM 1992. Evolution of sex determination and the Y chromosome: SRY-related sequences in marsupials. *Nature* **359:** 531–533.

Gartler SM, Dyer KA, Graves JAM and Rocchi M 1985. A two step model for mammalian X-chromosome inactivation. *Prog. Clin. Biol. Res.* **198:** 223–235.

Glas R, De Leo AA, Delbridge ML, Reid K, Ferguson-Smith MA, O'Brien PC, Westerman M and Graves JAM 1999. Chromosome painting in marsupials: genome conservation in the kangaroo family. *Chromosome Res.* **7:** 167–176.

Graves JAM 1995. The origin and function of the mammalian Y chromosome and Y-borne genes: an evolving understanding. *Bioessays* **17:** 311–320.

Graves JAM and Gartler SM 1986. Mammalian X chromosome inactivation: testing the hypothesis of transcriptional control. *Somat. Cell Mol. Genet.* **12:** 275–280.

Graves JAM and Schmidt MM 1992. Mammalian sex chromosomes: design or accident? *Curr. Opin. Genet. Dev.* **2:** 890–901.

Graves JAM, Disteche CM and Toder R 1998. Gene dosage in the evolution and function of mammalian sex chromosomes. *Cytogenet. Cell Genet.* **80:** 94–103.

Graves JAM, Koina E and Sankovic N 2006. How the gene content of human sex chromosomes evolved. *Curr. Opin. Genet. Dev.* **16:** 219–224.

Heard E 2004. Recent advances in X-chromosome inactivation. *Curr. Opin. Cell Biol.* **16:** 247–255.

Hore TA, Koina E, Wakefield MJ and Graves JAM 2007. The region homologous to the X-chromosome inactivation centre has been disrupted in marsupial and monotreme mammals. *Chromosome Res.* **15:** 147–161.

Hornecker JL, Samollow PB, Robinson ES, Vandeberg JL and McCarrey JR 2007. Meiotic sex chromosome inactivation in the marsupial *Monodelphis domestica*. *Genesis* **45:** 696–708.

Horsthemke B and Buiting K 2006. Imprinting defects on human chromosome 15. *Cytogenet. Genome Res.* **113:** 292–299.

Johnston CM, Lovell FL, Leongamornlert DA, Stranger BE, Dermitzakis ET and Ross MT 2008. Large-scale population study of human cell lines indicates that dosage compensation is virtually complete. *PLoS Genet.* **4:** e9.

Kaslow DC and Migeon BR 1987. DNA methylation stabilizes X chromosome inactivation in eutherians but not in marsupials: evidence for multistep maintenance of mammalian X dosage compensation. *Proc. Natl Acad. Sci. USA* **84:** 6210–6214.

Koina E, Wakefield MJ, Walcher C, Disteche CM, Whitehead S, Ross M and Graves JAM 2005. Isolation X location and activity of the marsupial homologue of SLC16A2, an XIST-flanking gene in eutherian mammals. *Chromosome Res.* **13:** 687–698.

Koina E, Chaumeil J, Greaves IK, Tremethick DJ and Graves JA 2009. Specific patterns of histone marks accompany X chromosome inactivation in a marsupial. *Chromosome Res.* **17:** 115–126.

Koopman P, Gubbay J, Vivian N, Goodfellow P and Lovell-Badge R 1991. Male development of chromosomally female mice transgenic for Sry. *Nature* **351:** 117–121.

Loebel DA and Johnston PG 1996. Methylation analysis of a marsupial X-linked CpG island by bisulfite genomic sequencing. *Genome Res.* **6:** 114–123.

Mikkelsen TS, Wakefield MJ, Aken B, Amemiya CT, Chang JL, Duke S, Garber M, Gentles AJ, Goodstadt L, Heger A, Jurka J, Kamal M, Mauceli E, Searle SM, Sharpe T, Baker ML, Batzer MA, Benos PV, Belov K, Clamp M, Cook A, Cuff J, Das R, Davidow L, Deakin JE, Fazzari MJ, Glass JL, Grabherr M, Greally JM, Gu W, Hore TA, Huttley GA, Kleber M, Jirtle RL, Koina E, Lee JT, Mahony S, Marra MA, Miller RD, Nicholls RD, Oda M, Papenfuss AT, Parra ZE, Pollock DD, Ray DA, Schein JE, Speed TP, Thompson K, VandeBerg JL, Wade CM, Walker JA, Waters PD, Webber C, Weidman JR, Xie X, Zody MC, Graves JAM, Ponting CP, Breen M, Samollow PB, Lander ES and Lindblad-Toh K 2007. Genome of the marsupial *Monodelphis domestica* reveals innovation in non-coding sequences. *Nature* **447:** 167–177.

Rapkins RW, Hore T, Smithwick M, Ager E, Pask AJ, Renfree MB, Kohn M, Hameister H, Nicholls RD, Deakin JE and Graves JAM 2006. Recent assembly of an imprinted domain from non-imprinted components. *PLoS Genet.* **2:** e182.

Reed ML and Leff SE 1994. Maternal imprinting of human SNRPN, a gene deleted in Prader-Willi syndrome. *Nat. Genet.* **6:** 163–167.

Rens W, O'Brien PC, Fairclough H, Harman L, Graves JAM and Ferguson-Smith MA 2003. Reversal and convergence in marsupial chromosome evolution. *Cytogenet. Genome Res.* **102:** 282–290.

Sharman GB 1971. Late DNA replication in the paternally derived X chromosome of female kangaroos. *Nature* **230:** 231–232.

Shevchenko AI, Zakharova IS, Elisaphenko EA, Kolesnikov NN, Whitehead S, Bird C, Ross M, Weidman JR, Jirtle RL, Karamysheva TV, Rubtsov NB, VandeBerg JL, Mazurok NA, Nesterova TB, Brockdorff N and Zakian SM 2007. Genes flanking Xist in mouse and human are separated on the X chromosome in American marsupials. *Chromosome Res.* **15:** 127–136.

Sinclair AH, Berta P, Palmer MS, Hawkins JR, Griffiths BL, Smith MJ, Foster JW, Frischauf AM, Lovell-Badge R and Goodfellow PN 1990. A gene from the human sex-determining region encodes a protein with homology to a conserved DNA-binding motif. *Nature* **346:** 240–244.

Sinclair AH, Foster JW, Spencer JA, Page DC, Palmer M, Goodfellow PN and Graves JAM 1988. Sequences homologous to ZFY, a candidate human sex-determining gene, are autosomal in marsupials. *Nature* **336:** 780–783.

Springer MS, Westerman M and Kirsch JAW 1994. Relationships among orders and families of marsupials based on 12S ribosomal DNA sequences and the timing of the marsupial radiation. *J. Mamm. Evol.* **2:** 85–115.

Suzuki S, Renfree MB, Pask AJ, Shaw G, Kobayashi S, Kohda T, Kaneko-Ishino T and Ishino F 2005. Genomic imprinting of IGF2, p57(KIP2) and PEG1/MEST in a marsupial, the tammar wallaby. *Mech. Dev.* **122:** 213–222.

Veyrunes F, Waters PD, Miethke P, Rens W, McMillan D, Alsop AE, Grützner F, Deakin JE, Whittington C, Schatzkamer K, Kremitzki CL, Graves T, Ferguson-Smith MA, Warren W and Graves JAM 2008. Bird-like sex chromosomes of platypus imply recent origin of mammal sex chromosomes. *Genome Res.* In press.

Vu TH and Hoffman AR 1997. Imprinting of the Angelman syndrome gene UBE3A, is restricted to brain. *Nat. Genet.* **17:** 12–13.

Wakefield MJ and Graves JA 2003. The kangaroo genome: leaps and bounds in comparative genomics. *EMBO Rep.* **4:** 143–147.

Wakefield MJ and Graves JA M 2005. Marsupials and monotremes sort genome treasures from junk. *Genome Biol.* **6:** 218.

Wakefield MJ, Keohane AM, Turner BM and Graves JAM 1997. Histone underacetylation is an ancient component of mammalian X chromosome inactivation. *Proc. Natl Acad. Sci. USA* **94:** 9665–9668.

Watson D, Jacombs AS, Loebel DA, Robinson ES and Johnston PG 2000. Single nucleotide primer extension (SNuPE) analysis of the G6PD gene in somatic cells and oocytes of a kangaroo (*Macropus robustus*). *Genet. Res.* **75:** 269–274.

Wilcox SA, Watson JM, Spencer JA and Graves JAM 1996. Comparative mapping identifies the fusion point of an ancient mammalian X-autosomal rearrangement. *Genomics* **35:** 66–70.

Zenger KR, McKenzie LM and Cooper DW 2002. The first comprehensive genetic linkage map of a marsupial: the tammar wallaby (*Macropus eugenii*). *Genetics* **162:** 321–330.

2 Genetic architecture of the macropodid immune system

H.V. Siddle, C.E. Sanderson, J.E. Deakin and K. Belov

SUMMARY

Macropodid marsupials are important models for developmental and comparative immunology studies. Early work on the marsupial immune system reported an unusual immunoglobulin isotype switch and weak mixed lymphocyte responses, leading to suggestions that the marsupial immune system is weaker and more primitive than that of eutherian mammals. More recent studies suggest that these results may be due to ineffective immunisation regimes and low levels of cross-reactivity between eutherian reagents and marsupial immune molecules. The availability of genomic resources for the tammar wallaby (*Macropus eugenii*) and other marsupial species has facilitated the isolation and characterisation of macropodid immune genes. These studies have demonstrated that the genetic architecture of the macropodid immune system is comparable in gene content and complexity to that of eutherians. Analysis of macropodid immune genes is providing insights into the evolution of mammalian immunity and how immune function is adapted to unique aspects of macropodid physiology, particularly reproduction.

INTRODUCTION

Recent advances in our understanding of the marsupial immune system are paving the way for the use of macropodids in developmental and comparative immunology studies, particularly the model macropodid species, the tammar wallaby (*Macropus eugenii*). Until recently, relatively little was known about the molecular genetics of the marsupial immune system, and our understanding of the vertebrate immune system and its genetic basis was confined to the well-studied eutherian (human and mouse) and non-mammalian (chicken and xenopus) model species. However, eutherians and non-mammals last shared a common ancestor ~315 million years ago (Bininda-Emonds *et al.* 2007) and this relationship can be too distant for effective comparisons, leaving a large gap in our understanding of the evolution of the vertebrate immune system (Wakefield and Graves 2003). Marsupials last shared a common ancestor with eutherians ~148 million years ago (Bininda-Emonds *et al.* 2007) and fill this gap in the vertebrate phylogeny.

Although macropodids share mammalian features with eutherians and monotremes they have a unique physiology and mode of reproduction, which makes their immune system of particular interest for comparative studies. For example, macropodid young are born without differentiated immune tissues and their immune system develops outside the sterile environment of the uterus, in the pathogen-laden pouch (Basden *et al.* 1997; Old and Deane 1998). An understanding of the macropodid immune system will provide insights into how marsupial species have adapted their immune response to their environment, and the challenges of their mode of reproduction.

Early investigations into macropodid immunology focused on humoral and cellular immune responses, but in many cases were hindered and confounded by a lack of marsupial-specific reagents and molecular knowledge of the marsupial immune system (Deakin *et al.* 2005; Young *et al.* 2003). Recently, the genome of the tammar wallaby was sequenced jointly by the Australian Genome Research Facility (AGRF) and Baylor College of Medicine (US) to provide a two-fold coverage of the genome (http://www.genome.gov/12512299). This resource has increased the speed and ease of gene discovery in the tammar wallaby and facilitated in-depth molecular studies on immune gene function. These studies suggest that the genetic architecture of the marsupial innate and adaptive immune responses is similar to that of eutherians (Belov *et al.* 2007; Wong *et al.* 2006). To date, studies on macropodids have focused on the quokka (*Setonix brachyurus*) and the tammar wallaby, with a recent emphasis on the tammar wallaby as the model Australian marsupial species. In this chapter we review the early functional studies on the humoral and cellular immune response in macropodids, but our main focus is on molecular developments and their application to our understanding of immune function in macropodids. Where relevant, we refer to studies of non-macropodid marsupial species.

EARLY WORK ON THE HUMORAL AND CELLULAR IMMUNE RESPONSE OF MACROPODIDS

The adaptive immune system is present in all jawed vertebrates and comprises two parts – humoral immunity and cell-mediated immunity. Both humoral and cell-mediated immunity need to recognise antigenic peptides produced by an invading pathogen. The humoral immune response primarily relies on the production of antibodies. Antibodies are immunoglobulins that recognise specific antigens and are secreted by B-lymphocytes. Immunoglobulins are classified into different isotypes according to function in the immune response. For example, IgM is the first isotype produced when a new pathogen is encountered; this stage of antibody production is referred to as the primary response. During the secondary immune response, which occurs when the same pathogen is re-encountered, in eutherians there is a rapid switch to a different isotype, IgG. Cell-mediated immunity involves the activation of macrophages and natural killer cells, the production of antigen-specific cytotoxic T-cells and the release of signalling molecules called cytokines. Cell-mediated immunity is triggered by the recognition of antigenic peptides by polymorphic cell-surface major histocompatibility complex (MHC) molecules, which present these peptides to T-lymphocytes.

Initial studies on the humoral immune response of macropodids (and other marsupial species) concluded that marsupials have a primitive immune response compared to eutherians (Stone *et al.* 1996; Wilkinson *et al.* 1994). One characteristic of macropodid humoral immunity is a prolonged primary response, reported to last as long as 26 weeks in the quokka (Yadav 1971) and at least nine weeks in the tammar wallaby (Deakin *et al.* 2005). A similar response has been reported in the South American gray short-tailed opossum (*Monodelphis domestica*) (Croix *et al.* 1989). The prolonged primary response of macropodids is very different from the short eutherian primary response, which typically peaks around two weeks after immunisation. This difference in primary response has made the true nature of the secondary response in macropodids, and marsupials in general, more difficult to characterise, with most immunisation trials in macropodids utilising a typical eutherian immunisation regime (Asquith *et al.* 2006; Kitchener *et al.* 2002). Deakin *et al.* (2005) argue that as the marsupial primary immune response is different from that of eutherians, immunisation regimes need to be adjusted to measure the secondary response accurately (Deakin *et al.* 2005). For example, a booster administered to quokkas 27 weeks after the initial immunisation resulted in increased antibody levels in line with a true secondary response (Yadav 1971).

The tammar wallaby and other marsupial species were thought to have an ineffective cellular immune response due to weak mixed lymphocyte reactions (MLRs) (Infante *et al.* 1991; Stone *et al.* 1998). MLRs test the ability of T-lymphocytes to proliferate in response to an antigen. The reaction is carried out between T-lymphocytes from two individuals, where the cells from one individual are treated to prevent proliferation but act as an antigen to the T-lymphocytes of the second individual. T-cell proliferation should occur in response to foreign MHC molecules on the surface of the T-cells of the second individual. Stone *et al.* (1998) proposed that low

MLRs observed in the opossum were a result of unusual T-cells or a lack of polymorphism between MHC genes in opossums (Stone *et al.* 1998). This hypothesis was supported by a report that the MHC class II genes of the tammar wallaby lack the level of polymorphism found in eutherian MHC genes (McKenzie and Cooper 1994).

RECENT MOLECULAR ADVANCES

Over the past 10 years, advances in molecular technologies have allowed us to begin characterising the macropodid immune system at the molecular level. An important step towards understanding immune gene function is to identify immune genes. In this section we review the current genetic knowledge of key molecules involved in the immune response of macropodids, including the immunoglobulins, T-cell receptors, MHC genes and cytokines (Table 2.1). We also summarise recent advances in characterising the immune tissues and immune response of adult and developing macropodids.

Immunoglobulins

All immunoglobulins have a similar structure consisting of four chains – two identical light chains and two identical heavy chains. In eutherians, immunoglobulins are encoded by three types of genes – a heavy chain gene (μ, δ, γ, α and ε) and two light chain genes (κ and λ), which can associate with any of the heavy chains. Both the heavy chain gene and light chain genes have a constant region at the C-terminal and a variable region at the N-terminal, where antigen binding occurs.

Marsupial immunoglobulins are more similar to those found in eutherians than non-mammals (Miller and Belov 2000). Eutherian mammals possess five classes of immunoglobulins – IgA, IgE, IgD, IgG and IgM, corresponding to the five heavy chain genes listed above. Four of the immunoglobulin classes found in eutherians – IgG, IgM, IgA and IgE – have been characterised at a genetic level in the tammar wallaby (Daly *et al.* 2007b), the common brushtail possum (*Trichosurus vulpecula*) and the gray short-tailed opossum (Miller and Belov 2000). Many non-mammalian species have IgA, IgD and IgM in common with mammals, but not IgG or IgE. Instead, they have IgY (Miller and Belov 2000). Phylogenetic comparison of eutherian, marsupial and non-mammalian immunoglobulin genes has provided insights into the evolution of mammals and their immune system. For example, duplication of the non-mammalian IgY gene gave rise to the IgG and IgE isotypes found in modern mammals, including marsupials (Aveskogh and Hellman 1998; Belov *et al.* 1998, 1999). Analysis of immunoglobulin heavy chains in eutherians, marsupials and monotremes provides support for the Therian hypothesis of mammalian evolution (Belov *et al.* 2002), that eutherians and marsupials are sister groups and are more closely related to each other than to monotremes.

Daly *et al.* (2007b) isolated the heavy chain constant regions for each class of immunoglobulin and the κ and λ immunoglobulin light chains from a tammar wallaby mammary gland cDNA library. This study indicated that the tammar wallaby preferentially utilises the λ light chain over κ and that the κ and λ immunoglobulin light chain diversity in the tammar wallaby is comparable to that seen in eutherians. A summary of light chain diversity in the common brushtail possum and gray short-tailed opossum is given in Miller and Belov (2000). Daly *et al.* (2007b) also isolated the immunoglobulin receptor (pIgR) and the J chain gene (the J chain facilitates the formation of polymeric IgM and IgA, important for mucosal immunity) to examine the role of these receptors in immunity of the tammar wallaby pouch young.

T-cell receptors

The weak MLR reported for some marsupial species led to speculation that marsupial T-cells do not recognise antigens effectively. T-cell receptors are heterodimers on the surface of T-cells, essential for antigen recognition. They consist of either α and ß polypeptide chains (αß TCR) or γ and δ polypeptide chains (γδ TCR) in addition to a CD3 complex that transduces signals from the TCR and bound antigen into intracellular signals. T-cell receptors (including alpha, beta, gamma and delta) have been isolated from the tammar wallaby, the common brushtail possum, northern brown bandicoot (*Isoodon macrourus*) and gray short-tailed opossum (Baker *et al.* 2001; Harrison *et al.* 2003a; Zuccolotto *et al.* 2000). A novel receptor, TCRμ (Parra *et al.* 2007), has been reported in a number of marsupial species including the tammar wallaby (Baker *et al.* 2005). This receptor has some similarities to a TCRδ isoform found in sharks and it has been proposed that TCRμ may correspond to a primordial TCR form (Parra *et al.* 2007).

Table 2.1: Comparison of eutherian and macropodid immune genes.

	Eutherians		Marsupials		
Components of immune system	**Human**	**Mouse**	**Tammar wallaby**	**Opossum**	**References (for opossum and tammar wallaby)**
Thymus	Single thoracic	Thoracic and cervical	Thoracic and cervical	Single thoracic	Stanley *et al.* 1972; Yadav *et al.* 1972b
Innate					
Cathelicidins	1	1	7	12	Belov *et al.* 2007; Daly *et al.* 2008a
Defensins	10 alpha, 39 beta, 1p theta	52 alpha, 6 beta	nd	1 alpha, 32 beta	Belov *et al.* 2007
Chemokines	47	45	1	31	Belov *et al.* 2007; Daly *et al.* 2008b
Natural killer complex receptors	15 C-type lectin receptors	28 C-type lectin receptors	nd	9 C-type lectin receptors	Belov *et al.* 2007
Leukocyte receptor complex receptors	~30 Ig-like receptors	10 Ig-like receptors	nd	~45 Ig-like receptors	Belov *et al.* 2007
Adaptive					
Immunoglobulins	IgA, IgE, IgD, IgG, IgM	IgA, IgE, IgD, IgG, IgM	IgA, IgE, IgG, IgM	IgA, IgE, IgG, IgM	Miller and Belov 2000
T-cell receptors	TCR α β δ, γ	TCR α β δ, γ	TCR α β δ γ, m	TCR α β δ γ, m	Deakin *et al.* 2006b; Parra *et al.* 2007; Zuccolotto *et al.* 2000
MHC class I	6 functional loci	~50 functional loci	11 functional loci	9 functional loci	Belov *et al.* 2006; Siddle *et al.* 2006
MHC class II	DO, DP, DQ, DR, DN, DM	DP(p), DO, DR, DQ, DN, DM	DA, DB, DC (nd), DM (nd)	DA, DB, DC, DM	Belov *et al.* 2006; Browning *et al.* 2004; Slade and Mayer 1995

nd = not determined, p = pseudogene, mouse MHC class II genes are named according to human orthologues.

A CD3 epsilon chain has been cloned in the tammar wallaby (Old *et al.* 2001). This gene has the required ITAM (immunoreceptor tyrosine-based activation motif) that is essential for T-cell activation. Notably, the extracellular domain of the protein showed the least conservation to eutherian CD3 peptide sequences. The availability of sequence for this gene has allowed further development of marsupial-specific reagents to improve the detection of T-cells.

MHC and comparative genomics

The MHC is a large multi-gene family found in all vertebrates. It is responsible for antigen recognition and presentation to T-lymphocytes. MHC genes are classified as class I, class II or class III; the class I and class II MHC genes are responsible for recognising and presenting antigens to T-lymphocytes, triggering a cellular immune response. Class I molecules are found on the surface of all nucleated cells and bind to endogenous antigens. The class II molecules are found on the surface of specialised immune cells and bind exogenously derived antigens. MHC class I and class II genes are usually highly polymorphic, facilitating the recognition of a broad range of antigenic peptides. In eutherians the different classes of MHC genes are genetically linked, although the organisation of the genes can vary between species (Kelley *et al.* 2005).

The MHC genes of macropodids have been investigated as a tool for understanding the evolution of MHC in vertebrates and for better understanding of the genetic mechanisms underlying the macropodid immune system. Initial investigation of macropodid MHC genes focused on characterising class I and class II genes in a piecemeal fashion; however, with the sequencing of the gray short-tailed opossum genome to a 6× coverage it has been possible to annotate the entire opossum MHC (Belov *et al.* 2006). This information is assisting in the construction and sequencing of a Bacterial Artificial Chromosome (BAC) contig of the tammar wallaby MHC (Siddle, unpubl. data).

While the grey short-tailed opossum MHC is as complex and gene-rich as the MHC of eutherians, the overall organisation is different (Belov *et al.* 2006). The opossum MHC contains class I, class II and class III genes, but has a class I/II region rather than separate class I and class II regions, as in eutherians. The class III region is highly conserved in gene content and order between the opossum (Belov *et al.* 2006), the tammar wallaby (Deakin *et al.* 2006a) and eutherians.

Isolation of class I MHC genes of macropodids indicates these genes have a unique organisation compared to eutherians and the gray short-tailed opossum. The opossum has 11 class I genes within the MHC and two class I genes (*Modo-UB* and *Modo-UC*) outside and distal to the MHC (Belov *et al.* 2006; Miska *et al.* 2004). It is estimated that the tammar wallaby has a similar number of class I genes to the gray short-tailed opossum, with approximately 12 loci (Siddle *et al.* 2006). However, the organisation of the tammar wallaby MHC is unique among vertebrates as the class II and class III genes are located on chromosome 2q, while the class I genes are found on chromosomes 1, 3, 4, 5, 6 and 7 (Deakin *et al.* 2007). In addition, analysis of MHC class I transcripts from the tammar wallaby (and other marsupial species) suggest that some structural differences in these genes may affect NK cell receptor binding (Daly *et al.* 2007a).

Class II MHC genes have been isolated in the tammar wallaby (Browning *et al.* 2004) and red-necked wallaby *(Macropus rufogriseus)* (Schneider *et al.* 1991) and in a number of other marsupial species (Lam *et al.* 2001; O'Huigin *et al.* 1998). Initially, marsupial class II genes were thought to be orthologous to eutherian class II gene families (Slade and Mayer 1995). However, a 2004 study concluded that polymorphic MHC class II genes of marsupials are not orthologous to the gene families of eutherians (Belov *et al.* 2004). The marsupial-specific class II families have been named DA, DB and DC. Only class II genes from the DA and DB gene families have been isolated from macropodids; the presence of the DC family has not been confirmed (Browning *et al.* 2004; Slade and Mayer 1995).

Initial studies of macropodid MHC genes found low levels of diversity at class II MHC genes and it was thought that this could explain weak MLRs in macropodids (McKenzie and Cooper 1994). However, Cheng *et al.* (2009a, b) have used MHC-linked microsatellite markers to demonstrate that class I and class II MHC loci of Kangaroo Island tammar wallabies are polymorphic. These genetic markers have been tested successfully on a range of macropod species, providing tools for studying the genetic diversity at functionally important immune genes in threatened macropodid species. In addition, studies on the MHC class II genes of the common brushtail possum (Holland *et al.* 2008) and a species of South American mouse opossum (*Gracilinanus microtarsus*) (Meyer-Lucht *et al.* 2008) demonstrate that other marsupial species have polymorphic MHC genes.

High levels of genetic diversity at MHC genes are believed to help populations and individuals respond to new disease threats (Amills *et al.* 2004). An understanding of MHC genes in a model macropodid, the tammar wallaby, has facilitated the development of markers for MHC genes in a range of threatened marsupial species, including the Tasmanian devil (*Sarcophilus harrisii*) (Siddle *et al.* 2007a, b), koala (*Phascolarctos cinereus*), western barred bandicoot (*Perameles bougainville*) and spotted tailed quoll (*Dasyurus maculatus*) (Sanderson *et al.* unpubl. data). The emergence and prevalence of diseases affecting Australian marsupial species, including devil facial tumour disease and chlamydia in koala populations, emphasises the importance of further understanding the immune system of the tammar wallaby, as a model for studying immune genes of other marsupial species.

Cytokines

Cytokines are proteins released by a cell that affect the growth and proliferation of immune cells. Cytokines form part of the innate immune system, the oldest form of protection against pathogens. Until recently, very little was known about marsupial cytokine genes as they were difficult to isolate in the laboratory, owing to low levels of sequence similarity with eutherian molecules (Harrison and Deane 1999). To date, lymphotoxin beta (*LTB*) (Harrison and Deane 1999) and lymphotoxin alpha (*LTA*) have been isolated from the tammar wallaby (Harrison and Deane 2000) and the genomic region encompassing these genes has been characterised to detect potentially important regulatory sequences (Cross *et al.* 2005; Deakin *et al.* 2006a). In addition, interleukin 5 has been isolated from the tammar wallaby and mapped to chromosome 1 (Hawken *et al.* 1999). Transcripts of this gene were detected in a range of marsupial species and the molecule was found to have a similar immune function to its homologue in eutherians. More recently, key cytokines including interferon gamma and interleukins 2, 4, 6, 12 and 13 have been isolated from the gray short-tailed opossum by mining the sequenced genome using bioinformatic techniques (Wong *et al.* 2006).

The type I interferon genes are the most studied innate immune components. They are a multi-gene family of glycoprotein mediators that have immune and developmental functions. It has been speculated that the interferon family of genes became more complex as mammals developed viviparity (Harrison *et al.* 2003b). Non-mammals, such as birds, have only one type I interferon gene; eutherians have many interferon genes, that are classified into four main subtypes (De Maeyer and De Maeyer-Guignard 1998). Harrison *et al.* (2003b) studied type I interferons in the common brushtail possum and tammar wallaby and found that while marsupials have α and β subtypes, monotremes appear to have only the β subtype. They have speculated that the limited interferon diversity found in monotremes is indicative of their egg-laying reproductive strategy. The emergence of additional diversity in marsupials and in eutherians supports this hypothesis. However, more interferon families present in the tammar wallaby may have been overlooked in the laboratory – the availability of a genomic sequence for the tammar wallaby will allow this question to be definitively answered.

MACROPODID THYMUS AND OTHER IMMUNOLOGICAL TISSUES

The identification of tammar wallaby immune genes has been used to develop antibodies that detect T- and B-lymphocytes of a number of macropodid species, allowing a more thorough characterisation of macropodid lymphoid tissues. The immunohistochemistry of the tammar wallaby spleen, thymus, gut-associated lympohoid tissue (GALT) and bronchus-associated lympohoid tissue (BALT) have been examined in detail using antibodies to cell surface markers for T-lymphocytes (CD3 and CD5) and B-lymphocytes (CD79b) (Old and Deane 2002). The structure of the lymphoid tissues and distribution lymphocytes were found to be similar to those of eutherians. An exception was the apparent absence of B-lymphocytes in the tammar wallaby GALT. However, the T- and B-lymphocyte distribution within the GALT of the eastern grey kangaroo (*Macropus giganteus*) is similar to that in eutherians, indicating that this macropodid species is capable of responding to pathogens entering via the mouth (Old and Deane 2001). Similarly, studies of GALT and BALT tissues of the rufous hare-wallaby (*Lagorchestes hirsutus*) documented the ability of this macropodid to mount an immune response to mycobacteria, a disease to which macropodids are susceptible (Young *et al.* 2003). Antigen-presenting cells (APCs), peripheral blood monocytes and monocyte-derived adherent cells and lymphocytes of the tammar wallaby have been successfully cultured in the laboratory and characterised (Young and Deane 2005, 2007). These studies have provided evidence that macropodids have an efficient immune system with a complexity similar to that of eutherians.

While many adult macropodid immunological tissues have a similar structure and function to those found in eutherians (Old and Deane 2000), a notable exception is the macropodid thymus. In the thymus T-lymphocytes develop and are positively selected if they can recognise antigen/MHC complexes and are destroyed if they are self-reactive. Thus, the thymus is the basis for the development of self/non-self recognition. Macropodids have both a cervical and thoracic thymus, a feature shared by a number of other diprotodont species (Yadav *et al.* 1972b). The presence of two thymuses was thought to be specific to some marsupial species, as the majority of eutherian species have only a thoracic thymus. However, a second cervical thymus was recently discovered in the mouse and was found to support T-lymphocyte development (Terszowski *et al.* 2006). Early work using thymectomy experiments in quokkas indicated that both the cervical and thoracic thymus play an important role in lymphoid tissue development (Ashman and Papadimitriou 1975) and the production of circulating T-lymphocytes (Yadav *et al.* 1972b). However, Stanley *et al.* (1972) hypothesised that the thymuses were programmed to operate at different stages of development. More recent investigations into the development of the cervical and thoracic thymuses in the tammar wallaby have shown that the cervical thymus develops more rapidly after birth and perhaps is more important for development of the immune system (Basden *et al.* 1997). We have investigated whether MHC class I transcripts are differentially expressed in the cervical and thoracic thymuses of tammar wallaby pouch young. Our results at a single stage of development found no differences in class I expression, supporting previous evidence that both thymuses have a role in T-lymphocyte selection and proliferation (Siddle, unpubl. data). We are currently sequencing all the genes expressed by the cervical and thoracic thymuses of tammar wallaby pouch young, the thymus transcriptome, to further investigate the roles of these tissues.

DEVELOPMENT OF IMMUNOCOMPETENCE IN POUCH YOUNG

All marsupials are born after a very short gestation period, which ranges from 10.5 days in the stripe-faced dunnart (*Sminthopsis macroura*) to 38 days in the eastern grey kangaroo (Tyndale-Biscoe and Renfree 1987). The result is a developmentally immature young, which undergoes immunological development in the pouch during lactation (Baker *et al.* 1999). At birth the immune tissues of macropodids are undifferentiated, but development of lymphoid tissues, particularly the cervical and thoracic thymuses, lymph nodes and spleen, occurs rapidly (Basden *et al.* 1997). Hassall's corpuscles, an indication of mature lymphoid tissue, are visible in the cervical and thoracic thymuses by day 30 in the tammar wallaby and day 60 in the quokka (Basden *et al.* 1997). Lymphocytes have been detected using immunohistological techniques from day 2–3 post-partum in the cervical thymus of the tammar wallaby and the quokka and day 5–7 in the thoracic thymus of the tammar wallaby (day 4 in the quokka) (Basden *et al.* 1997). While the tammar wallaby and quokka show signs of immunocompetence at day 21 and day 14 post-partum respectively, the tammar wallaby is not considered capable of adult humoral immune response until day 90–120, which coincides with release from the mother's teat (Basden *et al.* 1997).

Characterisation of the bacteria flora of the pouch of the tammar wallaby and the quokka (Deakin and Cooper 2004; Old and Deane 1998; Yadav *et al.* 1972a) detected potentially pathogenic bacteria and changes in bacterial composition throughout the reproductive cycle. Evidence suggests that immunological protection of developing pouch young against bacteria is provided by the mother in the form of antibodies and antimicrobials in the milk. Initial examination of quokka pouch young demonstrated an ability to absorb antibodies across the gut until 170 to 200 days post-partum (Yadav 1971). Further studies have shown that marsupial lactation is extremely specialised and can be divided into different phases as the composition of the milk changes (Cowan 1989). In the common brushtail possum, while the young is permanently attached to the teat (0–80 days) IgA and IgG are transferred from the mother to the young. When the young begins to suckle intermittently and to venture out of the pouch (the switch phase, 80–120 days) the levels of IgG in the milk increase and the young begins producing its own specific immunoglobulins (Adamski and Demmer 2000). Further studies have detected antibodies in common brushtail possum maternal milk and serum, specific for bacteria commonly found in the possum pouch (Deakin and Cooper 2004). A recent study of tammar wallaby lactation identified differential expression of immunoglobulins throughout the lactation cycle, corresponding to two phases (Daly *et al.* 2007b). These phases were found to correspond to times when the young was most at risk from pathogens in the environment – when they are first born and when they first leave their mother's pouch (Daly *et al.* 2007b).

It has been suggested that the macropodid pouch contains antimicrobial peptides, secreted by the mother or pouch young, which function by disrupting the cell membrane of pathogenic bacteria (Yadav *et al.* 1972a). Antimicrobial peptides from macropodids have proven difficult to isolate, but the recent availability of genomic resources for the tammar wallaby has led to the isolation of a cathelicidin-like transcript (*MaeuCath*) from the tammar wallaby (Daly *et al.* 2008a). Investigation of *MaeuCath* expression in tammar pouch young found that transcript expression increased during the first three to four weeks of life, particularly in the lung, gastrointestinal tract and skin, indicating a role in protection of the young before the adaptive immune system is fully developed (Daly *et al.* 2008a). In addition, a peptide isolated from the pouch of the tammar wallaby has been shown to act as an immunomodulator and increase proliferation of lymphocytes when cultured with a mitogen (Baudinette *et al.* 2005). We have recently identified key antimicrobial peptide genes, including defensins and cathelicidins in the tammar wallaby and opossum genomes, and are studying their expression in the pouch and pouch young (Whittington, Papenfuss and Belov, unpubl. data).

CONCLUSIONS AND FUTURE DIRECTIONS

Immunogenetic studies suggest that the macropodid immune system is not 'retarded' (Wilkinson *et al.* 1994). The macropodid immune gene repertoire is equal in complexity to that of eutherians and includes MHC genes, immunoglobulins, T-cell receptors, interferons, cytokines and antimicrobial peptides. The most recent studies have demonstrated that the macropodid immune system has similarities in structure and function to that of eutherians. However, it is perhaps the differences in immune genes and immune response between eutherians and macropodids which will prove most interesting, as these differences will demonstrate how the macropodid immune system has adapted to specific environmental pressures. Already, studies have shown that macropodid immune function is intricately related to their specialised mode of reproduction. They indicate the macropodid immune system has a novel T-cell receptor, MHC organisation and antimicrobial peptides. Sequencing of the tammar wallaby genome, the availability of genetic resources (including cDNA, Bacterial Artificial Chromosome libraries and expressed sequence tags) and several large breeding colonies have made the tammar wallaby an effective and convenient marsupial model for further comparative immunology studies focusing on macropodids. Australia's marsupial species are unique and valuable, both to Australia's diverse ecosystems and for comparative studies. Furthering our knowledge of the macropodid immune system is important for understanding the immune response of endangered and less-studied marsupial species threatened by disease, predation and habitat fragmentation.

ACKNOWLEDGMENTS

We thank the ARC and the University of Sydney for funding our immunogenetics work. HS is supported by a University of Sydney scholarship. CS is supported by a scholarship from the ARC Centre for Kangaroo Genomics and the Jean Walker Trust.

REFERENCES

Adamski FM and Demmer J 2000. Immunological protection of the vulnerable marsupial pouch young: two periods of immune transfer during lactation in *Trichosurus vulpecula* (brushtail possum). *Dev. Comp. Immunol.* **24:** 491–502.

Amills M, Jimenez N, Jordana J, Riccardi A, Fernandez-Arias A, Guiral J, Bouzat JL, Folch J and Sanchez A 2004. Low diversity in the major histocompatibility complex class II DRB1 gene of the Spanish ibex, *Capra pyrenaica*. *Heredity* **93:** 266–272.

Ashman RB and Papadimitriou JM 1975. The effect of thymectomy on the lymphoid tissue of the quokka (*Setonix brachyurus*). *Aust. J. Exp. Biol. Med. Sci.* **53:** 129–136.

Asquith KL, Kitchener AL and Kay DJ 2006. Immunisation of the male tammar wallaby (*Macropus eugenii*) with spermatozoa elicits epididymal antigen-specific antibody secretion and compromised fertilisation rate. *J. Reprod. Immunol.* **69:** 127–147.

Aveskogh M and Hellman L 1998. Evidence for an early appearance of modern post-switch isotypes in mammalian evolution; cloning of IgE, IgG and IgA from the marsupial *Monodelphis domestica. Eur. J. Immunol.* **28:** 2738–2750.

Baker ML, Gemmell E and Gemmell RT 1999. Ontogeny of the immune system of the brushtail possum, *Trichosurus vulpecula. Anat. Rec.* **256:** 354–365.

Baker ML, Rosenberg GH, Zuccolotto P, Harrison GA, Deane EM and Miller RD 2001. Further characterization of T cell receptor chains of marsupials. *Dev. Comp. Immunol.* **25:** 495–507.

Baker ML, Osterman AO and Brumburgh S 2005. Divergent T cell receptor delta chains from marsupials. *Immunogenetics* **57:** 665–673.

Basden K, Cooper DW and Deane EM 1997. Development of the lymphoid tissues of the tammar wallaby *Macropus eugenii. Reprod. Fertil. Dev.* **9:** 243–254.

Baudinette RV, Boontheung P, Musgrave IF, Wabnitz PA, Maselli VM, Skinner J, Alewood PF, Brinkworth CS and Bowie JH 2005. An immunomodulator used to protect young in the pouch of the tammar wallaby, *Macropus eugenii. Febs. J.* **272:** 433–443.

Belov K, Harrison GA and Cooper DW 1998. Molecular cloning of the cDNA encoding the constant region of the immunoglobulin A heavy chain (C alpha) from a marsupial: *Trichosurus vulpecula* (common brushtail possum). *Immunol. Lett.* **60:** 165–170.

Belov K, Harrison GA, Miller RD and Cooper DW 1999. Isolation and sequence of a cDNA coding for the heavy chain constant region of IgG from the Australian brushtail possum, *Trichosurus vulpecula. Mol. Immunol.* **36:** 535–541.

Belov K, Zenger KR, Hellman L and Cooper DW 2002. Echidna IgA supports mammalian unity and traditional Therian relationship. *Mamm. Genome* **13:** 656–663.

Belov K, Lam MK and Colgan DJ 2004. Marsupial MHC class II beta genes are not orthologous to the eutherian beta gene families. *J. Hered.* **95:** 338–345.

Belov K, Deakin JE, Papenfuss AT, Baker ML, Melman SD, Siddle HV, Gouin N, Goode DL, Sargeant TJ, Robinson MD, Wakefield MJ, Mahony S, Cross JG, Benos PV, Samollow PB, Speed TP, Graves JA and Miller RD 2006. Reconstructing an ancestral mammalian immune supercomplex from a marsupial major histocompatibility complex. *PLoS Biol.* **4:** e46.

Belov K, Sanderson CE, Deakin JE, Wong ES, Assange D, McColl KA, Gout A, de Bono B, Barrow AD, Speed TP, Trowsdale J and Papenfuss AT 2007. Characterization of the opossum immune genome provides insights into the evolution of the mammalian immune system. *Genome Res.* **17:** 982–991.

Bininda-Emonds OR, Cardillo M, Jones KE, MacPhee RD, Beck RM, Grenyer R, Price SA, Vos RA, Gittleman JL and Purvis A 2007. The delayed rise of present-day mammals. *Nature* **446:** 507–512.

Browning TL, Belov K, Miller RD and Eldridge MD 2004. Molecular cloning and characterization of the polymorphic MHC class II DBB from the tammar wallaby (*Macropus eugenii*). *Immunogenetics* **55:** 791–795.

Cheng Y, Wakefield M, Siddle HV, Coggill P, Herbert CA, Beck S, Belov K and Eldridge MDB 2009a. Isolation and characterization of 10 MHC Class I-associated microsatellite loci in the tammar wallaby (*Macropus eugenii*). *Mol. Ecol. Res.* **9:** 346–349.

Cheng Y, Siddle HV, Coggill P, Beck S, Eldridge MDB and Belov K 2009b. High levels of genetic variation at MHC class II DBB loci in the tammar wallaby (*Macropus eugenii*). *Immunogen.* **61:** 111–118.

Cowan PE 1989. Changes in milk composition during lactation in the common brushtail possum, *Trichosurus vulpecula* (Marsupialia: Phalangeridae). *Reprod. Fertil. Dev.* **1:** 325–335.

Croix DA, Samples NK, Vandeberg JL and Stone WH 1989. Immune response of a marsupial (*Monodelphis domestica*) to sheep red blood cells. *Dev. Comp. Immunol.* **13:** 73–78.

Cross JG, Harrison GA, Coggill P, Sims S, Beck S, Deakin JE and Graves JA 2005. Analysis of the genomic region containing the tammar wallaby (*Macropus eugenii*) orthologues of MHC class III genes. *Cytogenet. Genome Res.* **111:** 110–117.

Daly K, Church WB, Nicholas K and Williamson P 2007a. Comparative modeling of marsupial MHC class I molecules identifies structural polymorphisms affecting functional motifs. *J. Exp. Zool. A. Ecol. Genet. Physiol.* **307:** 611–624.

Daly KA, Digby M, Lefevre C, Mailer S, Thomson P, Nicholas K and Williamson P 2007b. Analysis of the expression of immunoglobulins throughout lactation suggests two periods of immune transfer in the tammar wallaby (*Macropus eugenii*). *Vet. Immunol. Immunopathol.* **120:** 187–200.

Daly KA, Digby MR, Lefevre C, Nicholas KR, Deane EM and Williamson P 2008a. Identification, characterization and expression of cathelicidin in the pouch young of tammar wallaby (*Macropus eugenii*). *Comp. Biochem. Physiol. B. Biochem. Mol. Biol.* **149:** 524–533.

Daly KA, Lefevre C, Nicholas K, Deane E and Williamson P 2008b. CD14 and TLR4 are expressed early in tammar (*Macropus eugenii*) neonate development. *J. Exp. Biol.* **211:** 1344–1351.

Deakin JE and Cooper DW 2004. Characterisation of and immunity to the aerobic bacteria found in the pouch of the brushtail possum *Trichosurus vulpecula. Comp. Immunol. Microbiol. Infect. Dis.* **27:** 33–46.

Deakin JE, Belov K, Curach NC, Green P and Cooper DW 2005. High levels of variability in immune response using antigens from two reproductive proteins in brushtail possums. *Wildl. Res.* **32:** 1–6.

Deakin JE, Papenfuss AT, Belov K, Cross JG, Coggill P, Palmer S, Sims S, Speed TP, Beck S and Graves JA 2006a. Evolution and comparative analysis of the MHC Class III inflammatory region. *BMC Genomics* **7:** 281.

Deakin JE, Parra ZE, Graves JA and Miller RD 2006b. Physical mapping of T cell receptor loci (TRA@, TRB@, TRD@ and TRG@) in the opossum (*Monodelphis domestica*). *Cytogenet. Genome. Res.* **112:** 342K.

Deakin JE, Siddle HV, Cross JG, Belov K and Graves JA 2007. Class I genes have split from the MHC in the tammar wallaby. *Cytogenet. Genome Res.* **116:** 205–211.

De Maeyer E and De Maeyer-Guignard J 1998. Type I interferons. *Int. Rev. Immunol.* **17:** 53–73.

Harrison GA and Deane EM 1999. cDNA sequence of the lymphotoxin beta chain from a marsupial, *Macropus eugenii* (tammar wallaby). *J. Interferon Cytokine Res.* **19:** 1099–1102.

Harrison GA and Deane EM 2000. cDNA cloning of lymphotoxin alpha (LT-alpha) from a marsupial, *Macropus eugenii. DNA Seq.* **10:** 399–403.

Harrison G, Taylor C, Miller R and Deane E 2003a. Primary structure and variation of the T-cell receptor delta-chain from a marsupial, *Macropus eugenii. Immunol. Lett.* **88:** 117–125.

Harrison GA, Young LJ, Watson CM, Miska KB, Miller RD and Deane EM 2003b. A survey of type I interferons from a marsupial and monotreme: implications for the evolution of the type I interferon gene family in mammals. *Cytokine* **21:** 105–119.

Hawken RJ, Maccarone P, Toder R, Graves JAM and Maddox JF 1999. Isolation and characterization of marsupial IL5 genes. *Immunogen.* **49:** 942–948.

Holland OJ, Cowan PE, Gleeson DM and Chamley LW 2008. High variability in the MHC class II DA beta chain of the brushtail possum (*Trichosurus vulpecula*). *Immunogen.* In press.

Infante AJ, Samples NK, Croix DA, Redding TS, VandeBerg JL and Stone WH 1991. Cellular immune response of a marsupial, *Monodelphis domestica. Dev. Comp. Immunol.* **15:** 189–199.

Kelley J, Walter L and Trowsdale J 2005. Comparative genomics of major histocompatibility complexes. *Immunogen.* **56:** 683–695.

Kitchener AL, Edds LM, Molinia FC and Kay DJ 2002. Porcine zonae pellucidae immunization of tammar wallabies (*Macropus eugenii*): fertility and immune responses. *Reprod. Fertil. Dev.* **14:** 215–223.

Lam MK, Belov K, Harrison GA and Cooper DW 2001. Cloning of the MHC class II DRB cDNA from the brushtail possum (*Trichosurus vulpecula*). *Immunol. Lett.* **76:** 31–36.

McKenzie LM and Cooper DW 1994. Low MHC class II variability in a marsupial. *Reprod. Fertil. Dev.* **6:** 721–726.

Meyer-Lucht Y, Otten C, Puttker T and Sommer S 2008. Selection, diversity and evolutionary patterns of the MHC class II DAB in free-ranging neotropical marsupials. *BMC Genet.* **9:** 39.

Miller RD and Belov K 2000. Immunoglobulin genetics of marsupials. *Dev. Comp. Immunol.* **24:** 485–490.

Miska KB, Wright AM, Lundgren R, Sasaki-McClees R, Osterman A, Gale JM and Miller RD 2004. Analysis of a marsupial MHC region containing two recently duplicated class I loci. *Mamm. Genome* **15:** 851–864.

O'Huigin C, Sultmann H, Tichy H and Murray BW 1998. Isolation of mhc class II DMA and DMB cDNA sequences in a marsupial: the gray short-tailed opossum (*Monodelphis domestica*). *J. Mol. Evol.* **47:** 578–585.

Old JM and Deane EM 1998. The effect of oestrus and the presence of pouch young on aerobic bacteria isolated from the pouch of the tammar wallaby, *Macropus eugenii. Comp. Immunol. Microbiol. Infect. Dis.* **21:** 237–245.

Old JM and Deane EM 2000. Development of the immune system and immunological protection in marsupial pouch young. *Dev. Comp. Immunol.* **24:** 445–454.

Old JM and Deane EM 2001. Histology and immunohistochemistry of the gut-associated lymphoid tissue of the eastern grey kangaroo, *Macropus giganteus. J. Anat.* **199:** 657–662.

Old JM and Deane EM 2002. Immunohistochemistry of the lymphoid tissues of the tammar wallaby, *Macropus eugenii. J. Anat.* **201:** 257–266.

Old JM, Deane EM and Harrison GA 2001. Molecular characterisation of the tammar wallaby (*Macropus eugenii*) CD3 epsilon chain cDNA *Mol. Immunol.* **38:** 359–364.

Parra ZE, Baker ML, Schwarz RS, Deakin JE, Lindblad-Toh K and Miller RD 2007. A unique T cell receptor discovered in marsupials. *Proc. Natl Acad. Sci. USA* **104:** 9776–9781.

Schneider S, Vincek V, Tichy H, Figueroa F and Klein J 1991. MHC class II genes of a marsupial, the red-necked wallaby (*Macropus rufogriseus*): identification of new gene families. *Mol. Biol. Evol.* **8:** 753–766.

Siddle HV, Deakin JE, Baker ML, Miller RD and Belov K 2006. Isolation of major histocompatibility complex Class I genes from the tammar wallaby (*Macropus eugenii*). *Immunogenetics* **58:** 487–493.

Siddle HV, Kreiss A, Eldridge MDB, Noonan E, Clarke CJ, Pyecroft S, Woods GM and Belov K 2007a. Transmission of a fatal clonal tumor by biting occurs due to depleted MHC diversity in a threatened carnivorous marsupial. *Proc. Natl Acad. Sci. USA* **104:** 16221–16226.

Siddle HV, Sanderson C and Belov K 2007b. Characterization of major histocompatibility complex class I and class II genes from the Tasmanian devil (*Sarcophilus harrisii*). *Immunogen.* **59:** 753–760.

Slade RW and Mayer WE 1995. The expressed class II alpha-chain genes of the marsupial major histocompatibility complex belong to eutherian mammal gene families. *Mol. Biol. Evol.* **12:** 441–450.

Stanley NF, Yadav M, Waring H and Eadie M 1972. The effect of thymectomy on response to various antigens of a marsupial *Setonix brachyurus* (Quokka). *Aust. J. Exp. Biol. Med. Sci.* **50:** 689–702.

Stone WH, Bruun DA, Manis GS, Holste SB, Hoffman ES, Spong KD and Malunas TL 1996. The immunobiology of the marsupial *Monodelphis domestica*. SOS Publications, Fairhaven NJ.

Stone WH, Brunn DA, Foster EB, Manis GS, Hoffman ES, Saphire DG, VandeBerg JL and Infante AJ 1998. Absence of a significant mixed lymphocyte reaction in a marsupial (*Monodelphis domestica*). *Lab. Anim. Sci.* **48:** 184–189.

Terszowski G, Muller SM, Bleul CC, Blum C, Schirmbeck R, Reimann J, Pasquier LD, Amagai T, Boehm T and Rodewald HR 2006. Evidence for a functional second thymus in mice. *Science* **312:** 284–287.

Tyndale-Biscoe CH and Renfree MB 1987. *Reproductive Physiology of Marsupials*. Cambridge University Press, Cambridge.

Wakefield MJ and Graves JA 2003. The kangaroo genome: leaps and bounds in comparative genomics. *EMBO Rep.* **4:** 143–147.

Wilkinson R, Kotlarski I and Barton M 1994. Further characterisation of the immune response of the koala. *Vet. Immunol. Immunopathol.* **40:** 325–339.

Wong ES, Young LJ, Papenfuss AT and Belov K 2006. In silico identification of opossum cytokine genes suggests the complexity of the marsupial immune system rivals that of eutherian mammals. *Immunome Res.* **2:** 4.

Yadav M 1971. The transmissions of antibodies across the gut of pouch-young marsupials. *Immunol.* **21:** 839–851.

Yadav M, Stanley NF and Waring H 1972a. The microbial flora of the gut of the pouch-young and the pouch of a marsupial, *Setonix brachyurus*. *J. Gen. Microbiol.* **70:** 437–442.

Yadav M, Stanley NF and Waring H 1972b. The thymus glands of a marsupial, *Setonix brachyurus* (quokka), and their role in immune responses. Structure and growth of the thymus glands. *Aust. J. Exp. Biol. Med. Sci.* **50:** 347–356.

Young LJ and Deane EM 2005. Culture and characterisation of peripheral blood monocytes and monocyte-derived adherent cells of the tammar wallaby, *Macropus eugenii*. *Immunol. Lett.* **96:** 253–259.

Young LJ and Deane EM 2007. Culture and stimulation of tammar wallaby lymphocytes. *Vet. Res. Commun.* **31:** 685–701.

Young LJ, McFarlane R, Slender AL and Deane EM 2003. Histological and immunohistological investigation of the lymphoid tissue in normal and mycobacteria-affected specimens of the rufous hare-wallaby (*Lagorchestes hirsutus*). *J. Anat.* **202:** 315–325.

Zuccolotto PD, Harrison GA and Deane EM 2000. Cloning of marsupial T cell receptor alpha and beta constant region cDNAs. *Immunol. Cell Biol.* **78:** 103–109.

3 Marsupials as models for mammalian pre-natal development: the tammar wallaby

D. Hickford

SUMMARY

Marsupials, as well as the mouse, provide a model for studies of developmental biology. Their embryos develop in a planar manner and are not covered by maternal or extraembryonic cell lineages until amnion formation. Attachment to the uterus occurs during mid-somitogenesis and gestation is usually short, with the birth of altricial young. Furthermore, the long period of divergence between marsupials and eutherians provides an opportunity to study the conservation of various developmental mechanisms and signals. Most studies of marsupial postnatal development have utilised the macropodid *Macropus eugenii* (tammar wallaby). *M. eugenii* has several features that make it an attractive model for studies of pre-natal development, particularly as the sequencing of its genome is now in progress. To be useful as a model for developmental biology, a species must be able to be cultured *in vitro* through much of gestation. Embryos or fetuses from a small number of marsupial species have been cultured, some with considerable success. This chapter outlines progress in the culture of marsupial embryos and fetuses, with particular emphasis on *M. eugenii*, and describes some of the experiments made possible by the ability to maintain pre-natal stages *in vitro*. Such experiments not only contribute to knowledge of marsupial development but have the potential to be applied to mammalian development in general.

INTRODUCTION

Most research on mammalian pre-natal development uses the mouse as a model because of the ease of labelling of cell lineages, micro-surgical ablation of cells or embryo tissues and transplantation of cells or tissues (Tam 1998). All these techniques require the ability to culture embryos *in vitro* for at least a short time. Culture methods for maintaining pre-implantation mouse and rat embryos *in vitro* are well established and mouse zygotes can be cultured until the blastocyst stage (Nagy *et al.* 2003). Culture techniques for human pre-implantation conceptuses are also well established (Pool 2004), a result of extensive research coinciding with the development of assisted reproductive techniques such as *in vitro* fertilisation (IVF) and intra-cytoplasmic sperm injection (ICSI).

Concerted attempts at the culture of pre-natal stages have really only been attempted for the mouse and rat (Cockroft 1990, 1997) and to a lesser extent the rabbit (Naya *et al.* 1991; Ninomiya *et al.* 1991).

Mouse embryos can routinely be cultured from early primitive streak formation until early somitogenesis (Quinlan *et al.* 1995; Tam 1998) and from early somitogenesis until the early hind limb bud stage (Tam 1998; Trainor *et al.* 1994), in equal quantities of complex medium (Dulbccco's Modified Eagle's Medium, DMEM) and rat serum in 5% CO_2 in air. However, the success rate for culturing mouse blastocysts through gastrulation to early somitogenesis is less than 50%, despite extensive testing (Hsu 1978; Libbus and Hsu 1980). Culturing rodent embryos through organogenesis is also difficult because of the failure of the chorioallantoic placenta to develop and survive well enough in culture to meet all the embryo's nutritional needs (Tam 1998).

With the established manipulative techniques, its well-characterised genetics and ready availability, the mouse is an obvious choice for studies of pre-natal mammalian development. However, the mouse embryo develops within a complicated egg cylinder that precludes viewing of the embryo itself. It also implants early in gestation, which can complicate separation of maternal and embryonic signalling during development. Finally, although the results of research in mouse embryos are applied to other mammals, it is often not known how applicable they are to other mammals, nor how conserved the various developmental pathways are among mammals.

Marsupials make excellent additional models for studying pre-natal mammalian development. In contrast to the mouse, the marsupial embryo develops in a planar manner on the surface of the yolk sac (Fig. 3.1) and is not covered by any extraembryonic cell lineages until amnion formation. Thus it is readily visible for most of its development, including the times at which the earliest cell lineages appear and axis formation and gastrulation occur. Another advantage of marsupials is that axis formation occurs well before attachment of the embryo to the uterus (Selwood 2001), making it easy to separate maternal and embryo-derived signals. Furthermore, because most marsupials do not develop a chorioallantoic placenta but obtain most of their nutrients through the yolk sac, they should be easier to maintain *in vitro*. Another advantage of marsupials is that they give birth to altricial young: marsupial neonates are developmentally equivalent to eutherian late fetuses (Nelson 1992) and therefore the post-natal stages provide a good opportunity to study processes that occur during late organogenesis. Finally, marsupials and eutherians diverged more than 125 million years ago (Luo *et al.* 2003), so any developmental pathways conserved between the two are likely to be conserved among mammals in general.

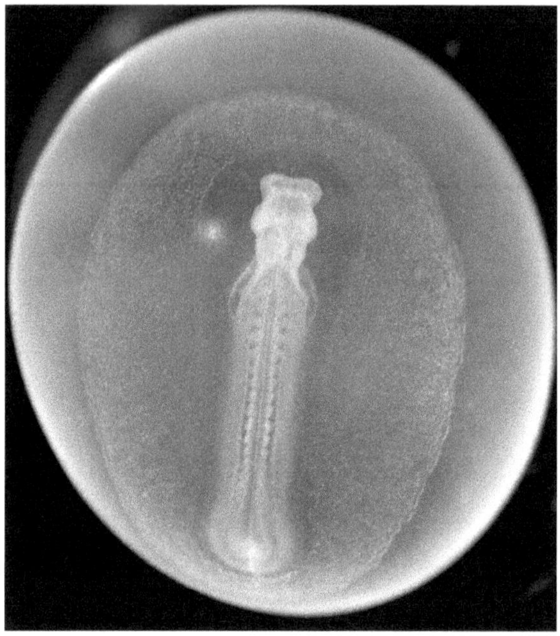

Figure 3.1: A tammar wallaby (*M. eugenii*) embryo with 15 somites and heart tubes, collected on day 16 after removal of pouch young. Total diameter of embryo and yolk sac is 1.26 cm.

MARSUPIAL PRE-NATAL DEVELOPMENT: AN OVERVIEW

The ovulated oocyte is encased within a zona pellucida. During its passage along the oviduct it acquires a mucoid coat and, at the utero–tubal junction and the uterus, a shell coat (Tyndale-Biscoe and Renfree 1987). The mucoid coat becomes attenuated during development, disappearing before gastrulation. The shell is lost just before attachment of the embryo to the uterus (Selwood 2000). Fertilisation occurs in the oviduct and cleavage begins a short time later, in the uterus.

In the species in which cleavage has been studied in detail – the dasyurids *Sminthopsis macroura* (stripe-faced dunnart), *Antechinus stuartii* (brown antechinus), *Dasyurus viverrinus* (eastern quoll) and *Sminthopsis crassicaudata* (fat-tailed dunnart), the didelphids *Didelphis virginiana* (Virginia opossum) and *Monodelphis domestica* (gray short-tailed opossum), the phalangerid *Trichosurus vulpecula* (common brushtail possum) and the peramelid *Perameles nasuta* (long-nosed bandicoot) – cell–zona adhesion precedes cell–cell adhesion (Hill 1910; Hartman 1916; Lyne and Hollis 1976; Selwood and Young 1983; Selwood 1987; Baggott and Moore 1990; Selwood and Smith 1990; Frankenberg and Selwood 1998). In contrast, in the macropodid *Macropus eugenii* (tammar wallaby) there is apparently no separation of blastomeres before the eight-cell stage (Renfree and Lewis 1996). In all species cleavage yields a unilaminar blastocyst – a hollow ball of cells with no inner cell mass (Tyndale Biscoe and Renfree 1987). The first two-cell populations formed are pluriblast, which gives rise to the embryo proper as well as to extraembryonic lineages, and trophoblast, which forms the outermost cell layer of the extraembryonic yolk sac. The unilaminar blastocyst becomes bilaminar with the appearance of extraembryonic hypoblast from the pluriblast. The hypoblast cells enter the blastocoele and proliferate until they totally line the blastocoele, forming the complete bilaminar blastocyst (Selwood 1992). With the appearance of the hypoblast the residual pluriblast becomes known as epiblast, which gives rise to all embryonic lineages as well as the extraembryonic mesoderm of the trilaminar yolk sac (Selwood 1992; Selwood and Hickford 1999).

The embryonic area (epiblast population) of the bilaminar blastocyst is round and uniform in appearance until immediately before primitive streak formation, when it elongates and becomes pyriform in shape. The most complete descriptive studies of marsupial peri-gastrulation development include those of *Bettongia cuniculus* (Tasmanian bettong) (Kerr 1936), *D. virginiana* (Hartman 1919, 1928; McCrady 1938) and *S. macroura* (Hickford and Selwood 2003). These studies and others of isolated gastrulation stages in various marsupial species suggest that the morphology and histology of early marsupial gastrulation is quite similar to that in the chick (Hickford and Selwood 2003; Selwood and Hickford 1999). Migration of extraembryonic mesoderm from the primitive streak out into the proximal yolk sac results in formation of the trilaminar (vascular) yolk sac. The extraembryonic mesoderm extends over at least half the yolk sac in all species examined (Freyer *et al.* 2003); the remaining (distal) yolk sac is therefore the bilaminar (avascular) yolk sac. During somitogenesis the embryo retains its planar arrangement, whereby it is continuous with the trilaminar yolk sac and thus readily visible. Amnion formation occurs during mid–late somitogenesis and shell loss and subsequent attachment to the uterus occurs not long after. Organogenesis in marsupials is rapid and neonates exhibit precocious development of anterior structures needed to reach the pouch. Many processes of later organogenesis, such as differentiation of the eye, hind limbs and gonads, occur after birth (Nelson 1992; Hughes and Hall 1988; Gemmell and Rose 1989).

CULTURE OF NON-MACROPODID MARSUPIAL EMBRYOS AND FETUSES

Attempts have been made to culture all stages of marsupial embryonic development, some with more success than others. Oocytes from only two species, *M. domestica* and *S. macroura*, have been matured *in vitro*. Oocytes from Graafian follicles from both species have been cultured through maturity to the release of the first, and in some cases second, polar body (Breed *et al.* 1994; Maleszewski and Selwood 2004; Selwood and VandeBerg 1992). *M. domestica* zygotes cultured *in vitro* can develop to the early expanding blastocyst stage (Baggott and Moore 1990; Moore and Taggart 1993; Selwood *et al.* 1997; Selwood and VandeBerg 1992). Three dasyurid species, *S. macroura*, *A. stuartii* and *S. crassicaudata*, have also been cultured during cleavage. All three species have a developmental block at the four-cell stage *in vitro* and only *S. macroura* can be cultured through this block from zygote to expanding blastocyst (Gardner *et al.* 1996; Selwood 1987; Selwood and Smith 1990). *A. stuartii* and *S. crassicaudata* can be cultured from the zygote to the four-cell stage and from the four-cell stage to early blastocysts (Selwood 1986, 1987; Selwood and Smith 1990; Selwood and Young 1983). Culture of all four species has been successful using DMEM with high glucose and 10% fetal calf serum (FCS) in 5% CO_2 in air. DMEM with 10% FCS has also been used to culture isolated cleavage stages of *T. vulpecula* from the late four-cell to the eight-cell and from the 16- to 32-cell stage (Casey and Selwood 2003).

Unilaminar blastocysts from *A. stuartii*, *S. macroura*, *M. domestica* and *T. vulpecula* have been cultured in various media including DMEM, Whittingham's T6 medium (WT6) and Monomed medium supplemented with 5–10% FCS in 5% CO_2 in air but none developed any further *in vitro*, although late unilaminar blastocysts of *T. vulpecula* did undergo some expansion (Casey and Selwood 2003; Selwood *et al.* 1997; Yousef and Selwood 1993).

Dasyurid unilaminar blastocysts last less than six hours before collapsing (Yousef and Selwood 1993). Early bilaminar blastocysts of *S. macroura* and *A. stuartii* do not develop *in vitro* either, suggesting that formation of the hypoblast lineage requires some as yet unidentified maternal signals (Selwood *et al.* 1997). Bilaminar blastocysts of *S. macroura* and *A. stuartii* do develop further *in vitro* once their future embryonic region has started to elongate, with most initiating primitive streak formation and some undergoing early somitogenesis (Yousef and Selwood 1993).

Primitive streak stage *S. macroura* and *A. stuartii* embryos can be cultured to fetal stages in DMEM with 10% FCS in 5% CO_2 in air using static culture. Late primitive streak stage *S. macroura* embryos can develop to C-shaped fetuses with cervical and thoracic flexures within 30 hours *in vitro*, developing at a rate approximately four hours slower than *in vivo* (Yousef and Selwood 1993). In a study by New *et al.* (1977), primitive streak stage embryos of *D. virginiana* did not fare as well *in vitro*, developing only three to five somites over 24 hours in '199' tissue culture medium with 20% opossum serum or straight opossum serum in 5% CO_2/10% oxygen using roller culture. Embryos of this stage have normally developed 12–13 somites within 24 hours *in vivo* (McCrady 1938). *D. virginiana* embryos with eight to 10 somites cultured in 95% oxygen developed beating hearts but showed little evidence of embryonic growth or yolk sac expansion (New and Mizell 1972). More advanced embryos (12–20 somites) fared better, forming beating hearts, fore limb buds and lumbar flexure. Amnion development continued and blood circulation was maintained for 12–18 hours (New *et al.* 1977).

D. virginiana fetuses have been maintained *in vitro* in various media with 20% opossum serum or in straight opossum serum, in either 20% or 95% oxygen using roller culture. Development in 95% oxygen was superior to that in 20% oxygen (New and Mizell 1972). The fetuses ranged in stage from those with ~28 somites, fore limb paddles, an almost complete amnion and an early allantois to those with fore limbs with claws and hind limb paddles. In most cases blood circulation was maintained for approximately 12 hours, fore and hind limbs continued to differentiate, the allantois expanded and somite number increased (New *et al.* 1977; New and Mizell 1972). *S. macroura* fetuses have also been maintained *in vitro* using roller culture, in 5% CO_2 in air in DMEM with 10–30% FCS (Cruz *et al.* 2000). Good development was attained with early fetuses that initially had fore limb ridges and an early amniotic head fold; these fetuses continued to develop, completing amnion formation and developing fore limb paddles after 24 hours *in vitro*. More advanced fetuses (those with a complete amnion and subsequent stages thereafter) progressed in culture but did not develop as well as the earlier stages, possibly because the later stages may require higher oxygen concentrations *in vitro*.

CULTURE OF MACROPODID EMBRYOS AND FETUSES

There is little published information on the culture of macropodid embryos and fetuses. *M. eugenii* oocytes can be cultured from the germinal vesicle stage to metaphase II (Mate and Buist 1999; Mate and Rodger 1993; Richings 2004). Mature oocytes collected from Graafian follicles have been fertilised *in vitro* and cultured through early cleavage in DMEM (Richings *et al.* 2004) (Fig. 3.2). *In vivo* fertilised *M. eugenii* oviductal or uterine oocytes cultured in DMEM initiated cleavage, with most reaching the four-cell stage within 24 hours (Renfree and Lewis 1996). Only a small proportion of these conceptuses cleaved to the eight-cell stage, suggesting a 'four-cell block' *in vitro*, similar to that in the dasyurids. However, like *S. macroura*, *M. eugenii* conceptuses can be cultured through this block, as two-cell but not one-cell stages can cleave to the eight-cell stage. All embryos initially at the four- or eight-cell stage formed early blastocysts *in vitro* within 24–48 hours (Renfree and Lewis 1996).

Macropus eugenii unilaminar blastocysts in diapause have been cultured in phosphate-buffered saline supplemented with glucose and bovine serum albumin for a few hours to investigate aspects of their metabolism (Spindler *et al.* 1995, 1998, 1999). Diapausing blastocysts of *M. eugenii* and *Macropus rufus* (red kangaroo) have also been maintained in 0.9% saline or Krebs-Ringer gassed with 95% oxygen/5% CO_2, in which they survive for up to a week without collapsing (Renfree and Tyndale-Biscoe 1978). Culture in various tissue culture media supplemented with serum accentuates blastocyst shrinkage (Renfree and Tyndale-Biscoe 1978); when cultured in DMEM with 10% FCS blastocysts shrink

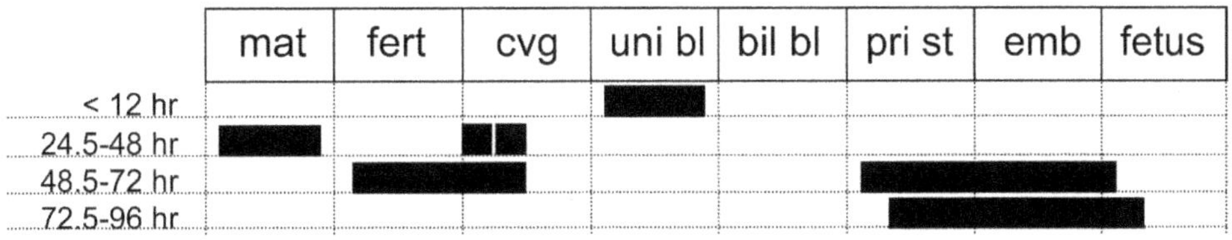

Figure 3.2: The developmental stages at which tammar (*M. eugenii*) conceptuses have been cultured. The time in culture is specified in the column on the left. Mat = oocyte maturation, fert = fertilisation, cvg = cleavage, uni bl = unilaminar blastocyst, bil bl = bilaminar blastocyst, pri st = primitive streak, emb = embryo. Sources: Mate and Buist (1999), Richings *et al.* (2004), Spindler *et al.* (1998, 1999), Renfree and Lewis (1996), Hickford *et al.* (2008) and unpublished information from Renfree and Trounson and from Tyndale-Biscoe and Hinds.

within two days (Renfree and Lewis 1996). Similar to dasyurid unilaminar blastocysts, diapausing blastocysts of *M. eugenii* do not progress in culture. However, once reactivated, unilaminar blastocysts of *M. eugenii* do expand significantly *in vitro* (M.B. Renfree, unpubl. obs.).

Early primitive streak and head fold stage *M. eugenii* embryos have been cultured using roller culture in Whittingham's T6 (WT6) medium supplemented with 10% wallaby serum (WS) and gassed with 95% oxygen. These embryos developed to the head fold and late somite stages (with heart formation and fore limb buds) respectively after two to three days *in vitro* (M.B. Renfree and A.O. Trounson, *pers. comm.*). Pre-somite to early somite stages have also been cultured in Medium 199 with 20% FCS gassed with 95% oxygen in roller culture for an average of four days. At the termination of culture these embryos had a partially closed amnion, numerous (20+) somites, fore limb buds and beating hearts (C.H. Tyndale-Biscoe and L. Hinds, unpubl. obs.; L. Hinds, *pers. comm.*), reaching a similar end stage to the embryos cultured in WT6 medium. Peri-gastrulation stage *M. eugenii* embryos also fare well in static culture. Primitive streak/regressing node stage embryos cultured in 95% oxygen/5% CO_2 in either DMEM with 10% WS or Medium 199 with 10% WS survived for up to 78 hours *in vitro*, with several developing numerous somites that had differentiated into sclerotome, myotome and dermamyotome, beating heart tubes, fore limb ridges and mesonephric ducts (Hickford *et al.* 2008).

MANIPULATIONS AND EXPERIMENTS INVOLVING MARSUPIAL EMBRYOS AND FETUSES

The experimental work on marsupial conceptuses can be divided into two categories. The first category includes the experimental techniques developed to provide a means by which specific questions can be addressed and answered. Such techniques include the development of methods for IVF, artificial insemination (AI), *in vitro* culture of various developmental stages (as outlined above) and blastocyst transfer. Once refined, these techniques can be utilised or modified to allow specific aspects of development to be investigated. The second category includes experiments undertaken to answer specific questions, and all experiments undertaken to date have utilised the ability to culture marsupial embryos *in vitro*.

IVF has been successfully achieved for two marsupial species. Oocytes of *M. domestica* were fertilised by culturing them together with epididymal sperm, and the resulting zygotes were cultured until they reached the 16- to 32-cell stage (Moore and Taggart 1993; Taggart *et al.* 1993). In contrast, oocytes of *M. eugenii* have been fertilised *in vitro* using ICSI (Magarey and Mate 2003; Richings *et al.* 2004) but not by co-culture with sperm (Renfree and Lewis 1996). Zygotes fertilised by ICSI have been cultured through to the eight-cell stage, although cleavage-stage embryos created from follicular oocytes exhibited greater separation between blastomeres than occurs *in vivo* (Richings *et al.* 2004). The ability to fertilise marsupial oocytes *in vitro* has applications for the management of endangered species. From an experimental point of view, the ability to successfully inject material into oocytes or embryos is the first step towards creating transgenic animals. To this end, preliminary efforts have been made to inject *M. eugenii* embryos with a lentivirus construct (M.B. Renfree, A.J. Pask and R.R. Behringher, *pers. comm.*). Artificial insemination has also been successful in *M. eugenii*, with two pouch young resulting from intrauterine AI using electro-ejaculated spermatozoa (Paris *et al.* 2005). Live births following AI have also been recorded in *Phascolarctos cinereus* (koala) (Johnston *et al.* 2003). Successful AI has not been reported in any other marsupial species.

Blastocyst transfers have been successfully carried out in two marsupial species – *M. eugenii* (Renfree 1972; Renfree and Tyndale-Biscoe 1978; Tyndale-Biscoe 1970) and *Setonix brachyurus* (quokka) (Tyndale-Biscoe 1963). The ability to manipulate embryos and transplant them to a surrogate mother to complete their development *in vivo* solves the problem of not being able to culture certain stages *in vitro*. For example, one of the first experimental manipulations undertaken on marsupial embryos *in vitro* was the destruction of some blastomeres of cleaving *A. stuartii* conceptuses (Selwood 1986). The aim was to ascertain whether the separation of embryonic and extraembryonic cell lineages was due to some cleavage divisions being determinant, as had been suggested by earlier investigators. The question is still unanswered because of the inability to culture *A. stuartii* conceptuses through the four-cell block or the unilaminar blastocyst stage.

A previously unresolved issue in marsupial developmental biology concerned the existence of a medullary plate. Based on examination of a very limited number of specimens of *D. virginiana*, McCrady (1938) stated that separate allocation events produce first the neuroectoderm and later the integumentary ectoderm, all before notochord formation. This issue was solved by marking various areas of primitive streak stage *S. macroura* embryos with the lipophilic cell surface marker DiI and following the fate of these cells after 24 hours *in vitro* – the first attempt at fate mapping a marsupial (Cruz *et al.* 1996). Different regions of the so-called medullary plate gave rise to both neuroectoderm and integumentary ectoderm, showing that at least in dasyurids (and probably in all marsupials) there is no medullary plate.

Marsupial embryos can also be dissociated into single cells or cut into fragments and then cultured. In most species the unilaminar blastocyst epithelium has a uniform appearance throughout, leading to it being regarded as a protoderm by many researchers (Selwood 1996). To show that dasyurid unilaminar blastocysts in fact consist of two cell populations (pluriblast and trophoblast), unilaminar and bilaminar blastocysts of *S. macroura* and *A. stuartii* were separated into embryonic and abembryonic hemispheres (based on the association of the yolk with the embryonic hemisphere) then dissociated. The dissociated cells were maintained *in vitro* for up to eight weeks. Pluriblast and trophoblast cells could be distinguished from each other and from hypoblast cells (from bilaminar blastocysts) *in vitro*. During the culture period the pluriblast cells and the majority of trophoblast cells remained undifferentiated and did not proliferate (Yousef and Selwood 1996). It would be interesting to expand this research to see if differentiation of specific cell lineages could be promoted by exposing the cells to various growth factors and signalling molecules.

To investigate whether the pre-gastrulation round embryonic area (epiblast) is pre-patterned before any overt morphological differences become obvious, embryonic areas of bilaminar blastocysts of *S. macroura* were cut into four to six pieces and cultured individually in DMEM with various exogenous growth factors and signalling molecules. Several explants responded differently to some of the treatments, suggesting that there is some degree of pre-patterning (Hickford and Selwood 2003).

The form of the peri-gastrulation stage marsupial embryo (a planar embryo developing on the surface of the extraembryonic lineages) is advantageous for observing development but severely limits the types of manipulations that can be performed. Primitive streak stage *S. macroura* embryos can sustain numerous injections of DiI through the shell coat as long as care is taken (Cruz *et al.* 1996; D. Hickford, pers. obs.). However, techniques such as cell or tissue transplantation or grafting are not possible because if the yolk sac or embryo is punctured carelessly the embryo collapses and does not reinflate. To overcome this problem, a method of culturing the embryonic area as an explant has been devised. The embryonic area is removed by cutting around it, leaving a border of yolk sac, and the explant is laid hypoblast side down on filter paper or a coverslip. In the original study using *S. macroura* embryos (Hickford and Selwood 2003), the explants were kept flat by using a piece of chick vitelline membrane weighed down with a stainless steel ring. This technique has been refined using *M. eugenii* embryos. The explants are now cultured just below the surface of the medium on collagen-coated membrane on top of a metal grid in an organ culture dish, using the resulting surface tension of the medium to keep the embryo flat. Embryos cultured this way develop much better, probably because of the higher oxygen concentration available to them (D. Hickford, unpubl. obs.). This culture technique means that researchers can undertake cell and tissue grafts, transplantations and the localised application of exogenous growth factors, signalling molecules or neutralising antibodies by implanting beads soaked in the relevant factor into specific regions of the embryo.

Isolated embryonic area explants of primitive streak stage *S. macroura* embryos have also been cultured within Matrigel, a synthetic basement membrane matrix. After 24–48 hours *in vitro* the explants resemble murine embryoid bodies (Hickford and Selwood 2003). The similarity of these explants to embryoid bodies suggests that they may be a possible source of marsupial embryonic stem cells.

CONCLUSIONS

Macropus eugenii has proved a valuable model for the study of post-natal development by contributing numerous valuable insights into mammalian sex determination and differentiation, sex reversal and virilisation (Renfree *et al.* 2006; Renfree 2006). *M. eugenii* also makes an excellent model for studies of pre-natal development, despite being monovular. It has the best characterised reproductive biology of any marsupial (Tyndale-Biscoe and Renfree 1987) and breeds readily in captivity. Embryos or fetuses of the desired stage can be obtained by removing the suckling pouch young, reactivating development of the diapausing blastocyst (Tyndale-Biscoe and Renfree 1987). The greatest advantage of *M. eugenii* is that its genome is being sequenced, which dramatically increases its value as a model for developmental biology.

Most of the investigative experiments performed on marsupial embryos have used the dasyurid *S. macroura*. The advantages of *S. macroura* include a short (10.7 day) gestation (Selwood and Woolley 1991), a well-characterised developmental timetable (Selwood and Hickford 1999) and the ability to obtain many embryos of the desired stage from a single animal. Several characteristics of *M. eugenii* complement some of the shortfalls of *S. macroura*. First, as noted above, the genome of *M. eugenii* is being sequenced. Second, *M. eugenii* has a longer gestation period than *S. macroura*, which has a highly compressed period of organogenesis. In *S. macroura*, development from the first somites through to heart tube formation, body flexures and fore limb paddles all occur on day 9 of gestation (Cruz *et al.* 2000). While this can be advantageous for some applications (e.g. culture), the extremely fast rate of development can make it difficult to separate the signals required for a particular developmental event within that period. In *M. eugenii* the same events occur over approximately two to three days, providing more time to separate any signals and to better characterise various events. Third, *M. eugenii* embryos are larger – early somite stages are approximately 11–14 mm in diameter (Tyndale-Biscoe and Renfree 1987), compared to 3.5–4.5 mm for *S. macroura* at the same stage (Selwood and Hickford 1999). The larger size of *M. eugenii* is advantageous for many experimental manipulations and means that all the embryo can be visualised at once; the smaller size and resulting greater axis of curvature of *S. macroura* embryos during peri-gastrulation prevents viewing of the entire embryo from any one angle.

In conclusion, marsupials make excellent additional models to the mouse for studies of mammalian pre-natal development, especially by combining information gained from the study of the two most suitable marsupial species – *M. eugenii* and *S. macroura*.

ACKNOWLEDGMENTS

Thank you to Marilyn Renfree for her helpful comments on the manuscript. Thank you also to Marilyn Renfree, Lyn Hinds, Hugh Tyndale-Biscoe and Alan Trounson for sharing their unpublished data.

REFERENCES

Baggott LM and Moore HDM 1990. Early embryonic development of the grey short-tailed opossum, *Monodelphis domestica, in vivo* and *in vitro*. *J. Zool., Lond.* **222:** 623–639.

Breed WG, Simerly C, Navara CS, VandeBerg JL and Schatten G 1994. Microtubule configurations in oocytes, zygotes, and early embryos of a marsupial, *Monodelphis domestica*. *Dev. Biol.* **164:** 230–240.

Casey N and Selwood L 2003. Removal of the shell coat affects maintenance of epithelia in blastocysts of the brushtail possum *in vitro*. *Cells Tissues Organs* **173:** 21–45.

Cockroft DL 1990. Dissection and culture of postimplantation mouse embryos. In *Postimplantation Mammalian Embryos: A Practical Approach*. (Eds AJ Copp and DL Cockroft) pp. 15–40. IRL Press, Oxford.

Cockroft DL 1997. A comparative and historical review of culture methods for vertebrates. *Int. J. Dev. Biol.* **41:** 127–137.

Cruz YP, Yousef A and Selwood L 1996. Fate map analysis of the epiblast of the dasyurid marsupial *Sminthopsis macroura* (Gould). *Reprod. Fertil. Dev.* **8:** 779–788.

Cruz YP, Hickford D and Selwood L 2000. A staging scheme for assessing development *in vitro* of organogenesis stage embryos of the stripe-faced dunnart, *Sminthopsis macroura* (Marsupialia: Dasyuridae). *J. Reprod. Fertil.* **120:** 99–108.

Frankenberg S and Selwood L 1998. An ultrastructural study of the role of an extracellular matrix during normal cleavage in a marsupial, the brushtail possum. *Molec. Reprod. Dev.* **50:** 420–433.

Freyer C, Zeller U and Renfree MB 2003. The marsupial placenta: a phylogenetic analysis. *J. Exp. Zool. A Comp. Exp. Biol.* **299:** 59–77.

Gardner DK, Selwood L and Lane M 1996. Nutrient uptake and culture of *Sminthopsis macroura* (stripe-faced dunnart) embryos. *Reprod. Fertil. Dev.* **8:** 685–690.

Gemmell RT and Rose RW 1989. Organ development in some newborn marsupials with particular reference to the rat kangaroos. In *Kangaroos, Wallabies and Rat-kangaroos.* (Eds I Hume and G Grigg) pp. 349–354. Australian Mammal Society/Surrey Beatty and Sons, Sydney.

Hartman CG 1916. Studies in the development of the opossum *Didelphis virginiana*. I. History of the early cleavage. II. Formation of the blastocyst. *J. Morphol.* **27:** 1–83.

Hartman CG 1919. Studies in the development of the opossum *Didelphis virginiana*. III. Description of new material on maturation, cleavage and endoderm formation. IV. The bilaminar blastocyst. *J. Morphol.* **32:** 1–139.

Hartman CG 1928. The breeding season of the opossum (*Didelphis virginiana*), and the rate of intrauterine and postnatal development. *J. Morphol.* **46:** 143–215.

Hickford D and Selwood L 2003. Peri-gastrulation development of the dasyurid marsupial *Sminthopsis macroura* (stripe-faced dunnart) *in vitro* and evidence for patterning of the epiblast prior to gastrulation. *Molec. Reprod. Dev.* **65:** 402–419.

Hickford D, Shaw G and Renfree MB 2008. *In vitro* culture of peri-gastrulation embryos of a macropodid marsupial. *J. Anat.* **212:** 180–191.

Hill JP 1910. The early development of the marsupialia, with special reference to the native cat (*Dasyurus viverrinus*). Contributions to the embryology of the marsupialia IV. *Q. J. Microsc. Sci.* **56:** 1–134.

Hsu YC 1978. *In vitro* development of whole mouse embryos beyond the implantation stage. In *Methods in Mammalian Reproduction.* (Ed JC Daniel Jr) pp. 229–245. Academic Press, New York.

Hughes RL and Hall LS 1988. Structural adaptations of the newborn marsupial. In *The Developing Marsupial.* (Eds CH Tyndale-Biscoe and PA Janssens) pp. 8–27. Springer, Berlin.

Johnston SD, McGowan MR, O'Callaghan P, Cox R, Houlden N, Haig S and Taddeo G 2003. Birth of koalas at the Lone Pine Koala Sanctuary following artificial insemination. *Int. Zoo. YB.* **38:** 160–172.

Kerr T 1936. On the primitive streak and associated structures in the marsupial *Bettongia cuniculus. Q. J. Microsc. Sci.* **78:** 687–715.

Libbus BL and Hsu YC 1980. Sequential development and tissue organization in whole mouse embryos cultured from blastocyst to early somite stage. *Anat. Rec.* **97:** 317–329.

Lyne AG and Hollis DE 1976. Early embryology of the marsupials *Isoodon marcourus* and *Perameles nasuta. Aust. J. Zool.* **24:** 361–382.

Luo Z-X, Ji Q, Wible JR and Yuan CX 2003. An early cretaceous tribosphenic mammal and metatherian evolution. *Science* **302:** 1934–1940.

Magarey GM and Mate KE 2003. Fertilization following intracytoplasmic sperm injection of *in vivo* and *in vitro* matured oocytes from an Australian marsupial, the tammar wallaby (*Macropus eugenii*). *Zygote* **11:** 339–346.

Maleszewski M and Selwood L 2004. Induced parthenogenetic activation of oocytes of the marsupial *Sminthopsis macroura. Reprod. Fertil. Dev.* **16:** 599–604.

Mate KE and Buist JM 1999. Timing and regulatory aspects of oocyte maturation *in vitro* in the tammar wallaby (*Macropus eugenii*). *Reprod. Fertil. Dev.* **11:** 247–254.

Mate KE and Rodger JC 1993. *In vitro* maturation of oocytes from a marsupial, the tammar wallaby (*Macropus eugenii*). *J. Molec. Reprod. Dev.* **34:** 329–336.

McCrady E Jr. 1938. The embryology of the opossum. *Am. Anat. Mem.* **16:** 1–233.

Moore HDM and Taggart DA 1993. *In vitro* fertilization and embryo culture in the grey short-tailed opossum, *Monodelphis domestica. J. Reprod. Fertil.* **98:** 267–274.

Nagy A, Gertsenstein M, Vintersten K and Behringer R 2003. *Manipulating the Mouse Embryo: A Laboratory Manual.* 3rd edn. Cold Spring Harbor Laboratory Press, New York.

Naya M, Kito Y, Eto K and Deguchi T 1991. Development of rabbit whole embryo culture during organogenesis. *Cong. Anom.* **31:** 153–156.

Nelson J 1992. Developmental staging in a marsupial *Dasyurus hallucatus. Anat. Embryol.* **185:** 335–354.

New DAT and Mizell M 1972. Opossum fetuses grown in culture. *Science* **175:** 533–536.

New DAT, Mizell M and Cockroft DL 1977. Growth of opossum embryos *in vitro* during organogenesis. *J. Exp. Morphol.* **41:** 111–123.

Ninomiya H, Kishida K, Ohno Y and Eto K 1991. Development of rabbit whole embryo culture techniques. *Teratology* **44:**14.

Paris DBBP, Taggart DA, Shaw G, Temple-Smith PD and Renfree MB 2005. Birth of pouch young after artificial insemination in the tammar wallaby (*Macropus eugenii*). *Biol. Reprod.* **72:** 451–459.

Pool TB 2004. Development of culture media for human assisted reproductive technology. *Fertil. Steril.* **81:** 287–289.

Quinlan GA, Williams EA, Tan S-S and Tam PPL 1995. Neuroectodermal fate of epiblast cells in the distal region of the egg cylinder: implication for body plan organization during early embryogenesis. *Development* **121:** 87–98.

Renfree MB 1972. Influence of embryo on marsupial uterus. *Nature* **240:** 475–477.

Renfree MB 2006. Life in the pouch: womb with a view. *Reprod. Fertil. Dev.* **18:** 721–734.

Renfree MB and Lewis AMcD 1996. Cleavage *in vivo* and *in vitro* in the marsupial *Macropus eugenii. Reprod. Fertil. Dev.* **8:** 725–742.

Renfree MB and Tyndale-Biscoe CH 1978. Manipulation of marsupial embryos and pouch young. In *Methods of Mammalian Reproduction.* (Ed JC Daniel) pp. 307–331. Academic Press, London.

Renfree MB, Pask AJ and Shaw G 2006. Sexual development of a model marsupial male. *Aust. J. Zool.* **54:** 151–158.

Richings N 2004. Growth, development and maturation of the marsupial follicle and oocyte. PhD thesis. University of Melbourne, Melbourne, Australia.

Richings NM, Shaw G, Temple-Smith PD and Renfree MB 2004. Intra-cytoplasmic sperm injection in a marsupial. *Reproduction* **128:** 595–605.

Selwood L 1986. Cleavage *in vitro* following destruction of some blastomeres in the marsupial *Antechinus stuartii* (Macleay). *J. Embryol. Exp. Morphol.* **92:** 71–84.

Selwood L 1987. Embryonic development in culture of two dasyurid marsupials, *Sminthopsis crassicaudata* (Gould) and *Sminthopsis macroura* (Spencer), during cleavage and blastocyst formation. *Gamete Res.* **16:** 355–370.

Selwood L 1992. Mechanisms underlying the development of pattern in marsupial embryos. *Curr. Topics Dev. Biol.* **27:** 175–233.

Selwood L 1994. Development of early cell lineages in marsupial embryos: an overview. *Reprod. Fertil. Dev.* **6:** 507–527.

Selwood L 1996. The blastocyst epithelium is not a protoderm in dasyurid marsupials: a review of the evidence. *Reprod. Fertil. Dev.* **8:** 711–723.

Selwood L 2000. Marsupial egg and embryo coats. *Cells Tissues Organs* **166:** 208–219.

Selwood L 2001. Mechanisms for pattern formation leading to axis formation and lineage allocation in mammals: a marsupial perspective. *Reproduction* **121:** 677–683.

Selwood L and Hickford D 1999. Early cell lineages in marsupial embryos. In *Cell Lineage and Fate Determination.* (Ed. SA Moody) pp. 505–519. Academic Press, San Diego.

Selwood L and Smith D 1990. Time lapse analysis and normal stages of development of cleavage and blastocyst formation in the marsupials the brown antechinus and the stripe-faced dunnart. *Molec. Reprod. Dev.* **26:** 53–62.

Selwood L and VandeBerg J 1992. The influence of incubation temperature on oocyte maturation, parthenogenetic and embryonic development *in vitro* of the marsupial *Monodelphis domestica. Anim. Reprod. Sci.* **29:** 99–116.

Selwood L and Woolley P 1991. A timetable of embryonic development and ovarian and uterine changes during pregnancy, in the stripe-faced dunnart, *Sminthopsis macroura* (Marsupialia: Dasyuridae). *J. Reprod. Fertil.* **91:** 213–227.

Selwood L and Young GJ 1983. Cleavage *in vivo* and culture in the dasyurid marsupial *Antechinus stuartii* (Macleay). *J. Morphol.* **176:** 43–60.

Selwood L, Robinson ES, Pedersen RA and VandeBerg JL 1997. Development *in vitro* of marsupials: a comparative review of species and a timetable of cleavage and early blastocyst stages of development in *Monodelphis domestica. Int. J. Dev. Biol.* **41:** 397–410.

Spindler RE, Renfree MB and Gardner DK 1995. Metabolic assessment of wallaby blastocysts during embryonic diapause and subsequent reactivation. *Reprod. Fertil. Dev.* **7:** 1157–1162.

Spindler RE, Renfree MB, Shaw G and Gardner DK 1998. Reactivating tammar wallaby blastocysts oxidize glucose. *Biol. Reprod.* **58:** 1425–1431.

Spindler RE, Renfree MB, Shaw G and Gardner DK 1999. Reactivating tammar wallaby blastocysts oxidize fatty acids and amino acids. *J. Reprod. Fertil.* **115:** 79–86.

Taggart DA, O'Brien HP and Moore HDM 1993. Ultrastructural characteristics of *in vivo* and *in vitro* fertilization in the grey short-tailed opossum, *Monodelphis domestica. Anat. Rec.* **237:** 21–37.

Tam PPL 1998. Postimplantation mouse development: whole embryo culture and micro-manipulation. *Int. J. Dev. Biol.* **42:** 895–902.

Trainor PA, Tan S-S and Tam PPL 1994. Cranial somitomeres: regionalisation of cell fate and impact on craniofacial development in mouse embryos. *Development* **120:** 2397–2408.

Tyndale-Biscoe CH 1963. Blastocyst transfer in the marsupial *Setonix brachyurus. J. Reprod. Fertil.* **6:** 41–48.

Tyndale-Biscoe CH 1970. Resumption of development by quiescent blastocysts transferred to primed, ovariectomized recipients in the marsupial, *Macropus eugenii. J. Reprod. Fertil.* **23:** 25–32.

Tyndale-Biscoe H and Renfree MB 1987. *Reproductive Physiology of Marsupials.* Cambridge University Press, Cambridge.

Yousef A and Selwood L 1993. Embryonic development in culture of the marsupials *Antechinus stuartii* (Macleay) and *Sminthopsis macroura* (Spencer) during preimplantation stages. *Reprod. Fertil. Dev.* **5:** 445–458.

Yousef A and Selwood L 1996. The type and differentiation of cells *in vitro* from unilaminar and bilaminar blastocysts of two marsupials, *Antechinus stuartii* and *Sminthopsis macroura. Reprod. Fertil. Dev.* **8:** 743–752.

4 Population genetic studies of the Macropodoidea: a review

M.D.B. Eldridge, M.P. Piggott and S.L. Hazlitt

SUMMARY

Recent advances have transformed the field of population genetics and led to a dramatic renaissance in the application of these techniques to the study of ecology and conservation. We review the progress made in applying these tools, notably hypervariable genetic markers (microsatellite loci and mitochondrial control region) and new statistical methods, to the study of wild macropodoid populations. Population genetic studies have been conducted on 20 macropodoid species, more species than any other marsupial group, although most studies have focused on threatened species or those with restricted distributions. A substantial number of microsatellite loci (n = 79) have been isolated from seven macropodoid species, and have been used to successfully amplify polymorphic loci in representatives of all three extant macropodoid families and most genera. Levels of microsatellite diversity identified within the 61 macropodoid populations so far examined have varied enormously, although the range of values is similar to those reported from eutherians and other vertebrates. Macropodoid populations with high levels of genetic diversity tended to be those from relatively abundant or widespread species, while those with lower levels were typically from species with restricted distributions, that are threatened or found on islands or have been established via translocation. The level of genetic structuring among macropodoid populations was variable, but consistent with known species biology. Limited structure has been documented in widespread vagile species such as kangaroos, while moderate or high amounts of genetic structure were found in habitat specialists, such as rock-wallabies. Studies to date suggest that dispersal for macropodoids is male-biased, which is broadly consistent with behavioural observations. Although the power to detect fine-scale genetic substructuring within a population has only recently become available, two studies have independently detected a pattern of differential spatial structuring between male and female rock-wallabies within single colonies. This suggests strong female philopatry and that within-colony dispersal is also male-biased. Data on the genetic mating systems of macropodoids in the wild are available for only four species. These studies have largely confirmed behavioural-based mating system hypotheses but have also shed light on alternative mating strategies. The potential of hypervariable genetic markers to investigate ecological patterns has been powerfully demonstrated within wild macropodoid populations. These studies have improved our understanding of macropodoid population biology, behaviour and reproduction, as well as informing conservation initiatives and management plans for many threatened macropodoid taxa. It is hoped that, in the future, population genetic studies can be expanded to include a larger number of widespread and abundant macropodoid species, especially those in northern Australia that have been less severely affected by European settlement.

INTRODUCTION

The unique biology and iconic status of the marsupial superfamily Macropodoidea (kangaroos, wallabies and rat-kangaroos) has made them of enduring international scientific interest and obvious candidates for ongoing research initiatives. Population genetic studies of the Macropodoidea have, however, traditionally been hampered, as have studies of other mammals, by the difficulty of detecting sufficient genetic variation. Throughout the 1970s and 1980s a series of allozyme electrophoresis studies revealed that macropodoids contained the relatively low levels of genetic diversity (average heterozygosity $He = 0.046 \pm 0.013$ from 14 species) typically detected with this technique (Sherwin and Murray 1990). The studies made some progress in clarifying species boundaries and determining broad-scale population structure (Richardson and Sharman 1976; Johnston and Sharman 1977, 1979; Seebeck and Johnston 1980).

During the late 1980s and early 1990s a series of discoveries and innovations transformed the field of molecular genetics and led to a dramatic renaissance in the study of population genetics. Of central importance was the development of the polymerase chain reaction (PCR) (Arnheim *et al.* 1990; Avise 1994) which enabled usable quantities of specifically targeted parts of the genome to be amplified from tiny amounts of starting DNA material. The second key development was the identification of hypervariable parts of the genome (both nuclear and mitochondrial) which could now be accessed, and variation assessed, via PCR (Sunnucks 2000). Coupled with this dramatic increase in the information content of genetic data were equally innovative developments in methods of data analysis (Luikart and England 1999; Excoffier and Heckel 2006; Marjoram and Tavare 2006), allowing the potential of these data to be more fully realised.

Our purpose in this chapter is to summarise the progress in applying the most commonly used of these new generation of hypervariable genetic markers (microsatellies and the mtDNA control region) to wild macropodoid populations and to survey how these studies have enhanced our knowledge of macropodoid biology, conservation and management.

MICROSATELLITES

Microsatellites, tandem repeats of DNA sequences two to five base pairs in length, have become the marker of choice in population genetics studies over the past decade. They are hypervariable in repeat length, widespread in the genome and mostly show codominant Mendelian inheritance and selective neutrality (Bruford *et al.* 1996; Jarne and Lagoda 1996; Oliveira *et al.* 2006). To date, 79 polymorphic microsatellite loci have been isolated from seven macropodoid species, including a total of nine from two potoroid species and 70 from five species of macropodid (Table 4.1). Over half (44/79, 56%) have been isolated from a single species, the tammar wallaby (*Macropus eugenii*). This bias is likely to intensify with *M. eugenii* being regarded as a model marsupial, as well as the first, and currently the only, Australasian marsupial to have its entire genome sequenced (Alsop *et al.* 2005). Of the polymorphic microsatellites isolated to date, the majority (65/79, 82%) have been autosomal (Zenger *et al.* 2002; MacDonald *et al.* 2007), eight have been located on the X chromosome (Zenger *et al.* 2002; MacDonald *et al.* 2007) and six on the Y chromosome (MacDonald *et al.* 2006). Of the loci identified, 44 have been integrated, with other markers, into nine linkage groups as part of efforts to establish a comprehensive genetic linkage map for the *M. eugenii* genome (Zenger *et al.* 2002). Another 21 loci are known from specific chromosomes (MacDonald *et al.* 2006, 2007).

Although microsatellite loci have been isolated from only seven of the ~70 macropodoid species, these loci have exhibited remarkable cross-species utility (Zenger *et al.* 2003b) and have successfully amplified homologous polymorphic loci in a wide range of macropodoid species, including representatives of all three extant families and 11 of the 15 extant macropodoid genera (Table 4.1). Even though a decrease in successful cross-species amplification with increasing genetic distance has been observed, as well as some reduction in diversity in non-source species (Zenger *et al.* 2003b), the loci that are currently available appear sufficiently robust and versatile to enable population genetic studies for most macropodoid species. Microsatellite-based population genetic studies have been conducted on only 18 macropodoid species (Table 4.2). Relatively few widespread or common species have been examined, with most studies focusing on threatened species or those with relatively restricted distributions. Despite these limitations, considerably more species of macropodoids have been examined with microsatellites than any other marsupial group.

Table 4.1: Polymorphic microsatellite loci isolated from macropodoids and their cross-species utility in other macropodoid taxa.

Family	Source taxon	No. of loci	Reference	Non-source taxa in which loci are polymorphic	Reference
Potoroidae	*Potorous longipes*	6	Luikart *et al.* 1997	*Bettongia*, other *Potorous*	Luikart *et al.* 1997; Donaldson and Vercoe 2008
	Bettongia tropica	3	Pope *et al.* 2000	*Aepyprymnus*, other *Bettongia*, *Hypsiprymnodon*	Johnson *et al.* 2005; Pope *et al.* 2005; Donaldson and Vercoe 2008
Macropodidae	*Macropus eugenii*	44[a]	Taylor and Cooper 1998; Hawken *et al.* 1999; Zenger and Cooper 2001a; MacDonald *et al.* 2006, 2007	*Bettongia, Dendrolagus, Hypsiprymnodon, Lagorchestes*, other *Macropus, Petrogale, Thylogale, Wallabia*	Heckenberg 1997; Taylor and Cooper 1998; Eldridge *et al.* 1999; Bowyer *et al.* 2002; Eldridge *et al.* 2004a; Johnson *et al.* 2005; Paplinska 2005; Donaldson and Vercoe 2008
	Macropus giganteus	9	Zenger and Cooper 2001b	*Hypsiprymnodon*, other *Macropus*, *Petrogale, Wallabia*	Eldridge *et al.* 2001b; Zenger *et al.* 2003a; Johnson *et al.* 2005; Paplinska 2005
	Petrogale xanthopus	7	Pope *et al.* 1996; Pope *pers. comm.*[b]	*Aepyprymnus, Bettongia, Dendrolagus, Lagorchestes, Macropus, Onychogalea*, other *Petrogale*	Eldridge *et al.* 1999; Pope *et al.* 2000; Bowyer *et al.* 2002; Zenger *et al.* 2003a; Eldridge *et al.* 2004a; Pope *et al.* 2005; Sigg *et al.* 2005; Donaldson and Vercoe 2008
	Petrogale assimilis	6	Spencer *et al.* 1995; Spencer 1996	*Aepyprymnus, Dendrolagus, Macropus*, other *Petrogale, Thylogale, Wallabia*	Heckenberg 1997; Eldridge *et al.* 1999; Bowyer *et al.* 2002; Zenger *et al.* 2003a; Paplinska 2005; Pope *et al.* 2005
	Onychogalea fraenata	4	Pope *et al.* 2000; Moritz *pers. comm.*[b]	*Aepyprymnus, Bettongia, Dendrolagus, Hypsiprymnodon, Macropus*	Pope *et al.* 2000; Bowyer *et al.* 2002; Zenger *et al.* 2003a; Johnson *et al.* 2005; Pope *et al.* 2005

a: An additional nine monomorphic loci have been described (MacDonald *et al.* 2006, 2007).

b: Details of loci given in Zenger *et al.* (2002) or Sigg *et al.* (2005).

Table 4.2: Allelic diversity (*AD*) and average heterozygosity (*He*) at polymorphic microsatellite loci, in 61 populations of 18 macropodoid taxa.

Taxon	*n*	No. of loci	*AD*	*He*	Reference
Macropus eugenii (Tutanning Nature Reserve)[a]	63	7	13.9	0.86	Eldridge *et al.* 2004a
Aepyprymnus rufescens (Mt Fox)	134	6	12.3	0.83	Pope *et al.* 2005
[T]*Onychogalea fraenata* (Taunton)[a]	100	7	11.7	0.81	Sigg 2006
Petrogale assimilis (Black Rock)	128	5	11.6	0.86	Spencer and Marsh 1997
Macropus eugenii (Kangaroo Island)	46	7	11.1	0.84	Taylor and Cooper 1999
Macropus giganteus (Tidbinbilla)	48	8	10.4	0.81	Zenger *et al.* 2003a
Macropus giganteus (Breadalbane)	29	8	10.1	0.80	Zenger *et al.* 2003a
Macropus rufogriseus (Gretna, Tasmania)	44	5	10.0	0.76	Le Page *et al.* 2000
Aepyprymnus rufescens (School House)	58	6	9.8	0.82	Pope *et al.* 2005
Aepyprymnus rufescens (Dingo Creek)	39	6	9.7	0.81	Pope *et al.* 2005
Macropus giganteus (South Bathurst)	35	8	9.6	0.81	Zenger *et al.* 2003a
Macropus rufogriseus (Deloraine, Tasmania)	26	5	9.2	0.79	Le Page *et al.* 2000
Macropus giganteus (Marulan)	15	8	8.9	0.85	Zenger *et al.* 2003a
Macropus rufogriseus (se mainland)[b]	15	5	8.8	0.87	Le Page *et al.* 2000
Thylogale s. stigmatica[b]	24	8	8.8	0.80	Heckenberg 1997
Thylogale thetis[b]	20	9	8.6	0.76	Heckenberg 1997
Petrogale assimilis (Little Black Rock)	15	5	8.6	0.85	Spencer *et al.* 1995
Aepyprymnus rufescens (Quartz Gum)	17	6	8.2	0.84	Pope *et al.* 2005
[T]*Petrogale x. xanthopus* (Flinders Ranges)[b]	69	7	7.8	0.75	Zenger *et al.* 2003b
[T]*Onychogalea fraenata* (Townsville, captive)[a]	34	7	7.6	0.79	Sigg 2006
Thylogale stigmatica wilcoxi[b]	18	8	7.6	0.76	Heckenberg 1997
Aepyprymnus rufescens (Loading Ramp)	21	6	7.5	0.78	Pope *et al.* 2005
Petrogale xanthopus celeris (Hill of Knowledge)	50	4	7.5	0.69	Pope *et al.* 1996
[T]*Bettongia tropica* (Davies Creek)	94	7	7.1	0.73	Pope *et al.* 2000
Macropus rufogriseus (Flinders Island)	28	5	7.0	0.71	Le Page *et al.* 2000
Aepyprymnus rufescens (Mt Claro)	17	6	6.7	0.82	Pope *et al.* 2005
Petrogale assimilis (Mt Stuart)	20	5	6.6	0.89	Spencer *et al.* 1995
Macropus eugenii (Kawau Is, New Zealand)[a]	37	7	6.3	0.62	Taylor and Cooper 1999
[T]*Onychogalea fraenata* (Idalia translocation)[a]	75	7	6.0	0.78	Sigg 2006
Macropus rufogriseus (Kempton, Tasmania)	16	5	6.0	0.74	Le Page *et al.* 2000
[T]*Bettongia tropica* (Emu Creek)	21	7	5.7	0.79	Pope *et al.* 2000
Dendrolagus lumholtzi[b]	45	10	5.3	0.60	Bowyer *et al.* 2002
[T]*Bettongia tropica* (Tinaroo)	20	7	5.2	0.68	Pope *et al.* 2000
[T]*Petrogale penicillata* (Ingles Rd)	28	11	5.2	0.70	Eldridge *et al.* 2004b
[T]*Petrogale penicillata* (Crocodile Rock)	15	7	5.1	0.65	Piggott *et al.* 2006b
[T]*Petrogale penicillata* (Rocky Creek)	17	7	5.1	0.65	Piggott *et al.* 2006b
[T]*Petrogale penicillata* (Bowmans Rd)	20	11	5.1	0.69	Eldridge *et al.* 2004b
[T]*Petrogale penicillata* (Wolgan, Main)	44	7	4.9	0.65	Piggott *et al.* 2006b

Taxon	n	No. of loci	AD	He	Reference
Macropus rufogriseus (New Zealand)[a]	44	5	4.6	0.59	Le Page *et al.* 2000
[T]*Petrogale penicillata* (Hurdle Creek)	54	12	4.5	0.67	Hazlitt *et al.* 2006a
[T]*Lagorchestes hirsutus* (Tanami Desert)[a]	15	8	4.3	0.61	Eldridge *et al.* 2004a
Macropus eugenii (Rotorua, New Zealand)[a]	61	7	4.1	0.60	Taylor and Cooper 1999
[T]*Petrogale penicillata* (Broke)	17	11	4.0	0.64	Eldridge *et al.* 2004b
Macropus parma (Kawau Island, New Zealand)[a]	51	2	4.0	0.55	Taylor *et al.* 1999
[T]*Petrogale l. lateralis* (Tutakin)[a]	19	11	4.0	0.62	Eldridge *et al.* 2004a
Macropus eugenii (Garden Island)	29	7	3.7	0.44	Eldridge *et al.* 2004a
[T]*Petrogale l. lateralis* (Sales Rock)[a]	31	11	3.5	0.55	Eldridge *et al.* 2004a
[T]*Petrogale penicillata* (Perseverance Dam)[a]	15	12	3.4	0.57	Hazlitt *et al.* 2006a
[T]*Petrogale l. lateralis* (Exmouth)[a]	15	10	3.4	0.62	Eldridge *et al.* 1999
[T]*Petrogale l. lateralis* (Mt Caroline East)[a]	32	11	3.3	0.51	Eldridge *et al.* 2004a
[T]*Petrogale penicillata* (Jenolan 1985, captive)[a]	20	11	3.2	0.50	Eldridge *et al.* 2004b
[T]*Petrogale l. lateralis* (Mt Caroline West)[a]	19	11	3.1	0.49	Eldridge *et al.* 2004a
[T]*Petrogale penicillata* (Kawau Is, New Zealand)[a]	18	11	2.6	0.44	Eldridge *et al.* 2001a
[T]*Potorous gilbertii* (Two Peoples Bay)[a]	25	5	2.6	0.46	Sinclair *et al.* 2002
[T]*Petrogale l. lateralis* (Nangeen Hill)[a]	19	11	2.5	0.41	Eldridge *et al.* 2004a
[T]*Petrogale penicillata* (Jenolan 1996, captive)[a]	10	11	2.5	0.45	Eldridge *et al.* 2004b
[T]*Petrogale penicillata* (Kangaroo Valley)[a]	19	12	2.4	0.37	Paplinska 2006
[T]*Petrogale l. lateralis* (Querekin)[c]	43	11	2.2	0.27	Eldridge *et al.* 2004a
[T]*Lagorchestes hirsutus* (Bernier Island)	17	8	1.8	0.25	Eldridge *et al.* 2004a
[T]*Petrogale penicillata* (Rocky Plains Creek)[a]	8	11	1.9	0.34	Browning *et al.* 2001
[T]*Petrogale l. lateralis* (Barrow Island)[a]	28	10	1.2	0.05	Eldridge *et al.* 1999

Data are presented only for studies where n > 15 or where the entire population was sampled. Populations are ordered by decreasing values of *AD*.

T: Threatened taxon (Maxwell *et al.* 1996).

a: Populations that have experienced a recent demographic bottleneck or a recent substantial range contraction (see Strahan 1995 or cited reference).

b: Samples pooled from multiple localities.

c: Population re-established in 1990 following the translocation of five animals from Nangeen Hill.

Genetic diversity at microsatellite loci

The levels of genetic diversity identified within the 61 macropodoid populations examined to date using microsatellite loci have varied enormously (Table 4.2). Allelic diversity (*AD*), the average numbers of alleles per locus, was found to vary from 1.2 to 13.9 (average ± se = 6.2 ± 0.3), while reported values of average heterozygosity (*He*) have ranged from 0.05 to 0.86 (average ± se = 0.66 ± 0.02). This range of values is similar to those reported from eutherians (Garner *et al.* 2005), non-macropodoid marsupials (Bowyer *et al.* 2002) and other vertebrates (DeWoody and Avise 2000). Garner *et al.* (2005) reported that on average marsupials had significantly higher heterozygosity values than eutherians (0.765 ± 0.002, n = 11 vs 0.677 ± 0.01, n = 74). However, the data presented here for macropodoids suggests that this relationship may not be sustained now that a larger number of marsupial populations have been studied.

As might be predicted, macropodoid populations with high levels of microsatellite diversity tend to be those sampled from relatively abundant or widespread species, while those with lower levels are typically populations of species with restricted distributions, that are threatened or found on islands or that have been established via translocation. Statistical analysis confirms that populations of threatened macropodoid species have significantly reduced levels of genetic diversity (*AD*; $t = 6.5$, $df = 57.3$, $p < 0.001$: *He*; $t = 5.1$, $df = 50.0$, $p < 0.001$), compared to non-threatened populations, while populations that have experienced a recent demographic bottleneck have significantly lower levels of genetic diversity (*AD*; $t = 3.9$, $df = 39.5$, $p < 0.001$: *He*; $t = 4.6$, $df = 34.3$, $p < 0.001$), compared to non-bottlenecked populations (Garner *et al.* 2005). Given the accumulating evidence that population bottlenecks and reductions in genetic diversity can significantly reduce population fitness (Frankham *et al.* 2002; Briskie and Mackingtosh 2004) and increase the likelihood of extinction (Saccheri *et al.* 1998; Frankham 2005), it is important that genetic factors be taken into account when managing threatened or bottlenecked macropodoid populations. The reduced diversity observed in these threatened species is most likely a direct consequence of their persistence at small population size for multiple generations, and is likely to make them more vulnerable to extinction in the future (Frankham *et al.* 2002). In addition, that populations of threatened macropodoids and other taxa (Spielman *et al.* 2004) show significantly reduced genetic diversity contradicts the proposition that environmental and demographic factors drive species to extinction before genetic factors can impact (Lande 1988; Caughley 1994).

Interestingly, the difference in levels of genetic diversity between bottlenecked and unbottlenecked macropodoid populations was highly significant ($p < 0.001$) despite bottlenecked populations of two species (bridled nailtail wallaby, *Onychogalea fraenata*, and *M. eugenii*) having levels of *AD* and *He* among the highest reported for macropodoids or other marsupials (Bowyer *et al.* 2002). These exceptions illustrate not only the strength of the underlying relationship but that, as population genetic theory predicts (Frankham *et al.* 2002), not all demographic bottlenecks are sufficiently severe or prolonged to result in a major loss of genetic diversity. Nevertheless, the overall pattern in the macropodoid (Table 4.2) and other data (Briskie and Mackingtosh 2004; Garner *et al.* 2005) is clear and it would be prudent to pay close attention to the genetic health of bottlenecked populations, which are often already of conservation concern.

Another trend evident in these data (Table 4.2) is that rock-wallaby (*Petrogale*) populations have significantly lower levels of genetic diversity (*AD*; $t = 4.7$, $df = 58.1$, $p < 0.001$: *He*; $t = 3.5$, $df = 43.7$, $p = 0.001$) than other macropodoids. This difference may relate to the highly patchy nature of preferred rock-wallaby habitat that naturally limits the size of individual populations, as well as restricted gene flow among habitat isolates (Hazlitt *et al.* 2006a; Piggott *et al.* 2006b). However, it could also be a sampling artefact since most rock-wallaby populations so far examined have been from southern Australia (*P. lateralis, P. penicillata, P. xanthopus*) and have been severely affected by hunting and by species introduced in the last 150 years (Maxwell *et al.* 1996). Further studies of rock-wallaby populations that have remained relatively unaffected by European settlement, such as in northern Australia, will be required to resolve this question.

MITOCHONDRIAL DNA

The advantages and limitations of mitochondrial DNA (mtDNA) for population and evolutionary studies have been extensively discussed (Moritz *et al.* 1987; Harrison 1989; Avise 1994; Moritz 1994a). Being maternally inherited, mtDNA allows the impact of female-mediated population processes to be investigated over both the short term (haplotype frequencies) and long term (sequence divergence and phylogeny) (Sunnucks 2000). Most population genetic studies that have utilised mtDNA have focused on the hypervariable, non-coding control region or D-loop (Taberlet 1996). In macropodoids the 5' end of this region (Domain 1) has typically been accessed using a set of conserved marsupial primers (developed by Fumagalli *et al.* 1997), with individual haplotypes being identified via direct sequencing or SSCP techniques (Sunnucks *et al.* 2000). These primers have been used successfully across a range of macropodoid species (six genera from two families; Table 4.3) and in other marsupials (Firestone *et al.* 1999; Fumagalli *et al.* 1999; Zenger *et al.* 2005). Compared to microsatellite studies, relatively few macropodoid species ($n = 18$ vs $n = 9$) or populations ($n = 61$ vs $n = 20$) have been examined for mtDNA variation, with most studies again focusing on populations of threatened or restricted range taxa rather than widespread or common species. The relatively small number of population genetic studies utilising mtDNA may be related to the greater difficulty and expense of assessing variation through sequencing a DNA fragment compared with the ease of high throughput microsatellite genotyping.

Table 4.3: Haplotype diversity (*h*) (Nei 1987) for mtDNA control region in 20 populations of nine macropodoid taxa.

Taxon	*n*	No. of haplotypes	*h*	Reference
[T]*Setonix brachyurus*[b]	21	21	1.00	Sinclair 2001
Macropus rufus[b]	34	31	0.99	Clegg *et al.* 1998
Macropus giganteus (South Bathurst)	35	8	0.91	Zenger *et al.* 2003a
Macropus giganteus (Tidbinbilla)	48	7	0.71	Zenger *et al.* 2003a
Macropus giganteus (Marulan)	15	7	0.76	Zenger *et al.* 2003a
Petrogale xanthopus celeris[b]	29	7	0.66	Pope *et al.* 1996
Macropus giganteus (Breadalbane)	29	5	0.60	Zenger *et al.* 2003a
[T]*Bettongia tropica* (Emu Creek)	20	4	0.63	Pope *et al.* 2000
[T]*Bettongia tropica* (Davies Creek)	20	3	0.53	Pope *et al.* 2000
Dendrolagus lumholtzi[b]	45	3	0.31	Bowyer *et al.* 2002
Macropus rufogriseus (New Zealand)[a]	40	3	-	Le Page *et al.* 2000
[T]*Potorous gilbertii* (Two Peoples Bay)[a]	23	3	0.24	Sinclair *et al.* 2002
[T]*Bettongia tropica* (Tinaroo)	20	1	0.0	Pope *et al.* 2000
[T]*Petrogale penicillata* (Ingles Rd)	28	1	0.0	Eldridge *et al.* 2004b
[T]*Petrogale penicillata* (Bowmans Rd)	20	1	0.0	Eldridge *et al.* 2004b
[T]*Petrogale penicillata* (Jenolan 1985, captive)	20	1	0.0	Eldridge *et al.* 2004b
[T]*Petrogale penicillata* (Kawau Is, New Zealand)[a]	18	1	0.0	Eldridge *et al.* 2001a
[T]*Petrogale penicillata* (Broke)	17	1	0.0	Eldridge *et al.* 2004b
[T]*Petrogale penicillata* (Jenolan 1996, captive)[a]	10	1	0.0	Eldridge *et al.* 2004b
[T]*Petrogale penicillata* (Rocky Plains Creek)[a]	8	1	0.0	Browning *et al.* 2001

All data were obtained from the 5′ end of the control region by sequencing and/or SSCP techniques (Sunnucks *et al.* 2000). Data are presented only for studies where $n \geq 15$ or where the entire population was sampled. Populations are ordered by decreasing values of *h*.

T: Threatened taxon (Maxwell *et al.* 1996).

a: Populations that have experienced a recent demographic bottleneck or a recent substantial range contraction (see Strahan 1995 or cited reference).

b: Samples pooled from multiple localities.

Genetic diversity at the mtDNA control region

Levels of mtDNA control region diversity identified within the 20 macropodoid populations examined to date vary considerably (Table 4.3). Haplotypic diversity (*h*) varies across its full range from 0.0 to 1.0 (average ± se = 0.39 ± 0.09), while the number of haplotypes identified ranged from one to 31 (average ± se = 5.5 ± 1.7). These values are similar to those reported in populations of other marsupials (Bowyer *et al.* 2002) and eutherians (Fernando *et al.* 2000; Nyakaana *et al.* 2002; Arrendal *et al.* 2004). The limited number of macropodoid population studies that have utilised the control region limits detailed and conclusive analysis. Analysis is further compromised by the fact that in four studies (including the two reporting the highest values) samples were pooled from multiple localities (Table 4.3), which is likely to inflate the value of diversity parameters. Although trends in the levels of genetic diversity appear similar to those observed for the microsatellite data, only differences between *h* for threatened vs non-threatened populations (t = 3.9, df = 16.6, p = 0.001), as well as differences between the number of haplotypes (t = 2.3, df = 12.8, p = 0.037) and *h* (t = 4.1, df = 17.0, p = 0.001) for rock-wallabies vs other macropodoids, were significant.

POPULATION STRUCTURE

Advances in sophisticated microsatellite and mtDNA analyses offer new insights into studying population structure and dispersal (Manel *et al.* 2005; Marjoram and Tavare 2006; Waples and Gaggiotti 2006). The ability to more accurately interpret population structure and dispersal via genetic methods or to incorporate genetic estimates with empirical data has greatly increased our understanding of macropodoid biology, as well as benefiting conservation and management.

With microsatellite data, a range of different genetic methods can be used to identify population boundaries. Traditional genetic approaches involve comparing the amount of allelic differentiation between pre-defined subpopulations, which may require prior basic knowledge about the species being studied. Recent developments in analytical tools use the individual as the sampling unit to detect genetic similarities between individuals so there is no pre-defined assumption of population structure (Pritchard *et al.* 2000; Waples and Gaggiotti 2006). Individual-based methods (e.g. assignment tests, see below) are also useful for detecting dispersal, as well as fine-scale population structure within a single population (Eldridge *et al.* 2001b; Berry *et al.* 2004; Manel *et al.* 2005).

In addition to microsatellite analysis, mtDNA can be used to interpret population structure. Haplotype frequency variation can be used to estimate gene flow and population subdivision over the short term, while haplotype sequence variation (and phylogeny) can elucidate long-term processes inferring the effects of historical events on population structure (Sunnucks 2000). The comparison of both mtDNA and microsatellite genotypes can provide valuable information which may not be detectable using a single marker (Sunnucks 2000).

Population structure (among populations)

The level of differentiation (F_{ST}) among populations is quite variable within macropodoids, ranging from weak (e.g. eastern grey kangaroo, *Macropus giganteus*) to strong (e.g. brush-tailed rock-wallaby, *Petrogale penicillata*) (Table 4.4). For *M. giganteus*, Zenger *et al.* (2003a) identified weak genetic structure (F_{ST}, Φ_{ST}) on a local (<50 km) and regional (50–230 km) scale in central New South Wales (NSW), suggesting high levels of gene flow and a panmictic population at those spatial scales. *M. giganteus* has a wide and almost continuous distribution down the east coast of Australia, which is likely to contribute to low genetic structure. However, there was evidence of significant differentiation (F_{ST}, Φ_{ST}) and mtDNA phylogenetic distinction between samples from NSW and Queensland, suggesting the existence of a contemporary or historical restriction to gene flow (Zenger *et al.* 2003a). The cause of this genetic differentiation is unclear, as there are few obvious physical barriers to dispersal. However, Zenger *et al.* (2003a) suggest that Queensland may have been colonised only recently by a limited number of southern animals, potentially contributing to the observed differentiation.

Using mtDNA sequence data, Clegg *et al.* (1998) found evidence for long-term connectivity among populations of the red kangaroo (*Macropus rufus*) across its distribution. *M. rufus* is a wide-ranging species distributed over large areas of Australia. Although phylogeographic structure of *M. rufus* was low across regions, there was evidence from mtDNA haplotype frequency data of significant population structure among populations within regions. Differences in the scale of population structuring were associated with rainfall variation and topographic and habitat complexity (Clegg *et al.* 1998).

Genetic differentiation among populations of the rufous bettong (*Aepyprymnus rufescens*) were also low but significant (F_{ST} = 0.016, Table 4.4), suggesting some gene flow up to 6.5 km (Pope *et al.* 2005). A study on the related but highly endangered northern bettong (*Bettongia tropica*) found evidence for significant genetic structuring (F_{ST} = 0.062) in populations with limited gene flow up to 9 km (Pope *et al.* 2000). Pope *et al.* (2005) suggested dispersal in *B. tropica* was one-dimensional compared to *A. rufescens* because *B. tropica* populations were present in narrow linear habitat, which would result in greater F_{ST} values. It should be noted that the comparison of F_{ST} values among such studies is potentially problematic, as values are dependent on the population's heterozygosity levels and populations in linear habitats are likely to violate some F_{ST} model assumptions (Hedrick 2005). In addition, divergent northern and southern mtDNA clades (4.5% sequence difference) have been identified within *B. tropica* (Pope et al. 2000). These clades meet within the Lamb Range population, suggesting recent secondary contact between previously isolated and genetically differentiated northern and southern populations of this species.

Studies on rock-wallabies have found moderate to high levels of genetic differentiation (F_{ST}, Φ_{ST}) among colonies (Table 4.4). Piggott *et al.* (2006a) found significant genetic substructuring (F_{ST} = 0.115) among four *P. penicillata* colonies that were less than 8 km

Table 4.4: Levels of population differentiation reported from population genetic studies of seven macropodoid species.

Species	n	No. of sites	Geographic scale	F_{ST} (microsatellites)	Φ_{ST} (mtDNA)	Reference
Macropus giganteus	171	8	<230 km	0.015[a]	0.185[a]	Zenger *et al.* 2003a
Macropus rufus	34	10	>1000 km	-	0.180Ψ	Clegg *et al.* 1998
Aepyprymnus rufescens	229	7	<12 km	0.016Ψ	-	Pope *et al.* 2005
Bettongia tropica	141	4	<10 km	0.062Ψ	0.561Ψ	Pope *et al.* 2000
Petrogale penicillata	85	4	<10 km	0.072[a]	-	Hazlitt *et al.* 2006a (within a single valley)
Petrogale penicillata	112	6	<100 km	0.139[a]	-	Hazlitt *et al.* 2006a (between valleys)
Petrogale penicillata	81	4	<10 km	0.115[a]	-	Piggott *et al.* 2006b
Petrogale xanthopus	74	3	10–70 km	0.107[a]	0.235[a]	Pope *et al.* 1996
Petrogale lateralis	181	7	<10 km	0.238[a]	-	Eldridge *et al.* 2001b

a: Mean pairwise.
Ψ: Overall.

apart. The results indicated that each colony should be regarded as an independent management unit (MU, *sensu* Moritz 1994b) and that the valley should not be managed as a single continuous population. Hazlitt *et al.* (2006b) detected moderate genetic differentiation (F_{ST} = 0.061 to 0.094) within valley systems and higher levels of differentiation among valley systems (F_{ST} = 0.169 to 0.230) of *P. penicillata* in south-east Queensland. Remnant Victorian *P. penicillata* populations in East Gippsland also show high levels of genetic structuring from both microsatellite and mtDNA analysis (Browning *et al.* 2001). However, the current level of genetic structuring in that region is likely to have been influenced by reduced gene flow, inbreeding and loss of genetic diversity via genetic drift as a result of population declines and habitat fragmentation since the 1900s (Browning *et al.* 2001).

Moderate and significant genetic differentiation has been detected among three Queensland colonies of the yellow-footed rock-wallaby (*Petrogale xanthopus*), 10–70 km apart, using four microsatellite loci (F_{ST} = 0.107; Pope *et al.* 1996). Comparison of mtDNA between the two subspecies in South Australia (*P. x. xanthopus*) and Qld (*P. x. celeris*) revealed an average sequence divergence of 5.3% (Pope *et al.* 1996), which suggests the populations have been isolated for an extended period of time and the subspecies may constitute separate evolutionary significant units (ESUs, *sensu* Ryder 1986; Moritz 1994b). Further genetic analysis (including microsatellite analysis) and larger sample sizes are required to resolve this issue.

There is also evidence of significant genetic substructuring (mean F_{ST} = 0.239, Table 4.4) among seven colonies of the black-footed rock-wallaby (*P. lateralis*) in Western Australia (WA) located 0.5–8 km apart (Eldridge *et al.* 2001b). Remnant populations of *P. lateralis* were once common and widespread in the wheatbelt of south-western WA and suffered serious declines due to predation by the red fox (Kinnear *et al.* 1988, 1998; Pearson and Kinnear 1997). The current high levels of genetic structuring in *P. lateralis* may also be influenced by the inhospitable surrounding rural landscape, which has a high density of predators (Kinnear *et al.* 1988).

The differences in population structure between macropodoid species are understandable when comparing the biology of these species (e.g. habitat requirements and species vagility). For example, *M. giganteus* has a wide and basically continuous distribution down the east coast of Australia in a range of habitats (Strahan 1995), whereas rock-wallaby habitat is patchy with unsuitable habitat between rocky outcrops. High levels of genetic structuring are common in species that occur in naturally fragmented habitat (Stow *et al.* 2001; Berry *et al.* 2004; Hazlitt *et al.* 2006a; Piggott *et al.* 2006b). In comparison, the widespread and largely continuous distribution of suitable habitat is likely to influence the low genetic structure occurring in *M. giganteus* (Zenger *et al.* 2003a).

Dispersal

Traditionally, gene flow and dispersal have been difficult to detect and measure either ecologically or genetically (Slatkin 1987; Waser and Strobeck 1998; Bohonak 1999). The level of gene flow can be inferred by indirect measures (e.g. *Nm* from F_{ST}). For example, Zenger *et al.* (2003) calculated the average number of migrants ($N_e m$) among NSW populations of *M. giganteus* as 22.61 individuals per generation. This conclusion of high levels of dispersal was supported by the lack of genetic differentiation over large distances (230 km). These findings contradicted earlier ecological studies which suggested strong site fidelity and limited dispersal (<20 km) (Zenger *et al.* 2003a). However, using F_{ST} to measure gene flow is problematic and it is rare that it can be used to accurately estimate *Nm* due to the unrealistic biological assumptions within the statistical model (Whitlock and McCauley 1999).

The ability to detect dispersal has improved significantly with the development of genetic assignment tests that can assign an individual to the population where its multi-locus genotype is most likely to occur (Paetkau *et al.* 1995; Rannala and Mountain 1997; Cornuet *et al.* 1999; Pritchard *et al.* 2000). Therefore, individuals that are assigned to a different population (the natal population) rather than the population in which they were sampled can be identified as dispersers (Eldridge *et al.* 2001b). Some assignment tests can detect not only first-generation immigrants but also their offspring (Rannala and Mountain 1997; Pritchard *et al.* 2000).

Assignment tests have been used to determine the source of a new population of rock-wallabies (*P. lateralis*) that recolonised habitat from which they had previously been extirpated (Eldridge *et al.* 2001b). The source population was found to be 8 km away and the recently established colony appeared to have been founded by a single female carrying a pouch young and quiescent embryo, as the new group was highly interrelated (Eldridge *et al.* 2001b). Piggott *et al.* (2006a) investigated dispersal between four *P. penicillata* colonies and found approximately 5% of individuals were inferred to be dispersers. Both male and female migrants were detected (three males and one female). Interestingly, three of these dispersers could not be assigned as originating from any of the four sampled colonies, suggesting they were from unsampled colonies in the area. Assignment tests were also used by Pope *et al.* (2000) to detect dispersal events in *B. tropica*. In this species, the majority of sampled individuals were assigned to the population from which they were sampled but there was an indication from misassignments (individuals sampled from one population but assigned to another) that dispersal was occurring mainly from one population (source) to most of the others (sinks).

Sex-biased dispersal can be measured by analysing genetic differences between males and females, for example ancestry (Hazlitt *et al.* 2006a), assignment indices (Hazlitt *et al.* 2006b; Piggott *et al.* 2006a), *Nm* (Zenger *et al.* 2003a) and F_{ST} (Pope *et al.* 2005). Although not conclusive, studies suggest that dispersal for macropodoids is usually male-biased (Table 4.4), which often supports behavioural observations. Zenger *et al.* (2003a) concluded there was evidence of male-biased dispersal in *M. giganteus*, based on comparisons between the average number of migrants ($N_e m$ = 22.61) and the average number of female migrants ($N_f m$ = 2.73) among NSW populations. This finding supported ecological studies that suggested greater male dispersal associated with breeding and social structure constraints in *M. giganteus* (Johnson 1989b; Jaremovic and Croft 1991). Pope *et al.* (2005) suggested that dispersal in *A. rufescens* may be male-biased, based on a marginally higher F_{ST} for adult females than for males. However, there was no difference in assignment indices between the sexes. Pope *et al.* (2000) also suggested that dispersal in *B. tropica* may be male-biased, based on the difference in the order of magnitude in estimates of F_{ST} between mtDNA (which is maternally inherited) and microsatellite markers. However, as noted by the authors, this is not a conclusive method of measuring sex-biased dispersal since differences in ploidy and modes of inheritance are not taken into account.

Although there is evidence of male-biased dispersal within *P. penicillata* colonies (see below), Hazlitt *et al.* (2006b) and Piggott *et al.* (2006a) could not detect significant sex-biased dispersal among colonies based on genetic differences between males and females. Piggott *et al.* (2006a), however, found that out of four dispersal events three were males, which suggests that intercolony dispersal may be male-biased in *P. penicillata*. The difficulties in detecting sex-biased dispersal in rock-wallabies suggest that either dispersal between colonies is not sex-biased or that it is male-biased but intercolony dispersal events are too infrequent for the pattern to be detected (Hazlitt *et al.* 2006a).

Fine-scale structure within a population

Detecting fine-scale genetic substructuring within a population has only been carried out in a few studies (Hazlitt *et al.* 2004; Piggott *et al.* 2006b). The power to detect fine-scale structure has only recently become available via the development of analyses such as multi-locus spatial autocorrelation (Peakall *et al.* 2003). Studies on *P. penicillata* found evidence of fine-scale genetic structuring within two different colonies with differences in the pattern of spatial structuring between males and females (Hazlitt *et al.* 2004; Piggott *et al.* 2006b). Female rock-wallabies in close spatial proximity were more genetically similar than expected, and females separated by more than 400 m (Piggott *et al.* 2006a) and more than 600 m (Hazlitt *et al.* 2004) were less genetically similar than expected. In contrast, in both studies there was no evidence of a significant clustering of related males. This suggests strong female philopatry and that within-colony dispersal is likely to be male-biased. As discussed in Hazlitt *et al.* (2004, 2006a), within-colony dispersal by males and female philopatry may assist in inbreeding avoidance in *P. penicillata*. Male-biased dispersal within colonies has been suggested via behavioural observations in *P. xanthopus* (Sharp 1997). Genetic analyses on the same *P. xanthopus* colony did not detect any genetic substructuring (Pope *et al.* 1996). However, only four microsatellite loci were used and, as spatial autocorrelation analyses had not been developed, there may have been insufficient power to detect within-colony substructuring in the species.

PATERNITY AND MATING SYSTEMS

Advances in the analytical tools to infer paternity (Marshall *et al.* 1998; Piggott and Taylor 2003b) have revolutionised studies of reproductive success and genetic mating systems in the wild. In particular, paternity analyses using biparentally inherited microsatellite markers have frequently revealed that behavioural observations of mating do not always reflect paternity, bringing to light alternative mating tactics and social systems (Dunn *et al.* 1994; Coltman *et al.* 1999). Data on the genetic mating systems of macropodoids in the wild are available for only four of the ~70 species – *O. fraenata* (Fisher and Lara 1999; Sigg *et al.* 2005), *A. rufescens* (Pope *et al.* 2005), *P. penicillata* (Hazlitt *et al.* 2006b) and the allied rock-wallaby (*P. assimilis*) (Spencer and Marsh 1997; Spencer *et al.* 1998) (see also the musky rat-kangaroo *Hypsprimnodon moschatus*, Johnson *et al.* 2005). These studies have confirmed behavioural-based mating system hypotheses for the Macropodoidae (Fisher and Owens 2000; Pope *et al.* 2005) and shed light on alternative patterns of mating (Spencer *et al.* 1998) in this large superfamily.

Macropodoids are similar to other mammals in that they typically exhibit polygynous mating systems, where males compete for access to receptive females. The social mating systems of several large gregarious macropodoids have been well studied in the wild, for example *M. rufus* (Croft 1981) and *M. giganteus* (Jarman and Southwell 1986). They have also been studied in a few of the medium-sized solitary wallabies, for example the whiptail wallaby *M. parryi* (Kaufmann 1974), red-necked wallaby *M. rufogriseus* (Johnson 1989a) and *O. fraenata* (Fisher and Lara 1999; Sigg and Goldizen 2006). These species are promiscuous and exhibit strong sexual size dimorphism. Males form size-based hierarchies in which older, and hence larger, males are more dominant and have priority access to females in oestrus (Jarman and Southwell 1986; Fisher and Lara 1999). These hierarchies can be stable, as is the case in the gregarious kangaroo species (Jarman and Southwell 1986), or 'instantaneously' resolved as in the more solitary *O. fraenata* (Fisher and Lara 1999; Sigg and Goldizen 2006). While behavioural studies have suggested a very high degree of male reproductive skew in the larger, more gregarious kangaroo species (e.g. 80% of behaviour-estimated matings attributed to a single alpha male), there is no information on how priority of access relates to genetic paternity in these species (Jarman and Southwell 1986). Genetic parentage analyses in the *O. fraenata* confirmed that larger, more dominant males with priority access to receptive females produced the greatest proportion of offspring. In addition, variance in male *O. fraenata* reproductive success was only moderately skewed, with 24–26% of males siring offspring (Fisher and Lara 1999; Sigg *et al.* 2005). This was predicted based on the solitary nature and wide-roaming behaviour of the species, which would make it difficult for a single male to dominate a large proportion of oestrous females (Fisher and Lara 1999). Paternity analyses also revealed that smaller males accounted for c. 10% of sirings in both studies (Fisher and Lara 1999; Sigg *et al.* 2005). Home-range size, which was not related to body size, was also a significant predictor of reproductive success in male *O. fraenata* (Fisher and Lara 1999). Thus, while large males successfully monopolise oestrous females and gain paternity through mate-guarding and male–male competition, combining behavioural observations with genetic data has revealed that small males may adopt a different mating strategy, gaining paternity by frequently abandoning mating groups to search for lone oestrous females (Sigg and Goldizen 2006).

Genetic estimates of paternity have shed light on the behaviour and mating systems of some of the smaller macropodoids, namely the rocky terrain-dependent rock-wallabies and the small homomorphic *A. rufescens* (Spencer *et al.* 1998; Pope *et al.* 2005; Hazlitt *et al.* 2006b). Like the solitary *O. fraenata*, monopolising oestrous females in these species may be difficult due to constraints on monopolising limited daytime rocky refuges for male rock-wallabies or homomorphy and large home-range sizes in *A. rufescens* (Jarman and Bayne 1997; Pope *et al.* 2005). This has been supported by behavioural evidence of persistent pairing in *A. rufescens* and *P. penicillata* and even social monogamy in *P. assimilis* (Dennis 1988; Frederick and Johnson 1996; Horsup 1996; Jarman and Bayne 1997). Genetic parentage data confirmed that persistent pairing occurs in *A. rufescens* and *P. penicillata*, although half of the breeding *A. rufescens* females were promiscuous and female *P. penicillata* switched breeding partners when their mates disappeared (Pope *et al.* 2005; Hazlitt *et al.* 2006b). Paternity data also clearly demonstrated that, although female *P. assimilis* were behaviourally monogamous, 24% bred with a male other than their social consort (Spencer *et al.* 1998).

CONSERVATION APPLICATIONS

Molecular genetic techniques are frequently utilised in the management and conservation of macropodoid populations. Measures of population genetic diversity have been used to compare the variation of small remnant populations with appropriate reference populations to elucidate pre-decline or 'healthy' levels of genetic diversity in three macropodid species, *M. eugenii*, the rufous hare-wallaby (*Lagorchestes hirsutus*) and *P. lateralis*, providing appropriate goals for management of declining populations or for monitoring the ongoing genetic health of refuge populations (Eldridge *et al.* 2004a). Genetic data have also been used to identify population structure and therefore key populations or subpopulations (ESUs, MUs) with unique diversity for conservation. For example, these applications have been used to identify priority populations for conservation in *P. xanthopus* and *P. penicillata* (Pope *et al.* 1996; Hazlitt *et al.* 2006a; Piggott *et al.* 2006b), where understanding the population structure was critical for defining MUs and targeting specific management actions.

Genetic information is useful for identifying the provenance of introduced species, either to aid in eradication plans or for translocations where exotic species have become threatened in their native range, such as *P. penicillata*, *M. eugenii* and *M. rufogriseus* (Taylor and Cooper 1999; Le Page *et al.* 2000; Eldridge *et al.* 2001a; Eldridge and Browning 2002). The same genetic techniques are used to aid in the selection of appropriate individuals for population translocations and for monitoring the genetic health of introduced populations, as demonstrated with the translocated *O. fraenata* population at Idalia National Park, Queensland (Sigg 2006).

Molecular genetic techniques offer opportunities to investigate numerous aspects of community and species biology, including landscape surveys for species presence (Alacs *et al.* 2003), estimating population size (Piggott *et al.* 2006a), providing information on mating systems and population structure and individual dispersal, as well as enabling the taxonomic identification of populations (Johnson *et al.* 2001).

FUTURE DIRECTIONS

Population genetic studies utilising microsatellites or the mtDNA control region have been conducted on less than 25% of macropodoid species. However, the potential of these hypervariable markers has been powerfully demonstrated – the studies have begun to transform our understanding of macropodoid population biology, behaviour and reproduction. They have also been instrumental in formulating and guiding conservation initiatives and management plans for many threatened macropodoid taxa. It is hoped that future population genetic studies can be expanded to include a larger number of widespread and abundant Australian species, especially those in northern Australia that have been less severely affected by European settlement. Another priority should be an examination of the unique macropodoid fauna of New Guinea, which remains completely unknown from a population genetics perspective.

For macropodoid species whose capture is often difficult or impractical (e.g. tree-kangaroos in New Guinea), non-invasive sampling such as identifying individuals or species via DNA extracted from faeces, hairs or other remotely collected samples (Piggott and Taylor 2003a, b) is a potentially important method for obtaining vital information on genetic variation, effective population size, population structure, recruitment and dispersal. A pilot study on genetic sampling of faeces from Lumholtz's tree-kangaroo (*Dendrolagus lumholtzi*) in northern Australia showed that reliable genotypic data could be obtained (Piggott and Coombes, unpubl. data). Although non-invasive methods are constantly improving (Bonin *et al.* 2004; Piggott *et al.* 2004; Seddon *et al.* 2005; Smith *et al.* 2006), the time and costs of laboratory analysis are high and need to be weighed against the alternative costs and constraints of traditional methods of capture and sampling. We therefore recommend that non-invasive sampling of macropodoids be considered only where existing methods of sample collection are impractical or provide suboptimal results.

The scale and scope of macropodoid population genetic studies is being made increasingly easy through the continuing development of novel genetic markers, methods of high-throughput genotypic screening and the increasing reliability of non-invasive sampling. In particular, the availability of the DNA sequence for an entire macropodoid genome (Alsop *et al.* 2005) will dramatically increase the ease with which additional microsatellite loci, and other genetic markers, can be located and developed. It will also enable loci to be accurately targeted to link to specific genes or genome regions of interest (e.g. MHC), allowing the detailed exploration of life history, behavioural and reproductive strategies (e.g. mate choice, kin recognition, sociality, disease/parasite susceptibility). The recently developed X and Y chromosome-specific microsatellite loci (MacDonald *et al.* 2006, 2007) are an example of the evolving molecular tools available to researchers and managers. These markers have yet to be used for macropodoid population studies but will undoubtedly be instrumental in significantly advancing our understanding of sex differences in macropodoid behaviour, patterns of gene flow and dispersal, as well as the propensity to hybridise. The scientific and conservation community can expect to see continued growth in the number of population genetics studies of macropodoids and a corresponding increase in our detailed knowledge of the biology of these unique marsupials.

REFERENCES

Alacs E, Alpers D, de Tores PJ, Dillon MJ and Spencer PBS 2003. Identifying the presence of quokkas (*Setonix brachyurus*) and other macropods using cytochrome *b* analyses from faeces. *Wildl. Res.* **30:** 41–47.

Alsop AE, Miethke P, Rofe RH, Koina E, Sankovic N, Deakin JE, Haines H, Rapkins RW and Graves JAM 2005. Characterising the chromosomes of the Australian model marsupial *Macropus eugenii* (tammar wallaby). *Chrom. Res.* **13:** 627–636.

Arnheim N, White T and Rainey W 1990. Application of the polymerase chain reaction to organismal and population biology. *BioScience* **40:** 174–182.

Arrendal J, Walker CW, Sundqvist A-K, Hellborg L and Vila C 2004. Genetic evaluation of an otter program. *Conserv. Genet.* **5:** 79–88.

Avise JC 1994. *Molecular Markers, Natural History and Evolution*. Chapman and Hall, New York.

Berry O, Tocher MD and Sarre SD 2004. Can assignment tests measure dispersal? *Mol. Ecol.* **13:** 551–561.

Bohonak AJ 1999. Dispersal, gene flow and population structure. *Q. Rev. Biol.* **74:** 21–45.

Bonin A, Bellemain E, Bronken Eidesen P, Pompanon F, Brochmann C and Taberlet P 2004. How to track and assess genotyping errors in population genetics studies. *Mol. Ecol.* **13:** 3261–3273.

Bowyer JC, Newell GR and Eldridge MDB 2002. Genetic effects of habitat contraction on Lumholtz's tree-kangaroo (*Dendrolagus lumholtzi*) in the Australian Wet Tropics. *Conserv. Genet.* **3:** 61–69.

Briskie JV and Mackingtosh M 2004. Hatching failure increases with severity of population bottlenecks in birds. *Proc. Natl Acad. Sci. USA* **101:** 558–561.

Browning TL, Taggart DA, Rummery C, Close RL and Eldridge MDB 2001. Multifaceted genetic analysis of the 'critically endangered' brush-tailed rock-wallaby *Petrogale penicillata* in Victoria, Australia: implications for management. *Conserv. Genet.* **2:** 145–156.

Bruford MW, Cheesman DJ, Coote T, Green HAA, Haines SA, O'Ryan C and Williams TR 1996. Microsatellites and their application to conservation genetics. In *Molecular Genetic Approaches in Conservation*. (Eds TB Smith and RK Wayne) pp. 278–297. Oxford University Press, New York.

Caughley G 1994. Directions in conservation biology. *J. Anim. Ecol.* **63:** 215–244.

Clegg SM, Hale P and Moritz C 1998. Molecular population genetics of the red kangaroo (*Macropus rufus*): mtDNA variation. *Mol. Ecol.* **7:** 679–686.

Coltman DW, Bancroft DR, Robertson A, Smith JA, Clutton-Brock TH and Pemberton JM 1999. Male reproductive success in a promiscuous mammal: behavioural estimates compared with genetic paternity. *Mol. Ecol.* **8:** 1199–1209.

Cornuet J-M, Piry S, Luikart G, Estoup A and Solignac M 1999. New methods employing multilocus genotypes to select or exclude populations as origins of individuals. *Genetics* **153:** 1989–2000.

Croft DB 1981. Behaviour of red kangaroos, *Macropus rufus* (Desmarest, 1822) in northwestern New South Wales, Australia. *Aust. Mammal.* **4:** 5–58.

Dennis AJ 1988. Behaviour and movements of rufous bettongs (*Aepyprymnus rufescens*) at Black Rock, North Queensland. Honours thesis. James Cook University, Townsville, Australia.

DeWoody JA and Avise JC 2000. Microsatellite variation in marine, freshwater and anadromous fishes compared with other animals. *J. Fish Biol.* **56:** 461–473.

Donaldson FR and Vercoe PE 2008. Cross-family amplification: microsatellites isolated from Macropodidae are polymorphic in Potoroidae. *Mol. Ecol. Resour.* **8:** 452–454.

Dunn PO, Robertson RJ, Michaud-Freeman D and Boag PT 1994. Extra-pair paternity in tree-swallows: why do females mate with more than one male? *Behav. Ecol. Sociobiol.* **35:** 273–281.

Eldridge MDB and Browning TL 2002. Molecular genetic analysis of the naturalized Hawaiian population of the brush-tailed rock-wallaby, *Petrogale penicillata* (Marsupialia: Macropodidae). *J. Mammal.* **83:** 437–444.

Eldridge MDB, King JM, Loupis AK, Spencer PBS, Taylor AC, Pope LC and Hall GP 1999. Unprecedented low levels of genetic variation and inbreeding depression in an island population of the black-footed rock-wallaby. *Conserv. Biol.* **13:** 531–541.

Eldridge MDB, Browning TL and Close RL 2001a. Provenance of a New Zealand brush-tailed rock-wallaby (*Petrogale penicillata*) population determined by mitochondrial DNA sequence analysis. *Mol. Ecol.* **10:** 2561–2567.

Eldridge MDB, Kinnear JE and Onus ML 2001b. Source population of dispersing rock-wallabies (*Petrogale lateralis*) identified by assignment tests on multilocus genotypic data. *Mol. Ecol.* **10:** 2867–2876.

Eldridge MDB, Kinnear JE, Zenger KR, McKenzie LM and Spencer PBS 2004a. Genetic diversity in remnant mainland and 'pristine' island populations of three endemic Australian macropodids (Marsupialia): *Macropus eugenii, Lagorchestes hirsutus* and *Petrogale lateralis*. *Conserv. Genet.* **5:** 325–338.

Eldridge MDB, Rummery C, Bray C, Zenger K, Browning TL and Close RL 2004b. Genetic consequences of a population crash in brush-tailed rock-wallabies (*Petrogale penicillata*) from Jenolan Caves, south-eastern Australia. *Wildl. Res.* **31:** 229–240.

Excoffier L and Heckel G 2006. Computer programs for population genetics data analysis. *Nature Rev. Genet.* **7:** 745–748.

Fernando P, Pfrender ME, Encalada SE and Lande R 2000. Mitochondrial DNA variation, phylogeography and population structure of the Asian elephant. *Heredity* **84:** 362–372.

Firestone KB, Elphinstone MS, Sherwin WB and Houlden BA 1999. Phylogeographical population structure of tiger quolls *Dasyurus maculatus* (Dasyuridae: Marsupialia), an endangered carnivorous marsupial. *Mol. Ecol.* **8:** 1613–1625.

Fisher DO and Lara MC 1999. Effects of body size and home range on access to mates and paternity in male bridled nailtail wallabies. *Anim. Behav.* **58:** 121–130.

Fisher DO and Owens IPF 2000. Female home range size and the evolution of social organization in macropod marsupials. *J. Anim. Ecol.* **69:** 1083–1098.

Frankham R 2005. Genetics and extinction. *Biol. Conserv.* **126:** 131–140.

Frankham R, Ballou JD and Briscoe DA 2002. *Introduction to Conservation Genetics*. Cambridge University Press, Cambridge, UK.

Frederick H and Johnson CN 1996. Social organisation in the rufous bettongs (*Aepyprymnus rufescens*). *Aust. J. Zool.* **44:** 9–17.

Fumagalli L, Pope LC, Taberlet P and Moritz C 1997. Versatile primers for the amplification of the mitochondrial DNA control region in marsupials. *Mol. Ecol.* **6:** 1199–1201.

Fumagalli L, Moritz C, Taberlet P and Friend JA 1999. Mitochondrial DNA sequence variation within the remnant populations of the endangered (Marsupialia: Myrmecobiidae: *Myrmecobius fasciatus*). *Mol. Ecol.* **8:** 1545–1549.

Garner A, Rachlow JL and Hicks JF 2005. Patterns of genetic diversity and its loss in mammalian populations. *Conserv. Biol.* **19:** 1215–1221.

Harrison RG 1989. Animal mitochondrial DNA as a genetic marker in population and evolutionary biology. *Trends Ecol. Evol.* **4:** 6–11.

Hawken RJ, Maccarone P, Toder R, Graves JAM and Maddox JF 1999. Isolation and characterisation of marsupial IL5 genes. *Immunogen.* **49:** 942–948.

Hazlitt SL, Eldridge MDB and Goldizen AW 2004. Fine-scale spatial genetic correlation analyses reveal strong female philopatry within a brush-tailed rock-wallaby colony in southeast Queensland. *Mol. Ecol.* **13:** 3621–3632.

Hazlitt SL, Goldizen AW and Eldridge MDB 2006a. Significant patterns of population genetic structure and limited gene flow in a threatened macropodid marsupial despite continuous habitat in southeast Queensland, Australia. *Conserv. Genet.* **7:** 675–689.

Hazlitt SL, Sigg DP, Eldridge MDB and Goldizen AW 2006b. Restricted mating dispersal and strong breeding group structure in a mid-sized marsupial mammal (*Petrogale penicillata*). *Mol. Ecol.* **15:** 2997–3007.

Heckenberg KA 1997. DNA studies of *Thylogale*: relationships and potential hybridisation of species and subspecies. BSc Hons thesis. University of Western Sydney, Australia.

Hedrick PW 2005. A standardized genetic differentiation measure. *Evolution* **59:** 1633–1638.

Horsup AB 1996. The behavioural ecology of the allied rock-wallaby, *Petrogale assimilis.* PhD thesis. James Cook University, Townsville, Australia.

Jaremovic RV and Croft DB 1991. Social organisation of the eastern grey kangaroo (Macropodidae, Marsupialia) in southeastern New South Wales. I. Groups and group home ranges. *Mammalia* **55:** 169–185.

Jarman PJ and Southwell CJ 1986. Grouping, association and reproductive strategies in eastern grey kangaroos. In *Ecological Aspects of Social Evolution.* (Eds DI Rubenstein and RW Wrangham) pp. 399–428. Princeton University Press, Princeton.

Jarman PJ and Bayne P 1997. Behavioural ecology of *Petrogale penicillata* in relation to conservation. *Aust. Mammal.* **19:** 219–228.

Jarne P and Lagoda PJL 1996. Microsatellites, from molecules to populations and back. *Trends Ecol. Evol.* **11:** 424–429.

Johnson CN 1989a. Social interactions and reproductive tactics in red-necked wallabies (*Macropus rufogriseus banksianus*). *J. Zool., Lond.* **217:** 267–280.

Johnson CN 1989b. Dispersal and philopatry in macropodoids. In *Kangaroos, Wallabies and Rat-kangaroos.* (Eds G Grigg, P Jarman and I Hume) pp. 593–601. Surrey Beatty and Sons, Sydney.

Johnson PM, Eldridge MDB, Kiernan V and Cupitt RJ 2001. A significant range extension of the purple-necked rock-wallaby *Petrogale purpureicollis. Aust. Mammal.* **23:** 71–73.

Johnson PM, Lloyd S, Vallance T and Eldridge MDB 2005. First record of quadruplets in the musky rat-kangaroo *Hypsiprymnodon moschatus. Aust. Mammal.* **27:** 95–97.

Johnston PG and Sharman GB 1977. Studies on populations of *Potorous* Desmarest (Marsupialia). II. Electrophoretic, chromosomal and breeding studies. *Aust. J. Zool.* **25:** 733–747.

Johnston PG and Sharman GB 1979. Electrophoretic, chromosomal and morphometric studies on the red-necked wallaby *Macropus rufogriseus* (Desmarest). *Aust. J. Zool.* **27:** 433–441.

Kaufmann JH 1974. Social ethology of the whiptail wallaby (*Macropus parryi*) in northwestern New South Wales. *Anim. Behav.* **22:** 281–369.

Kinnear JE, Onus ML and Bromilow RN 1988. Fox control and rock-wallaby population dynamics. *Aust. Wildl. Res.* **15:** 435–450.

Kinnear JE, Onus ML and Sumner NR 1998. Fox control and rock-wallaby population dynamics. II. An update. *Wildl. Res.* **25:** 81–88.

Lande R 1988. Genetics and demography in biological conservation. *Science* **241:** 1455–1460.

Le Page SL, Livermore RA, Cooper DW and Taylor AC 2000. Genetic analysis of a documented population bottleneck: introduced Bennett's wallabies (*Macropus rufogriseus rufogriseus*) in New Zealand. *Mol. Ecol.* **9:** 753–763.

Luikart G and England PR 1999. Statistical analysis of microsatellite DNA data. *Trends Ecol. Evol.* **14:** 253–256.

Luikart G, Painter J, Crozier RH, Westerman M and Sherwin WB 1997. Characterization of microsatellite loci in the endangered long-footed potoroo *Potorous longipes. Mol. Ecol.* **6:** 497–498.

MacDonald AJ, Sankovic N, Sarre SD, Fitzsimmons NN, Wakefield MJ, Graves JAM and Zenger KR 2006. Y chromosome microsatellite markers identified from the tammar wallaby (*Macropus eugenii*) and their amplification in three other macropod species. *Mol. Ecol. Notes* **6:** 1202–1204.

MacDonald AJ, Sarre SD, Fitzsimmons NN and Graves JAM 2007. Chromosome-specific microsatellites from the tammar wallaby X chromosome and chromosome 2. *Mol. Ecol. Notes* **7:** 1063–1066.

Manel S, Gaggiotti OE and Waples RS 2005. Assignment methods: matching biological questions with appropriate techniques. *Trends Ecol. Evol.* **20:** 136–142.

Marjoram P and Tavare S 2006. Modern computational approaches for analysing molecular genetic variation data. *Nature Rev. Genet.* **7:** 759–770.

Marshall TC, Slate J, Kruuk LEB and Pemberton JM 1998. Statistical confidence for likelihood-based paternity inference in natural populations. *Mol. Ecol.* **7:** 639–655.

Maxwell S, Burbidge AA and Morris K (Eds) 1996. *The 1996 Action Plan for Australian Marsupials and Monotremes.* Wildlife Australia, Canberra.

Moritz C, Dowling TE and Brown WM 1987. Evolution of animal mitochondrial DNA: relevance for population biology and systematics. *Ann. Rev. Ecol. Syst.* **18:** 269–292.

Moritz C 1994a. Applications of mitochondrial DNA analysis in conservation: a critical review. *Mol. Ecol.* **3:** 401–411.

Moritz C 1994b. Defining 'evolutionary significant units' for conservation. *Trends Ecol. Evol.* **9:** 373–375.

Nyakaana S, Arctander P and Siegismund HR 2002. Population structure of the African savannah elephant inferred from mitochondrial control region sequences and nuclear microsatellite loci. *Heredity* **89:** 90–98.

Oliveira EJ, Padua JG, Zucchi MI, Vencovsky R and Vieira MLC 2006. Origin, evolution and genome distribution of microsatellites. *Genet. Mol. Biol.* **29:** 294–307.

Paetkau D, Calvert W, Stirling I and Strobeck C 1995. Microsatellite analysis of population structure in Canadian polar bears. *Mol. Ecol.* **4:** 347–354.

Paplinska JZ 2005. Seasonal breeding, puberty and population genetics of the swamp wallaby (*Wallabia bicolor*). PhD thesis. University of Melbourne, Australia.

Paplinska JZ 2006. Faecal DNA analysis of *Petrogale penicillata* in Kangaroo Valley. Final report for NSW National Parks and Wildlife Service. New South Wales National Parks and Wildlife Service, Sydney.

Peakall R, Ruibal M and Lindenmayer DB 2003. Spatial autocorrelation and analysis offers new insights into gene flow in the Australian bush rat, *Rattus fuscipes*. *Evolution* **57:** 1182–1195.

Pearson DJ and Kinnear JE 1997. A review of the distribution, status and conservation of rock-wallabies in Western Australia. *Aust. Mammal.* **19:** 137–152.

Piggott MP and Taylor AC 2003a. Extensive evaluation of faecal preservation and DNA extraction methods in Australian native and introduced species. *Aust. J. Zool.* **51:** 341–355.

Piggott MP and Taylor AC 2003b. Remote collection of animal DNA and its application in conservation management and understanding the population biology of rare and cryptic species. *Wildl. Res.* **30:** 1–13.

Piggott MP, Bellemain E, Taberlet P and Taylor AC 2004. A multiplex pre-amplification method that significantly improves microsatellite amplification and error rates for faecal DNA in limiting conditions. *Conserv. Genet.* **5:** 417–420.

Piggott MP, Banks SC, Stone N, Banffy C and Taylor AC 2006a. Estimating population size of endangered brush-tailed rock-wallaby (*Petrogale penicillata*) colonies using faecal DNA. *Mol. Ecol.* **15:** 81–91.

Piggott MP, Banks SC and Taylor AC 2006b. Population structure of brush-tailed rock-wallaby (*Petrogale penicillata*) colonies inferred from analysis of faecal DNA. *Mol. Ecol.* **15:** 93–105.

Pope LC, Sharp A and Moritz C 1996. Population structure of the yellow-footed rock-wallaby *Petrogale xanthopus* (Gray, 1854) inferred from mtDNA sequences and microsatellite loci. *Mol. Ecol.* **5:** 629–640.

Pope LC, Estoup A and Moritz C 2000. Phylogeography and population structure of an ectonal marsupial, *Bettongia tropica,* determined from mtDNA and microsatellites. *Mol. Ecol.* **9:** 2041–2053.

Pope LC, Blair D and Johnson CN 2005. Dispersal and population structure of the rufous bettong, *Aepyprymnus rufescens* (Marsupialia: Potoroidae). *Austral Ecol.* **30:** 572–580.

Pritchard JK, Stephens M and Donnelly P 2000. Inference of population structure using multilocus genotype data. *Genetics* **155:** 945–959.

Rannala B and Mountain JL 1997. Detecting immigration by using multilocus genotypes. *Proc. Natl Acad. Sci. USA* **94:** 9197–9201.

Richardson BJ and Sharman GB 1976. Biochemical and morphological observations on the wallaroos (Macropodidae: Marsupialia) with a suggested new taxonomy. *J. Zool., Lond.* **179:** 499–513.

Ryder OA 1986. Species conservation and systematics: the dilemma of subspecies. *Trends Ecol. Evol.* **1:** 9–10.

Saccheri I, Kuussaari M, Kankare M, Vikman P, Fortelius W and Hanski I 1998. Inbreeding and extinction in a butterfly metapopulation. *Nature* **392:** 491–494.

Seddon JM, Parker HG, Ostrander EA and Ellegren H 2005. SNPs in ecological and conservation studies: a test in the Scandinavian wolf population. *Mol. Ecol.* **14:** 503–511.

Seebeck JH and Johnston PG 1980. *Potorous longipes* (Marsupialia: Macropodidae): a new species from eastern Victoria. *Aust. J. Zool.* **28:** 119–134.

Sharp A 1997. Insights into the dispersal patterns of yellow-footed rock-wallabies, *Petrogale xanthopus*. *Aust. Mammal.* **19:** 229–238.

Sherwin WB and Murray ND 1990. Population and conservation genetics of marsupials. *Aust. J. Zool.* **37:** 161–180.

Sigg DP 2006. Reduced genetic diversity and significant genetic differentiation after translocation: comparison of the remnant and translocated populations of bridled nailtail wallabies (*Onychogalea fraenata*). *Conserv. Genet.* **7:** 577–589.

Sigg DP and Goldizen AW 2006. Male reproductive tactics and female choice in the solitary, promiscuous bridled nailtail wallaby (*Onychogalea fraenata*). *J. Mammal.* **87:** 461–469.

Sigg DP, Goldizen AW and Pople AR 2005. The importance of mating system in translocation programs: reproductive success of released male bridled nailtail wallabies. *Biol. Conserv.* **123:** 289–300.

Sinclair AH 2001. Phylogeographic variation in the quokka, *Setonix brachyurus* (Marsupialia: Macropodidae): implications for management. *Anim. Conserv.* **4:** 325–333.

Sinclair EA, Costello B, Courtenay JM and Crandall KA 2002. Detecting a genetic bottleneck in Gilbert's potoroo, *Potorous gilbertii* (Marsupialia: Potoroidae), inferred from microsatellite and mitochondrial DNA sequence data. *Conserv. Genet.* **3:** 191–196.

Slatkin M 1987. Gene flow and the geographic structure of natural populations. *Science* **236:** 787–792.

Smith DA, Ralls K, Hurt A, Adams B, Parker M and Maldonado JE 2006. Assessing reliability of microsatellite genotypes from kit fox faecal samples using genetic and GIS analyses. *Mol. Ecol.* **15:** 387–406.

Spencer PBS 1996. Coping with a naturally fragmented environment: a genetic and ecological study of the allied rock-wallaby, *Petrogale assimilis*. PhD thesis. James Cook University, Townsville, Australia.

Spencer PBS and Marsh HD 1997. Microsatellite DNA fingerprinting confirms dizygotic twinning and paternity in the allied rock-wallaby, *Petrogale assimilis*

(Marsupialia: Macropodidae). *Aust. Mammal.* **19:** 279–280.

Spencer PBS, Odorico DM, Jones SJ, Marsh HD and Miller DJ 1995. Highly variable microsatellites in isolated colonies of the rock- wallaby (*Petrogale assimilis*). *Mol. Ecol.* **4:** 523–525.

Spencer PBS, Horsup AB and Marsh HD 1998. Enhancement of reproductive success through mate choice in a social rock-wallaby, *Petrogale assimilis* (Macropodidae) as revealed by microsatellite markers. *Behav. Ecol. Sociobiol.* **43:** 1–9.

Spielman D, Brook BW and Frankham R 2004. Most species are not driven to extinction before genetic factors impact them. *Proc. Natl Acad. Sci. USA* **101:** 15261–15264.

Stow AJ, Sunnucks P, Briscoe DA and Gardner MG 2001. The impact of habitat fragmentation on dispersal of Cunningham's skink (*Egernia cunninghami*): evidence from alleleic and genotypic analyses of microsatellites. *Mol. Ecol.* **10:** 867–878.

Strahan R (Ed.) 1995. *The Mammals of Australia*. Reed Books, Sydney.

Sunnucks P 2000. Efficient genetic markers for population biology. *Trends Ecol. Evol.* **15:** 199–203.

Sunnucks P, Wilson AC, Beheregaray LB, Zenger K, French J and Taylor AC 2000. SSCP is not so difficult: the application and utility of single-stranded conformation polymorphism in evolutionary biology and molecular ecology. *Mol. Ecol.* **9:** 1699–1710.

Taberlet P 1996. The use of mitochondrial DNA control region sequencing in conservation genetics. In *Molecular Genetic Approaches in Conservation.* (Eds TB Smith and RK Wayne) pp. 125–142. Oxford University Press, New York.

Taylor AC and Cooper DW 1998. A set of tammar wallaby (*Macropus eugenii*) microsatellites tested for genetic linkage. *Mol. Ecol.* **7:** 925–926.

Taylor AC and Cooper DW 1999. Microsatellites identify introduced New Zealand tammar wallabies (*Macropus eugenii*) as an 'extinct' taxon. *Anim. Conserv.* **2:** 41–49.

Taylor AC, Sunnucks P and Cooper DW 1999. Retention of reproductive barriers and ecological differences between two introduced sympatric *Macropus* spp. in New Zealand. *Anim. Conserv.* **2:** 195–202.

Waples RS and Gaggiotti OE 2006. What is a population? An empirical evaluation of some genetic methods for identifying the number of gene pools and their degree of connectivity. *Mol. Ecol.* **15:** 1419–1439.

Waser PM and Strobeck C 1998. Genetic signatures of interpopulation dispersal. *Trends Ecol. Evol.* **13:** 43–44.

Whitlock MC and McCauley DE 1999. Indirect measures of gene flow and migration: $F_{ST} \neq 1/(4N_m + 1)$. *Heredity* **82:** 117–125.

Zenger KR and Cooper DW 2001a. Characterization of 14 macropod microsatellite genetic markers. *Anim. Genet.* **32:** 160–167.

Zenger KR and Cooper DW 2001b. A set of highly polymorphic microsatellite markers developed for the eastern grey kangaroo (*Macropus giganteus*). *Mol. Ecol. Notes* **1:** 98–100.

Zenger KR, McKenzie LM and Cooper DW 2002. The first comprehensive genetic linkage map for a marsupial: the tammar wallaby (*Macropus eugenii*). *Genetics* **162:** 321–330.

Zenger KR, Eldridge MD B and Cooper DW, 2003a. Intraspecific variation, sex-biased dispersal and phylogeography of the eastern grey kangaroo (*Macropus giganteus*). *Heredity* **92:** 153–162.

Zenger KR, Eldridge MDB, Pope LC and Cooper DW 2003b. Characterisation and cross-species utility of microsatellite markers within kangaroos, wallabies and rat-kangaroos (Macropodoidae: Marsupialia). *Aust. J. Zool.* **51:** 587–596.

Zenger KR, Eldridge MDB and Johnston PG 2005. Phylogenetics, population structure and genetic diversity of the endangered southern brown bandicoot (*Isoodon obesulus*) in south-eastern Australia. *Conserv. Genet.* **6:** 193–204.

5 Sex chromosome microsatellites: new tools for macropodoid population ecology

A.J. MacDonald, S.D. Sarre and N.N. FitzSimmons

SUMMARY

Microsatellites are important genetic markers that have been applied to answer questions at the population level in a range of macropodoid species. Most studies to date have been restricted to analyses of variation at autosomal microsatellite loci, sometimes in conjunction with mitochondrial DNA (mtDNA) sequence data. The recent development of microsatellite markers from the tammar wallaby (*Macropus eugenii*) X and Y chromosomes provides new opportunities to investigate macropodoid population genetic questions from different evolutionary perspectives. Sex chromosome markers can be used to shed light on sex-specific genetic processes. They may also provide different levels of resolution for a range of questions as a consequence of their smaller effective population sizes relative to the autosomes. In this chapter we detail the use of autosomal microsatellites in studies of macropodoids and use examples from studies of eutherian mammals to demonstrate the utility of sex chromosome microsatellites in areas such as population origins, sex-biased dispersal, hybridisation, mating system analysis and conservation genetics.

INTRODUCTION

The development of sophisticated laboratory analysis systems over the last 20 years, combined with the sheer number of genetic markers now available, has put population genetics within the reach of an ever-increasing number of ecologists. Each genome sequencing project brings in its wake a host of benefits for studies of related taxa. In particular, genome sequencing provides opportunities for new marker development because genetic markers developed from one species can often be applied to close relatives, as has been shown for microsatellites in a range of taxa including plants (Tuskan *et al.* 2004), insects (Wilson *et al.* 2004), birds (Primmer *et al.* 1996), turtles (FitzSimmons *et al.* 1995) and mammals (Bradley *et al.* 2000).

Autosomal microsatellites have been developed for a range of species from the marsupial superfamily Macropodoidea (reviewed in Ch. 4, this volume). These markers have been applied to a wide variety of macropodoid population genetic studies, including analyses of population structure and gene flow (Zenger *et al.* 2003a; Pope *et al.* 2005; Hazlitt *et al.* 2006a), genetic diversity of isolated populations (Eldridge *et al.* 1999, 2004a), mating systems (Sigg *et al.* 2005; Hazlitt *et al.* 2006b) and studies directly relevant to macropodoid conservation (Browning *et al.* 2001; Bowyer *et al.* 2002; Piggott *et al.* 2006a). It has been demonstrated that microsatellite markers developed for a particular macropodoid can often be optimised for use in other macropodoid species (see Ch. 4) and that this approach may be more likely to succeed where the two species are closely related (Zenger *et al.* 2003b). This means that microsatellite analyses are now within the reach of researchers working on most, if not all, species of macropodoids.

To date most macropodoid microsatellites have been isolated using laboratory-based techniques, including standard bacteria screening and microsatellite enrichment libraries (Karagyozov *et al.* 1993; Hakki and Akkaya 2000). These methods can be time-consuming and unpredictable, with no guarantees of obtaining the numbers or types of markers desired. In addition, these approaches do not permit the targeting of markers from particular chromosomes, or even the identification of the chromosomes of origin of known markers. Consequently, the availability of DNA sequence from the tammar wallaby (*Macropus eugenii*) (NHGRI 2008) is now providing unprecedented opportunities to identify novel genetic markers for use in the tammar wallaby and closely related species (MacDonald *et al.* 2006, 2007).

The ability to target the development of markers from particular regions of the genome will provide a new range of population genetic tools and will improve our understanding of the factors affecting genetic variability in macropodoid populations. For example, the development of markers from the sex chromosomes will provide new evolutionary perspectives from which to investigate topics such as population histories and interpopulation gene flow (Hurles and Jobling 2001; Petit *et al.* 2002; Schaffner 2004). The combination of data from maternal, paternal and biparentally inherited markers will add to analyses a male-specific component that has been utilised in studies of humans and other species (Mesa *et al.* 2000; Prugnolle and de Meeus 2002; Vilà *et al.* 2003b) but has been conspicuously absent from macropodoid studies. This will provide a much greater understanding of genetic diversity in natural populations than could be obtained from any of these classes of marker alone.

In this chapter we introduce microsatellite markers and review the potential application of sex chromosome microsatellites to various questions in macropodoid population ecology. In particular, we focus on areas of research in which sex chromosome microsatellites have already proven their value, and highlight case studies from eutherian mammals to illustrate the utility of these markers. Table 5.1 provides a summary of the characteristics of the genetic markers discussed and the applications to which each is suited.

MICROSATELLITE MARKERS

Microsatellites are a type of molecular marker used widely in the biological sciences, with applications ranging from medicine and forensics to agriculture and conservation genetics. These repetitive sequences consist of a simple DNA motif of one to six bases in length that is repeated in tandem at a particular locus. Microsatellites are classified according to the length of the DNA motif they contain, for example $[AC]_n$ is known as a dinucleotide repeat, $[GAT]_n$ a trinucleotide repeat and so on. Individual microsatellite loci are identified by the unique sequences that flank either side of the repetitive region of the microsatellite.

The potential use of microsatellites as population genetic markers was identified in the early 1990s (Bruford and Wayne 1993; Wright and Bentzen 1994). Microsatellites are particularly useful to population geneticists for several reasons, including their relative ease of use in the laboratory and their high levels of variation in natural populations (Schlötterer 2000; Sunnucks 2000). This variation is mostly the result of differences in the number of tandem repeats present at different alleles, seen as polymorphism in the overall length of the repeat region (Fig. 5.1). Microsatellite variation can be analysed in the laboratory with the use of the polymerase chain reaction (PCR) and gel electrophoresis (typically on laser-based sequencing machines) to indicate the number of repeat units present in each sample. The large numbers of alleles found at microsatellite loci make it possible to use them to investigate genetic variation even between closely related individuals, providing insights into population genetic processes at much greater resolution than might be possible with other classes of genetic marker.

SEX CHROMOSOME MICROSATELLITES

With only a few exceptions (e.g. Zenger and Cooper 2001), most of the microsatellites that have been developed from marsupials originate from the autosomes, i.e. the non-sex chromosomes. In comparison to autosomal microsatellites, it is much more difficult to identify microsatellites from the sex chromosomes

Table 5.1: Mammalian population genetic markers – the characteristics and potential applications of microsatellites from the sex chromosomes and autosomes, in comparison with mtDNA.

Application	mtDNA	Autosomal microsatellites	Y chromosome microsatellites	X chromosome microsatellites	Examples
Population origins	Regional haplotype variations indicate origins	Regional allele frequency variations indicate origins	Regional haplotype variations indicate origins – males only	Regional allele frequency variations indicate origins	Vilà *et al.* 2003a (wolves)
Sex-biased dispersal	Haplotype diversity reflects female-specific gene flow	Used to identify specific individuals from trace samples	Haplotype diversity reflects male-specific gene flow – males only. Limited by reduced Y chromosome variation	Used to identify specific individuals from trace samples	Eriksson *et al.* 2006 (bonobos)
Hybrid detection	Species-specific haplotypes determine mother's origin	Species-specific allele frequencies detect hybrids	Species-specific haplotypes determine father's origin – males only	Species-specific haplotypes determine mother's origin in males	Vilà *et al.* 2003b (dogs/wolves)
Admixture analysis	Limited: no recombination	Difficult: haplotypes ambiguous in heterozygotes	Limited: no recombination	Haplotypes unambiguous in males, reflect history of recombination in females	Freeman *et al.* 2006 (cattle)
Parentage analysis	Maternal relatives share same haplotype. Can exclude maternal non-matches	Can reconstruct parental genotypes from alleles present in offspring. Best when one parent already known	Paternal relatives share same haplotype. Can exclude paternal non-matches – males only	Maternal allele in males easily determined. Greater power to detect paternity when potential fathers are closely related: X haplotypes more different than autosomal genotypes	Vonhof *et al.* 2006 (big brown bats); Waser *et al.* 2006 (kangaroo rats)
Kinship analysis	Maternal relatives share same haplotype	Haplotypes not necessarily shared between relatives	Paternal relatives share same haplotype – males only	Paternal sisters/half-sisters share same haplotype	Smith *et al.* 2003 (baboons)

(Hurles and Jobling 2001). In part this is because microsatellite enrichment methods tend to identify microsatellites from different chromosomes in proportion to the percentage of the genome each chromosome represents. This is especially problematic for finding

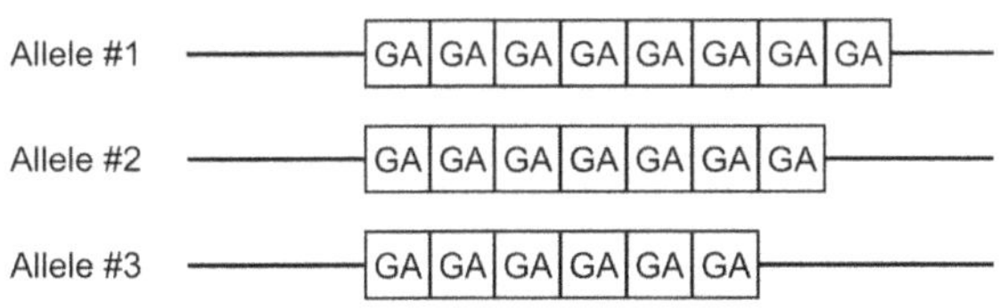

Figure 5.1: A hypothetical dinucleotide $[GA]_n$ microsatellite. Allele #1 has eight repeat units, allele #2 has seven repeat units and allele #3 has six repeat units present. Non-repetitive DNA sequence flanking the microsatellite, denoted by solid lines, is identical at each allele.

microsatellites on the Y chromosome, which tends to be the smallest mammalian chromosome. In the tammar wallaby the Y chromosome represents <3% of the haploid genome (Alsop *et al.* 2005), making it much less likely to be represented in enrichment libraries. It is also difficult to target microsatellite isolation towards a particular chromosome. Doing so requires access to resources that are not typically available for non-model species, for example flow-sorted chromosomes (Dickens *et al.* 1999; Korstanje *et al.* 2001) or DNA sequence known to originate from the target chromosome (Ayub *et al.* 2000). If discovered unwittingly, it is possible that some sex chromosome markers have been discarded from further analyses because they did not conform to Hardy-Weinberg equilibrium.

Sex chromosome microsatellites are now recognised as desirable genetic markers that can be used to provide a variety of evolutionary perspectives to population genetic studies. Sequencing of maternally inherited mtDNA is used widely in conjunction with autosomal markers to provide a more complete picture of population structures and histories. Now sex chromosomes can be added to the molecular ecology toolbox, with Y chromosome markers useful for elucidating male-specific genetic processes and X chromosome markers providing a unique perspective by virtue of their haplo-diploid inheritance in males versus females (Hurles and Jobling 2001; Schaffner 2004). The potential of sex chromosome microsatellites as population genetic markers has been demonstrated, for example where Y chromosome markers have been used to infer human population histories (Pérez-Lezaun *et al.* 1997; Jorde *et al.* 2000). However, the application of sex chromosome microsatellites to non-model species has yet to reach its full potential, as the identification and development of suitable markers lags behind.

Sex chromosome microsatellites in the tammar wallaby

Microsatellites have recently been identified from DNA sequence originating from the tammar wallaby X and Y chromosomes (MacDonald *et al.* 2006, 2007). These markers include seven microsatellites identified from the X chromosome and four from the Y chromosome that are polymorphic in Kangaroo Island tammar wallabies (two to eight alleles per locus). It is apparent that these markers can be adapted for use in other macropodid species, especially those closely related to the tammar wallaby. Preliminary work has shown that PCR amplification of some Y chromosome markers is possible from Bennett's wallaby (*Macropus rufogriseus rufogriseus*), eastern grey kangaroo (*Macropus giganteus*) and yellow-footed rock-wallaby (*Petrogale xanthopus*), although further optimisation of PCR conditions is needed in most cases (MacDonald *et al.* 2006). The development of these markers brings an opportunity to incorporate them into the suite of genetic tools currently applied to macropodoid population genetics. Consequently, it is timely to review the ways in which sex chromosome microsatellites can be applied to mammalian studies, in conjunction with autosomal and mitochondrial markers, to tackle questions of interest to ecologists and managers.

SEX CHROMOSOME MICROSATELLITES IN POPULATION ECOLOGY

Population origins and histories

Knowledge of a population's origins and of historical events such as expansions or bottlenecks can be crucial to understanding the current biological processes affecting genetic variation. Information on the genetic histories of populations can be obtained from analysis of regional allele frequency differences at genetic markers from all chromosomes (Table 5.1). Using paternally inherited Y chromosome microsatellites to address such questions has provided striking comparisons to analyses of maternally inherited mtDNA, and important insights to the study of human population histories across the globe (Hammer and Horai 1995; Ruiz-Linares *et al.* 1999; Seielstad *et al.* 1999; Zegura *et al.* 2004). One of the few wildlife examples of the application of sex chromosome microsatellites to this topic comes from a study of the Scandinavian wolf (*Canis lupus*) population, which was believed to be extinct in the 1970s but by 2002 had recovered to around 100 individuals. Combined data from mtDNA and autosomal, X and Y chromosome microsatellites suggest that the new population was initially founded by only two individuals. Subsequent to this founder event the arrival of a second male wolf in the early 1990s was identified: this male contributed additional genetic variation, including a new Y chromosome haplotype, and his arrival seems to have been the catalyst for a population expansion. Comparison of allele frequencies between populations also enabled identification of the population of origin of these colonists (Sundqvist *et al.* 2001; Vilà *et al.* 2003a).

Population origins, colonisation events, bottlenecks and inbreeding have been investigated in a range of macropodoid species using microsatellite markers and mtDNA analysis (Bowyer *et al.* 2002; Eldridge *et al.* 1999, 2004a; Le Page *et al.* 2000; Sinclair *et al.* 2002). In future macropodoid studies it may be possible to investigate the sex-specific components of genetic diversity with the inclusion of sex chromosome microsatellites. For example, microsatellites were used to identify, as descendants of a now-extinct Australian population, tammar wallabies that had been introduced to New Zealand during the second half of the 19th century (Taylor and Cooper 1999), leading to a recent translocation of New

Zealand wallabies to the South Australian mainland (DEH 2006). In cases such as this, the additional information available from Y chromosome markers may allow managers to identify different male lineages and maximise the genetic diversity of the translocated males. The number of founders of the New Zealand population of Bennett's wallaby, another species introduced to New Zealand in the 1870s, and the impacts of a post-translocation bottleneck have been investigated by analysis of microsatellites and mtDNA (Le Page *et al.* 2000). In future studies of macropodoid colonisation events, analysis of Y chromosome haplotype diversity could help to identify the number of male founders of a population. In addition, the small effective population sizes of the X and Y chromosomes relative to the autosomes mean that genetic drift will occur more quickly on the sex chromosomes (Schaffner 2004). Thus, sex chromosome markers may be particularly effective tools for investigating bottlenecks in threatened macropodoid species that cannot be detected with autosomal markers, for example as a consequence of relatively recent reductions in population size.

Sex-biased dispersal

An understanding of sex-biased dispersal is of great interest to many researchers and managers as it can be informative about numerous aspects of a species' ecology, from mating systems and habitat use to factors affecting differential mortality of males and females. Sex chromosome markers, especially Y chromosome markers when used in conjunction with mtDNA, provide a sex-specific perspective to a population's genetics (Table 5.1) and seem an obvious choice for the study of sex-biased dispersal (Prugnolle and de Meeus 2002). Yet most examples of the application of sex chromosome microsatellites to this topic come from studies of human migrations (Mesa *et al.* 2000; Pérez-Lezaun *et al.* 1999; Seielstad *et al.* 1998). One example of the use of Y chromosome microsatellites to investigate sex-biased dispersal in wildlife comes from a study of the bonobo (*Pan paniscus*), in which haplotype diversity was found to be significantly greater for mtDNA than for the Y chromosome. Additionally, 88% of mitochondrial variation, but only 2.6% of Y chromosome variation, was shared between populations, suggesting that dispersal in this species is highly female-biased. These results also hint at a low male effective population size (Eriksson *et al.* 2006). The paucity of further examples of the use of Y chromosome microsatellites to investigate sex-biased dispersal may result, at least in part, from a lack of sufficiently variable Y chromosome markers in other mammal species (Prugnolle and de Meeus 2002).

Sex-biased dispersal has been investigated in several species of macropodoid, including eastern grey kangaroos (Zenger *et al.* 2003a), brush-tailed rock-wallabies (*Petrogale penicillata*) (Hazlitt *et al.* 2004; Piggott *et al.* 2006b) and rufous bettongs (*Aepyprymnus rufescens*) (Pope *et al.* 2005). The future application of Y chromosome microsatellites may provide a valuable male-specific complement to recent macropodoid studies that have used autosomal microsatellites, sometimes in conjunction with mtDNA haplotypes, to compare dispersal from a female perspective with that seen in the population as a whole. For example, autosomal microsatellite data have shown that females within a colony of brush-tailed rock-wallabies were more genetically similar to their close female neighbours than expected. The same trend was not observed between males and their close male neighbours, suggesting that dispersal within the colony is male-biased (Hazlitt *et al.* 2006a; Piggott *et al.* 2006b). The lack of recombination on the Y chromosome means that multiple Y chromosome microsatellites, considered as a haplotype, provide more information on paternal lineages than do autosomal markers. Thus, in future studies, Y chromosome markers may provide greater power to detect movements of discrete male pedigrees within and between isolated colonies of rock-wallabies.

Hybridisation and admixture

Interspecies hybridisation is a complex issue with important implications for conservation management. The presence of naturally occurring hybrids may raise questions about the definitions of species boundaries, while both natural and anthropogenically assisted hybridisation can lead to the extinction of threatened taxa at the population and species levels (Rhymer and Simberloff 1996; Allendorf *et al.* 2001). In the past, morphological characteristics were used to identify hybrid individuals but the application of more informative genetic markers to these questions is now standard (Boecklen and Howard 1997; Allendorf *et al.* 2001).

In species where hybridisation is uncommon, it may be sufficient to simply identify hybrid individuals and determine the origins of each parent. Autosomal markers can indicate the presence of alleles specific to two different species, but not the contribution of each parent individually. In contrast, species-specific mitochondrial or Y chromosome haplotypes are particularly useful because their sex-specific inheritance and minimal recombination can unambiguously indicate the direction of hybridisation (Table 5.1), although the Y chromosome can only be informative in males. Autosomal microsatellites have been applied to the detection of hybrids in a range of mammalian taxa, in some cases in conjunction with other classes of Y chromosome marker (Culver *et al.* 2000; Evans *et al.* 2001; Nijman *et al.* 2003; Ermakov *et al.* 2006; Lancaster *et al.* 2006). In addition, Y chromosome microsatellites have been incorporated into at least one study, to identify a male wolf–dog hybrid as the product of a female wolf and a male dog (Vilà *et al.* 2003b). It is likely that Y chromosome markers will be applied to hybrid studies in a wider range of taxa as more markers and species-specific polymorphisms are identified from non-model species.

Where hybridisation occurs frequently, it can be difficult to interpret the effects on population genetic variation of many generations of backcrossing between hybrids and parent populations. Genetic markers can be used in this context to determine the degree of admixture between the groups, as indicated by the decay of species-specific haplotypes over time (Allendorf *et al.* 2001). Hybridisation between species creates individuals with chromosomes originating from each species. When these chromosomes undergo recombination, hybrid chromosomes are created, containing alleles from both the parent taxa. Admixture over the generations leads to more recombination events between the parental haplotypes, creating chromosomes that are mosaics of alleles from both the parent species (Fig. 5.2). Analysis of linked genetic markers from the same chromosome can show how much admixture has occurred between species and how long it has persisted (Lecis *et al.* 2006; Verardi *et al.* 2006). Here, the X chromosome is particularly useful for investigating admixture (Table 5.1). Recombination in females acts as a measure of admixture, but X chromosome haplotypes can be unambiguously determined by restricting genetic analyses to haploid males (Schaffner 2004). This approach has been used to investigate admixture between cattle lineages (*Bos taurus* and *B. indicus*) in Africa (Freeman *et al.* 2006). A Y chromosome microsatellite (Ward *et al.* 2001) and groups

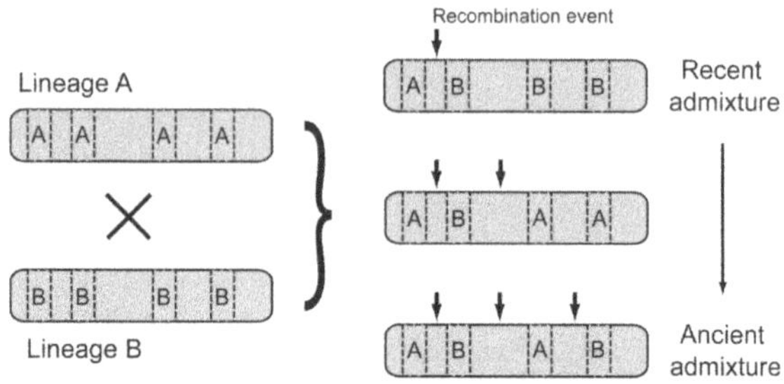

Figure 5.2: Using X chromosome haplotypes to investigate admixture between different lineages. Shaded boxes represent a section of the X chromosome in different individuals. Four hypothetical microsatellite loci are denoted by dashed boxes, with A and B representing lineage-specific alleles at each locus. Admixture between lineage A and lineage B leads to recombination between lineage-specific alleles, with the degree of mosaicism representative of the amount and duration of admixture.

of linked microsatellites from the autosomes and the X chromosome (Halbert *et al.* 2005) have also been used to investigate cattle (*B. taurus*) introgression into bison (*Bison bison*) populations in North America, which were driven to the brink of extinction in the 19th century.

Hybridisation has been demonstrated in captivity between a range of macropodoid species in which it is either unlikely or poorly understood in natural populations (Close and Lowry 1990). However, hybrid animals are used as models in studies of reproduction and genetics (Close and Bell 1997; Zenger *et al.* 2002), so an understanding of hybridisation may be important to managers of captive research colonies. If wild hybrids are identified in the future, species-specific genetic markers will be invaluable in determining how the hybridisation occurred. For example, hybridisation in captivity has been observed between eastern grey kangaroo females and western grey kangaroo (*Macropus fuliginosus*) males, but not the reverse (Poole and Catling 1974). Tammar wallaby Y chromosome microsatellites have recently been shown to be variable in both species of grey kangaroo (L. Neaves, *pers. comm.*), thus it may be possible to identify the paternal species should hybrid eastern–western grey kangaroos or their descendants be identified in the wild. The application of admixture analysis to macropods such as these using X chromosome haplotypes may also provide a better understanding of patterns of interspecific hybridisation, particularly in macropodoids with overlapping species boundaries where hybrids may be infrequent and difficult to detect morphologically.

Mating systems and kinship

Genetic analysis of parentage and the degree of relatedness between individuals can tell us much about mating systems and the social structure of a population. Molecular markers have been used to infer genetic relationships for some decades, but the advent of microsatellite genotyping and related statistical analyses enabled the widespread application of molecular techniques to wildlife parentage studies (Jones and Ardren 2003). Juveniles and adults can be genotyped across a range of microsatellite loci, allowing the exclusion of putative parents who do not carry alleles observed in the juveniles, and the estimation or reconstruction of the most likely parental genotypes for each individual. The analysis of markers from different chromosomes, with different modes of inheritance, can provide direct information on alleles inherited from a specific parent (Table 5.1).

Although Y chromosome microsatellites can only be informative for father–son analyses in parentage studies, their utility has been demonstrated in humans, especially when used in conjunction with autosomal markers. Y chromosome microsatellites are now a standard tool in human forensics and paternity analyses (Jobling *et al.* 1997). Markers from the Y chromosome may be particularly useful when they can exclude the possibility of paternity, or when a sample is not available from the putative father but is available from his paternal male relatives who share the same Y chromosome haplotype. X chromosome microsatellites may be even more useful, as they are informative for both males and females. They may also have advantages over autosomal markers in certain circumstances. In parent–offspring cases involving a daughter, where multiple putative fathers are themselves close blood relatives (e.g. father and son), X chromosome markers can provide greater power than autosomal markers to exclude paternity. Each male inherits his single X chromosome from his mother, so in male relatives with different mothers the X chromosome haplotypes will not be identical by descent, making the X chromosome the region in which their genetic differences are most obvious (Szibor *et al.* 2003). X chromosome microsatellites are also more powerful for determining paternal sisters, as all daughters of a single male inherit the same paternal X chromosome haplotype. This approach has been used in baboons to identify paternal sisters and half-sisters (Smith *et al.* 2003), while in big brown bats (*Eptesicus fuscus*) a single X-linked microsatellite was used to identify multiple paternity in sets of female twins (Vonhof *et al.* 2006). In banner-tailed kangaroo rats (*Dipodomys spectabilis*), X chromosome microsatellites have been used to exclude potential parents (Waser *et al.* 2006; Winters and Waser 2003).

Parentage analyses have been conducted in several macropodoid species using autosomal microsatellites. In rufous bettong populations there was evidence of both monogamous and promiscuous mating (Pope *et al.* 2005), while brush-tailed rock-wallabies had a polygynous mating system with restricted mating dispersal and relatively low levels of skew in reproductive success between individual males (Hazlitt *et al.* 2006b). In contrast, both sexes were promiscuous in the bridled nailtail wallaby (*Onychogalea fraenata*) (Fisher and Lara 1999) although, in common with the brush-tailed rock-wallaby, this species shows a male size threshold above which access to females is more likely. The value of mating system analysis to wildlife management has been demonstrated in the bridled nailtail wallaby, with an investigation into reproductive success in captive-bred versus wild-caught translocated animals (Sigg *et al.* 2005). As described above, the particular characteristics of X and Y chromosome microsatellites may make them valuable new tools for the analysis of macropodoid mating systems, particularly where they can easily exclude potential parents. For example, in populations with restricted dispersal ability or some degree of inbreeding (characteristics that are common to many macropodoid species), the X and Y chromosomes may provide greater resolution for parentage analysis than autosomal markers alone. The potential for sex chromosome markers to assign fathers to greater proportions of offspring than in previous studies may be of particular benefit to studies of macropods, as maternal identity can be determined more easily for these species if DNA samples are collected from mothers and pouch young. However, this approach is still limited by dependence on researchers' ability to sample all or most of the potential fathers in each population (Pope *et al.* 2005).

Conservation genetics of mammal populations

We have discussed the potential for sex chromosome microsatellite markers to improve our understanding of macropodoid population histories and structures, but this new class of marker may be of even greater importance to macropodoid conservation genetics. Many macropodoid species have suffered serious population declines over the last 200 years (Tyndale-Biscoe 2005) and 22 species or subspecies are now listed as 'critically endangered', 'endangered' or 'vulnerable' under Australia's *Environment Protection and Biodiversity Conservation Act 1999* (DEWHA 2008). The most threatened macropodoid species is the 'critically endangered' Gilbert's potoroo (*Potorous gilbertii*). A genetic bottleneck has been detected in this species using mtDNA and autosomal microsatellites (Sinclair *et al.* 2002). In situations such as this, where low genetic diversity is observed in a threatened species, analysis of Y chromosome haplotypes of different males might enhance genetic management of captive breeding colonies and improve our ability to maximise the genetic diversity of individuals reintroduced to the wild. As demonstrated with the Scandinavian wolf population, the introduction of a single genetically distinct male into a small population, originally founded by just two individuals, was sufficient catalyst for a population expansion and increase in genetic diversity in subsequent generations (Vilà *et al.* 2003a).

Sex chromosome microsatellites may also be valuable in determining the conservation status of small, fragmented macropodoid populations. Genetic studies of the brush-tailed rock-wallaby (Browning *et al.* 2001; Hazlitt *et al.* 2006a; Piggott *et al.* 2006b) and the yellow-footed rock-wallaby (Pope *et al.* 1996) have demonstrated that neighbouring colonies represent discrete genetic units and should be managed independently to maintain genetic diversity. The addition of X and Y chromosome microsatellite data to those obtained from existing autosomal and mitochondrial markers may further resolve patterns of gene flow between colonies, particularly male-specific movements, and assist in selection of individuals for captive breeding and relocation programs. Historical DNA samples are available for some populations, and the genetic effects of at least one population decline have been documented in brush-tailed rock-wallabies using autosomal and maternally inherited markers (Eldridge *et al.* 2004b). Future research may also investigate Y chromosome haplotypes in such instances, to determine the effects of contemporary population bottlenecks on the male-specific components of genetic diversity.

SUMMARY

The recent development of sex chromosome microsatellites from the tammar wallaby is an important step for macropodoid biologists. While the resources available to human geneticists are still far beyond our reach, it is becoming possible to select a 'designer suite' of markers to tackle some of the important questions in macropodoid population genetics. The combined use of markers from the autosomes, the sex chromosomes and the mitochondrial genome, selected as appropriate to the research question, provides the opportunity for a comparative approach: concordance between chromosomes gives additional support to conclusions whereas discrepancies may point to more complex factors affected by differing patterns of marker inheritance. However, sex chromosome microsatellites can provide more than just another line of evidence. Their sex-specific inheritance permits the elucidation of sex-linked gene flow and, relative to the autosomes, their lower effective population sizes provide opportunities to study population histories from a different evolutionary timescale.

We have reviewed the utility of sex chromosome microsatellites in population genetics of eutherian mammals and outlined some of the study areas to which their application is particularly suited. In particular, the ability to detect and study processes such as bottlenecks, dispersal patterns, introgression and mating system structures are of great importance to conservation biology. Many macropodoid species are currently subject to conservation management plans, while others are increasingly threatened by habitat destruction and invasive predators. Any improvement in our ability to understand macropodoid population genetic processes is welcome, including the development and application of new markers providing alternative evolutionary perspectives. Sex chromosome microsatellites have become an integral component of the human population genetics toolbox and we are optimistic that they will be of equal or greater value to macropodoid conservation and population biology.

ACKNOWLEDGMENTS

This work is supported by an Australian Research Council Discovery Grant (DP0211687) awarded to SS and NF. Thanks to two anonymous reviewers for their comments and helpful suggestions.

REFERENCES

Allendorf FW, Leary RF, Spruell P and Wenburg JK 2001. The problem with hybrids: setting conservation guidelines. *Trends Ecol. Evol.* **16:** 613–621.

Alsop A, Miethke P, Rofe R, Koina E, Sankovic N, Deakin J, Haines H, Rapkins R and Graves JAM 2005. Characterizing the chromosomes of the Australian model marsupial *Macropus eugenii* (tammar wallaby). *Chromosome Res.* **13:** 627–636.

Ayub Q, Mohyuddin A, Qamar R, Mazhar K, Zerjal T, Mehdi SQ and Tyler-Smith C 2000. Identification and characterisation of novel human Y-chromosomal microsatellites from sequence database information. *Nucleic Acids Res.* **28:** e8.

Boecklen WJ and Howard DJ 1997. Genetic analysis of hybrid zones: numbers of markers and power of resolution. *Ecology* **78:** 2611–2616.

Bowyer JC, Newell GR and Eldridge MDB 2002. Genetic effects of habitat contraction on Lumholtz's tree-kangaroo (*Dendrolagus lumholtzi*) in the Australian Wet Tropics. *Conserv. Genet.* **3:** 59–67.

Bradley BJ, Boesch C and Vigilant L 2000. Identification and redesign of human microsatellite markers for genotyping wild chimpanzee (*Pan troglodytes verus*) and gorilla (*Gorilla gorilla gorilla*) DNA from faeces. *Conserv. Genet.* **1:** 289–292.

Browning TL, Taggart DA, Rummery C, Close RL and Eldridge MDB 2001. Multifaceted genetic analysis of the 'critically endangered' brush-tailed rock-wallaby *Petrogale penicillata* in Victoria, Australia: implications for management. *Conserv. Genet.* **2:** 145–156.

Bruford MW and Wayne RK 1993. Microsatellites and their application to population genetic studies. *Curr. Opin. Genet. Dev.* **3:** 939–943.

Close RL and Bell JN 1997. Fertile hybrids in two genera of wallabies: *Petrogale* and *Thylogale*. *J. Hered.* **88:** 393–397.

Close RL and Lowry PS 1990. Hybrids in marsupial research. *Aust. J. Zool.* **37:** 259–267.

Culver M, Johnson WE, Pecon-Slattery J and O'Brien SJ 2000. Genomic ancestry of the American puma (*Puma concolor*). *J. Hered.* **91:** 186–197.

DEH (Department for Environment and Heritage) 2006. Reintroduction of mainland SA tammar wallaby to Innes National Park. Progress report. Adelaide, South Australia.

DEWHA (Department of Environment, Water, Heritage and Arts) 2008. EPBC Act List of Threatened Fauna. http://www.environmnt.gov.au/cgi-bin/sprat/public/publicthreatenedlist.pl?wanted=fauna. Accessed 10 October 2008.

Dickens HF, Holmes NG, Ryder E, Breen M, Thomas R, Suter N, Sampson J, Langford CF, Ross M, Carter NP and Binns MM1999. Use of cosmid-derived and chromosome-specific canine microsatellites. *J. Hered.* **90:** 52–54.

Eldridge MDB, King JM, Loupis AK, Spencer PBS, Taylor AC, Pope LC and Hall GP 1999. Unprecedented low levels of genetic variation and inbreeding depression in an island population of the black-footed rock-wallaby. *Conserv. Biol.* **13:** 531–541.

Eldridge MDB, Kinnear JE, Zenger KR, McKenzie LM and Spencer PBS 2004a. Genetic diversity in remnant mainland and 'pristine' island populations of three endemic Australian macropodids (Marsupialia): *Macropus eugenii*, *Lagorchestes hirsutus* and *Petrogale lateralis*. *Conserv. Genet.* **5:** 325–338.

Eldridge MDB, Rummery C, Bray C, Zenger KR, Browning TL and Close RL 2004b. Genetic analysis of a population crash in brush-tailed rock-wallabies (*Petrogale penicillata*) from Jenolan Caves, south-eastern Australia. *Wildl. Res.* **31:** 229–240.

Eriksson J, Siedel H, Lukas D, Kayser M, Erler A, Hashimoto C, Hohmann G, Boesch C and Vigilant L 2006. Y-chromosome analysis confirms highly sex-biased dispersal and suggests a low male effective population size in bonobos (*Pan paniscus*). *Mol. Ecol.* **15:** 939–949.

Ermakov OA, Surin VL, Titov SV, Zborovsky SS and Formozov NA 2006. A search for Y-chromosomal species-specific markers and their use for hybridization analysis in ground squirrels (Spermophilus: Rodentia, Sciuridae). *Russ. J. Genet.* **42:** 429–438.

Evans BJ, Supriatna J and Melnick DJ 2001. Hybridization and population genetics of two macaque species in Sulawesi, Indonesia. *Evolution* **55:** 1686–1702.

Fisher DO and Lara MC 1999. Effects of body size and home range on access to mates and paternity in male bridled nailtail wallabies. *Anim. Behav.* **58:** 121–130.

FitzSimmons NN, Moritz C and Moore SS 1995. Conservation and dynamics of microsatellite loci over 300 million years of marine turtle evolution. *Mol. Biol. Evol.* **12:** 432–440.

Freeman AR, Hoggart CJ, Hanotte O and Bradley DG 2006. Assessing the relative ages of admixture in the bovine hybrid zones of Africa and the Near East using X chromosome haplotype mosaicism. *Genetics* **173:** 1503–1510.

Hakki EE and Akkaya MS 2000. Microsatellite isolation using amplified fragment length polymorphism markers: no cloning, no screening. *Mol. Ecol.* **9:** 2152–2154.

Halbert ND, Ward TJ, Schnabel RD, Taylor JF and Derr JN 2005. Conservation genomics: disequilibrium mapping of domestic cattle chromosomal segments in North American bison populations. *Mol. Ecol.* **14:** 2343–2362.

Hammer MF and Horai S 1995. Y chromosomal DNA variation and the peopling of Japan. *Am. J. Hum. Genet.* **56:** 951–962.

Hazlitt SL, Eldridge MDB and Goldizen AW 2004. Fine-scale spatial genetic correlation analyses reveal strong female philopatry within a brush-tailed rock-wallaby colony in southeast Queensland. *Mol. Ecol.* **13:** 3621–3632.

Hazlitt SL, Goldizen AR and Eldridge MDB 2006a. Significant patterns of population genetic structure and limited gene flow in a threatened macropodid marsupial despite continuous habitat in southeast Queensland, Australia. *Conserv. Genet.* **7:** 675–689.

Hazlitt SL, Sigg DP, Eldridge MDB and Goldizen AW 2006b. Restricted mating dispersal and strong breeding group structure in a mid-sized marsupial mammal (*Petrogale penicillata*). *Mol. Ecol.* **15:** 2997–3007.

Hurles ME and Jobling MA 2001. Haploid chromosomes in molecular ecology: lessons from the human Y. *Mol. Ecol.* **10:** 1599–1613.

Jobling MA, Pandya A and Tyler-Smith C 1997. The Y chromosome in forensic analysis and paternity testing. *Int. J. Leg. Med.* **110:** 118–124.

Jones AG and Ardren WR 2003. Methods of parentage analysis in natural populations. *Mol. Ecol.* **12:** 2511–2523.

Jorde LB, Watkins WS, Bamshad MJ, Dixon ME, Ricker CE, Seielstad MT and Batzer MA 2000. The distribution of human genetic diversity: a comparison of mitochondrial, autosomal and Y-chromosome data. *Am. J. Hum. Genet.* **66:** 979–988.

Karagyozov L, Kalcheva ID and Chapman VM 1993. Construction of random small-insert genomic libraries highly enriched for simple sequence repeats. *Nucleic Acids Res.* **21:** 3911–3912.

Korstanje R, Gillissen GF, den Bieman MG, Versteeg SA, van Oost B, Fox RR, van Lith HA and van Zutphen LFM 2001. Mapping of rabbit chromosome 1 markers generated from a microsatellite-enriched chromosome-specific library. *Anim. Genet.* **32:** 308–312.

Lancaster ML, Gemmell NJ, Negro S, Goldsworthy S and Sunnucks P 2006. Ménage à trois on Macquarie Island: hybridization among three species of fur seal (*Arctocephalus* spp.) following historical population extinction. *Mol. Ecol.* **15:** 3681–3692.

Le Page SL, Livermore RA, Cooper DW and Taylor AC 2000. Genetic analysis of a documented population bottleneck: introduced Bennett's wallabies (*Macropus rufogriseus rufogriseus*) in New Zealand. *Mol. Ecol.* **9:** 753–763.

Lecis R, Pierpaoli M, Biro ZS, Szemethy L, Ragni B, Vercillo F and Randi E 2006. Bayesian analyses of admixture in wild and domestic cats (*Felis silvestris*) using linked microsatellite loci. *Mol. Ecol.* **15:** 119–131.

MacDonald AJ, Sankovic N, Sarre SD, FitzSimmons NN, Wakefield MJ, Graves JAM and Zenger KR 2006. Y chromosome microsatellite markers identified from the tammar wallaby (*Macropus eugenii*) and their amplification in three other macropod species. *Mol. Ecol. Notes* **6:** 1202–1204.

MacDonald AJ, Sarre SD, FitzSimmons NN and Graves JAM 2007. Chromosome-specific microsatellites from the tammar wallaby X chromosome and chromosome 2. *Mol. Ecol. Notes* **7:** 1063–1066.

Mesa NR, Mondragón MC, Soto ID, Parra MV, Duque C, Ortíz-Barrientos D, García LF, Velez ID, Bravo ML, Múnera JG, Bedoya G, Bortolini MC and Ruiz-Linares A 2000. Autosomal, mtDNA and Y-chromosome diversity in Amerinds: pre- and post-Columbian patterns of gene flow in South America. *Am. J. Hum. Genet.* **67:** 1277–1286.

NHGRI (National Human Genome Research Institute) 2008. http://www.genome. gov/12512299. Accessed 16 October 2008.

Nijman J, Otsen M, Verkaar ELC, de Ruijter C, Hanekamp E, Ochieng JW, Shamshad S, Rege JEO, Hanotte O, Barwegen MW, Sulawati T and Lenstra JA 2003. Hybridization of banteng (*Bos javanicus*) and zebu (*Bos indicus*) revealed by mitochondrial DNA, satellite DNA, AFLP and microsatellites. *Heredity* **90:** 10–16.

Pérez-Lezaun A, Calafell F, Seielstad M, Mateu E, Comas D, Bosch E and Bertranpetit J 1997. Population genetics of Y-chromosome short tandem repeats in humans. *J. Mol. Evol.* **45:** 265–270.

Pérez-Lezaun A, Calafell F, Comas D, Mateu E, Bosch E, Martínez-Arias R, Clarimón J, Fiori G, Luiselli D, Facchini F, Pettener D and Bertranpetit J 1999. Sex-specific migration patterns in central Asian populations, revealed by analysis of Y-chromosome short tandem repeats and mtDNA *Am. J. Hum. Genet.* **65:** 208–219.

Petit E, Balloux F and Excoffier L 2002. Mammalian population genetics: why not Y? *Trends Ecol. Evol.* **17:** 28–33.

Piggott MP, Banks SC, Stone N, Banffy C and Taylor AC 2006a. Estimating population size of endangered brush-tailed rock-wallaby (*Petrogale penicillata*) colonies using faecal DNA *Mol. Ecol.* **15:** 81–91.

Piggott MP, Banks SC and Taylor AC 2006b. Population structure of brush-tailed rock-wallaby (*Petrogale penicillata*) colonies inferred from analysis of faecal DNA *Mol. Ecol.* **15:** 93–105.

Poole W and Catling P 1974. Reproduction in the two species of grey kangaroos, *Macropus giganteus* Shaw and *M. fuligimsus* (Desmarest). I. Sexual maturity and oestrus. *Aust. J. Zool.* **22:** 277–302.

Pope LC, Sharp A and Moritz C 1996. Population structure of the yellow-footed rock-wallaby *Petrogale xanthopus* (Gray, 1854) inferred from mtDNA sequences and microsatellite loci. *Mol. Ecol.* **5:** 629–640.

Pope LC, Blair D and Johnson CN 2005. Dispersal and population structure of the rufous bettong, *Aepyprymnus rufescens* (Marsupialia: Potoroidae). *Austral Ecol.* **30:** 572–580.

Primmer CR, Moller AP and Ellegren H 1996. A wide-range survey of cross-species microsatellite amplification in birds. *Mol. Ecol.* **5:** 365–378.

Prugnolle F and de Meeus T 2002. Inferring sex-biased dispersal from population genetic tools: a review. *Heredity* **88:** 161–165.

Rhymer JM and Simberloff D 1996. Extinction by hybridization and introgression. *Ann. Rev. Ecol. Syst.* **27:** 83–109.

Ruiz-Linares A, Ortiz-Barrientos D, Figueroa M, Mesa N, Múnera JG, Bedoya G, Vélez ID, García LF, Pérez-Lezaun A, Bertranpetit J, Feldman MW and Goldstein DB 1999. Microsatellites provide evidence for Y chromosome diversity among the founders of the New World. *Proc. Natl Acad. Sci. USA* **96:** 6312–6317.

Schaffner SF 2004. The X chromosome in population genetics. *Nat. Rev. Genet.* **5:** 43–51.

Schlötterer C 2000. Evolutionary dynamics of microsatellite DNA. *Chromosoma* **109:** 365–371.

Seielstad MT, Minch E and Cavalli-Sforza LL 1998. Genetic evidence for a higher female migration rate in humans. *Nat. Genet.* **20:** 278–280.

Seielstad M, Bekele E, Ibrahim M, Toure A and Traore M 1999. A view of modern human origins from Y chromosome microsatellite variation. *Genome Res.* **9:** 558–567.

Sigg DP, Goldizen AW and Pople AR 2005. The importance of mating system in translocation programs: reproductive success of released male bridled nailtail wallabies. *Biol. Conserv.* **123:** 289–300.

Sinclair EA, Costello B, Courtenay JM and Crandall KA 2002. Detecting a genetic bottleneck in Gilbert's potoroo (*Potorous gilbertii*) (Marsupialia: Potoroidae), inferred from microsatellite and mitochondrial DNA sequence data. *Conserv. Genet.* **3:** 191–196.

Smith K, Alberts SC and Altmann J 2003. Wild female baboons bias their social behaviour towards paternal half-sisters. *Proc. R. Soc. Lond. B Biol. Sci.* **270:** 503–510.

Sundqvist AK, Ellegren H, Olivier M and Vila C 2001. Y chromosome haplotyping in Scandinavian wolves (*Canis lupus*) based on microsatellite markers. *Mol. Ecol.* **10:** 1959–1966.

Sunnucks P 2000. Efficient genetic markers for population biology. *Trends Ecol. Evol.* **15:** 199–203.

Szibor R, Krawczak M, Hering S, Edelmann J, Kuhlisch E and Krause D 2003. Use of X-linked markers for forensic purposes. *Int. J. Leg. Med.* **117:** 67–74.

Taylor AC and Cooper DW 1999. Microsatellites identify introduced New Zealand tammar wallabies (*Macropus eugenii*) as an 'extinct' taxon. *Anim. Conserv.* **2:** 41–49.

Tuskan GA, Gunter LE, Yang ZK, Yin T, Sewell MM and DiFazio SP 2004. Characterization of microsatellites revealed by genomic sequencing of *Populus trichocarpa*. *Can. J. For. Res.* **34:** 85–93.

Tyndale-Biscoe H 2005. *Life of Marsupials*. CSIRO Publishing, Melbourne.

Verardi A, Lucchini V and Randi E 2006. Detecting introgressive hybridization between free-ranging domestic dogs and wild wolves (*Canis lupus*) by admixture linkage disequilibrium analysis. *Mol. Ecol.* **15:** 2845–2855.

Vilà C, Sundqvist AK, Flagstad Ø, Seddon J, Bjornerfeldt S, Kojola I, Casulli A, Sand H, Wabakken P and Ellegren H 2003a. Rescue of a severely bottlenecked wolf (*Canis lupus*) population by a single immigrant. *Proc. R. Soc. Lond. B Biol. Sci.* **270:** 91–97.

Vilà C, Walker C, Sundqvist AK, Flagstad Ø, Andersone Z, Casulli A, Kojola I, Valdmann H, Halverson J and Ellegren H 2003b. Combined use of maternal, paternal and bi-parental genetic markers for the identification of wolf–dog hybrids. *Heredity* **90:** 17–24.

Vonhof MJ, Barber D, Fenton MB and Strobeck C 2006. A tale of two siblings: multiple paternity in big brown bats (*Eptesicus fuscus*) demonstrated using microsatellite markers. *Mol. Ecol.* **15:** 241–247.

Ward TJ, Skow LC, Gallagher DS, Schnabel RD, Nall CA, Kolenda CE, Davis SK, Taylor JF and Derr JN 2001. Differential introgression of uniparentally inherited markers in bison populations with hybrid ancestries. *Anim. Genet.* **32:** 89–91.

Waser PM, Busch JD, McCormick CR and Dewoody JA 2006. Parentage analysis detects cryptic precapture dispersal in a philopatric rodent. *Mol. Ecol.* **15:** 1929–1937.

Wilson ACC, Massonnet B, Simon J-C, Prunier-Leterme N, Dolatti L, Llewellyn KS, Figueroa CC, Ramirez CC, Blackman RL, Estoup A and Sunnucks P 2004. Cross-species amplification of microsatellite loci in aphids: assessment and application. *Mol. Ecol. Notes* **4:** 104–109.

Winters JB and Waser PM 2003. Gene dispersal and outbreeding in a philopatric mammal. *Mol. Ecol.* **12:** 2251–2259.

Wright JM and Bentzen P 1994. Microsatellites: genetic markers for the future. *Rev. Fish Biol. Fish.* **4:** 384–388.

Zegura SL, Karafet TM, Zhivotovsky LA and Hammer MF 2004. High-resolution SNPs and microsatellite haplotypes point to a single, recent entry of Native American Y chromosomes into the Americas. *Mol. Biol. Evol.* **21:** 164–175.

Zenger KR and Cooper DW 2001. Characterization of 14 macropod microsatellite genetic markers. *Anim. Genet.* **32:** 166–167.

Zenger KR, McKenzie LM and Cooper DW 2002. The first comprehensive genetic linkage map of a marsupial: the tammar wallaby (*Macropus eugenii*). *Genetics* **162:** 321–330.

Zenger KR, Eldridge MDB and Cooper DW 2003a. Intraspecific variation, sex-biased dispersal and phylogeography of the eastern grey kangaroo (*Macropus giganteus*). *Heredity* **91:** 153–162.

Zenger KR, Eldridge MDB, Pope LC and Cooper DW 2003b. Characterisation and cross-species utility of microsatellite markers within kangaroos, wallabies and rat-kangaroos (Macropodoidea : Marsupialia). *Aust. J. Zool.* **51:** 587–596.

6 Sperm competition in the Macropodoidea: a review of evidence

J.Z. Paplinska, R. Bencini, D.O. Fisher, G.R. Newell, A.W. Goldizen, S.L. Hazlitt, D.P. Sigg, G.R. Finlayson, A.J. Munn, B.K. Chambers and D.A. Taggart

SUMMARY

This chapter reviews the evidence for sperm competition in the marsupial superfamily Macropodoidea. The opportunity for sperm competition arises when a female is mated by more than one male during a single oestrous period. Previous studies suggest that the male that inseminates the female with the greatest number of sperm, or the highest-quality sperm, will have the highest probability of successfully fertilising the female's egg(s). If sperm competition is operating, selection should favour males with larger testes that produce more sperm, or males capable of producing higher-quality sperm. In a 'raffle' system, in which sperm numbers are crucial to success, males of species with polyandrous mating systems are expected to have larger testes relative to their body weight than males of species with monandrous mating systems. This relationship holds true for many taxa including marsupials. Hence, testes:body weight ratios have been used in conjunction with other morphological and behavioural data to predict the likelihood of sperm competition and the possible mating system. We have assembled testis data from numerous studies, both published and unpublished, resulting in testis weights for 17 species in eight genera, and data on scrotum width for 17 species in eight genera. The literature has been examined for information about the mating system and other correlates of sperm competition to determine if the expected relationships between testis weight, body weight and mating system hold true within the Macropodoidea. The paucity of data about mating systems, especially with respect to the degree of polyandry in females, prevents us from clearly stating that there is a positive relationship between the degree of polyandry and testis weight although none of our data contradict this relationship. There is some indication that seasonal breeding may increase the probability of sperm competition.

INTRODUCTION

Intrasexual and intersexual selection are the two main components of sexual selection, which act to maximise individual reproductive success (Darwin 1871). In most species with two sexes, males compete to maximise their reproductive success by increasing the number of females with which they mate, and females seek to maximise their reproductive success by choosing high-quality males (Qvarnstrom and Forsgren 1998). In addition to these pre-copulatory mechanisms of sexual selection, in situations where a female is mated by more than one male during a single fertile period there is also the opportunity for post-copulatory sexual selection in the form of sperm competition (Parker 1970; Birkhead 1995) and/or sperm selection (Thornhill 1983; Birkhead 1995; Gomendio *et al.* 1998; Snook 2005). Sperm competition is often visualised as a raffle, in which individual males are more likely to succeed with fertilisation if they contribute more sperm to the competition occurring within the female reproductive tract (Parker 1970; Taborsky 1998). Alternatively, sperm quality can influence the outcome of competition between males for fertilisation, independently of the number of sperm inseminated (a good-sperm model, Yasui 1997). Both these mechanisms have been demonstrated in marsupials (Fisher *et al.* 2006; Holleley *et al.* 2006). Sperm choice by females occurs when females bias fertilisation success towards certain males after mating (Thornhill 1983; Birkhead 1995; Snook 2005). It is near-impossible to distinguish between sperm competition and sperm selection in free-living animals, but large testis size is predicted by the raffle model of sperm competition and not necessarily the good sperm or sperm choice mechanisms (Birkhead 1995). We review the variation in testis weight relative to body weight in macropodoids, and its association with mating system.

Sperm competition has the potential to occur in all species where females are mated by multiple males. In vertebrates its existence has been confirmed in fish (Petersen and Warner 1998; Taborsky 1998), amphibians (Jennions and Passmore 1993; Halliday 1998), reptiles (Olsson and Madsen 1998), birds (Birkhead 1995, 1998) and mammals (Tyndale-Biscoe and Renfree 1987; Harcourt *et al.* 1995; Gomendio *et al.* 1998; Kraaijeveld-Smit *et al.* 2002; Gage and Freckleton 2003; Preston *et al.* 2003; Fisher *et al.* 2006; Holleley *et al.* 2006). Males may maximise the probability of outcompeting rival males and successfully fertilising a female by increasing the number of spermatozoa with which they inseminate the female (Martin *et al.* 1974; Martin and Dziuk 1977). The number of spermatozoa produced is related to the amount of seminiferous tissue in the testes and the weight of the testes (adjusted for body weight) is positively correlated with ejaculate volume and sperm counts in mammals (Moller 1989; Simmons *et al.* 2004). Therefore, males of polyandrous species are expected to have larger testes and produce more sperm than males of monandrous species (Gage and Freckleton 2003). This relationship between testis size and behaviour holds true for many mammals, including marsupials (Short 1977, 1979, 1997; Harcourt *et al.* 1981; Kenagy and Trombulak 1986; Tyndale-Biscoe and Renfree 1987; Gomendio and Roldan 1991; Rose 1997; Taggart *et al.* 1998; Sacher *et al.* 1999; Ramm *et al.* 2005).

Sperm morphological characteristics can also play a part in increasing male reproductive success during sperm competition (Taborsky 1998; Snook 2005). Characteristics such as midpiece volume, tail length and tail straightness play a large role in determining how fast a sperm can move and whether it has progressive or non-progressive motility (Barth and Oko 1989; Gomendio and Roldan 1991; Anderson and Dixson 2002), and sperm morphology is the best parameter to predict sperm fertilising capacity (Ombelet *et al.* 1997; Kubo-Irie *et al.* 2004). These features of the ejaculate in mammals may be positively correlated with each other and with sperm number and testes size (Gomendio *et al.* 1998), although recent findings show that sperm morphology is not associated with relative testis size in mammals once phylogenetic associations are controlled for (Gage and Freckleton 2003).

The marsupial superfamily Macropodoidea is composed of 73 species in 16 genera (Appendix this volume), but the majority of our knowledge comes from a handful of species. Previous reviews of sperm competition in marsupials included information about only 11 species of macropodoid (Rose 1997; Taggart *et al.* 1998). The macropodoids pose special challenges to the study of sperm competition and the evolution of male reproductive anatomy because the best evidence of viable multiple mating of females by males is multiple paternity of litters and, with the exception of the musky rat-kangaroo (*Hypsiprymnodon moschatus*) which regularly gives birth to twins or triplets (Johnson and Strahan 1982; Dennis and Marsh 1997), all macropodoids are monovular (Tyndale-Biscoe and Renfree 1987). Data about the mating systems of macropodoid species are very scant and have generally been inferred from their social organisation (Jarman 1983, 1991), which has repeatedly been shown to be a poor indicator of the mating system (Kempenaers *et al.* 1992; Lifjeld *et al.* 1992; Dixon *et al.* 1994; Hughes 1998; Spencer *et al.* 1998; Coltman *et al.* 1999; Hynes *et al.* 2005). The purpose of this review is to assemble the available data about testes size and behavioural and genetic mating systems in the Macropodoidea and interpret the relationships between these, including any new data which have accumulated since the previous reviews (Rose 1997; Taggart *et al.* 1998).

COMPARISON OF RELATIVE TESTES SIZE

Testes weights and body weights were collected from a total of 17 macropodoid species (Table 6.1) and linear dimensions of testes measured within the scrotum and body weights were collected for a total of 17 species (Table 6.2). While it is possible to convert linear dimensions to weights, all three radii of the testis must be measured and these measures were only available for three species – bridled nailtail wallaby (*Onychogalea fraenata*), allied rock-wallaby (*Petrogale assimilis*) and brush-tailed rock-wallaby (*Petrogale penicillata*). For these species the linear dimensions were converted to weights using the formula (Harcourt *et al.* 1995):

$$\text{weight (g)} = 2 \times \text{volume of a single testis (cm}^3\text{)} \times 1.1$$

where the volume of a single testis is the volume of an ellipsoid.

Due to the large variation in sample sizes (n = 1 to n = 265) a stepwise multiple linear regression was carried out in *STATISTIXL* (StatistiXL, Nedlands, Australia) to test whether sample size affected the relationship between body size and testis size. Where no sample size was given the sample size was assumed to be one. There was no significant relationship found between sample size and body weight (r^2 = 0.039, p = 0.878) or testis weight (r^2 = 0.087, p = 0.731) or between sample size and body weight (r^2 = 0.067, p = 0.780) or scrotum width (r^2 = 0.100, p = 0.676). All measurements were from adult males assumed to be sexually mature based on visual assessment of testis size. Seasonal information was not collected for any

Table 6.1: Residuals of log testes weight (g) regressed against log body weight (g) and likely mating system in 17 macropodoid species.

Species	Code	n	Residual	Mating
Brush-tailed rock-wallaby (*Petrogale penicillata*)	Ppen	61	–0.26[1]	M[8, 9]
Brush-tailed bettong (*Bettongia penicillata*)	Bpen	1	–0.24[2]	?
Long-nosed potoroo (*Potorous tridactylus*)	Ptri	2	–0.24[2]	?
Red kangaroo (*Macropus rufus*)	Mruf	4	–0.24[3]	?
Eastern grey kangaroo (*Macropus giganteus*)	Mgig	9	–0.20[3]	P
Rufous bettong (*Aepyprymnus rufescens*)	Aruf	1	–0.17[2]	P[10, 11]
Common wallaroo (*Macropus robustus*)	Mrob	?	–0.12[4]	?
Allied rock-wallaby (*Petrogale assimilis*)	Pass	17	–0.11	M[13]
Swamp wallaby (*Wallabia bicolor*)	Wbic	149	–0.09[5]	?
Western grey kangaroo (*Macropus fuliginosus*)	Mful	10	–0.05[6]	?
Red-necked wallaby (*Macropus rufogriseus banksianus*)	Mrfgb	6	–0.01[1]	P[13]
Agile wallaby (*Macropus agilis*)	Magi	6	0.02[3]	?
Tasmanian bettong (*Bettongia gaimardi*)	Bgai	3	0.04[6]	?
Black-footed rock-wallaby (*Petrogale lateralis*)	Plat	3	0.15[7]	?
Bridled nailtail wallaby (*Onychogalea fraenata*)	Ofra	265	0.17[1]	P[14, 15, 16]
Bennett's wallaby (*Macropus rufogriseus rufogriseus*)	Mrrfg	9	0.19[3]	?
Tasmanian pademelon (*Thylogale billardierii*)	Tbil	12	0.41[6]	?
Tammar wallaby (*Macropus eugenii*)	Meug	17	0.64[3]	P[17, 18]

P = polyandry, M = monandry.

1 This study. **2** Taggart *et al.* (1998). **3** Tyndale-Biscoe and Renfree (1987). **4** Setchell (1977). **5** Paplinska (2005). **6** Rose (1997). **7** Jones (2002). **8** Jarman and Bayne (1997). **9** Hazlitt *et al.* (2006). **10** Frederick and Johnson (1996). **11** Pope *et al.* (2005). **12** Spencer *et al.* (1998). **13** Watson *et al.* (1992). **14** Sigg and Goldizen (2006). **15** Sigg *et al.* (2005). **16** Fisher and Lara (1999). **17** Hynes *et al.* (2005). **18** Ewen *et al.* (1993).

Table 6.2: Residuals of log scrotum width (cm) regressed against the cube root of body weight (g) and likely mating system in 17 macropodoid species.

Species	Code	n	Residual	Mating
Lumholtz's tree-kangaroo (*Dendrolagus lumholtzi*)	Dlum	13	–0.16[1]	?
Doria's tree-kangaroo (*Dendrolagus dorianus notatus*)	Ddno	?	–0.13[2]	?
Tenkile (*Dendrolagus scottae*)	Dsco	?	–0.09[2]	?
Brush-tailed rock-wallaby (*Petrogale penicillata*)	Ppen	78	–0.07[1]	M[4, 5]
Brush-tailed bettong (*Bettongia penicillata*)	Bpen	1	–0.07[1]	?
Rufous bettong (*Aepyprymnus rufescens*)	Aruf	1	–0.05[1]	P[6, 7]
Goodfellow's tree-kangaroo (*Dendrolagus goodfellowi buergersi*)	Dgbu	?	–0.03[2]	?
Grizzled tree-kangaroo (*Dendrolagus inustus finschi*)	Difi	?	–0.02[2]	?
Red-necked wallaby (*Macropus rufogriseus banksianus*)	Mrfgb	9	–0.01[1]	P[8]
Bridled nailtail wallaby (*Onychogalea fraenata*)	Ofra	306	–0.01[1]	P[9, 10, 11]
Doria's tree-kangaroo (*Dendrolagus dorianus dorianus*)	Dddo	?	–0.01[2]	?
Allied rock-wallaby (*Petrogale assimilis*)	Pass	17	0.02	M[12]
Doria's tree-kangaroo (*Dendrolagus dorianus stellarum*)	Ddst	?	0.03[2]	?
Goodfellow's tree-kangaroo (*Dendrolagus goodfellowi goodfellowi*)	Dggo	?	0.04[2]	?
Swamp wallaby (*Wallabia bicolor*)	Wbic	146	0.05[3]	?
Matschei's tree-kangaroo (*Dendrolagus matschei*)	Dmat	2	0.06[1]	?
Long-nosed potoroo (*Potorous tridactylus*)	Ptri	2	0.10[1]	?
Lowland tree-kangaroo (*Dendrolagus spadix*)	Dspa	?	0.11[2]	?
Yellow-footed rock-wallaby (*Petrogale xanthopus*)	Pxan	1	0.12[1]	?
Tammar wallaby (*Macropus eugenii*)	Meug	88	0.14[1]	P[13, 14]

P = polyandry, M = monandry.

1 This study. **2** Flannery *et al.* (1996). **3** Paplinska (2005). **4** Jarman and Bayne (1997). **5** Hazlitt *et al.* (2006). **6** Frederick and Johnson (1996). **7** Pope *et al.* (2005). **8** Watson *et al.* (1992). **9** Sigg and Goldizen (2006). **10** Sigg *et al.* (2005). **11** Fisher and Lara (1999). **12** Spencer *et al.* (1998). **13** Hynes *et al.* (2005). **14** Ewen *et al.* (1993).

species, but seasonal regression and recrudescence of the testes has not been reported in any male macropodoids (Poole and Pilton 1964; Poole 1973, 1975; Paplinska *et al.* 2007) except for the strictly seasonal tammar wallaby (*Macropus eugenii*) which has an approximately 20% decrease in testis weight outside the breeding season (Paris *et al.* 2005). Since there is only one other strictly seasonal breeding macropodoid (Bennett's wallaby, *Macropus rufogriseus rufogriseus*; Catt 1977; Merchant and Calaby 1981) the season of collection is not likely to have affected testis size for the majority of species.

There is usually an allometric relationship between the size of an organ and body size. If body size has a significant effect on testes size then its effects must be excluded in order to compare testes sizes between differently sized species. Simple linear regressions were carried out in *SYSTAT* on log body weight (g) and log testis weight (g) (Fig. 6.1) and on log scrotum width (cm) and the cube root of body weight (g) (Fig. 6.2). Body weight had a significant effect on both testis weight ($r = 0.86$, $p = 0.00$) and scrotum width ($r = 0.77$, $p = 0.00$). Traditional statistical analyses assume independence of data points but, because more closely related species tend to be more similar than distantly related species, this assumption does not hold true (Felsenstein 1985). Since there are only limited data about phylogenetic relationships among New Guinean tree-kangaroos (*Dendrolagus* spp.) the following analyses were only carried out on the testis weight group; standard residual values were used as the measure of testis size independent of body size for the scrotum width group. We used *COMPARE* (Martins 2004) to investigate if phylogeny had a significant effect on the relationship between body size and testis size by

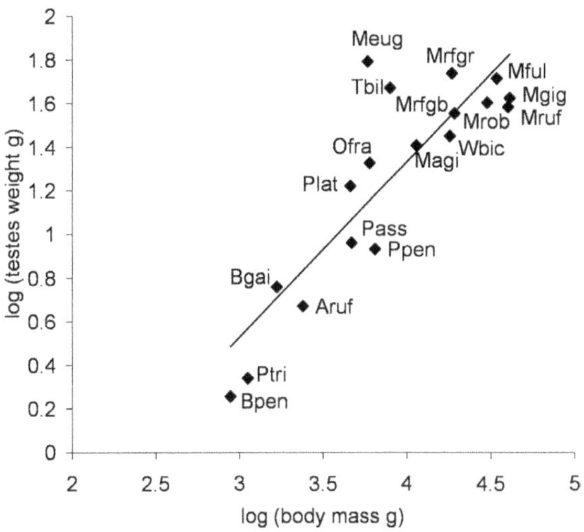

Figure 6.1: Log testes weight (g) plotted against log body weight (g) showing the least-squares regression line. The relationship between testes weight and body weight is highly significant (r = 0.86, p = 0.00). See Table 6.1 for species codes.

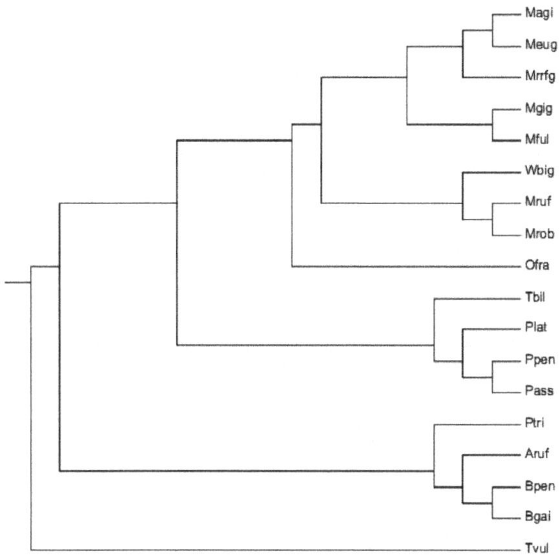

Figure 6.3: Phylogeny used for *CONTINUOUS* analysis, constructed by combining the phylogenies of Kirsch *et al.* (1997) and Cardillo *et al.* (2004), using the common brushtail possum, *Trichosurus vulpecula* (Tvul) as the outgroup. See Table 6.1 for species codes.

using the phylogenetically mixed model (PMM) to calculate the phylogenetic heritability (Phy h^2, the relative importance of phylogeny in explaining phenotypic variation). We then tested whether the significant association between testis weight and body weight still existed if phylogeny was accounted for, using an analysis of Felsenstein's Independent Contrasts (IC, Felsenstein 1985). The phylogeny used was a composite of the two most extensive phylogenies available (Kirsch *et al.* 1997; Cardillo *et al.* 2004) and used the common brushtail possum (*Trichosurus vulpecula*) as an outgroup (Fig. 6.3). Because branch lengths were not known for all the species we assumed equal branch lengths, as recommended by Purvis and Rambaut (1995). The IC analysis has been shown to be very robust to even large errors in branch length (Diaz-Uriarte and Garland 1998). The PMM analysis showed that there was a strong effect of phylogeny on the positive association between body weight and testis weight (Phy h^2 = 1.0). The IC analysis showed that the association still existed when phylogeny was accounted for (r = 0.49, p = 0.039, Fig. 6.3), so residual values were used as the measure of testis size independent of body size for the testis weight group.

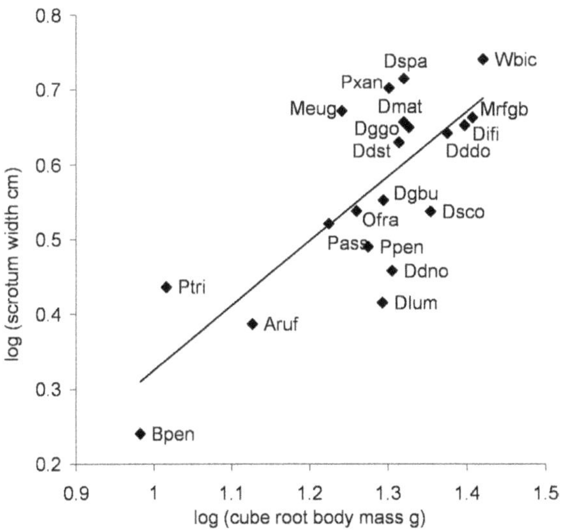

Figure 6.2: Log scrotum width (cm) plotted against the log of the cube root of body weight (g) showing the least-squares regression line. The relationship between scrotum width and body weight is highly significant (r = 0.77, p = 0.00). See Table 6.1 for species codes.

DOES TESTIS SIZE INCREASE WITH INCREASING POLYANDRY?

There was a wide range of residual values for both the testis weight and scrotum width groups (Figs 6.1 and 6.2, Tables 6.1 and 6.2). If testis weight is indicative of mating system, this points to variation in mating strategies within the Macropodoidea. Very little reliable information about genetic mating systems, especially with respect to the occurrence of polyandry, is available for any species of macropodoid. Of the species included in this review, there are only seven for which the mating system has been explicitly studied: brush-tailed rock-wallabies (Jarman and Bayne 1997; Hazlitt *et al.* 2006), allied rock-wallabies (Spencer *et al.* 1998), bridled nailtail wallabies (Fisher and Lara 1999; Sigg *et al.* 2005; Sigg and Goldizen 2006), rufous bettongs (*Aepyprymnus rufescens*) (Frederick and Johnson 1996; Pope *et al.* 2005), red-necked wallabies (*M. rufogriseus banksianus*) (Watson *et al.* 1992), eastern grey kangaroos (*Macropus giganteus*) (Jarman and Southwell 1986; Walker 1996) and tammar wallabies (Ewen *et al.* 1993; Rudd 1994; Paris *et al.* 2004; Hynes *et al.* 2005).

Brush-tailed rock-wallabies are polygynous, with one adult male sharing a rocky refuge and siring young with one to five adult females (Jarman and Bayne 1997; Hazlitt *et al.* 2006). Males defend access to the refuge and the females and generally sire young with multiple females during a single oestrous period and across successive oestrous periods (Hazlitt *et al.* 2006). Over successive oestrous periods females usually conceive by a single male, whose home range overlaps significantly with their home range; they will conceive by a different male only when their current partner disappears from the colony (Hazlitt *et al.* 2006). This indicates that the probability of multiple mating within a single oestrous period is low for female brush-tailed rock-wallabies. This is consistent with males of the species having the lightest and almost the smallest testes in this review (Table 6.1).

Allied rock-wallabies have a mating system similar to that of brush-tailed rock-wallabies. Adult males often share a home range and refuge site with two or three females, so the species has been described as facultatively monogamous (Horsup 1996). Spencer *et al.* (1998) showed that about 50% of females have a monogamous mating system, based on genetic data, and are behaviourally pair-bonded to the father of all their young. These females shared a refuge with one adult male. The other half of females were polygynous, either sharing a refuge with the father of some of their young, or mating with different males outside their partner's territory in successive oestrous periods but not conceiving with their social mate. Thus the probability of multiple mating within a single oestrous period is low for female allied rock-wallabies, as most females are either faithful to a single male to which they are pair-bonded, or appeared to mate with only one male outside their partner's territory (Spencer *et al.* 1998). The smaller than average testes of males of this species (Table 6.1) are consistent with a lack of polyandry.

Behavioural observations have shown eastern grey kangaroos to have a stable dominance hierarchy, with the largest male defending access to the female while she is in oestrus (Lee and Cockburn 1985; Walker 1996) and the majority of females mating with the most dominant male (Jarman and Southwell 1986). Multiple mating of females by males has been observed in this species (Walker 1996) but it is not clear how often this happens or whether mating of females by subordinate males happens during physiological oestrus, so we are unable to say whether the relationship between polyandry and testis size in macropodoids is supported by the data for this species. If the relationship does hold true, then the relatively low testis weights in this species would indicate that dominant males are usually successful at preventing mating by subordinates or that mating by subordinates occurs when females are not fertile.

Observations of rufous bettongs in the wild show that most males do not associate with a particular female outside of oestrus but rather try to maintain contact with as many different females as possible. Females approaching oestrus are continually followed by several males, suggesting a promiscuous mating system for this species (Frederick and Johnson 1996). Microsatellite analysis of paternity shows that, over successive oestrous periods, some females conceive young by the same male whereas others conceive young by several males (Pope *et al.* 2005). The frequency of multiple mating at a single oestrus is not known in this species and the testis weight is based on a small sample size ($n = 1$) so we cannot be sure if the data for this species support the relationship between polyandry and testis size in macropodoids. If the relationship holds, however, and if the testis weight measured in this study is representative of males of this species, then the low testis weights of rufous bettongs indicate an absence of multiple mating of females by males.

Captive red-necked wallabies appear to be polygynous with a dominance hierarchy where the position of alpha male alternates between the two largest males (Watson *et al.* 1992). Mating by subordinates can be successful at least 30% of the time although being the alpha male may slightly enhance reproductive success (Watson *et al.* 1992). We do not know whether both dominant red-necked wallaby males and subordinate red-necked wallaby males mate with the female during a single oestrus, and so do not know whether the data for this species support a positive relationship between polyandry and testis weight in macropodoids. However, if testis weight is indicative of the degree of polyandry in macropodoids, then the smaller than average testes of males of this species suggest that multiple mating during a single oestrus is rare or absent.

Bridled nailtail wallabies are a solitary species with a promiscuous mating system (Fisher and Lara 1999; Sigg *et al.* 2005). Observations of wild bridled nailtail wallabies show that males do not form a predetermined, stable dominance hierarchy, but do establish temporary dominance hierarchies during an oestrous event (Sigg *et al.* 2005). Sigg and Goldizen (2006) found that in 44% of sightings oestrous females were seen associating with a single male, and in 56% of sightings they associated with two to four males. Fisher and Lara (1999) saw females known to be in oestrus associating with one male at a time in 84% of observations, with a maximum of six males at once, and nine males altogether during the week of oestrus. Males sire the young of several females, and females often conceive by a different male at each mating (Fisher and Lara 1999). Some males sired a disproportionate number of offspring within their home ranges, although a relatively low male reproductive skew was observed in this species (Fisher and Lara 1999). There is some evidence of multiple mating – two females were observed to mate with two males that were later excluded as the sires of the resulting young (Fisher and Lara 1999) – but the frequency of multiple mating is not known. However, males engage in post-copulatory guarding of females for up to 48 hours, suggesting that males perceive a high risk of polyandry (Fisher and Lara 1999). Based on testis size, bridled nailtail wallabies are likely to have a high degree of polyandry.

Tammar wallabies have a promiscuous mating system (*sensu* Wittenberger 1979). In captivity, male fighting establishes a dominance hierarchy among males and oestrous females are mate-guarded by the dominant male for around eight hours, thus preventing the majority of subordinates from mating (Rudd 1994). After this time the alpha male ceases to guard the female and she is mated by subordinate males (Rudd 1994). The alpha male generally mates before subordinate males (Hynes *et al.* 2005) but mate-guarding by the alpha male does not prevent subordinates from mating with a female and siring a young (Ewen *et al.* 1993; Hynes *et al.* 2005). Second and third matings result in the same number of young as first matings (Hynes *et al.* 2005). It is uncertain how closely the captive studies resemble the situation in the wild. In all three of the studies mentioned, oestrus was induced in females by the removal of pouch young, leading to a number of females undergoing oestrus simultaneously. Dominant males are unable to guard more than one receptive female at once, resulting in increased mating and siring success by subordinates. Breeding in the wild is also highly synchronous (Tyndale-Biscoe and Renfree 1987), so multiple mating of females by many males is not likely to be an artefact of captivity. Multiple mating in the wild has been inferred from high ejaculate weights within female tracts collected from shot specimens during the major breeding season on Kangaroo Island (Paris *et al.* 2004).

The highly synchronous breeding season may be partly responsible for tammar wallabies having the largest and heaviest testes of any macropodoid included in this review. A highly synchronous breeding season with large numbers of females coming into oestrus at the same time is likely to increase competition among males and decrease their ability to defend a female (Taggart *et al.* 1998). This will increase the degree of polyandry and the intensity of sperm competition. Although there are insufficient quantitative data about seasonal breeding for formal statistical analysis, there is some information which indicates that increased seasonality increases the intensity of sperm competition. First, the only two macropodoid species which are strictly seasonal breeders (tammar wallabies and Bennett's wallabies), have the largest and third-largest testes, respectively, of the species in this review. However, the species with the second-largest testes (the Tasmanian pademelon, *Thylogale billardierii*) breeds year-round (Strahan 2002). The measurements of Bennett's wallaby testes may not have all been taken during the breeding season; if they had been, perhaps the average testis size would be much larger than that of the Tasmanian pademelon. More convincingly, the mainland subspecies of red-necked wallabies are continuous breeders whereas the Tasmanian subspecies (Bennett's wallabies) are highly synchronous breeders with a strict seasonal cycle as the result of a photoperiod-controlled seasonal quiescence (Catt 1977; Merchant and Calaby 1981). While we do not know anything else about their mating system,

Bennett's wallabies have testes which are significantly larger than those of the red-necked wallaby (t_{13} = 4.46, p = 0.0006) although their body weights do not differ significantly (t_{13} = 0.52, p = 0.61).

Data about the mating systems of the remainder of the species in this review are unavailable and have generally been inferred from the social system which, as mentioned, is rarely a good predictor of the mating system. The remaining large kangaroos (red (*Macropus rufus*) and western grey (*Macropus fuliginosus*) kangaroos) are known to establish dominance hierarchies like the eastern grey kangaroo (Lee and Cockburn 1985), but we do not know the extent to which this allows a male to monopolise mating. Perhaps the larger testes size of western grey kangaroos is an indication that dominant males are less successful in preventing sneak mating than in the other two species, although the difference may be due to small sample sizes (Table 6.1). Black-footed rock-wallabies (*Petrogale lateralis*) occupy a similar habitat niche to brush-tailed and allied rock-wallabies (Strahan 2002) and may be similar to those species in terms of behaviour and social system, but have larger testes than males of either species. This could indicate that the opportunity for polyandry in black-footed rock-wallabies is higher, but because their testes measures are based on a very small sample size (n = 3) we cannot be sure.

In mammals, territory exclusivity in conjunction with small home range size in females allows males to most effectively monopolise female territories, promoting the evolution of monogamy (Komers and Brotherton 1997) or polygyny, with single males controlling small groups of territorial females (Hixon 1987). We would therefore expect males of macropodoid species in which females have small defended territories to have the smallest testis sizes, if testis size is an indicator of the degree of polyandry. Tree-kangaroos, potoroos and rock-wallabies are the only macropodoids in which males defend territories or shelter sites of females. These macropodoids have small female home ranges, particularly the long-nosed potoroo (*Potorous tridactylus*; Fisher and Owens 2000). In a radio-tracking study of long-nosed potoroos (Long 2001), multiple squats (resting areas belonging to females and presumed to be where males associate with females for sexual checking) were sometimes overlapped by the home ranges of more than one male, but some males were seen to only associate with one female, suggesting social monogamy in some females. Long-footed potoroos (*Potorous longipes*) appear to be monogamous, and pairs share and defend a small territory (Green *et al.* 1998). Brush-tailed and allied rock-wallabies both have a low probability of polyandry (see above). Accordingly, long-footed potoroos and brush-tailed and allied rock-wallabies have very low relative testes weights. Both Lumholtz's (*Dendrolagus lumholtzi*) and Bennett's (*Dendrolagus bennettianus*) tree-kangaroos are territorial (Martin 1992). The species included in this review (*D. lumholtzi*) has the smallest testes of the scrotum width group (Table 6.2), implying that males successfully defend females against sneak mating by other males. If there is a positive relationship between the degree of polyandry and scrotum width, it is likely that the other species of tree-kangaroo have varying degrees of polyandry with at least some having levels of polyandry comparable to that found in tammar wallabies – there is a wide range of scrotum widths in this genus (Table 6.2) and males of at least one species (lowland tree-kangaroo, *Dendrolagus spadix*) have scrota of a similar width to those of male tammar wallabies.

The linear testes measurements are a less reliable measure of testes size than testes weight or volume by displacement (only available post-mortem), so the calculated residuals for the scrotum width group may be inaccurate. An illustration is that long-nosed potoroos, allied rock-wallabies and swamp wallabies have smaller than average testes if weight is used to calculate a residual (Table 6.1) but larger than average testes if width is used to calculate a residual (Table 6.2). Bridled nailtail wallabies have larger than average testes if weight is used to calculate a residual (Table 6.1) but smaller than average testes if width is used (Table 6.2). Despite this, the other four species in both the scrotum width and testes weight groups (red-necked wallabies, brush-tailed rock-wallabies, rufous bettongs and tammar wallabies) have broadly similar residuals in both analyses. It is not clear whether the differences in residuals between the testis weight and scrotum width groups are due to measurement errors or to the inclusion of different species in each analysis, which have altered the intercept and gradient of the regression line.

CONCLUSIONS

Due to the paucity of data about mating systems, especially with respect to multiple mating of females by males, we cannot be sure whether the relationship between testis size and mating system holds true within the Macropodoidea. However, the four species of macropodid for which we do have an indication of the degree of polyandry all have the predicted residuals of the testes weight–mating system hypothesis. Brush-tailed and allied rock-wallaby females are highly unlikely to have the opportunity for multiple mating; brush-tailed rock-wallaby males have the smallest testes of any species in this review. Allied rock-wallabies also have smaller than average testes. Both bridled nailtail and tammar wallabies are promiscuous breeders with confirmed occurrences of multiple mating; in both cases residual testis values are positive. Although multiple mating has been observed in eastern grey kangaroos, males of this species have smaller than average testes. This could be because the majority of females mate with only the dominant male or because sneak mating by subordinates occurs only once physiological oestrus has passed, eliminating the possibility of sperm competition.

There is some suggestion that highly synchronous seasonal breeding may increase the probability of sperm competition, as the only two strictly seasonally breeding macropodoids have the largest and third-largest testes of any of the species reviewed. In addition, the continuously breeding mainland subspecies of *M. rufogriseus* has significantly smaller testes than the Tasmanian subspecies, which has a photoperiodically controlled breeding season.

There are other correlates of sperm competition not discussed here (Gomendio *et al.* 1998; Taggart *et al.* 1998). Sperm quality measures such as sperm length, percentage motile sperm, percentage normal sperm and total sperm counts are unavailable for the majority of macropodoids. To expand our understanding of how sperm quality affects macropodoid male reproductive success, ejaculates could be collected from males and scored for quality measures. The males could then be divided into males with low-, medium- and high-quality ejaculate and be used in mating trials where oestrous females are mated sequentially to a low-, high- and medium-quality male in a random order. Paternity of young could be used to investigate how sperm quality affects male reproductive success.

A combination of field and captive observational and molecular studies to elucidate the genetic mating systems of a wider range of macropodoid species should be carried out to examine the strength of the relationship between the degree of polyandry and testis size in this group. Captive mating observations of females housed with a group of sexually mature males during oestrus could be carried out to ascertain whether females will mate with more than one male if more than one male is present, and the frequency of the behaviour. In addition, females of different species could be trapped in the field and have their vaginae flushed at times when mating is likely to be occurring. However, because few of the macropodoids are seasonal breeders, times when many females are likely to be mated are hard to predict – the data would be difficult to collect. If sperm is found in the flushed fluid, microsatellite analysis could be used to ascertain the minimum number of males which had contributed the sperm and the degree of polyandry in different species could thus be ascertained. This chapter has highlighted the paucity of reliable information about the evolutionary ecology of sperm competition in the Macropodoidea.

ACKNOWLEDGMENTS

Many thanks to Chris Mayberry for access to data and Matt Symonds for answering comparative analysis questions, providing helpful literature and explaining how to use *COMPARE*.

REFERENCES

Anderson MJ and Dixson AF 2002. Motility and the midpiece in primates. *Nature* **416:** 496.

Barth AD and Oko RJ 1989. *Abnormal Morphology of Bovine Spermatozoa*. Iowa State University Press, Ames.

Birkhead TR 1995. Sperm competition: evolutionary causes and consequences. *Reprod. Fertil. Dev.* **7:** 755–775.

Birkhead TR 1998. Sperm competition in birds. *Rev. Reprod.* **3:** 123–129.

Cardillo M, Bininda-Emonds ORP, Boakes E and Purvis A 2004. A species-level phylogenetic supertree of marsupials. *J. Zool. (Lond.)* **264:** 11–31.

Catt DC 1977. The breeding biology of Bennett's wallaby (*Macropus rufogriseus fruticus*) in South Canterbury, New Zealand. *N.Z. J. Zool.* **4:** 401–411.

Coltman DW, Bancroft DR, Robertson A, Smith JA, Clutton-Brock TH and Pemberton JM 1999. Male reproductive success in a promiscuous mammal: behavioural estimates compared with genetic paternity. *Mol. Ecol.* **8:** 1199.

Darwin C 1871. *The Descent of Man: And Selection in Relation to Sex.* John Murray, London.

Dennis AJ and Marsh H 1997. Seasonal reproduction in musky rat-kangaroos, *Hypsiprymnodon moschatus*: a response to changes in resource availability. *Wildl. Res.* **24:** 561–578.

Diaz-Uriarte R and Garland T Jr 1998. Effects of branch length errors on the performance of phylogenetically independent contrasts. *Syst. Biol.* **47:** 654–672.

Dixon A, Ross D, O'Malley SLC and Burke T 1994. Paternal investment inversely related to degree of extra-pair paternity in the reed bunting. *Nature* **371:** 698–700.

Ewen K, Temple-Smith P, Bowden D, Marinopoulos J, Renfree M and Yan H 1993. DNA fingerprinting in relation to male dominance and paternity in a captive colony of tammar wallabies (*Macropus eugenii*). *J. Reprod. Fertil.* **99:** 33–37.

Felsenstein J 1985. Phylogenies and the comparative method. *Am. Nat.* **125:** 1–15.

Fisher DO and Lara MC 1999. Effects of body size and home range on access to mates and paternity in male bridled nailtail wallabies. *Anim. Behav.* **58:** 121–130.

Fisher DO and Owens IPF 2000. Female home range size and the evolution of social organisation in macropod marsupials. *J. Anim. Ecol.* **69:** 1083–1089.

Fisher DO, Double MC, Blomberg SP, Jennions MD and Cockburn A 2006. Post-mating sexual selection increases lifetime fitness of polyandrous females in the wild. *Nature* **444:** 89–92.

Flannery TF, Martin RW and Szalay A 1996. *Tree-kangaroos: A Curious Natural History.* Reed Books, Melbourne.

Frederick H and Johnson CN 1996. Social organisation in the rufous bettong, *Aepyprymnus rufescens*. *Aust. J. Zool.* **44:** 9–17.

Gage MJG and Freckleton RP 2003. Relative testis size and sperm morphometry across mammals: no evidence for an association between sperm competition and sperm length. *Proc. R. Soc. Lond. B Biol. Sci.* **270:** 625–632.

Gomendio M and Roldan ERS 1991. Sperm competition influences sperm size in mammals. *Proc. R. Soc. Lond. B Biol. Sci.* **243:** 181–185.

Gomendio M, Harcourt AH and Roldan ERS 1998. Sperm competition in mammals. In *Sperm Competition and Sexual Selection.* (Eds TR Birkhead and AP Moller) pp. 667–756. Academic Press, New York.

Green K, Mitchell AT and Tennant P 1998. Home range and microhabitat use by the long-footed potoroo, *Potorous longipes*. *Wildl. Res.* **25:** 357–372.

Halliday T 1998. Sperm competition in amphibians. In *Sperm Competition and Sexual Selection.* (Eds TR Birkhead and AP Moller) pp. 465–502. Academic Press, New York.

Harcourt AH, Harvey PH, Larson SG and Short RV 1981. Testis weight, body weight and breeding system in primates. *Nature* **293:** 55–57.

Harcourt AH, Purvis A and Liles L 1995. Sperm competition: mating system, not breeding season, affects testes size of primates. *Funct. Ecol.* **9:** 468–476.

Hazlitt SL, Sigg DP, Eldridge MDB and Goldizen AW 2006. Restricted mating dispersal and strong breeding group structure in a mid-sized marsupial mammal (*Petrogale penicillata*). *Mol. Ecol.* **15:** 2997–3007.

Hixon MA 1987. Territory area as a determinant of mating system. *Am. Zool.* **27:** 229–247.

Holleley CE, Dickman CR, Crowther MS and Oldroyd BP 2006. Size breeds success: multiple paternity, multivariate selection and male semelparity in a small marsupial, *Antechinus stuartii*. *Mol. Ecol.* **15:** 3439–3448.

Horsup AB 1996. The behavioural ecology of the allied rock- wallaby, *Petrogale assimilis*. PhD thesis. Dept of Zoology, James Cook University, Townsville, Australia.

Hughes C 1998. Integrating molecular techniques with field methods in studies of social behavior: a revolution results. *Ecology* **79:** 383–399.

Hynes EF, Rudd CD, Temple-Smith PDM, Sofronidis G, Paris D, Shaw G and Renfree MB 2005. Mating sequence, dominance and paternity success in captive male tammar wallabies. *Reproduction* **130:** 123–130.

Jarman P 1983. Mating system and sexual dimorphism in large, terrestrial, mammalian herbivores. *Biol. Rev.* **58:** 485–520.

Jarman P 1991. Social behavior and organization in the Macropodoidea. *Adv. Study Behav.* **20:** 1–50.

Jarman P and Bayne P 1997. Behavioural ecology of *Petrogale penicillata* in relation to conservation. *Aust. Mammal.* **19:** 219–228.

Jarman PJ and Southwell CJ 1986. Grouping, associations and reproductive strategies in eastern grey kangaroos. In *Ecological Aspects of Social Evolution.* (Eds DI Rubenstein and RW Wrangham) pp. 399–428. Princeton University Press, Princeton.

Jennions MD and Passmore NI 1993. Sperm competition in frogs: testis size and a 'sterile male' experiment on *Chiromantis xerampelina* (Rhacophoridae). *Biol. J. Linn. Soc.* **50:** 211–220.

Johnson PM and Strahan R 1982. A further description of the musky rat-kangaroo *Hypsiprymnodon moschatus* Ramsey, 1876 (Marsupialia, Potoroidae), with notes on its biology. *Aust. Zool.* **21:** 27–46.

Jones M 2002. The black-footed rock-wallaby, *Petrogale lateralis*: reproductive ecology and the effect of population bottlenecks on genetic diversity. PhD thesis. Dept of Zoology, University of Melbourne, Australia.

Kempenaers B, Verheyen GR, Van den Broeck M, Burke T, Van Broeckhoven C and Dhondt AA 1992. Extra-pair paternity results from female preference for high quality males in the blue tit. *Nature* **357:** 494–496.

Kenagy GJ and Trombulak C 1986. Size and function of mammalian testes in relation to body size. *J. Mammal.* **67:** 1–22.

Kirsch JAW, Lapointe FJ and Springer MS 1997. DNA-hybridisation studies of marsupials and their implications for metatherian classification. *Aust. J. Zool.* **45:** 211–280.

Komers PE and Brotherton PNM 1997. Female space use is the best predictor of monogamy in mammals. *Proc. R. Soc. Lond. B Biol. Sci.* **264:** 1261–1270.

Kraaijeveld-Smit FJL, Ward SJ, Temple-Smith PDM and Paetkau D 2002. Factors influencing paternity success in *Antechinus agilis*: last-male sperm precedence, timing of mating and genetic compatibility. *J. Evol. Biol.* **15:** 100–107.

Kubo-Irie M, Matsumiya K, Iwamoto T, Kaneko S and Ishijima S 2004. Morphological abnormalities in the spermatozoa of fertile and infertile men. *Mol. Reprod. Dev.* **70:** 70–81.

Lee AK and Cockburn A 1985. *Evolutionary Ecology of Marsupials*. Cambridge University Press, Cambridge.

Lifjeld JT, Dunn PO, Robertson RJ and Boag PT 1992. Extra-pair paternity in monogamous tree swallows. *Anim. Behav.* **45:** 213–229.

Long KI 2001. Spatio-temporal interactions among male and female long-nosed potoroos, *Potorous tridactylus* (Marsupialia: Macropodoidea): mating system implications. *Aust. J. Zool.* **49:** 17–26.

Martin R 1992. An ecological study of Bennett's tree-kangaroo (*Dendrolagus bennettianus*). WWF Project 116. Final report. World Wide Fund for Nature, Sydney.

Martin PA and Dziuk PJ 1977. Assessment of relative fertility of males (cockerels and boars) by competitive mating. *J. Reprod. Fertil.* **49:** 323–329.

Martin PA, Reimers TJ, Lodge JR and Dziuk PJ 1974. The effect of ratios and numbers of spermatozoa mixed from two males on proportions of offspring. *J. Reprod. Fertil.* **39:** 251–258.

Martins EP 2004. Compare, version 4.6b. Computer programs for the statistical analysis of comparative data. http://compare.bio.indiana.edu/.

Merchant JC and Calaby JH 1981. Reproductive biology of the red-necked wallaby (*Macropus rufogriseus banksianus*) and the Bennett's wallaby (*M. r. rufogriseus*) in captivity. *J. Zool.* **194:** 203–217.

Moller AP 1989. Ejaculate quality, testes size and sperm production in mammals. *Funct. Ecol.* **3:** 91–96.

Olsson M and Madsen T 1998. Sexual selection and sperm competition in reptiles. In *Sperm Competition and Sexual Selection*. (Eds TR Birkhead and AP Moller) pp. 503–564. Academic Press, New York.

Ombelet W, Bosmans E, Janssen M, Cox A, Vlasselaer J, Gyselaers W, Vandeput H, Gielen J, Pollet H, Maes M, Steeno O and Kruger T 1997. Semen parameters in a fertile versus subfertile population: a need for change in the interpretation of semen testing. *Hum. Reprod.* **12:** 987–993.

Paplinska JZ 2005. Seasonal breeding, puberty and population genetics of the swamp wallaby (*Wallabia bicolor*). PhD thesis. Dept of Zoology, University of Melbourne, Australia.

Paplinska JZ, Moyle RLC, Wreford NG, Temple-Smith PDM and Renfree MB 2007. Reproduction in male swamp wallabies (*Wallabia bicolor*): puberty and the effects of season. *J. Anat.* **211:** 518–533.

Paris DBBP, Taggart DA, Paris MCJ, Temple-Smith PDM and Renfree MB 2004. Sperm transport, size of the seminal plug and the timing of ovulation after natural mating in the female tammar wallaby *Macropus eugenii*. *Reprod. Fertil. Dev.* **16:** 811–822.

Paris DBBP, Taggart DA, Shaw G, Temple-Smith PDM and Renfree MB 2005. Changes in semen quality and morphology of the reproductive tract of the male tammar wallaby parallel seasonal breeding in the female. *Reproduction* **130:** 367–378.

Parker GA 1970. Sperm competition and its evolutionary consequences in the insects. *Biol. Rev.* **45:** 525–567.

Petersen CW and Warner RR 1998. Sperm competition in fishes. In *Sperm Competition and Sexual Selection*. (Eds TR Birkhead and AP Moller) pp. 435–464. Academic Press, New York.

Poole WE 1973. A study of breeding in grey kangaroos, *Macropus giganteus* (Shaw) and *M. fuliginosus* (Desmarest), in central New South Wales. *Aust. J. Zool.* **21:** 183–212.

Poole WE 1975. Reproduction in two species of grey kangaroos, *Macropus giganteus* (Shaw) and *M. fuliginosus* (Desmarest). II. Gestation, parturition and pouch life. *Aust. J. Zool.* **23:** 333–353.

Poole WE and Pilton PE 1964. Reproduction in the grey kangaroo, *Macropus canguru*, in captivity. *CSIRO Wildl. Res.* **9:** 218–234.

Pope LC, Blair D and Johnson CN 2005. Dispersal and population structure of the rufous bettong, *Aepyprymnus rufescens* (Marsupialia: Potoroidae). *Austral Ecol.* **30:** 572–580.

Preston BT, Stevenson IR, Pemberton JM, Coltman DW and Wilson K 2003. Overt and covert competition in a promiscuous mammal: the importance of weaponry and testes size to male reproductive success. *Proc. R. Soc. Lond. B Biol. Sci.* **270:** 633–640.

Purvis A and Rambaut A 1995. Comparative analysis by independent contrasts (CAIC): an Apple Macintosh application for analysing comparative data. *Comput. Appl. Biosci.* **11:** 247–251.

Qvarnstrom A and Forsgren E 1998. Should females prefer dominant males? *Trends Ecol. Evol.* **13:** 498–501.

Ramm SA, Parker GA and Stockley P 2005. Sperm competition and the evolution of male reproductive anatomy in rodents. *Proc. R. Soc. Lond. B Biol. Sci.* **272:** 949–955.

Rose RW 1997. Testes weight, body weight and mating systems in marsupials and monotremes. *J. Zool. (Lond.)* **243:** 523–531.

Rudd CD 1994. Sexual behaviour of male and female tammar wallabies (*Macropus eugenii*) at post-partum oestrus. *J. Zool. (Lond.)* **232:** 151–162.

Sacher N, Schwarz-Weig E, Keil A and Epplen JT 1999. Behavioural strategies, testis size, and reproductive success in two cavimorph rodents with different mating strategies. *Behaviour* **136:** 1203–1217.

Setchell BP 1977. Reproduction in male marsupials. In *The Biology of Marsupials.* (Eds B Stonehouse and D Gilmore) pp. 411–457. University Park Press, Baltimore.

Short RV 1977. Sexual selection and the descent of man. In *Reproduction and Evolution.* (Eds JH Calaby and CH Tyndale-Biscoe) pp. 3–19. Australian Academy of Science, Canberra.

Short RV 1979. Sexual selection and its component parts, somatic and genital selection, as illustrated by man and the great apes. *Adv. Study Behav.* **9:** 131–158.

Short RV 1997. The testis: the witness of the mating system, the site of mutation and the engine of desire. *Acta Paediatr. Suppl.* **422:** 3–7.

Sigg DP and Goldizen AW 2006. Male reproductive tactics and female choice in the solitary, promiscuous bridled nailtail wallaby (*Onychogalea fraenata*). *J. Mammal.* **87:** 461–496.

Sigg DP, Goldizen AW and Pople AR 2005. The importance of mating system in translocation programs: reproductive success of released bridled nailtail wallabies. *Biol. Conserv.* **123:** 289–300.

Simmons LW, Firman RC, Rhodes G and Peters M 2004. Human sperm competition: testis size, sperm production and rates of extrapair copulations. *Anim. Behav.* **68:** 297–302.

Snook RR 2005. Sperm in competition: not playing by the numbers. *Trends Ecol. Evol.* **20:** 46–53.

Spencer PBS, Horsup AB and Marsh HD 1998. Enhancement of reproductive success through mate choice in a social rock-wallaby, *Petrogale assimilis* (Macropodidae) as revealed by microsatellite markers. *Behav. Ecol. Sociobiol.* **43:** 1–9.

Strahan R 2002. *The Mammals of Australia.* Reed New Holland, Sydney.

Taborsky M 1998. Sperm competition in fish: bourgeois males and parasitic spawning. *Trends Ecol. Evol.* **13:** 222–227.

Taggart DA, Breed WG, Temple-Smith PDM, Purvis A and Shimmin G 1998. Reproduction, mating strategies and sperm competition in marsupials and monotremes. In *Sperm Competition and Sexual Selection.* (Eds TR Birkhead and AP Moller) pp. 623–666. Academic Press, New York.

Thornhill R 1983. Cryptic female choice and its implications in the scorpionfly *Harpobittacus nigriceps. Am. Nat.* **122:** 765–788.

Tyndale-Biscoe CH and Renfree MB 1987. *Reproductive Physiology of Marsupials.* Cambridge University Press, Cambridge.

Walker LV 1996. Female mate-choice: are marsupials really so different? In *Comparison of Marsupial and Placental Behaviour.* (Eds DB Croft and U Ganslosser) pp. 208–225. Filander Press, Fuerth.

Watson DM, Croft DB and Crozier RH 1992. Paternity exclusion and dominance in captive red-necked wallabies, *Macropus rufogriseus* (Marsupialia: Macropodidae). *Aust. Mammal.* **15:** 31–36.

Wittenberger JF 1979. The evolution of mating systems in birds and mammals. In *Handbook of Behavioural Neurobiology.* (Eds P Marler and JC Vandenbergh) pp. 271–349. Plenum Press, New York.

Yasui Y 1997. A 'good-sperm' model can explain the evolution of costly multiple mating by females. *Am. Nat.* **149:** 573–584.

7 Dominance and paternity in the tammar wallaby

E.J. Miller, M.D.B. Eldridge and C.A. Herbert

SUMMARY

In polygynous mating systems, it is often difficult to identify factors contributing to male reproductive success. We sought to determine if particular behavioural, morphological or physiological traits were correlated with male dominance and paternity in a captive population of tammar wallabies (*Macropus eugenii*), and whether male dominance conferred a reproductive advantage. Four replicate groups were held at a sex ratio of four males to seven females in large outdoor enclosures. A range of behavioural (agonistic and sexual) and morphological (body weight, pes length, testis volume and body condition) measurements were collected and serum testosterone concentrations were determined. Paternity of pouch young was determined using microsatellite loci. This study found that alpha males could be clearly distinguished from lower-ranked males based on the frequency of sexual behaviour (mate-guarding, checking females and disrupting consorts). Of the five non-behavioural variables, only pes length differed significantly between males of different ranks. Dominance rank was significantly correlated with reproductive success and alpha males that maintained post-copulatory guarding had higher reproductive success than males who did not. Alpha males secured 55% of first copulations and the sire corresponded with the first male to mate with the female 69% of the time. Sires could be clearly distinguished from non-sires based on dominance rank, weight and pes length, and they also tended to be in better condition. These data suggest that the physical size of a male is an important determinant of dominance status and reproductive success, probably by enhancing a male's competitiveness in combat.

INTRODUCTION

Two processes are central to Darwin's (1871) theory of sexual selection: male–male competition and female mate choice. The male-biased sexual size dimorphism that exists in many mammalian taxa is often attributed to competition between individuals for reproductive opportunities, encouraging the evolution of secondary sexual traits, for example larger body size, bright colours and weaponry. Such traits are thought to have evolved as indicators of quality, strength or fighting ability via sexual selection (Andersson 1994).

The tammar wallaby (*Macropus eugenii*) is a sexually dimorphic macropodid marsupial with a polygynous mating system, in which females usually mate with more than one male (Rudd 1994). Breeding is strictly seasonal with a photoperiodically controlled quiescence of pregnancy and a highly synchronised period of mating, with a large proportion of births occurring in late January and early February (Tyndale-Biscoe and Renfree 1987). Females exhibit a post-partum oestrus within an hour or two of giving birth and all mating activity usually concludes within four hours of parturition in undisturbed animals (Renfree *et al.* 1989; Rudd 1994). The resultant conceptus grows to a 100-cell blastocyst before entering embryonic diapause (Tyndale-Biscoe and Renfree 1987). The loss or removal of a pouch young during the breeding season stimulates the quiescent blastocyst to reactivate. Birth and a new post-partum oestrus then occurs approximately 26.5 ± 0.4 days later (Renfree *et al.* 1989).

In male tammars there is little seasonal variation in testis size (Inns 1982). Serum testosterone concentrations peak at the onset of the breeding season, coincident with increases in the weight of accessory reproductive organs, but remain low for most of the year (Catling and Sutherland 1980; Inns 1982). This seasonal variation in testosterone does not occur in the absence of female tammars (Catling and Sutherland 1980). The expression of male-type sexual behaviour in the tammar wallaby appears to be determined by the activating effects of testosterone during adulthood, as these behaviours are significantly reduced in castrated males and can be induced in females by treatment with testosterone implants (Rudd *et al.* 1996).

Approximately two weeks prior to the first female giving birth, male tammar wallabies begin to display sexual checking behaviour, a characteristic of male macropods. 'Checking' involves regular sniffing of the urogenital opening (UGO) and pouch of females (Rudd 1994). This is presumably to evaluate their reproductive state via pheromonal cues (Coulson 1989). Checking behaviour intensifies around the time of birth, as the female is approaching oestrus. Copulation starts with a brief period of mounting and intromission by the male. In undisturbed animals, following ejaculation, males may remain *in copulo* for six to eight minutes (Rudd 1994). However, because of intense intermale aggression, copulation is often disrupted. Males rigorously guard oestrous females from subordinate males' advances both pre- and post-copulation, which limits the number of subordinate males gaining access to receptive females and successfully copulating but does not entirely prevent it (Rudd 1994).

In macropodids, as with other species, dominance 'relates to a situation in which individuals can be ranked so that a higher-ranking individual may attack or show aggressive threat, while the subordinate avoids conflict ... A dominance hierarchy is commonly established through fighting ... or through differences in age' (Poole 1985). Dominance is exerted by an individual via active or passive supplanting of another individual (Croft 1981). Previous studies on the reproductive success of male tammars have given conflicting results. Ewen *et al.* (1993) found that there was no significant difference in the number of offspring sired by each of four males that were paired with 13 females in captivity. Conversely, a second study on captive animals held at varying sex ratios found that the dominant male sired 50% of the young (Hynes *et al.* 2005).

In this study, we investigated the predicted relationships between body size, dominance rank and reproductive success in the tammar wallaby. Three interrelated questions were addressed.

1. Is dominance correlated with male sexual behaviours, morphological variables (weight, pes length, testis size and body condition) or endocrine (testosterone) concentrations?
2. Does behavioural dominance translate into higher reproductive success?
3. What traits distinguish sires from non-sires?

MATERIALS AND METHODS

Study subjects and site

This study was conducted at the Macquarie University Fauna Park in north-west Sydney, Australia (33°45'S, 151°06'E). The original source population of the breeding colony was Kangaroo Island, South Australia. All animals (n = 37) were sexually mature and identified by a unique numeric code. The animals were maintained in outdoor enclosures with trees and shelter, supplied with water and kangaroo pellets *ad libitum*, as well as being supplemented with lucerne three times a week. Animals were randomly allocated to one of three groups to avoid bias and held at a sex ratio of four males to seven females. The sex ratio selected was believed to approach male:female ratios recorded in the wild (Inns 1982). Males were placed in their enclosure four weeks prior to females to allow time for them to settle and establish a dominance hierarchy.

Behavioural data collection

Behavioural observations were conducted during the 2004–05 breeding season by a single observer (E.J.M.). All animals were collared with unique colours for observational purposes. Observations were made during daylight hours behind shade-cloth where possible, but often with the observer in full view of the animals as they habituated quickly to the observer's presence. Animals were identified using binoculars (Pentax 10 × 24 UCF WR). The animals were observed daily at dawn and/or dusk for a set duration of one hour using scan sampling (Martin and Bateson 1993), recording all agonistic or sexual behaviours observed. Behavioural categories were adapted from Coulson (1989), Ganslosser (1989) and Renfree *et al.* (1989). Male position and proximity to one another and to females within a 5 m radius was recorded at two-minute intervals. Additional behaviour sampling (Martin and Bateson 1993) of whole groups was conducted, looking specifically for fighting, birth behaviour and copulations beginning two days prior to and during the estimated time of oestrus. Any female showing characteristic birth behaviour was observed until mating was complete (complete mating sequence) and the sequence of males observed copulating with the female was recorded (Hynes *et al.* 2005). The sexual period in each yard began when male sexual behaviour was first observed and concluded when guarding ceased for the last female in oestrus.

Agonistic interactions between males were arranged into a dominance matrix (Martin and Bateson 1993). For each male a dominance index (I) $= N_a/(N_a + N_b)$ was calculated, where N_a is the number of encounters won and N_b is the number of encounters lost. Males with the highest I values (>0.7) were considered dominant. Landau's linearity index (h) $= 12/(n^3 - n)^* \sum(v_a - ½\,(n - 1))^2$ was used to estimate the linearity of the dominance hierarchy, where n is the number of animals and v_a is the number of individuals whom a has dominated. The h values range from zero (non-linear) to 1.0 (perfect linearity), with values >0.9 considered significantly linear (Martin and Bateson 1993).

Sampling of males

The length and breadth (excluding the epididymis) of the left testis of each male was measured weekly by the same person (E.J.M.) using digital Vernier callipers. Testicular volume was calculated using the formula for an oblate sphere, $V = (\pi/6).B^2.L$, where V = volume, B = breadth of testis and L = length of testis (Williamson *et al.* 1990).

A condition index was calculated for each male using the ratios from a regression of body mass and a skeletal measurement (pes). We follow accepted practice in assuming that a higher ratio (a lower mass at a given body size) indicates an individual in poorer condition. Every animal was weighed weekly with a 10 kg Salter suspended dial scale (SA-235/6, Wedderburn, accuracy 0.05 kg). Pes was measured from the tip of the toe (excluding the nail) to the base of the heel using digital Vernier callipers to the nearest 0.01 mm three times during the study for each male.

Blood was collected at weekly intervals from the lateral tail vein of males using a 21 gauge winged infusion set (Surflo, Terumo Corp.) and a 5 mL syringe. Blood was transferred immediately into serum separation tubes (Vacuette, Greiner Labortechnik, Austria). The serum was separated and stored in duplicate aliquots at –20°C until assayed to determine the concentrations of testosterone.

Sampling of females

Females were checked for the presence of pouch young (PY) at weekly intervals. Removal of PY (RPY) enabled data collection from repeated cycles by stimulating reactivation of the quiescent blastocyst (Tyndale-Biscoe and Renfree 1987). The first PY removed (late Jan.–early Feb.) was conceived during the previous breeding season (2003–04). All young were removed and euthanased at approximately seven days of age to maintain natural variation between cycles of individual females. Head length (tip of nose to back of head) was measured using digital Vernier callipers to the nearest 0.01 mm to determine age from established growth curves (Poole *et al.* 1991). The interval between RPY and oestrus is shorter for a pregnant (26.2 ± 0.67 days) than a non-pregnant (29.3 ± 1.06 days) cycle (Tyndale-Biscoe and Renfree 1987). This information facilitated the estimation of the next date of birth and oestrus for females, assisting in observing copulations.

Genetic assignment of paternity

Ear biopsies were collected from all adult (n = 37, 16 males and 21 females) tammar wallabies. A sterile leather punch was used to remove a small (5 mm diam.) sample from the outermost region of an ear and the samples were fixed in 80% ethanol. Tissue samples were also collected from all euthanased PY (n = 27).

DNA extraction was performed using a salting-out method (Sunnucks and Hales 1996). Three females produced no PY and were excluded from paternity analysis. Adults (n = 34) and PY (n = 27) were typed for five microsatellite loci (Me2, Me14, G19-1, G26-4 and Pa595) with high allelic diversity, to determine paternity. PCR reactions were as previously described (Spencer *et al.* 1995; Taylor and Cooper 1998; Zenger and Cooper 2001). The PCR products were resolved on 6% polyacrylamide gels and visualised by autoradiography. After genotyping, the data were checked to ensure Mendelian inheritance between mothers and their offspring.

Since maternity was known, paternally derived alleles were readily identified. Paternity was then determined by exclusion, since each PY could only have been sired by one of four fathers. Males were excluded as potential fathers if they did not possess the identified paternal allele at one or more loci. Alternatively, males were assigned paternity if they possessed the identified paternal alleles at all loci.

Testosterone radioimmunoassay

Testosterone concentrations were determined using the automated Immulite® 2000 immunoassay analyser with a solid-phase competitive chemiluminescent enzyme immunoassay. Single 20 ml serum samples were assayed using the Immulite® total testosterone kit (L2KTW2, Diagnostics Product Corp., Los Angeles).

Statistical analyses

Sexual behaviours (guarding, checking and disrupting consort) were recorded as counts and converted to a frequency (number of interactions per hour) to compare the rate of occurrence between individuals. Sexual behaviour for each male was not normally distributed, so the relationship between dominance rank and number of interactions per hour was analysed using a non-parametric Kruskal-Wallis ANOVA test in SPSS 12.0.1. To test whether sexual behaviours (guarding, checking and disrupting consort) were useful predictors of male dominance rank, a Linear Discriminant Analysis (LDA) in SYSTAT® 10 was used.

Temporal variation in male morphological measurements (weight, condition index, testis volume and pes length) or testosterone concentrations were analysed using a GLM repeated-measures ANOVA to ensure that any changes over time were similar between animals of different ranks, before averaging the values.

Dominance effects were determined using the males' nominal rank positions, i.e. the alpha male was allocated a rank of 1 and the most subordinate male a rank of 4. Comparisons of each trait for each male were compared using one-way ANOVA. Multiple comparisons of group means were conducted using the Tukey's HSD test for any variables that were significantly different between ranks.

To determine whether dominance rank was a useful predictor for the number of offspring sired, a Spearman's rank order correlation was performed. As the hierarchy is defined by a nominal rank position (alpha male rank = 1), any dominance-related effects would be reflected in a significant negative correlation. The relative importance of the various traits for paternity success were determined by dividing the males into two subsets – sires and non-sires – and comparing them using one-way ANOVA. Any variables that were not normally distributed were analysed using a non-parametric Kruskal-Wallis ANOVA test.

RESULTS

Dominance hierarchies and linearity

Significant linear dominance hierarchies were observed in three of four yards when females were in oestrus ($h = 0.9–1.0$; yards 2, 3 and 4) (Table 7.1). The alpha male maintained his rank in both yards that continued into a second cycle (yards 3 and 4).

Sexual behaviour

The number of sexual interactions with females differed significantly between males of different ranks. Alpha males participated in significantly more mate-guarding ($H_{11} = 9.963$, df = 3, $p < 0.05$, Fig. 7.1a), checking ($H_{11} = 9.684$, df = 3, $p < 0.05$, Fig. 7.1b) and disrupting consort ($H_{11} = 8.634$, df = 3, $p < 0.05$, Fig. 7.1c) behaviours than did lower-ranked males. Pre-copulatory guarding behaviour by alpha males was similar among yards, but post-copulatory behaviour differed between yards (Fig. 7.2). The alpha male in yard 4 appeared to perform less post-copulatory guarding than alpha males in other yards, but there were insufficient data for statistical analysis.

Non-behavioural factors associated with male rank

Of the five variables tested, only pes length varied significantly between the ranks ($F_{3,12} = 5.439$, $p < 0.05$). The alpha male had the largest pes, and pes

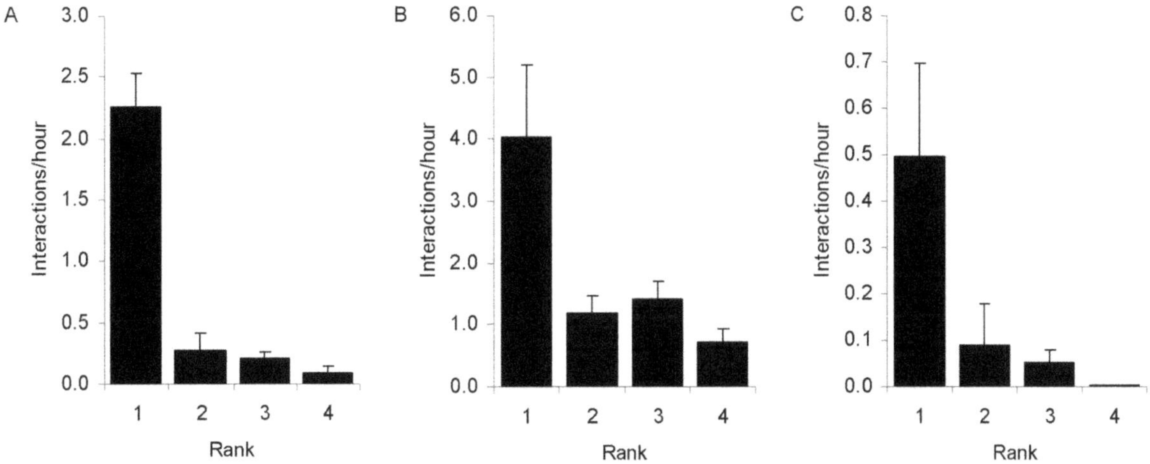

Figure 7.1: The mean (± s.e.m.) frequency of sexual interactions with females (interactions per hour) for male tammar wallabies (*M. eugenii*) of each dominance rank. (a) Mate guarding. (b) Checking females. (c) Disrupting consort behaviours.

length decreased with decreasing rank. Alpha males tended to have higher weight, higher testosterone concentrations and better body condition than lower-ranked males but these differences were not significant ($F_{3,12} = 2.301$, $p = 0.13$; $F_{3,12} = 1.428$, $p = 0.28$; $F_{3,12} = 1.940$, $p = 0.18$, respectively). Dominance status was independent of testis volume ($F_{3,12} = 0.246$, $p = 0.86$).

Table 7.1: Summary of dominance hierarchy rank (*I*) and linearity (*h*) in each yard for male tammar wallabies (*M. eugenii*) during the first oestrous cycle.

Yard	Male ID	Rank[a]	*I*	*h*
1	678	1	0.972	0.7
	933T	2	0.200	
	FICT	3	0.182	
	1152	4	0.071	
2	13DT	1	0.983	1.0
	965T	2	0.348	
	18ET	3	0.190	
	B2OT	4	0.076	
3	669	1	0.960	1.0
	4E1T	2	0.290	
	A2ET	3	0.227	
	351T	4	0.111	
4	19FT	1	0.950	0.9
	1179	2	0.318	
	1931	3	0.111	
	8500	4	0.091	

a: Dominance rank (highest to lowest) is represented by 1, 2, 3 and 4.

Male dominance and reproductive success

Paternity was assigned to each PY using five microsatellite loci. The data were for two oestrous cycles (n = 27). The first cycle included four replicates (yards 1, 2, 3 and 4). The second cycle included two replicates (yards 3 and 4) due to removal of males from other yards. During the first oestrous cycle, dominance rank and the number of offspring sired were significantly negatively correlated ($r_s = -0.542$, $n = 16$, $p < 0.05$). As mate-guarding was a significant contributor to predicting male dominance rank, a Spearman's rank order correlation was calculated for guarding and number of offspring sired for the first oestrous cycle for all yards. A positive and significant correlation ($r_s = 0.519$, $n = 16$, $p < 0.05$) exists between guarding and reproductive success.

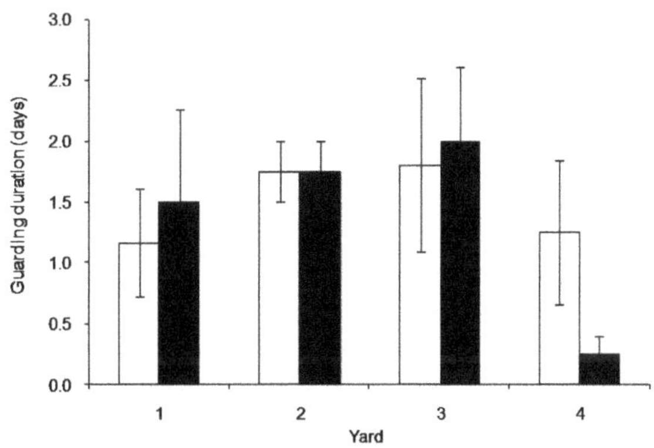

Figure 7.2: Differences in the duration (mean ± s.e.m.) of pre-copulatory (open bars) and post-copulatory (closed bars) guarding behaviour by alpha male tammar wallabies (*M. eugenii*) from different yards during the first oestrous cycle.

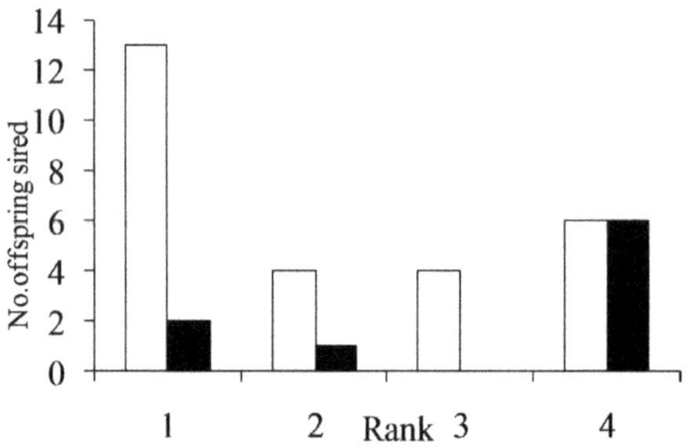

Figure 7.3: Number of offspring sired by male tammar wallabies (*M. eugenii*) according to rank for all yards (open bars) and only yard 4 (closed bars), illustrating the high reproductive success observed for the lowest-ranked male in that yard.

The overall trend for paternity was that the alpha male sired the most offspring (48%, n = 27), but two trends were observed. First, in three of four yards (1, 2 and 3) for the first cycle and in one of two yards (yard 3) for the second cycle, the alpha male sired the most offspring, 57% (n = 14) and 75% (n = 4) respectively (Fig. 7.3). Second, for two consecutive cycles (both cycle one and two), yard 4 behaved differently, with the lowest-ranked male (rank 4) siring the most offspring, 75% (n = 4) and 80% (n = 5) respectively (Fig. 7.3).

Mating sequences

A total of 13 complete mating sequences were observed during the first oestrous cycle and six during the second oestrus cycle. Of the observed matings, 37% (n = 7) did not produce an offspring. On the eight occasions where multiple males mated with a female and a young was produced, the sire corresponded with the first male to mate 63% of the time (n = 5). From the 19 observed matings, the dominant male secured the first copulation 58% of the time but lower-ranking males copulated first on several occasions: rank 2 = 16%, rank 3 = 10% and rank 4 = 16%.

Comparison of sires and non-sires

Eleven of the 16 males successfully sired offspring. Sires were characterised by a range of traits that distinguished them from non-sires (Table 7.2). There was a significant difference in the dominance rank, weight and pes length of sires and non-sires, and sires also tended to be in better condition. Sires and non-sires could not be distinguished based on the frequency of the three categories of sexual behaviour (mate-guarding, checking and disrupting consorts).

DISCUSSION

This study has shown that sexual behaviour was a superior predictor for alpha male dominance status, but an inferior predictor among subordinate male tammar wallabies. Of the five non-behavioural variables, only pes length differed significantly with rank, suggesting that body size or age may be important determinants of dominance rank. Sires could

Table 7.2: Differences in socioendocrine and morphological traits of sire and non-sire tammar wallabies (*M. eugenii*).

Variable	Sire (n = 11)	Non-sire (n = 5)	$F_{(1,14)}$	*P* value
Dominance rank	2.09 ± 0.29	3.40 ± 0.40	5.845	0.030[b]
Behavioural traits				
Mate-guarding	0.93 ± 0.33	0.21 ± 0.12	2.363[a]	0.124
Checking females	2.18 ± 0.60	1.09 ± 0.33	1.388	0.258
Disrupting consorts	0.22 ± 0.10	0.02 ± 0.02	2.248[a]	0.134
Non-behavioural traits				
Testicular volume (cm^3)	14.69 ± 0.69	15.41 ± 1.21	0.308	0.587
Condition index	2.01 ± 0.44	2.44 ± 0.17	3.494[a]	0.062
Testosterone (ng/mL)	4.40 ± 0.61	3.47 ± 0.71	0.810	0.383
Weight (kg)	7.48 ± 0.15	6.01 ± 0.50	14.137	0.002[c]
Pes (mm)	14.94 ± 0.12	14.34 ±0.22	1.249	0.018[b]

Values reported are means ± s.e.m.

a: Data were not normally distributed, Kruskal-Wallis test (χ^2 values).

b: Significant at $p < 0.05$.

c: Significant at $p < 0.005$.

be clearly distinguished from non-sires based on dominance rank, weight and pes length. Dominance rank was significantly correlated with paternity, with the alpha male siring approximately 50% of offspring. Mate-guarding was an important contributor to male reproductive success. Failure to mate-guard resulted in poor reproductive success. For an individual to successfully guard a female, they had to hold a position of dominance. This suggests dominance and mate-guarding are important contributors to male reproductive success. The results also suggest that the attainment of a minimum weight, size and dominance rank are important for males to effectively compete for and gain access to receptive females.

A clear dominance hierarchy was evident during the oestrous period. Three of the four yards exhibited a significantly linear dominance hierarchy, indicating the alpha male dominated all other males in the group. Sexual behaviour significantly separated alpha males from subordinates based on their dominance rank. The alpha male in each yard spent significantly more time guarding, checking and disrupting consorts of the oestrous females. Guarding emerged as the most significant sexual behaviour correlated with alpha male status. Mate-guarding is one of the two main behavioural adaptations to sperm competition (Parker 1998). It is one of the most effective ways a male can ensure he sires a female's offspring. Guarding prevents other males from copulating with a female while she is fertile, to protect his paternity. Our finding of a significant correlation between guarding and reproductive success is consistent with Rudd's (1994) hypothesis that guarding of oestrous females would increase the dominant male's reproductive success. In yards 1–3, the alpha male maintained pre- and post-copulatory guarding for a similar length of time. In yard 4, the alpha male failed to maintain post-copulatory guarding and this corresponded with lower reproductive success. This variation in behaviour is of interest as it potentially reflects different reproductive strategies being utilised by different males.

In polygynous species, males have to balance the trade-off between paternity assurance (guarding) and acquiring multiple mates (Alcock 1993). This decision will affect other males' behaviour and their consequent attempts to mate with receptive females. In this study, restricted utilisation of the space within enclosures was observed for both males and females. In all yards, the alpha males were observed defending (guarding and displacing lower-ranked males) a specific shelter site where four to seven females tended to congregate (≥76% of observations). The alpha males in yards 1–3 spent most of their time (average 60 ± 12%) within 9 m of this shelter. However, these males also guarded females elsewhere in the enclosure (Miller 2004, unpubl. data). In yard 4 the alpha male remained within 9 m of the shelter 84% of the observation time and was not seen to actively guard oestrous females elsewhere in the enclosure, a likely factor in his lower reproductive success.

Why females should show a preference for a particular shelter when similar others were available is unclear. However, the females may have preferentially congregated to minimise harassment from males. For example, Carranza (1999) observed that female red deer (*Cervus elephaus*) within harems suffered a lower harassment rate than females outside the harem. The selective advantage for males to utilise such an area would be to maximise their access to females and enhance their ability to exclude other males. As female oestrus is highly synchronised, territorial behaviour would be beneficial for increasing the probability of males copulating (Clutton-Brock 1989). Similar territorial behaviour in wild tammar populations is yet to be reported.

In this study, pes length increased with increasing dominance rank, supporting the notion that body size is influential in determining dominance status. These results are consistent with the findings of Hynes *et al.* (2005), that male weight differed significantly between male tammar wallabies of different ranks (Table 7.3). The sexual dimorphism of body size in the tammar and other medium and large macropodids arises from differential growth rates and persistent growth in the male (Jarman 1983). Assuming that the foot grows at the same relative rate as the rest of the body, pes length is likely to be indicative of the overall skeletal size of the animal and/or its age. Supporting this conclusion is the fact that the same relationship was observed for weight. Although the relationship between weight and dominance status was not significant, males that successfully sired offspring were significantly heavier than non-sires. No male less than 6.2 kg successfully sired an offspring, suggesting that surpassing a minimum body mass may enable males to enter the competitive arena and successfully gain access to receptive females.

There is evidence in numerous mammalian taxa that a larger body size is advantageous when

Table 7.3: Comparisons of the traits analysed to determine their influence on dominance status and reproductive success in male tammar wallabies (*M. eugenii*).

	Current study					
	Dominance		**Paternity**		**Hynes *et al.* 2005**	
Body weight	1 > 2 > 3 > 4	n.s.	sire > n. sire	c	1 > 2 > 3 > 4	b
Pes length	1 > 2 > 3 > 4	a	sire > n. sire	a	–	
Body condition	1, 2, 3 > 4	n.s.	sire > n. sire	n.s.	–	
Testosterone	1 > 2, 3, 4	n.s.	sire > n. sire	n.s.	NR	
Testis size	NR		NR		–	
1st mating	1 = 55%	(n = 20)	sire = 69%	(n = 13)	1 = 60% sire = 44%	(n = 40) (n = 9)
Paternity	1 = 48% 2 = 15% 3 = 15% 4 = 22%	(n = 27)	N/A		1 = 50% 2 = 35% 3 = 15% 4 = 0%	(n = 20)

NR = no relationship, N/A = not applicable, – = no data, n. sire = non-sire.
Dominance rank (highest to lowest) is represented by 1, 2, 3 and 4.
Each relationship is reported as n.s. = not significant at $p < 0.05$, or significant as shown below.
a: $p < 0.05$.
b: $p < 0.01$.
c: $p < 0.005$.

competing for access to receptive females, providing further support for the idea that contests between males have driven the evolution of sexual size dimorphism (Andersson 1994; Birkhead 2000). By winning male–male conflicts, the dominant male wallaby earns the right to guard and monopolise access to that female. This is reflected in the fact that dominant males in the current study spent significantly more time mate-guarding than did males of any other rank. In this study, and that of Hynes *et al.* (2005), the alpha male secured 55% and 60% of first matings respectively (Table 7.3), presumably as a result of his ability to successfully compete with other males and guard females during the pre-copulatory period. Circulating testosterone concentrations may also be important for securing a position of dominance, particularly at the start of the breeding season, but the results from different studies are unequivocal and probably reflect the need for a larger sample size.

Sperm competition theory predicts that males with larger testes may have a reproductive advantage (Short 1979) as the critical determinant of testis size is the number of Sertoli cells, which is proportional to sperm production (Bardin *et al.* 1994). Recent empirical evidence suggests that this may be the case both between and within mammalian species (Preston *et al.* 2003). However, despite the high relative testes size of male tammars (Taggart *et al.* 1998) and previous hypotheses that dominant males may have larger epididymal sperm reserves and may therefore be able to produce a larger number of effective ejaculates (Hynes *et al.* 2005), there was no relationship between testis volume and dominance status or reproductive success in the current study.

The tammar has a range of characteristics that are often associated with a first-mating advantage, such as pre-copulatory mate-guarding and the formation of a large, firm copulatory plug (Rudd 1994; Hynes *et al.* 2005). In this study, the male who mated first corresponded with the sire 63% of the time, whereas in the study by Hynes *et al.* (2005) the figure was only 44% (Table 7.3). These variations may be related to differences in the extent of post-copulatory guarding, the varying sex ratios between the two studies or the small number of observations in both studies. The precise function of copulatory plugs has not been determined in tammar wallabies, nor in any other macropodid.

The findings of this experiment indicate alpha males had significantly higher reproductive success than subordinate males, with the dominant male successfully siring 48% of the offspring, which is remarkably similar to the 50% observed by Hynes *et al.* (2005). In an earlier study of tammar wallaby paternity, Ewen *et al.* (1993) found that no one male sired significantly more offspring than any other. This may have been influenced by the high degree of oestrus synchronisation of the females and/or the low ratio of males to females (4M:13F) in Ewen *et al.*'s study. Where oestrus was naturally synchronised in this study, alpha males increased their reproductive success by guarding a female post-copulation.

In conclusion, this study of captive tammar wallabies has demonstrated that dominance confers a reproductive advantage. Dominant males are larger (and/or older) than subordinates and this size advantage allows them to gain greater access to females by winning aggressive encounters with other males. These findings support the hypothesis that male-biased sexual size dimorphism in the tammar wallaby has evolved as a result of competition between males for access to receptive females. Male dominance rank and siring success was independent of testis size in this study, but post-copulatory guarding appeared to be important. Failure to mate-guard post-copulation corresponded with poor reproductive success. In the wild, female tammars have been observed to mate with between five and seven males (Renfree and Shaw, unpubl. obs. cited in Hynes *et al.* 2005). Under such circumstances, males may find it difficult to successfully guard females post-copulation. Captive tammar wallaby populations may represent a good model system to investigate the relative importance of some variables, such as operational sex ratio and the functional significance of the copulatory plug, in a controlled environment.

ACKNOWLEDGMENTS

Special thanks to Rob Harcourt, Mariella Herberstein, Macquarie University Fauna Park staff and volunteers, Teena Browning for previously collected ear biopsies, Richard Peters, Matthew Vidler, Paul Herbert and the team at the Endocrinology lab at Royal North Shore Hospital, especially Cameron Wood and Margaret Wilkinson. All experimental work was approved by the Macquarie University Animal Ethics Committee (approval no. 2003/032). We thank the two anonymous referees for their constructive comments on the earlier draft of this manuscript.

REFERENCES

Alcock J 1993. *Animal Behaviour: An Evolutionary Approach.* Sinauer Associates, Massachusetts.

Andersson M 1994. *Sexual Selection.* Princeton University Press, New Jersey.

Bardin CW, Cheung CY, Mustow NA and Gunsalus GL 1994. The sertoli cell. In *The Physiology of Reproduction.* (Eds E Knobil and JD Neill) pp. 1291–1333. Raven Press, New York.

Birkhead TR 2000. *Promiscuity: An Evolutionary History of Sperm Competition and Sexual Conflict.* Academic Press, London.

Carranza J and Valencia J 1999. Red deer females collect on male clumps at mating areas. *Behav. Ecol.* **10**: 525–532.

Catling PC and Sutherland RL 1980. Effect of gonadectomy, season and the presence of female tammar wallabies (*Macropus eugenii*) on concentrations of testosterone, luteinising hormone and follicle-stimulating hormone in the plasma of male tammar wallabies. *J. Endocrinol.* **86:** 25–33.

Clutton-Brock TH 1989. Mammalian mating systems. *Proc. Roy. Soc. Lond. B Biol. Sci.* **236**: 339–372.

Coulson G 1989. Repertoires of social behaviour in the Macropodoidea. In *Kangaroos, Wallabies and Rat-kangaroos.* (Eds G Grigg, P Jarman and I Hume) pp. 457–473. Surrey Beatty and Sons, Sydney.

Croft DB 1981. Behaviour of the red kangaroo, *Macropus rufus* (Desmarest, 1822) in north-western New South Wales. *Aust. Mammal.* **5:** 5–13.

Ewen KR, Temple-Smith PD, Bowden DK, Marinopoulos J, Renfree MB and Yan H 1993. DNA fingerprinting in relation to male dominance and paternity in a captive colony of tammar wallabies (*Macropus eugenii*). *J. Reprod. Fertil.* **99:** 33–37.

Ganslosser U 1989. Agonistic behaviour in Macropodoids: a review. In *Kangaroos, Wallabies and Rat-kangaroos.* (Eds G Grigg, P Jarman and I Hume) pp. 475–503. Surrey Beatty and Sons, Sydney.

Hynes EF, Rudd CD, Temple-Smith PD, Sofronidis G, Paris D, Shaw G and Renfree MB 2005. Mating sequence, dominance and paternity success in captive male tammar wallabies. *Reproduction* **130:** 123–130.

Inns RW 1982. Seasonal changes in the accessory reproductive system and plasma testosterone levels of the male tammar wallaby, *Macropus eugenii*, in the wild. *J. Reprod. Fertil.* **66:** 675–680.

Jarman PJ 1983. Mating system and sexual dimorphism in large, terrestrial, mammalian herbivores. *Biol. Rev.* **58:** 485–520.

Martin P and Bateson P 1993. *Measuring Behaviour: An Introductory Guide.* Cambridge University Press, Cambridge.

Parker GA 1998. Sperm competition and the evolution of ejaculates: towards a theory base. In *Sperm Competition and Sexual Selection.* (Eds TR Birkhead and AP Moller) pp. 3–54. Academic Press, London.

Poole TB 1985. *Social Behaviour in Mammals.* Blackie, Oxford, London.

Poole WE, Simms NG, Wood JT and Lubulwa M 1991. Tables for age determination of the Kangaroo Island wallaby (tammar), *Macropus eugenii*, from body measurements. Technical Memorandum No. 32, CSIRO Division of Wildlife and Ecology, Canberra.

Preston BT, Stevenson IR, Pemberton JM, Coltman DW and Wilson K 2003. Overt and covert competition in a promiscuous mammal: the importance of weaponry and testes size to male reproductive success. *Proc. Roy. Soc. Lond. B Biol. Sci.* **270:** 633–640.

Renfree MB, Fletcher TP, Blanden DR, Lewis PR, Shaw G, Gordon K, Short RV, Parer-Cook E and Parer D 1989. Physiological and behavioural events around the time of birth in macropod marsupials. In *Kangaroos, Wallabies and Rat-kangaroos.* (Eds G Grigg, P Jarman and I Hume) pp. 323–337. Surrey Beatty and Sons, Sydney.

Rudd CD 1994. Sexual behaviour of male and female tammar wallabies (*Macropus eugenii*) at post-partum oestrus. *J. Zool.* **232:** 151–162.

Rudd CD, Short RV, Shaw G and Renfree MB 1996. Testosterone control of male-type sexual behaviour in the tammar wallaby (*Macropus eugenii*). *Horm. Behav.* **30:** 446–454.

Short RV 1979. Sexual selection and its component parts, somatic and genital selection, as illustrated by man and the great apes. *Adv. Stud. Behav.* **9:** 131–158.

Spencer PBS, Odorico DM, Jones J, Marsh HD and Miller DJ 1995. Highly variable microsatellites in isolated colonies of the rock-wallaby (*Petrogale assimilis*). *Mol. Ecol.* **4:** 523–525.

Sunnucks P and Hales D 1996. Numerous transposed sequences of mitochondrial cytochrome oxidase I–II in aphids of the genus *Sitobion* (Hemiptera: Aphidae). *Mol. Ecol.* **13:** 510–524.

Taggart DA, Breed WA, Temple-Smith PD, Purvis A and Shimmin G 1998. Reproduction, mating strategies and sperm competition in marsupials and monotremes. In *Sperm Competition and Sexual Selection.* (Eds TR Birkhead and AP Moller) pp. 623–666. Academic Press, London.

Taylor AC and Cooper DW 1998. A set of tammar wallaby (*Macropus eugenii*) microsatellites tested for genetic linkage. *Mol. Ecol.* **7:** 925–931.

Tyndale-Biscoe CH and Renfree MB 1987. *Reproductive Physiology of Marsupials.* Cambridge University Press, Cambridge.

Williamson P, Fletcher TP and Renfree MB 1990. Testicular development and maturation of the hypothalamic-pituitary-testicular axis in the male tammar, *Macropus eugenii. J. Reprod. Fertil.* **88:** 549–557.

Zenger KR and Cooper DW 2001. A set of highly polymorphic microsatellite markers developed for the eastern grey kangaroo (*Macropus giganteus*). *Mol. Ecol.* **1:** 98–100.

8 Strong matrilineal structuring in the brush-tailed rock-wallaby confirmed by spatial patterns of mitochondrial DNA

S.L. Hazlitt, M.D.B. Eldridge and A.W. Goldizen

SUMMARY

Matrilineal population structure, or male-biased dispersal among female kin groups, is typical of many eutherian mammals. In contrast, matrilineal genetic structuring has not been well documented in marsupials, although recent studies on the brush-tailed rock-wallaby (*Petrogale penicillata*) are suggestive of the presence of female kin groups within colonies. We used spatial patterns of mitochondrial DNA (mtDNA) variation and long-term mark–recapture records to examine patterns of population structure and dispersal within and among brush-tailed rock-wallaby colonies in south-east Queensland. We test two hypotheses: that strong female philopatry has led to lineage sorting among neighbouring brush-tailed rock-wallaby colonies, and that female rock-wallabies form kin groups within a single colony. We documented unique haplotypes within most colonies, with strong and significant population structuring (high Φ_{ST} values) among colonies within two neighbouring valleys. In addition, females sharing mtDNA haplotypes formed tight and significant spatial clusters along only 800 m of cliff line within a single colony; no such spatial clustering was found for males. These findings further demonstrate that female brush-tailed rock-wallabies are faithful to their natal colony and provide support for the presence of kin groups in a marsupial.

INTRODUCTION

Most mammal species are characterised by male-biased dispersal, where males tend to disperse further or more frequently than females and females are often philopatric (Greenwood 1980). Theory suggests that in the typical mammalian polygynous mating system, where females invest heavily in offspring, females may benefit more from philopatry than males as a result of experience in local habitats (Greenwood 1980). In addition, limited habitat availability or benefits to survival or reproduction from social aggregations of kin may further promote selection for strong female philopatry (Greenwood 1980; Kerth *et al.* 2000). When strong female philopatry persists over long time scales, females in an area become highly related and can form closed kin groups or matrilines within populations (Hoelzer *et al.* 1994). Strong spatial patterns of female kinship and matrilineal social groups have been demonstrated in many eutherian mammals, including primates (Kappeler *et al.* 2002), elephants (Vidya and

Sukumar 2005), bats (Kerth *et al.* 2000) and wolves (Girman *et al.* 1997). This pattern of matrilineal genetic structuring has, through kin selection theory (Hamilton 1963; Maynard Smith 1964), formed one of the foundations for the explanation of social behaviours in mammals, with recent molecular studies providing support for the influence of kinship on patterns of behaviour and reproductive success (Morin *et al.* 1994; Packer *et al.* 1991; Parsons *et al.* 2003; Pope 2000).

Data on spatial patterns of female kinship and social structuring for marsupials are extremely limited, although such patterns are demonstrated widely in eutherian taxa (Johnson 1989a). The social structure of macropodid marsupials, a large family of primarily terrestrial herbivores (Macropodidae, the kangaroos and wallabies), ranges from solitary living to open-membership mobs (Jarman 1991). Many marsupial species show a pattern of female philopatry and male-biased dispersal at the population level (Pope *et al.* 2005; Zenger *et al.* 2003) or within populations (Hazlitt *et al.* 2004; Piggott *et al.* 2006). Behavioural work has shown that adult female red-necked wallabies (*Macropus rufogriseus*) and eastern grey kangaroos (*M. giganteus*) associate with older daughters more often than expected by chance (Jarman 1991; Johnson 1989b). Moreover, recent studies on the brush-tailed rock-wallaby are suggestive of the presence of female kin groups within colonies (Hazlitt *et al.* 2004, 2006b; Piggott *et al.* 2006).

Rock-wallabies are a marsupial species in which more complex social structuring is likely to have evolved (Jarman 1991; Johnson 1989a). Rock-wallabies (genus *Petrogale*) are medium-sized (1–10 kg) macropodids dependent on heterogeneous and limited rocky habitats for daytime shelter from the environment and predators (Strahan 1995). Strong philopatry may be particularly adaptive in these species, in which home-range familiarity and alertness could decrease the risk of predation (Jarman and Bayne 1997; Johnson 1989a). Dispersal events among rock-wallaby colonies are relatively limited (Eldridge *et al.* 2001; Pope *et al.* 1996; Sharp 1997), making it difficult to assess patterns in sex-biased dispersal at the among-colony scale (Hazlitt *et al.* 2006a; Piggott *et al.* 2006). However, two independent studies have used nuclear DNA markers to demonstrate a pattern of male-biased dispersal and strong female philopatry within single, relatively large brush-tailed rock-wallaby (*P. penicillata*) colonies (Hazlitt *et al.* 2004; Piggott *et al.* 2006). Females showed much stronger levels of philopatry compared to males within these colonies, but both studies also indicated that males in close proximity were more genetically similar than males further apart, suggesting that male dispersal is not random within a colony (Hazlitt *et al.* 2004; Piggott *et al.* 2006).

Colonies of the brush-tailed rock-wallaby, one of the better-studied rock-wallaby species, often comprise several distinct social groups (Jarman and Bayne 1997; Laws and Goldizen 2003). Each group typically consists of a single male and multiple females sharing the same rocky caves during the day (Jarman and Bayne 1997). Group members that emerge from the same region of cliff line also forage together in the same area of the colony (Laws and Goldizen 2003) and recent work has demonstrated that these groups represent discrete breeding groups, at least in one well-studied colony (Hazlitt *et al.* 2006b). The pattern of within-colony sex-biased dispersal, strong female philopatry and discrete breeding groups is suggestive of the presence of female kin groups or matrilines in brush-tailed rock-wallabies (Hazlitt *et al.* 2006b; Piggott *et al.* 2006).

For this study we used spatial patterns of mitochondrial DNA (mtDNA) control region haplotypes, which reflect female-mediated gene flow, to examine patterns of population structure and dispersal within and among brush-tailed rock-wallaby colonies in south-east Queensland. First, we present a fine-scale phylogeographic analysis using mtDNA from four neighbouring rock-wallaby colonies to test the hypothesis that strong female philopatry has led to lineage sorting (no female migration) among colonies. Second, we use spatial data based on trapping locations of 40 breeding individuals in the well-studied Hurdle Creek colony, in combination with mtDNA control region haplotype diversity, to confirm the hypothesis that female rock-wallabies form kin groups (Kappeler *et al.* 2002) and to examine patterns of male dispersal within colonies. Finally, we use long-term mark–recapture records to examine the contemporary pattern of female and male natal recruitment within the Hurdle Creek colony.

MATERIALS AND METHODS

Study sites and sample collection

The study sites and field methods are described in detail in Hazlitt *et al.* (2004, 2006a). Briefly,

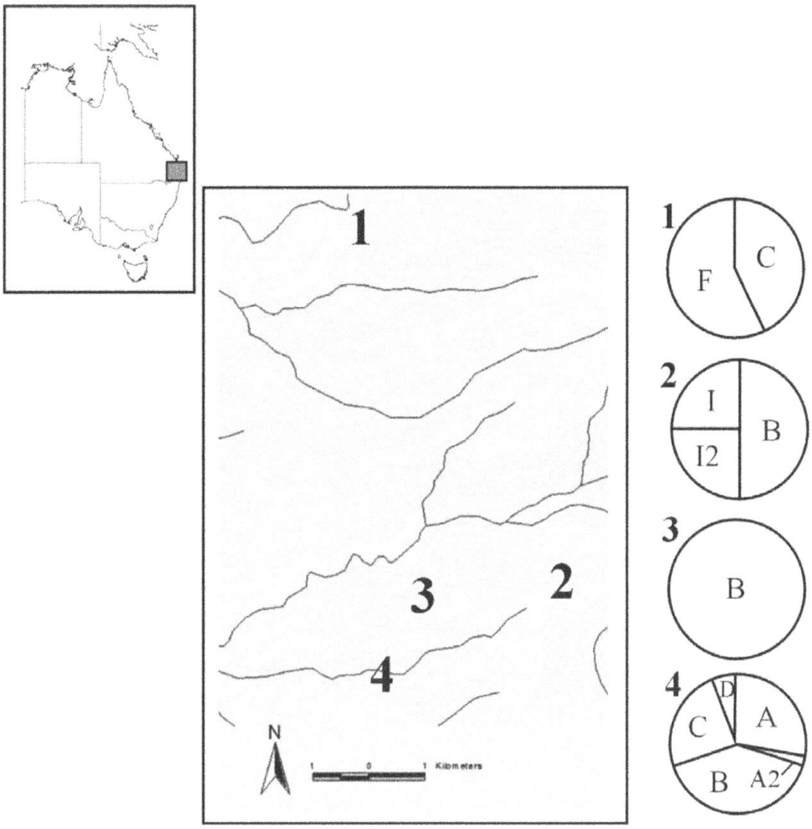

Figure 8.1: Location of the four sampled brush-tailed rock-wallaby (*P. penicillata*) colonies in south-east Queensland – 1 Emu Creek (n = 7), 2 Farm Creek East (n = 4), 3 Farm Creek (n = 9), 4 Hurdle Creek (n = 47) – and the frequencies and distribution of eight mtDNA control region haplotypes.

we sampled brush-tailed rock-wallabies from four colonies – Hurdle Creek (n = 47), Farm Creek (n = 9), Farm Creek East (n = 4) and Emu Creek (n = 7) – located within 1–8 km of each other in the Condamine Valley region of south-east Queensland (28°17'S, 152°19'E; Fig. 8.1). For the fine-scale phylogeographic mtDNA analysis, data were available from 67 individuals trapped between 2001 and 2004, except for two samples from Emu Creek which were collected by the Queensland Parks and Wildlife Service in 1997. The small sample sizes were reflective of the small population sizes of most brush-tailed rock-wallaby colonies (Jarman and Bayne 1997) and the difficulty of accessing and trapping rock-wallabies within colonies (Eldridge *et al.* 2001). However, our Hurdle Creek sample represented the census size of the colony (Hazlitt *et al.* 2004). Females were classified as adults if they had a pouch young or one or more teats had been used for suckling (Poole and Catling 1974), males were classified as adults based on size (breeding size ≥4 kg) and subadults were animals trapped independently and ≥2 kg but less than 4 kg (Hazlitt *et al.* 2006b; Johnson and Delean 2002). The 67 individuals trapped included 41 females (seven subadults), 25 males (two subadults) and one individual of unknown gender.

MtDNA sequence data and phylogeographic analyses

Ear biopsies of 3 mm diameter were collected from all animals and stored in 90% ethanol before total cellular DNA was extracted using the salting-out methods of Sunnucks and Hales (1996) or Sigg *et al.* (2005; for samples collected after March 2003). The mtDNA control region was amplified using conserved marsupial primers (L15999M and H16498M, Fumagalli *et al.* 1997), and mtDNA control region variation among all sampled rock-wallabies was assessed using the SSCP technique (Sunnucks *et al.* 2000) as previously described (Browning *et al.* 2001). Sequence data were obtained for each unique haplotype using BigDye termination chemistry and resolved using an ABI 3730 capillary sequencer. Approximately 630 base pairs (bp) of control region sequence was obtained from three to 11 individuals of each SSCP haplotype except where a unique haplotype was identified in only one or two individuals (e.g. haplotype D). Sequences were checked and aligned by eye and using MEGA3 (Kumar *et al.* 2004).

Phylogenetic patterns in the sequence data were examined using the neighbour-joining (NJ) analysis in MEGA3 with distances among sequences modelled using the Kimura 2-parameter (K2P) genetic distance (Kimura 1980) with a gamma distribution value of $\alpha = 0.5$ (Zenger *et al.* 2003). Support for branching topology was assessed using bootstrap analysis with 10 000 replicates. Sequence data obtained from a closely related taxon, *P. herberti* (Genbank no. AY040892; Herbert's rock-wallaby), were used as an outgroup. Spatial patterns of mtDNA structure among colonies were examined using pairwise Φ_{ST} values (K2P with gamma at 0.5), a measure of genetic differentiation based on the number and distribution of haplotypes among colonies. Pairwise Φ_{ST} and associated significance levels (based on 10 000 permutations) were calculated using ARLEQUIN version 2.0 (Schneider *et al.* 2000).

Within-colony spatial distribution of mtDNA haplotypes and natal recruitment

The spatial distribution of individuals with different mtDNA haplotypes within a colony was investigated in the Hurdle Creek colony. Spatial analyses were based on centres of cliff emergence ranges for each animal, which were calculated using cliff line trap records from the four-year mark–recapture study, described in detail in Hazlitt *et al.* (2004). These spatial data are a consistent aspect of each animal's home range, because we trapped where individuals emerge from their sheltering locations along cliff faces to forage on the grassy slopes above (Laws and Goldizen 2003). The spatial data utilised here have been shown to reflect other aspects of home range patterns such as nocturnal foraging areas (Laws and Goldizen 2003) and breeding groups (Hazlitt *et al.* 2006b). The genetic data presented here include only breeding rock-wallabies first trapped as subadults or adults, therefore excluding any known relatives. In addition, since the age of first breeding ranges from two to three years for females and older for males (Delaney 1997; Hazlitt *et al.* 2006b; Johnson and Delean 2002), using only adults or subadults minimised the overlap of generations. The data included 11 successfully breeding males, determined using molecular techniques (Hazlitt *et al.* 2006b), and 29 breeding females (mean ± SE number of trapping records per animal 15 ± 2).

We compared mean distances among the centres of cliff emergence ranges of both breeding males and breeding females with identical and different haplotypes, using the two-sample randomisation test with 10 000 iterations in POPTOOLS 2.6.2 (Hood 2003), to test the assumption that the observed mean distances did not differ significantly from a random assortment of values among the categories (Taylor *et al.* 1997). Four animals caught as pouch young in the first six months of the Hurdle Creek mark–recapture study (Nov. 2000–April 2001) were still present in the population in 2004. Two females born into the colony recruited into the breeding population approximately three years later, and two males were still present in the population well after reaching breeding size (>5 kg) (Johnson and Delean 2002). We compared the distance between the centres of cliff emergence ranges of these second-generation individuals (mean ± SE number of independent trapping records 18 ± 3) to the centres of the cliff emergence ranges of their mothers. We tested whether settlement distances from their mothers were different from expected from random for settling females and males, by comparing the observed distance values with distances between the settling recruits and all other breeding females in the colony, using the two-sample randomisation test with 10 000 iterations in POPTOOLS 2.6.2 (Hood 2003).

RESULTS

Eight mtDNA haplotypes were found among the 67 rock-wallabies sampled (GenBank Accession Nos EU887004-EU887011; Table 8.1, Fig. 8.1). The 628 bp of control region sequence aligned unambiguously, exhibiting 13 (2.1%) variable sites, including nine transitions, a single transversion, two single nucleotide insertions and a 26 bp insertion in haplotype C, which was a repeat of the previous 26 bp therefore treated as a single mutation (Table 8.1). The mean sequence divergence was 0.0086 (range 0.0016–0.016) among all haplotypes. The sequences clustered into three groups, two that were well supported by high bootstrap values (>85%) and a third with reasonable bootstrap support (70%; Fig. 8.2). Two common

Table 8.1: Sequence variation among 628 bp of mtDNA control region haplotypes found in 67 brush-tailed rock-wallabies (*P. penicillata*).

				1	1	1	1	2	3	3	3	3	4	4	4	4	5
	7	7	8	0	2	5	7	9	2	5	6	7	0	3	3	8	4
	8	9	1	3	9	6	3	2	9	9	7	0	5	2	4	2	6
A	–	–	C	C	G	T	T	C	T	G	C	T	A	A	–	C	C
A2	.	.	.	.	.	.	.	.	C	.	.	.	.	.	.	.	.
B	.	.	.	.	.	.	C	.	C	.	.	C	C	G	.	.	.
C	.	.	.	.	.	.	.	.	C	.	.	.	.	.	*	.	.
D	T	T	.	.	.	C	C	.	C	A	T	.	C	.	.	T	T
I	.	.	.	.	.	.	C	.	C	.	.	C	C	.	.	.	.
I2	T	T	.	.	.	C	C	.	C	.	T	.	C	.	.	T	T
F	T	.	.	.	.	.	.	.	C	.	.	.	.	.	.	.	.

Only variable sites are shown, with nucleotide positions indicated above. Letters indicate nucleotide substitutions; · = identity with the A haplotype; – = indels; * = 26 bp insertion (AGTACATAAAGACATTAAACTCAAAC).

haplotypes (B and C) were shared among two or three colonies; however, less common and unique haplotypes were present in three of the four colonies (Fig. 8.1). Some of the unique haplotypes supported a pattern of sequence divergence within colonies with no dispersion of those haplotypes among neighbouring colonies (Figs 8.1, 8.2). The presence of unique haplotypes within most colonies was reflected by very high pairwise differentiation values among colonies, with a mean Φ_{ST} value of 0.45 (range 0.14–1.00; Table 8.2). Most among-colony Φ_{ST} comparisons were significantly differentiated except comparisons with Farm Creek East, which was likely due to the small sample size (n = 4).

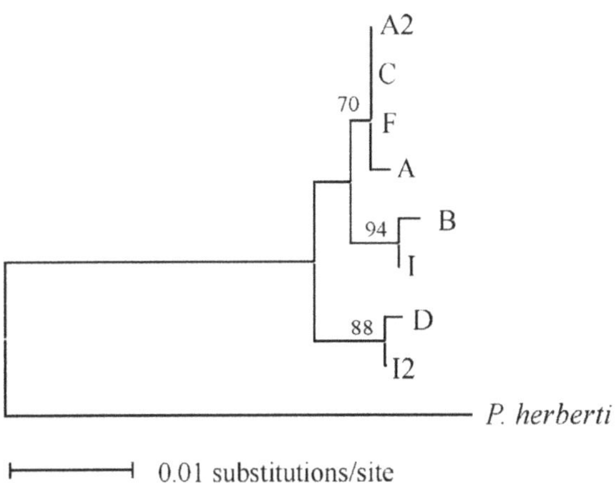

Figure 8.2: Neighbour-joining tree displaying the relationship between brush-tailed rock-wallaby (*P. penicillata*) mtDNA control region haplotypes, with *P. herberti* (Herbert's rock-wallaby) as an outgroup. Values at nodes indicate the support obtained from 10 000 bootstrap replicates.

Spatial patterns of mtDNA haplotypes and natal recruitment within a colony

Breeding females sharing the same mtDNA haplotype formed spatially distinct clusters along the colony's 800 m cliff line (Fig. 8.3a). Females with the two most common haplotypes, A and B, formed distinct clusters at either end of the colony (Fig. 8.3a). The mean distance between the centres of cliff emergence ranges of pairs of females sharing mtDNA haplotypes was significantly less than expected from random (mean ± SE same haplotypes = 97 ± 7 m, n = 99 comparisons, P < 0.0001), while females with different mtDNA haplotypes were separated by distances significantly greater than expected from random within the colony (mean ± SE different haplotypes = 288 ± 9 m, n = 302 comparisons, P < 0.0001). In contrast, the mean distances between males with the same and different haplotypes were similar and did not differ

Table 8.2: Pairwise differentiation (Φ_{ST}) among brush-tailed rock-wallaby (*P. penicillata*) colonies (below diagonal) and significance based on 10 000 permutations (above diagonal).

	Hurdle Creek	Farm Creek	Farm Creek East	Emu Creek
Hurdle Creek		b	–	a
Farm Creek	0.318		–	b
Farm Creek East	0.142	0.321		b
Emu Creek	0.191	1.000	0.728	

a: Indicates significance prior to Bonferroni correction at P < 0.05.
b: Values were significant after Bonferroni correction.

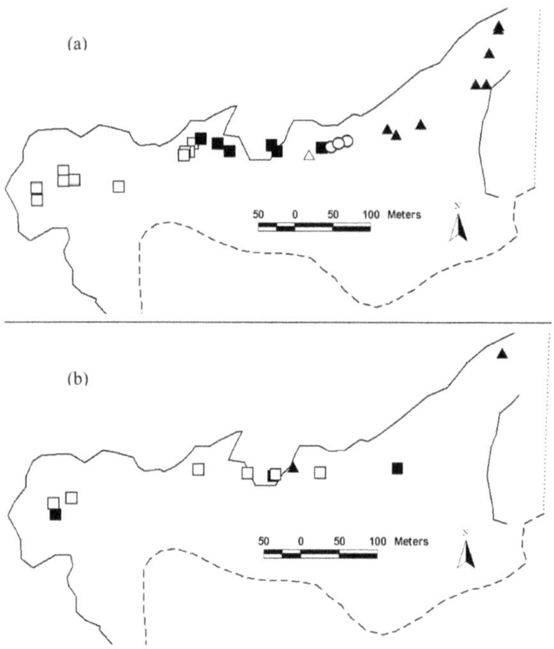

Figure 8.3: Spatial patterning of different mtDNA haplotypes in the Hurdle Creek brush-tailed rock-wallaby (*P. penicillata*) colony. The centres of cliff emergence ranges of (a) breeding females and (b) breeding males are depicted with mtDNA haplotypes denoted by symbol shape/colour (black triangle = haplotype A, white square = haplotype B, black square = haplotype C, white circle = haplotype D, white triangle = haplotype A2). Solid black lines depict the cliff edge, dotted lines denote dirt road and track.

from random (means ± SE: same haplotypes = 198 ± 26 m, n = 22 comparisons; different haplotypes = 227 ± 26 m, n= 44 comparisons, both P > 0.05) (Fig. 8.3b). In addition to the spatial pattern of individuals with similar mtDNA haplotypes, four years of mark–recapture data provided contemporary data on the natal recruitment of four individuals (two females and two males) in the Hurdle Creek colony. Breeding daughters settled significantly closer to their mothers than expected from random (mean ± SE mother–daughter distance = 16 ± 1 m, n = 2 comparisons, P < 0.01), while the distances that adult sons settled from their mothers did not differ from random (mean ± SE mother–son distance = 154 ± 85 m, n = 2 comparisons, P > 0.05).

DISCUSSION

We used mtDNA haplotypes and spatial data to examine patterns of population structure and dispersal within and among brush-tailed rock-wallaby colonies in south-east Queensland. We tested the hypotheses that strong female philopatry has led to lineage sorting among neighbouring brush-tailed rock-wallaby colonies and that female rock-wallabies form kin groups within a single colony. Finally, we used long-term mark–recapture records to examine spatial patterns of female and male natal recruitment within the Hurdle Creek colony.

We documented significant population structuring among neighbouring brush-tailed rock-wallaby colonies, as shown in earlier studies that utilised both mitochondrial and nuclear variation to examine patterns of dispersal among colonies of brush-tailed and other rock-wallaby species (Browning *et al.* 2001; Eldridge *et al.* 2001; Hazlitt *et al.* 2004, 2006a; Piggott *et al.* 2006; Pope *et al.* 1996). Most of the sampled brush-tailed rock-wallaby colonies in this study shared one or two common mtDNA haplotypes, but neighbouring colonies also contained unique mtDNA haplotypes. These findings suggest some level of among-colony female dispersal. Although small sample sizes from three of the four colonies make definitive conclusions difficult (as haplotypes may have been present within a colony but not sampled), the presence of novel haplotypes that appear diverged from more common haplotypes but have not been redistributed among colonies suggests that female dispersal among these colonies is low, despite the colonies being separated by only 1–2 km of suitable habitat.

Our data suggest that female brush-tailed rock-wallabies form matrilineal or kin groups within the Hurdle Creek colony, similar in scale to the tight kin clusters found in eutherian species, such as bats (Kerth *et al.* 2000) and non-gregarious primates (Kappeler *et al.* 2002), as well as kin groups found in obligately social species like long-tailed macaques (de Ruiter and Geffen 1998) and red howler monkeys (Pope 2000). An earlier study of our brush-tailed rock-wallaby colony was strongly suggestive of the presence of matrilineal structuring based on patterns of nuclear DNA variation, but was only able to demonstrate two significant clusters of related females within the colony (Hazlitt *et al.* 2004). This study was able to demonstrate that the multiple, spatially distinct female groups within the colony are each represented by unique mtDNA haplotypes. In addition, paternity analysis showed that these same female groups represent discrete breeding groups (Hazlitt *et al.* 2006b). These findings are further supported by the mark–recapture data for two females born into the colony who later recruited into the breeding population in the same spatial location and breeding group as their mothers (Hazlitt *et al.* 2006b).

No significant pattern of structure was found using mtDNA haplotypes of breeding male brush-tailed rock-wallabies in the Hurdle Creek colony. Additionally, two males born during the study and still alive in Hurdle Creek at breeding age did not remain in their natal areas but moved into other regions of the colony. Although the sample size for breeding males within the colony was small, our data support the notion that within-colony male dispersal among breeding groups (and high male turnover), coupled with infrequent among-colony dispersal events, facilitates inbreeding avoidance in this colony-dwelling species (Hazlitt *et al.* 2004, 2006b; Piggott *et al.* 2006). Two studies detected weak signals of male brush-tailed rock-wallaby population structure within colonies (Hazlitt *et al.* 2004; Piggott *et al.* 2006), which could be the result of sons moving only short distances to neighbouring breeding groups within colonies.

The among-colony patterns of mtDNA variation documented in south-east Queensland provide some insight into patterns of mtDNA variation among brush-tailed rock-wallaby colonies in the southern part of the species' distribution, where populations have undergone major declines in abundance and where the species is listed as endangered. Studies on these remnant colonies found a pattern of haplotype uniformity within colonies with no sharing of haplotypes among colonies (Browning *et al.* 2001; Eldridge *et al.* 2004). This lack of within-colony mtDNA variation and lack of shared haplotypes among colonies in the south of the species' distribution is likely the result of major declines in population sizes over the last century, as well as the absence of contemporary female-mediated gene flow among colonies (Browning *et al.* 2001).

It has been suggested that, because of their strong dependency on complex rocky habitats, rock-wallabies are a taxon that might especially benefit from strong philopatry (Johnson 1989a). Rock-wallaby refuges often have multiple entrances and pathways to and from foraging areas are consistently utilised by individuals (Jarman and Bayne 1997; Laws and Goldizen 2003), which supports the idea that habitat familiarity is an important element to survival. In addition, the sharing of limited cliff refuges by related females could provide direct and indirect reproductive fitness benefits (Hamilton 1963). Although we have no data on the reproductive fitness of female lineages, female brush-tailed rock-wallabies sharing the same cliff region form discrete breeding groups, usually breeding with a single unrelated male (Hazlitt *et al.* 2006b), which is similar to the pattern of breeding in matrilineal groups in the social primates (de Ruiter and Geffen 1998). We propose that rock-wallabies would be appropriate marsupial species for further investigations of the relationship between kinship and social behaviour in marsupials.

ACKNOWLEDGMENTS

We are indebted to the private land owners for access to their properties and we thank numerous researchers, field assistants and volunteers for help with field work. We are grateful to Queensland Parks and Wildlife Service and the Toowoomba City Council for in-kind support and samples. We are indebted to D. Sigg and J. Nicholls for help with the molecular work and to J. Nicholls, A. Lowe, D. Putland and D. Sigg for comments and discussion. We thank the anonymous reviewers who provided suggestions and comments that improved this manuscript. This study was approved by the University of Queensland Animal Ethics Committee and the Queensland Parks and Wildlife Service (Scientific Purposes Permit# W4/002560/00/SAA). The research was funded by the Australian Research Council, National Geographic Society, Norman Wettenhall Foundation, University of Queensland and Australian Commonwealth.

REFERENCES

Browning TL, Taggart DA, Rummery C, Close RL and Eldridge MDB 2001. Multifaceted genetic analysis of the 'critically endangered' brush-tailed rock-wallaby *Petrogale penicillata* in Victoria, Australia: implications for management. *Conserv. Genet.* **2:** 145–156.

Delaney R 1997. Reproductive ecology of the allied rock-wallaby, *Petrogale assimilis*. *Aust. Mammal.* **19:** 209–218.

de Ruiter JR and Geffen E 1998. Relatedness of matrilines, dispersing males and social groups in long-tailed macaques (*Macaca fascicularis*). *Proc. Roy. Soc. Lond. B Biol. Sci.* **265:** 79–87.

Eldridge MDB, Kinnear JE and Onus ML 2001. Source population of dispersing rock-wallabies (*Petrogale lateralis*) identified by assignment tests on multilocus genotypic data. *Mol. Ecol.* **10:** 2847–2876.

Eldridge MDB, Rummery C, Bray C, Zenger KR, Browning TL and Close RL 2004. Genetic analysis of a population crash in brush-tailed rock-wallabies (*Petrogale penicillata*) from Jenolan Caves, south-eastern Australia. *Wildl. Res.* **31:** 229–240.

Fumagalli L, Pope LC, Taberlet P and Moritz C 1997. Versatile primers for the amplification of the mitochondrial DNA control region in marsupials. *Mol. Ecol.* **6:** 1199–1201.

Girman DJ, Mills MGL, Geffen E and Wayne RK 1997. A molecular genetic analysis of social structure, dispersal, and interpack relationships of the African wild dog (*Lycaon pictus*). *Behav. Ecol. Sociobiol.* **40:** 187–198.

Greenwood PJ 1980. Mating systems, philopatry and dispersal in birds and mammals. *Anim. Behav.* **28:** 1140–1162.

Hamilton WD 1963. Evolution of altruistic behavior. *Am. Nat.* **97:** 354–356.

Hazlitt SL, Eldridge MDB and Goldizen AW 2004. Fine-scale spatial genetic correlation analyses reveal strong female philopatry within a brush-tailed rock-wallaby colony in southeast Queensland. *Mol. Ecol.* **13:** 3621–3632.

Hazlitt SL, Goldizen AW and Eldridge MDB 2006a. Significant patterns of population genetic structure and limited gene flow in a threatened macropodid marsupial despite continuous habitat in southeast Queensland, Australia. *Conserv. Genet.* **7:** 675–689.

Hazlitt SL, Sigg DP, Eldridge MDB and Goldizen AW 2006b. Restricted mating dispersal and strong breeding group structure in a mid-sized marsupial mammal (*Petrogale penicillata*). *Mol. Ecol.* **15:** 2997–3007.

Hoelzer GA, Dittus WPJ, Ashley MV and Melnick DJ 1994. The local distribution of highly divergent mitochondrial DNA haplotypes in toque macaques *Macaca sinica* at Polonnaruwa, Sri Lanka. *Mol. Ecol.* **3:** 451–458.

Hood GM 2003. PopTools version 2.6.2. http://www.cse.csiro.au/poptools/.

Jarman PJ 1991. Social behavior and organization in the Macropodoidea. *Adv. Stud. Behav.* **20:** 1–50.

Jarman PJ and Bayne P 1997. Behavioural ecology of *Petrogale pencillata* in relation to conservation. *Aust. Mammal.* **19:** 219–228.

Johnson CN 1989a. Dispersal and philopatry in the Macropodoids. In *Kangaroos, Wallabies and Rat-kangaroos.* (Eds G Grigg, P Jarman and I Hume) pp. 593–601. Surrey Beatty and Sons, Sydney.

Johnson CN 1989b. Grouping and the structure of association in the red-necked wallaby. *J. Mammal.* **70:** 18–26.

Johnson PM and Delean S 2002. Reproduction of the purple-necked rock-wallaby, *Petrogale purpureicollis* Le Souef (Marsupialia: Macropodidae) in captivity, with age estimation and development of the pouch young. *Wildl. Res.* **29:** 463–468.

Kappeler PM, Wimmer B, Zinner D and Tautz D 2002. The hidden matrilineal structure of a solitary lemur: implications for primate social evolution. *Proc. Roy. Soc. Lond. B Biol. Sci.* **269:** 1755–1763.

Kerth G, Mayer F and Konig B 2000. Mitochondrial DNA (mtDNA) reveals that female Bechstein's bats live in closed societies. *Mol. Ecol.* **9:** 793–800.

Kimura M 1980. A simple method for estimating evolutionary rates of base substitutions through comparative studies of nucleotide sequences. *J. Mol. Evol.* **16:** 111–120.

Kumar S, Tamura K and Nei M 2004. MEGA3: Integrated software for molecular evolutionary genetics analysis and sequence alignment. *Brief. Bioinform.* **5:** 150–163.

Laws RJ and Goldizen AW 2003. Nocturnal home ranges and social interactions of the brush-tailed rock-wallaby *Petrogale penicillata* at Hurdle Creek, Queensland. *Aust. Mammal.* **25:** 169–176.

Maynard Smith J 1964. Group selection and kin selection. *Nature* **201:** 1145–1147.

Morin PA, Moore JJ, Chakraborty R, Jin L, Goodall J and Woodruff DS 1994. Kin selection, social structure, gene flow, and the evolution of chimpanzees. *Science* **265:** 1193–1201.

Packer C, Gilbert DA, Pusey AE and Obrien SJ 1991. A molecular genetic analysis of kinship and cooperation in African lions. *Nature* **351:** 562–565.

Parsons KM, Durban JW, Claridge DE, Balcomb KC, Noble LR and Thompson PM 2003. Kinship as a basis for alliance formation between male bottlenose dolphins, *Tursiops truncatus*, in the Bahamas. *Anim. Behav.* **66:** 185–194.

Piggott MP, Banks SC and Taylor AC 2006. Population structure of brush-tailed rock-wallaby (*Petrogale penicillata*) colonies inferred from analysis of faecal DNA *Mol. Ecol.* **15:** 93–105.

Poole WE and Catling PC 1974. Reproduction in two species of grey kangaroos, *Macropus giganteus* (Shaw) and *Macropus fuliginosus* (Desmarest). I. Sexual maturity and oestrus. *Aust. J. Zool.* **22:** 277–302.

Pope TR 2000. Reproductive success increases with degree of kinship in cooperative coalitions of female red howler monkeys (*Alouatta seniculus*). *Behav. Ecol. Sociobiol.* **48:** 253–267.

Pope LC, Sharp A and Moritz C 1996. Population structure of the yellow-footed rock-wallaby *Petrogale xanthopus* (Gray, 1854) inferred from mtDNA sequences and microsatellite loci. *Mol. Ecol.* **5:** 629–640.

Pope LC, Blair D and Johnson CN 2005. Dispersal and population structure of the rufous bettong, *Aepyprymnus rufescens* (Marsupialia: Potoroidae). *Austral Ecol.* **30:** 572–580.

Schneider S, Roessli D and Excoffier L 2000. Arlequin (version 2.0). A software for population genetic data analysis. Genetics and Biometry Laboratory, University of Geneva, Switzerland.

Sharp A 1997. Insights into the dispersal patterns of yellow-footed rock-wallabies, *Petrogale xanthopus*. *Aust. Mammal.* **19:** 229–238.

Strahan R (Ed.) 1995. *The Mammals of Australia*. Reed Books, Sydney.

Sunnucks P and Hales DF 1996. Numerous transposed sequences of mitochondrial cytochrome oxidase I–II in aphids of the genus *Sitobion* (Hemiptera: Aphididae). *Mol. Biol. Evol.* **13:** 510–524.

Sunnucks P, Wilson ACC, Beheregaray LB, French J, Zenger KR and Taylor AC 2000. SSCP is not so difficult: the application and utility of single-stranded conformation polymorphism in evolutionary biology and molecular ecology. *Mol. Ecol.* **9:** 1699–1710.

Taylor AC, Horsup CN, Johnson CN, Sunnucks P and Sherwin WB 1997. Relatedness structure detected by microsatellite analysis and attempted pedigree reconstruction in an endangered marsupial, the northern hairy-nosed wombat *Lasiorhinus krefftii*. *Mol. Ecol.* **6:** 9–19.

Vidya TNC and Sukumar R 2005. Social organization of the Asian elephant (*Elephas maximus*) in southern India inferred from microsatellite DNA. *J. Ethol.* **23:** 205–210.

Zenger KR, Eldridge MDB and Cooper DW 2003. Intraspecific variation, sex-biased dispersal and phylogeography of the eastern grey kangaroo (*Macropus giganteus*). *Heredity* **91:** 153–162.

PART II

MORPHOLOGY AND PHYSIOLOGY

9 Energetic characteristics of macropodoid locomotion

T.J. Dawson and K.N. Webster

SUMMARY

Hopping defines the macropodoids. Uniquely, the group contains large species (>5 kg); no other large vertebrates use or have used this mode of locomotion. Our current understanding of this unique association follows 30+ years of study. We have focused recently on species of disparate mass, the red kangaroo (*Macropus rufus*) and the brush-tailed bettong (*Bettongia penicillata*), to appreciate mass relationships within the group. Treadmill experiments were used to understand the relationships between speed and energy use. Detailed analysis of muscle, including electron microscopy, facilitated insights into total power use and its distribution among muscle groups. Hopping is characterised by basic biomechanical features over wide speed ranges, notably a relatively fixed stride frequency with speed increasing via an expanding stride length. Energy use during hopping is partly entrained with these features and is economical in comparison to quadrupedal running. Disadvantages associated with hopping include an increased cost of slow locomotion and reduced manoeuvrability at slow speeds, particularly as body mass increases. This is especially the case for larger macropodoids when slow walking progression involves use of the tail. Overall energy use patterns vary with the body mass of different species. Although the hopping gait is associated with energy efficiency, this does not mean that macropodoids are low-energy use mammals. Their efficient gait is linked with a high aerobic capacity. This capacity is due to a large mass of skeletal muscle that has relatively high densities of mitochondria and capillaries. Macropodoids are among the most athletic of mammals. The combination of locomotor efficiency and muscle power enables macropodoids to characteristically sustain high speeds aerobically.

INTRODUCTION

Continuous locomotion using the bipedal hop, also known as saltatory locomotion, is an unusual mammalian gait. It, however, defines the macropodoids, a single superfamily of marsupials, which includes the kangaroos, wallabies and rat-kangaroos (superfamily Macropodoidea). Among this group are the only large vertebrates (>5 kg) that use or have used this mode of locomotion. There is no evidence from the fossil record that other large vertebrates have hopped. Saltatory locomotion is also found in four families of rodents, animals often associated with arid habitats (Webster and Dawson 2004). An African hopping rodent *Pedetes capensis* (springhare) reaches 3–4 kg, but other species are small, with body masses typically below 200 g (Vaughan *et al.* 2000). Another marsupial, the kultarr (*Antechinomys laniger*), a mouse-sized member of the family Dasyuridae, also probably hops. Limited laboratory studies suggest otherwise (Ride 1965) but clear field observations from inland Australia are hard to refute (Troughton 1962). Macropodoid hopping is unique for large vertebrates. How did this unusual gait develop?

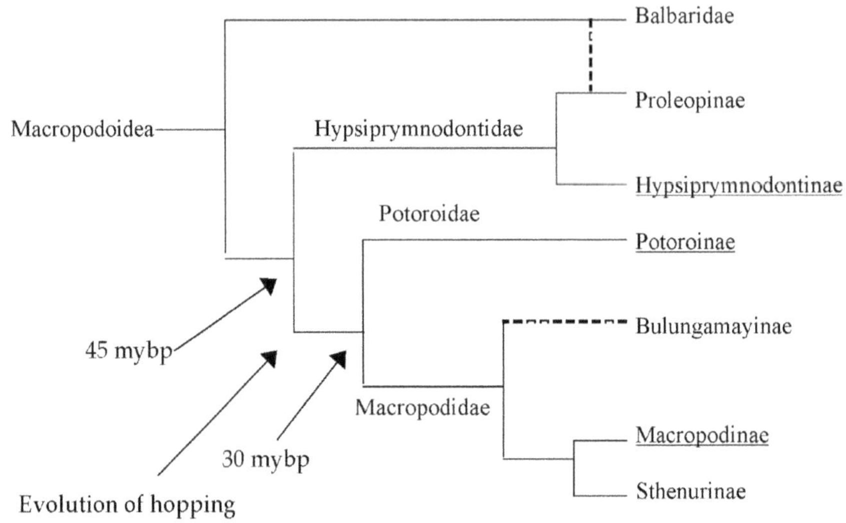

Figure 9.1: Probable relationships within the superfamily Macropodoidea (Archer *et al.* 1999). Dotted lines indicate uncertain relationships, extant subfamilies are underlined. Some likely divergence dates and the probable point at which bipedal hopping evolved are shown, based on the gaits of the extant groups and on molecular data (Burk *et al.* 1998). After Webster and Dawson (2004).

ORIGIN

Most evidence indicates a monophyletic origin for hopping in the superfamily Macropodoidea. It appears to be a phylogenetically old trait that arose in pre-arid Australia. Given the gaits used by the extant species of superfamily Macropodoidea (i.e. the musky rat-kangaroo (*Hypsiprymnodon moschatus*) does not hop, though this is not certain), it could be that bipedal hopping arose after the divergence of the common ancestor of the families Macropodidae and Potoroidae from the Hypsiprymnodontidae, but before the macropodids and potoroids diverged (Fig. 9.1). All three families are represented in late Oligocene fossil deposits (23–26 million years old), and so must have diverged earlier (Archer *et al.* 1999). Clear insight into this divergence is lacking, because of the paucity of Australia's fossil record for terrestrial mammals between the middle Eocene and middle Miocene. Using molecular techniques, Burk *et al.* (1998) estimated that the divergence of family Hypsiprymnodontidae from the common ancestor of families Potoroidae and Macropodidae occurred around 45 mybp, with the subsequent divergence of potoroids from macropodids occurring around 30 mybp, during the mid to late Oligocene. If these divergence dates are correct, then hopping probably developed during the mid to late Oligocene (Fig. 9.1) when Australia was largely covered by temperate rainforests. However, the situation is still unclear because of the post-cranial fossil elements of the Oligo-Miocene macropodoid *Nambaroo gillespieae*, a member of the extinct family Balbaridae (Kear *et al.* 2007). The Balbaridae are considered to be basal to the Macropodoidea, i.e. they predated the Hypsiprymnodontidae (Fig. 9.1). While Kear *et al.* (2007) concluded that *N. gillespieae* likely focused on a quadrupedal gait (as do some potoroids when moving at slower speeds), it shows some distinct adaptations towards hopping locomotion in its pelvis and upper hind limb (Kear *et al.* 2007).

By the mid to late Miocene (15–25 mybp) macropodid kangaroos had an obvious bipedal morphology (Archer *et al.* 1999). With the drying out of the continent and the spread of grasslands during the Miocene, the Macropodoidea underwent a major radiation (Archer *et al.* 1999). The dominance of this group of hopping herbivores continued until European times, with both Potoroidae and Macropodidae families having many 'athletic' species. Among the Macropodidae the kangaroos are obvious examples, and in the potoroids the bettongs tend to be the more specialised hoppers. Other features pertinent to the evolution of macropodoid locomotion are that the original ancestor must have been small and terrestrial and, while the evolution of hopping is often associated with arid environments, this does not appear to be so for the macropodoids (Webster and Dawson 2004).

HOPPING BY KANGAROOS

The hopping of macropodoids has been regarded as most unusual since the earliest encounters by

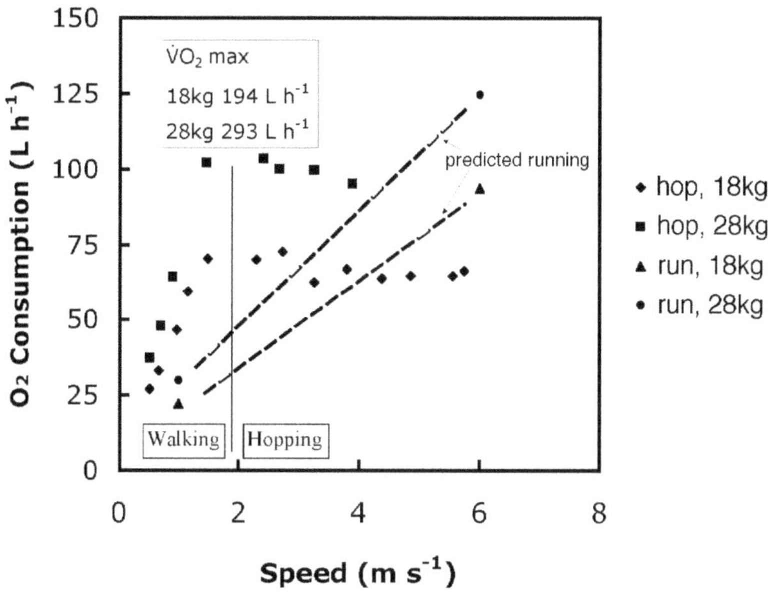

Figure 9.2: Relationship between steady-state speed of locomotion and energetic cost, measured via oxygen consumption, for two *M. rufus* of differing mass (Dawson and Taylor 1973). The vertical solid line approximates the transition between walking, in which the tail is used, and bipedal hopping. Dashed lines indicate the predicted cost of locomotion for quadrupedal mammals of equivalent masses (Taylor *et al.* 1982). The maximum oxygen consumption values ($\dot{V}O_2max$; Taylor *et al.* 1981) are for mammals generally.

European scientists (Sir Joseph Banks, *Journal of Voyage of the Endeavour 1768–1771*, cited in Dawson 1995) and initial examination of its energetic characteristics (Dawson and Taylor 1973) reinforced that opinion. This study of two red kangaroos (*Macropus rufus*) on a treadmill focused attention on hopping because it showed that the energetic cost (as measured by steady-state oxygen consumption) of travelling at increasing speed differed markedly from that of mammalian quadrupeds (Fig. 9.2). *M. rufus* did not significantly increase its energy expenditure over a large range of hopping speeds, compared with quadrupedal running mammals in which energetic costs tend to increase linearly with increasing speed (Taylor *et al.* 1970, 1982). Notably, at moderately fast speeds the metabolic cost of hopping by *M. rufus* was less than that of running quadrupeds of the same mass. However, slow hopping and 'walking' (slow progression involving the tail) were more costly than quadrupedal running. *M. rufus* begins to hop at around 2 m s^{-1} but does not travel inexpensively until speeds above 4 m s^{-1} (Dawson and Taylor 1973). The tammar wallaby (*Macropus eugenii*) exhibits similar energetic characteristics of locomotion but uses less energy than a running mammal at almost all hopping speeds (above 3 m s^{-1}; Baudinette *et al.* 1992). This suggests that body size affects hopping, as it does quadrupedal running. Do larger hoppers have to go faster before hopping becomes energetically economical? Data from two female *M. rufus* of different mass do not give a clear answer for kangaroos (Fig. 9.2) and information is required from mature males, that may be three times larger than females.

Since the initial findings of Dawson and Taylor (1973) we have struggled to understand the mechanisms behind the economy of hopping. Initially we considered that a high level of elastic conservation of energy was the distinctive feature of kangaroo hopping, but this does not seem to be so. The spring-like nature of hopping focused interest more generally on elastic energy storage during locomotion and it appears significant in both hopping and galloping mammals. In hopping and galloping, energy is stored in the tendons and muscles during the first part of a stride and recovered during the subsequent phase (Alexander and Vernon 1975; Ker *et al.* 1986; Bennett and Taylor 1995). This energy recovery is substantial in cursorial quadrupeds such as the ungulates (Dimery *et al.* 1986; Ker *et al.* 1988; Biewener 1998) and may be greater than in kangaroos. Ker *et al.* (1986) indicate that, for optimal storage and recovery of elastic energy, the muscle fibres must be exceedingly short and the tendons very long – these are anatomical features of the ungulates but not of kangaroos in general. Thus, while elastic energy effects are important in hopping, they do not explain its low metabolic costs relative to the galloping of ungulates. The results of Biewener *et al.* (2004) on *M. eugenii* hopping on inclines reinforced this point.

Hopping allows for the specialised concentration of locomotory muscles around the upper hind leg and the reduction in mass of the fore legs (Dawson *et al.* 2004), features which appear to improve locomotory efficiency. However, the key difference between the kangaroos and ungulates appears to be in gait characteristics as typified by the relationships between stride length and stride frequency. Hopping enables the marked extension of the stride length, thereby reducing the frequency that the limbs are required to accelerate and decelerate. In kangaroos and wallabies, stride frequency is relatively constant except at the highest speeds (Dawson and Taylor 1973; Dawson 1977; Baudinette *et al.* 1987). Quadrupeds, however, increase their speed by increasing both the frequency and length of their strides, with the rate of increase in stride frequency highest when walking and lowest when galloping (Heglund and Taylor 1988). From this, Heglund and Taylor proposed that the cost of locomotion is partially determined by stride frequency, i.e. there are costs associated with accelerating and decelerating limbs. In *M. rufus,* hopping frequency and the rate of oxygen consumption are strongly linked over the speed range of 2 m s^{-1} to 6 m s^{-1}. From field observations at speeds greater than 10 m s^{-1}, stride frequency of hopping *M. rufus* begins to increase (Fig. 9.3a) and Dawson (1977) considered that costs rise. Using these results, Dawson (1977) predicted the pattern of energy use at speeds higher than reasonably obtainable on a treadmill (Fig. 9.3b).

At low speeds in their walking gait kangaroos actually move pentapedally, with the tail acting like a fifth limb. When walking, kangaroos increase speed by increasing both stride length and frequency, with frequency predominating. This is associated with a large increase in energetic costs (Fig. 9.3a, b). Thus the notion was accepted in our laboratory that hopping was a mechanism to extend stride length, thereby increasing maximum speeds relative to metabolic capability. This idea appealed during our early experiments, when it was considered (erroneously!) that generally marsupials, including kangaroos, had a relatively low metabolic capability (Dawson 1977; Hinds and MacMillen 1984).

A range of models relating biomechanical parameters to the metabolic cost of terrestrial locomotion have been put forward but the mechanism underlying this relationship is still debated (Alexander 2005). One well-regarded model was that of Kram and Taylor (1990). They found that in several species the metabolic rate was inversely proportional to the foot contact time (t_c) over a large speed range. Kram and

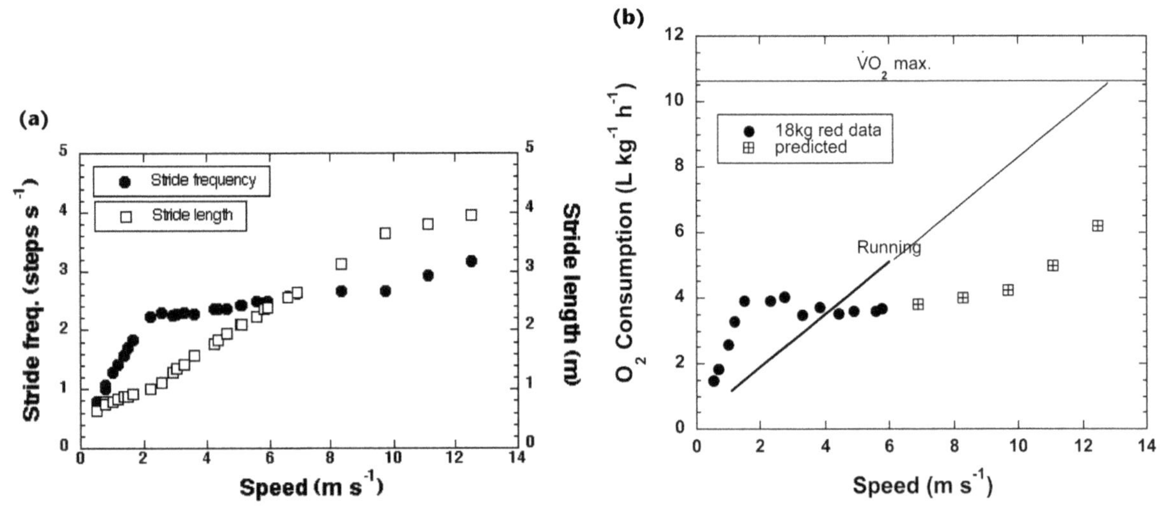

Figure 9.3: (a) Changes in stride frequency and length across a wide range of speeds for *M. rufus* females. Speeds up to and including 6 m s^{-1} are from treadmill studies (Dawson and Taylor 1973); above 6 m s^{-1} data are from field observations (Dawson 1977). (b) Costs of locomotion for a *M. rufus* female (18 kg) at differing speeds. The data below 6 m s^{-1} are from treadmill studies (Dawson and Taylor 1973); above 6 m s^{-1} values are those predicted by Dawson (1977) based on changes in gait. Solid lines indicate the predicted cost of locomotion for quadrupedal mammals of equivalent masses (Taylor *et al.* 1982). Maximum oxygen consumption values ($\dot{V}O_2$max) are from Taylor *et al.* (1981). *M. rufus* should achieve higher aerobically sustained speeds than quadrupeds.

Taylor (1990) suggested that at fast speeds the decrease in t_c necessitates rapid development of muscle forces, requiring the recruitment of faster but less energy-efficient muscle fibres. However, this model fails to explain locomotory costs in *M. rufus* and thus probably fails generally; at faster hopping speeds, t_c decreases, but metabolic rate remains nearly constant and is related to stride frequency (Kram and Dawson 1998).

A recent analysis by Pontzer (2005, 2007) has potential to provide insight into hopping. It suggests that variation in locomotor costs between species is dependent on differences in leg length, notably 'effective' limb length. Pontzer produced a model (LiMb model) that calculated the rate of force production during terrestrial locomotion as (mean force per step) × (step frequency), which was a much better predictor of cost than was body mass. In physiological terms, this model indicates that the volume of muscle activated per step, and step frequency, are sufficient to predict the cost of locomotion with changing speed, gait and body size. In hoppers, such as kangaroos, step frequency is relatively constant. Therefore, if they increase 'effective' leg length as hopping speed increases then mean force per step, and thus locomotor costs, would be reduced and perhaps held constant over some speed ranges. Initial viewing of films of hopping kangaroos suggests changes in effective leg length occur with changes in speed of travel. This model may hold the key to our understanding of hopping, but further detailed assessment is required. Interestingly, the model is also consistent with findings from load-carrying experiments in which costs do not rise proportionally with added mass. The cheap carriage of pouch young in macropodids (Baudinette 1998) is perhaps explained.

HOPPING IN RAT-KANGAROOS AND KANGAROO RATS/MICE

To examine the broad, size-related energetic patterns of hopping, Dawson (1976) investigated a small Australian hopping rodent, *Notomys cervinus* (fawn hopping mouse). The species was selected because it was able to hop at low speed on a treadmill, unlike other related species such as *N. alexis* (Webster and Dawson 2004). With *N. cervinus*, the variation in gait characteristics and energetic responses to changing speeds generally followed those of kangaroos. Although the inflection in the curve showing the relationship between energetic cost of locomotion and speed was not as extreme as that of *M. rufus*, the costs fell well below those predicted for a running quadruped, particularly at higher speeds (Fig. 9.4a). Notably, the

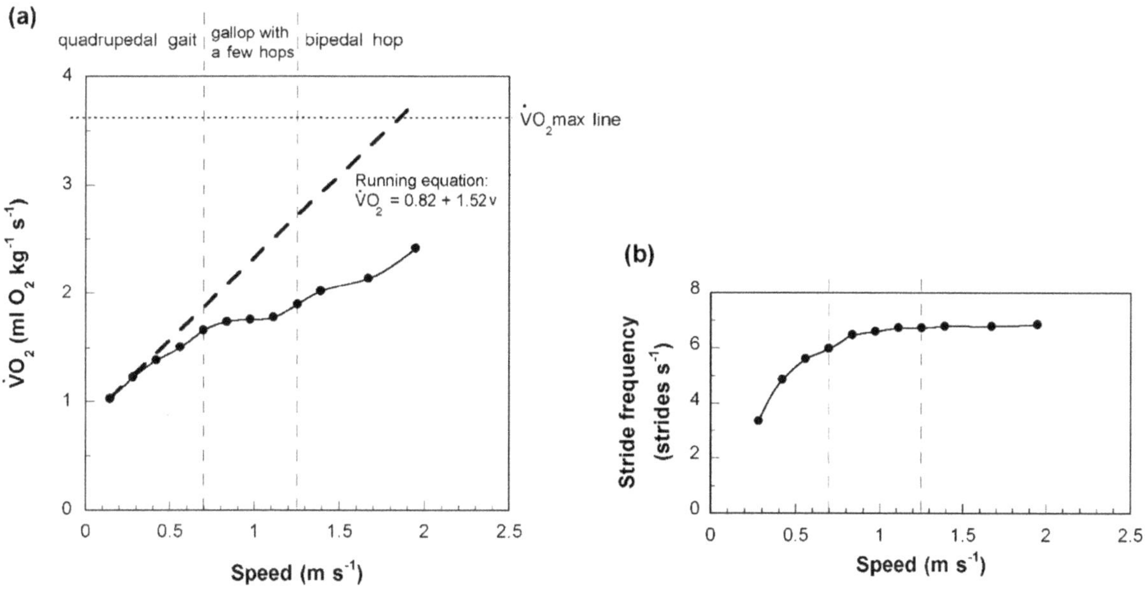

Figure 9.4: (a) Rate of oxygen consumption ($\dot{V}O_2$) vs speed for *N. cervinus*. Data are means (n = 3), connected by a cubic spline curve of best fit (Dawson 1976). The heavy dashed line indicates the expected $\dot{V}O_2$ for a 37 g running mammal (equation derived from Taylor *et al.* 1982). Light dashed lines denote gait transition speeds. Dotted $\dot{V}O_2$max line indicates the expected $\dot{V}O_2$max of a 37 g mammal (3.6 mL s^{-1} kg^{-1}; from Taylor *et al.* 1981); a running mammal reaches this level at about 1.8 m s^{-1}. (b) Mean stride frequency (n =3) vs speed for *N. cervinus* (Dawson 1976).

stride frequency response to speed increases showed a kangaroo-like pattern, with frequency remaining stable at higher speeds (Fig. 9.4b). Subsequent investigation (Webster and Dawson 2004) revealed that, in addition to economy of travel, *N. cervinus* appeared able to reach significantly higher speeds than comparable quadrupeds with similar maximum metabolic capacities (Fig. 9.4a).

The energetic characteristics of hopping in mammals, particularly in marsupials, seemed clarified at the end of the 1970s but uncertainty arose with studies suggesting that small hoppers with masses below 4–5 kg were not different from quadrupeds in their locomotor energetics. The studies were by Baudinette (1977) on the quokka (*Setonix brachyurus*), a 3 kg hopping macropodid, and by Thompson *et al.* (1980) on several small hopping rodents and a potoroid marsupial, the brush-tailed bettong (*Bettongia penicillata*, mass ~1 kg). Some researchers consequently accepted that this pattern was general for small hoppers (Baudinette 1989). Additional studies on the long-nosed potoroo (*Potorous tridactylus*, mass ~1 kg), which indicated that its locomotory energetics were quadrupedal-like, appeared to further support this conclusion (Baudinette *et al.* 1993).

The issue, however, is not clear-cut. The data for *S. brachyurus* are inconclusive because energy costs were measured only below hopping speeds. However, the reasons for differences between the data of Dawson (1976) and of Thompson *et al.* (1980) on hopping rodents were difficult to unravel, particularly because gait characteristics were not measured in Thompson *et al.* (1980). The data on *P. tridactylus* are also circumspect in regard to the energetics of hopping in small mammals. Among the rat-kangaroos (family Potoroidae) propensity to hop varies, with the potoroos being less obligatory hoppers than the bettongs. *P. tridactylus* is primarily quadrupedal but can hop at fast speeds (Buchmann and Guiler 1974). However, bipedal hopping was rare during the treadmill study by Baudinette *et al.* (1993) and, like *S. brachyurus*, it exhibited a quadrupedal energetics pattern. Although the stride frequency of *P. tridactylus* did not stabilise during quadrupedal bounding, the rate of increase in stride frequency declined at the higher speeds where hopping occasionally occurred (Baudinette *et al.* 1993).

While the potoroos may not have kangaroo-like energetics, what of the bettongs? A recently extinct species, the desert rat-kangaroo (*Caloprymnus campestris*), was probably the most accomplished potoroid hopper. Despite weighing less than a kilogram, it could outhop galloping horses for up to c. 20 km (Finlayson 1932, 1943). These characteristics urged us to reinvestigate the locomotion of this group of rat-kangaroos and we undertook a further examination of the locomotion energetics and gait characteristics of *B. penicillata* (Webster and Dawson 2003). We showed that this small marsupial has significantly lower energetic costs than a quadruped of the same size when hopping at speeds above 2 m s^{-1} (Fig. 9.5a). Unlike the larger hopping species, *B. penicillata* generally exhibits an increasing cost of locomotion with increasing speed, but the rate of increase is significantly lower than that of quadrupeds. This differed from the results of Thompson *et al.* (1980), who found no significant difference from quadrupedal running energetics for this species. The divergence between the two *B. penicillata* sets of data appears due to different measurements at speeds when the gait was quadrupedal and to the generally higher energetics measured by Thompson *et al.* (1980) at all speeds. We found that, like kangaroos and wallabies, *B. penicillata* increased its hopping speed by lengthening stride; stride lengths of up to 2 m were measured. There was effectively no increase in stride frequency over a nearly six-fold increase in hopping speed (Fig. 9.5b).

By hopping, *B. penicillata* can reach higher speeds than quadrupedal mammals, due to the locomotory economy at speeds when hopping is well established. Because of the limitation on aerobic capacity in both groups, *B. penicillata* has an apparent maximum speed more than twice that predicted for a similar-sized quadruped (Fig. 9.5a). Predation risk would presumably be reduced by the ability to reach relatively high speeds. Also, if not hopping at high speeds, *B. penicillata* moves with lower costs than a quadruped, which probably facilitates its foraging over a wide area. There are trade-offs associated with hopping. At lower speeds locomotory costs are actually higher than those for quadrupedal movement, presumably because of the anatomical specialisations associated with bipedality. These characteristics are most definitely kangaroo-like!

HOPPING AND AEROBIC METABOLISM

During our studies it became apparent that the high speeds usually attained by hoppers relative to those

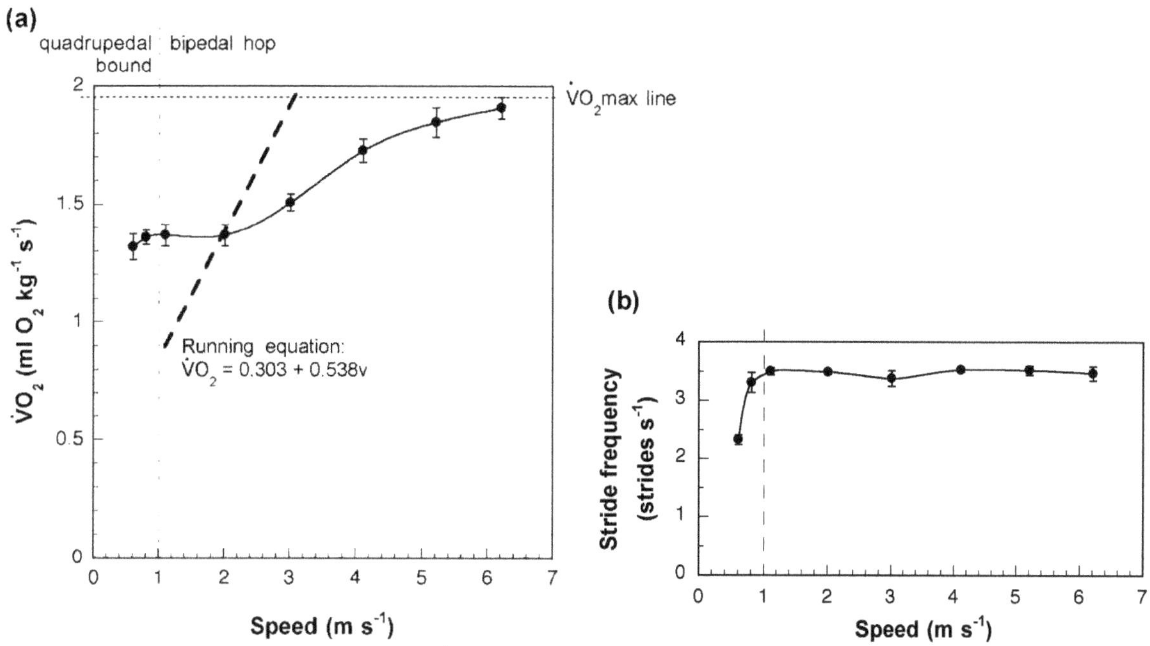

Figure 9.5: (a) Rate of oxygen consumption ($\dot{V}O_2$) vs speed for *B. penicillata*. Data are means ± SE (n = 5), connected by a cubic curve of best fit (Webster and Dawson 2003). Heavy dashed line indicates the expected $\dot{V}O_2$ for a 1 kg running mammal (equation derived from Taylor *et al.* 1982). Light dashed line denotes gait transition speed. Dotted $\dot{V}O_2$max line indicates the expected $\dot{V}O_2$max of a 1 kg mammal (1.94 mL s^{-1} kg^{-1}; Taylor *et al.* 1981). *B. penicillata* reaches this level at 6.2 m s^{-1}, a running mammal would reach it at only 3 m s^{-1}.(b) Stride frequency vs speed for *B. penicillata*. Data are means ± SE (n = 5) (Webster and Dawson 2003).

of most runners (Webster and Dawson 2004) are associated with high energy output capabilities in addition to apparent energetic efficiency. This pattern is directly opposite to that suggested by Dawson (1977), who considered that the efficiency of hopping at higher speeds might compensate for lower metabolic capabilities associated with the lower basal metabolism of marsupials. Contrary to these opinions, marsupials have been found to have generally high metabolic capabilities, at least comparable with those of placental mammals (Dawson and Dawson 1982; Dawson and Olson 1988; Hinds *et al.* 1993). During exercise, *M. rufus* has a maximum aerobic capacity equivalent to the most athletic of placental mammals (Kram and Dawson 1998). This achievement in relation to hopping was examined in a study of muscle structure and muscle mitochondrial and capillary morphology (Dawson *et al.* 2004).

Macropus rufus females shot in the field comprised 47% skeletal muscle (on a full gut basis). Values of 47–52% have been reported previously for red kangaroos; variability can occur due to variable fill of the large foregut and/or sexual dimorphism (Grand 1990; Hopwood 1981; Hopwood and Griffiths 1984; Tribe and Peel 1963). However, a value near 50% places kangaroos among the most muscular of mammals (Table 9.1) (Grand 1990). Most muscle mass in *M. rufus* is concentrated around the pelvis, particularly in the upper hind leg. Not only does the upper hind leg make up 44.3% of all skeletal muscle – when the trunk and back, which have their bulk posteriorly located (Fig. 9.6), are included, some 80% of skeletal muscle is so positioned.

The underlying units of aerobic metabolism, mitochondria, have a high density in kangaroo muscle, comparable with the densities in pronghorn antelope and dogs (Dawson *et al.* 2004). Of the total body mitochondrial volume in kangaroos, 86% is positioned around the upper hind leg and pelvis. Blood capillary densities in the muscles of *M. rufus* match their high mitochondrial densities. The lower hind leg, with its large Achilles tendon, has been the focus of much interest in relation to the elastic conservation of energy during hopping, but as it contains only 7% of total muscle mitochondria its role needs re-evaluation. Overall, muscle characteristics of kangaroos were matched to aerobic capacity similarly to those in placental mammals and were equivalent to those in the most aerobic (athletic) of placentals (Fig. 9.7).

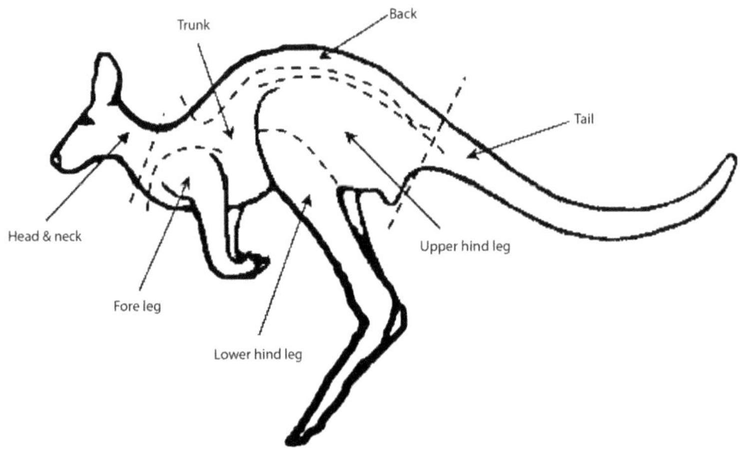

Figure 9.6: Body regions of a red kangaroo (*M. rufus*) from which muscle samples were taken, to provide the distribution data in Table 9.1.

In summary, hopping in mammals is a mechanism that enables stride length to be extended so that relatively high speeds can be achieved at comparative low energy costs. A reduction in metabolic costs appears to be achieved by focusing muscle mass around the pelvis, limiting increases in stride frequencies and capitalising on elastic energy conservation. These specialisations come with disadvantages at slower speeds, principally during 'walking'; these impacts are body size related. For example, a reduction in nimbleness occurs at low speeds, particularly in larger species such as the kangaroos. Metabolic costs of slow walking and slow hopping are not lower than in quadrupeds and as body mass increases these costs surpass quadrupedal costs; again this is most noticeable in kangaroos. On balance, hopping or saltatory locomotion is a mechanism to achieve high aerobic speeds, i.e. speeds that can be sustained. While specialisations result in efficiencies in such conditions, full use of the benefits of hopping, particularly in large mammals, are achieved by also having a high metabolic capability. Dawson *et al.* (2004) show that kangaroos have very powerful 'motors' in their muscles!

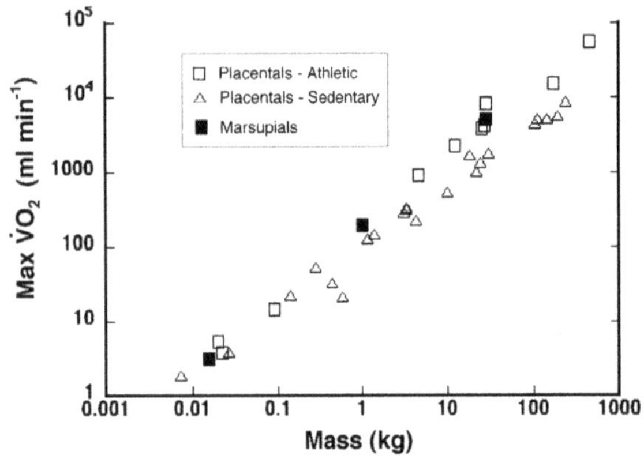

Figure 9.7: Relationship between maximum aerobic capability and body mass in mammals. Data for 'athletic' and 'sedentary' placental mammals come from Weibel *et al.* (2004). Data for the three marsupials – *M. rufus, B. penicillata* and *Sminthopsis crassicaudatus* – come from Dawson *et al.* (2009, in prep.). Note that the marsupial values fall at the upper range of the athletic placentals.

REFERENCES

Alexander RM 2005. Models and the scaling of energy costs for locomotion. *J. Exp. Biol.* **208:** 1645–1652.

Alexander RM and Vernon A 1975. The mechanics of hopping by kangaroos (Macropodidae). *J. Zool. Lond.* **177:** 265–303.

Archer M, Arena R, Bassarova M, Black K, Brammall J, Cooke B, Creaser P, Crosby K, Gillespie A, Godthelp H, Gott M, Hand S, Kear B, Krikmann A, Mackness B, Muirhead J, Musser A, Myers T, Pledge N, Wang Y and

Table 9.1: Distribution of muscle and muscle mitochondria in the body of female *M. rufus*; total muscle mass was 47% of body mass.

Body section	Total muscle mass (%)	Total mitochondrial volume (%)
Head and neck	2.5	1.1
Fore leg	3.9	1.7
Trunk	25.6	26.9
Back	10.1	12.9
Upper hind leg	44.3	45.7
Lower hind leg	9.1	7.0
Tail	7.0	4.8

See Dawson *et al.* (2004) for detailed data.

Wroe A 1999. The evolutionary history and diversity of Australian mammals. *Aust. Mammal.* **21:** 1–45.

Baudinette RV 1977. Locomotion energetics in a marsupial, *Setonix brachyurus. Aust. J. Zool.* **25:** 423–428.

Baudinette RV 1989. The biomechanics and energetics of locomotion in Macropodoidea. In *Kangaroos, Wallabies and Rat-kangaroos.* (Eds G Grigg, P Jarman and I Hume) pp. 245–253. Surrey Beatty and Sons, Sydney.

Baudinette RV 1998. Young wallabies get a free ride. *Nature* **395:** 653.

Baudinette RV, Gannon BJ, Runciman WB and Wells S 1987. Do cardiorespiratory frequencies show entrainment with hopping in the tammar wallaby? *J. Exp. Biol.* **129:** 251–263.

Baudinette RV, Snyder GK and Frappell PB 1992. Energetic cost of locomotion in the tammar wallaby. *Am. J. Physiol.* **262:** R771–R778.

Baudinette RV, Halpern EA and Hinds DS 1993. Energetic cost of locomotion as a function of ambient temperature and during growth in the marsupial, *Potorous tridactylus. J. Exp. Biol.* **174:** 81–95.

Bennett MB and Taylor GB 1995. Scaling of elastic strain energy in kangaroos and the benefits of being big. *Nature* **378:** 56–59.

Biewener AA 1998. Muscle–tendon stresses and elastic energy storage during locomotion in the horse. *Comp. Biochem. Physiol. B* **120:** 73–87.

Biewener AA, McGowan G, Card GM and Baudinette RV 2004. Dynamics of leg muscle function in tammar wallabies (*M. eugenii*) during level *versus* incline hopping. *J. Exp. Biol.* **207:** 211–223.

Buchmann OLK and Guiler ER 1974. Locomotion in the potoroo. *J. Mammal.* **55:** 203–206.

Burk A, Westerman M and Springer MS 1998. The phylogenetic postion of the musky rat-kangaroo and the evolution of bipedal hopping in kangaroos (Macropodidae: Diprotodontia). *Syst. Biol.* **47:** 437–474.

Dawson TJ 1976. Energetic cost of locomotion in Australian hopping mice. *Nature* **259:** 305–307.

Dawson TJ 1977. Kangaroos. *Sci. Am.* **237:** 78–89.

Dawson TJ 1995. *Kangaroos: Biology of the Largest Marsupials.* University of New South Wales Press, Sydney.

Dawson TJ and Dawson WR 1982. Metabolic scope and conductance in response to cold of some dasyurid marsupials and Australian rodents. *Comp. Biochem. Physiol.* **71A:** 59–64.

Dawson TJ and Olson JM 1988. Thermogenic capabilities of the opossum *Monodelphis domestica* when warm and cold acclimated: similarities between American and Australian marsupials. *Comp. Biochem. Physiol.* **89A:** 85–91.

Dawson TJ and Taylor CR 1973. Energetic cost of locomotion in kangaroos. *Nature* **246:** 313–314.

Dawson TJ, Mifsud B, Raad MC and Webster KN 2004. Aerobic characteristics of red kangaroo skeletal muscles: is a high aerobic capacity matched by muscle mitochondrial and capillary morphology as in eutherian mammals? *J. Exp. Biol.* **207:** 2811–2821.

Dimery NJ, Ker RF and Alexander RM 1986. Elastic properties of the feet of deer (Cervidae). *J. Zool. Lond.* **208:** 161–169.

Finlayson HH 1932. *Caloprymnus campestris*, its recurrence and characters. *T. Roy. Soc. South Aust.* **56:** 148–1676.

Finlayson HH 1943. *The Red Centre: Man and Beast in the Heart of Australia.* Halstead Press, Sydney.

Grand TI 1990. Body composition and the evolution of the Macropodidae (*Potorous, Dendrolagus, and Macropus*). *Anat. Embryol.* **182:** 85–92.

Heglund NC and Taylor CR 1988. Speed, stride frequency and energy cost per stride: how do they change with body size and gait? *J. Exp. Biol.* **138:** 301–318.

Hinds DS and MacMillen RE 1984. Energy scaling in marsupials and eutherians. *Science* **225:** 335–337.

Hinds DS, Baudinette RV, MacMillen RE and Halpern EA 1993. Maximal metabolism and the aerobic factorial scope of endotherms. *J. Exp. Biol.* **182:** 41–56.

Hopwood PR 1981. Carcass muscle weight distribution and yield: a comparison between the grey kangaroo *Macropus giganteus*, and red kangaroos, *M. rufus. Aust. Wildl. Res.* **8:** 263–268.

Hopwood PR and Griffiths DA 1984. Carcass muscle weight distribution and yield: a comparison between male and female grey kangaroos, *Macropus giganteus. Aust. Wildl. Res.* **11:** 299–302.

Kear BP, Cooke BN, Archer M and Flannery TF 2007. Implications of a new species of the Oligo-Miocene kangaroo (Marsupialia: Macropodoidea) *Nambaroo*, from the Riversleigh World Heritage Area, Queensland, Australia. *J. Paleontol.* **81:** 1147–1167.

Ker RF, Dimery NJ and Alexander RM 1986. The role of tendon elasticity in hopping in a wallaby (*Macropus rufogriseus*). *J. Zool. Lond.* **208:** 417–428.

Ker RF, Alexander RM and Bennett MB 1988. Why are mammalian tendons so thick? *J. Zool. Lond.* **216:** 309–324.

Kram R and Dawson TJ 1998. Energetics and biomechanics of locomotion by red kangaroos (*Macropus rufus*). *Comp. Biochem. Physiol. B* **120:** 41–49.

Kram R and Taylor CR 1990. Energetics of running: a new perspective. *Nature* **346:** 265–267.

Pontzer H 2005. A new model predicting locomotor costs from limb length *via* force production. *J. Exp. Biol.* **208:** 1513–1524.

Pontzer H 2007. Predicting the energy cost of terrestrial locomotion: a test of the LiMb model in humans and quadrupeds. *J. Exp. Biol.* **210:** 484–494.

Ride WDL 1965. Locomotion in the Australian marsupial *Antechinomys. Nature* **205:** 199.

Taylor CR, Schmidt-Nielsen K and Raab JL 1970. Scaling of the energetic cost of running to body size in mammals. *Am. J. Physiol.* **219:** 1104–1107.

Taylor CR, Maloiy GMO, Weibel ER, Langman VA, Kamau JMZ, Seeherman HJ and Heglund NC 1981.

Design of the mammalian respiratory system. III. Scaling maximum aerobic capacity to body mass: wild and domestic mammals. *Respir. Physiol.* **44:** 25–38.

Taylor CR, Heglund NC and Maloiy GMO 1982. Energetics and mechanics of terrestrial locomotion. I. Metabolic energy consumption as a function of speed and body size in birds and mammals. *J. Exp. Biol.* **97:** 1–21.

Thompson SD, MacMillen RE, Burke EM and Taylor CR 1980. The energetic cost of bipedal hopping in small mammals. *Nature* **287:** 223–224.

Tribe DE and Peel L 1963. Body composition of the kangaroo (*Macropus* sp.). *Aust. J. Zool.* **11:** 273–289.

Troughton ELeG 1962. *Furred Animals of Australia.* 7th edn. Angus and Robertson, Sydney.

Vaughan TA, Ryan JM and Czaplewski NJ 2000. *Mammalogy.* 4th edn. Saunders College Publishing, Fort Worth.

Webster KN and Dawson TJ 2003. Locomotion energetics and gait characteristics of a rat-kangaroo, *Bettongia penicillata*, have some kangaroo-like features. *J. Comp. Physiol. B* **173:** 549–557.

Webster KN and Dawson TJ 2004. Is the energetics of mammalian hopping locomotion advantageous in arid environments? *Aust. Mammal.* **26:** 153–160.

Weibel ER, Bacigalupe LD, Schmitt B and Hoppeler H 2004. Allometric scaling of maximal metabolic rate in mammals: muscle aerobic capacity as a determinant factor. *Respir. Physiol. Neurobiol.* **140:** 115–132.

10 Development of endothermy and non-shivering thermogenesis in the Tasmanian bettong: a review

R.W. Rose

SUMMARY

The Tasmanian bettong (*Bettongia gaimardi*) is a small marsupial able to increase its heat production (and thus its metabolic rate) by non-shivering thermogenesis. Oxygen consumption increases in response to various stimuli: vasopressin, nor-epinephrine, both α_1 and β adrenergic stimulation, thyroid hormone and cold acclimation (CA) (4–5°C for 14 days). Moreover, the metabolic rate response to nor-epinephrine is enhanced after CA (71% increase 'v' 47% prior to CA). The ability to respond to these stimuli appears only after week 10 of the 15-week pouch life. We have attempted to identify the source and cause of the metabolic response. Brown adipose tissue and the uncoupling protein are totally absent in this species, as in most marsupials. Our research led us to investigate skeletal muscle as a source of heat production in this species and further research into this tissue's involvement in non-shivering thermogenesis seems appropriate.

INTRODUCTION

Although Haeckel's theory that 'Ontogeny recapitulates phylogeny' has long been questioned (Gould 1977), normal usage sees it applied to morphological/embryological features; however, it is still tempting to discuss physiological development in evolutionary terms. Whittow (1973) and Ruben (1995) have attempted to discuss the development of endothermy in these terms. Indeed, Hulbert (1988) stated that there is a remarkable similarity in the sequence of events in the development of endothermy in extant animals and those independently deduced and proposed for the evolution of endothermy.

It is clearly accepted that endothermy evolved separately in mammals and birds (Ruben 1995). It therefore requires no great leap of faith to accept that endothermy evolved somewhat differently in the two major groups of mammals, the eutherians and marsupials. Unfortunately, potential signals of endothermy are not preserved well in fossils and in their absence other means are required.

According to McNab (1978) and Hulbert (1987), there have been three major innovations during the acquisition of mammalian endothermy: the development of insulation in the form of fur and subcutaneous fat, a three- to five-fold jump in resting metabolism and an increase in the ability to raise metabolic heat production in response to cold.

Changes in all these parameters can be seen during pouch occupation by macropodid young.

This chapter will present data on the attainment of homeothermy and endothermy in a rat-kangaroo, the Tasmanian bettong (*Bettongia gaimardi*), with reference to previously published data from other macropodoid species.

Many eutherian mammals are born naked and susceptible to cold but can produce body heat in response to nor-adrenaline by metabolising brown adipose tissue (BAT). This is accomplished via an uncoupling protein (UCP1) which 'short-circuits' ATP production, leading to heat liberation – a process which is usually referred to as non-shivering thermogenesis. At birth, however, marsupials are reptile-like and only gradually become able to produce metabolic heat. No marsupial possesses appreciable BAT, yet a number can produce body heat in response to nor-adrenaline. Hence, it seems apparent that heat production has evolved somewhat differently in the two major mammalian groups; one group relies on BAT and marsupials use a different mechanism.

Previous work showed that Tasmanian macropodoids demonstrate nor-adrenaline induced non-shivering thermogenesis (the long-nosed potoroo, *Potorous tridactylus* (Nicol 1978), Bennett's wallaby, *Macropus rufogriseus rufogriseus* (Loudon *et al.* 1987) and Tasmanian bettong (Rose *et al.* 1998)) but most other marsupials do not (Wallis 1979; Dawson and Dawson 1982; Dawson and Olson 1988; Nicol *et al.* 1997; Reynolds and Hulbert 1982; Wunder, *pers. comm.*). However, more recent information on the Tasmanian devil (*Sarcophilus harrisii*) (Kabat *et al.* 2003a) and eastern barred bandicoot (*Perameles gunnii*) (Rose and Ikonomopoulou 2005) shows that both these species from different taxonomic groups are also able to respond to nor-adrenaline by increasing metabolic rate.

TASMANIAN BETTONG: A HOMEOTHERMIC ENDOTHERM

Adult Tasmanian bettongs (a rat-kangaroo) are typically homeothermic (Rose *et al.* 1990) and endothermic mammals (Rose 1997). They are able to maintain a relatively stable and high body temperature over a wide range of ambient temperatures and increase their metabolic rate at temperatures outside their thermoneutral zone (10–20°C)(Figs 10.1, 10.1). The body temperature of this species is relatively high and undermines the prevailing idea that marsupial body temperatures are lower than those of similar sized eutherians.

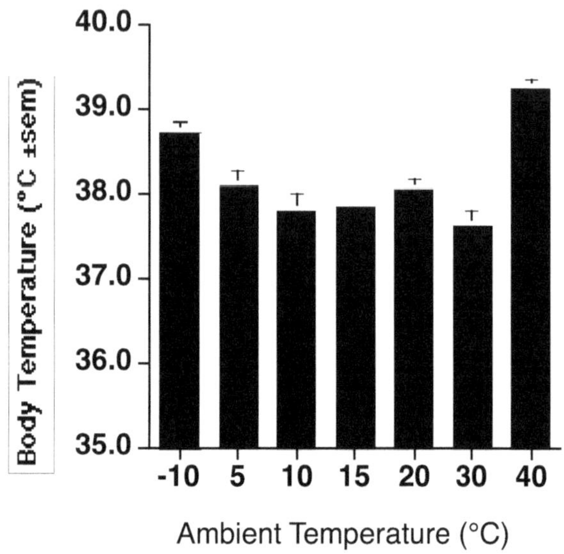

Figure 10.1: Body temperature ± sem of five adult Tasmanian bettongs (*B. gaimardi*) over a range of ambient temperatures (Rose *et al.* 1990).

Bartholomew (1956) and Wallis and Maynes (1973) produced 'cooling curves' for the quokka (*Setonix brachyurus*) and parma wallaby (*Macropus parma*) during the developmental periods before and after homeothermy occurred. In the bettong, the development of a stable body temperature (homeothermy) occurs slowly during the 15-week pouch life. At six weeks of age the young is unable to maintain

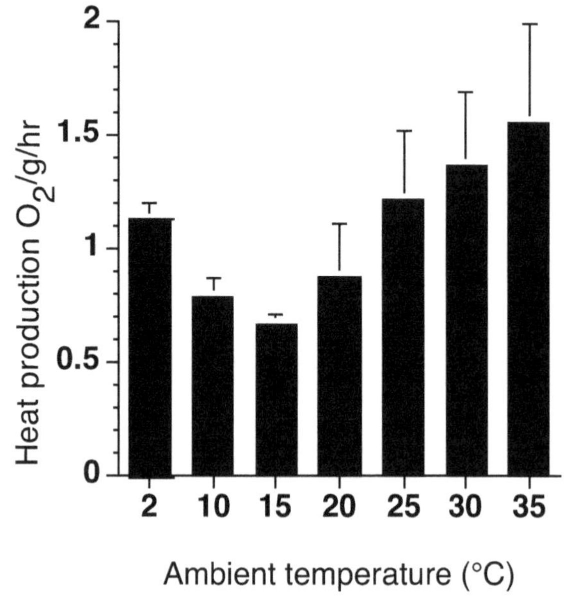

Figure 10.2: Mean resting metabolic rate (± sem) of five adult Tasmanian bettongs (*B. gaimardi*) at ambient temperatures between 2°C and 35°C. A thermoneutral zone appears to exist at 10–20°C (Rose 1997).

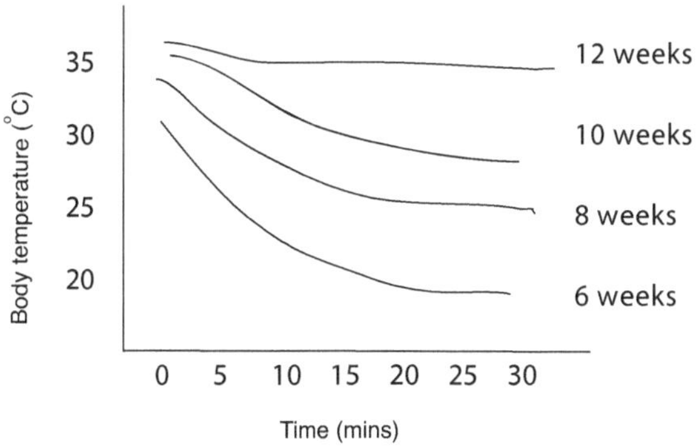

Figure 10.3: Representative set of cooling curves (reduction in body temperature at an ambient temperature of 20°C) from single young at four different ages during pouch life in the Tasmanian bettong (*B. gaimardi*). Homeothermy is achieved by age 12 weeks, three weeks before final pouch vacation (Rose 1987).

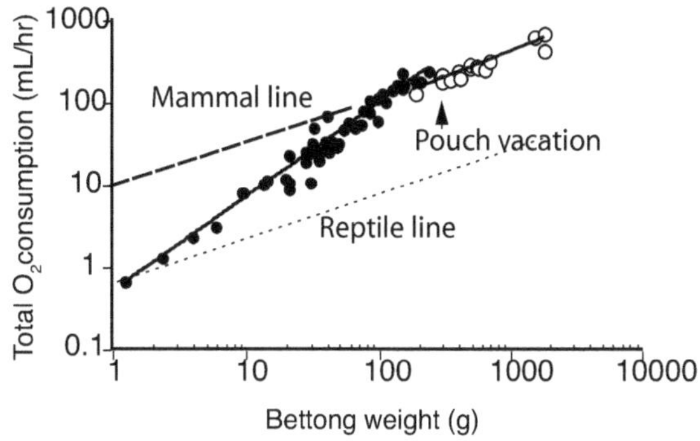

Figure 10.5: Log-log plot of the change in total oxygen consumption as Tasmanian bettong (*B. gaimardi*) weight (g) increases, measured at 35°C. There is a significant change in the slope near pouch vacation (Rose *et al.* 1998).

its body temperature above that of ambient, and not until 12 weeks is it able to maintain a Tb above 30°C for an hour at room temperature (Fig. 10.3). This development is due to reduction in the surface area:volume ratios, changes in skin thickness, changes in insulation/ thermal conductance (Fig. 10.4) and control of peripheral blood flow, plus behavioural (postural) changes and the physiological ability to produce heat.

Shield (1966) noted that quokka pouch young are able to increase metabolic rate in response to cold after age 100 days but that prior to this, metabolic rate varies directly with ambient temperature (ectothermy). Initially, bettong young are also ectothermic with the ability to produce little if any internal heat; this changes as the young transforms from a reptilian thermoregulatory mode to that of a marsupial/mammal. A steep slope is necessary to accomplish the change (Fig. 10.5), the cause of which is unknown. It has been speculated that it may be due to changes in cell membrane leakiness (Hulbert 1987; G. Grigg, *pers. comm.*). As physiological maturity occurs the ability to produce heat develops as does the ability to increase heat production in response to

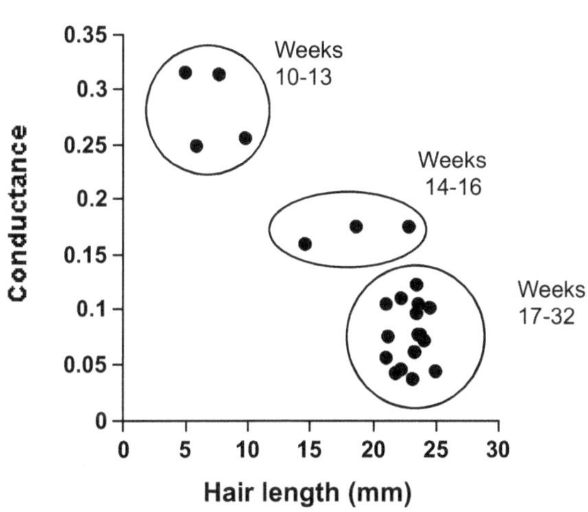

Figure 10.4: The change in mean conductance measured at 35°C during different weeks of pouch life plotted against mean fur length in the Tasmanian bettong (*B. gaimardi*) (Rose *et al.* 1998).

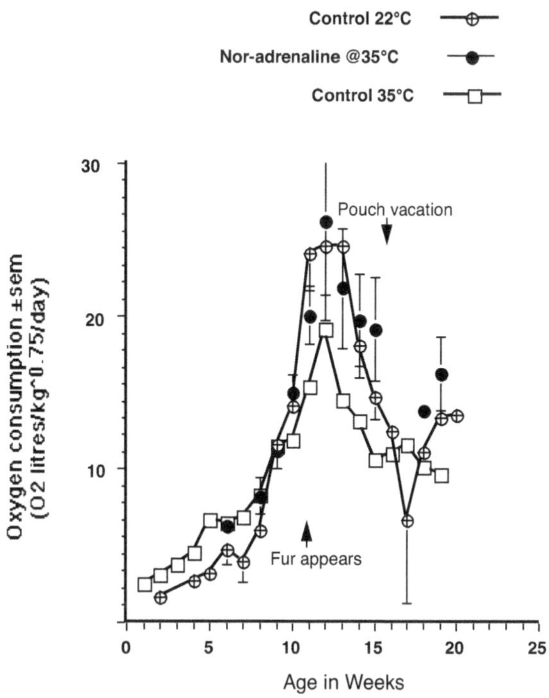

Figure 10.6: Mean change in oxygen consumption (mL/g/hr ± sem) with Tasmanian bettong (*B. gaimardi*) age (weeks) measured at 35°C, 22°C and after injection of nor-adrenaline (Rose *et al.* 1998).

cold, first present after week 10. This is a primary requirement of endothermy. Maximal heat production in response to a cold challenge occurs around the time of pouch vacation.

Nicol (1978) was the first to demonstrate that an adult macropodoid marsupial – the long-nosed potoroo – could respond to nor-adrenaline injection with an increase in metabolic rate. Loudon *et al.* (1985) found that Bennett's wallaby pouch young, originating from Tasmania, demonstrated a positive response to nor-adrenaline by increasing metabolic rate when approximately 450 g, but prior to its first emergence from the pouch. Rose *et al.* (1998) found that a response to nor-adrenaline (0.4 mg/kg) first appears at week 10 in the bettong and the response at 35°C mirrors the metabolic rate changes in response to cold in the absence of injected nor-adrenaline (Fig. 10.6). This suggested that the bettong's natural response to cold was mediated via nor-adrenaline, as occurs in many eutherians.

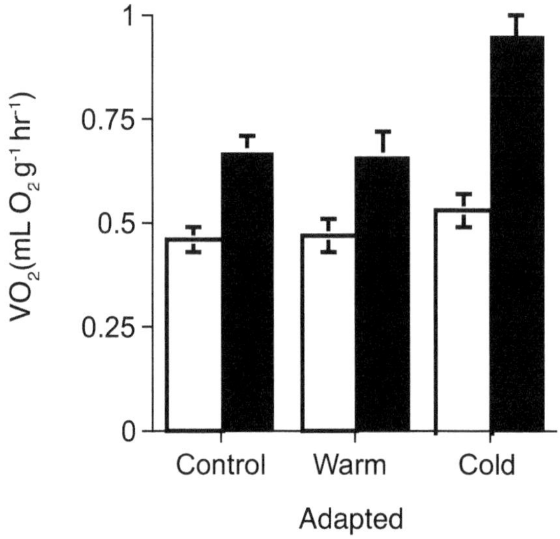

Figure 10.7: Effect of temperature acclimation on Tasmanian bettong (*B. gaimardi*) whole body metabolic rate. The metabolic rate, expressed as VO_2 (mL O_2/g/hr) was measured at 20°C before (open bars) and after nor-adrenaline (dark bars). Animals (n = 5) were subsequently acclimated for two weeks at 25°C (warm) then 5°C (cold). Modified from Rose *et al.* (1999).

NON-SHIVERING THERMOGENESIS

An ability to increase heat production (metabolic rate) by mechanisms other than shivering is termed 'non-shivering thermogenesis'. Although there is some confusion over the exact meaning as well as the usage of this term, the Commission for Thermal Physiology of the International Union of Physiological Sciences (IUPS Thermal Commission 2003) states that the principal effector, particularly in small mammals, is brown adipose tissue. Usually only animals that hibernate, are cold-adapted or are newborn mammals possess brown fat. Hayward and Lisson (1992) failed to observe the presence of BAT in a large number of marsupial species; however, insignificant amounts of BAT have been reported in the pouch young of Bennett's wallaby (in a convertible form of adipose tissue, similar to white adipose tissue) (Loudon *et al.* 1985). Very small amounts of BAT were noted in the small marsupial *Sminthopsis crassicaudata* (fat-tailed dunnart) after cold acclimation, although its presence could not be related to thermogenesis (Hope *et al.* 1997). BAT produces heat due to the presence of the uncoupling protein UCP1 but this protein is absent in marsupials (Kabat *et al.* 2003a, b). As marsupials do not posses BAT/UCP1 but are able to increase their heat production (thermogenesis) by means other than shivering, the above definition seems far too narrow. A definition for non-shivering thermogenesis that encompasses marsupial physiology in addition to that of eutherians should include the ability to increase heat production in response to cold in the presence/absence of BAT (Grigg *et al.* 2004).

Nor-adrenaline led to an increase in heat production of up to 50% in adult bettongs, but was this related to cold acclimation? We subsequently cold-acclimated bettongs and re-investigated the role of nor-adrenaline. After two weeks at 5°C, the metabolic rate of bettongs increased by 71% after nor-adrenaline injection (Fig. 10.7) (Rose *et al.* 1999). This enhanced response is indicative of 'true' non-shivering thermogenesis in the Tasmanian bettong, in the absence of BAT and UCP1 (Rose *et al.* 1999; Kabat *et al.* 2003b).

In a major study, Golozoubova *et al.* (2001) stated that 'only UCP1 can mediate adaptive non-shivering thermogenesis in the cold'. Although this may be well be the case for rodents (the animals studied) it cannot be the explanation for the occurrence of non-shivering thermogenesis in marsupials! Other studies with nor-adrenaline on the isolated hind limb of the bettong demonstrated metabolic and physiological effects resulting in increased blood pressure and metabolism in the limb's musculature. This led to the consideration that skeletal muscle held the key to heat production in marsupials without BAT (Ye *et al.* 1995, 1996). Heat production in skeletal muscle could be by friction due to blood vessel smooth muscle contraction (Clark *et al.* 1995), a proposal thought 'attractive' by Grigg *et al.* (2004), or due to a change in muscle tone (Kabat 2003). The site and mode of action of nor-adrenaline remains unclear. Both alpha$_1$ and beta$_3$ adrenergic receptors are involved (Ye *et al.* 1995, 1996; Rose *et al.* 1999), suggesting action in blood vessels and muscle. At present, we can only speculate about the source and site of non-shivering thermogenesis in marsupials (Grigg *et al.* 2004).

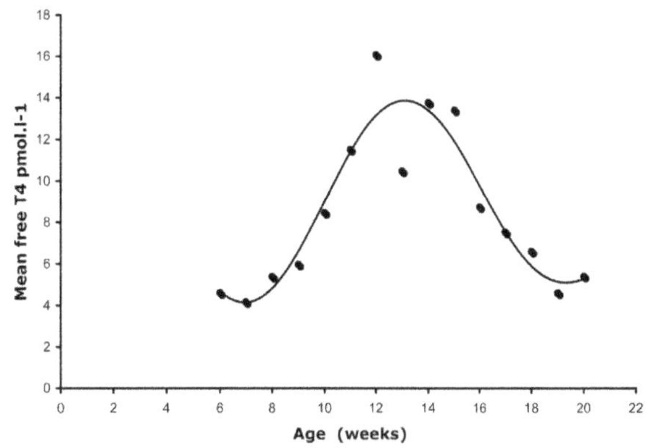

Figure 10.8: The change in estimated free thyroxine (T4 pmol/L^{-1}) until week 20 in Tasmanian bettongs (*B. gaimardi*) (Rose and Kuswanti 2004).

THYROID GLAND

The thyroid gland is intimately involved in mammalian thermogenesis (Hulbert 1987). It has a role in the development of thermoregulation and the ability to increase metabolic rate in marsupials. Levels of thyroid hormone (T4) increase during pouch life of the Tasmanian bettong, peaking at week 12 (Fig. 10.8). This is 1–2 weeks after the maximal responses to nor-adrenaline. Metabolic responses to thyroxine injection only become apparent from week 13–14 and reach a maximum at week 14–15, just before final vacation of the pouch (Rose and Kuswanti 2004). The peak of T4 prior to its maximal effect on metabolic rate may result in maturation of nor-adrenaline receptors, as shown by Rozanov and Dratman (1996). Hence nor-adrenaline and thyroid hormones may act synergistically in the development of endothermy.

Not all eutherians are born with BAT or substantial insulation; an example is the southern elephant seal (*Mirounga leonine*). In this species there is an eight-fold increase in thyroid gland activity within 24 hours of birth that maintains body temperature (Little 1991). However, the thyroid gland could not account for any major heat production in newborn bettongs or any other newborn marsupial, as the thyroid is not fully functional until mid-pouch life (Buaboocha and Gemmell 1995; Rose *et al.* 1998) and results in no metabolic heat production until week 14 of pouch life in the bettong.

SHIVERING AND THE EMG

Kabat (2003) has measured the electromyogram (EMG) from developing bettongs. He showed that the EMG changed little in bettong pouch young between the age of 6–14 weeks when exposed to 35°C (= pouch temperature) but that obvious shivering started at week 8 and was fully developed by week 10 when measured at an ambient temperature of 20°C. Of interest is his finding that nor-adrenaline resulted in increased muscle activity/tone even though the ambient temperature was 35°C (Fig. 10.9). Kabat (2003) stated that this increase in tone differed from muscle activity seen during shivering. Rose and Ikonomopoulou (2005) carried out similar experiments with the eastern barred bandicoot and, in the clearest demonstration of non-shivering thermogenesis in a marsupial to date, noted that after cold acclimation this species increased its metabolic rate in response to nor-adrenaline, in the absence of an increase in shivering as measured by EMG.

DEVELOPMENT OF ENDOTHERMY

Table 10.1 describes the ontogeny of homeothermy/endothermy in the Tasmanian bettong based on data in Rose (1987, 1997) and Rose *et al.* (1998). From an ectothermic neonate the animal remains attached to the teat for the first five weeks of life in the pouch. Subsequently, exposure to cold results in peripheral vasoconstriction (skin blanching) and postural changes (curling up). Shortly thereafter (week 8) shivering can be noted. From week 10 onwards the metabolic rate can be increased in response to cold and/or nor-adrenaline injection and by week 12, shortly after fur develops, the bettong is homeothermic, develops a maximal BMR and response to cold and nor-adrenaline; T4 levels are also maximal. This period coincides with that when the young becomes able to leave and return to the pouch (Rose 1986) and the energy supplied by the milk is at a high level (Rose 1989).

Marsupials after birth remain in a warm and stable environment – the pouch. Exposure to cold

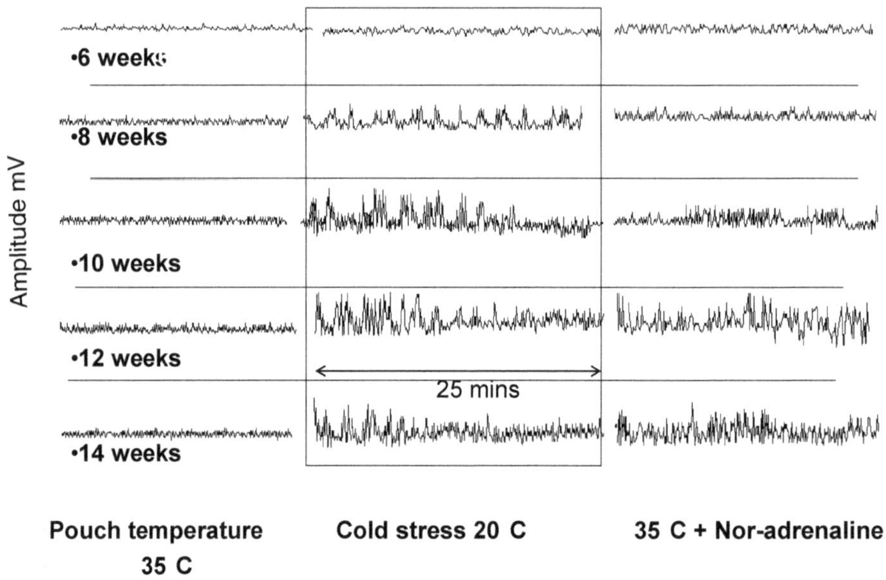

Figure 10.9: Electromyograms from Tasmanian bettong (*B. gaimardi*) pouch young of different ages, measured at 35°C, 20°C and at 35°C after injection of nor-adrenaline.

during much of pouch life is unlikely. The brown adipose tissue found in many eutherian neonates is just not required in marsupials and perhaps it never evolved or has been lost over time.

ACKNOWLEDGMENT

I thank Professor G. Grigg for his constructive comments on this chapter.

Table 10.1: Ontogeny of homeothermy and endothermy in the Tasmanian bettong (*B. gaimardi*).

Age (weeks)	First appearance of parameter
0	Birth (0.3 g)
1–5	Firmly attached to teat in warm pouch
6	Peripheral vasoconstriction and behavioural regulation
8	Shivering first appears
10	Increased metabolic response (MR) to cold and nor-adrenaline, fur first appears
12	Homeothermic, peak MR and T4 levels, peak response to cold and nor-adrenaline
14	Increased MR response to T4
15	Final pouch vacation

Based on data from Rose (1987), Rose *et al.* (1998, 1999) and unpublished data.

REFERENCES

Bartholomew G 1956. Temperature regulation in the macropod marsupial, *Setonix brachyurus*. *Physiol. Zool.* **29:** 26–40.

Buaboocha W and Gemmell RT 1995. Thyroid gland development in the brushtail possum, *Trichosurus vulpecula*. *Anat. Rec.* **243:** 254–260.

Clark MG, Colquhoun EQ, Rattigan S, Dora KA, Eldershaw TPD, Hall JL and Ye JM 1995. Vascular and endocrine control of muscle metabolism. *Am. J. Physiol.* **268:** E797–E812.

Dawson TJ and Dawson WR 1982. Metabolic scope and conductance in response to cold of some dasyurid marsupials and Australian rodents. *Comp. Biochem. Physiol.* **71A:** 59–64.

Dawson TJ and Olson JM 1988. Thermogenic capabilities of the opossum *Monodelphis domestica* when warm and cold acclimated: similarities between American and Australian marsupials. *Comp. Biochem. Physiol.* **89:** 85–91.

Golozoubova V, Hohtola E, Matthias A, Jacobsson A, Cannon B and Nedergaard J 2001. Only UCP1 can mediate adaptive nonshivering thermogenesis in the cold. *FASEB* **10:** 1096.

Gould SJ 1977. *Ontogeny and Phylogeny.* Harvard University Press, Cambridge, Massachusetts.

Grigg G, Beard L and Augee M 2004. The evolution of endothermy and its diversity in mammals and birds. *Physiol. Biochem. Zool.* **77:** 982–997.

Hayward JS and Lisson PA 1992. Evolution of brown fat: its absence in marsupials and monotremes. *Can. J. Zool.* **70:** 171–179.

Hope PJ, Pyle D, Daniels CB, Chapman I, Horowitz M, Morley JE, Trayhurn P, Kumaratilake J and Wittert G 1997. Identification of brown fat and mechanisms for energy balance in the marsupial, *Sminthopsis crassicaudata*. *Am. J. Physiol.* **273:** R161–167.

Hulbert AJ 1987. Thyroid hormones, membranes and the evolution of endothermy. In *Advances in Physiological Research*. (Eds CHS McLennan and DR Jones) pp. 305-319. Plenum Publishing, New York.

Hulbert AJ 1988. Metabolism and the development of endothermy. In *The Developing Marsupial: Models for Biomedical Research*. (Eds CH Tyndale-Biscoe and PA Janssens) pp. 148–161. Springer-Verlag, Berlin.

IUPS Thermal Commission 2003. Glossary of terms for thermal physiology. *J. Therm. Biol.* **28:** 75–106.

Kabat AP 2003. Thermogenesis and uncoupling proteins in marsupials. PhD thesis. University of Tasmania, Australia.

Kabat AP, Rose RW and West AK 2003a. Non-shivering thermogenesis in a carnivorous marsupial *Sarcophilus harrisii*, in the absence of UCP1. *J. Therm. Biol.* **28:** 413–420.

Kabat AP, Rose RW, Harris T and West AK 2003b. Molecular identification of uncoupling proteins (UCP2 andUCP3) and absence of UCP1 in the marsupial Tasmanian bettong, *Bettongia gaimardi*. *Comp. Biochem. Physiol.* **134B:** 71–77.

Little GJ 1991. Thyroid morphology and function and its role in thermoregulation in the newborn southern elephant seal (*Mirounga leonina*) at Macquarie Island. *J. Anat.* **176**: 55–69.

Loudon ASI 1987. The reproductive energetics of lactation in a seasonal macropodid marsupial. *Symp. Zool. Soc. (Lond.)* **57**: 127–147.

Loudon A, Rothwell N and Stock M 1985. Brown fat, thermogenesis and physiological birth in a marsupial. *Comp. Biochem. Physiol.* **81A:** 815–819.

McNab BK 1978. The evolution of homeothermy in the phylogeny of mammals. *Am. Nat.* **112:** 1–21.

Nicol SC 1978. Non-shivering thermogenesis in the potoroo, *Potorous tridactylus* (Kerr). *Comp. Biochem. Physiol.* **59C:** 33–37.

Nicol SC, Pavlides D and Andersen NA 1997. Nonshivering thermogenesis in marsupials: absence of thermogenic response to beta 3-adrenergic agonists. *Comp. Biochem. Physiol.* **117A:** 399–405.

Reynolds W and Hulbert AJ 1982. Cold acclimation in a small dasyurids. In *Carnivorous Marsupials*. (Ed. M Archer) pp. 279–283. Surrey Beatty and Sons, Sydney.

Rose RW 1986. Pouch vacation in the Tasmanian bettong. *Aust. J. Zool.* **34:** 485–491.

Rose RW 1987. Reproductive energetics of two Tasmanian rat-kangaroos. *Symp. Zool. Soc. (Lond)*. **57:** 149–165.

Rose RW 1989. Reproductive biology of rat-kangaroos. In *Kangaroos, Wallabies and Rat-kangaroos*. (Eds G Grigg, P Jarman and I Hume) pp. 307–315. Surrey Beatty and Sons, Sydney.

Rose RW 1997. The metabolic rate of a mycophagous marsupial: *Bettongia gaimardi*. *J. Biol.* **2:** 7.

Rose RW and Ikonomopoulou MP 2005. Shivering and non-shivering thermogenesis in a marsupial, the eastern barred bandicoot (*Perameles gunnii*). *J. Therm. Biol.* **30**: 85–92.

Rose RW and Kuswanti N 2004. Thyroid function and the development of endothermy in a marsupial, the Tasmanian bettong *Bettongia gaimardi* (Desmarest 1822). *Gen. Comp. Endocr.* **136:** 17–22.

Rose RW, Swain R and Bryant SL 1990. Body temperature; rhythm and regulation in the Tasmanian bettong (*Bettongia gaimardi*). *Comp. Biochem. Physiol.* **97A:** 573–576.

Rose RW, Kuswanti N and Colquhoun EQ 1998. The development of endothermy in the Tasmanian bettong and the effects of cold and nor-adrenaline. *J. Comp. Physiol.* **168:** 359–363.

Rose RW, West AK, Ye JM, McCormack GH and Colquhoun EQ 1999. Non-shivering thermogenesis in the marsupial (*Bettongia gaimardi*) is not attributable to brown adipose tissue. *Physiol. Biochem. Zool.* **72:** 699–704.

Rozanov CB and Dratman MB 1996. Immunohistochemical mapping of brain triiodothyronine reveals prominent localization in central nor-adrenergic systems. *Neurosci.* **74:** 897–915.

Ruben J 1995. The evolution of endothermy in mammals and birds: from physiology to fossils. *Ann. Rev. Physiol.* **57:** 69–95.

Shield J 1966. Oxygen consumption during pouch development of the macropod marsupial *Setonix brachyurus*. *J. Physiol.* **187:** 257–270.

Wallis RL 1979. Responses to low temperature in small marsupial mammals. *J. Therm. Biol.* **4:** 105–111.

Wallis RL and Maynes GM 1973.Ontogeny of thermoregulation in *Macropus parma*. *J. Mammal.* **54:** 278–281.

Whittow GC 1973. Evolution of thermoregulation. In *Comparative Physiology of Thermoregulation*, vol 3. (Ed. GC Whittow) pp. 202–258. Academic Press, New York.

Ye JM, Edwards SJ, Rose RW, Rattigan S, Clark MG and Colquhoun EQ 1995. Vasoconstrictors alter oxygen, lactate and glycerol metabolism in the perfused hindlimb of a rat-kangaroo. *Am. J. Physiol.* **268:** R1217–R1223.

Ye JM, Edwards SJ, Steen JT, Rose RW, Rattigan S, Clark MG and Colquhoun EQ, 1996. α-adrenergic stimulation of thermogenesis in a rat-kangaroo (Marsupialia: *Bettongia gaimardi*). *Am. J. Physiol.* **271:** R586–R592.

11 Mechanistic explanations for drought-related mortality of juvenile red kangaroos: implications for population dynamics and modelling

A.J. Munn and T.J. Dawson

SUMMARY

Why do juvenile red kangaroos (*Macropus rufus*) die during drought? Answering this question has important implications for understanding the population dynamics not only of red kangaroos, but of large macropodids generally. Highly variable and extensive juvenile mortalities are a feature in populations of large herbivorous mammals worldwide. Population growth and persistence in many species is largely dependent on juvenile recruitment through favourable environmental conditions. Understanding juvenile kangaroo survival in relation to environmental conditions is therefore crucial for accurately modelling their population dynamics. This chapter reviews recent studies that have compared the physiological ecology of juvenile and adult *M. rufus*, providing the first mechanistic explanations for the vulnerability of young kangaroos to drought conditions.

INTRODUCTION

Red kangaroos (*Macropus rufus*: superfamily Macropodoidea; family Macropodidae) are recognisable and iconic Australian marsupials. They inhabit much of Australia's arid and semi-arid interior, where they contend with unpredictable climates that are typified by extremes in daily temperature (–5–>45°C), scarce free water and frequent drought. The abilities of red kangaroos to deal with such environments have enabled them to prosper even in the face of domestic stock, in contrast to many other herbivorous marsupials, which have declined (Dawson 1995; McCullough and McCullough 2000). Recent estimates indicate that there are around 15 million red kangaroos in the arid rangelands, mostly in areas where dingos are controlled (Hacker and McLeod 2003). Concerns about the population dynamics of red kangaroos, as well as other large arid and semi-arid kangaroos (the eastern *Macropus giganteus* and western *Macropus fuliginosus* grey kangaroos and the euro *Macropus robustus*), arise because of their potential impact on agriculture and their importance as harvestable species (Grigg 2002; Dawson and Munn 2006).

While adult red kangaroos possess a suite of behavioural and physiological features for surviving drought, young red kangaroos are notably vulnerable to dry conditions. Juvenile red kangaroos typically have the highest mortality rates of any age cohort during drought (Bayliss 1987; Shepherd 1987). Such high rates of mortality contrast with the much better survival of prime-aged females, that usually make up the bulk of red kangaroo populations and are the main source of population fecundity. Such patterns of juvenile mortality are seen in other large mammalian herbivores worldwide and the phenomenon appears tightly linked with environmental factors (Clutton-Brock *et al.* 1992; Owen-Smith 1990; Sæther 1997; Gaillard *et al.* 1998, 2000). Offspring survival varies considerably between years, with survival being high during favourable environmental conditions but typically poor during lean years (Sæther 1997; Gaillard *et al.* 1998, 2000; Forchhammer *et al.* 2001). Population recruitment via juvenile survival is therefore a major limiting factor for many large herbivores, irrespective of climate/habitat (e.g. temperate vs tropical, forest vs grassland) or of the proximate causes of mortality (e.g. predation, disease, resource limitation). Such patterns of variable annual recruitment are also seen in other species of kangaroo (Kirkpatrick and McEvoy 1966; Bayliss 1987; Shepherd 1987; Arnold *et al.* 1991; Dawson 1995; Bilton and Croft 2004; Fisher *et al.* 2000).

In arid and semi-arid Australia rainfall is the notable environmental factor affecting kangaroo populations (Newsome *et al.* 1967; Bayliss 1985a, b; Robertson 1986; Shepherd 1987; Caughley *et al.* 1984, 1985; Dawson 1995; McCullough and McCullough 2000). The bulk of the population recruitment for red kangaroos occurs after periods of substantial rain and population growth is markedly truncated by drought (Shepherd 1987; Cairns and Grigg 1993; Fig. 11.1). Although numerous mathematical models have attempted to explain the dynamics of kangaroo populations in relation to rainfall (Caughley *et al.* 1984; Bayliss 1987; Caughley 1987; Cairns 1989; Caughley and Gunn 1993), more recent studies have indicated that the rainfall–population dynamic in red kangaroos is complex and that rainfall alone is insufficient to accurately predict population changes (McCarthy 1996; Jonzen *et al.* 2005). Importantly, few models of kangaroo population dynamics incorporate fundamental differences in the physiological ecology of differently aged animals (McLeod *et al.* 2004). Clearly, a mechanistic understanding of the survival of juvenile red kangaroos is essential for improving population models and therefore evaluating sustainable land and animal management strategies.

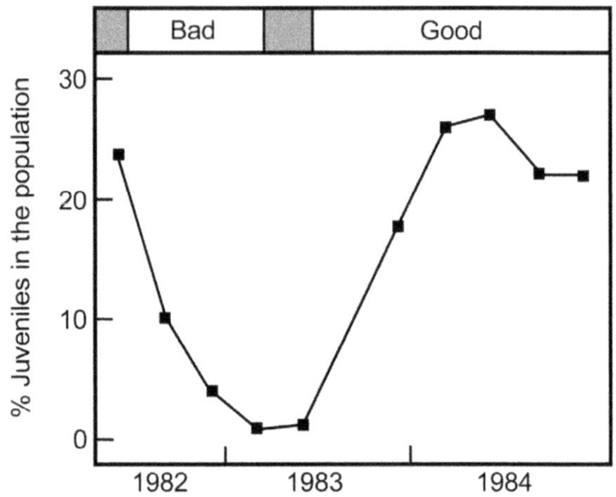

Figure 11.1: Proportion of juvenile red kangaroos (*M. rufus*) within a population in western New South Wales throughout a severe drought (1982–83, = bad times) and after drought-breaking rains (good times). Redrawn from Shepherd (1987).

The basic question arises, why do juvenile red kangaroos die during drought? Is their mortality related to lack of water or lack of food, or both? If food is involved, is it simply related to the quantity of forage available, or is quality important? We summarise recent studies that have identified causal mechanisms underlying the vulnerability of juvenile red kangaroos to drought. This approach provides an ecophysiological framework for understanding the mortality patterns of red kangaroos, and for other kangaroo species and mammalian herbivores generally.

ENERGY NEEDS AND THE BURDEN OF GROWTH

An animal's basal metabolic rate (BMR) describes its minimal energy requirements and to a large extent correlates with feed requirements (Hume 1999). BMR is defined as the minimal resting metabolism of mature (i.e. not growing), non-reproductive, post-absorptive, non-stressed wakeful animals in a thermoneutral environment and with minimal activity (Withers 1992). In common with most marsupials, adult red kangaroos have BMRs around 70% of those of comparable eutherians (Dawson and Hulbert 1970; Dawson *et al.* 2000). Similarly, mature red kangaroos have lower maintenance energy needs than eutherians (e.g. sheep). This may have profound impacts on their ability to survive arid conditions, particularly during droughts (Hume 1974; Munn and Dawson 2003b).

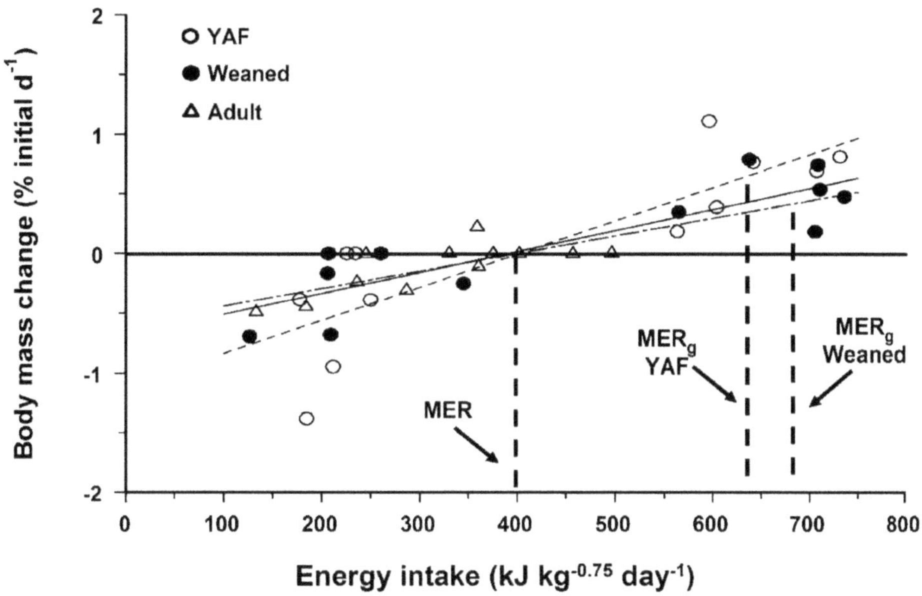

Figure 11.2: Maintenance energy requirements (MER) and MER plus growth-energy requirements (MER_g) of YAF, weaned and non-lactating adult female red kangaroos (*M. rufus*) measured as digestible energy intake ($kJ/kg^{-0.75}/d^{-1}$). Modified from Munn and Dawson (2003b).

How do these features affect juvenile survivorship? Unlike adults, juvenile animals have additional energy turnovers associated with growth. In red kangaroos, growth rate peaks between the time when the young permanently leaves the mother's pouch (c. 250 days since birth) and shortly after weaning (c. 360 days). These young-at-foot (YAF) animals are the focus of juvenile mortality in red kangaroos during drought. Because of the growth-related metabolism of juveniles, it is difficult to determine their underlying energy needs compared with those of adults, or to quantify the impact of growth on their daily energy/food requirements. The closest analogue to BMR in young mammals is a fasted resting metabolic rate (RMR) in a thermoneutral environment. The small body size of juveniles relative to adult size also presents a problem when comparing their fundamental energy needs. In mammals, metabolism does not scale directly (isometrically) with body mass (i.e. per kg), but allometrically; it usually is accepted to scale with body mass raised to a power of 0.75 (Kleiber 1975; Schmidt-Nielsen 1984; West *et al.* 1997). To compare juvenile kangaroos with adults we examined juvenile RMRs per unit of body $mass^{0.75}$ (Munn and Dawson 2001). On this basis, the RMR of YAF red kangaroos was 1.5 times that expected for an adult marsupial of equivalent body mass (Dawson and Hulbert 1970; Hayssen and Lacy 1985; Hinds *et al.* 1993). That is, an adult kangaroo of the same body mass as a YAF (7.4 kg) is predicted to have an RMR of approximately 200 $kJ/kg^{-0.75}/d^{-1}$ (Dawson and Hulbert 1970; Hayssen and Lacy 1985; Hinds *et al.* 1993) but we found that YAF and weaned red kangaroos have RMRs of near 300 $kJ/kg^{-0.75}/d^{-1}$ (Munn and Dawson 2001).

Interestingly, the higher RMR of juvenile red kangaroos, relative to adults, did not translate into higher maintenance energy requirements (MER). Basically, MER is the daily energy required by an animal to maintain their body mass: contrary to findings for RMR, we found no significant differences in the MERs of YAF, weaned and adult red kangaroos – all were 384–390 $kJ/kg^{-0.75}/d^{-1}$(Munn and Dawson 2003a). These estimates do not include the contribution of growth to daily energy requirements or intakes. When the cost of maintaining ideal growth rates was taken into account, juveniles, as expected, had markedly higher food requirements than adults (Fig. 11.2) (Munn and Dawson 2003a).

On average, YAF and recently weaned red kangaroos have maximal (or ideal) growth rates of 50–60 g/d^{-1} (Frith and Sharman 1964; Sharman *et al.* 1964; Munn and Dawson 2003a). To sustain this level of growth, YAF and weaned red kangaroos had daily energy requirements of 639 kJ and 68 $kJ/kg^{-0.75}/d–^{1}$ respectively, i.e. 1.7–1.8 times the MER values given above. Therefore, because there was no difference in the MERs of YAF, weaned and adult red kangaroos, the higher resting metabolic requirements of the juveniles are explained by their additional energy requirements for growth.

The burden of growth for juvenile red kangaroos has a major impact on their daily food requirements and hence their ecology. For example, to sustain ideal growth rates even within their thermoneutral zone, YAF red kangaroos would require total daily digestible energy intakes (kJ/d^{-1}) approaching 70% that required by mature non-lactating females, despite being only 20–25% of the adult's body mass. Moreover, during dry conditions, especially drought, the mother's milk output usually declines, rendering the juveniles more dependent on forage. By time of weaning the young red kangaroos require as much as 95% of the daily energy intake of mature females in order to maintain optimal growth, though they are still only half the body mass of mature females (Munn and Dawson 2003a).

THERMOREGULATION: A ROLE FOR WATER IN LIMITING JUVENILE KANGAROO SURVIVAL?

The additional energy and water costs associated with free living must also be taken into account when comparing the ecology of juvenile and adult animals. In particular, the costs of thermoregulation may be expected to be higher for smaller juvenile red kangaroos than for adults. By the time the young kangaroo exits the mother's pouch permanently it is less than a third of the mature female body mass, but must contend with the same environmental conditions as larger animals. A small body size has distinct disadvantages with regard to thermoregulation because of a higher surface area to volume ratio. Smaller animals lose relatively more heat to the environment under cold conditions and this further intensifies the differentially higher food requirements of juveniles (Munn and Dawson 2001). They also potentially suffer greater external heat loads at ambient temperatures above deep body temperature. In their arid environment, red kangaroos routinely experience environmental temperatures well in excess of 37°C and adult animals are known to adjust their behaviour to reduce water requirements for thermoregulation.

Red kangaroos including juveniles are largely active nocturnally, foraging at night and at dusk and dawn (Watson and Dawson 1993; Dawson *et al.* 2006). By limiting activity and seeking shade during the day, red kangaroos minimise their overall water use. Despite these behavioural adjustments, juvenile red kangaroos may be expected to have a greater water debt than adults during dry conditions. Under warm conditions, for example, when ambient temperatures approach deep body temperature (c. T_a 33°C), both YAF and weaned red kangaroos are unable to maintain their body temperature at basal levels as seen in adults (Munn and Dawson 2001). The T_b of juveniles increased from basal levels of 36.9 ± 0.2°C at T_a 25°C to 37.7 ± 0.2°C at T_a 33°C. To prevent further overheating under these warm conditions, YAF and weaned kangaroos evaporated around 14 mL H_2O/hr^{-1} by panting and/or licking the body surface (mainly the fore limbs, legs, belly and tail). Under similarly hot conditions, adult kangaroos maintained T_b at basal levels (range 36.3 ± 0.2° to 36.5 ± 0.2°C) by evaporating some 25 mL H_2O/hr^{-1}. This level of water loss by adults was less than twice that of juveniles, despite the adults being more than three times larger than YAF. Under hot conditions (T_a 45°C), both YAF and weaned red kangaroos evaporated as much water as adults, 79–84 mL H_2O/hr^{-1}. This level of thermoregulatory water loss by juvenile kangaroos could have profound implications for their abilities to survive during dry conditions.

Adult red kangaroos do not drink every day, even during summer (Dawson *et al.* 1975). However, being smaller than adults, juvenile kangaroos have smaller absolute body-water reserves and under hot conditions should need to drink more frequently than adults, especially in light of their high water use for thermoregulation. While still sucking, the water in their mother's milk could ameliorate the water losses by YAF kangaroos, but lactation declines during hot conditions and may prematurely shut down during drought (Newsome 1964a, b; Frith and Sharman 1964; Newsome 1965). This means that YAF and weaned red kangaroos may be more dependent on water points, where predators focus and food is sparse. Shepherd (1981) found that 97% of the red kangaroos killed by dingoes around watering holes were juveniles. Also, adult red kangaroos, and presumably YAF, do not show water-focused grazing. They must travel between water sources and their distant home ranges (Montague-Drake and Croft 2004). If juveniles access water more frequently than adults they will incur higher travelling costs (Baudinette *et al.* 1993).

FOOD LIMITATION AND NUTRITION: QUANTITY OR QUALITY?

Primarily grass-eaters, adult kangaroos are able to utilise fibrous vegetation by fermentative digestion in a large forestomach, in a manner analogous to digestion in ruminants (Moir *et al.* 1956; Foot and Romberg 1965; McIntosh 1966; Forbes and Tribe 1970; Hume 1974). However, in form and function, the tubiform foregut of kangaroos is more like that of the equine colon than the vat-like structure of ruminants (Hume 1999). It has been suggested that a tubiform foregut contributes to the superior ability of kangaroos to persist on fibrous grasses compared with ruminants (Freudenberger *et al.* 1989; Hume 1999; but see Munn *et al.* 2008). However, red kangaroos prefer fresh green growth and forage on this when possible (Dawson and Ellis 1994).

Young, fresh plant growth is generally more nutritious than mature dry forage; it contains higher proportions of easily digestible cell contents relative to the less digestible structural carbohydrates of cellulose and hemi-cellulose, i.e. fibre (Terry and Tilley 1964; Short *et al.* 1974). Fresh plant growth is also higher in protein (nitrogen) than mature forage. Juvenile red kangaroos appear particularly reliant on fresh high-quality forage (Moss and Croft 1999). For example, measured as an index of plant greenness and hence age (Moss and Croft 1999), or directly as nitrogen or fibre content (Shepherd 1987), forage quality was found to be the best predictor of body condition in red kangaroos in arid regions of western New South Wales. This was especially so for juvenile red kangaroos, particularly young males. During dry conditions, production of new herbage declines and fresh green herbage is quickly eaten out. In drought, mature fibrous grasses can be over 90% of the adult red kangaroos' diet (Dawson and Ellis 1994). The limited extent to which juvenile red kangaroos appear able to utilise fibrous grasses suggests that food quality, not just quantity, is important (Munn *et al.* 2006).

Forage quality is generally discussed in terms of the level of plant fibre because mammalian enzymes cannot readily break down these fibre fractions. The digestion of fibre requires fermentation by specialised microbes, usually in expanded regions of the fore- or hindgut (Stevens and Hume 1995). Small herbivores, however, have smaller absolute gut sizes relative to their metabolic needs and necessarily have faster rates of feed passage (Parra 1978; Mould and Robbins 1982; Demment and Van Soest 1985; Freudenberger *et al.* 1989). Compared with larger animals, material ingested by smaller herbivores is exposed to microbial action for a shorter time, thereby reducing digestive efficiency (i.e. nutrient extracted per unit feed ingested). Consequently, fibre digestion is usually less efficient among smaller herbivores (Demment and Van Soest 1985; Freudenberger *et al.* 1989). It was considered possible that such a scenario might explain the apparently lower ability of juvenile red kangaroos to cope with the fibrous forage available during droughts. However, using the most reliable inert markers to measure food passage in juvenile and adult kangaroos, Munn and Dawson (2006) have shown that the situation is more complex.

When offered high-quality chopped lucerne hay containing about 44% neutral-detergent fibre (NDF; see Munn and Dawson 2006), overall food passage rates were actually similar for YAF, weaned and mature female kangaroos (Fig. 11.3). Although the mean retention time (MRT) of particles in the gastrointestinal tract, for example, varied between 25 hours and 22 hours for YAF and weaned kangaroos respectively, and 29 hours for mature females, the differences were not significant (Munn and Dawson 2006). As expected, digestion of NDF (% digestibility) was lowest for the smaller YAF (27%), followed by weaned (34%) and then mature female (40%) kangaroos, but total dry matter digestibility of chopped lucerne was similar for all age classes (55–59%). This level of dry matter digestibility was comparable to that seen in mature male red kangaroos fed similar forage (McIntosh 1966). Therefore, on high-quality forage, YAF and weaned kangaroos had food passage rates comparable with larger adults and were capable of maintaining ideal growth rates, even when the YAF did not receive any milk that they would normally be taking at this stage. However, on a poor-quality diet of chopped oaten hay (*Avena sativa*), containing c. 68% NDF, neither the YAF nor weaned kangaroos could maintain growth, and both age classes lost body mass. The smaller YAF kangaroos were least capable of digesting NDF from oaten hay, at only c. 18%, followed by weaned (23%) and mature female (27%) kangaroos. The digestibility of dry matter from the oaten hay was lowest for the YAF (34%), significantly less than the 42% digested by weaned and mature female kangaroos. That the weaned kangaroos could digest the fibrous oaten hay as well as mature females could is important because it indicates that extraction of nutrients by the smaller juveniles is comparable to that by adults despite large differences in their body masses; weaned kangaroos are around 10–12 kg compared with mature females of 25–30 kg and males of 60–70 kg. This result is somewhat contrary to that expected, based on the smaller body size and apparently faster feed passage in smaller herbivores. Indeed, on the chopped oaten hay the MRT of particles in YAF, weaned and mature female kangaroos was not significantly different, ranging from 30 hours to 35 hours. Why then do recently weaned animals, as well as YAF whose milk has probably been cut off, still show high mortalities during drought? The answer lies with total gastrointestinal capacity, or gut fill.

Using calculations of gut fill based on forage digestibility and particle MRTs (Holleman and White 1989), we found that smaller juveniles were limited in their capacity to expand the gut to accommodate poorer-quality forage (Munn and Dawson 2006). On both poor (oaten hay) and high (chopped lucerne) quality forages, YAF and weaned kangaroos had levels of gut fill of 25–30 g DM/kg$^{-0.75}$ (Table 11.1). This level of fill was similar to that of mature females on chopped lucerne hay at 24 g DM/kg$^{-0.75}$. However, on the poor-quality forage the mature female kangaroos were capable of expanding total gut fill by 1.6 times, to 38 g DM/kg$^{-0.75}$ (Table 11.1). Adult female red kangaroos, therefore, were able to respond to the decline in forage quality by increasing the total nutrient load within the gut. Juveniles, on the other hand, appear limited in their absolute gut capacity. Even on high-quality forages the juvenile red kangaroos apparently maximised gut fill in order to extract sufficient nutrients to maintain growth. This is particularly important in light of the fact that, to sustain growth, the juvenile red kangaroos required around 1.8 times as much energy each day per unit body mass$^{0.75}$ as did mature females. Even on the poor-quality forage the mature females maintained body mass.

At the scale of the individual, juvenile red kangaroos therefore may be expected to employ different foraging strategies from adults. At the least, juvenile red kangaroos require forage with less than 40–50%

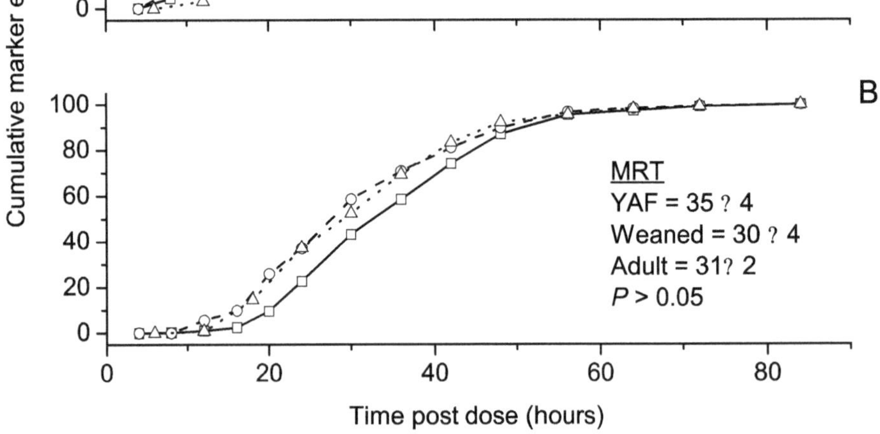

Figure 11.3: Mean cumulative elimination (% total excretion) of particle marker (size > 500 μm, < 1 mm) and mean retention time (MRT; hr) for particles from YAF, weaned and non-lactating adult red kangaroos (*M. rufus*) following a single oral dose at time zero (hr) and on a diet of (a) chopped lucerne or (b) oaten hays. Summarised from Munn and Dawson (2006).

Table 11.1: Total dry matter gastrointestinal tract fill (gut fill) in young-at-foot (YAF), weaned and adult red kangaroos (*M. rufus*) fed chopped lucerne and oaten hays.

	YAF	Weaned	Adult
Gut fill (g DM)			
Lucerne hay	121.6 ± 20.8^{b}	158.6 ± 23.6^{b}	$264.0 \pm 23.7^{a*}$
Oaten hay	113.9 ± 22.9^{b}	150.7 ± 43.3^{b}	426.5 ± 25.5^{a}
Gut fill (g DM/kg^{-1})			
Lucerne hay	18.4 ± 2.8^{b}	$14.8 \pm 2.0^{a,b}$	$10.6 \pm 1.5^{a*}$
Oaten hay	18.0 ± 3.4^{a}	13.5 ± 3.2^{a}	16.8 ± 0.8^{a}
Gut fill (g DM/kg$^{-0.75}$)			
Lucerne hay	29.4 ± 4.5^{a}	26.7 ± 3.7^{a}	$23.7 \pm 2.8^{a*}$
Oaten hay	28.5 ± 5.4^{a}	24.6 ± 6.2^{a}	37.7 ± 1.5^{a}

Means ± SEM bearing different superscripts are significantly different within diets between ages ($^{a,b,c}P < 0.05$). Asterisks (*) associated with lucerne hay diets indicate significant differences between diets within ages ($p < 0.01$). After Munn and Dawson (2006).

fibre (NDF) to sustain growth. Under good environmental conditions this may not be a problem. Juveniles could satisfy their requirements by feeding on highly nutritious, fresh plant growth. However, during dry conditions, when feed quantity and quality fall, juveniles may need to spend more time searching for high-quality feed, thereby increasing predation and thermal risks. It is not known whether YAF and adult red kangaroos differ in behavioural responses to changes in vegetation, but such differences could be important for understanding the vulnerability of juveniles to drought. Watson and Dawson (1993), for example, observed that subadult red kangaroos spent more time searching for feed and less time chewing than did mature males and females. Nonetheless, the main difference between the ability of juveniles and adults to utilise fibrous forage was the juvenile's inability to further expand gut fill as forage quality declined. Juvenile red kangaroos appear limited not by how well they can process poor-quality forage, as predicted from theory (Demment and Van Soest 1985), but by how much they can process.

CONCLUSIONS: FUTURE DIRECTIONS FROM A MECHANISTIC APPROACH

Speare *et al.* (1989) suggested that the most likely causes of juvenile kangaroo mortality during drought were dehydration and malnutrition. This chapter has presented findings that have teased apart some of these broad-brush ideas and provided specific information about how juvenile kangaroos' physiology interacts with their environment to affect their survival. This information is crucial for modelling juvenile red kangaroo survival in relation to food requirements and their ecology as a whole. Nonetheless, more information is needed to fully understand the role of juveniles in the population dynamics not only of red kangaroos, but of large macropodids generally. Several pieces of the puzzle are still lacking. In particular, we need data on the costs of locomotion in juvenile kangaroos. Can juveniles make the long-distance movements for food or water that are occasionally reported for adults (Priddel *et al.* 1988)? Is it relatively more costly for young kangaroos to travel even short distances from water sources to grazing sites? What are the short-term impacts of water restriction on juveniles, and how well do they cope with dehydration? Information on the water turnovers of juvenile kangaroos in the field is lacking. Their apparently greater need for water, however, suggests that research on the survival patterns of red kangaroos in relation to the distance between water and grazing sites could prove very useful for refining population models at local and regional scales.

ACKNOWLEDGMENT

A modified version of this chapter was published as part of the conference proceedings for the 4th International Conference in Africa for Comparative Physiology and Biochemistry, 2008.

REFERENCES

Arnold GW, Grassia A, Steven DE and Weeldenburg JR 1991. Population ecology of western grey kangaroos in a remnant of wandoo woodland at Bakers Hill, southern Western Australia. *Wildl. Res.* **1:** 561–575.

Baudinette RV, Halpern EA and Hinds DS 1993. Energetic cost of locomotion as a function of ambient temperature and during growth in the marsupial *Potorous tridactylus*. *J. Exp. Biol.* **174:** 81–95.

Bayliss P 1985a. The population dynamics of red and western grey kangaroos in arid New South Wales, Australia. I. Population trends and rainfall. *J. Animal Ecol.* **54:** 111–125.

Bayliss P 1985b. The population dynamics of red and western grey kangaroos in arid New South Wales, Australia. II. The numerical response function. *J. Animal Ecol.* **54:** 127–135.

Bayliss P 1987. Kangaroo dynamics. In *Kangaroos: Their Ecology and Management in Sheep Rangelands of Australia*. (Eds G Caughley, N Shepherd and J Short) pp. 119–134. Cambridge University Press, Cambridge.

Bilton AD and Croft DB 2004. Lifetime reproductive success in a population of female red kangaroos *Macropus rufus* in the sheep rangelands of western New South Wales: environmental effects and population dynamics. *Aust. Mammal.* **26:** 45–60.

Cairns SC 1989. Models of macropodid populations. In *Kangaroos, Wallabies and Rat-kangaroos*. (Eds G Grigg, P Jarman and I Hume) pp. 695–704. Surrey Beatty and Sons, Sydney.

Cairns SC and Grigg GC 1993. Population dynamics of red kangaroos (*Macropus rufus*) in relation to rainfall in South Australian pastoral zone. *J. Appl. Ecol.* **30:** 444–458.

Caughley G 1987. Ecological relationships. In *Kangaroos: Their Ecology and Management in Sheep Rangelands of Australia*. (Eds G Caughley, N Shepherd and J Short) pp. 159–187. Cambridge University Press, Cambridge.

Caughley G and Gunn A 1993. Dynamics of large herbivores in deserts: kangaroos and caribou. *Oikos* **67:** 47–55.

Caughley J, Bayliss P and Giles J 1984. Trends in kangaroo numbers in western New South Wales and their relation to rainfall. *Aust. Wildl. Res.* **11:** 415–422.

Caughley G, Grigg GC and Smith L 1985. The effect of drought on kangaroo populations. *J. Wildl. Manag.* **49:** 679–685.

Clutton-Brock TH, Price OF, Albon SD and Jewel PA 1992. Early development and population fluctuations in Soay sheep. *J. Appl. Ecol.* **61:** 381–396.

Dawson TJ 1995. *Kangaroos: Biology of the Largest Marsupials*. UNSW Press, Sydney.

Dawson TJ and Ellis BA 1994. Diets of mammalian herbivores in Australian arid shrublands: seasonal effects on overlap between red kangaroos, sheep and rabbits and on dietary niche breadths and electivities. *J. Arid Environ.* **26:** 257–271.

Dawson TJ and Hulbert AJ 1970. Standard metabolism, body temperature, and surface areas of Australian marsupials. *Am. J. Physiol.* **218:** 1233–1238.

Dawson TJ and Munn AJ 2006. How much do kangaroos of differing age and size eat relative to domestic stock? Implications for the arid rangelands. In *Animals of Arid Australia*. (Eds C Dickman, D Lunney and S Burgin) pp. 99–101. Royal Zoological Society of New South Wales, Sydney.

Dawson TJ, Denny MJS, Russell EM and Ellis B 1975. Water usage and diet preferences of free-ranging kangaroos, sheep and feral goats in the Australian arid zone during summer. *J. Zool. Lond.* **177:** 1–23.

Dawson TJ, Blaney CE, Munn AJ, Krockenberger A and Maloney SK 2000. Thermoregulation by kangaroos from mesic and arid habitats: influence of temperature on routes of heat loss in eastern grey kangaroos (*Macropus giganteus*) and red kangaroos (*Macropus rufus*). *Physiol. Biochem. Zool.* **73:** 374–381.

Dawson TJ, McTavish KJ, Munn AJ and Holloway J 2006. Water use and the thermoregulatory behaviour of kangaroos in arid regions: insights into the colonisation of arid rangelands in Australia by the eastern grey kangaroo (*Macropus giganteus*). *J. Comp. Physiol. B* **176:** 45–53.

Demment MW and Van Soest PJ 1985. A nutritional explanation for body-size patterns of ruminant and nonruminant herbivores. *Am. Nat.* **125:** 641–672.

Fisher DO, Hoyle SD and Blomberg SP 2000. Population dynamics and survival of an endangered wallaby: a comparison of four methods. *Ecol. Applic.* **10:** 901–910.

Foot JZ and Romberg B 1965. The utilisation of roughage by sheep and the red kangaroo, *Macropus rufus* (Desmarest). *Aust. J. Agric. Res.* **16:** 429–435.

Forbes DK and Tribe DE 1970. The utilization of roughages by sheep and kangaroos. *Aust. J. Zool.* **18:** 247–256.

Forchhammer MC, Clutton-Brock TH, Lindstrom J and Albon SD 2001. Climate and population density induce long-term cohort variation in a northern ungulate. *J. Animal Ecol.* **70:** 721–729.

Freudenberger DO, Wallis IR and Hume ID 1989. Digestive adaptations of kangaroos, wallabies and rat-kangaroos. In *Kangaroos, Wallabies and Rat-kangaroos*. (Eds G Grigg, P Jarman and I Hume) pp. 179–187. Surrey Beatty and Sons, Sydney.

Frith HJ and Sharman GB 1964. Breeding in wild populations of the red kangaroo, *Megaleia rufa*. *CSIRO Wildl. Res.* **9:** 86–114.

Gaillard J, Festa-Bianchet M and Yoccoz N 1998. Population dynamics of large herbivores: variable recruitment with constant adult survival. *Trends Ecol. Evol.* **13:** 58–63.

Gaillard JM, Festa-Bianchet M, Yoccoz NG, Loison A and Toïgo C 2000. Temporal variation in fitness and population dynamics of large herbivores. *Ann. Rev. Ecol. Syst.* **31:** 367–393.

Grigg G 2002. Using native fauna to assist in its own survival. In *A Zoological Revolution*. (Eds D Lunney and C Dickman) pp. 53–76. Royal Zoological Society of New South Wales/Australian Museum, Sydney.

Hacker R and McLeod S 2003. *Living with Kangaroos: A Guide to Kangaroos and their Management in the Murray-Darling Basin*. NSW Agriculture, Sydney.

Hayssen V and Lacy RC 1985. Basal metabolic rates in mammals: taxonomic differences in the allometry of

BMR and body mass. *Comp. Biochem. Phys. A* **81:** 741–754.

Hinds DS, Baudinette RV, Macmillen RE and Halpern EA 1993. Maximum metabolism and the aerobic factorial scope of endotherms. *J. Exp. Biol.* **182:** 41–56.

Holleman DF and White RG 1989. Determination of digesta fill and passage rate from non-absorbed particulate phase markers using the single dosing method. *Can. J. Zool.* **67:** 488–494.

Hume ID 1974. Nitrogen and sulphur retention and fibre digestion by euros, red kangaroos and sheep. *Aust. J. Zool.* **22:** 13–23.

Hume ID 1999. *Marsupial Nutrition*. Cambridge University Press, Cambridge.

Jonzén N, Pople AR, Grigg GC and Possingham HP 2005. Of sheep and rain: large-scale population dynamics of the red kangaroo. *J. Anim. Ecol.* **74:** 22–30.

Kirkpatrick TH and McEvoy JS 1966. Studies of Macropodidae in Queensland. V. Effects of drought on reproduction in the grey kangaroo (*Macropus giganteus*). *Qld J. Agric. Anim. Sci.* **23:** 439–442.

Kleiber M 1975. *The Fire of Life*. Krieger Publishing, New York.

McCarthy MA 1996. Red kangaroo (*Macropus rufus*) dynamics: effects of rainfall, density dependence, harvesting and environmental stochasticity. *J. Appl. Ecol.* **33:** 45–53.

McCullough DR and McCullough Y 2000. *Kangaroos in Outback Australia: Comparative Ecology and Behavior of Three Coexisting Species*. Columbia University Press, New York.

McIntosh DL 1966. The digestibility of two roughages and the rates of passage of their residues by the red kangaroo, *Megaleia rufa* (Desmarest), and the merino sheep. *CSIRO Wild. Res.* **11:** 125–135.

McLeod SR, Hacker RB and Druhan JP 2004. Managing the commercial harvest of kangaroos in the Murray-Darling Basin. *Aust. Mammal.* **26:** 9–22.

Moir RJ, Somers M and Waring H 1956. Studies in marsupial nutrition. I. Ruminant-like digestion in an herbivorous marsupial (*Setonix brachyurus*). *Aust. J. Biol. Sci.* **9:** 293–304.

Montague-Drake R and Croft DB 2004. Do kangaroos exhibit water-focussed grazing patterns in arid New South Wales? A case study in Sturt National Park. *Aust. Mammal.* **26:** 87–100.

Moss GL and Croft DB 1999. Body condition of the red kangaroo (*Macropus rufus*) in arid Australia: the effect of environmental condition, sex and reproduction. *Aust. J. Ecol.* **24:** 97–109.

Mould ED and Robbins CT 1982. Digestive capabilities in elk compared to white-tailed deer. *J. Wildl. Man.* **46:** 22–29.

Munn AJ and Dawson TJ 2001. Thermoregulation in juvenile red kangaroo (*Macropus rufus*) after pouch exit: higher metabolism and evaporative water requirements. *Phys. Biochem. Zool.* **74:** 917–927.

Munn AJ and Dawson TJ 2003a. How important is milk for near-weaned red kangaroos (*Macropus rufus*) fed different forages? *J. Comp. Phys. B* **173:** 141–148.

Munn AJ and Dawson TJ 2003b. Energy requirements of the red kangaroos (*Macropus rufus*): impacts of age, growth and body size in a large desert-dwelling herbivore. *J. Comp. Phys. B* **173:** 575–582.

Munn AJ and Dawson TJ 2006. Forage fibre digestion, rates of feed passage and gut fill in juvenile and adult red kangaroos *Macropus rufus* (Desmarest): why body size matters. *J. Exp. Biol.* **209:** 1535–1547.

Munn AJ, Dawson TJ and Hume ID 2006. Endogenous nitrogen excretion by red kangaroos (*Macropus rufus*): effects of animal age and forage quality. *Phys. Biochem. Zool.* **79:** 424–436.

Munn AJ, Streich WJ, Hummel J and Clauss M 2008. Modelling digestive constraints in non-ruminant and ruminant foregut-fermenting mammals. *Comp. Biochem. Physiol. A* **151:** 78–84.

Newsome AE 1964a. Anoestrus in the red kangaroo, *Megaleia rufa* (Desmarest). *Aust. J. Zool.* **12:** 6–17.

Newsome AE 1964b. Oestrus in the lactating red kangaroo, *Megaleia rufa* (Desmarest). *Aust. J. Zool.* **12:** 315–321.

Newsome AE 1965. Reproduction in natural populations of the red kangaroo *Megaleia rufa* (Desmarest), in central Australia. *Aust. J. Zool.* **13:** 303–311.

Newsome AE, Stephens DR and Shipway AK 1967. Effect of a long drought on the abundance of red kangaroos in central Australia. *CSIRO Wildl. Res.* **12:** 1–8.

Parra R 1978. Comparison of foregut and hindgut fermentation in herbivores. In *The Ecology of Arboreal Folivores*. (Ed. GG Montgomery) pp. 205–230. Smithsonian Institution Press, Washington DC.

Priddel D, Wellard G and Shepherd N 1988. Movements of sympatric red kangaroos, *Macropus rufus*, and western grey kangaroos, *M. fuliginosus*, in western New South Wales. *Aust. Wildl. Res.* **15:** 339–346.

Owen-Smith N 1990. Demography of a large herbivore, the greater kudu *Tragelaphus strepsiceros*, in relation to rainfall. *J. Anim. Ecol.* **59:** 893–913.

Robertson G 1986. The mortality of kangaroos in drought. *Aust. Wildl. Res.* **13:** 349–354.

Sæther B 1997. Environmental stochasticity and population dynamics of large herbivores: a search for mechanisms. *Trends Ecol. Evol.* **12:** 143–149.

Schmidt-Nielsen K 1984. *Scaling: Why is Animal Size so Important?* Cambridge University Press, Cambridge.

Sharman GB, Frith HJ and Calaby JH 1964. Growth of the pouch young, tooth eruption and age determination in the red kangaroo, *Megaleia rufa*. *CSIRO Wildl. Res.* **9:** 20–49.

Shepherd N 1981. Predation of red kangaroos, *Macropus rufus*, by the dingo, *Canis familiaris dingo* (Blumenbach), in north-western New South Wales. *Aust. Wildl. Res.* **8:** 255–262.

Shepherd N 1987. Condition and recruitment of kangaroos. In *Kangaroos: Their Ecology and*

Management in Sheep Rangelands of Australia. (Eds G Caughley, N Shepherd and J Short) pp. 135–158. Cambridge University Press, Cambridge.
Short HL, Blair RM and Segelquist CA 1974. Fibre composition and forage digestibility by small ruminants. *J. Wildl. Man.* **38:** 197–209.
Speare R, Donovan JA, Thomas AD and Speare PJ 1989. Diseases of free-ranging Macropodoidea. In *Kangaroos, Wallabies and Rat-kangaroos*. (Eds G Grigg, P Jarman and I Hume) pp. 705–734. Surrey Beatty and Sons, Sydney.
Stevens CE and Hume ID 1995. *ComparativePphysiology of the Vertebrate Digestive System*. 2nd edn. Cambridge University Press, Cambridge.
Terry RA and Tilley JMA 1964. The digestibility of leaves and stems of perennial ryegrass, cocksfoot, timothy, tall fescue, lucerne and sainfoin, as measured by an in vitro procedure. *J. British Grassl. Soc.* **19:** 363–372.
Watson DM and Dawson TJ 1993. The effects of age, sex, reproductive status and temporal factors on the time use of free-ranging red kangaroos (*Macropus rufus*) in western New South Wales. *Wildl. Res.* **20:** 785–801.
West GB, Brown JH and Enquist BJ 1997. A general model for the origin of allometric scaling laws in biology. *Science* **276:** 122–126.
Withers PC 1992. *Comparative Animal Physiology*. Saunders College Publishing, Sydney.

12 Mesial drift and mesial shift in the molars of four species of wallaby: the influence of chewing mechanics on tooth movement in a group of species with an unusual mode of jaw action

R.G. Lentle and I.D. Hume

SUMMARY

The mode of action of the lower macropodid jaw is unusual. On occlusion, each lower hemijaw rotates medially, causing the transversely oriented system of molar lophs on the lower molars to be drawn across those of the upper jaw, producing medial rotational shear in the plane of occlusion. Such occlusal forces cause molars to move within the bony substance of the jaw ('drift') in the bucco-lingual rather than the mesio-distal plane (i.e. sideways rather than forward and back). Thus the lower molars would be expected to exhibit lingual drift while those in the upper jaw would exhibit buccal drift. However, macropodid molars are known to move mesially with age in respect to a reference point on the viscerocranium (i.e. to exhibit molar progression). We discuss evidence that molar progression in four species of wallaby results from mesial shift (differential bony growth in the rear of the upper and lower molar arrays that causes their progressive forward displacement) rather than from mesial drift (movement of the molars within the bony substance of the jaw). In the absence of any stress-sensitive augmentation of bony growth in the rear of the molar array, mesial shift may cause molar progression to be more sensitive to nutritional factors that affect bony growth than to mechanical factors that promote occlusal pressure. It is also possible that mesial shift is more important in younger and smaller animals, and that mesial drift is more important in larger-bodied species.

MOLAR PROGRESSION

Anterior (mesial) movement of erupted molars in the jaw has been recorded in macropodid marsupials and a number of eutherians, such as crab-eating macaques (*Macaco fasicularis*) (Moss and Picton 1982), brown rats (*Rattus norvegicus*) (King *et al.* 1991) and humans (*Homo sapiens*) (Moss and Picton 1975), in which it is known as mesial drift. Hypotheses for the mechanisms of molar progression (Sanson 1989) are strikingly similar to those for mesial drift (Moss and Picton 1982). In both cases the forces that induce tooth movement are thought to be transmitted from tooth to tooth by transseptal fibres within the peridontal ligament (Sampson and Sims 1977). This produces forward drift of distal molars into gaps (Papandreas *et al.* 1993) such as the leeway space (Araya and Savara 1977) left by deciduous teeth.

Rates of mesial drift obtained with continuously applied displacement forces are known to approach the maximum turnover rate for alveolar bone. The maximal rates of absorption of bone matrix at the front and deposition of bone matrix at the rear of the direction of drift in rats are approximately 6.7 μm per day, or 2.5 mm per year (King *et al.* 1991). It is likely that these rates are similar among species as they result from common cellular physiological processes. Thus the reported absolute rate of mesial movement of first upper molars of tammar wallabies (*Macropus eugenii*) (2.2–2.7 mm/yr) (Lentle *et al.* 1998) is likely to lie at the upper limit of mesial drift, even though displacement forces were applied only intermittently during occlusion. This finding suggests that mesial molar movement (molar progression) in the tammar results in part from mechanisms other than those involving the direct application of displacement forces to teeth.

The overall length of the jaw and corresponding viscerocranium is known to increase through early life in a number of eutherian herbivores, such as fallow deer (*Dama dama*) (Frampton and Nugent 1992), red deer (*Cervis elaphus*) (Challies 1978), mustelids (He and Killaridis 2004) and rodents (Provencal 1993). A similar increase occurs in the macropodid jaw (Lentle *et al.* 1998). This is likely to be due to deposition of bone in the rear of the dentary, which we call 'mesial shift' (Lentle *et al.* 1998, 2003a; Fig. 12.1), and may be the main contributor to the high rates of mesial movement in the four species of wallaby studied (*M. eugenii*, *M. parma* (parma wallaby), *Wallabia bicolor* (swamp wallaby) and *Petrogale penicillata* (brush-tailed rock-wallaby). This continuing addition of bone to the rear of the dental array may reflect the general ontogenic order of ossification of the macropodid mandible from front to rear (Clark and Smith 1993).

The existence of such a process may be detected by its relationship to the growth of the anterior viscerocranium. Hence, if molar movement occurs by mesial shift alone then the absolute distance of the leading edge of the molar dentary will be tightly related to the growth of the anterior viscerocranium. On the other hand, if there is mesial drift, this relationship will be changed as forward progression will consistently exceed mesial shift from growth of the anterior viscerocranium. Thus, if two closely related species with differing dietary habits are compared we would expect to see differences between their rates of mesial drift because the forces on the teeth would differ in response to different chewing actions. Such a comparison was made by Lentle *et al.* (2003a) on Kawau Island, New Zealand, between introduced populations of tammar and parma wallabies that were known to live and feed exclusively in areas with different vegetation (Lentle *et al.* 2003d). Results showed that while there were significant differences in the rates of skull growth between the two species there were no significant differences in the relationship between the length of the viscerocranium and the horizontal distance between the front of the molar array and the temporomandibular joint. It is likely that the bulk of the forward movement of the molar array resulted from mesial shift rather than from mesial drift.

The fact that a number of browsing species of macropodid have low rates of molar progression compared with those of grazing species has been used to support the hypothesis that molar progression is due to mesial drift, and that mesial drift is blocked in browsers by the presence of large permanent sectorial premolars (Sanson 1989). However, when measured in relation to body mass, the absolute rates of molar progression in browsing species do not differ significantly from those of smaller-bodied grazing species with deciduous P4 premolars (Lentle *et al.* 2003b), a finding that fits more readily with the hypothesis of mesial shift.

The finding that mesial shift may contribute to molar progression is important from an ecological viewpoint. The rate of mesial drift is likely to be influenced by forces on the periodontal ligament, which may vary with the physical characteristics of the diet, such as chewing requirement (Sanson 1989). In contrast, the rate of mesial shift is related to the rate of bone growth in the rear of the dentary, which may be similarly stimulated by forces resulting from occlusive pressure, such as tension at the rear of the viscerocranium (Yamamoto 1996), and the anterior

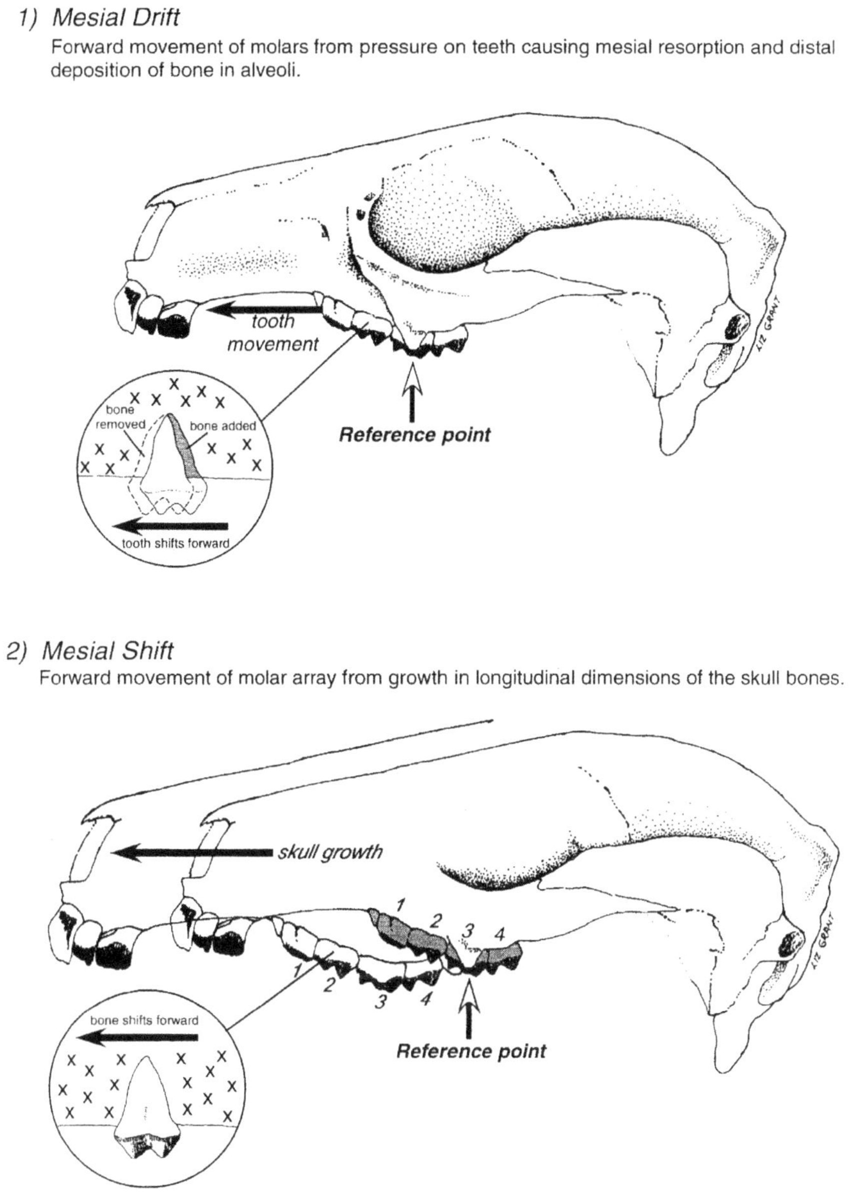

Figure 12.1: The two possible mechanisms for molar progression in macropodids.

limit of the junction of the ramus with the body of the mandible. Mesial shift may also be affected by nutritional status and by genetic differences, as these are known to influence the rate of jaw growth (Frampton and Nugent 1992).

PROBLEMS RELATING TO THE MACROPODID JAW ACTION MECHANISM

Our current understanding of the action mode of the macropod jaw raises further problems with the hypothesis that molar progression is largely due to mesial drift. This action, which involves independent medial rotation of the two halves of the lower jaw (Ride 1959; Lentle *et al.* 1998, 2003a), appears to be more marked than in other herbivore species with unfused lower anterior symphyses (Lieberman and Crompton 2000).

Ontogeny of the development and action of the macropodid jaw

The lack of fusion of the mandibular symphysis and accentuation of the rotatory action of the two lower hemijaws of macropods (Lentle *et al.* 2003a) may result from fundamental differences between macropodids and eutherian herbivores in the chronology of intrauterine development of the skull (Maunz and

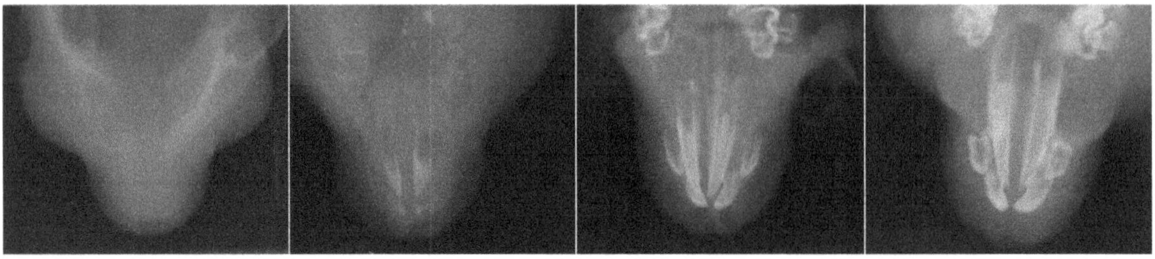

Figure 12.2: Developing dentition and associated structures in tammar wallaby (*M. eugenii*) pouch young of various ages. Radiographs showing the anterior viscero-cranium following dissection of the roof of the nasal cavity and lateral reflection of the flaps containing the nasal bones (lying lateral to the shafts of the procumbent incisors in the two older specimens).

German 1996; Nunn and Smith 1998). That is, the short gestation and long lactation of marsupials may lead to continual separation of the two halves of the lower jaw by the maternal nipple during the first 100 days of in-pouch development (Fig. 12.2).

Comparing the morphological development of the skulls of marsupials with those of eutherians indicates fundamental differences in the timing of the development of the viscero- and neuro-cranium (Maunz and German 1996; Nunn and Smith 1998). In marsupials, there is heterochronic delay in the timing of the development of elements of the central nervous system relative to many elements of the somatic and muscular systems. This may result from constraints on embryonic tissue development and energy resources during the very short period of intrauterine organogenesis, leading to postponement of the energetically expensive development of the central nervous system (Smith 1997). One outcome is that the bulk of the marsupial viscero-cranium commences ossification before the development of the telencephalon (Maier 1987; Clark and Smith 1993), whereas in eutherians the same bones commence ossification only after evagination of the telencephalon (Smith 1997).

Adequate differentiation of the viscero-cranium and associated structures is necessary for successful post-natal nutrition, i.e. the attachment and retention of the maternal nipple soon after birth (Lillegraven 1975). Ossification commences in the anterior of the viscero-cranium and mandible, and travels caudally with increasing age (Clark and Smith 1993). However, radiographs of young tammar wallabies taken from above (Fig. 12.2) show that calcification of the anterior hemijaws is sparse and occurs tardily, so that the principal elements of longitudinal jaw strength during early in-pouch development are the developing procumbent incisors. Moreover, these elements are separated medially by an area of radiolucency, indicating little if any bony union between the two hemijaws. This finding fits with the observation that the significantly wider gape of neonatal macropodids leads to deficiencies in obturation around the nipple and the secretion of a cornified cutaneous seal (Lillegraven 1974). Thus, it seems likely that the continuing presence of the maternal nipple in the anterior part of the buccal cavity contributes to the separation of the two hemijaws during their early development. However, it is noteworthy that many eutherian species also have flexible mandibular symphyses and exhibit mesial rotation during occlusion, without the continual presence of the maternal nipple in the anterior buccal cavity (Lieberman and Compton 2000).

The marsupial tongue and mandibular musculature are well developed in the neonate (Tyndale-Biscoe 2005), as is the palate (Clark and Smith 1993). Within the buccal cavity the maternal nipple lies between the concave groove in the dorsum of the tongue and the developing palate (Lillegraven 1975). The mandible moves little during suckling (Lillegraven 1975; German and Crompton 1998), the negative pressure for 'pump-sucking' (Lillegraven 1975) being generated by periodic contractions of the hyoid muscles which depress the floor of the buccal cavity (Griffiths and Slater 1988; Filan 1991; Tyndale-Biscoe 2005).

In some species, torsional forces around the relatively insubstantial portion of the body of the mandible (Clark and Smith 1993) generated during the prolonged period of suckling may cause a spiral distortion of the developing hemijaws bearing the lower

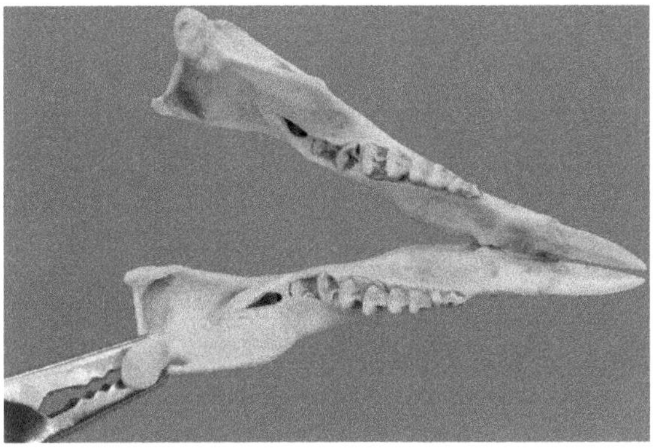

Figure 12.3: Air-dried lower jaw of a yearling tammar wallaby (*M. eugenii*). The anterior diaphysis is seen as a dark line between the anterior limits of the two hemijaws. Note the torsion in the line of the molar array with the distal-most, non-erupted molar lying behind the anterior limit of the mandibular ramus, as well as the characteristic inflections of the sites of attachment of the medial pterygoid and masseter.

molar arrays (Fig. 12.3), which subsequently aids their medial rotation during occlusion.

Movements of the lower jaw

The dP4 premolars and molars of macropodids are characterised by transverse clefts between the mesial and distal lophs (Sanson 1989). The initial occlusion of the lophs generates shear on the lophine crests (Sanson 1989). Subsequently there is significant interlocking of alternate lophs of the upper and lower molar arrays, although the presence of an elevated longitudinal link between the mesial and distal lophs of grazing species may prevent complete intercuspation (Sanson 1989). This interlocking precludes any subsequent mesio-distal movement between the upper and lower jaw but does not prevent rotation in the mesio-lingual plane. Lophine morphology restricts mesio-distal sliding of ingesta from molar to molar during occlusion, which limits the wear to the teeth that are effectively occluded.

It has been proposed that each lower hemijaw rotates around a longitudinally oriented axis of rotation immediately following intercuspation (Ride 1959; Lentle *et al.* 2003a). Thus, prior to intercuspation, there is eversion of the upper surface of each hemijaw as described by Leibermann and Crompton (2000) in the Virginia opossum (*Didelphis virginiana*). This results from the action of the medial pterygoid (Tomo *et al.* 2007). The pronounced lingual inflection of the attachment of this muscle on the medial surface of each hemijaw (Fig. 12.2) allows it to draw the ventral surface of the hemijaw jaw lingually (Leiberman and Crompton 2000), pulling the lower borders of each hemijaw towards each other and causing the upper borders to evert, i.e. move buccally. In macropods, this leads to the occlusion of the buccal edge of the procumbent incisor with the buccal edge of the conjoined concave facets of the occlusal surfaces of the upper incisors. At the same time, the lingual limits of the occluding surfaces of the lower molars occlude with the buccal limits of the occluding surfaces on the upper molars.

Immediately after intercuspation and the initial phase of lophine cutting (Sanson 1989), contraction of the outermost fibres of the masseter that are attached to the laterally extended portion of the rear of each hemijaw causes the upper surface to be rotated medially, as has been described in the opossum (Leibermann and Crompton 2000). In macropods, this action causes the buccal edge of the procumbent incisor to be drawn lingually across the concave surface of the conjoined groove on the occlusal surface of upper incisors at the same time as the lingual limit of the occlusal facet on the lower molars is drawn lingually across the surface of the opposing facet on the upper molars. Thus, in unworn molars of grazing macropod species, the prominent links between the anterior and posterior molar lophs (Sanson 1989) shear past the opposing occlusal surfaces on the corresponding upper lophs in the bucco-lingual plane. In older animals, the links are worn down and the bulk of the opposing surfaces on the corresponding lophs of upper and lower molars are worn to form opposing planes that are inclined laterally in the lower molars and medially in the upper molars. Thus, on occlusion, these opposing surfaces will cause some of the force applied to the lower molars to be vectored medially, assisting lingual rotation of the lower hemijaws (Figs 12.4, 12.5).

This mode of jaw action requires hypertrophy of the medial pterygoids for efficient initial eversion of the upper surfaces of the lower hemijaws. Conversely, the limited antero-posterior movement that occurs during the initial translation of the (limited) occlusal surfaces on the opposing anterior and posterior lophs (Sanson 1989) requires little development of the lateral pterygoid. These mechanical requirements are reflected in the greater development of the medial than the lateral pterygoid in marsupial species (Abbie 1939; Ride 1959; Lentle *et al.* 1998; Tomo *et al.* 2007).

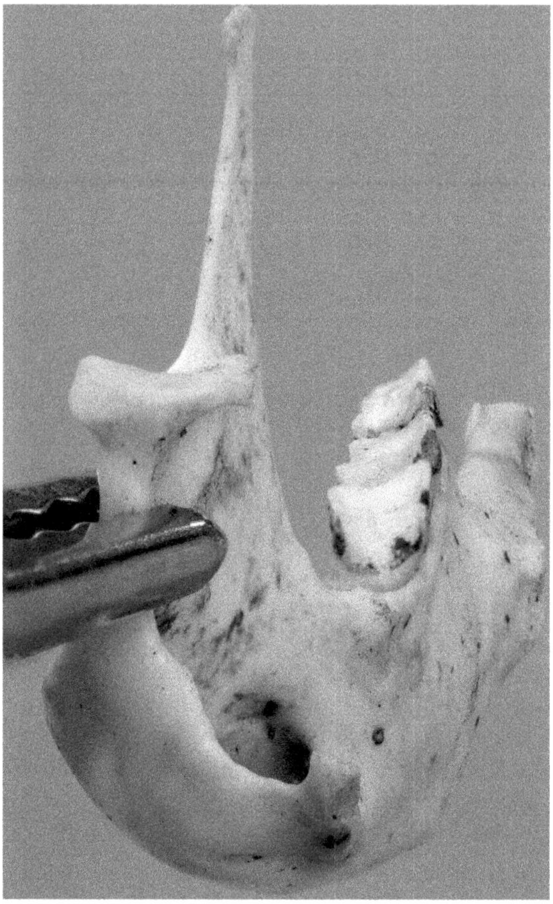

Figure 12.4: Hemijaw of a mature tammar wallaby (*M. eugenii*) viewed from the rear. The laterally inclined molar occlusal facets of the worn molars and the characteristic depression in the rear of the lateral leading edge of the occlusal facet on the procumbent incisor are clear. Note the flattened concave surface of the condyle, permitting inward rotation of the hemijaw at the temporomandibular joint.

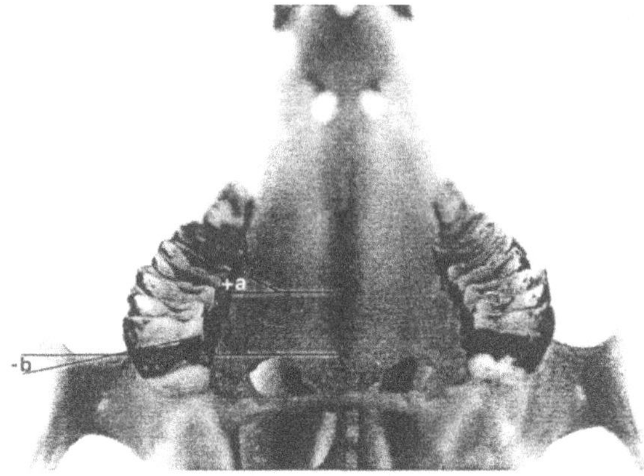

Figure 12.5: Viscero-cranium of a mature tammar wallaby (*M. eugenii*) showing upper molar arrays. The lophine structure is seen in the rearward molars. The occlusal surfaces of the molars are worn and the interlophine links are no longer visible. Note the angulation of the palatine bone of the viscero-cranuim with respect to the base of the neurocranium (in the foreground).

Synchronous occlusion of the ipsilateral incisor arcade and molar array during the medial rotation of a lower hemijaw requires a bony architecture that aligns both these surfaces with the longitudinal axis of rotation. The angle of the viscero-cranium with the neuro-cranium varies between macropod species (Milne and O'Higgins 2002), thus the vertical distance between the longitudinal axis of rotation of the lower hemijaw and the occlusal planes of the upper molar arrays also varies. Compensation for this variation in order to align the occlusal surfaces of the incisors may involve changes in both the length and angle of the mandibular ramus as well as the angle at which the procumbent incisors emerge from the lower jaw. The outcome of such complex geometry may require local and angulated increases in the vertical depth of the body of each hemijaw to align the occlusal surfaces of the upper and lower molar arrays. The extent of the latter may vary with species and with dietary habit (Sanson 1989).

EVIDENCE FOR ROTATIONAL MOVEMENT OF THE HEMIJAWS

Recent evidence supporting the hypothesis of rotational movement of the hemijaws following intercuspation comes from the disposition of microstriae and from X-ray cinematography.

Disposition of microstriae

The orientation of microstriae on dental enamel formed by the abrasive action of siliceous components of grasses can be used to determine the direction of translational movement between the occluding surfaces (Rensberger 1978).

In four species of macropod, including a browser, the conjoined occlusion facets of the upper incisors are bucco-lingually concave and bear transversely oriented striae (Lentle *et al.* 1998, 2003c). The surface of the dentine lingual to the labial enamel on the procumbent incisors is worn below that of the enamel (Fig. 12.4). These findings are commensurate with the labial surface of each procumbent incisor being rotated and drawn supero-lingually over that of the upper incisors during occlusion.

In tammar, parma and swamp wallabies, as well as brush-tailed rock-wallabies, the occlusal facets of the lophine and interlophine areas of the upper and

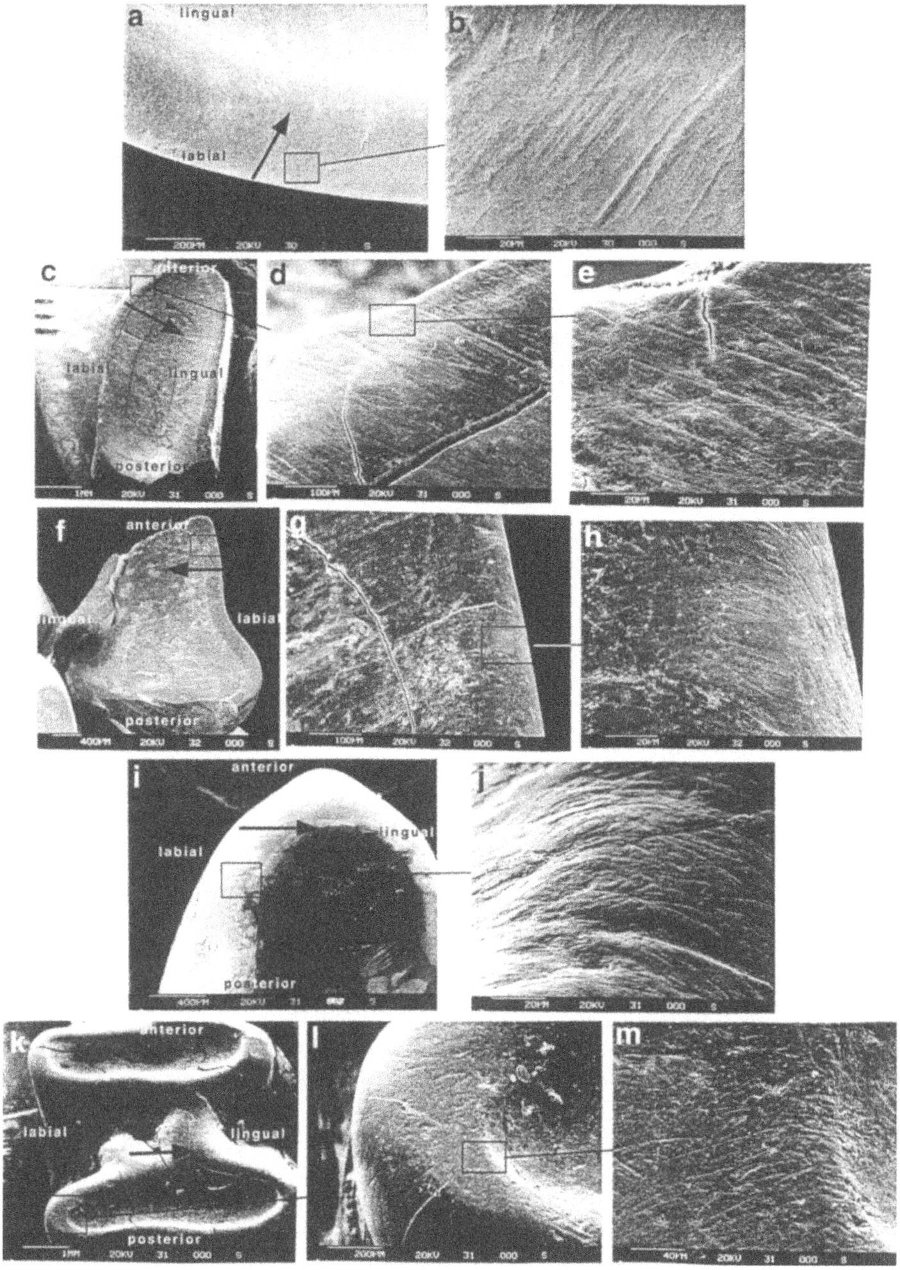

Figure 12.6: Selected scanning electron micrographs showing the orientation of macroscopic and microscopic striae on the dentition of the swamp wallaby (*W. bicolor*). (a, b) Labial edge of lower (procumbent I1 incisors. (c–e). Right upper I1 incisor. (f–h) Right upper I2 incisor. (i, j) Labial edge on mesial end of lower left P4 premolar. (k–m) Lower left M2 molar. All photographs are oriented identically. Arrows showing the direction of striae from labial to lingual are shown on the low-magnification views (a, e, f, i and k).

lower molars also bear transversely oriented striae (Fig. 12.6), indicating a similar mode of action.

X-ray cinematography

X-ray cinematography of the jaws of an adult female tammar wallaby trained to eat when confined in a box, confirmed that rotation of each of the two hemijaws around its longitudinal (mesio-distal) axis took place during occlusion (Lentle *et al.* 2003a). Following an initial phase of eversion, when the superior (cranial) margin of each of the two lower hemijaws was everted (rotated bucally), each lower hemijaw then rotated medially so that the superior margins of the two lower hemijaws were each rotated lingually (toward each other) while the lower surface was abducted, causing the tip of the lateral inflexion for the attachment of the masseter to be thrown into greater lateral relief (Fig. 12.7) (Lentle *et al.* 2003a).

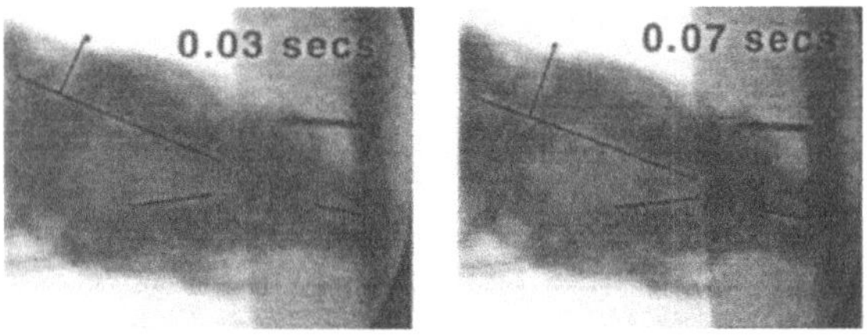

Figure 12.7: Selected frames from X-ray cinematography of a tammar wallaby (*M. eugenii*) chewing carrot, viewed from above. In the first photograph the left hemijaws are fully everted. In the second the hemijaws are commencing inversion, which causes their superior edges to approximate more closely and the outer limit of the lateral bony inflection (black circle) in the rear of the jaw to be thrown into greater relief.

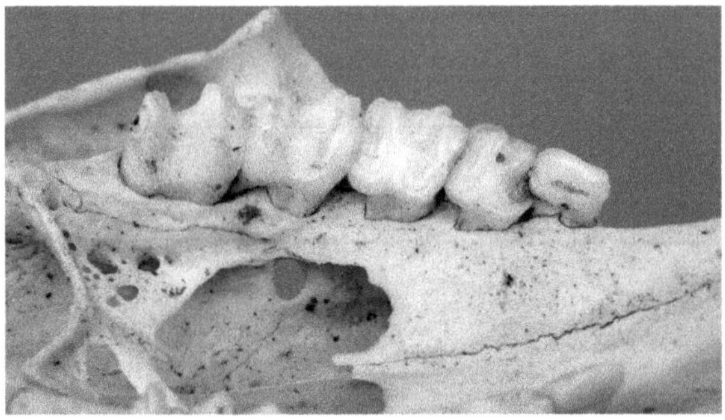

Figure 12.8: Skull of a parma wallaby (*M. parma*) showing buccal drift and tip of the anterior teeth in the molar array. This is best seen in P4. The lingual alveolar margin lies buccal to that of M1 and the adjacent surface of M1 is furrowed with increasing depth buccally. This is commensurate with P4 having undergone buccal drift with increasing buccal tip.

RELATING ROTATIONAL MOVEMENT TO MOLAR PROGRESSION

If rotation of the two hemijaws causes the occlusal surfaces of the lower molars to be drawn bucco-lingually across those of the upper jaw during occlusion, forces arising from such movement will be directed buccally in the upper molar array and lingually in the lower molar array. This will cause the molars to drift buccally in the upper jaw and lingually in the lower jaw. Relative lingual movement is difficult to discern in the lower molar array owing to twisting of the molar array across the upper surface of the body of the mandible. Relative buccal movement is more readily seen in the upper molar array by comparing the points of emergence of successive molars from the corresponding bone (Fig. 12.8). Apart from causing the bucco-lingual displacement of the tooth within the substance of the bone, the vectoring of forces arising from the angled occlusion facets can bring about a change in the angle of emergence of the tooth from the bone, that is it will induce progressive 'tipping' of the long axis of the tooth (Kaifu *et al.* 2003). Thus, the vertical axes of the upper molars will become increasingly angled outwards with advancing age. Figure 12.8 shows that the angle of emergence of P4 has been progressively tipped onto the mesial surface of the adjacent molar, causing it to be increasingly and more laterally furrowed as angulation increases. A combination of buccal or medial drift and tip is likely to account for the observation of teeth being 'squeezed out' from the front of the upper and lower molar array (Sanson 1989).

Clearly, if occlusal forces are to some extent vectored bucco-lingually, causing teeth to drift and tip in the manner shown in the specimen, mesial drift alone cannot account for molar progression. However, it is noteworthy that the wear progression from P4 across M1 is somewhat oblique, indicating that it results from a combination of considerable buccal and smaller mesial movements.

WHY SOME MACROPODID SPECIES DON'T EXHIBIT MOLAR PROGRESSION

It is evident that the rate of molar progression with age is greater in grazing than in browsing species, a phenomenon that has previously been attributed to differences in diet and propensity to tooth wear (Sanson 1989). Further, it has been proposed that persistence of the sectorial premolars in browsing species, such as the brush-tailed rock-wallaby, acts to arrest mesial drift and thus to slow the rate of molar progression. If, in the four species of wallaby discussed in this chapter, molar progression does not result principally from mesial drift but from mesial shift, this explanation cannot hold. It is unlikely that the presence of sectorial premolars would have any effect on bony deposition in the rear of the molar array. How, then, can the rate of molar progression differ among these species? Is there any unrelated variation between species in the rate of mesial shift? In fact, when molar index is regressed against body mass for a grazer (tammar wallaby), a grazer-browser (parma wallaby) and a browser-grazer species (brush-tailed rock-wallaby), the mean rates of increase in molar index are the same for all three species. The differences in the rate of molar progression turn out to be due to differences in tissue growth rates (Lentle *et al.* 2003b). The slower rate of molar progression of swamp wallabies results from slower ongoing growth and consequent slower ongoing bony apposition to secure mesial shift.

CONCLUSION

It is probable, given the evidence presented in this chapter, that, at least in the four species examined, molar progression results largely from differential bony growth in the rear of the upper and lower molar arrays (mesial shift) that causes their progressive forward displacement, rather than from movement of the molars within the bony substance of the jaw (mesial drift). However, the relative importance of mesial drift and mesial shift may differ according to the pattern of vectoring of occlusive forces and thus among species and with animal age. It is possible that molar shift (differential bony growth in the rear of the molar array) is more prominent in younger, faster-growing animals and more important in species with continuous growth curves, such as macropodid marsupials. On the other hand, molar drift (forward movement of molars though the jaw bones) might be more important in older, slower-growing animals and in species with arrested (or climax) growth curves, such as humans. In this respect it is noteworthy that molar progression may be relatively accelerated in large-bodied grazing macropodids compared to the group of species examined here. Future studies should thus include large-bodied macropodids, with particular attention to their relative growth rates as well as body mass and age.

REFERENCES

Abbie AA 1939. A masticatory adaptation peculiar to some diprotodont marsupials. *Proc. Zool. Soc. Lond.* **109:** 261–279.

Araya BS and Savara BS 1977. Analysis of leeway space. *J. Indian Orthod. Soc.* **9:** 1–10.

Challies CN 1978. An assessment of the physical well-being of red deer (*Cervus elaphus*) populations in South Westland, New Zealand. PhD thesis. University of Canterbury, Christchurch, New Zealand.

Clark CT and Smith KK 1993. Cranial osteogenesis in *Monodelphyis domestica* (Didelphidae) and *Macropus eugenii* (Macropodidae). *J. Morphol.* **215:** 119–149.

Filan SL 1991. Developments of the middle ear region in *Monodelphys domestica* (Marsupialia, Didelphidae): marsupial solutions to early birth. *J. Zool. Lond.* **225:** 577–588.

Frampton CM and Nugent G 1992. Age and sex-independent comparisons of morphological measurements. *Growth Dev. Ageing* **56:** 45–52.

German RZ and Crompton AW 1996. Ontogeny of suckling mechanisms in opossums (*Didelphis virginiana*). *Brain Behav. Evol.* **48:** 157–164.

Griffiths M and Slater E 1988. The significance of striated muscle in the mammary glands of marsupials. *J. Anat.* **156:** 141–156.

He TL and Killaridis S 2004. Craniofacial growth in the ferret (*Mustela putorius furo*): a cephalometric study. *Arch. Oral Biol.* **49:** 837–848.

Kaifu Y, Kasai K, Townsend GC and Richards LC 2003. Tooth wear and the 'design' of the human dentition: a perspective from evolutionary medicine. *Yearbook Physical Anthropol.* **46:** 47–61.

King GL, Keeling SD, McCoy EA and Ward TH 1991. Measuring dental drift and orthodontic movement in response to various initial forces in adult rats. *Am. J. Orthodont. Dentofacial Orthoped.* **99:** 456–465.

Lentle RG, Stafford KJ, Potter MA, Springett BP and Haslett S 1998. Incisor and molar wear in the tammar wallaby (*Macropus eugenii* Desmarest). *Aust. J. Zool.* **46:** 509–527.

Lentle RG, Hume ID, Stafford KJ, Kennedy M, Potter MA, Springett BP and Haslett S 2003a. Molar progression and tooth wear in tammar (*Macropus eugeni*) and parma (*Macropus parma*) wallabies. *Aust. J. Zool.* **51**: 137–151.

Lentle RG, Hume ID, Stafford KJ, Kennedy M, Potter MA, Springett BP and Haslett S 2003b. Comparison of indices of molar progression and dental function of brush-tailed rock-wallabies (*Petrogale penicillata*) with those of tammar (*Macropus eugenii*) and parma wallabies (*Macropus parma*). *Aust. J. Zool.* **51:** 259–269.

Lentle RG, Hume ID, Stafford KJ, Kennedy M, Potter MA, Springett BP and Haslett S 2003c. Comparison of tooth morphology and wear pattern of four species of wallabies. *Aust. J. Zool.* **51:** 61–79.

Lentle RG, Hume ID, Stafford KJ, Kennedy M, Potter MA, Springett BP and Haslett S (2003d). Differences in renal and alimentary water conservation capacity account for different distributions of tammar and parma wallabies on Kawau Island, New Zealand. *Aust. J. Zool.* **51:** 371–385.

Liebermann DE and Crompton AW 2000. Why fuse the mandibular symphysis? A comparative analysis. *Am. J. Anthropol.* **112:** 517–540.

Lillegraven JA 1975. Biological considerations of the marsupial placenatal dichotomy. *Evolution* **29:** 702–722.

Maier W 1987. The ontogenic development of the orbitotemporal region in the skull of *Monodelphis domestica* (Didelphidae, Marsupialia) and the problem of the mammalian alisphenoid. In *Morphogenesis of the Mammalian Skull.* (Eds HJ Zeller and U Kuhn) pp. 71–90. Verlag/Paul Pary, Hamburg.

Maunz M and German RZ 1996. Craniofacial hererochrony and sexual diamorphism in the short-tailed opossum (*Mondelphis domestica*). *J. Mammal.* **77:** 992–1005.

Milne N and O'Higgins P 2002. Inter-specific variation in *Macropus* crania: form, function and phylogeny. *J. Zool.* **256:** 523–535.

Moss JP and Picton DC 1975. The causes of migration of teeth. In *Transactions of the 3rd International Orthodontic Congress*. (Ed. JT Cook) pp. 356–343. Staples, London.

Moss JP and Picton DC 1982. Short-term changes in the mesio-distal position of teeth following removal of approximal contacts in the monkey *Macaca fasicularis*. *Arch. Oral Biol.* **27:** 273–278.

Nunn CL and Smith KK 1998. Statistical analyses of developmental sequences: the craniofacial region in marsupial and placental mammals. *Am. Nat.* **152:** 82–101.

Papandreas SG, Buschang PH, Alexander RG, Kennedy DB and Koyama I 1993. Physiologic drift of the mandibualr dentition following first premolar extractions. *Angle Orthodont.* **63:** 127–134.

Provensal MC 1993. Growth and determination of age in *Calomys musculinus* (Rodentia: Cricetidae). *Mammalia* **57:** 245–254.

Rensberger JM 1978. Scanning electron microscopy of wear and occlusal events in some small herbivores. In *Development Function and Evolution of Teeth*. (Eds PM Butler and KA Joysey) pp. 415–438. Academic Press, London.

Ride WDL 1959. Mastication and taxonomy in the macropodine skull. In *Function and Taxonomic Importance*. (Ed. AJ Cain) pp. 33–55. Publication No. 3, Systematics Association, London.

Sampson WJ and Simms MR 1977. Oxytalan fibre organisation in marsupial mandibular periodontal tissues. *J. Morphol.* **154:** 115–132.

Sanson GD 1980. The morphology and occlusion of the molariform cheek teeth in some Macropodinae (Marsupialia: Macropodidae). *Aust. J. Zool.* **28:** 341–356.

Sanson GD 1989. Morphological adaptations of teeth to diets and feeding in the Macropodidea. In *Kangaroos, Wallabies and Rat-kangaroos*. (Eds G Grigg, P Jarman and I Hume) pp. 151–168. Surrey Beatty and Sons, Sydney.

Smith KK 1997. Comparative patterns of craniofacial development in eutherian and metatherian mammals. *Evolution* **51:** 1663–1678.

Tomo S, Tomo I, Townsend GC and Hirata K 2007. Masticatory muscles of the great grey kangaroo (*Macropus giganteus*). *Anat. Rec.* **290:** 382–388.

Tyndale-Biscoe H 2005. *Life of Marsupials*. CSIRO Publishing, Melbourne.

Yamamoto S 1996. The effects of food consistency on maxillary growth in rats. *Eur. J. Orthod.* **18:** 601–615.

13 Functional pedal morphology of the extinct tree-kangaroo *Bohra* (Diprotodontia: Macropodidae)

N.M. Warburton and G.J. Prideaux

SUMMARY

Species in the extinct genus *Bohra* have been allied with living tree-kangaroos (*Dendrolagus*) on the basis of marked similarities in isolated cranio-dental and hind limb elements. Prompted by recent discoveries of the first near-complete *Bohra* skeletons in Pleistocene deposits in caves beneath the central Nullarbor Plain, south-central Australia, we compared the pedal morphology of tree-kangaroos with a range of terrestrial macropodines. Our objectives were to more clearly detail the key functional attributes of the *Dendrolagus* pes and to assess the likely arboreal adeptness of *Bohra*. Overall proportions of the calcaneus, talus, cuboid and metatarsals, as well as the morphology of their articular facets, suggest that the pes of *Bohra* was specialised for enhanced mobility and flexibility, and thus well adapted to the functional demands of an arboreal environment. The presence of these Pleistocene species on the Nullarbor Plain indicates that tree-kangaroos had far broader geographic and climatic ranges than hitherto anticipated, and also demonstrates that the 'treeless' plain was not always treeless.

INTRODUCTION

Macropodoids (kangaroos and their kin) descended from an arboreal, possum-like ancestor in the Palaeogene period, possibly during the Eocene epoch (Burk *et al.* 1998). Following a postulated origin in the middle Miocene (16–11 million years ago), members of the family Macropodidae (kangaroos and wallabies) radiated into a broad array of terrestrial herbivore niches. The hind limb morphology of macropodines is among the most specialised of all mammals, reflecting their unique method of saltatorial locomotion (Alexander and Vernon 1975; Flannery 1982). The pes (hind foot) is characterised by loss of the hallux, marked reduction of the syndactylous metatarsals II–III, and enlargement and elongation of metatarsals IV–V. Articulations between tarsal elements have been modified to enhance the hinge action at the ankle while restricting rotational movements (Szalay 1994). Given this evolutionary trajectory and the functional–adaptive constraints posed by the terrestrial macropodine bauplan, it is remarkable that one lineage (tree-kangaroos) returned to the arboreal realm during the late Miocene (11–5 million years ago), dispersing into New Guinea before its isolation from the Australian mainland (Hall 2001).

Compared with terrestrial macropodines, extant tree-kangaroos (species of *Dendrolagus*) have a shorter broader pes characterised by stout tarsal and metatarsal bones (Fig. 13.1). Articular surfaces have been modified for increased flexibility, allowing greater balance and mobility in an arboreal environment (Flannery 1982; Flannery and Szalay 1982). Of

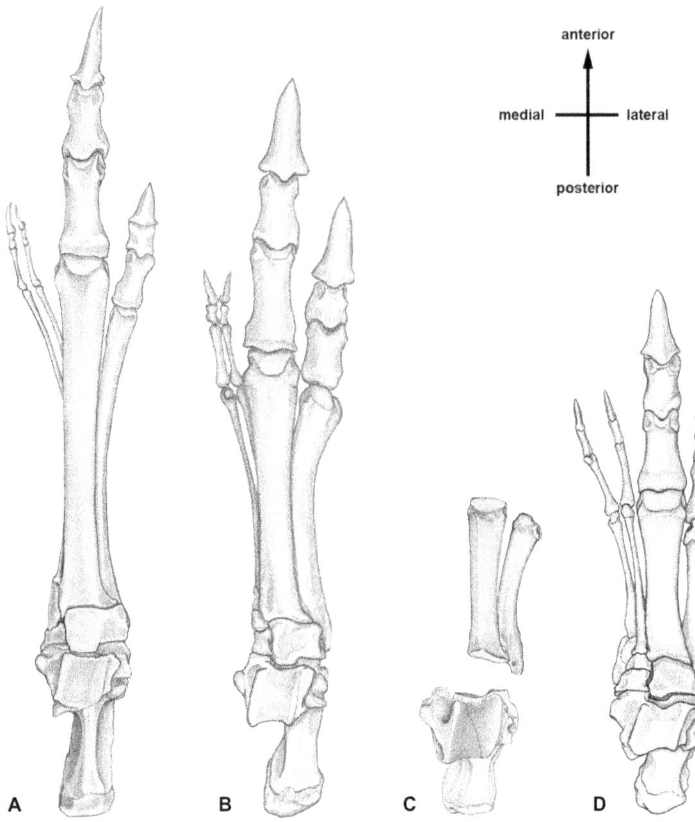

Figure 13.1: Right pes of four macropodines in dorsal view. (a) *M. fuliginosus*. (b) *Thylogale billardierii*. (c) *Bohra nullarbora* (Western Australian Museum 05.4.70, reversed, incomplete). (d) *D. bennettianus*. Scaled to same talus width.

the 10 extant species of *Dendrolagus*, two inhabit tropical forest in north-east Queensland, while the other eight occupy generally limited distributions in different parts of New Guinea (Flannery *et al.* 1996). *Dendrolagus* has been divided into a long-footed group, comprising both Australian species (*D. bennettianus*, *D. lumholtzi*) plus *D. inustus*, and a short-footed group including the remaining New Guinea species (Groves 1982). The long-footed group is apparently the more plesiomorphic (Groves 1982; Flannery *et al.* 1996).

Bohra paulae was raised by Flannery and Szalay (1982) to describe a large fossil tree-kangaroo represented by a few isolated hind limb elements from Wellington Caves in central eastern New South Wales. The fossils were collected in the late 19th century, but retained no detailed provenance data. Their age is thus uncertain; deposits in these caves range from early Pliocene to late Pleistocene (Dawson and Augee 1997; Dawson *et al.* 1999). Dawson (2004) described a second species, *B. wilkinsonorum*, on the basis of a partial maxilla from the late Pliocene Chinchilla Local Fauna (LF) of south-east Queensland, while Hocknull (2005) identified a calcaneus attributable to the genus from both the Chinchilla LF and the middle Pleistocene Mount Etna cave fills. Two new Pleistocene species of *Bohra* have recently been identified on the basis of more complete specimens, including cranial and post-cranial elements, collected from caves beneath the Nullarbor Plain of south-central Australia (Prideaux *et al.* 2007; Prideaux and Warburton 2008, 2009).

Flannery and Szalay (1982) identified several features in common between *Bohra* and *Dendrolagus*, including a relatively smoothed calcaneal–cuboid articulation, the transverse orientation of the talar head, the transversely expanded tuber calcanei and the reduced length of the fibular crest of the tibia. This suggested at least a partially arboreal lifestyle for *B. paulae* and a phylogenetic link with *Dendrolagus*, views confirmed by the analyses of *B. wilkinsonorum* (Dawson 2004) and the new Nullarbor species, *B. illuminata* and *B. nullarbora*.

In this chapter we present a detailed functional analysis of the pedal elements of *Bohra*, prompted by the well-preserved and more complete Nullarbor

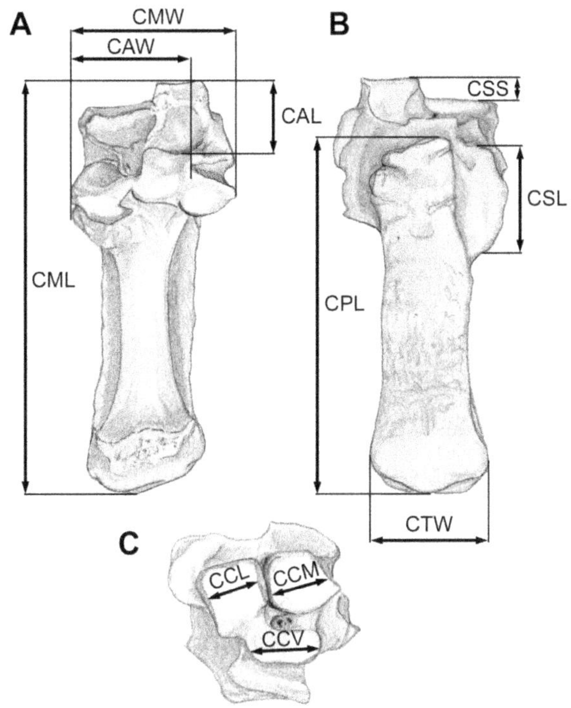

Figure 13.2: Right calcaneus of *M. fuliginosus*, illustrating features that reflect functional attributes and dimensions measured. (a) Dorsal view. (b) Plantar view. (c) Anterior view.

specimens. Our aims are to establish what the morphology tells us about its arboreal aptitude relative to *Dendrolagus*. We expect this to provide part of the foundation required for the development of an empirically grounded hypothesis for the evolutionary transition from the terrestrial to the arboreal realm. We identify the unique pedal features of *Dendrolagus* and *Bohra* compared with terrestrial macropodoids. We then employ multivariate analyses to establish whether these differences can be quantified morphometrically, and consider the functional significance of different pedal morphologies.

MATERIALS AND METHODS

Taxonomic comparisons

Specimens of an ecomorphologically diverse range of extant macropodines and potoroines (21 species) were examined and measured (Table 13.1), along with those of three species of *Bohra* (Table 13.2). Our analysis focused on the three major elements of the tarsus (calcaneus, talus [=astragalus] and cuboid) but also included observations on the tibia and metatarsals. Taxonomic comparisons focused on the relative size and shape of musculo-tendon attachment sites and articular surfaces. As far as possible, descriptive terminology follows the Nomina Anatomica Veterinaria (2005). For annotated illustrations of morphological features referred to, see Prideaux and Warburton (2008).

Linear measurements (recorded to 0.1 mm) were made using fine-point callipers for 36 tarsal dimensions (Table 13.3; Figs 13.2–13.4). Measurements included those used in previous studies of the tarsus and those that incorporate features with proposed functional significance (Bishop 1997; Carrano 1997; Argot 2002; Sargis 2002; De Gusta and Vrba 2003; Salton and Szalay 2004). Due to the unequal number of individuals measured for each species (Table 13.1), we analysed mean measurements. Some sexual dimorphism in body size is apparent but because of the small sample sizes available we combined female and male specimens within groups. Because shape changes with ontogeny, measurements were restricted to subadult and adult specimens (denoted by fused epiphyses). Raw numerical data is available from NMW.

The three species of *Bohra* represented by tarsal elements were examined: *B. paulae* (AM F62099, AM F62100), *B. illuminata* (WAM 03.5.10) and *B. nullarbora.* (WAM 05.4.70). The complete set of tarsal

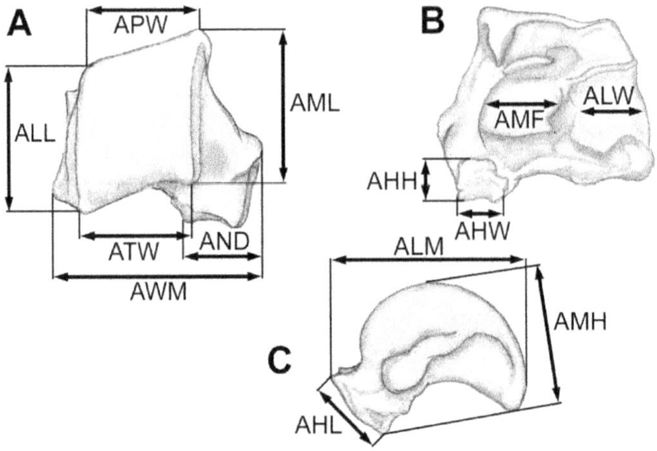

Figure 13.3: Right talus of *M. fuliginosus*, illustrating features that reflect functional attributes and dimensions measured. (a) Dorsal view. (b) Ventral view. (c) Medial view.

Table 13.1: Species used to investigate pedal morphology and their preferred terrain.

Species	Preferred habitat	Specimens examined	n
Aepyprymnus rufescens	Mixed	WAM M26119	1
Bettongia lesueur	Mixed	WAM M3644, M4392, M15488, M19232	3
Bettongia penicillata	Woodlands	WAM M13405, M14961, M19064, M19521, M21028	5
Potorous tridactylus	Thickets	WAM M9867	1
Dendrolagus bennettianus	Arboreal	NMV C7125, C7123; SAM M5530	3
Dendrolagus dorianus	Arboreal	AM M9109	1
Dendrolagus inustus	Arboreal	AM M25390	1
Dendrolagus lumholtzi	Arboreal	QM M9333, JM13924, JM3860, JM6667; AMNH M65257, M65258	6
Dendrolagus matschiei	Arboreal	WAM M21013, M2252	2
Lagorchestes conspicillatus	Grasslands	WAM M7032, M19854, M19855, M6733	4
Lagorchestes hirsutus	Grasslands	WAM M7965, M7968	2
Lagostrophus fasciatus	Mixed	WAM M3625, M3626; M4393, M5658, M6303, M8006, M16285	5
Macropus eugenii	Thickets	WAM M6573, M6737, M7654, M8357, M7093, M4162, M16726	6
Macropus fuliginosus	Woodlands	WAM M6750, M3328, M13015, M13451, M13586	5
Macropus irma	Woodlands	WAM M19864, M19581, M19869, M13027	4
Macropus robustus	Woodlands	WAM M14955, M15848, M19223, M19554, M34125	5
Macropus rufus	Grasslands	WAM M4953, M6749, M19099	1
Onychogalea unguifera	Grasslands	WAM M6942, M11622, M13969, M17142, M17147	3
Petrogale brachyotis	Rocky outcrops	WAM M19066, M19540, M19865	3
Setonix brachyurus	Thickets	WAM M6792, M6795, M13404, M19866, M13936, M51202	6
Thylogale billardierii	Thickets	Prideaux private collection	2

Source: Strahan (1995).

n = number of specimens used for principal components and discriminant analyses.

AM M = Mammal Section, Australian Museum, Sydney; AMNH M = Dept of Mammalogy, American Museum of Natural History, New York; QM M and QM JM = Mammal Section, Queensland Museum, Brisbane; SAM M = Dept of Mammals, South Australian Museum, Adelaide; NMV C = Mammal Section, Museum Victoria, Melbourne (formerly National Museum of Victoria); WAM M = Mammal Collection, Dept of Terrestrial Vertebrates, Western Australian Museum, Perth.

Table 13.2: Observed range of calcaneal and metatarsal ratios (mean and sample size in parentheses).

Ratio	*Dendrolagus*	Terrestrial species	*Bohra*
CMW:CML	0.62–0.88 (0.69; 13)	0.37–0.59 (0.50; 70)	0.61–0.73 (0.66; 4)
CAW:CML	0.45–0.57 (0.54; 13)	0.33–0.47 (0.40; 70)	0.46–0.66 (0.53; 4)
CTW:CML	0.33–0.48 (0.41; 13)	0.29–0.31 (0.30; 70)	0.38–0.45 (0.43; 4)
CTM:CML	0.29–0.36 (0.32; 13)	0.17–0.20 (0.18; 70)	0.33–0.34 (0.33; 4)
CPL:CML	0.60–0.79 (0.71; 13)	0.54–0.78 (0.63; 70)	0.69–0.80 (0.74; 4)
CCM:CML	0.18–0.27 (0.21; 13)	0.17–0.21 (0.18; 70)	0.21–0.29 (0.25; 4)
CCS:CML	0.05–0.09 (0.06; 13)	0.08–0.17(0.11; 70)	0.04–0.11 (0.07; 4)
Length metatarsal IV:metatarsal V	1.03–1.10 (1.07; 9)	1.07–1.18 (1.12; 15)	1.11 (1)
Metatarsal IV distal epiphysis width:length	0.22–0.27 (0.25; 9)	0.11–0.21 (0.15; 16)	0.22 (2)

measurements were only available for *B. illuminata*. A second analysis using a reduced dataset including only calcaneal measurements was undertaken to investigate the morphometric affinities of all three species.

Data analysis

Preliminary univariate analysis was used to determine basic descriptors: mean, standard deviation and coefficients of variation. Scatter plots demonstrated a strong positive regression between means and their standard deviations, and data were subsequently $\log_{10}$-transformed to accommodate this size-dependent variance. All measurements were divided by calcaneal length (CML) to standardise for size, thus, data are indices of relative size. Nine indices proved useful in discriminating taxa (Table 13.2).

Principal components analysis (PCA) was conducted using 36 morphometric variables (Table 13.3) incorporating all extant taxa and *B. illuminata*. Additional analyses on reduced datasets investigated variation within macropodines alone, and calcaneal measurements only. The first three principal components were extracted in each case as determined by their respective eigenvalues, by the percentage of variation explained in each case and by scree plot interpretation.

Discriminant grouping analysis was used to determine whether grouping by habitat preference yielded obvious trends in morphological data, and to reveal the likely affinities of the species of *Bohra*. Habitat preference was classified according to Strahan (1995). The dataset was reduced to a set of six uncorrelated (as determined by PCA) variables to meet the assumptions of the analysis. These variables were CAW, CTM, CPL, CCH, AWM and QTL (Table 13.3). Set tolerance limits were 0.005. Analyses were conducted in Microsoft Excel using statistiXL (http://www.statistixl.com/).

RESULTS

Pedal morphology

Calcaneus

Species of *Dendrolagus* are united by the following calcaneal attributes compared to terrestrial macropodines: a short, very broad calcaneus; medial

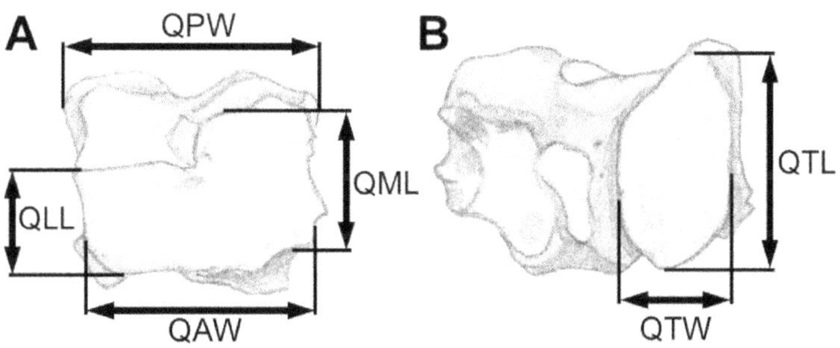

Figure 13.4: Right cuboid of *M. fuliginosus*, illustrating features that reflect functional attributes and dimensions measured. (a) Dorsal view. (b) Plantar view.

Table 13.3: Abbreviations and definitions.

Abbreviation	Definition
AHH	Anteroposterior height of talar head measured at 90° to mediolateral length
AHL	Dorsoventral length of talar head measured on medial aspect from anterior margin to posterior margin of navicular facet
AHW	Maximum mediolateral width of talar head
ALH	Maximum height of lateral trochlear crest of talus from basal plane to crest apex
ALL	Maximum length of lateral trochlear crest of talus in sagittal plane
ALM	Maximum medial anteroposterior length measured from posterior plantar tubercle to anterior margin of talar head
ALW	Width of lateral facet measured parallel to coronal plane of talus
AMF	Width of medial facet measured parallel to coronal plane of talus
AMH	Maximum height of medial trochlear crest of talus from basal plane to crest apex
AMM	Minimum median height of talus from basal plane to lowest point on trochlea
AML	Maximum length of medial trochlear crest of talus in sagittal plane
APW	Distance between posterior medial and lateral margins of the trochlear facet taken parallel to the coronal talar plane
ATW	Maximum distance from anterior medial and lateral margins of trochlear facets measured parallel to coronal plane of talus
AWM	Maximum talar width in coronal plane from lateral tubercle to medial margin
CAL	Distance between anteriormost margin of dorsal calcaneal surface measured from level with anterior facet for talus to anterior margin of posterolateral facet
CAW	Maximum linear distance on calcaneus between medial and lateral margins of posterior talar articular facets measured along long axis of articulation
CCH	Height of cuboid articulation on calcaneus measured from dorsal to plantar margin
CCM	Maximum width of dorsomedial facet of cuboid articulation on calcaneus
CCL	Maximum width of dorsolateral facet of cuboid articulation on calcaneus
CCV	Maximum width of ventromedian facet of cuboid articulation on calcaneus
CCS	Projected height of cuboid step on calcaneus measured in sagittal plane
CML	Maximum linear length of calcaneus in sagittal plane measured from posteriormost tip of tuber calcanei epiphysis to anteriormost point of cuboid articular facet
CMW	Maximum linear calcaneal width in transverse plane
CPL	Maximum length of rugose plantar surface of calcaneus measured parallel to sagittal plane
CSL	Length of sustentaculum tali of calcaneus
CTD	Minimum dorsoventral height of tuber calcanei measured at 90° to sagittal plane (perpendicular to CTM)
CTH	Maximum dorsoventral height of tuber calcanei measured at 90° to sagittal plane (perpendicular to CTW)
CTM	Minimum mediolateral width of tuber calcanei
CTW	Maximum mediolateral width of tuber calcanei measured as maximum width of epiphysis
QAW	Maximum width of dorsal anterior aspect of cuboid
QLL	Distance on cuboid from most dorsal lateral point of metatarsal facet to most dorsal lateral point of calcaneal facet
QML	Distance on cuboid from most dorsal medial point of metatarsal facet to most dorsal medial point of calcaneal facet
QPH	Maximum height of posterior aspect of cuboid corresponding to dorsoventral plane of calcaneal facet, perpendicular to QPW
QPW	Maximum width of posterior aspect of cuboid corresponding to transverse plane of calcaneal facet
QTL	Maximum length of posterior plantar process of cuboid measured parallel to sagittal plane
QTW	Maximum width of posterior plantar process of cuboid measured parallel to transverse plane

expansion and dorsoventral compression of the tuber calcanei; a short, broad plantar surface; increased size of the sulcus anterior to the rugose area on the plantar surface (Fig. 13.5). The talocalcaneal articulation is the main component lower ankle joint of macropods. It comprises an oval-shaped, concave posteromedial facet, a convex posterolateral facet and usually a small anterior facet. In *Dendrolagus*, this articulation is markedly broadened and equivalent to at least half the maximum length of the calcaneus. In addition, the lateral and medial facets are confluent and disproportionate in size, with the lateral facet larger (Fig. 13.5b). The lateral facet is medially constricted and laterally expanded producing a conical rather than cylindrical convexity when viewed dorsally. The medial facet is flatter, more obliquely aligned, and less pronounced posteriorly than in terrestrial species. In all species of *Dendrolagus* except *D. inustus*, the anterior talocalcaneal facet is lost due to the more medial placement of talar head. *D. inustus* has a large anterior facet; dorsomedial orientation, separated a short distance from the postero-medial facet.

Expansion (transversely) of the sustentaculum tali in *Dendrolagus* is correlated with the medial extension of the medial talar facet, which results in enlargement of the entire anteromedial portion of the calcaneus (Fig. 13.5d). Ventrally, the sustentaculum tali is convex, with a very shallow flexor groove.

The anterior end of the calcaneus is dominated by the cuboid articular facets. The calcaneal–cuboid articulation is the major component of the transverse tarsal joint. In macropods, this articulation is divided into two major facets separated by a 'step', which results from the medial facet extending much more anteriorly than the lateral facet. A third, ventromedian facet is variably developed across the family. In *Dendrolagus*, the calcaneal–cuboid articulation is relatively expanded and flattened, the cuboid step is smoothed and curved in profile and the ventromedian facet is indistinct, i.e. completely merged with dorsolateral facet (Fig. 13.5f).

The calcaneus of *Bohra* is very similar in overall appearance to that of *Dendrolagus*, in particular *D. inustus* and the Australian species, *D. bennettianus* and *D. lumholtzi*. It is stout and very broad, with enlarged articular facets and a robust, medially expanded tuber calcanei (Fig. 13.5). The anteromedial aspect, including the medial talar facet, is very broad and centrally expanded, as in *Dendrolagus*.

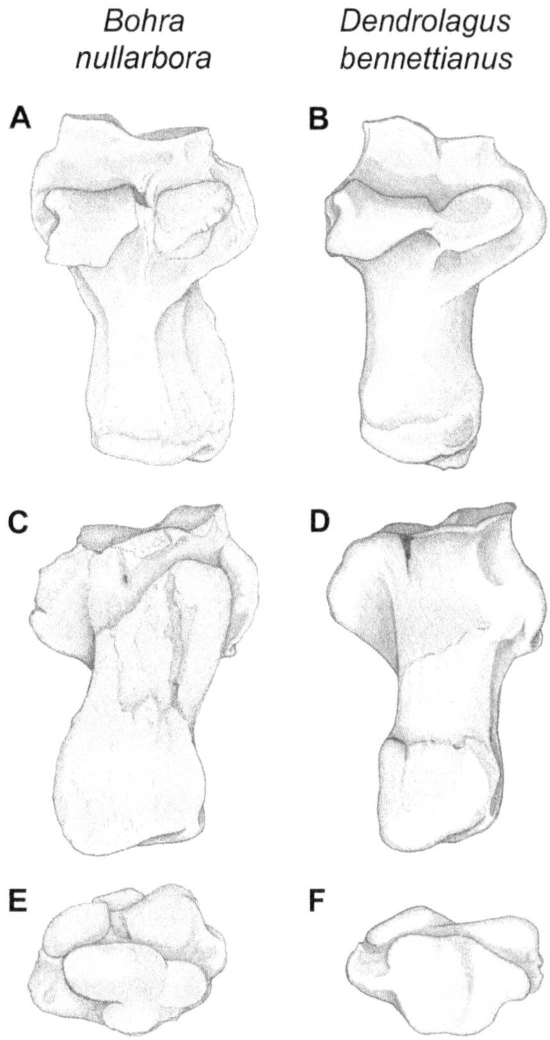

Figure 13.5: Left calcanea of *B. nullarbora* and *D. bennettianus*. (a–b) Dorsal view. (c–d) Plantar view. (e–f) Anterior view. a–d scaled to same length, e–f scaled to same width.

Overall, the calcaneal–talar articulation of *Bohra* most closely resembles that of *D. inustus*, with a large medial facet and a strongly convex, conical lateral facet (Fig. 13.5). The articular surface is continuous in *B. paulae* and *B. nullarbora*, but there is a partial separation in *B. illuminata*. In this species, the lateral facet is relatively narrow and less tapered mesially. The dorsal anteromedial facet, for articulation with the talar head, is reduced but not lost entirely. The cuboid step of *Bohra* is smoothed, as in *Dendrolagus*, and the ventromedian facet is broad and partially integrated with the dorsal facets (Fig. 13.5e). The medial surface of the lateral facet is medially concave, i.e. the boundary between the medial and lateral facets is arcuate.

Talus

The upper ankle joint is fundamentally a hinge that joins the pes to the crus (tibia and fibula). The main component of the joint is the talotibial articulation. The talus of *Dendrolagus* is much broader relative to its length than it is in terrestrial kangaroos and wallabies (Fig. 13.1). The increase in overall width is primarily due to the expansion of the talar head, neck and medial malleolus and, to a lesser extent, the expanded width of the articular trochlea. The talar head is ovoid and wider than it is long (although the orientation of the long axis varies among the species; oblique in *D. inustus* and *D. bennettianus*, and roughly transverse in other species). In terrestrial macropodines, the articular surface for the navicular is dorsoventrally long and narrow and rectangular in shape (Fig. 13.1). The trochlear crests are low and more gently sloped laterally in *Dendrolagus* than in terrestrial macropodines, where the crests are high with near-vertical sides. The lateral crest is also higher than the medial crest. The ventral articular facets for the calcaneus, in particular the ovoid medial facet, is lengthened in *Dendrolagus*, and the two facets are confluent, as in the corresponding calcaneal features. The talar neck is very short and broad. The talotibial pit on the dorsal surface of the talar neck is broad and shallow, in contrast to the deep morphology observed in terrestrial macropodines.

The talus of *Bohra* (Fig. 13.6a–d) is intermediate in general proportion between those of terrestrial macropodines and *Dendrolagus*, i.e. relatively wider than that of terrestrial macropods and relatively longer than that of *Dendrolagus* (Fig. 13.1). As in *Dendrolagus*, increased width of the talus in *Bohra* has been primarily achieved by mesial expansion of the medial malleolus and the talar head (Fig. 13.6b). The head is obliquely orientated, as in *D. inustus* and *D. bennettianus* (Fig. 13.6d). Unlike *Dendrolagus*, however, the head is relatively anteriorly placed due to the longer neck region and is reminiscent of the morphology observed in terrestrial macropodines. The trochlear crests are long, and the trochlear fossa is relatively narrow relative to *Dendrolagus* (Fig. 13.1). The height of the crests is intermediate between *Dendrolagus* and terrestrial species, and the lateral crest is more flattened than the medial (Fig. 13.6d). The flattened lateral crest is similar to *Dendrolagus* while the higher medial crest is a feature retained from terrestrial species. The ventral facets for articulation with the calcaneus are greatly expanded (Fig. 13.6c), and show a tendency toward the confluence seen in *Dendrolagus*.

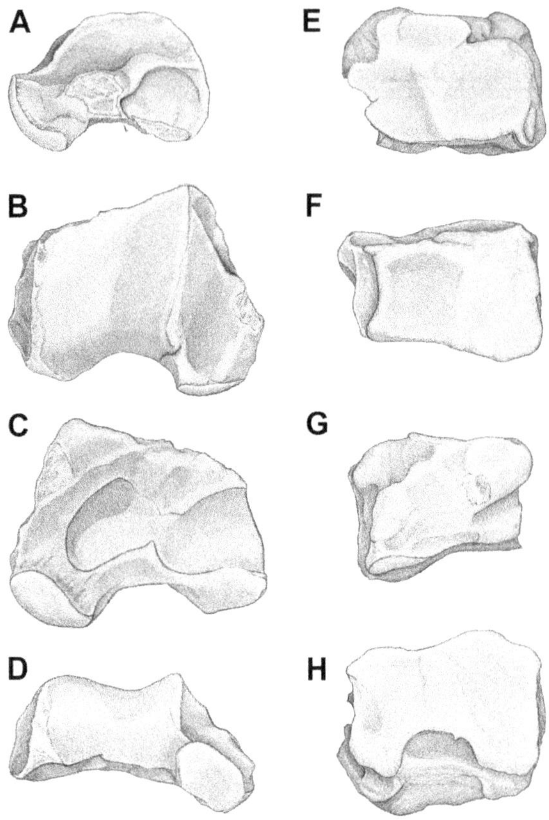

Figure 13.6: Right tarsal bones of *B. illuminata*. (a) Talus in medial view. (b) Talus in dorsal view. (c) Talus in ventral view. (d) Talus in anterior view. (e) Cuboid in posterior view. (f) Cuboid in dorsal view. (g) Cuboid in ventral view. (h) Cuboid in anterior view.

Cuboid

In comparison with the very cubic morphology of terrestrial macropodines (see Fig. 13.4, *M. fuliginosus*), the cuboid of *Dendrolagus* is distinctly compressed dorsoventrally and anteroposteriorly (Fig. 13.1d). This is emphasised in the short-footed species group. The step on the posterior surface of the cuboid is smoothly contoured, reflecting the morphology of the calcaneus. The lateral plantar process is much more elongate and obliquely oriented in *Dendrolagus* than in the terrestrial taxa. It forms a broad shelf across the posterolateral quadrant of the plantar surface, and is bounded by the oblique peroneal sulcus anteriorly and by the deep flexor sulcus medially. The medial plantar process is flattened in *D. lumholtzi* and *D. bennettianus*, and is barely distinguishable in *D. inustus*, *D. dorianus* and *D. matschiei*.

The cuboid of *Bohra* (Fig. 13.6e–h) is intermediate in morphology between those of *Dendrolagus* and terrestrial macropodines. It retains a primarily cubic appearance, but the lateral plantar process is laterally expanded (Fig. 13.6g) and its anterior border, together with the peroneal sulcus, is oblique in orientation. In addition, the lateral portion of the cuboid is slightly compressed when viewed dorsally (Fig. 13.6f), and the step between the medial and lateral calcaneal facets is smoothed.

Metatarsals

The metatarsals are short and robust in all species of *Dendrolagus* relative to those of terrestrial macropodines (Fig. 13.1). They are relatively longest in *D. inustus* (mean metatarsal IV length to proximal width: 3.9) and relatively shortest in *D. matschiei* (length to width: 2.8). Typically, metatarsal IV in *Macropus* is six to seven times longer than its proximal width. This ratio varies in other terrestrial macropodine genera from approximately four (*Setonix, Petrogale*) to eight times (*Lagorchestes*). Metatarsals IV and V of *Dendrolagus* also differ by being subequal in length and more ventrally convex in the sagittal plane, and by having relatively much broader, deeper distal epiphyses and expanded articular facets for the interosseous articulation (Table 13.2, Fig. 13.1). The proximolateral process of metatarsal V is also expanded. The head is asymmetrical. Among terrestrial macropodines, the interosseous facets of metatarsals IV and V are elongate and variably constricted mesially. In *Setonix* and some species of *Macropus* this mesial constriction is complete, but in other taxa (e.g. *Petrogale, Thylogale*) the two portions are contiguous. In *Dendrolagus*, the facet for metatarsal V is relatively large and continuous.

Metatarsals IV–V in *Bohra* are large and relatively broad in cross-section in comparison to *Macropus* (Fig. 13.1). The articulation between them resembles that of *Dendrolagus* in general shape and the confluence of the two facets. The shaft length of metatarsal IV is intermediate between those of *Dendrolagus* and *Macropus*, and most similar to those of *Setonix* and *Petrogale*. The *Bohra* metatarsal V is particularly robust compared with that of both *Macropus* and *Petrogale*. Metatarsal IV differs from that of terrestrial macropodines in the greater expansion of the cuboid articulation, the plantar tuberosity medially and the articulation for metatarsal V laterally. Similarly, the proximolateral expansion of metatarsal V and the expansion of the articular facets closely resemble the conditions expressed in *Dendrolagus*.

STATISTICAL ANALYSES

Principal components analysis

Analysis of the dataset comprising calcaneal, talar and cuboid indices of *B. illuminata* and 21 extant macropodines, including five species of *Dendrolagus* (Fig. 13.7), demonstrated strong separation of the tree-kangaroos from the terrestrial species in the first principal component (PC1). This accounts for 57.9% of the variance. Nearly all component loadings for PC1 are positive but not all the positive loadings are close to unity, and PC1 is not correlated with size (CML; $R^2 = 0.28$). Even after the removal of the tree-kangaroos and potoroines, no clear relationship between this axis and body size is evident. The components that contribute to the strong positive placement of *Dendrolagus* and *Bohra* on this axis, in decreasing order of significance, include AWM, CAW, ATW, ALH, CMW, QPW, APW, AMT, AMF, AHW, ALL, QML, ALM and CTW. The most significant of these measures relate to the relative widths of the tarsal elements, especially those of the articular surfaces, as well as the expansion of the lateral aspect of the trochlear articulation of the talus. The strong negative placement of the large species of *Macropus* is attributed to, in decreasing order of significance, QTL, CPL, QPH and CSL. These correspond to the relatively large size of the plantar features of the cuboid and calcaneus, and the length of the sustentaculum tali.

The second principal component (PC2) accounts for 10.7% of the variation and separates the potoroines and *Lagostrophus* from the remainder of the species (Fig. 13.7a). Their positive placement reflects the contribution of, in decreasing order of significance, CCS, CCH, CCL, QPH, CTD, AHL, CAL and AMF. The first four of these dimensions relate to the shape of the calcaneal–cuboid articulation. The remaining four dimensions relate to the depth of the tuber calcanei, the relative shape of the talar head, the talocalcaneal articular length and the relative proportions of the medial talar facet. *Dendrolagus*, *Bohra* and the remaining macropodines are negatively placed on PC2, reflecting, in decreasing order of significance, CTM, AML, CPL, CTW and ALM. These relate to features including the relative shape of the tuber calcanei and length of the talus. PC2 is not correlated with size (CML; $R^2 = 0.37$).

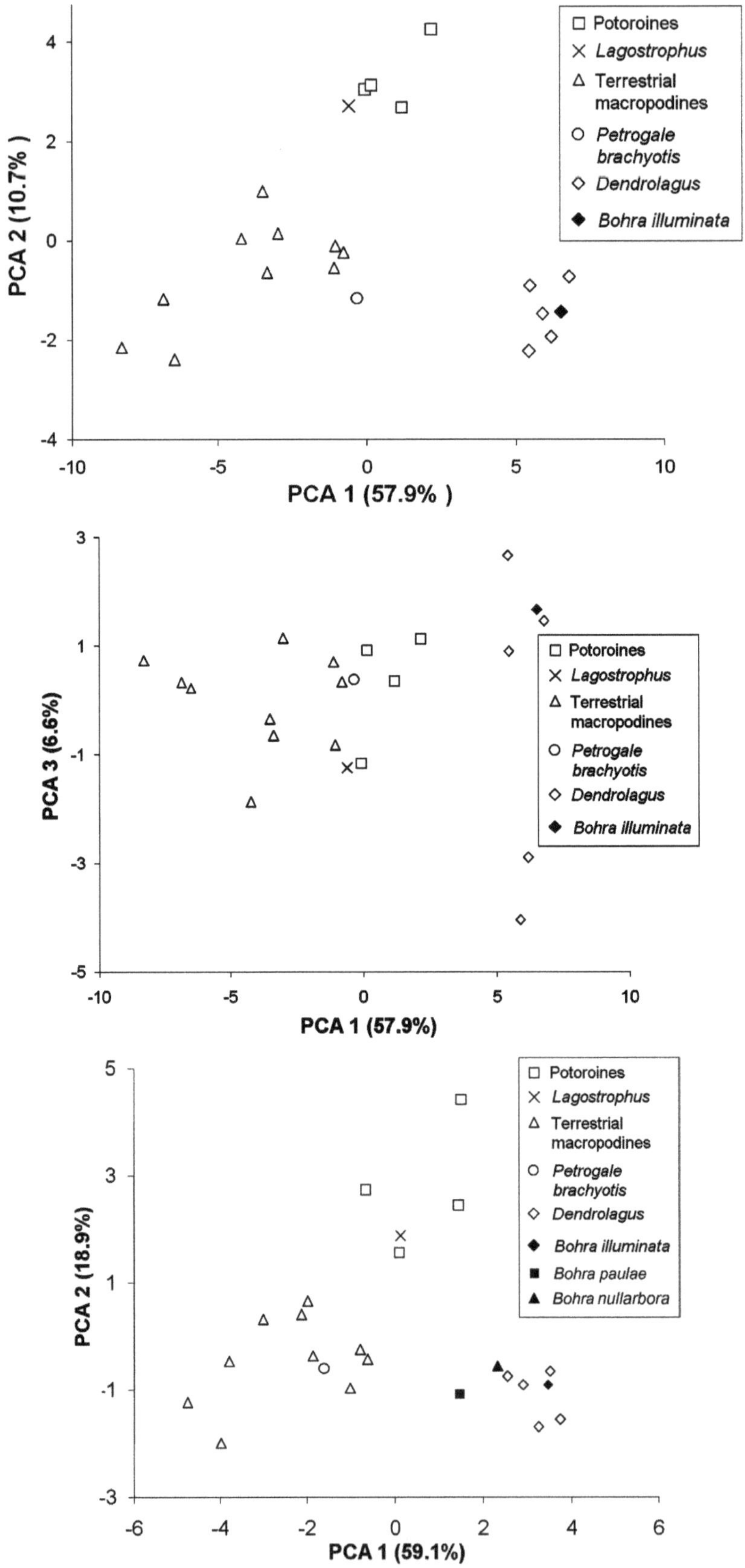

Figure 13.7: Principal components analysis of tarsal morphological variables – calcaneus, talus and cuboid dimensions. (a) PC2 vs PC1. (b) PC3 vs PC1. (c) Calcaneus dimensions, PC2 vs PC1.

PC3 accounts for 6.6% of the variation in the dataset but does not show any obvious differences between the main species groups (Fig. 13.7b). Tree-kangaroos demonstrate the greatest range and include both the highest (*D. inustus*; 2.667) and lowest (*D. dorianus*; –4.035) score in PC3. *Bohra* has an intermediate score (1.668). The strongly positive components include QLL, AHL, CSL, QPH, QTW and CCV. The most significant are the relative length of the cuboid, the length of the talar head and the length of the sustentaculum tali; the remaining three correspond to the relative size of the cuboid facets for the calcaneus. The strongly negative components are CCS, QAW and ALW.

Removing potoroines and *Lagostrophus* from the dataset has little effect on the manner in which the macropodine species are differentiated for PC1. This accounted for 65.9% of the variation and is not strongly correlated with size (CML; R^2 = 0.31) although nearly all the components had positive loadings, the majority close to unity. Strongly negative components, in decreasing order of significance, were CPL, QTL, QPH, CCH and CSL. PC2 (7.8% of the variation) corresponded strongly with PC3 from the previous analysis, and resulted in scattering of the tree-kangaroo species. PC3 accounts for 7.0% of the variation and separates taxa into a strongly positive pair of *Lagorchestes conspicillatus* and *M. irma*, a strongly negative pair, comprising two of the large species of *Macropus*, and an intermediate group, made up of *Thylogale, Petrogale, Onychogalea, Lagorchestes hirsutus* and the remaining species of *Macropus*. The species of *Dendrolagus* are widely spread in PC3 while *Bohra* is placed at the positive end of the axis. The strongly positive components include, in decreasing order of significance, CCH, QPH and CTD. The most strongly negative components include CSL, AHL, CTH, QTW and CTM.

Analysis of calcaneal dimensions alone for all species, plus *B. nullarbora* and *B. paulae*, showed a very similar distribution of the taxa in the first three principal components (Fig. 13.7c), accounting for 51.9%, 18.9% and 10.4% of the data, respectively. All three species of *Bohra* cluster with *Dendrolagus*, being strongly positive in PC1 and negative in PC2. Among them, *B. paulae* and *B. nullarbora* lie closer to the terrestrial macropodines than any of the species of *Dendrolagus*. The most strongly positive loadings in PC1 are CAW, CWS, CTH, CAL, CCM, CTW and CTM. All reflect the relative width of the calcaneal features. The strongest negative loading is for CPL. There is no correlation between PC1 and size (CML; $R^2 = 0.21$).

Discriminant analysis

Discriminant analysis of the dataset comprising calcaneal, talar and cuboid indices of the extant species plus *B. illuminata* utilised six uncorrelated measures as determined from the principal components analysis (CAW, CTM, CPL, CCH, AWM, QTL; n = 17; tolerance = 0.005) to plot groups based on habitat preference (Table 13.1). The first function explains 94.5% of the variation between the groups and demonstrates a clear separation of the groups, ranging from the terrestrial species adapted to the most open terrain, through inhabitants of denser understorey habitats, to rocky habitats and finally to arboreal specialists (Fig. 13.8). *B. illuminata* clusters with the species of *Dendrolagus*. The variable with the strongest positive loadings for function 1 is CTM, followed by CCH and AWM; the strongest negative loading is for CAW.

DISCUSSION

Functional morphology

The short, wide tuber calcanei of *Dendrolagus* and *Bohra* provides for less torque in the sagittal plane than in terrestrial macropodines, but provides greater mechanical advantage over a wider range of

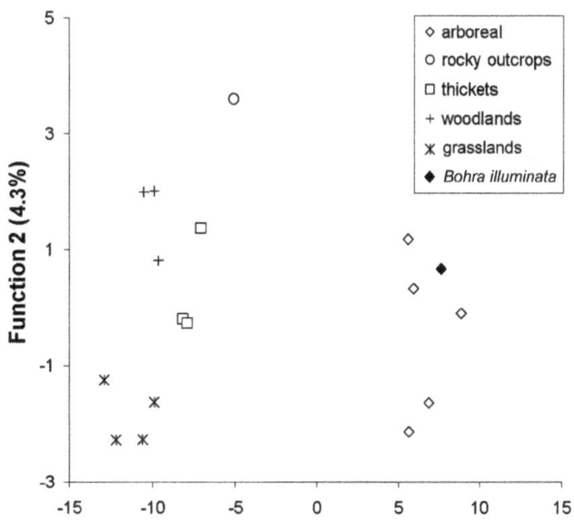

Figure 13.8: Discriminant analysis of tarsal variables within macropodines (excludes potoroines and *Lagostrophus*).

positions. The disproportionate development of the lateral facet relative to the medial facet suggests a less stable joint in which unbalanced loads, experienced during variable inversion and eversion of the pes, are a dominant pattern during locomotion. A similar pattern of modification characterises other arboreal marsupials (Argot 2002). Furthermore, the anterior articulation between the calcaneus and the talar head, which helps to stabilise the lower ankle joint in terrestrial kangaroos, has been entirely lost in tree-kangaroos, again suggesting a wider range in rotational movements.

High crests and a deep trochlea groove of the talus typically correlate with strong flexion–extension movements and restricted rotation at the upper ankle joint in mammals (Szalay 1994; Argot 2002). The vertical sides of the crests form a strong locking mechanism with the tibial and fibular malleoli, restricting movement to the sagittal plane. This morphology is exemplified in terrestrial macropodines. Reduced height of trochlear crests on the talus of *Dendrolagus* highlights greater mobility at the joint. The unequal height of the crests, and their smoother, flatter-sloping external faces resembles the morphology of other arboreal mammals (Argot 2002) and facilitates a comparatively wider range of rotational movements. The form of the lateral crest facilitates a more mobile fibular articulation. The relatively short and broad trochlear surface of *Dendrolagus* further serves to enhance mobility at the talotibial articulation; the reduced length of the trochlea highlights a shift away from restricted fore-and-aft hinge movements, and the increased width of the surface suggests a spreading of forces acting at the joint over a wider range of postures. *Bohra* retains a more intermediate morphology of the talotibial articular surfaces, suggesting that major changes at this joint occurred later in the specialisation to an arboreal lifestyle. Presence of a deep talotibial pit on the dorsal surface of the talar neck denotes a large tibial malleolus, which prevents extreme dorsoventral flexion in terrestrial mammals. The feature is usually reduced or lost in arboreal forms (Salton and Szalay 2004). In *Dendrolagus* and *Bohra*, the talotibial pit is broad, but not as deep or steeply sided as in terrestrial kangaroos. The tree-kangaroo morphology reflects enhanced dorsoventral flexion and a greater emphasis on rotation. The more anteroventrally rounded head of *Dendrolagus* facilitates a wider range of possible movements and the short robust talar neck suggests a wider range of pedal positions (Szalay 1994; Argot 2002; Salton and Szalay 2004). *Bohra* is less specialised in this regard, which suggests a more generalised function than in *Dendrolagus*, and that specialisation of this tarsal region occurred subsequent to those relating to enhanced mobility at the talocalcaneal and calcaneal–cuboid joints.

The stepped morphology of the transverse tarsal joint restricts rotational movements, thereby stabilising the pes of terrestrial macropodines during bipedal saltation (Bishop 1997). Strong development of a ventromedian facet and plantar tuberosity of the cuboid further restricts movement at the joint. *Dendrolagus*, and to a lesser extent *Bohra*, have modified the calcaneal–cuboid articulation by smoothing the step, increasing the curvature of the articular surface and coalescing the ventromedian and dorsal facets to facilitate greater rotation (inversion/eversion) and translation. The more oblique orientation and elongate morphology of the cuboid plantar tuberosity, and the flexor tendons that pass around it, aid in this movement (Gebo 1987; Argot 2002), helping overcome the phylogenetic constraints of the stepped calcaneal–cuboid articulation. While the morphology of the cuboid of *Bohra* is less specialised than that of *Dendrolagus*, the expansion of the lateral plantar process and oblique peroneal sulcus are well-developed. Overall expansion of the articular surface, particularly of the lateral facet, has been shown to facilitate a wider range of inversion movements in other mammals (Salton and Szalay 2004). The smoothed arcuate step in the calcaneal–cuboid articulation of *Bohra* reflects an enhanced range of rotational movements compared with terrestrial kangaroos, but less translation than in *Dendrolagus*, as revealed by its more completely smoothed articulation. The functional modification of the transverse tarsal joint represents perhaps the key step in the evolution of tree-kangaroo pes.

The short robust metatarsals and phalanges of *Dendrolagus* reflect selection for increased manoeuvrability and balance in an uneven three-dimensional environment (Flannery 1982). *Bohra* also possesses relatively short and robust metatarsals, but the phalanges are not as stout proportionally as those of *Dendrolagus*. The enlarged distal epiphyses and expanded interosseous articular facets of the metatarsals of *Dendrolagus* and *Bohra* facilitate a greater range of movement in the metatarsal and phalangeal region than is possible in terrestrial macropodines. Similar observations have been made in other arboreal marsupials (Argot 2002). Reduced length and increased flexibility of the pes is a critical adaptation in the assumption of an arboreal habitat.

MORPHOMETRY

Multivariate analyses of linear landmark data provide an opportunity to test whether morphological differences observed through visual comparison also separate species statistically. They allow morphological variation to be plotted graphically to help discern both large-scale and fine-scale patterns within and between taxa. The interaction between PC1 and PC2 spreads the terrestrial groups along a broad gradient from large species occupying open areas through to thicket inhabitants (Fig. 13.7). The first discriminant function accounts for 96.9% of the variation between species and also clearly distinguishes, irrespective of body size, species adapted to open terrain, inhabitants of denser understorey habitats and rocky habitats, and tree-kangaroos (Fig. 13.8). Within the PCA, the potoroines and *Lagostrophus* are distinct from the terrestrial macropodines, suggesting a clear deviation in tarsal morphology between them and the macropodines. Further quantitative analyses incorporating larger sample sizes and a wider spread of terrestrial taxa may help to resolve the fine-scale shape changes of the tarsus and its relationship to the functional and ecological evolution of terrestrial macropodines.

Principal components analysis confirms that the three species of *Bohra* represented by pedal elements have a quantitatively distinctive calcaneal form. The species plot closer to those of *Dendrolagus* than to any of the terrestrial macropodines, although their exact positions differ slightly. *B. paulae* plots relatively closer to the terrestrial kangaroos in calcaneal morphometry than the other two species, *B. nullarbora* is intermediate between *B. paulae* and *Dendrolagus*, and *B. illuminata* lies within the bounds of the *Dendrolagus* cluster (Fig. 13.7c). When measurements of the calcaneus, talus and cuboid are analysed together, *B. illuminata* remains clustered with *Dendrolagus* (Fig. 13.7a–b). Thus, PCA strongly confirms the qualitative comparisons: the tarsus of *B. illuminata* was as well-adapted to the arboreal realm as the extant, rainforest-dwelling tree-kangaroos. *B. nullarbora* and *B. paulae* differ only by being displaced along one axis of the PC1 plot due to a slightly longer tuber calcanei, but the similarities with *B. illuminata* strongly outweigh the differences.

By contrast, substantially more variation in tarsal morphology is observed between tree-kangaroo species for PC3, which accounts for 6.6 % of the total variation in the entire group (Fig. 13.7b). Whether this reflects functional or phylogenetic patterns within the tree-kangaroo group remains unresolved. There is strong separation between the short-footed New Guinea species (*D. dorianus*, *D. matschiei*) and the long-footed Australian species (*D. bennettianus*, *D. lumholtzi*, *B. illuminata*). There are also more subtle morphometric differences between species within these groups. Understanding the functional or phylogenetic basis for this variation requires a more detailed analysis incorporating a broader range of extant species.

PALAEOECOLOGY

The concept of tree-kangaroos as prior inhabitants of the now treeless Nullarbor Plain initially seems to beggar belief. However, given that most other taxa in the late Cenozoic faunal assemblages of Chinchilla (south-east Queensland) and Wellington Caves (central eastern New South Wales) are more suggestive of open woodland and grassland than forest, it is perhaps not quite as astonishing as it first seems. Flannery and Szalay (1982) and Dawson (2004) were well aware that *B. paulae* and *B. wilkinsonorum* may not have inhabited forest, although they justifiably left open the possibility of a preference for remnant riparian forest. The flat limestone karst terrain of the Nullarbor Plain is too permeable to support surface streams and during the middle Pleistocene, as now, it is likely that animals incapable of surviving without free water utilised ephemeral rock holes. Palaeoecological and stable isotopic analyses of the diverse Nullarbor fossil fauna (70 species, including 23 kangaroos) point to a similar climate to that of today (230–260 mm mean annual rainfall) but a much more diverse vegetation structure than the modern shrub steppe (Prideaux *et al.* 2007).

Our analyses show that species of *Bohra* had a very similar level of pedal mobility and hind limb strength as *Dendrolagus* species. Even though they may have weighed up to twice as much as their extant relatives, the species of *Bohra* were clearly well-adapted to an arboreal realm. *Bohra* differed from *Dendrolagus* in overall build, possessing a larger body relative to the size of the head, and longer hind limbs relative to both fore limb length and body size (Prideaux and Warburton 2008, 2009). While the species of *Bohra* were taxonomically and functionally tree-kangaroos, their body proportions were not quite as derived relative to their terrestrial ancestor as modern *Dendrolagus*.

Within the scleromorphic woodland/shrubland mosaic inferred for the middle Pleistocene environment of the Nullarbor Plain (Prideaux *et al.* 2007), the two species of *Bohra* may have utilised a range of small tree species now largely restricted in this region to remnant stands on the periphery of the plain. These include *Myoporum platycarpum* (sugarwood), *Pittosporum phylliraeoides* (native apricot), *Santalum acuminatum* (sweet quandong), *S. spicatum* (sandalwood), *Alectryon oleifolius* (bullock bush) and *Acacia papyrocarpa* (western myall) – plants with palatable leaves and/or fleshy fruit now highly sought after by introduced ungulates and rabbits (Mitchell and Wilcox 1998).

CONCLUSIONS

Pedal attributes shared by *Bohra* and *Dendrolagus* to the exclusion of all other macropodines reflect enhanced mobility of the pes for greater flexibility and balance in a three-dimensional environment. These include marked expansion of the tuber calcanei, enlargement and confluence of the posterior facets for the calcaneal–talar articulation, expansion and smoothing of the facets of the calcaneal–cuboid articulation, broadening of the talus, deepening of the talar head, flattening of the cuboid, and subequal length and stoutness of metatarsals IV–V. Together, these facilitate an expanded range of inversion/eversion and abduction/adduction movements. Based on this evidence, the species of *Bohra* were adapted to moving in an arboreal environment. Still, the pes of *Bohra* was not quite as specialised as that of *Dendrolagus*, which suggests that *Bohra* is the more plesiomorphic taxon. Species of *Bohra* bear closer resemblance to the long-footed species of *Dendrolagus* than to the short-footed species and it is probable that the similarities represent symplesiomorphies, although confirmation awaits completion of our phylogenetic analysis of the group.

ACKNOWLEDGMENTS

We thank the following individuals and institutions for the loan of or access to comparative specimens: Norah Cooper, Western Australian Museum; Robert Jones, Australian Museum; Scott Hocknull, Steve Van Dyck, Heather Janetski and Kristen Spring, Queensland Museum; Robert Palmer, Australian National Wildlife Collection; David Stemmer, South Australian Museum; Robert Voss, American Museum of Natural History. Jane Prince and Phil Withers advised on statistical analyses. This study was supported by the Rio Tinto WA Future Fund and an anonymous donation to the Western Australian Museum Foundation.

REFERENCES

Alexander RM and Vernon A 1975. The mechanics of hopping by kangaroos. *J. Zool.* **177:** 265–303.

Argot C 2002. Functional–adaptive analysis of the hindlimb anatomy of extant marsupials and the paleobiology of the Paleocene marsupials *Mayulestes ferox* and *Pucadelphys andinus*. *J. Morphol.* **253:** 76–108.

Bishop NJ 1997. Functional anatomy of the macropodid pes. *Proc. Linn. Soc. NSW* **117:** 17–50.

Burk A, Westerman M and Springer MS 1998. The phylogenetic position of the musky rat-kangaroo and the evolution of bipedal hopping in kangaroos (Macropodidae: Diprotodontia). *Syst. Biol.* **47:** 457–474.

Carrano MT 1997. Morphological indicators of foot posture in mammals: a statistical and biomechanical analysis. *Zool. J. Linn. Soc.* **121:** 77–104.

Dawson L 2004. A new Pliocene tree kangaroo species (Marsupialia, Macropodinae) from the Chinchilla Local Fauna, southeastern Queensland. *Alcheringa* **28:** 267–273.

Dawson L and Augee ML 1997. The late Quaternary sediments and fossil vertebrate fauna from Cathedral Cave, Wellington Caves, New South Wales. *Proc. Linn. Soc. NSW* **117:** 51–78.

Dawson L, Muirhead J and Wroe S 1999. The Big Sink Local Fauna: a lower Pliocene mammalian fauna from the Wellington Caves complex, Wellington, New South Wales. *Rec. West. Aust. Mus., Suppl.* **57:** 265–290.

DeGusta D and Vrba ES 2003. A method for inferring paleohabitats from the functional morphology of bovid astragali. *J. Archaeol. Sci.* **30:** 1009–1022.

Flannery TF 1982. Hindlimb structure and evolution in the kangaroos (Marsupialia: Macropodoidea). In *The Fossil Vertebrate Record of Australasia*. (Eds PV Rich and EM Thompson) pp. 508–524. Monash University Press, Melbourne.

Flannery TF and Szalay FS 1982. *Bohra paulae*, a new giant fossil tree kangaroo (Marsupialia: Macropodidae) from New South Wales, Australia. *Aust. Mammal.* **5:** 83–94.

Flannery TF, Martin R and Szalay A 1996. *Tree-kangaroos: A Curious Natural History*. Reed Books, Melbourne.

Gebo DL 1987. Functional anatomy of the tarsier foot. *Am. J. Phys. Anthrop.* **73:** 9–31.

Groves CP 1982. The systematics of tree kangaroos (*Dendrolagus*; Marsupialia, Macropodidae). *Aust. Mammal.* **5:** 157–186.

Hall R 2001. Cenozoic reconstructions of SE Asia and the SW Pacific: changing patterns of land and sea. In *Faunal and Floral Migration and Evolution in SE Asia–Australasia*. (Eds I Metcalfe, JMB Smith, M Morwood and IDAA Davidson) pp. 35–56. Balkema, Lisse.

Hocknull SA 2005. Additional specimens of *Bohra* (Marsupialia: Macropodidae) from the Pliocene of Queensland. *Mem. Qld Mus.* **51:** 26.

Mitchell AA and Wilcox DG 1998. *Arid Shrubland Plants of Western Australia*. University of Western Australia Press, Perth.

Nomina Anatomica Veterinaria 2005. International Committee on Veterinary Gross Anatomical Nomenclature, 5th edn. http://www.wava-amav.org/Downloads/nav_2005.pdf.

Prideaux GJ and Warburton NM 2008. A new Pleistocene tree-kangaroo (Diprotodontia: Macropodidae) from the Nullarbor Plain of south-central Australia. *J. Vert. Paleo.* **28:** 463–478.

Prideaux GJ and Warburton NM 2009. *Bohra nullarbora* sp. nov., a second tree-kangaroo (Marsupialia: Macropodidae) from the Pleistocene of the Nullarbor Plain, Western Australia. *Rec. West. Aust. Mus.* **25:** 165–178.

Prideaux GJ, Long JA, Ayliffe LK, Hellstrom JC, Pillans B, Boles WE, Hutchinson MN, Roberts RG, Cupper ML, Arnold LJ, Devine PD and Warburton NM 2007. An arid-adapted middle Pleistocene vertebrate fauna from south-central Australia. *Nature* **445**: 422–425.

Salton JA and Szalay FS 2004. The tarsal complex of Afro-Malagasy Tenrecoidea: a search for phylogenetically meaningful characters. *J. Mammal. Evol.* **11:** 73–104.

Sargis EJ 2002. Functional morphology of the hindlimb of tupaiids (Mammalia, Scandentia) and its phylogenetic implications. *J. Morphol.* **254:** 149–185.

Strahan R (Ed.) 1995. *The Mammals of Australia*. Reed Books, Sydney.

Szalay FS 1994. *Evolutionary History of the Marsupials and an Analysis of Osteological Characters*. Cambridge University Press, New York.

PART III

ECOLOGY

14 Mycophagy in a community of macropodoid species

K. Vernes

SUMMARY

Mycophagy (fungus-eating) by mammals is likely to be an important ecosystem process in Australian forests and woodlands, because the hypogeous ectomycorrhizal fungi (truffles) that are dispersed by mycophagous mammals are symbiotic with trees, aiding them in healthy growth. Studies of mammals that consume and disperse these fungi have generally focused on potoroid marsupials, or more occasionally bandicoots or rodents. A common and diverse component of Australian forest and woodlands, the macropodids, have largely been overlooked as mycophagists. This chapter presents data on the occurrence of fungi in the diets of a community of six macropodoid species in north-east New South Wales, including four species of macropodid and two species of potoroid. Diets of red-necked pademelons (*Thylogale thetis*), swamp wallabies (*Wallabia bicolor*), parma wallabies (*Macropus parma*), brush-tailed rock-wallabies (*Petrogale penicillata*), rufous bettongs (*Aepyprymnus rufescens*) and long-nosed potoroos (*Potorous tridactylus*) were examined. All these macropodoids consumed hypogeous fungi to various degrees, the diversity of spore types in the diet ranging from just a few taxa that were detected occasionally (e.g. brush-tailed rock-wallaby diets) to more than 30 taxa that were detected frequently (e.g. swamp wallaby diets). Fungal diets of the different macropodoids were significantly different from one another in the types of fungi eaten and the relative contribution of the different fungal types to the diet. Diets also changed seasonally. The ecological implications of fungal consumption and dispersal by such a diversity of macropodoids, some of which are widely distributed in eastern Australia, are discussed.

INTRODUCTION

Importance of mammalian mycophagy

It is generally assumed that mammalian mycophagists are important in the maintenance of forest ecosystems as they disperse the spores of mycorrhizal fungi (Johnson 1996). Mycorrhizal fungi promote the healthy growth of vascular plants by enhancing mineral nutrient uptake and increasing drought tolerance and disease resistance (Claridge and May 1994; Claridge 2002). Mature hypogeous (below-ground) fruiting bodies (truffles) encourage consumption, and thus spore dispersal, by emitting odours (Maser *et al.* 1978) that attract mycophagous mammals (Donaldson and Stoddart 1994; Pyare and Longland 2001). Spores remain viable after passage through the gut of mammals and resultant spore-laden faeces are typically deposited on the forest floor, often close to established host root systems (Claridge 2002). Experiments have shown that spore dispersal by mammals enhances the likelihood of mycorrhiza formation (Claridge *et al.* 1992; Reddell *et al.* 1997; Caldwell *et al.* 2005). Furthermore, because mammals may travel relatively long distances before defecation, the ingestion and passing of faecal pellets probably represents an efficient mechanism for the widespread dispersal of spores by the fungus (Claridge 2002; Johnson 1996). Australian mycophagous mammals are therefore likely to contribute disproportionately more towards the maintenance of forest ecosystem health than other vertebrates (Johnson 1996).

In a little over 200 years since European settlement, the Australian continent has lost 22 terrestrial mammal species to extinction, with a further 47 listed nationally as threatened (EPBC Act 1999). Many of these extinct or threatened mammals are small, critical-weight range (35–5500 g) terrestrial species (Burbidge and McKenzie 1989; Dickman 1994), some of which (e.g. potoroids and bandicoots) have traditionally been thought to consume the greatest diversity and abundance of hypogeous mycorrhizal fungi (Claridge 2002). However, despite a broad knowledge of mammal–fungal relationships, understanding of community-level interactions between different species of mycophagous mammals remains poor, and in most ecosystems it is unknown which species of mammal consume fungi (Claridge and May 1994). This is not merely because the basic ecology of some mammals is unknown (Claridge 2002), but also because the consumption of fungus by mammals has historically been overlooked in dietary studies.

In global terms, the diversity of Australian hypogeous fungi is prodigious, estimated at well over 1000 species (Bougher and Lebel 2001). This diversity has been variously attributed to nutrient-poor soils that promote mycorrhizal associations (Claridge 2002), climatic fluctuations that favoured multiple speciation events (Johnson 1996), Australia's past and present arid climate that favoured subterranean fungi (Trappe 1988) and the presence of several obligate mycophagous mammals in most Australian mammal communities that depended on these fungi for food and subsequently dispersed their spores (Claridge 2002). A rich and diverse mycorrhizal fungal community probably equates to a healthy and resilient forest ecosystem, which may influence its ability to respond to environmental change (Allen *et al.* 1995). It follows that long-term management of fungal communities requires a good understanding of how mammals influence fungal diversity and which mammal species, or combinations of species, are the key to maintaining or enhancing fungal diversity. Even a basic understanding of ecosystem processes between forest trees and fungi in Australian forests and woodland requires a thorough understanding of which species of mammals actually consume fungi.

Macropodoids as spore dispersers

Reviews by Claridge and May (1994) and Claridge (2002) of mycophagy by Australian mammals listed 41 species of mammal recorded to consume fungi, of which only six were macropodids. Christensen (1980) showed that, in Western Australia, tammar wallabies (*Macropus eugenii*), western grey kangaroos (*M. fuliginosus*) and quokkas (*Setonix brachyurus*) all consumed fungi, albeit in small quantities. On the New England Tablelands in New South Wales, Redenbach (1982) showed that pademelon (*Thylogale thetis* and *T. stigmatica*) diets could contain up to 8% fungus although, as with most early studies, no attempt was made to describe the types of fungi consumed. More recently, Vernes and Trappe (2007) showed that the red-legged pademelon (*T. stigmatica*) in the north Queensland Wet Tropics ate at least 15 species of fungi, many of them hypogeous. Redenbach (1982) also examined stomachs from three parma wallabies (*M. parma*) from the New England Tableland and found them to contain up to 11% fungal matter; no identification of fungal type was attempted. Swamp wallabies (*Wallabia bicolor*) have long been known to consume fungi as part of their mixed diet (Harrington 1976; Hollis *et al.* 1986; Osawa 1990). Claridge *et al.* (2001) confirmed that they consumed hypogeous fungi, finding at least 20 species were consumed across a range of sites in south-east Australia and postulating that mycophagy was a common feeding habit for the species.

Consumption of fungi by the potoroids (bettongs, potoroos and rat-kangaroos) has been well studied. There is general agreement in the literature that potoroids are obligate feeders on a great diversity of hypogeous fungi, and that fungus forms the majority of their diet (Hill and Triggs 1985; Taylor 1992; Bennett and Baxter 1989; Claridge *et al.* 1993; Tory *et al.* 1997; Green *et al.* 1999; Vernes *et al.* 2001; Nguyen *et al.* 2005).

This chapter presents data on fungal consumption by four macropodid and two potoroid species on the New England Tableland in north-east New South Wales. Presented patterns in consumption between species and across seasons demonstrate that mycophagy is common in this group of mammals. The main conclusion is that forest-dwelling macropodids eat more fungi than previously recognised, and consequently that mammal–fungal interactions at a community level are probably more complex than has traditionally been portrayed.

METHODS

Study area and collection of faecal samples

All collection sites were on the eastern edge of New England Tableland (New England Bioregion) in north-east New South Wales (Fig. 14.1). A diversity of vegetation types occurs at these sites, but an element common to all is the dominance of woody vegetation (primarily *Eucalyptus* spp.), which supports communities of ectomycorrhizal fungi. Detailed descriptions of vegetation for the study region can be found in Hunter and Clarke (1998) and Sheringham and Hunter (2002).

At the main site, Gibraltar Range National Park (29°31'S, 152°18'E) and adjacent Washpool National Park (29°28'S, 152°20'E), faecal samples were collected from swamp wallabies, parma wallabies, red-necked pademelons (*T. thetis*) and a single long-nosed potoroo (*Potorous tridactylus*). In addition, samples from

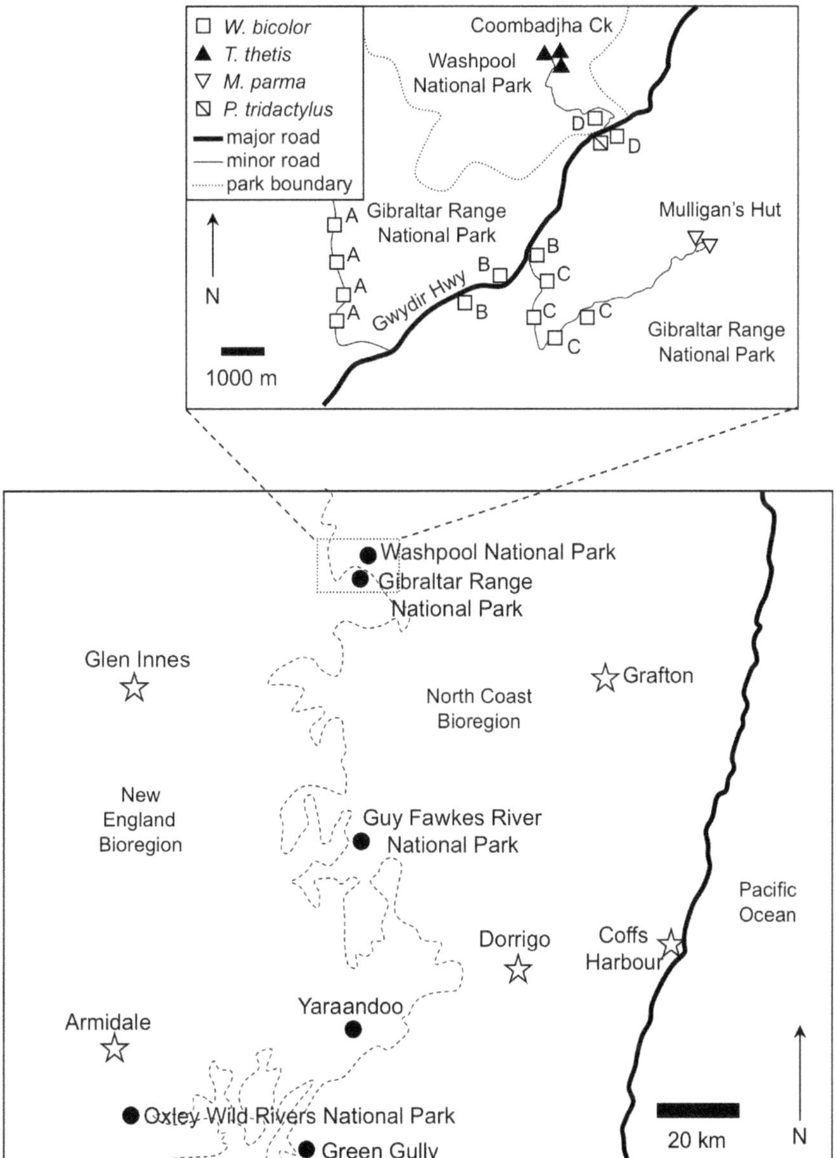

Figure 14.1: The study region in north-east New South Wales showing location of sites (solid circles), New England Bioregion boundary (dashed line) and major townships (open stars). The inset shows the main study sites at Gibraltar Ranges and Washpool National Parks in detail, with collection sites for four macropodoid species marked. Letters (A–D) associated with swamp wallaby sites denote clusters of sites.

brush-tailed rock-wallabies (*Petrogale penicillata*) were collected from sites along the edge of gorges in different parts of Oxley Wild Rivers National Park (30°40'S, 151°23'E and 30°51'S, 152°08'E) and Guy Fawkes River National Park (30°05'S, 152°18'E). I also gathered samples from a semi-wild colony of rufous bettongs (*Aepyprymnus rufescens*) that were maintained within a 7 ha area of native bush enclosed by a predator-proof fence at 'Yaraandoo' (30°29'S, 152°18'E) adjacent to New England National Park. Although the diets of these bettongs were supplemented *ad libitum* with kangaroo pellets and fresh water, the animals foraged nocturnally for wild food and sheltered by day in nests typical of the species (Youman 2003) and generally avoided human contact.

All samples were collected from scats deposited on the ground by macropodoids, with the exception of the single long-nosed potoroo sample that came from a cage-trapped individual. At Gibraltar Range National Park, a maximum of six swamp wallaby scat samples were collected from each of 12 areas (representing four sites, A–D; see Fig. 14.1) per sampling period in order to maximise the numbers of animals sampled across the study area. The location of these sites at least 1 km apart ensured sites were independent, based upon the average size (16 ha) of a swamp wallaby home range (Troy and Coulson 1993). For all other animals, collection was necessarily driven by our knowledge of where animals congregated (e.g. red-necked pademelons on the grassy edges of rainforest), partly due to the difficulty of randomly locating samples from these mammals and partly to ensure that we were collecting from the intended species. Samples were collected fresh (estimated to be deposited within the last 7 days based upon colour, odour and intactness) and dried immediately, or frozen until they could be dried later. In all cases, I attempted to gather samples as far from one another as possible to avoid sampling the same individual during the same sampling period, although this was not always possible for species that congregated in small clearings (e.g. pademelons and parma wallabies). If there were any doubt over the species of mammal that produced the scat, it was not collected.

Faecal samples were briefly softened in 5% KOH, macerated, then rinsed with distilled water through a 125 × 125 μm mesh. Two to three drops of the mixture that passed through the mesh were mounted on a glass slide with glycerine jelly and topped with a glass cover slip. Slides were systematically scanned at ×400 under a light microscope; ×1000 magnification was used to define morphologic characters (size, shape, ornamentation, wall thickness, symmetry) of each spore type encountered and to identify them to the lowest taxonomic level possible. I used scanning electron microscopy (SEM) to further resolve spore identification. For SEM, a subsample of faecal suspension (stored in 70% ethanol), typically 0.5–1.5 mL depending on density, was withdrawn and washed with 250 mL distilled water in a vacuum filtration apparatus containing a 25 mm diameter, 3 μm pore size polytetrafluoroethylene (PTFE) membrane. Filters containing the samples were dried (1 h at 45°C), and portions of the filters were mounted on 10 mm diameter aluminium support stubs using double-sided tape, rimmed with colloidal carbon and coated with c. 15 nm gold. SEM was performed using a JEOL JSM-5600 scanning electron microscope (JEOL USA, Peabody, MA) operating at 10 kV and 8–48 mm working distance at the Digital Microscopy Facility, Mount Allison University, New Brunswick. I photographed all spore types detected, and identified them using available guides (primarily Castellano *et al.* 1989).

Data analyses

Two-way unreplicated analysis of variance (ANOVA) tests were performed on percentage occurrence of hypogeous and epigeous fungi in mammal diets using season and mammal species as independent variables. For each mammal species, paired *t*-tests were also used to detect significant differences in the percentage of samples within each season that contained hypogeous versus epigeous fungi. ANOVA tests were used to detect significant differences in the number of unique spore types seen within each sample across mammals and seasons.

Multivariate analyses available within the PRIMER 6 software package (Clarke and Gorley 2006) were used to examine seasonal and interspecies trends in the diversity of fungal taxa consumed by the different mammal species. To determine whether all data for a species gathered across sites and years could be pooled by season (summer, autumn, winter or spring), I calculated between-sample similarities based upon the composition of each individual sample (presence/absence of each fungal taxon within each scat) for each mammal species separately by constructing a Bray-Curtis similarity matrix. Significant patterns in sample composition were then analysed using analysis of similarity (ANOSIM) tests. For each mammal, the resultant global *R*-values and corresponding *P*-values for each test supported the pooling of data by season (swamp wallaby $R = 0.28$, $P < 0.001$; red-necked pademelon $R = 0.23$, $P < 0.001$; parma wallaby $R = 0.14$, $P < 0.001$).

Based upon preliminary analyses that supported the grouping of data by season, I constructed a Bray-Curtis similarity matrix using log-transformed percentage occurrence data for each species × season combination for each of the 56 fungal taxa identified across the diets of swamp wallabies, parma wallabies, red-necked pademelons, brush-tailed rock-wallabies and rufous bettongs. Patterns in diets were then displayed using non-metric multi-dimensional scaling (MDS), and differences in the groups that were identified visually from the MDS plot were tested for significance using an ANOSIM test. The similarity percentages routine (SIMPER) was then used to identify the fungal taxa that contributed most to the different dietary groups that the ANOSIM statistically supported.

Finally, to obtain an estimate of how well my sampling of macropodoid diets captured the predicted diversity of fungal spore types likely to actually occur in the diet of each species, I constructed species accumulation curves for swamp wallabies, red-necked pademelons and parma wallabies by calculating the number of species that are returned in successively pooled randomly selected samples. This process was repeated 999 times, with the resultant curve for each species representing a smoothed average of the iterations (Clarke and Gorley 2006). This curve was compared with the predicted true total of taxa that would be observed as the number of samples tended towards infinity using the Bootstrap model in PRIMER 6 (Clarke and Gorley 2006).

RESULTS

Seasonal consumption of fungi

The three mammals for which I obtained the full range of seasonal data (swamp wallaby, red-necked pademelon and parma wallaby) showed high fungal consumption in all seasons (Table 14.1). The percentage of swamp wallaby samples that contained fungus ranged from a low of about 53% in spring to a high of about 95% in autumn and winter. Red-necked pademelon and parma wallaby samples also had the lowest occurrence of fungus in spring (67% for red-necked pademelons, 79% for parma wallabies), with the peak occurrence (92–100%) from summer through to winter. Other mammals for which I had limited data also showed degrees of mycophagy. Brush-tailed rock-wallabies were sampled only in summer and spring; fungus was detected in both seasons, but in fewer samples relative to the other macropodoids. Samples from semi-wild rufous bettongs were collected only in autumn and winter, and the proportion of these samples containing fungus was high (84 – 95%). I captured only one long-nosed potoroo at Gibraltar Range (in winter 2004), and this adult male had 10 different spore types in its scat (Table 14.2).

Across seasons, the percentage occurrence of hypogeous versus epigeous fungi in swamp wallaby samples did not differ (paired t-test; $t_3 = 0.31$, $P = 0.8$; Fig. 14.2); however, for both red-necked pademelons and parma wallabies, more samples contained hypogeous fungi than epigeous fungi (red-necked pademelon $t_3 = 10.1$, $P = 0.002$; parma wallaby $t_3 = 5.0$, $P = 0.02$; Table 14.1). Mammal species did not differ in the percentage of samples in each season that contained hypogeous fungi ($F_{2,11} = 4.5$, $P = 0.06$), but differed significantly in the percentage of samples in each season that contained epigeous fungi ($F_{2,11} = 7.5$, $P = 0.02$), the difference being driven by more swamp wallaby than red-necked pademelon samples containing epigeous fungi (Tukey-Kramer $q = 5.49$, $P < 0.05$). Occurrence of epigeous fungus was not different between the other species pairwise comparisons.

Number of taxa per sample

Mammal species differed significantly in the numbers of spore types per sample for both hypogeous ($F_{2,377} = 33.6$, $P < 0.0001$) and epigeous ($F_{2,377} = 49.1$, $P < 0.0001$) fungi. For both kinds of fungi, differences reflected fewer spore types in swamp wallaby than in red-necked pademelon and parma wallaby samples (Tukey HSD; $P < 0.0001$ in each case; Fig. 14.2). Seasonal differences detected for both hypogeous ($F_{2,377} = 11.2$, $P < 0.0001$) and epigeous ($F_{2,377} = 19.5$, $P < 0.0001$) fungi related to more hypogeous fungal taxa occurring in autumn and winter than in spring samples (Fig. 14.2). Summer was a transitional season, having few spore types per sample than winter, but not autumn. Similarly, the number of epigeous spore types per sample was greater in autumn and winter than in spring and summer (Fig. 14.2).

Patterns in fungal taxa consumed

Analysis of Similarity (ANOSIM) revealed a significant difference between Bray-Curtis similarity coefficients coded by mammal species ($R = 0.667$,

Table 14.1: Occurrence of fungi in the diets of five macropodoids from the New England Bioregion.

Species	Season	No. of samples	% with hypogeous fungi	% with epigeous fungi	% with any fungus
Swamp wallaby (*Wallabia bicolor*)	Summer	27	59.3	40.7	59.3
	Autumn	47	59.6	93.6	95.7
	Winter	91	91.2	92.3	94.5
	Spring	34	50.0	47.1	52.9
	All samples	199	72.4	77.9	82.9
Red-necked pademelon (*Thylogale thetis*)	Summer	17	94.1	23.5	94.1
	Autumn	17	100.0	52.9	100.0
	Winter	48	91.7	14.6	91.7
	Spring	42	66.7	2.4	66.7
	All samples	124	84.7	16.9	84.7
Parma wallaby (*Macropus parma*)	Summer	18	88.9	44.4	100.0
	Autumn	7	100.0	57.1	100.0
	Winter	11	90.9	72.7	100.0
	Spring	28	75.0	17.9	78.6
	All samples	45	84.4	39.1	90.6
Brush-tailed rock-wallaby (*Petrogale penicillata*)	Summer	13	15.4	38.5	46.2
	Autumn	0	–	–	–
	Winter	0	–	–	–
	Spring	23	17.4	4.3	17.4
	All samples	36	16.7	16.7	27.8
Rufous bettong (*Aepyprymnus rufescens*)	Summer	15	6.7	6.7	6.7
	Autumn	19	84.2	36.8	84.2
	Winter	20	95.0	35.0	95.0
	Spring	–	–	–	–
	All samples	54	66.7	27.8	66.7

The long-nosed potoroo (*P. tridactylus*) is not shown, because diet of that species was based on only one sample.

$P < 0.001$; Fig. 14.3), but not by season ($R = -0.047$, $P = 0.66$), indicating that species' diets differed distinctly from one another regardless of seasonal shifts in the consumption of some fungal taxa. Furthermore, R-values were high for all mammal species pairs (Table 14.3), suggesting that the diet of each mammal was considerably different from the diets of all others. However, the small sample size for brush-tailed rock-wallabies, which contributed to high R-values and paradoxically non-significant P-values, means that the statistical power of comparisons involving this species is low.

Similarity analysis (SIMPER) identified the fungal taxa in the diet of each mammal that explained the greatest between-species differences. Two epigeous (*Boletellus* and Boletoid 1) and two hypogeous (Russuloid 4 and *Labyrinthomyces* 2) taxa contributed most to the differences between diets of swamp wallabies and the diets of other mammals (Table 14.2). Taxa that distinguished red-necked pademelon diets from other mammal diets were *Chamonixia*, Russuloid 4 and *Hysterogaster* 1, while parma wallaby diets were characterised by the taxa *Chamonixia*, *Hysterangium* and *Thaxterogaster*. Taxa that together contributed more than 50% to the between-species differences are indicated in Table 14.2.

Seasonal and site differences in swamp wallaby diet

Swamp wallabies were sampled from five sites (total of 12 discrete collection locations) at Gibraltar Range National Park over different seasons in order to

Table 14.2: Fungal taxa detected in the diets of six macropodoids.

	Mammal species					
	W. bicolor	***T. thetis***	***M. parma***	***P. penicillata***	***A. rufescens***	***P. tridactylus***
Fungal taxa	**n = 199**	**n = 124**	**n = 45**	**n = 36**	**n = 54**	**n = 1**
Epigeous (above-ground or mushroom-like) taxa						
Agaricus 1	9.0	0.8	1.6	5.7	5.6	
Agaricus 2	0.5	0.0	0.0	0.0	1.9	
Boletellus	**50.3**	1.6	3.1	0.0	3.7	
Boletoid 1	**38.7**	4.0	14.1	**8.6**	1.9	
Boletoid 2	22.6	0.8	9.4	0.0	0.0	
Mushroom 1	37.2	7.3	20.3	0.0	24.1	
Mushroom 2	16.1	3.2	1.6	**2.9**	7.4	
Copriniaceae	2.0	1.6	0.0	0.0	0.0	
Hypogeous (below-ground or truffle-like) taxa						
Andebbia/ Gummiglobus	1.5	0.0	4.7	0.0	0.0	
Aroramyces 1	0.5	0.0	3.1	0.0	13.0	Present
Aroramyces 2	0.0	0.0	1.6	0.0	11.1	
Ascomycete (*Dicena*-like)	0.0	0.0	0.0	0.0	1.9	
Austrogautieria	2.5	0.8	28.1	0.0	18.5	
Chamonixia	8.0	**41.1**	**53.1**	0.0	**59.3**	
*Descomyces stolatus**	0.0	7.3	0.0	0.0	0.0	
Descomyces 2	0.0	12.9	0.0	0.0	13.0	
Descomyces 3	0.0	18.5	0.0	0.0	0.0	
Dingleya	0.0	0.0	1.6	0.0	3.7	Present
Elaphomyces	2.0	2.4	1.6	0.0	13.0	Present
Gautieria	0.5	0.0	1.6	0.0	0.0	
Glomus	0.0	0.8	0.0	2.9	0.0	
Hydnoplicata convoluta	0.0	0.8	0.0	0.0	1.9	
Hymenogaster group	6.5	0.0	1.6	0.0	0.0	
Hysterangium spp.	4.0	3.2	**37.5**	0.0	44.4	Present
Hysterangium inflatum	0.0	0.0	0.0	0.0	1.9	
Hysterogaster 1	1.0	**34.7**	0.0	0.0	0.0	
Hysterogaster 2	5.5	0.0	1.6	0.0	0.0	
Labyrinthomyces 1	1.0	0.0	4.7	0.0	0.0	Present
Labyrinthomyces 2	**31.7**	0.8	3.1	0.0	0.0	
Labyrinthomyces 3	1.0	0.0	0.0	0.0	0.0	
Leucogaster meridionalis	0.0	0.0	0.0	0.0	5.6	Present
Mesophellia	1.5	0.0	3.1	0.0	0.0	
Octaviania	2.5	0.8	4.7	0.0	3.7	
Protoglossum	0.5	0.0	0.0	0.0	13.0	
Quadrispora oblongispora	0.0	0.0	0.0	0.0	1.9	

Table 14.2: Fungal taxa detected in the diets of six macropodoids (Continued).

	Mammal species					
	W. bicolor	*T. thetis*	*M. parma*	*P. penicillata*	*A. rufescens*	*P. tridactylus*
Fungal taxa	**n = 199**	**n = 124**	**n = 45**	**n = 36**	**n = 54**	**n = 1**
Russuloid 1	1.5	0.0	0.0	0.0	0.0	
Russuloid 2	4.0	8.9	10.9	0.0	**53.7**	
Russuloid 3	0.5	4.8	4.7	0.0	0.0	
Russuloid 4	**19.6**	**37.1**	**43.8**	8.6	0.0	Present
Russuloid 5	0.0	0.0	1.6	0.0	0.0	
Russuloid 6	0.0	0.0	0.0	0.0	25.9	
Scleroderma 1	2.0	1.6	1.6	0.0	0.0	
Scleroderma 2	2.5	4.0	3.1	**2.9**	3.7	
Scleroderma 3	4.5	8.9	9.4	0.0	11.1	
Scleroderma tommayi	1.5	0.0	1.6	0.0	1.9	
Sclerogaster	4.0	4.0	0.0	0.0	0.0	
Sphaerosoma	0.5	0.0	0.0	0.0	0.0	
Thaxterogaster	32.2	26.6	**43.8**	5.7	**53.7**	Present
Timgrovea	0.0	0.0	0.0	0.0	1.9	
Unknown 1	2.0	0.0	0.0	0.0	0.0	
Unknown 2	0.0	0.0	0.0	0.0	0.0	Present
Unknown 3	0.5	0.0	0.0	0.0	0.0	
Unknown 4	0.5	0.0	0.0	0.0	0.0	
Unknown 5	0.0	0.0	0.0	0.0	7.4	
Unknown 6	0.0	0.0	0.0	0.0	3.7	
Unknown 7	0.0	0.0	0.0	0.0	3.7	
Total taxa consumed	37	27	30	7	31	10

*Trappe and Claridge nom. prov.
The long-nosed potoroo (*P. tridactylus*) is not shown, because diet of that species was based on only one sample.
Values in bold represent the fungal taxa identified as having contributed most to the differences found between species.

examine the influence of site differences on diet. In all respects, these data were analysed as for the full interspecies comparisons described above. ANOSIM revealed a significant difference between Bray-Curtis similarity coefficients coded by season ($R = 0.677$, $P < 0.001$; Fig. 14.3) but not by site ($R = -0.168$, $P = 0.78$), demonstrating that season, not site, influenced patterns in spore consumption (Fig. 14.4).

Predicted dietary diversity

Curves describing the number of fungal taxa detected in the diets of swamp wallabies, red-necked pademelons and parma wallabies with increasing numbers of samples show that sampling captured a considerable amount of the diversity of fungal spore types in the diet of these mammals (Fig. 14.5). However, no species' diet was described fully in terms of its fungal component; bootstrapped (predicted) values suggested that at least three or four additional taxa would need to have been detected for diversity to reach an asymptote in each case.

DISCUSSION

Macropodid mycophagy

This study greatly extends the knowledge of mycophagy among macropodids in three main ways. First, the numbers of Australian mammal species known to include fungi in their diet has been expanded, and the research shows that consumption and dispersal of fungi by macropodids has probably been underestimated or overlooked in past studies. Mycophagy appears widespread among the macropodoids in the region where this work was conducted. Assuming the

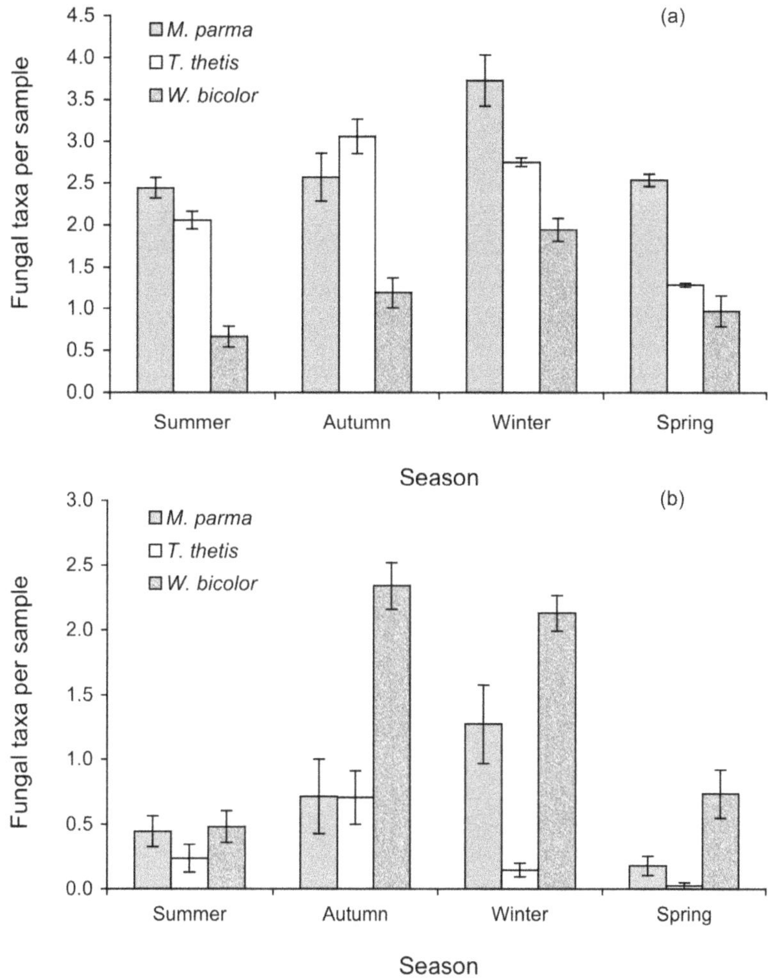

Figure 14.2: Mean number (±SE) of different spore types of (a) hypogeous fungi and (b) epigeous fungi found in each scat sample during different seasons.

animals sampled and the study areas are not unique, it is likely that most medium-sized macropodids inhabiting forests and woodlands in Australia are mycophagous to some extent. Second, the results show that, far from being an occasional component of the diet, fungi are regularly consumed by small and medium-sized macropodids, with a diversity of fungi consumed in all seasons. Third, the work has highlighted that most of the fungi consumed are hypogeous ectomycorrhizal (truffle-like) species, suggesting that current understanding of mammal-mediated spore dispersal in Australian forested habitats – where a few specialist mycophagous mammals consume and disperse most of the fungi – is probably a greatly simplified view of mammal–fungal community ecology.

All six macropodoids examined in this study consumed fungi. As expected, the two potoroid marsupials in the study region – the rufous bettong and long-nosed potoroo – consumed a great diversity of fungi, although for the latter this assumption was based upon a single scat sample containing 10 fungal taxa (Table 14.2) and on evidence from past studies showing the species to be heavily mycophagous (Bennett and Baxter 1989; Claridge *et al.* 1993). Of greater interest was the discovery that species not previously reported in the literature as consuming hypogeous fungi, principally the red-necked pademelon and parma wallaby, regularly consumed a great diversity (27–30 taxa) of such fungi. Previous work by Redenbach (1982) identified that both species were mycophagous, but there was no attempt to determine the kind of fungi eaten. For both species, Redenbach's data suggested that fungus was only a minor component of the diet. Site differences may play a part in the small amount of fungus reported as consumed by pademelons, because most of Redenbach's animals

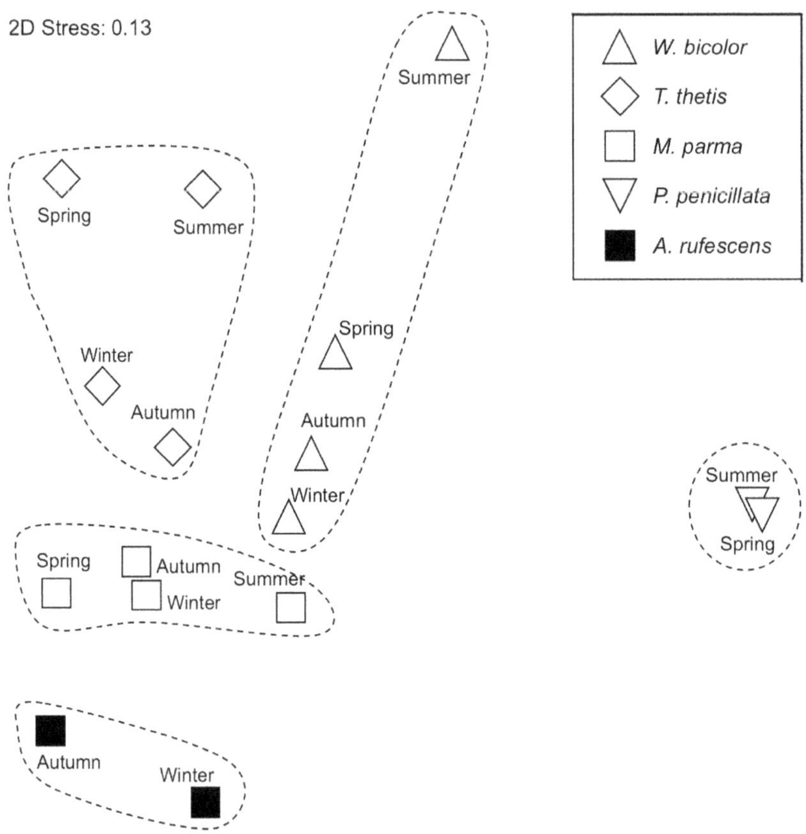

Figure 14.3: MDS plot of Bray-Curtis similarity coefficients for the seasonal diets of five macropodoid species, according to the percentage occurrence of different fungal taxa in each species diet. Dashed lines indicate the statistically supported grouping of diets by species.

Table 14.3: Species pairs compared using similarity analysis and the *R*-statistic and corresponding *P*-values for each comparison.

Species being compared	R-statistic	Significance level
W. bicolor – T. thetis	0.594	*
W. bicolor – M. parma	0.448	*
W. bicolor – P. penicillata	1.000	n.s.
W. bicolor – A. rufescens	0.667	*
T. thetis – M. parma	0.646	*
T. thetis – P. penicillata	1.000	n.s.
T. thetis – A. rufescens	0.630	*
M. parma – P. penicillata	1.000	*
M. parma – A. rufescens	0.481	*
P. penicillata – A. rufescens	0.917	n.s.

R-statistics are interpreted in this context as proportion of the diet of each species that is similar, with 0 being completely the same and 1 being completely different. Comparisons involving brush-tailed rock-wallabies (*P. penicillata*) are based on small sample sizes, so despite diets being highly dissimilar the power of the analysis was low and *P*-values were not significant.

came from Dorrigo where rainforest abuts cleared pasture. Under these conditions, hypogeous fungal diversity would be expected to be low or absent, but at Gibraltar Range, rainforest abuts open eucalypt forest where the diversity of hypogeous fungi would be expected to be high (Claridge 2002).

Despite small sample sizes and patchy seasonal sampling, I was able to demonstrate that brush-tailed rock-wallabies are also mycophagous, consuming at least seven fungal taxa (some of which are hypogeous) in and around the gorges that flank the eastern edge of the New England Tableland. Brush-tailed rock-wallabies have not previously been recorded as eating fungus, although they have been seen to dig for unidentified food items on the New England Tableland (Paul Bayne, NSW NPWS, pers. obs.).

The swamp wallaby has previously been described as mycophagous (Hollis *et al.* 1986; Osawa 1990; Claridge *et al.* 2001) and work in the New England region has suggested that fungus makes up about 15–20% of its diet (Harrington 1976; Hollis *et al.* 1986). In my study, swamp wallabies consumed more

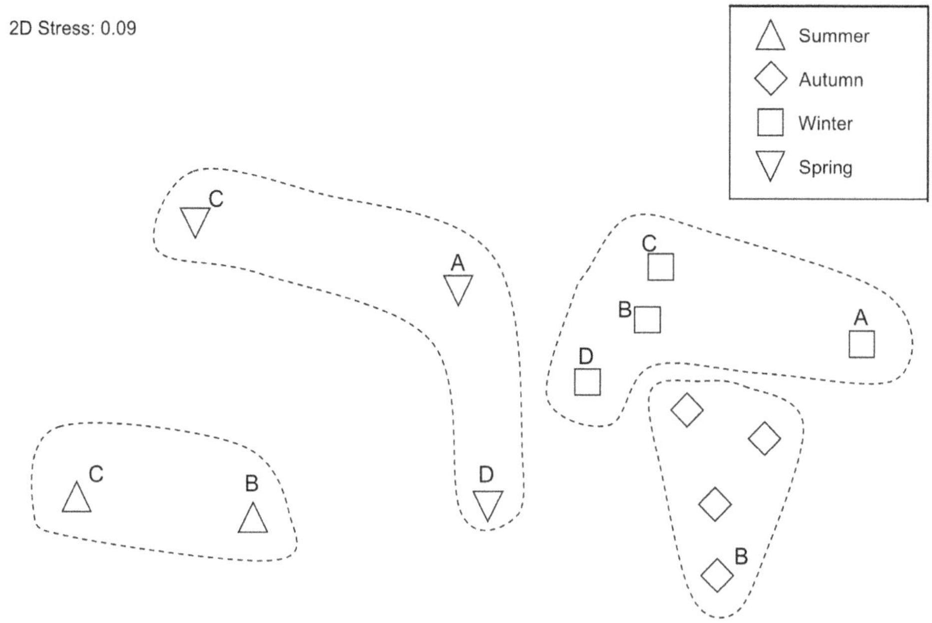

Figure 14.4: MDS plot of Bray-Curtis similarity coefficients for the seasonal diets of swamp wallabies from different sites at Gibraltar Ranges National Park, according to the percentage occurrence of different fungal taxa in the diet. Dashed lines indicate the statistically supported grouping of diets by season. Letters (A–D) refer to the swamp wallaby collection sites in Figure 14.1.

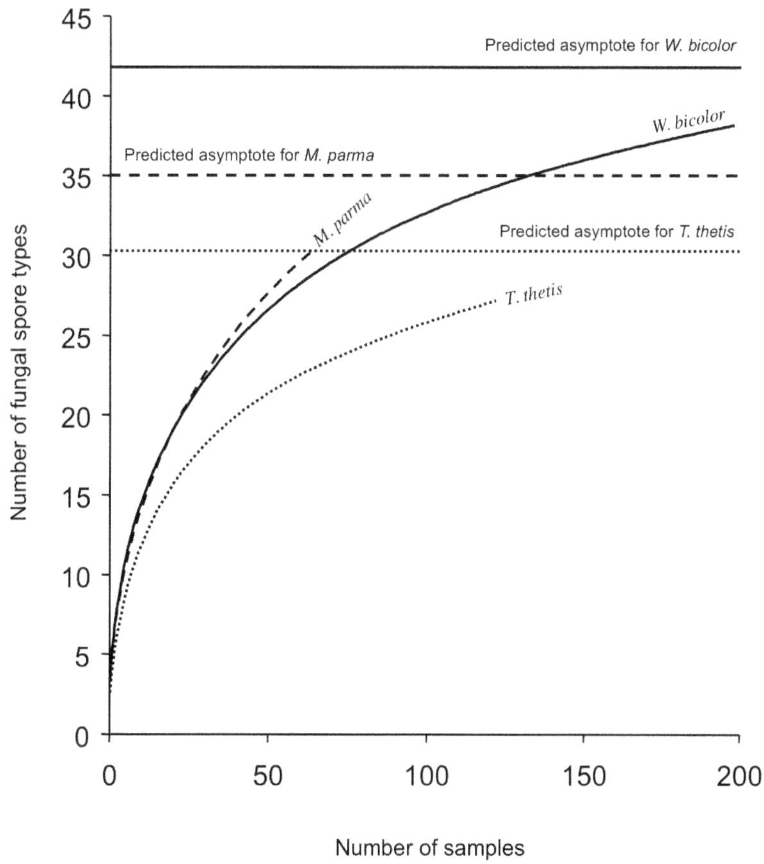

Figure 14.5: Species accumulation curves describing the number of fungal taxa returned for swamp wallabies, red-necked pademelons and parma wallabies with successively pooled randomly selected samples (999 iterations). The predicted true totals of taxa that would be observed for each species are shown as horizontal lines.

epigeous fungi than the other species, but most of the 37 spore types consumed by swamp wallabies were of hypogeous taxa and were consumed across a range of habitats and during all seasons at Gibraltar Range. My results therefore support the observation of Claridge *et al.* (2001), who demonstrated that swamp wallabies consumed a diversity of epigeous and hypogeous fungi, and show unequivocally that these fungi are consumed as part of the normal diet of the species.

Other forest-dwelling macropodids that occur along the Great Escarpment in north-east New South Wales (Jarman and Phillips 1989; Vernes *et al.* 2006) were not included in this study, mainly because I could not sample them indirectly (i.e. by collection of scats) with confidence. For example, red-necked wallabies (*M. rufogriseus*) are widespread in New England but I could not separate the scats of this mammal from other *Macropus* species in the region. Although the red-legged pademelon occurs at Washpool and Gibraltar Range, I did not sample its diet because the animal typically resides deep within the rainforest and scats are difficult to find. However, Vernes and Trappe (2007) have recently shown that this species consumes hypogeous fungi at rainforest/sclerophyll forest boundaries in north-east Queensland, and Redenbach (1982) detected fungi in its diet at Wallaby Creek (Jarman and Phillips 1989).

Seasonality in fungal consumption

The three macropodids for which I have sufficient data (swamp wallaby, red-necked pademelon and parma wallaby) consumed fungi in all seasons. All species, however, clearly differed seasonally in the proportions of samples containing fungus and the numbers of spore types recorded per sample. All species consumed less fungus of fewer types in spring than in other seasons. These trends are consistent with the results of Hollis *et al.* (1986) for swamp wallabies, where percentage of fungus in the diet was high in summer (20%), autumn (23%) and winter (17%), but low in spring (4%).

Hollis *et al.* (1986) suggested that considerable fungal consumption by swamp wallabies in all seasons except spring might be related to nutritional demands for protein. Mesic-adapted forest-dwelling macropodids have high nitrogen maintenance requirements (Hume 1977, 1986), so fungi may provide macropodids with a readily available source of protein in autumn and winter when plant proteins are low, but not in spring, when plant protein levels are high (Hollis *et al.* 1986). Young and Hume (2005) determined that another partially mycophagous marsupial in mesic forests and woodlands, the northern brown bandicoot (*Isoodon macrourus*), also has unusually high nitrogen maintenance requirements.

Like the potoroid marsupials, macropodids are foregut fermenters, using pre-gastric fermentation in the forestomach to begin the breakdown of ingested food (Hume 1982). This strategy allows potoroids to better utilise the protein in fungus that is largely inaccessible to hindgut fermenters, because much of the protein in fungus is tied up in cell-wall tissue that is difficult to digest (Claridge and Cork 1994). Macropodid forestomachs are less developed than potoroid forestomachs (Freudenberger *et al.* 1989). However, there is a grade in forestomach volume as a percentage of total gut volume, with forest-dwelling macropodids intermediate between the small potoroids and the larger and more arid-adapted macropodids (Dellow 1979; Hume 1982, Freudenberger *et al.* 1989). Considering their high nitrogen maintenance requirements (Hume 1977, 1986) and the advantage of a relatively large forestomach, mycophagy by small and medium-sized forest-dwelling macropodids may represent an efficient nutritional strategy at certain times of the year.

Fungal consumption and potential for dispersal by a community of macropodoids

There is general agreement in the literature that potoroid marsupials rely heavily on fungi as a food source, but that most macropodids are primarily browsers and grazers on a range of plant foods (Hume 1982; Norbury *et al.* 1989; Claridge 2002). I did not measure the amount of fungus in the diet relative to other food types, as my primary interest was not to determine how much fungus was consumed but the regularity of fungal consumption and how many different kinds of fungi were eaten. For the macropodids I studied, previous estimates of fungi in the diet ranged from 0–11% for red-necked pademelons, red-legged pademelons and parma wallabies (Redenbach 1982) to around 20% for swamp wallabies (Harrington 1976; Hollis *et al.* 1986). This indicates that fungus is consumed in considerably smaller amounts than is seen in potoroids, where the diet can be as much as 90% fungus (Nguyen *et al.* 2005). Importantly, though, fungus is a food item regularly consumed by the small and medium-sized macropodids

I examined, and the diversity of fungus consumed is great. In the New England Bioregion there are about 43 extant species of mammal, and of these 19 (44%) are mycophagous to some degree (Vernes 2007). Of the 10 species of macropodid that occur there, at least six are mycophagous, including the four that I examined. Mycophagy among macropodoids is clearly widespread in this region, which has ecological implications for the dispersal of fungal spores in local forests and woodlands.

The macropodids in my study area had different fungal diets, with all species significantly different in terms of what taxa of fungi they consumed, the proportions in which they consumed each taxa and the time of year when they consumed them. Each macropodoid consumed at least a few fungal taxa that were not eaten by any other (Table 14.2). Habitat undoubtedly played some role in these differences. The red-necked pademelon is a rainforest/open forest ecotone species, the parma wallaby at my study area lived in woodland with a dense heath understorey and swamp wallabies occupied a broad range of habitats, including those occupied by the other macropodoid species. Furthermore, most sampling locations sampled just one of the species examined (Fig. 14.1), so site differences might also play a part in the differences. Nevertheless, such differences indicate that mammals in different habitats are consuming and dispersing unique suites of fungi, and might be playing an important local role in fungal community dynamics.

Most mycophagous mammals probably consume palatable fungi in proportions similar to their local availability, so it is likely that the contents of the macropodid diets examined loosely reflected what was available to each species at the time, although this remains to be tested. Nevertheless, the samples collected from six macropodoid species collectively contained a far greater diversity of fungal taxa than would have been sampled from any one of these mammals alone. Dispersal of the rich diversity of fungi typical of forested habitats is thus accomplished by a range of mammals, not simply by a few specialist mycophagous species, and community interactions between mammals and fungi is complex. From an ecological standpoint, the dispersal of fungi by a wide range of mammal species is encouraging for the long-term maintenance of fungal communities because mammals traditionally thought of as important mycophagists, such as the potoroids, are among those most affected by human-induced changes to the New England Bioregion, and Australia in general, in the last 150 years. New England has lost two mycophagous bettongs (the Tasmanian bettong, *Bettongia gaimardi*, and brush-tailed bettong, *B. penicillata*), and the two remaining potoroids (the rufous bettong and long-nosed potoroo) are much reduced in range and numbers since European settlement (Jarman and Vernes 2006). The swamp wallaby, which remains widespread and abundant in eastern Australian forests, may therefore be fulfilling an important ecological role that was previously performed to a greater degree by the potoroids. Claridge *et al.* (2001) suggested that swamp wallabies might be performing the same general spore dispersal role as the more heavily mycophagous marsupials. My work shows that statement could be expanded to include many of the forest-dwelling macropodids such as the pademelons, parma wallaby and brush-tailed rock-wallaby, which occupy different types of habitats and are consuming and dispersing different fungal taxa.

ACKNOWLEDGMENTS

Stuart Green (University of New England) was instrumental in the selection of study sites, collection of samples and organisation of field trips. Without his input, much of this work could not have been achieved. James Trappe (Oregon State University) and Michael Castellano (USDA Forest Service) gave great guidance in the identification of fungal spores and shared their rich knowledge of fungal taxonomy and mycophagy. Dean Carter, Pam Brook and Tani Cooper prepared some of the mammal samples for analysis. Tani Cooper also provided valuable comments on an earlier draft of this chapter, that greatly improved its clarity. Peter Jarman provided some of the rock-wallaby samples, provided astute feedback on an earlier draft of the chapter and was a source of scholarly advice and encouragement throughout. This research was supported under the Australian Research Council's Discovery Projects funding scheme (DP0557022).

REFERENCES

Allen EB, Allen MF, Helm DJ, Trappe JM, Molina R and Rincon E 1995. Patterns and regulation of mycorrhizal plant and fungal diversity. *Plant Soil* **170:** 47–62.

Bennett AF and Baxter BJ 1989. Diet of the long-nosed potoroo, *Potorous tridactylus* (Marsupialia: Potoroidae), in south-western Victoria. *Aust. Wildl. Res.* **16:** 263–271.

Bougher NL and Lebel T 2001. Sequestrate (truffle-like) fungi of Australia and New Zealand. *Aust. Syst. Bot.* **14:** 1–47.

Burbidge AA and McKenzie NL 1989. Patterns in the modern decline of Western Australia's vertebrate fauna: causes and conservation implications. *Biol. Conserv.* **50:** 143–198.

Caldwell IR, Vernes K and Bärlocher F 2005. The northern flying squirrel as a vector for inoculation of red spruce (*Picea rubens*) seedlings with ectomycorrhizal fungi. *Sydowia* **57:** 166–178.

Castellano MA, Trappe JM, Maser Z and Maser C 1989. *Key to the Spores of the Genera of Hypogeous Fungi of North Temperate Forests.* Mad River Press, California.

Christensen PES 1980. The biology of *Bettongia penicillata* (Gray, 1837) and *Macropus eugenii* (Desmarest, 1817) in relation to fire. Report No. 91. Forests Department of Western Australia.

Claridge AW 2002. Ecological role of hypogeous ectomycorrhizal fungi in Australian forests and woodlands. *Plant Soil* **244:** 291–305.

Claridge AW and Cork SJ 1994. Nutritional value of hypogeal fungal sporocarps for the long-nosed potoroo (*Potorous tridactylus*), a forest-dwelling mycophagous marsupial. *Aust. J. Zool.* **42:** 701–710.

Claridge AW and May TW 1994. Mycophagy among Australian mammals. *Aust. J. Ecol.* **19:** 251–275.

Claridge AW, Tanton MT, Seebeck JH, Cork SJ and Cunningham RB 1992. Establishment of ectomycorrhizae on the roots of two species of *Eucalyptus* from fungal spores contained in the faeces of the long-nosed potoroo (*Potorous tridactylus*). *Aust. J. Ecol.* **17:** 207–217.

Claridge AW, Tanton MT and Cunningham RB 1993. Hypogeal fungi in the diet of the long-nosed potoroo (*Potorous tridactylus*) in mixed-species and regrowth eucalypt forest stands in south-eastern Australia. *Wildl. Res.* **20:** 321–337.

Claridge AW, Trappe JM and Claridge DL 2001. Mycophagy by the swamp wallaby (*Wallabia bicolor*). *Wildl. Res.* **28:** 643–645.

Clarke KR and Gorley RN 2006. PRIMER v6: user manual/tutorial. PRIMER-E, Plymouth.

Dellow DW 1979. Physiology of digestion in the macropodine marsupials. PhD thesis. University of New England, Armidale, Australia.

Dickman CR 1994. Mammals of New South Wales: past, present and future. *Aust. Zool.* **29:** 158–165.

Donaldson R and Stoddart M 1994. Detection of hypogeous fungi by the Tasmanian bettong (*Bettongia gaimardi*: Marsupialia; Macropodoidea). *J. Chem. Ecol.* **20:** 1201–1207.

EPBC (Environment Protection and Biodiversity Conservation) Act 1999. Australian Government, Canberra.

Freudenberger DO, Wallis IR and Hume ID 1989. Digestive adaptations of kangaroos, wallabies and rat-kangaroos. In *Kangaroos, Wallabies and Rat-kangaroos.* (Eds G Grigg, P Jarman and I Hume) pp. 179–187. Surrey Beatty and Sons, Sydney.

Green K, Tory MK, Mitchell AT, Tennant P and May TW 1999. The diet of the long-footed potoroo (*Potorous longipes*). *Aust. J. Ecol.* **24:** 151–156.

Harrington JR 1976. The diet of the swamp wallaby (*Wallabia bicolor*) at Diamond Flat, New South Wales. Dip.Natural Resources thesis. University of New England, Armidale, Australia.

Hill FAR and Triggs BE 1985. Ecology and distribution of the long-footed potoroo (*Potorous longipes*): a secondary preliminary examination. Report No. 310. Dept of Conservation, Forests and Lands, State Forests and Lands Service, Research Branch.

Hollis CJ, Robertshaw JD and Harden RH 1986. Ecology of the swamp wallaby (*Wallabia bicolor*) in New South Wales. I. Diet. *Aust. Wildl. Res.* **13:** 355–365.

Hume ID 1977. Maintenance nitrogen requirements of the macropod marsupials *Thylogale thetis*, red-necked pademelon, and *Macropus eugenii*, tammar wallaby. *Aust. J. Zool.* **25:** 407–417.

Hume ID 1982. *Digestive Physiology and Nutrition of Marsupials.* Cambridge University Press, Sydney.

Hume ID 1986. Nitrogen metabolism in the parma wallaby, *Macropus parma. Aust. J. Zool.* **34:** 147–155.

Hunter JT and Clarke PJ 1998. The vegetation of granite outcrop communities on the New England Batholith of eastern Australia. *Cunninghamia* **5:** 547–618.

Jarman PJ and Phillips CM 1989. Diets in a community of macropod species. In *Kangaroos, Wallabies and Rat-kangaroos.* (Eds G Grigg, P Jarman and I Hume) pp. 143–149. Surrey Beatty and Sons, Sydney.

Jarman P and Vernes K 2006. Wildlife. In *High Lean Country: Land, People and Memory in New England.* (Eds A Atkinson, JS Ryan, I Davidson and A Piper) pp. 44–56. Allen and Unwin, Sydney.

Johnson CN 1996. Interactions between mammals and ectomycorrhizal fungi. *Trends Ecol. Evol.* **11:** 503–507.

Maser C, Trappe JM and Nussbaum RA 1978. Fungal–small mammal interrelationships with emphasis on Oregon coniferous forests. *Ecology* **59:** 799–809.

Nguyen V, Needham A and Friend JA 2005. A quantitative dietary study of the 'critically endangered' Gilbert's potoroo, *Potorous gilberti. Aust. Mammal.* **27:** 1–6.

Norbury GL, Sanson GD and Lee AK 1989. Feeding ecology of the Macropodoidea. In *Kangaroos, Wallabies and Rat-kangaroos.* (Eds G Grigg, P Jarman and I Hume) pp. 169–178. Surrey Beatty and Sons, Sydney.

Osawa R 1990. Feeding strategies of the swamp wallaby, *Wallabia bicolor*, on North Stradbroke Island, Queensland. I. Composition of diets. *Aust. Wildl. Res.* **17:** 615–621.

Pyare S and Longland WS 2001. Mechanisms of truffle detection by northern flying squirrels. *Can. J. Zool.* **79:** 1007–1015.

Reddell P, Spain AV and Hopkins M 1997. Dispersal of spores of mycorrhizal fungi in scats of native mammals in tropical forests of northeastern Australia. *Biotropica* **29:** 184–192.

Redenbach CJ 1982. A dietary comparison of three sympatric forest-dwelling macropodids. B.Natural Resources thesis. University of New England, Armidale, Australia.

Sheringham P and Hunter JT 2002. Vegetation and floristics of Gibraltar Range National Park. Report to the NSW National Parks and Wildlife Service, Northern Tablelands Region.

Taylor RJ 1992. Seasonal changes in the diet of the Tasmanian bettong (*Bettongia gaimardi*), a mycophagous marsupial. *J. Mammal.* **73:** 408–414.

Tory MK, May TW, Keane PJ and Bennett AF 1997. Mycophagy in small mammals: a comparison of the occurrence and diversity of hypogeal fungi in the diet of the long-nosed potoroo *Potorous tridactylus* and the bush rat *Rattus fuscipes* from southwestern Victoria, Australia. *Aust. J. Ecol.* **22:** 460–470.

Trappe JM 1988. Lessons from alpine fungi. *Mycologia* **80:** 1–10.

Troy S and Coulson G 1993. Home range of the swamp wallaby, *Wallabia bicolor. Wildl. Res.* **20:** 571–577.

Vernes K 2007. Are diverse mammal communities important for maintaining plant–fungal associations and ecosystem health? *Aust. Plant Conserv.* **15:** 16–18.

Vernes K and Trappe JM 2007. Hypogeous fungi in the diet of the red-legged pademelon (*Thylogale stigmatica*) from a rainforest–open forest interface in northeastern Australia. *Aust. Zool.* **34:** 203–208.

Vernes K, Castellano M and Johnson CN 2001. Effects of season and fire on the diversity of hypogeous fungi consumed by a tropical mycophagous marsupial. *J. Anim. Ecol.* **70:** 945–954.

Vernes K, Green S, Howes A and Dunn L 2006. Species richness and habitat associations of non-flying mammals in Gibraltar Range National Park. *Proc. Linn. Soc. NSW* **127:** 93–105.

Young V and Hume ID 2005. Nitrogen requirements and urea recycling in an omnivorous marsupial, the northern brown bandicoot, *Isoodon macrourus. Physiol. Biochem. Zool.* **78:** 456–467.

15 Determining the diet of tammar wallabies on Garden Island, Western Australia, using stable isotope analysis

A. McMillan, G. Coupland, B.K. Chambers, H.R. Mills and R. Bencini

SUMMARY

The ratios of the stable isotopes of carbon and nitrogen ($^{13}C/^{12}C$ and $^{15}N/^{14}N$) were measured in samples of hair collected from tammar wallabies (*Macropus eugenii*) between July 2005 and February 2006 to determine the long-term diet of three subpopulations from Garden Island, Western Australia. These populations were living on HMAS Stirling Naval Base, located at the southern end of the island, and in native bushland at the northern and southern ends of the island. The stable isotopic signatures of hair samples were compared with those of 12 plant species known to be consumed by tammar wallabies from previous dietary studies to test the hypothesis that the subpopulation living on the base would be relying heavily on the grasses from its ovals and lawns, resulting in higher $\delta^{13}C$ and $\delta^{15}N$ values in their tissues than in those of wallabies living in native bushland. We also expected that the $\delta^{13}C$ and $\delta^{15}N$ value of hair collected in summer, the period of reduced food availability, would be higher than that of hair collected during winter/spring. The $\delta^{13}C$ value of hair from tammar wallabies caught on the base was higher than that of wallabies from the northern and southern subpopulations ($p < 0.05$). In addition, tammar wallabies from the northern subpopulation had higher $\delta^{15}N$ values than those from the southern subpopulation ($p < 0.05$). These differences reflected the consumption of irrigated C_4 grasses (e.g. *Cynodon dactylon*) found on the base and demonstrate the long-term reliance of these tammar wallabies on this artificial source of food. The $\delta^{15}N$ value of hair did not differ between seasons ($p = 0.861$) but hair samples collected in December were lower in ^{13}C than hair collected in July and September ($p < 0.05$). Our findings demonstrate that tammar wallabies living on the base rely mainly on introduced grasses as a source of food. Limiting access to this source of food might be adopted to manage this population.

INTRODUCTION

Since 1972, Garden Island has been the location of the largest naval base in Australia, HMAS Stirling. The infrastructure of the naval base is primarily in the southern part of the island, including a causeway to facilitate road access from the mainland (Brooker 1992). The introduction to the island of irrigated lawns and ovals associated with the base has provided additional food sources for the tammar wallabies

(*Macropus eugenii*) that live in the area (Moredoundt 1983; Bell *et al.* 1987) and their numbers are believed to have increased.

It has been suggested that Garden Island supports three subpopulations of tammar wallabies: the northern and southern animals that consume native and introduced vegetation and the base 'roadside' or 'urban' animals that can access the grasses on the ovals and lawn areas (Brooker *et al.* 1992). There is variation in the quantity and floristic composition of the vegetation available between the locations of these subpopulations (Moredoundt 1983; Bell *et al.* 1987; Brooker *et al.* 1992). The accidental introduction of some edible plant species and the deliberate establishment of irrigated grass fields on the base are possibly responsible for increased numbers of tammar wallabies in this subpopulation (Moredoundt 1983; Bell *et al.* 1987). In comparison, the resources available to tammar wallabies in the northern and southern subpopulations are limited by the abundance of naturally occurring vegetation (Bell *et al.* 1987; Brooker *et al.* 1992).

Plants differ in their abundance of $\delta^{13}C$ depending upon the photosynthetic pathway used: C_3 plants ($\delta^{13}C < -28‰$) have relatively low values relative to C_4/CAM plants ($\delta^{13}C$ −12 to −13‰; Stapp and Polis 2003). The majority of introduced grasses on Garden Island are C_4. Therefore it was expected the $\delta^{13}C$ isotopic signature of the vegetation on the base would be sufficiently distinct to determine if it forms a large part of the long-term diet of the tammar wallabies living in this area. Stable isotope ratios can be measured in the hair of animals to gain an understanding of their diet (Gannes *et al.* 1998).

The primary objective of this work was to investigate the long-term diet of tammar wallabies in the northern, southern and base subpopulations by determining the stable isotope signatures of their hair and to study seasonal differences within each subpopulation. It was expected that tammar wallabies from the base subpopulation would rely more on irrigated grasses and would therefore show elevated (more positive) $\delta^{13}C$ and $\delta^{15}N$ values than the northern subpopulation, while the southern subpopulation would have intermediate values because members can access the irrigated grasses on the base but also consume native vegetation.

During periods of limited resource availability animals experience enriched $\delta^{15}N$ values, as animals utilise tissue protein for energy and essential amino acids (Gannes *et al.* 1998). Therefore it was expected that the $\delta^{13}C$ and $\delta^{15}N$ value of tammar wallaby hair collected in summer, the period of reduced food availability, would be higher than that of hair collected during winter/spring.

METHODS

To test our hypotheses, we collected samples of hair from tammar wallabies living in the three subpopulations and determined their isotopic signatures at different times of the year. We also used stomach content analysis of tammar wallabies killed by vehicles to confirm the percentage composition of plant species consumed.

Study site

Garden Island is located 5 km off the Western Australian (WA) coast, approximately 35 km south-west of the city of Perth (115°40'E, 32°16'S, Fig. 15.1). The island hosts the HMAS Stirling Naval Base, which was built during the 1970s. The infrastructure of the base is primarily in the southern part of the island and includes two sports ovals and several road verges

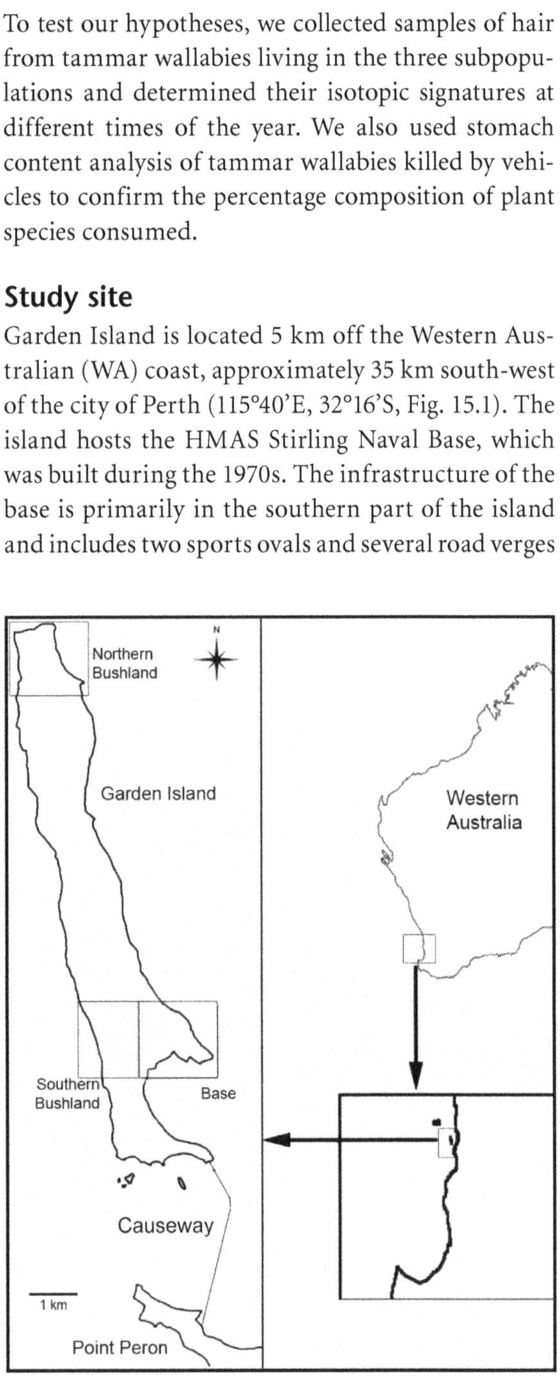

Figure 15.1: Location of Garden Island, Western Australia, showing the naval base and northern and southern bushland study sites.

Table 15.1: Carbon ($\delta^{13}C$) and nitrogen ($\delta^{15}N$) isotopic means (±SE) of plant species collected from Garden Island, Western Australia.

Species	$\delta^{13}C$	±SE	$\delta^{15}N$	±SE	No. of samples
Acacia rostellifera	–28.77	0.61	2.74	0.15	3
Asparagus asparagoides	–25.51	0.66	0.71	0.68	3
Asphodelus fistulosus	–28.03	0.49	2.04	0.50	3
Austrostipa flavescens	–28.03	0.07	2.86	0.23	3
Eremophila glabra	–26.13	0.39	6.35	0.21	3
Rhagodia baccata	–27.40	0.09	4.78	0.02	3
Solanum symonii	–26.93	0.23	2.59	0.36	3
Trachyandra divaricata	–29.28	0.40	2.86	0.61	3
Cynodon dactylon (oval)	–12.15	0.28	0.12	0.52	3
Cynodon dactylon (verge)	–13.28	0.08	-0.15	0.10	3

vegetated with introduced grasses, mainly *Cynodon dactylon*. The rest of the island is largely undisturbed and includes dense bushland dominated by the native *Acacia rostellifera, Callitris preissii* and *Melaleuca lanceolata* as well as the introduced weeds *Asparagus asparagoides, Asphodelus fistulosus* and *Trachyandra divaricata.*

Collection of plant material

Three replicate samples of species known to be consumed by tammar wallabies, based on stomach content analysis, were collected. Species collected included the native plants *Acacia rostellifera, Callitris preissii, Rhagodia baccata* and *Austrostipa flavescens* and the introduced weeds *Asparagus asparagoides, Asphodelus fistulosus* and *Trachyandra divaricata* as well as the *Cynodon dactylon* growing on the base's ovals and road verges (Table 15.1).

Collection of hair samples from tammar wallabies

Tammar wallabies from the northern, southern and base subpopulations on Garden Island (Fig. 15.1) were trapped using Thomas soft-wall traps made with shade-cloth stretched over a steel frame (450 × 450 × 800 mm, Sheffield Wire Works, Welshpool, WA). Traps were baited with a mixture of rolled lupins, rolled oats and sunflower seeds, flavoured with molasses (Kangaroo Muesli, Thompson & Redwood Produce Supplies, Upper Swan, WA). Thirty traps were set over the study areas in a grid at intervals of approximately 200 m during July, September and December 2005 and February 2006. Although the southern and base areas were separated by a 3 m high chain link fence, animals could move to and from the base through gaps under the fence. The northern subpopulation is located approximately 5.6 km to the north, a distance that effectively precluded the movement of animals between that and the other two locations (Chambers 2009).

Captured wallabies were weighed, measured, sexed, aged and tagged. Hair was collected from the right thigh of each trapped individual by trimming the region with surgical scissors. As the patch remained visible between successive trapping trips, we were able to assign the period in which the hair had been grown to each successive sample. This allowed us to determine seasonal differences in the isotopic signature of the hair. Hair samples were frozen within 1 hour at –20°C, and kept frozen until preparation.

Determination of stable isotope profiles of plant and animal material

Samples of plants were first washed in deionised water and dried at 60°C to a constant weight, crumbled and ground to a homogeneous particle size. Three replicate dried samples of leaves and flowers of plants were weighed into 4 × 6 mm tin cups. Samples of hair from each tammar wallaby were placed into individual petri dishes, then soaked and agitated in a 2:1 chloroform:methanol solution. Samples were left to dry in a fume hood for 24 hours, then cut manually into fine homogenous pieces using surgical scissors. Approximately 1.4–1.7 mg subsamples of hair were weighed into 4 × 6 mm tin cups.

Samples of plant material and hair were analysed using an automated coupling of a RoboPrep gas chromatograph with a Tracermass mass spectrometer (Europa Scientific, Crewe, UK). Gases from the combusted samples were separated in the GC and bled directly into the MS. Isotope signatures were expressed in the δ-notation, as ratios ($^{13}C/^{12}C$, $^{15}N/^{14}N$) of the sample relative to PeeDee Belemnite (PDB) (carbon) and atmospheric N_2 (air) standards, in parts per 1000 (‰) (Farquhar *et al.* 1982):

$$\delta X\% = \left(\left(R_{sample}/R_{standard}\right)\right) - 1 \times 100$$

where X is ^{13}C or ^{15}N and R is the ratio of heavy to light isotope ($^{13}C/^{12}C$, $^{15}N/^{14}N$). The laboratory standard for estimating $\delta^{13}C$ was Radish (Anca 54) and for $\delta^{15}N$, L Glutamic acid (Anca 5). The analytical error associated with $\delta^{13}C$ was 0.1‰ and $\delta^{15}N$ was 0.3‰.

Stomach content analysis

Samples of stomach contents were collected from the forestomach of road-killed tammar wallabies from the base (n = 3) and northern (n = 3) subpopulations to determine the percentage composition of plant species consumed. Five 5 g samples were removed from the stomach of each individual, placed in separated 100 mL beakers and 20 mL of 10% nitric acid was added. The contents of each beaker were boiled for 3 min. A further 20 mL of water was added and the contents left to simmer for a further 15 min. The contents were then left to cool, and continually flushed through a 500 μm sieve with water to remove residual acidity. A fraction of digesta was taken from each subsample and spread evenly on a microscope slide with fine forceps. Slides were first scanned quantitatively under a dissecting microscope (×4 magnification) to determine the type of food present. Ten sites on each slide were then photographed under a compound microscope at ×10 magnification along a line transect. Plant species in each photograph were identified and the percentage composition of each field of view tallied. Stomach particles were identified to species by comparison with a reference slide herbarium prepared from plant species collected from Garden Island.

Data analysis

Differences in the carbon and nitrogen isotopic signature were identified using multivariate Analysis of Similarity (ANOSIM), based on the Euclidean distance between mean $\delta^{13}C$ and $\delta^{15}N$ values of species from the different dietary sources. The similarity in the isotopic signature (carbon and nitrogen) of plant species within dietary sources was analysed using a multivariate link tree analysis.

Specific differences in the mean natural abundance of $\delta^{13}C$ and $\delta^{15}N$ between terrestrial plant species found on Garden Island were analysed using a univariate one-way analysis of variance (ANOVA). All means were of equal variance according to Cochran's test for homogeneity of variance (Snedecor and Cochran 1989).

A two-way ANOVA was used to determine equality of carbon and nitrogen isotope signatures of hair samples collected from different trapping periods using as fixed factors season (July, September and December 2005, February 2006) and location (north, base, south) and the interaction between the two factors. Tukey's multiple comparison tests were used to identify significant differences between samples from different locations and trapping times where results were significant (Crichton 1999). Prior to analysis, all mean values were tested for equal variance using the Cochran's test for homogeneity of variance (Snedecor and Cochran 1989).

RESULTS

Stable isotope profiles

The $\delta^{13}C$ value of hair from tammar wallabies caught on the base was higher than that of individuals from the northern and southern subpopulations ($p < 0.05$, Fig. 15.2a). Individuals from the northern subpopulation had higher $\delta^{15}N$ values relative to individuals from the southern subpopulation ($p < 0.05$, Fig. 15.2b). The $\delta^{15}N$ value of hair from tammar wallabies did not differ between seasons ($p = 0.861$, Table 15.2). By contrast, hair samples collected in December were significantly more depleted in ^{13}C than hair collected in July and September ($p < 0.05$, Table 15.2).

Stomach content analysis

Microscope analysis of the stomach contents of three tammar wallabies from both the northern and base subpopulations revealed a clear distinction in ingested plant material. Road verge and oval grasses were the main contributors to the diet of wallabies from the base, whereas native and introduced shrubs and trees (*Asphodelus fistulosus, Austrostipa flavescens, Asparagus asparagoides*) made a greater contribution to the diet of northern animals (Fig. 15.3).

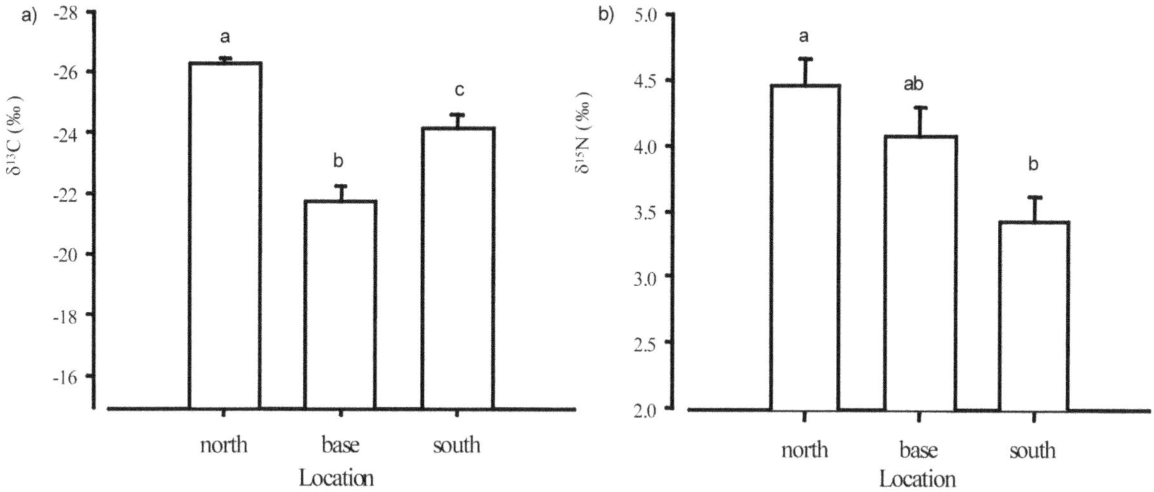

Figure 15.2: Mean (a) δ^{13}C and (b) δ^{15}N values of hair collected from tammar wallabies (*M. eugenii*) from different subpopulations on Garden Island, Western Australia (north, n = 30; base, n = 20; south, n = 24). Significant differences are indicated by different letters ($p < 0.05$).

DISCUSSION

The δ^{13}C value of hair from tammar wallabies sampled from the base was higher than that of animals living in the southern and northern bushlands, supporting our hypothesis that wallabies on the base rely heavily on introduced C_4 grasses as their source of food. Tammar wallabies from the base have an almost unlimited supply of grasses from irrigated road verges and ovals. As the δ^{13}C value of grass from the base (–12 to –13‰) is higher in δ^{13}C than any other terrestrial plant species, ingestion and reliance on this source of food best explains the elevated value of δ^{13}C in hair from tammars sampled from the base. This conclusion is also supported by the high proportion of grass fragments found in the stomach content of wallabies from the base.

Hair sampled from animals living in the northern bushland had higher δ^{15}N values than hair sampled from the base and southern bushland subpopulations. Tammar wallabies from the northern subpopulation may experience food shortages to a greater extent than those living on the base and southern bushland and

Table 15.2: δ^{13}C and δ^{15}N (mean and SE) values of hair from tammar wallabies (*M. eugenii*) collected in July, September and December 2005 and February 2006 from different subpopulations on Garden Island, Western Australia.

				δ^{13}C			δ^{15}N	
Location	**Month**	**n**	**mean**	**SE**	**Significance**	**mean**	**SE**	**Significance**
North	July	8	–25.48	0.38	a	4.37	0.34	a
	September	8	–25.92	0.43	a	4.34	0.36	a
	December	8	–27.29	0.28	b	4.41	0.41	a
	February	6	–26.26	0.47	b	4.76	0.73	a
South	July	8	–23.69	0.14	c	3.13	0.28	b
	September	4	–21.41	1.16	d	3.49	0.40	b
	December	8	–25.63	0.49	a	3.44	0.33	b
	February	4	–24.91	0.93	a	3.92	0.73	a
Base	July	8	–21.09	0.95	d	4.42	0.34	a
	September	4	–21.24	0.25	d	4.21	0.48	a
	December	4	–23.76	1.06	c	3.71	0.72	b
	February	4	–21.50	1.23	d	3.58	0.23	b

For each isotope ratio, different letters indicate significant differences ($p < 0.05$).

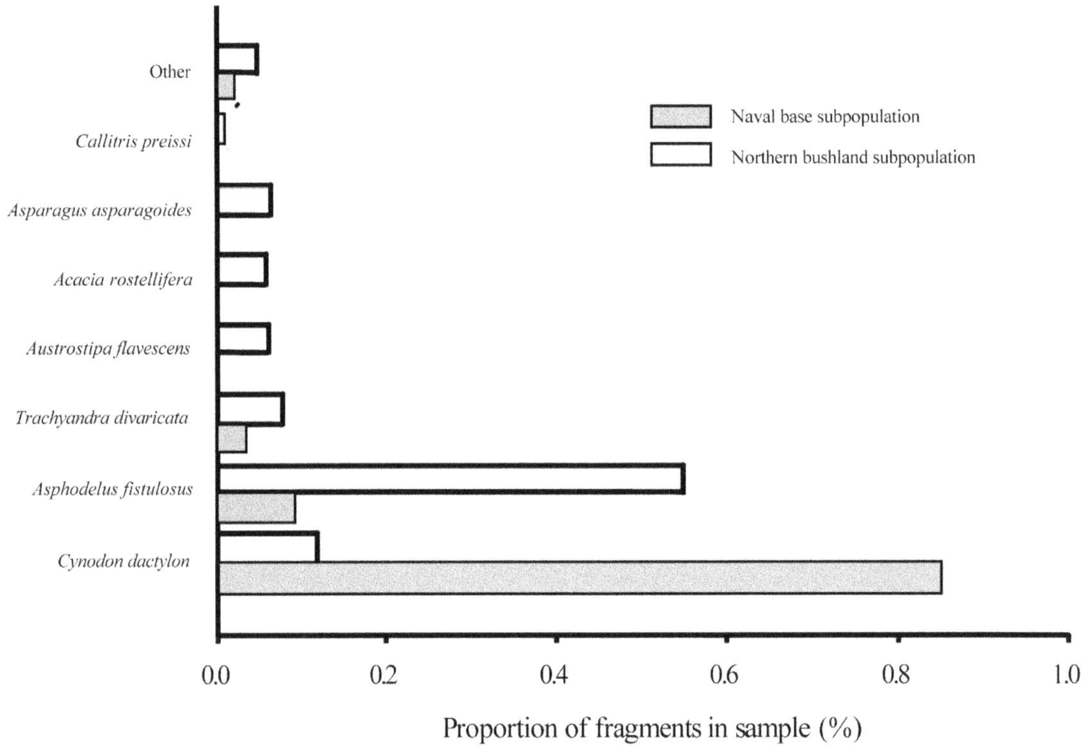

Figure 15.3: Relative proportion of fragments of plant material observed in the stomach contents of tammar wallabies (*M. eugenii*) from northern (n = 3) and base (n = 3) subpopulations on Garden Island, Western Australia.

may therefore resort to utilising tissue protein for energy, resulting in higher $\delta^{15}N$ values (Gannes *et al.* 1998). More likely, the higher $\delta^{15}N$ values of hair from the northern and base tammar wallabies reflect the consumption of a different diet. Animals from the south share access to road verge grasses with animals from the base. As road verge grasses have low $\delta^{15}N$ relative to native vegetation, this dietary source may be contributing to the low $\delta^{15}N$ value observed in these animals. However, these grasses could not be solely responsible for the lower $\delta^{15}N$ value of southern animals. Stomach content analysis of base tammar wallabies indicated that a high proportion of their diet was composed of grasses, yet their $\delta^{15}N$ value was higher than that of animals from the southern bushland. The southern bushland wallabies also have access to a selection of native and introduced plant species. Therefore it is likely that tammar wallabies from the southern bushland subpopulation are ingesting a species of plant found in the southern bushland with a lower $\delta^{15}N$ value, as well as consuming grasses.

Stable isotope analysis on the hair of tammar wallabies allowed us to establish the reliance on imported grasses over the long time span entailed in the growth of hair. Scat analysis and direct observation, by contrast, would have simply revealed what the wallabies were eating 24–48 hours prior to the collection of samples, in the case of scats, and would only have allowed us to infer the consumption of grass at the moment of observation in the case of direct observation. Additionally, only indigestible material ends up in the scats and therefore highly digestible items often go undetected when using scat analysis.

Stable isotope analysis indicated significantly lower $\delta^{13}C$ in hair from tammars captured in December compared with those captured in July and September. In agreement with the hypothesis, in summer tammar wallabies are utilising a greater proportion of dietary sources that are low in $\delta^{13}C$. This also suggests that tammars from Garden Island are changing their diet during the year.

MANAGEMENT IMPLICATIONS

Our results support the notion that the recent increase in numbers of tammars living on the base and southern part of the island is due to their reliance on grasses that represent a practically unlimited source of food throughout the year. Fencing off ovals or removing the grass from verges would deprive tammars of this resource and probably result in a decline in the population.

The determination of the stable isotopic signatures of hair offers wildlife managers a simple and reliable technique to examine long-term dietary preferences of wildlife. It could be an important tool to determine dietary preferences in populations of macropodids, especially when it is suspected that they damage crops, as the isotopic signatures of the fur would reflect their reliance on crops rather than native species. This method is also preferable to the use of scat analysis, which only allows the detection of indigestible material recently consumed by the animals.

In the case of Garden Island, our findings confirm that tammars living on the base consume mainly introduced C_4 grasses. Limiting access to this source of food might be adopted by the Navy to manage this population of tammars.

ACKNOWLEDGMENTS

We thank the Australian Defence Force and the School of Animal Biology at the University of Western Australia for funding this research. Our thanks also go to Trevor Smith (Defence Public Area Ranger, Garden Island), Joanne Wann (Defence Regional Environmental Officer) and the security and grounds and gardens staff of HMAS Stirling. Experimental protocols were approved by the University of Western Australia's Animal Ethics Committee (approval no. 04/100/376) and followed the recommendations of the Australian National Health and Medical Research Council.

REFERENCES

Bell DT, Moredoundt JC and Loneragan A 1987. Grazing pressure by the tammar (*Macropus eugenii* Desm.) on the vegetation of Garden Island, Western Australia, and the potential impact on food reserves of a controlled burning regime. *J. Roy. Soc. West. Aust.* **69:** 89–94.

Brooker M 1992. Biological survey of Garden Island. Final report. CSIRO Division of Wildlife and Ecology, Perth.

Brooker MG, Smith GT, Leone J and Ingram JA 1992. A biological survey of Garden Island, Western Australia. II. Terrestrial mammals. CSIRO Division of Wildlife and Ecology, Perth.

Chambers BK 2009. Human disturbance affects the ecology and population dynamics of tammar wallabies (*Macropus eugenii*) on Garden Island. PhD thesis, The University of Western Australia.

Crichton N 1999. Tukey multiple comparison test. *J. Clinic. Nur.* **8:** 304.

Farquhar GD, O'Leary MH and Berry JA 1982. On the relationship between carbon isotope discrimination and the intercellular carbon dioxide concentration in leaves. *Aust. J. Plant Physiol.* **9:** 121–137.

Gannes LZ, del Rio CM and Koch P 1998. Natural abundance variations in stable isotopes and their potential uses in animal physiological ecology. *Comp. Biochem. Physiol. Part A Mol. Integr. Physiol.* **119:** 725–737.

Moredoundt JC 1983. The ecological management of Garden Island. BSc Hons thesis. University of Western Australia, Perth, Australia.

Snedecor GW and Cochran WG 1989. *Statistical Methods.* 8th edn. Blackwell Publishing.

Stapp P and Polis GA 2003. Marine resources subsidize insular rodent populations in the Gulf of California, Mexico. *Oecologia* **134:** 496–504.

16 Ecology and conservation of the antilopine wallaroo: an overview of current knowledge

E.G. Ritchie

SUMMARY

The antilopine wallaroo (*Macropus antilopinus*) is restricted entirely to the tropical savannas of northern Australia and, along with the black wallaroo (*Macropus bernardus*), remains the least studied of the large macropodids. There is currently great concern that many of the endemic mammals of this region are undergoing declines in their distribution and population size; without detailed ecological information, proactive management to prevent further declines will be difficult. Between 2002 and 2005, I conducted an extensive study of the ecology and conservation of the antilopine wallaroo at 50 sites across the species' entire distribution. I calculated relative density estimates for each site using line transects and recorded the vegetation characteristics, fire history and resource availability at each site. I collected detailed information on the socioecology and behaviour of antilopine wallaroos and other sympatric macropodid species opportunistically. This information, together with that of previously published studies, is presented in the form of a general overview of the ecology and conservation of the antilopine wallaroo.

The antilopine wallaroo has a wide but patchy distribution across northern Australia, with a mean abundance of three animals per square kilometre (range 0.33–31). I identified the availability of permanent water, frequency of fire and the density of eastern grey kangaroos (*Macropus giganteus*) as the factors that most strongly influenced antilopine wallaroo abundance. In the tropics, the antilopine wallaroo is the most gregarious of the macropodid species, with an average group size of three. Group composition changes from predominantly mixed in the wet season, the time of mating, to largely all-male and all-female groups in the dry season (non-breeding period). At one site in Queensland, antilopine wallaroos fed predominantly on grass, particularly black speargrass (*Heteropogon contortus*), but forbs constituted 5–6% of the diet, which varied seasonally and between sexes. Under climate change scenarios predicted for northern Australia, the antilopine wallaroo may experience a severe range contraction, with possible extinction by 2070.

INTRODUCTION

The antilopine wallaroo (*Macropus antilopinus*; Fig. 16.1) is a large macropodid (18–51 kg) restricted to the savannas of northern Australia (Ritchie 2008), ranging from the Kimberley region of Western Australia, across the top end of the Northern Territory and to the Cape York and Einasleigh Uplands bioregions of Queensland. Tropical savannas comprise approximately one quarter of mainland Australia (~1.9 million square kilometres) and support the highest diversity of large macropodids (Johnson

Figure 16.1: (a) Medium male, (b) large male and (c) adult female antilopine wallaroos (*M. antilopinus*). Photos ©David Webb.

2003). Macropodid ecology has been studied extensively in temperate Australia (Grigg *et al.* 1989), but to date few data exist for tropical species (Croft 1987; Press 1988; Stirrat 2003), in particular the antilopine wallaroo and black wallaroo (*Macropus bernardus*). This lack of information is of great concern because many of northern Australia's mammal species are thought to be undergoing extensive population and distribution declines (Woinarski *et al.* 2001), and a lack of detailed ecological information prevents the proactive management of biodiversity within the region. Anecdotal reports of antilopine wallaroo population declines exist across the species' range, including the Mt Surprise area in Queensland (Peter Johnson, *pers. comm.*), Coomalie Farm in the Northern Territory (Richard Noske, *pers. comm.*) and across the Kimberley region in Western Australia (Tim Willing, *pers. comm.*). The antilopine wallaroo is a vital species to conserve, for a number of reasons. First, being a large grazing herbivore, it is likely to profoundly shape savanna ecosystem function (Gordon *et al.* 2004). Second, it is the only large-bodied macropodid restricted entirely to the tropics, and is therefore critical to preserve from a biodiversity perspective (Van Dyck and Strahan 2008). Third, it has strong cultural significance for indigenous people, and in many local communities is an important food resource (Yibarbuk *et al.* 2001). Finally, it is highly valued by nature-based ecotourism businesses.

To date, the only ecological studies of the antilopine wallaroo have occurred in the Northern Territory (Croft 1982, 1987; Russell and Richardson 1971). These studies investigated the species' socioecology, behaviour, breeding, age structure and habitat use. However, there is no long-term, systematic or broad-scale information available for the species. Between 2002 and 2005, I studied the ecology and conservation of the antilopine wallaroo across its entire distribution. This chapter provides an overview of the major results from that study and, where available, presents other data from previous studies for comparison.

STUDY SITE AND SURVEY METHODS

I surveyed 50 sites across the entire known distribution of the antilopine wallaroo (Fig. 16.2): 29 sites were in Queensland (Cape York and Einasleigh Uplands bioregions), 11 sites in the Northern Territory and 10 in the Kimberley region of Western Australia. I surveyed sites using driven line transects (5 km) conducted twice in the morning (5–8.30 am) and twice in the afternoon (4.30–6.30 pm) over two days, during the late wet season (April–May) and late dry season (October–November), totalling eight sampling occasions for each site. Along each transect I recorded all macropodid species observed, their group size and composition, mapped the vegetation and assessed water availability. I compiled fire histories for all sites using satellite images of fire scars.

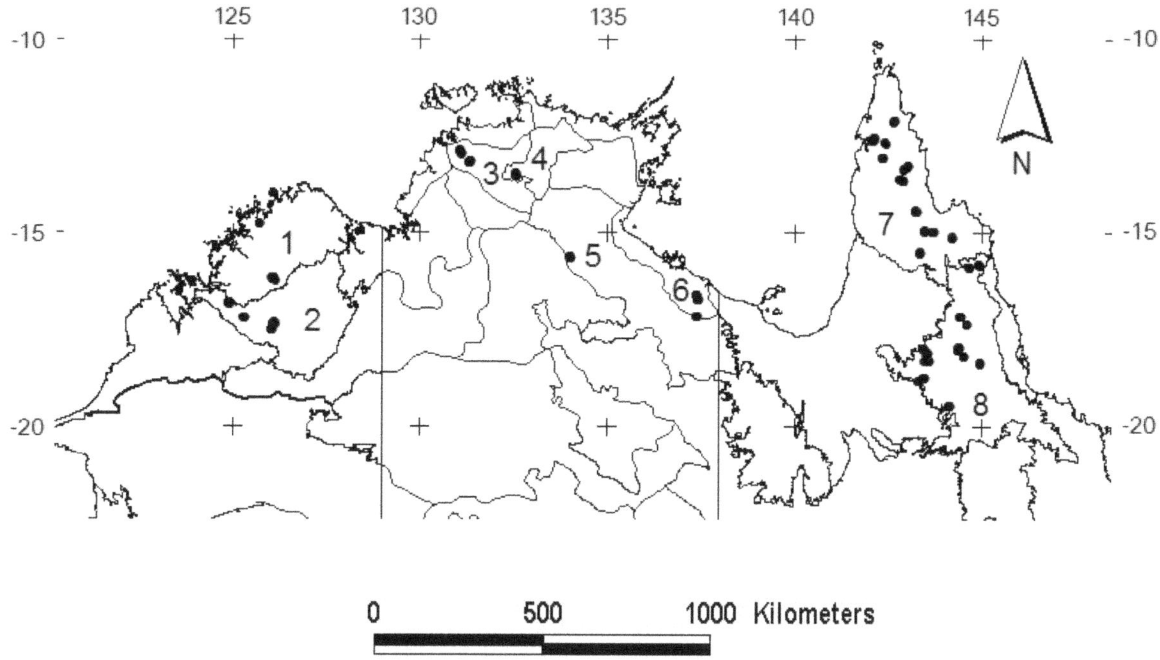

Figure 16.2: The study region in northern Australia. Bioregion boundaries are represented by freeform lines, state boundaries are shown with straight lines (left to right = Western Australia, Northern Territory and Queensland). Solid circles are study sites. Bioregions are (1) Northern Kimberley, (2) Central Kimberley, (3) Pine Creek, (4) Arnhem Plateau, (5) Gulf Fall and Uplands, (6) Gulf Coastal, (7) Cape York and (8) Einasleigh Uplands.

The information was used to derive habitat models for the antilopine wallaroo at varying spatial scales. For more detailed methodologies see Ritchie (2007) and Ritchie *et al.* (2008).

DISTRIBUTION AND ABUNDANCE

The antilopine wallaroo has a broad but patchy distribution across northern Australia (Fig. 16.3). Few records for the species exist from the Arnhem Land Plateau in the Northern Territory, which most likely reflects survey bias rather than the true absence of the species from this region. I recorded antilopine wallaroos at 34 of the 50 sites surveyed (68%). Antilopine wallaroos have not been recorded from the gulf plains at the base of the Gulf of Carpentaria, an area characterised by a change from open savanna woodland to semi-arid grasslands. This area is where red kangaroos (*Macropus rufus*) are most common and may represent an ecological barrier for dispersal between antilopine wallaroo populations in Queensland and the Northern Territory. During the study period, I observed a total of 5084 antilopine wallaroos; common wallaroos (*Macropus robustus*) were also widespread and common (644 individuals). I frequently observed eastern grey kangaroos (*Macropus giganteus*) within north Queensland (364 individuals), which included a significant range extension for this species into northern Cape York (Ritchie 2005). Other species recorded were the agile wallaby (*Macropus agilis*) (218), whiptail wallaby (*Macropus parryi*) (70), red kangaroo (29) and northern nailtail wallaby (*Onychogalea unguifera*) (4). On average, antilopine wallaroo density was three animals per square kilometre. Croft (1987) recorded densities ranging between 8.8 and 30.9 animals per square kilometre at two sites in the Northern Territory, and densities were higher in the wet season than the dry season. My study (Ritchie *et al.* 2008) and those of Croft (1987) and Russell and Richardson (1971) found antilopine wallaroos most frequently occurred in sympatry with the common wallaroo.

Across the entire range of the antilopine wallaroo, the amount of permanent water and the frequency of late dry season fires (August–November) were the most significant influences on antilopine wallaroo abundance (Ritchie *et al.* 2008), as identified by generalised linear models using the information theoretic approach (see Burnham and Anderson 2002). Despite rainfall being higher in the tropics than in temperate areas, the availability of water is largely ephemeral in most of the tropical savannas and becomes less plentiful with the advancement of the

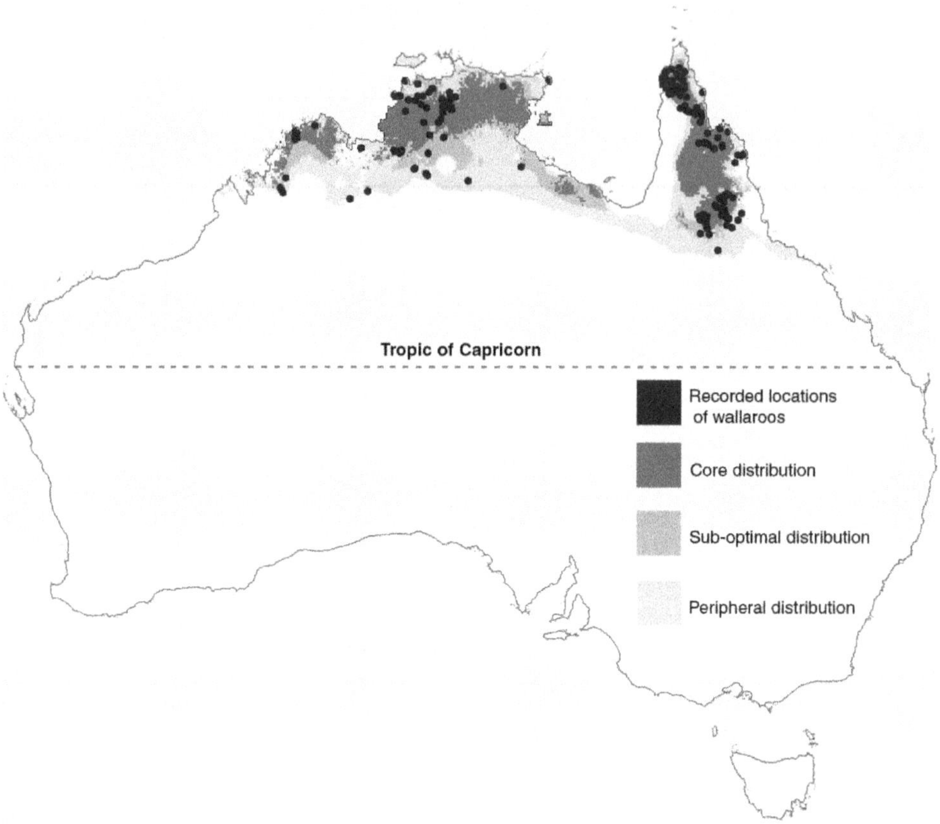

Figure 16.3: Distribution of the antilopine wallaroo (*M. antilopinus*) as modelled by BIOCLIM. Modified from Ritchie and Bolitho (2008).

extended dry season (May–November). Permanent natural water bodies (e.g. springs and water holes) as well as artificial sources (e.g. bores and dams) may therefore be critical in supporting local populations of antilopine wallaroos. Late-season fires, which occur frequently in northern Australia (Williams *et al.* 2003), may benefit antilopine wallaroos by opening up the understorey and stimulating new grass growth. However, this fire regime is known to negatively affect other species (e.g. the Gouldian finch, *Erythrura gouldiae*, and the black-footed tree-rat, *Mesembriomys gouldii*; Woinarski *et al.* 2005), which highlights the challenges of fire management in the north. In Queensland, in addition to water availability and fire, the increased density of eastern grey kangaroos had a negative influence on antilopine wallaroo abundance. The association of these two species occurred significantly less than expected (C score = 112, $p = 0.006$, $n = 29$), which is suggestive of competition (Ritchie *et al.* 2008). A further unpublished study by Ritchie has determined that eastern grey kangaroos exert a stronger negative effect on antilopine wallaroo abundance than the reverse, and this appears related to the larger size and more specialised habitat preferences of the eastern grey kangaroo. My results highlight that generalisations about the factors which influence a species' distribution and abundance can only be made when reliable data at varying spatial scales is available.

REPRODUCTIVE PATTERNS

The reproductive pattern of the antilopine wallaroo has been subject to debate (Croft 1987; Russell and Richardson 1971). Seasonal data on the sizes of pouch young and the frequency of mothers with dependent young observed in my study populations support Russell and Richardson (1971), who suggested that the species mates towards the end of the wet season (February–April). In my study, in the mid to late wet season (February–April), 40% of all females were observed with young at foot and 4% had pouch young. In contrast, in the late dry season (September–November), only 6% of females had young at foot but 80% had large pouch young. This observation appears to be best explained in relation to seasonality of rainfall and pasture condition. By mating in the late wet season, mothers can carry young over the extended dry season when forage abundance and quality is typically lowest (Norman 1963). Young then exit the pouch at the beginning of the subsequent wet season, when grass is highest in nutrients. This strategy would increase the likelihood of young being able to find nutrient-rich forage and to establish digestive bacteria in their gut, which is thought to be critical to their survival (Marilyn Renfree, *pers. comm.*). Second, it would buffer females against the energetic demands associated with peak lactation (Merchant 1989), also due to an increase in available nutrients. It remains unclear why antilopine wallaroos have a strongly seasonal breeding pattern and other sympatric macropodid species (eastern grey kangaroos, common wallaroos and agile wallabies) do not. It is unknown whether antilopine wallaroos possess the adaptation known as embryonic diapause but anecdotal evidence suggests not (Poole and Merchant 1987; Sarah Hirst and Euan Ritchie, unpubl. data). Embryonic diapause is lacking in another highly seasonally breeding macropodid, the western grey kangaroo, *Macropus fuliginosus*; Tyndale-Biscoe and Renfree 1987). Further investigation of this may provide useful insights into the evolution of mating strategies within the family Macropodidae.

The only study to record the ages and longevity of antilopine wallaroos (Russell and Richardson 1971) found that individuals (n = 18) may live up to 10 years. Additional samples are required from across the species' range for a better understanding of the life history and demography of this species.

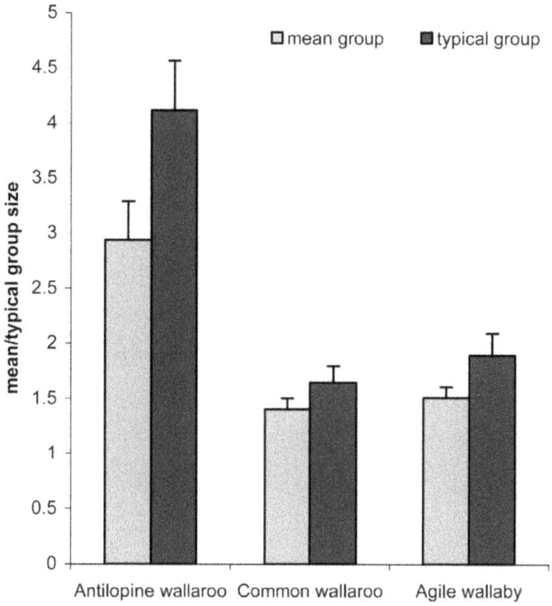

Figure 16.4: Mean and typical group size of the antilopine wallaroo (*M. antilopinus*), common wallaroo (*M. robustus*) and agile wallaby (*M. agilis*) across northern Australia (n = 50 sites). Error bars show standard error.

SOCIOECOLOGY

The socioecology and behaviour of the antilopine wallaroo has been studied in the Northern Territory by Croft (1982, 1987) and Russell and Richardson (1971). These studies concluded that the species was among the most gregarious of the large macropodids and that social and habitat segregation between the sexes commonly occurred. My study concurs with those findings (Fig. 16.4): based on 34 sites, the antilopine wallaroo had a mean group size of 2.95 (SE ± 0.15) and a mean typical group size (as defined by Jarman 1974) of 4.2 (SE ± 0.25), with groups of one to three the most commonly observed. Russell and Richardson (1971) found the majority of individuals were solitary or occurred in pairs but that, on average, more individuals were found in groups than alone. In contrast, sympatric common wallaroos and agile wallabies typically occur alone (Croft 1987; Ritchie 2007; Russell and Richardson 1971). In agreement with Croft (1987), I found a significant relationship between density and group size (Ritchie 2007).

I recorded moderate to high levels of social segregation of antilopine wallaroos, which varied considerably between sites. There was a consistent trend across populations of increasing segregation from the end of the wet season (April), through the dry season (May–November). Mixed-sex groups were most common in the wet season, and all-male and all-female groups were most common in the dry season (Fig. 16.5); this difference was statistically significant. Contrastingly, Croft (1987) found most individuals occurred in mixed-sex groups throughout seasons. My study and Croft (1987) both found that medium and large males are often alone, presumably due to their movement – this is the class most likely to be involved in mating, due to their larger size and therefore dominance – between groups seeking mating opportunities.

Little is known of the movements and home range of antilopine wallaroos. Croft (1987) collared one large male and calculated a home range estimate of

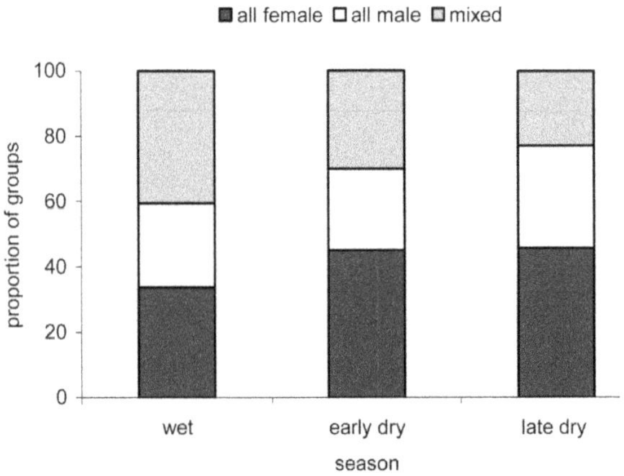

Figure 16.5: Seasonal group composition of the antilopine wallaroo (*M. antilopinus*) across northern Australia (n = 34 sites).

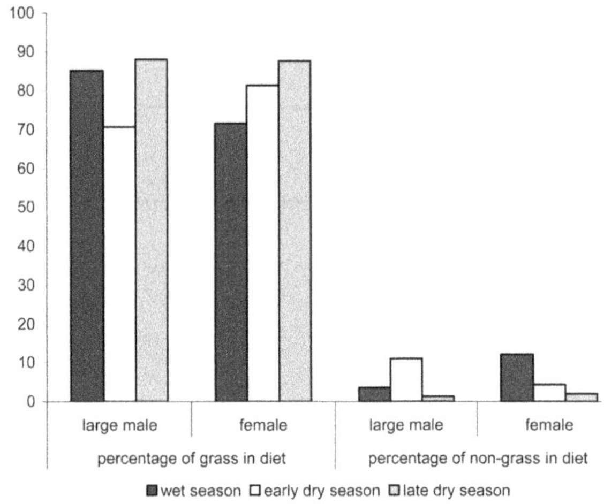

Figure 16.6: Seasonal diet of the antilopine wallaroo (*M. antilopinus*) at Rocky Springs cattle station in north Queensland. Missing proportion of diet is unidentified material.

76.1 ha. He observed two other individuals frequently and calculated estimates of 14.2 ha (female) and 102 ha (male). In comparison, a northern wallaroo at the same site had a home range of 9.9 ha. Further studies are required to understand how the seasonal environment of the tropics influences resource availability and movement patterns in the antilopine wallaroo.

DIET

In collaboration with Peter Fossan, Dr Jenny Martin and Prof. Chris Johnson, I studied the diet of the antilopine wallaroo using both microscopic analyses and near infrared spectroscopy (see Foley *et al.* 1998) over the course of one year at Rocky Springs cattle station in north Queensland (–18.09, 144.38). We examined faecal pellets from both sexes and from three different male body-size classes. On average, grass made up 80% of the diet of both males and females; non-grass (forbs) comprised 5% and 6% respectively (Fig. 16.6). There was variation in these relative proportions between seasons. Black speargrass (*Heteropogon contortus*) was the most common grass species recorded in the diet (for more detailed information see Fossan 2005). Our results contrast with those of Croft (1987), who concluded that antilopine wallaroos graze only on grasses.

CLIMATE CHANGE

Climate change is predicted to have significant effects on species globally (Hughes 2003; Walther *et al.* 2005), and the antilopine wallaroo has been identified as a species at particular risk due to its restricted range within the tropics (Busby 1988). In collaboration with Emily Bolitho I modelled the current and future distributions of the antilopine wallaroo, common wallaroo, eastern grey kangaroo and red kangaroo under varying combinations of increased and decreased rainfall. With as little as a 0.4°C increase in temperature the antilopine wallaroo's core distribution is predicted to decrease by 20%, and the remaining three species on average would decrease by 9%. With an increase of 6°C (possible by 2070; Hughes 2003) the antilopine wallaroo's distribution (climate envelope) disappears, equating to bioclimatic extinction, while the other three species' distributions are reduced by an average of 96% (Ritchie and Bolitho 2008). These results suggest that although large macropodids have wide geographic distributions, are relatively mobile and have good dispersal capabilities in comparison to other species, they may still be at significant risk from the effects of climate change.

CONCLUSIONS AND FUTURE DIRECTIONS

The antilopine wallaroo and black wallaroo remain the least studied of Australia's large macropodids. The results of my study provide the most detailed account of aspects of the ecology and conservation of the antilopine wallaroo across its entire distribution, the first study of its kind for a large free-ranging herbivore. Within northern Australia, the antilopine wallaroo has a broad distribution but on average is found at densities much lower than those of temperate macropodids. Fire regimes, water availability and potential competition with the eastern grey kangaroo appear to be key factors influencing variation in patterns of abundance. Details on the socioecology and behaviour of the antilopine wallaroo from my study build upon work by Croft (1982, 1987) and Russell and Richardson (1971), and provide great opportunity for comparative studies and synthesis within the family Macropodidae.

The current decline of mammals in northern Australia is of great concern as many of the species, including the antilopine wallaroo, are endemic. Their conservation is crucial to preserving Australia's biodiversity. My study suggests that the antilopine wallaroo is currently locally common across its range and at no immediate risk; however, the intensification of grazing, changes to fire regimes and climate change have the potential to adversely affect the species' survival and therefore require consideration by management agencies. To monitor the antilopine wallaroo effectively, I strongly advocate for the repeated and systematic resurveying of the sites established by my study.

ACKNOWLEDGMENTS

I would like to thank Jenny Martin, Chris Johnson, Stephen Garnett, Andrew Krockenberger, Peter Fossan and Emily Bolitho who all collaborated on aspects of this research. I was funded in parts by the Australia and Pacific Science Foundation, Australian Geographic, Cape York Peninsula Development Association, Cooperative Research Centre for Tropical Savannas, Ecological Society of Australia, James Cook University, Linnean Society of New South Wales and Royal Zoological Society of New South Wales. I sincerely thank all the traditional owners, cattle station owners, national parks staff, government agencies and the Australian Wildlife Conservancy who provided access to field sites and logistical support. Last I thank the many volunteers who assisted with the c. 150 000 km of field surveys!

REFERENCES

Burnham KP and Anderson DR 2002. Model selection and multimodel inference: a practice information-theoretic approach. Springer-Verlag, New York.

Busby GW 1988. Possible impacts of climate change on Australia's flora and fauna. In *Greenhouse: Planning for Climate Change.* (Ed. GI Pearman) pp. 375–386. CSIRO Division of Atmospheric Research, Melbourne.

Croft DB 1982. Some observations on the behavior of the antilopine wallaroo *Macropus antilopinus* (Marsupialia: Macropodidae). *Aust. Mammal.* **5:** 5–14.

Croft DB 1987. Socio-ecology of the antilopine wallaroo, *Macropus antilopinus*, in the Northern Territory, with observations on sympatric *Macropus robustus woodwardii* and *Macropus agilis. Aust. Wildl. Res.* **14:** 243–255.

Foley WJ, McIlwee A, Lawler I, Aragones L, Woolnough AP and Berding N 1998. Ecological applications of near infrared reflectance spectroscopy: a tool for rapid, cost-effective prediction of the composition of plant and animal tissues and aspects of animal performance. *Oecologia* **116:** 293–305.

Fossan PCG 2005. Intraspecific differences in diet selection in a population of antilopine wallaroos, *Macropus antilopinus* (Marsupialia: Macropodidae) in North Queensland, Australia. MAppSci thesis. James Cook University, Townsville, Australia.

Gordon IJ, Hester AJ and Festa-Bianchet M 2004. The management of wild large herbivores to meet economic, conservation and environmental objectives. *J. Appl. Ecol.* **41:** 1021–1031.

Grigg G, Hume I and Jarman P (Eds), 1989. *Kangaroos, Wallabies and Rat-Kangaroos.* Surrey Beatty and Sons, Sydney.

Hughes L 2003. Climate change and Australia: trends, projections and impacts. *Austral Ecol.* **28:** 423–443.

Jarman PJ 1974. Social organization of antelope in relation to their ecology. *Behaviour* **48:** 215–267.

Johnson PM 2003. *Kangaroos of Queensland.* Queensland Museum, Brisbane.

Merchant JC 1989. Lactation in macropodoid marsupials. In *Kangaroos, Wallabies and Rat-kangaroos.* (Eds G Grigg, P Jarman and I Hume) pp. 355–366. Surrey Beatty and Sons, Sydney.

Norman MJT 1963. The pattern of dry matter and nutrient content changes in native pastures at Katherine NT. *Aust. J. Exp. Agric. Anim. Husb.* **3:** 119–124.

Poole WE and Merchant JC 1987. Reproduction in captive wallaroos: the eastern wallaroo *Macropus robustus robustus*, the euro *Macropus robustus erubescens*, and the antilopine wallaroo *Macropus antilopinus. Aust. Wildl. Res.* **14:** 225–242.

Press AJ 1988. The distribution and status of macropods (Marsupialia: Macropodidae) in Kakadu National Park, Northern Territory. *Aust. Mammal.* **11:** 103–108.

Ritchie EG 2005. An extension to the known range of the eastern grey kangaroo (*Macropus giganteus*) on Cape York Peninsula. *Aust. Mammal.* **27:** 225–226.

Ritchie EG 2007. The ecology and conservation of the antilopine wallaroo (*Macropus antilopinus*). PhD thesis. James Cook University, Townsville, Australia.

Ritchie EG 2008. Antilopine wallaroo, *Macropus antilopinus*. In *The Mammals of Australia*, 3rd edn. (Eds S Van Dyck and R Strahan) pp. 325–326. New Holland Publishers, Sydney.

Ritchie EG and Bolitho EE 2008. Australia's savanna herbivores: bioclimatic distributions and an assessment of the potential impact of regional climate change. *Physiol. Biochem. Zool.* **81:** 880–890.

Ritchie EG, Martin JK, Krockenberger AK, Garnett S and Johnson CN 2008. Large-herbivore distribution and abundance: intra-and interspecific niche variation in the tropics. *Ecol. Monogr.* **78:** 105–122.

Russell EM and Richardson BJ 1971. Some observations on the breeding age structure dispersion and habitat of populations of *Macropus robustus* and *Macropus antilopinus* (Marsupialia). *J. Zool. Lond.* **65:** 131–142.

Stirrat SC 2003. Seasonal changes in home-range area and habitat use by the agile wallaby (*Macropus agilis*). *Wildl. Res.* **30:** 593–600.

Tyndale-Biscoe H and Renfree MB 1987. *Reproductive Physiology of Marsupials*. Cambridge University Press, Cambridge.

Van Dyck S and Strahan R (Eds) 2008. *The Mammals of Australia*. 3rd edn. New Holland Publishers, Sydney.

Walther G-R, Hughes L, Vitousek P and Stenseth NC 2005. Consensus on climate change. *Trends Ecol. Evol.* **20:** 648–649.

Williams RJ, Woinarski JCZ and Andersen AN 2003. Fire experiments in northern Australia: contributions to ecological understanding and biodiversity conservation in tropical savannas. *Int. J. Wildland Fire* **12:** 391–402.

Woinarski JCZ, Milne DJ and Wanganeen G 2001. Changes in mammal populations in relatively intact landscapes of Kakadu National Park, Northern Territory, Australia. *Austral Ecol.* **26:** 360–370.

Woinarski JCZ, Williams RJ, Price O and Rankmore B 2005. Landscapes without boundaries: wildlife and their environments in northern Australia. *Wildl. Res.* **32:** 377–388.

Yibarbuk D, Whitehead PJ, Russell-Smith J, Jackson D, Godjuwa C, Fisher A, Cooke P, Choquenot D and Bowman D 2001. Fire ecology and Aboriginal land management in central Arnhem Land, northern Australia: a tradition of ecosystem management. *J. Biogeogr.* **28:** 325–343.

17 Effect of habitat type, sex and time of day on space use by the swamp wallaby

J. Di Stefano, M. Swan, A. Greenfield and G. Coulson

SUMMARY

Quantifying patterns of animal movement in habitats with different characteristics, and at a range of temporal and spatial scales, will help determine the range of habitats and resources required for survival and reproduction. In this study we investigated the effect of habitat type, sex and time of day on space use by the swamp wallaby (*Wallabia bicolor*) within four habitat types in a landscape modified by timber harvesting. The difference between diurnal and nocturnal space use was quantified by calculating the change in home range size attributable to night locations, and we predicted that nocturnal and diurnal space use would differ only in habitats where food and shelter resources were spatially separated. We recorded total home ranges between 3.4 and 82.6 ha. On average, the addition of nocturnal data to diurnal home ranges caused female ranges to increase by 28.9% ± 9.7% (mean ± 95% CI) while male home ranges increased by 15.2% ± 12.4%. Female home ranges increased by 39.2% ± 17.7% at five-year-old regenerating sites and by 41.9% ± 20.9% at 10-year-old sites, while much smaller increases were observed at unharvested and recently harvested (<12 months old) locations. Males demonstrated a similar pattern, although the effects were generally smaller and had larger errors. For males, we observed a negative relationship between range size increase and body weight (adj. $r^2 = 53.4\%$, $P < 0.001$), although the relationship was substantially reduced when two outliers were removed (adj. $r^2 = 11.3\%$, $P = 0.11$). There was no similar relationship for females (adj. $r^2 = 0$, $P = 0.52$). Our prediction was largely supported by the data, which indicated that the degree to which swamp wallabies used different spaces during diurnal and nocturnal periods was influenced by specific characteristics of the habitat and sex. We suggest that quantifying animal movements during different diel or seasonal periods will further conservation objectives, and propose an experimental framework within which the impact of resources on the movement of relatively large forest-dwelling species can be studied.

INTRODUCTION

Mobile animals often use multiple habitat patches at a variety of spatial and temporal scales (Orians and Wittenberger 1991; Law and Dickman 1998), and using different areas during diurnal and nocturnal periods is one example of this phenomenon. Quantifying differences between diurnal and nocturnal space use provides information about the range of habitats and resources required for survival and reproduction, which, in many cases, can be used to inform programs for conservation or management

(Finlayson *et al.* 2008). Nevertheless, most descriptions of animal movements are derived from diurnal data, which may not adequately represent the use of habitats or resources throughout the 24-hour cycle (Beyer and Haufler 1994; Moe *et al.* 2007).

The swamp wallaby (*Wallabia bicolor*) is a medium-sized macropodid marsupial that occurs predominantly along Australia's east coast. With the exception of a recent study indicating that habitat selection differed between diurnal and nocturnal periods (Di Stefano *et al.* 2009), descriptions of its movements and habitat use have been based on diurnal observations (Kaufmann 1974), counts of faecal pellets (Floyd 1980; Lunney and O'Connell 1988; Hill and Phinn 1993; Ramsey and Wilson 1997; de Munk 1999) and diurnal and crepuscular radio location data (Edwards and Ealey 1975; Troy *et al.* 1992; Troy and Coulson 1993; Wood 2002; Ben-Ami 2005). The radio location studies have reported relatively small home ranges (mostly 10–40 ha but up to 80 ha), but the degree to which nocturnal data influence home range size is unknown.

As part of a study investigating the impact of timber harvesting on resource use by swamp wallabies, we collected nocturnal locations from radio-collared wallabies living in different parts of the harvested landscape. In this chapter, we use these data to quantify the effect of nocturnal locations on home range size for wallabies with access to five- and 10-year-old regenerating eucalypt forest, recently harvested sites (<12 months old) and unharvested control locations. Data from macropods and other mammals (Johnson 1980; Beyer and Haufler 1994; Fisher 2000; Ager *et al.* 2003; le Mar and McArthur 2005; Moe *et al.* 2007) indicate that differential space use during diurnal and nocturnal periods is relatively common. This behaviour is frequently interpreted as the result of a trade-off between avoiding predators and acquiring food, with individuals often occupying sheltered habitats during the day and moving into more open, forage-rich areas at night. We believe that swamp wallabies probably respond to predation risk at our study site in a similar way (Di Stefano *et al.* 2009), thus we expected them to use different areas during diurnal and nocturnal periods in habitats where food and shelter resources were spatially separated. In habitats where food and shelter resources occurred together, we expected swamp wallabies to use similar areas throughout the 24-hour cycle.

METHODS

Study site

The data were collected from the Pyrenees State Forest in west central Victoria (704000 E, 5892000 N; Datum = WGS 1984, Projection = UTM Zone 54S; Fig. 17.1). This 20 000 ha forest is underlain by stony red duplex soils and dominated by a single range running approximately east–west and rising to 750 m above sea level. Records from nearby Avoca (10 km off the mountain and at 250 m above sea level) show the mean annual rainfall to be 540 mm, and the mean winter (July) minimum and summer (January) maximum temperatures to be 3°C and 29°C respectively (BOM 2006).

This dry sclerophyll landscape is dominated by two ecological vegetation classes: grassy dry forest on the northerly aspects and herb-rich foothill forest on the southerly slopes. The two classes are broadly similar, although the wetter southerly slopes support a more productive and diverse forb community. The dominant overstorey species is messmate (*Eucalyptus obliqua*), but substantial stands of southern blue gum (*Eucalyptus globulus bicostata*) are also found. The sparse middle storey consists of silver wattle (*Acacia dealbata*) and cherry ballart (*Exocarpos cupressiformis*), while the shrub layer includes common heath (*Epacris impressa*), gorse bitter-pea (*Daviesia ulicifolia*), common cassinia (*Cassinia aculeata*) and prickly wattle (*Acacia paradoxa*). Structurally, the ground layer is dominated by austral bracken (*Pteridium esculentum*), grasses and forbs (LCC 1978; DSE 2006).

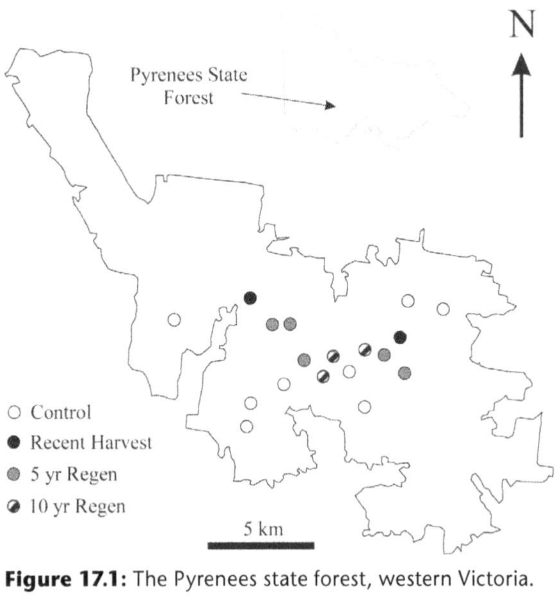

Figure 17.1: The Pyrenees state forest, western Victoria.

Since 1990, timber harvesting in the Pyrenees has created about 25 regenerating blocks (coupes) surrounded by unharvested forest. We used unharvested forest and three age classes of regenerating forest – recently harvested (data collected three to 10 months post-harvest), about five years post-harvest and about 10 years post-harvest – to define four habitat types that differed most obviously in regard to vegetation cover and visibility. Except for retained seed trees and some remaining ground vegetation, the recently harvested areas were initially devoid of live plant biomass, although some growth occurred during the monitoring period. The five-year-old sites were dominated by 1–3 m tall, densely regenerating eucalyptus seedlings, but also had substantial quantities of silver wattle, grass and forbs. Ten-year-old sites supported dense closed stands of 3–6 m tall eucalypt regeneration and consequently had low levels of forb, grass and shrub cover. The cover of bracken, however, remained relatively high. Unharvested sites were relatively open and thus supported substantial quantities of forbs, grass and occasional patches of shrub. They tended to be further from major ridge lines and on steeper slopes than harvested sites and the density and size of mature trees varied depending on local harvesting history.

Experimental design

We used timber harvesting operations as an experimental treatment to investigate the effect of habitat type, sex and time of day on wallaby movements. The location of unharvested control sites was limited by road access and proximity to each other and to non-control areas. Initially we identified around 15 potential control locations within the area used for timber harvesting (above c. 500 m elevation) and attempted to trap wallabies at 14 of these. We subsequently rejected six sites due to unsuccessful trapping regimes (four), and wallabies using other (non-control) areas (two). Swamp wallabies caught at the remaining eight control sites were used in the analysis. We used all available recently harvested sites (n = 2) and five-year-old sites (n = 5), and selected three 10-year-old sites randomly from six that we deemed appropriate. A number of 10-year-old sites were rejected *a priori* due to their close proximity to sites where wallabies had already been captured.

With two exceptions, sites were at least 1.5 km apart to increase the likelihood of spatial independence. In one case, a five-year-old and 10-year-old site were separated by about 500 m and in another, two five-year-old sites were 300 m apart at their closest point. Of the 13 wallabies tracked at these adjacent sites, two moved between them. On two additional occasions, minor home range overlap (<10%) occurred between animals at other adjacent sites. At each site we obtained location data (see below) for between one and four wallabies, and sampled both sexes whenever possible. In all, the data represented 42 wallabies (24 females and 18 males) from 18 sites.

Swamp wallaby capture and radio-tracking

We trapped swamp wallabies from March 2004 to October 2005 using double-layered traps designed for the purpose (Di Stefano *et al.* 2005). We 'free-fed' with carrots up to four weeks prior to trapping, then used carrots as bait, setting traps in the late afternoon and checking them early the following morning. Our total trapping effort constituted 897 trap nights for 102 captures (11.4% trap success), which included 94 individuals with eight recaptures.

Once caught, we sedated wallabies with an intramuscular injection of 5 mg/kg of Zoletil 100 (Virbac Australia Ltd) and fitted them with a Sirtrack radio-collar (approximately 30 g) and two reflective eartags. We also recorded the body weight, crus and pes length of adults, and the pes length of pouch young.

We radio-tracked wallabies by homing in on foot using a Telonics TR2 portable receiver and RA-14 Yagi antenna and recorded positions using a hand-held Garmin 12 GPS unit, which reported an estimated error of <15 m in 94.8% of cases. We tracked in all weather and at all times of the day and night (night was defined as completely dark), and scheduled tracking so that nocturnal and diurnal locations for each wallaby were spread approximately evenly throughout these periods. In order to minimise disturbance to individuals we left at least six hours between tracking events and obtained no more than two fixes per individual in a 24-hour period. The final data set contained 2163 positions, 78.9% ± 0.9 (mean ± 95% CI) collected during the day and 21.1% ± 0.9 collected at night. The mean number of positions per individual was 51.5 ± 3.9, 40.6 ± 3.1 of these representing diurnal locations.

Because we did not always see the wallabies before disturbing them, we used a 1–5 rating system to quantify location accuracy: 1 = within 5 m of exact location, 2 = 6–25 m, 3 = 26–100 m, 4 = 101–200 m

Table 17.1: Subjectively estimated accuracy of radio-tracking locations.

	% of locations given each accuracy rating				
Time period	**1**	**2**	**3**	**4**	**5**
Diurnal	42.2 ± 7.8	25.2 ± 4.1	26.4 ± 5.3	4.9 ± 1.5	1.3 ± 0.7
Nocturnal	34.7 ± 7.3	27.7 ± 4.9	30.5 ± 5.4	4.4 ± 2.0	2.7 ± 1.8

Values are the percentage of locations (± 95% confidence interval) relating to a 1–5 scale of spatial accuracy. 1 = within 5 m of exact location, 2 = 6–25 m, 3 = 26–100 m, 4 = 101–200 m and 5 = >200 m, or when a signal could not be detected.

and 5 = >200 m, or when a signal could not be detected (we did not record locations for rating 5s). The percentage of locations given each rating in both diurnal and nocturnal periods is shown in Table 17.1. In cases where we saw or heard the wallaby (1s, 2s and most 3s), ratings were based on visual and aural cues. In cases when we did not see or hear the wallaby (some 3s and all 4s), its position was estimated on the basis of aural feedback from the tracking equipment in relation to the local terrain. Relative to the spatial scale of the analysis, we considered ratings of 1 and 2 to be highly precise locations, 3 to be moderately precise and 4 to be imprecise. We removed a single rating 4 location from 5 individuals as these data resulted in large range size increases, which we defined as ≥10%.

Data screening and analysis

We generated location data in short bursts (field trips) with periods of relative inactivity in between, resulting in blocks of autocorrelated data that were likely to be relatively independent from one another (Katajisto and Moilanen 2006). As animal locations will always be autocorrelated to a greater or lesser extent and methods designed to derive a statistically independent sample (Swihart and Slade 1985) may reduce biological relevance (Reynolds and Laundre 1990; Rooney *et al.* 1998), we did not consider autocorrelation in the analysis. Our objectives required the estimation of home range boundaries and size, so we focused on acquiring a representative sample of positions within a period sufficient for the wallabies to adequately describe their home range (Otis and White 1999).

Our analysis depended on the accurate generation of diurnal ranges, so, with three exceptions, we excluded wallabies with <30 diurnal positions (Seaman *et al.* 1999). We included three males with smaller diurnal samples (n = 29, 26 and 26) to boost male sample size, accepting the trade-off between more replication and less subsampling precision.

We defined home range size and shape with fixed 95% kernels (Worton 1989) using Ranges VI (Anatrack Ltd). To differentiate between diurnal and nocturnal space use we calculated the change in home range size attributable to night locations using the following formula:

$$\%\text{ change in home range size} = (HR_T - HR_D)/HR_D \times 100$$

where HR_T is total range size and HR_D is diurnal range size. The change in range size was expressed as a percentage of the diurnal range to standardise results across animals that had markedly different ranging areas.

To determine an appropriate smoothing factor, we initially used the median multiplier from the sample of Least Squares Cross-Validated results (Kenward 2001) but this value (0.5) resulted in some home range outlines that were fractured into multiple segments and biologically nonsensical. We computed the analysis (see below) using 0.5, 0.8 and 1.0 as smoothing factor multipliers and in each case obtained similar statistical outputs. Results were reported based on a multiplier of 0.8, as the home range outlines generated were most consistent with the perception of space use we acquired throughout the tracking regime.

We used Restricted Maximum Likelihood (REML) in GenStat 9 to model the effect of treatment (unharvested control, recently harvested, five-year-old sites and 10-year-old sites) and sex on the percentage change in home range size, using site as a random factor in the analysis. REML provides efficient estimates of treatment effects in unbalanced designs (GenStat Committee 2006), and thus was appropriate for our asymmetrical dataset. We explored the use of body weight and sample size as covariates, but used neither as the relationship between these variables and the dependent variable was either small or highly influenced by extreme data points. We explored the relationship between the change in home range size and body weight further using linear regression, also in GenStat 9. Assumptions of normality and homogeneity of variance were tested for both the REML and regression analyses with a half-normal plot and a fitted-value plot respectively, and data transformation was deemed unnecessary.

RESULTS

For male and female swamp wallabies respectively we recorded total home ranges of 16.0–82.6 and 3.4–37.5 ha, and diurnal home ranges of 12.0–85.9 and 2.8–33.3 ha (Table 17.2). Five-year-old sites were characterised by relatively small home ranges, and females at these locations had some of the smallest ranges ever recorded for the species.

The degree to which diurnal home range size increased with the addition of nocturnal data differed statistically between treatments (P = 0.04) and sex (P = 0.03), but without an interaction between the two (P = 0.76). Averaging across treatments, the addition of nocturnal data caused female home ranges to increase by 28.9% ± 9.7% (mean ± 95% CI) while male home ranges increased by 15.2% ± 12.4%. For both sexes, the addition of nocturnal data caused home range size to increase more at five- and 10-year-old regenerating sites than it did at unharvested control and recently harvested sites (Fig. 17.2). This effect was much clearer for females than for males, with a small sample of males at recently harvested and 10-year-old sites (Fig. 17.2) adding substantial uncertainty to the corresponding estimates. The largest increases were for females at five- and 10-year-old regenerating sites, where the addition of nocturnal data caused diurnal ranges to increase by 39.2% ± 17.7% and 41.9% ± 20.9% respectively.

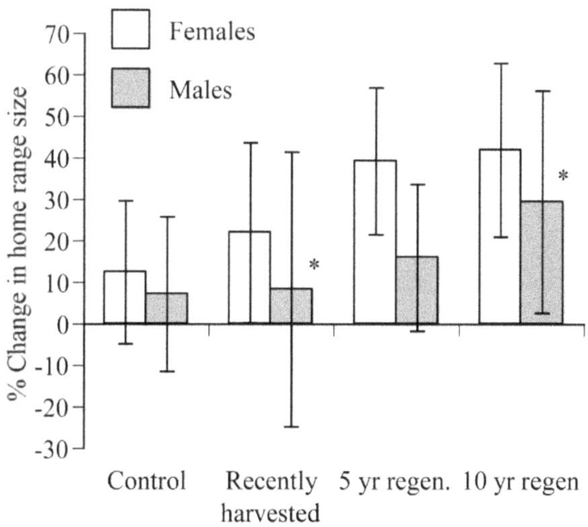

Figure 17.2: The degree to which swamp wallaby (*W. bicolor*) home range size increases with the addition of nocturnal data for each of four habitat types. Increases are expressed as a percentage of diurnal range size to standardise results across individuals that had markedly different ranging areas. Error bars are 95% CI and intervals that do not overlap zero indicate that range size increases are statistically different from this value at the 5% level. Values marked * may be inaccurate due to small sample sizes (n = 2 in one case, n = 3 in the other).

When all data were considered, there was a negative linear relationship between body weight and home range size increase for males (adj. r^2 = 53.4%, P < 0.001, Fig. 17.3a), indicating that, with the addition of nocturnal data, range sizes increased more for lighter individuals. However, this relationship was strongly influenced by two data points (open circles in Fig. 17.3a), and the exclusion of these data greatly reduced the negative trend (adj. r^2 = 11.3%, P = 0.11). There was no relationship between body weight and home range size increase for females (adj. r^2 = 0, P = 0.52, Fig. 17.3b). There was very little overlap between the body weight of males and females (Fig. 17.3a, b), demonstrating the sexually dimorphic nature of the species.

DISCUSSION

Our initial prediction – that diurnal and nocturnal space use would differ only for individuals living in habitat types where food and shelter resources were spatially separated – was generally supported by the data. At unharvested forest sites, food (mainly forbs and shrubs; Di Stefano and Newell 2008), and shelter (large trees, patches of austral bracken, gullies) were both well distributed within home range areas. As expected, the home range size of swamp wallabies living at these sites changed little when nocturnal data was added to diurnal ranges, presumably because both food and shelter resources could be accessed in similar locations, rendering diel movements between patches of shelter and food unnecessary. Recently harvested patches had few resources and swamp wallabies living at these sites established home ranges in the surrounding unharvested forest (Di Stefano *et al.* 2007). Thus, wallabies at recently harvested sites behaved similarly to their counterparts in the unharvested forest, although the addition of nocturnal data increased female range size to a greater degree. This was due to occasional

Table 17.2: Raw home range data used in the analysis.

Site type	Sex	Weight (kg)	n (total)	Total 95% home range (ha)	n (diurnal)	Diurnal 95% home range (ha)
Recent harvest	F	14.0	56	24.29	43	16.78
	F	12.0	54	16.77	43	14.90
	F	15.5	38	35.08	31	33.26
	F	15.0	55	22.00	44	16.66
	F	13.0	55	25.83	43	20.81
	M	20.0	41	35.01	33	27.84
	M	24.0	53	77.44	42	85.93
5 yr old	F	14.0	56	20.04	44	14.48
	F	14.0	62	6.03	48	3.36
	F	15.0	37	3.39	30	2.80
	F	14.0	49	9.78	37	6.17
	F	14.0	47	24.95	34	16.16
	F	14.0	64	14.27	50	12.58
	F	13.5	59	19.76	48	17.85
	M	21.5	46	42.77	35	42.78
	M	17.5	44	57.07	33	53.89
	M	14.5	37	16.04	30	11.95
	M	19.0	39	23.86	32	17.24
	M	22.0	47	30.18	35	29.45
	M	18.0	48	29.47	36	24.71
	M	22.5	41	40.91	33	36.30
10 yr old	F	15.0	47	32.79	37	24.37
	F	13.5	47	29.03	36	18.00
	F	13.5	58	20.30	43	14.83
	F	10.5	54	18.18	41	10.70
	F	14.0	59	18.37	45	16.38
	M	17.0	50	82.57	39	74.75
	M	17.5	47	71.28	36	77.90
	M	11.0	42	33.75	34	18.78
Control	F	12.5	63	13.48	52	10.76
	F	17.0	47	18.09	39	15.81
	F	11.0	63	22.26	49	21.37
	F	14.5	67	18.36	55	17.95
	F	13.5	38	15.29	32	16.62
	F	15.0	87	29.23	68	23.14
	F	12.0	64	37.53	54	30.58
	M	21.0	32	46.93	26	50.70
	M	23.0	52	78.11	41	82.13
	M	20.0	36	33.80	29	30.80
	M	18.5	33	46.45	26	46.00
	M	18.0	62	27.86	49	19.97
	M	21.5	89	16.59	71	15.12

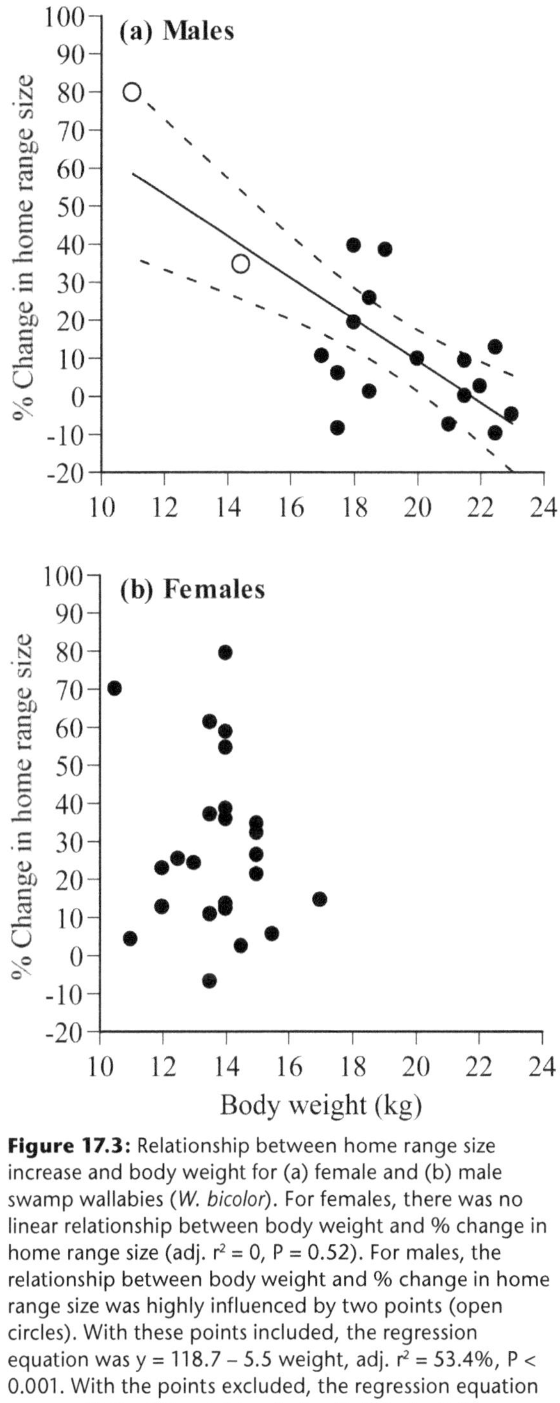

Figure 17.3: Relationship between home range size increase and body weight for (a) female and (b) male swamp wallabies (*W. bicolor*). For females, there was no linear relationship between body weight and % change in home range size (adj. $r^2 = 0$, P = 0.52). For males, the relationship between body weight and % change in home range size was highly influenced by two points (open circles). With these points included, the regression equation was y = 118.7 – 5.5 weight, adj. $r^2 = 53.4\%$, P < 0.001. With the points excluded, the regression equation was y = 69.2 – 3.0 weight, adj. $r^2 = 11.3\%$, P = 0.11.

nocturnal visits to the harvested area, presumably for individuals to test the suitability of this new and changing habitat.

In contrast, the addition of nocturnal data to diurnal ranges increased range size substantially for females at both five- and 10-year-old regenerating sites. In the case of 10-year-old sites this was consistent with our expectations, as food and shelter resources were spatially separated (Di Stefano *et al.* 2009), resulting in a marked temporal transition between sheltered diurnal and relatively forage-rich nocturnal habitats. At five-year-old sites, however, we have shown that regenerating areas contained a mix of food and shelter resources, and that female wallabies selected five-year-old regenerating patches during both diurnal and nocturnal periods (Di Stefano *et al.* 2009). As such, we did not expect the addition of nocturnal data to result in a home range size increase as large as we observed. To some extent, this effect is probably related to our use of percentages to quantify the change in home range size. Although percentages were necessary to standardise results across wallabies with markedly different range sizes, many individuals at five-year-old sites had very small diurnal home ranges (Table 17.2), and thus small differences between diurnal and nocturnal space use resulted in a relatively large percentage increase in range area. Although diurnal and nocturnal space use by females at five-year-old sites was somewhat different, the effect shown in Figure 17.2 probably overestimates the difference to some degree.

Not withstanding the uncertainty associated with male responses at recently harvested and 10-year-old sites, the addition of nocturnal data to diurnal ranges resulted in smaller range size increases for males than for females. Although the effect of sex on the movements of sexually dimorphic species may sometimes be confounded by differences in body weight, our analysis demonstrated that body weight had little influence on the observed change in home range size, indicating that the effect was primarily due to sex.

One plausible explanation for the observed sex effect is that males and females use different criteria to select habitat and thus are expected to demonstrate different patterns of space use. For example, it has been proposed (Main and Coblentz 1990; Main *et al.* 1996) that females will maximise their fitness by raising their offspring to sexual maturity and thus should select habitats that both reduce predation risk and provide adequate resources for rearing young. Males, on the other hand, maximise fitness by successfully competing for mates and so should select habitats that help meet this objective. We have discussed this issue in relation to our study species elsewhere (Di Stefano *et al.* 2009) but, in summary, male swamp wallabies are solitary and non-territorial and thus are expected to maximise their reproductive success by searching widely for mates. Such behaviour is consistent with the substantially larger male home ranges we observed (Table 17.2) and may have modulated the influence of food and shelter resources on their patterns of movement.

Our finding that space use differed between diurnal and nocturnal periods is consistent with previous work on macropodid marsupials (Johnson 1980; Fisher 2000; le Mar and McArthur 2005) and other species (Law 1993; Beyer and Haufler 1994; Ager *et al.* 2003; Moe *et al.* 2007) and suggests that diurnally based estimates of home range size for swamp wallabies (Edwards and Ealey 1975; Troy and Coulson 1993; Wood 2002; Ben-Ami 2005) may not adequately represent total ranging area or the diversity of habitats used. Nevertheless, we demonstrated that diurnal home ranges adequately represented total ranging area at unharvested control sites, and it has been shown that diurnal ranges can be adequate descriptors of total space use for other mammalian herbivores (Hayes and Krausman 1993; Perelberg *et al.* 2003). In general, we suggest patterns of space use during diurnal and nocturnal periods will be influenced by the spatial arrangement of multiple resources (food, shelter, den sites, water), thus the degree to which space use differs during the 24-hour cycle is likely to depend on specific characteristics of the landscape within and around individual home ranges. The concurrent influence of multiple resources on animal movement is an area that has received little attention in the ecological literature, and deserves an increased focus.

Implications for conservation and management

The use of different habitats during diurnal and nocturnal periods has implications for species conservation and management and for the design of studies attempting to quantify important habitats or resources. If important habitats are not identified due, for example, to diurnal-only sampling (Beyer and Haufler 1994), the full range of habitats required for the maintenance of viable populations may not be detected. This is clearly linked to the design of conservation areas and reserves, as the appropriate size, location and composition of such areas should reflect species habitat requirements at a variety of spatial and temporal scales (Noss *et al.* 2002).

A more complete knowledge of species habitat requirements may also enable habitat maintenance or enhancement using planned human disturbances such as timber harvesting or fire (Law and Dickman 1998). For example, the nationally endangered south-eastern red-tailed black cockatoo (*Calyptorhynchus banksii graptogyne*) has a specialised diet, depending on the seeds of a limited number of species in the stringybark woodlands of south-west Victoria (Joseph 1982). Using this knowledge, local authorities have implemented a prescribed burning regime that maintains patches of long-unburnt woodland in an attempt to increase the cockatoo's food supply (DSE 2004). More generally, a number of mammal species show strong preferences for early successional habitats (Fuller and DeStefano 2003) and are likely to respond positively to disturbance events that reset the plant successional process. Although not of current conservation concern, swamp wallabies exemplify this kind of response, showing marked density increases within early successional forest patches resulting from timber harvesting operations (Di Stefano *et al.* 2007).

Linking home range size to resources: the use of planned disturbances as an experimental framework

In this study we have shown that, for swamp wallabies, the addition of nocturnal data to diurnal home ranges resulted in substantial range size increases in some cases. Further, we suggested that these changes were likely due to the spatial arrangement of food and shelter resources and to the propensity of wallabies to use these during different parts of the diel cycle. Future work should aim to establish relationships between home range size and specific resources such as food and shelter and investigate these during both diurnal and nocturnal periods.

For medium to large mammals, such relationships have often been investigated by determining correlations between various resources and home range size using observational data sets (Gompper and Gittleman 1991; Tufto *et al.* 1996; Fisher 2000; Dahle and Swenson 2003). Manipulative experiments, however, are a better way to infer causal effects and examine interactions (Johnson 2002), and would provide strong evidence of relationships between home range size and the distribution and abundance of important resources. While the addition of food resources has been used to study the movement patterns of larger animals (Cooper *et al.* 2006), shelter is more difficult to manipulate, particularly at relevant spatial scales, and the manipulation of shelter in studies of medium and larger sized mammals is rare.

We suggest that the practical problems associated with manipulating resources at scales relevant to medium and large mammals can be reduced by using planned disturbances such as timber harvesting, plantation establishment and prescribed fire. These activities constitute shelter manipulations at large spatial scales, as they result in patches of dense regenerating vegetation that are often surrounded by more open, mature forest stands. Manipulating food resources within or adjacent to regenerating patches provides an opportunity to establish causal links between animal behaviour (home range size, habitat use etc.) and limiting resources such as food and shelter. We believe there is considerable scope for the development of such experiments in the future.

ACKNOWLEDGMENTS

We thank Graham Hepworth for statistical advice and Merv Flett for facilitating field work. Funding was provided by the Holsworth Wildlife Research Endowment, the Victorian Department of Sustainability and Environment and the University of Melbourne (Departments of Forest and Ecosystem Science and Zoology). This research was conducted in conjunction with appropriate permits (DSE permit no. 10002779; Melbourne University Faculty of Science Animal Experimentation Committee permit no. 03249).

REFERENCES

Ager AA, Johnson BK, Kern JW and Kie JG 2003. Daily and seasonal movements and habitat use by female rocky mountain elk and mule deer. *J. Mammal.* **84:** 1076–1088.

Ben-Ami D 2005. The behavioural ecology of the swamp wallaby, *Wallabia bicolor*, and its response to human induced disturbance. PhD thesis. University of New South Wales, Sydney.

Beyer DE and Haufler JB 1994. Diurnal versus 24-hour sampling of habitat use. *J. Wildl. Man.* **58:** 178–180.

BOM 2006. Bureau of Meteorology (Australia) website: http://www.bom.gov.au/.

Cooper SM, Owens MK, Cooper RM and Ginnett TF 2006. Effect of supplemental feeding on spatial distribution and browse utilization by white-tailed deer in semi-arid rangeland. *J. Arid. Environ.* **66:** 716–726.

Dahle B and Swenson JE 2003. Home ranges in adult Scandinavian brown bears (*Ursus arctos*): effect of mass, sex, reproductive category, population density and habitat type. *J. Zool.* **260:** 329–335.

de Munk FG 1999. Resource use by the eastern grey kangaroo and the black wallaby in a managed remnant woodland community. PhD thesis. Deakin University, Geelong, Australia.

Di Stefano J and Newell GR 2008. Diet selection by the swamp wallaby (*Wallabia bicolor*): feeding strategies under conditions of changed food availability. *J. Mammal.* **89:** 1540–1549.

Di Stefano J, Moyle R and Coulson G 2005. A soft-walled, double-layered trap for capture of swamp wallabies *Wallabia bicolor. Aust. Mammal.* **27:** 235–238.

Di Stefano J, Anson JA, York A, Greenfield A, Coulson G, Berman A and Bladen M 2007. Interactions between timber harvesting and swamp wallabies (*Wallabia bicolor*): space use, density and browsing impact. *For. Ecol. Man.* **253:** 128–137.

Di Stefano J, York A, Swan M, Greenfield A and Coulson G 2009. Habitat selection by the swamp wallaby (*Wallabia bicolor*) in relation to diel period, food and shelter. *Austral Ecol.* **34:** 143–155.

DSE 2004. Portand fire protection plan. Dept of Sustainability and Environment, Melbourne.

DSE 2006. Forest explorer map. Website: http://www.dse.vic.gov.au/dse/dsencor.nsf/ LinkView/836EE128E54D861FCA256DA200208B945FD09CE028D6AA58CA256 DAC0029FA1A.

Edwards GP and Ealey EHM 1975. Aspects of the ecology of the swamp wallaby *Wallabia bicolor* (Marsupialia: Macropodidae). *Aust. Mammal.* **1:** 307–317.

Finlayson GR, Vieira EM, Priddel D, Wheeler R, Bentley J and Dickman CR 2008. Multi-scale patterns of habitat use by re-introduced mammals: a case study using medium-sized marsupials. *Biol. Conserv.* **141:** 320–331.

Fisher DO 2000. Effects of vegetation structure, food and shelter on the home range and habitat use of an endangered wallaby. *J. Appl. Ecol.* **37:** 660–671.

Floyd RB 1980. Density of *Wallabia bicolor* (Desmarest) (Marsupialia: Macropodidae) in eucalypt plantations of different ages. *Aust. Wildl. Res.* **7:** 333–337.

Fuller TK and DeStefano S 2003. Relative importance of early-successional forests and shrubland habitats to mammals in the northeastern United States. *For. Ecol. Man.* **185:** 75–79.

GenStat Committee 2006. Guide to GenStat Release 9. Part 2: Statistics. VSN International, Hertfordshire.

Gompper ME and Gittleman JL 1991. Home range scaling: intraspecific and comparative trends. *Oecologia* **87:** 343–348.

Hayes CL and Krausman PR 1993. Nocturnal activity of female desert mule deer. *J. Wildl. Man.* **57:** 897–904.

Hill GJE and Phinn SR 1993. Revegetated sand mining areas, swamp wallabies and remote sensing: North Stradbroke Island, Queensland. *Aust. Geog. Stud.* **31:** 3–13.

Johnson KA 1980. Spatial and temporal use of habitat by the red-necked pademelon, *Thylogale thetis*

(Marsupialia: Macropodidae). *Aust. Wildl. Res.* **7:** 157–166.

Johnson DH 2002. The importance of replication in wildlife research. *J. Wildl. Man.* **66:** 919–932.

Joseph L 1982. The red-tailed black cockatoo in south-eastern Australia. *Emu* **82**:42–45.

Katajisto J and Moilanen A 2006. Kernel-based home range method for data with irregular sampling intervals. *Ecol. Model.* **194:** 405–413.

Kaufmann JH 1974. Habitat use and social organization of nine sympatric species of macropodid marsupials. *J. Mammal.* **55:** 66–80.

Kenward RE 2001. *A Manual for Wildlife Radio Tagging.* Academic Press, London.

Law BS 1993. Roosting and foraging ecology of the Queensland blossom bat (*Syconycteris australis*) in north-eastern New South Wales: flexibility in response to seasonal variation. *Wildl. Res.* **20:** 419–431.

Law BS and Dickman CR 1998. The use of habitat mosaics by terrestrial vertebrate fauna: implications for conservation and management. *Biodivers. Conserv.* **7:** 323–333.

LCC 1978. Report on the North Central Study Area. Land Conservation Council, Melbourne.

le Mar K and McArthur C 2005. Comparison of habitat selection by two sympatric macropods, *Thylogale billardierii* and *Macropus rufogriseus rufogriseus*, in a patchy eucalypt–forestry environment. *Austral Ecol.* **30:** 674–683.

Lunney D and O'Connell M 1988. Habitat selection by the swamp wallaby, *Wallabia bicolor*, the red-necked wallaby, *Macropus rufogriseus*, and the common wombat, *Vombatus ursinus*, in logged, burnt forest near Bega, New South Wales. *Aust. Wildl. Res.* **15:** 695–706.

Main MB and Coblentz BE 1990. Sexual segregation among ungulates: a critique. *Wildl. Soc. Bull.* **18:** 204–210.

Main MB, Weckerly FW and Bleich VC 1996. Sexual segregation in ungulates: new directions for research. *J. Mammal.* **77:** 449–461.

Moe TF, Kindberg J, Jansson I and Swenson JE 2007. Importance of diel behaviour when studying habitat selection: examples from female Scandinavian brown bears (*Ursus arctos*). *Can. J. Zool.* **85:** 518–525.

Noss RF, Carroll C, Vance-Borland K and Wuerthner G 2002. A multicriteria assessment of the irreplaceability and vulnerability of sites in the greater Yellowstone ecosystem. *Conserv. Biol.* **16:** 895–908.

Orians GH and Wittenberger JF 1991. Spatial and temporal scales in habitat selection. *Am. Nat.* **137:** S29–S49.

Otis DL and White GC 1999. Autocorrelation of location estimates and the analysis of radiotracking data. *J. Wildl. Man.* **63:** 1039–1044.

Perelberg A, Saltz D, Bar-David S, Dolev A and Yom-Tov Y 2003. Seasonal and circadian changes in the home ranges of reintroduced Persian fallow deer. *J. Wildl. Man.* **67:** 485–492.

Ramsey DSL and Wilson JC 1997. The impact of grazing by macropods on coastal foredune vegetation in southeast Queensland. *Aust. J. Ecol.* **22:** 288–297.

Reynolds TD and Laundre JW 1990. Time intervals for estimating pronghorn and coyote home ranges and daily movements. *J. Wildl. Man.* **54:** 316–322.

Rooney SM, Wolfe A and Hayden TJ 1998. Autocorrelated data in telemetry studies: time to independence and the problem of behavioural effects. *Mammal Rev.* **28:** 89–98.

Seaman DE, Millspaugh JJ, Kernohan BJ, Brundige GC, Raedeke KJ and Gitzen RA 1999. Effects of sample size on kernel home range estimates. *J. Wildl. Man.* **63:** 739–747.

Swihart RK and Slade NA 1985. Testing for independence of observations in animal movements. *Ecology* **66:** 1176–1184.

Troy S and Coulson G 1993. Home range of the swamp wallaby, *Wallabia bicolor. Wildl. Res.* **20:** 571–577.

Troy S, Coulson G and Middleton D 1992. A comparison of radio-tracking and line transect techniques to determine habitat preferences in the swamp wallaby (*Wallabia bicolor*) in south-eastern Australia. In *Wildlife Telemetry: Remote Monitoring and Tracking of Animals.* (Eds IG Priede and SM Swift) pp. 651–660. Ellis Horwood, Chichester.

Tufto J, Andersen R and Linnell J 1996. Habitat use and ecological correlates of home range size in a small cervid: the roe deer. *J. Anim. Ecol.* **65:** 715–724.

Wood MS 2002. Resource partitioning in sympatric populations of red-necked and black wallabies. PhD thesis. Deakin University, Geelong, Australia.

Worton BJ 1989. Kernel methods for estimating the utilization distribution in home-range studies. *Ecology* **70:** 164–168.

18 Reassessing the spatial and temporal dynamics of kangaroo populations

A.R. Pople, G.C. Grigg, S.R. Phinn, N. Menke, C. McAlpine and H.P. Possingham

SUMMARY

Long-running datasets from aerial surveys of kangaroos (*Macropus giganteus*, *Macropus fuliginosus*, *Macropus robustus* and *Macropus rufus*) across Queensland, New South Wales and South Australia have been analysed, seeking better predictors of rates of increase which would allow aerial surveys to be undertaken less frequently than annually. Early models of changes in kangaroo numbers in response to rainfall had shown great promise, but much variability. We used normalised difference vegetation index (NDVI) instead, reasoning that changes in pasture condition would provide a better predictor than rainfall. However, except at a fine scale, NDVI proved no better; although two linked periods of rainfall proved useful predictors of rates of increase, this was only in some areas for some species. The good correlations reported in earlier studies were a consequence of data dominated by large drought-induced adult mortality, whereas over a longer time frame and where changes between years are less dramatic, juvenile survival has the strongest influence on dynamics. Further, harvesting, density dependence and competition with domestic stock are additional and important influences and it is now clear that kangaroo movement has a greater influence on population dynamics than had been assumed. Accordingly, previous conclusions about kangaroo populations as simple systems driven by rainfall need to be reassessed. Examination of this large dataset has permitted descriptions of shifts in distribution of three species across eastern Australia, changes in dispersion in response to rainfall, and an evaluation of using harvest statistics as an index of density and harvest rate. These results have been combined into a risk assessment and decision theory framework to identify optimal monitoring strategies.

INTRODUCTION

This chapter describes summary results from a project that had three broad aims, as follows (see Pople 2006 for details):

- prediction of short-term changes in kangaroo (*Macropus giganteus*, *Macropus fuliginosus*, *Macropus robustus* and *Macropus rufus*) distribution and abundance using rainfall or satellite imagery and other environmental data;
- evaluation of the practicality of using harvest statistics (e.g. sex ratio, carcass weight, catch-per-unit-effort) for indirect monitoring of kangaroo density and harvest rate;
- optimisation of survey methods, frequency and design.

The need for better information about the short-term changes in distribution and abundance of kangaroos arises partly because kangaroo harvest quotas are set each December on the basis of surveys conducted the preceding winter or earlier, by which time the population estimates will be somewhat out of date. These surveys, usually aerial surveys, are expensive, large-scale and labour-intensive and there is an urgent need to develop supplementary or ancillary techniques which are both cost-effective and able to monitor population changes in a timelier manner. Our study explored two approaches. The first, reported here in summary form, was to combine extensive survey data on population size with rainfall, remotely sensed estimates of pasture condition (NDVI, available Australia-wide on a monthly basis) and other environmental variables (soil, vegetation and rainfall) with a view to determining predictive models. The second, reported in Chapter 32 this volume, was to examine the relationship between aerial survey data (i.e. direct monitoring) and the harvest data which are collected routinely (on a monthly basis) as an indication of population density or harvest rate (i.e. to explore the potential usefulness of indirect monitoring as a supplement or even a replacement for direct monitoring in kangaroo management).

Finally, we considered how these modelled relationships could fit into a monitoring program. Such a program needed to combine predictions of population size with direct survey estimates at some frequency and intensity. In turn, these predictions and estimates of population size need to be considered, not as en end in themselves, but as a part of the harvest strategy of allowing a set proportion of the population to be harvested each year.

HISTORICAL BACKGROUND

Aerial survey of kangaroos in Australia to monitor population size began with a series of experiments in New South Wales in the early 1970s (Caughley *et al.* 1976). The methodology (Caughley *et al.* 1976; Caughley and Grigg 1981) has been re-evaluated and improved over a number of years, including the development and validation of helicopter techniques (Clancy *et al.* 1997) and even an ultralight (Grigg *et al.* 1997). The first of what would become extensive and regular surveys was that of the kangaroo management zone in New South Wales in 1975, followed by the pastoral zone in South Australia in 1978 and the harvested areas in Queensland in 1980. Surveys have continued until the present in these areas, annually in most cases. In Western Australia, annual surveys are conducted of one-third of rangeland areas, on a rotational basis. The stimulus to extensive aerial surveys was the closure in 1973 of the export of kangaroo hides and meat, in response to concerns that kangaroo harvesting was putting at risk the long-term survival of kangaroo populations. Much of that concern was generated from the US, which subsequently implemented an import ban, in 1974. Responses by relevant state and Commonwealth governments involved implementing scientifically based management plans and funding extensive aerial surveys which assessed populations in each state and, subsequently, assessed for the first time the sizes and distributions of the populations Australia-wide (Caughley *et al.* 1983). The Commonwealth export ban was lifted in 1975 once satisfactory management plans were in place, but the US import ban remained in force until 1981.

Since the late 1970s and early 1980s, therefore, kangaroo harvests have been regulated by quotas set primarily on data gained from aerial surveys undertaken the previous winter using fixed-wing aircraft (New South Wales and South Australia) or, in Queensland, primarily a helicopter (see Table 18.1). While the surveys in other states have changed methodologies and areas covered from time to time, the South Australian survey has been undertaken in the same manner and on the same flight lines throughout (Grigg *et al.* 1999). There has been a strong emphasis on ensuring comparability between surveys, with great care to ensure thorough observer training (Beard 1999).

These aerial surveys have resulted in the accumulation of a number of very large databases tracking changes in populations of red kangaroos (*M. rufus*), eastern grey kangaroos (*M. giganteus*) and western grey kangaroos (*M. fuliginosus*) over a 25–30-year period involving several droughts and a diversity of climate histories.

Early attempts to model kangaroo population dynamics in New South Wales and South Australia

Work by Caughley *et al.* (1984), Bayliss (1985a, b) and Caughley *et al.* (1987) in western New South Wales largely confirmed the prevailing opinion that these rangeland-herbivore systems were comparatively simple and that rainfall is a fairly good predictor of population changes, especially if the time of year that the rain fell is taken adequately into account. Similar analyses of data from 1978–85 in the South Australian Pastoral Zone (SAPZ) led to similar conclusions but, with the longer run of data, the outcomes were less satisfying than hoped. When the survey period was broken into two, the first half was a fairly poor predictor of the events in the second half. In particular, the modelling was not successful at predicting events after conditions improved following the very severe drought in 1982–83 (Cairns 1989; Cairns and Grigg 1993).

It was thought that these poor predictions could be attributed to the sparse availability of rainfall recording sites, the lack of age-structured data and, particularly, the run of data still being too short. A longer run of data was expected to support models to provide more accurate predictions and so be more useful to managers.

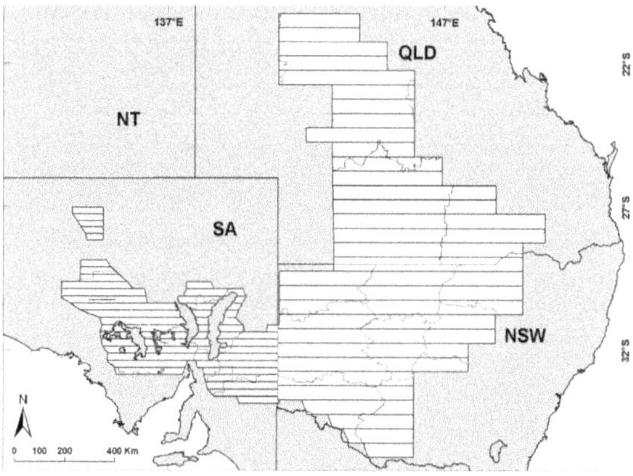

Figure 18.1: The study area within the sheep rangelands of eastern Australia. In New South Wales and South Australia, the survey area has been broken up into regions in which kangaroo harvests are managed. Biogeographic boundaries (Environment Australia 2000), adjusted to coincide with possible future administrative (shire) boundaries, are shown in the Queensland survey area. Transect lines flown in the 2001 aerial survey are shown as east–west lines.

Australian Research Council Linkage Grant 2001–03

During 1999–2000 our interest in further analysis over a longer time frame stirred again with two important additions, the inclusion of data from Queensland and New South Wales and, bearing in mind that the sparse distribution of rainfall observations had been a serious limitation in interpretation of the South Australian data, the use of remotely sensed NDVI data (the 'greenness index'). An Australian Research Council (ARC) Linkage application was developed, with new colleagues and new techniques. This chapter presents an overview of some outcomes from that study.

STUDY AREA

The study area spans three states across a region where sheep grazing has been the predominant land use (Fig. 18.1). The relatively flat terrain and open vegetation makes the area suitable for aerial survey of large wildlife such as kangaroos. The vast size of the study area (>1.2 million km^2) makes aerial survey the only feasible monitoring method. In Queensland, surveys have been conducted annually by helicopter, in blocks (Ch. 32 this volume).

DATASETS

The aerial survey and harvest data used in this report are summarised in Tables 18.1 and 18.2. Some data have been lost (e.g. burnt in a fire) but most have been maintained by individuals who were involved in their collection and recognised their value beyond the immediate management needs.

Harvest data required editing, presumably due to transcription errors during data entry (e.g. implausible weights or nightly harvest). The problem was not with the total harvest for the state, but the total harvest and associated statistics at finer scales such as a region or property. This unreliability cautions against the use of harvest statistics in management without thoughtful scrutiny. On a broader scale, small errors are likely to be overshadowed by the sheer volume of data. The approach taken in this project was to remove or correct obvious errors, but to generally accept that the harvest data were imperfect. The question was whether the harvest data, in their present form, could be useful to management.

MODELLING TEMPORAL CHANGES

Background

The dynamics of kangaroo populations have come to be been regarded as well understood, with the numerical response of populations to a variable, rainfall-driven food supply often providing a textbook example (Caughley and Sinclair 1994; Bayliss and Choquenot 2002; Sibly and Hone 2002; Sinclair

Table 18.1: Data on kangaroo density used in this study.

State	Area	Year(s)	Platform	Scale
Queensland	Core	1980,1984–92	Fixed-wing	2 km^2
	Core	2001	Fixed-wing	1 km^2
	Core	1991	HLT	5 blocks
	Core	1992–94	HLT	9 blocks
	Core	1995–2001	HLT	10 blocks
	>Core	2002–03	HLT	19 blocks
	Core	1980,1984–2003	Fixed-wing, HLT	Region
New South Wales	WP	1975–76	Fixed-wing	1:250 000 map sheets
	WP	1977–83	Fixed-wing	Monitor blocks
	WP	1984–2003	Fixed-wing	1 block, KMZ
	WP	1993–2000	Fixed-wing	2 km^2
	WP	2001–03	Fixed-wing	1 km^2
South Australia	PZ	1978–2003	Fixed-wing	2 km^2, SCB

Scale refers to the resolution of data available for analysis.
HLT = helicopter line transect; WP = western plains; PZ = pastoral zone; KMZ = kangaroo management zone; SCB = soil conservation board; region = areas based approximately on biogeographic boundaries within the core area (Fig. 18.1).

and Krebs 2002). This picture of kangaroo population dynamics falls within the mechanistic paradigm described by Krebs (1995, 2002) and has been used to contrast that approach with that of the density paradigm, with an emphasis on density and equilibria rather than on interactions with resources (Caughley and Sinclair 1994).

Recent analyses of aerial survey data, particularly longer time series, suggest that the dynamics of some kangaroo populations do not fall so neatly into the mechanistic paradigm (McCarthy 1996; Jonzen *et al.* 2005). Cairns *et al.* (2000) also found the dynamics of western grey kangaroos in South Australia could not be as well predicted as populations in western New South Wales. It is worth noting that, while populations of four species have been surveyed throughout semi-arid and arid Australia, our understanding has largely come from surveys of red and western grey kangaroos in western New South Wales and northern South Australia (Caughley *et al.* 1984b; Bayliss 1985a, b, 1987; Caughley 1987; Cairns and Grigg 1993). Data from elsewhere in these species' ranges, particularly more mesic areas and northern Australia, and on eastern

Table 18.2: Harvest statistics used in this study.

State	Area	Year(s)	Statistics	Spatial scale	Temporal scale
Queensland	Harvest zone	1975–2003	H	Harvest zone	Annual
	Harvest zone	1986, 1988–90	H	Grid square (½ block)	Annual
	Harvest zone	1991–2003	H, SR, WT, SS, CPUE	Grid square (½ block)	Month
New South Wales	Harvest zone	1975–2003	H	Harvest zone	Annual
	Harvest zone	1992	H	KMZ	Annual
	Harvest zone	1993–2004	H, SR, WT	KMZ	Month
	Harvest zone	1997–2004	CPUE	KMZ	3 months
South Australia	Harvest zone	1975–2003	H	Harvest zone	Annual
	Harvest zone	1980–2002	H, SR, WT, CPUE	Property, SCB	Month

Scale refers to the resolution of data available for analysis.
SR = sex ratio; H = harvest; WT = weight; SS = skin size; CPUE = catch-per-unit-effort; KMZ = kangaroo management zone; SCB = soil conservation board.

grey kangaroos and common wallaroos (*M. robustus*), have had limited assessment.

Ideally, numerical response models for kangaroos should be based directly on food supply such as pasture biomass (Bayliss 1987). However, this has rarely been available and rainfall has been used as a surrogate. Early assessments indicated recent rainfall was a useful predictor of rate of increase in red and western grey kangaroo populations (Bayliss 1985a; Cairns and Grigg 1993), but more recent investigations with a longer time series showed that rainfall with a longer time-lag gave a stronger correlation with rate of increase (McCarthy 1996). An alternative is to use satellite imagery, which has the advantage of being related directly to actual pasture response (Hobbs 1995) and is recorded throughout remote arid areas where rainfall recording stations are sparsely located and rainfall can be patchy.

Numerical response

For each region (Fig. 18.1 and survey blocks in Queensland, Ch. 32 this volume), rainfall was determined from interpolated rainfall surfaces for the six months and 12 months prior to the second of two consecutive surveys and for 12 months prior to the first survey. These periods have been found to correlate best with kangaroo rates of increase in northern South Australia (Cairns and Grigg 1993), western New South Wales (Bayliss 1987) and the entire SAPZ (McCarthy 1996), respectively. Dry- and wet-season rainfalls were considered as candidates for predicting rates of increase in Queensland. The advanced very high resolution NDVI (Tucker *et al.* 1985; Hobbs 1995), available from mid 1981, was error-corrected and transformed into growth flush layers (Cridland and Fitzgerald 2001). The growth flush layers with the best correlation with a time series of pasture-monitoring photographs in the South Australian sheep rangelands were determined (Menke *et al.* 2005). NDVI data were standardised over time for each pixel. Together with rainfall data for similarly lagged periods, they were averaged across pixels within each South Australian district soil conservation board (SCB) and averaged for six- and 12-month periods at increasing time-lags from each aerial survey. These proxies for food supply were then assessed in regression models as possible determinants of kangaroo rates of increase, adjusted for harvest rate. Explanatory variables included one or two periods of rainfall, curvilinear terms and interactions.

NDVI proved no better than rainfall for predicting rates of increase, at least in South Australia. However, the NDVI flush data are a spatially explicit direct representation of landscape pattern and may prove useful for predicting smaller-scale fluctuations in kangaroo numbers in response to patchy rainfall and movement, as well as spatially variable survival rates (see below).

For all three kangaroo species at relatively high densities in regions in South Australia (adjusted R^2 ranged 0.28–0.44) and New South Wales (adjusted R^2 ranged 0.22–0.70), piecewise regression models (Crawley 2002) comprising two rainfall terms were generally the most parsimonious explanations of the data. These models included 12 months rainfall with a 12-month lag, and usually a recent rainfall term only when it was below a certain amount. The rationale is that it is really survival of juveniles rather than mortality of adults which determines fluctuations in kangaroo numbers, as for large mammals in general (Gaillard *et al.* 1998). This is determined by rainfall, with a substantial lag because juveniles are not detected readily from the air (Bayliss 1985b; McCarthy 1996). In contrast, adult mortality can be substantial during drought (Robertson 1986) and influences population change immediately (Caughley *et al.* 1985), hence the requirement for a recent rainfall term. For species in regions with low average densities, relationships were poor (adjusted $R^2 < 0.20$).

In Queensland, relationships between rates of increase and rainfall were generally weak (adjusted R^2 ranged 0.01–0.37), particularly in the north, and were more variable among regions. Calendar year rainfall and summer rainfall had the best associations with rate of increase for red and eastern grey kangaroos and for common wallaroos; piecewise regression models offered no improvement. Notably, kangaroo populations are harvested at higher rates in Queensland than in the southern states. This was taken into account by considering models with either additive or partially compensatory harvest mortality.

There was often a strong correlation between past density and rate of increase and this has led to the inclusion of strong, direct density dependence in simple models describing kangaroo population dynamics, such as those by McCarthy (1996) and Jonzen *et al.* (2005). The inclusion of lagged density to quantify density dependence is unsatisfactory because density is only a surrogate for processes regulating the population (Krebs 2002). Furthermore, density dependence will be overestimated in regression analyses because past density is included in both response and explanatory variables (Burgman *et al.* 1993; McCarthy 1996). McCarthy (1996) also recognised that the apparent strong density dependence may be an artefact of uncertainty in the abundance estimates. We explored this by simulation, using Caughley's (1987) interactive model of an unstructured red kangaroo population.

In the interactive model, kangaroo rate of increase is a function of pasture biomass which, in turn, is determined by recent rainfall, past pasture biomass and the density of kangaroos (and livestock) consuming the pasture. The density dependence is indirect and weak. Rainfall drives pasture biomass, which in turn determines the rate of increase of the kangaroo population. There are two negative feedback loops. The first is pasture biomass reducing pasture growth. The second is kangaroo density reducing kangaroo rate of increase by reducing the available biomass by eating it. The interactive model can be described by the following equations:

$$r = -1.6 + 2(1 - e^{-0.007V})$$

$$\Delta V = -55.12 - 0.01535V - 0.00056V^2 + 2.5RT$$

$$I = 86(1 - e^{-V/34}) \times W^{0.75} \times (1 + S)$$

where r is the annual exponential rate of increase of kangaroos, V is pasture biomass (kg/ha^{-1}) at the start of the interval, ΔV is the growth increment in pasture biomass over three months and RT is rainfall (mm) over three months drawn from a lognormal distribution using the mean and standard deviation from Menindee in western New South Wales. I is intake (g) of pasture by kangaroos and sheep per day, S is the amount of pasture eaten by sheep relative to kangaroos, assumed here to be 1.5 (Caughley 1987), and W is the average body weight (kg) of a kangaroo, assumed here to be 26 kg. Pasture grew and was grazed down in weekly time-steps. Pasture growth was drawn from a normal distribution with a standard deviation of 52 kg/ha^{-1} as estimated by Robertson (1987). Kangaroo density was recalculated in quarterly time-steps.

The population was harvested at an annual instantaneous rate of 5–20% drawn from a uniform distribution. Harvesting was introduced in quarterly time-steps prior to reproduction and natural mortality. To mimic current management practice in setting quotas, the annual harvest was calculated as a percentage of the previous winter's population estimate. That density estimate was drawn from a lognormal distribution with a mean being the modelled population density (= 'true' density) and a standard deviation ranging 0 – 0.6 × density. Rate of increase was determined from these density estimates. The population was modelled over 100 years. For each level of precision in the density estimate 1000 iterations were performed, each providing correlations between the rate of increase of kangaroos and past rainfall and past density. These simulations recorded a declining correlation between rate of increase and rainfall and an increasing correlation between rate of increase and past density, with increasing uncertainty in density estimates. The strength of the correlation with density surpassed that with rainfall around a precision (SE/mean) of 0.2. By the time precision had worsened to 0.4, >5% of simulations recorded correlations < 0 with rainfall. In summary, measurement error in estimates of density, which usually have a precision > 0.2, can mask the link between food supply and population dynamics and give the appearance of strong density dependence.

This is an important point. Ricker (logistic) and ratio-dependent models with strong density dependence fit these data well (McCarthy 1996; Jonzen *et al.* 2005). From the results described above, a good fit should not be interpreted as an adequate representation of the system. These models can give markedly different management recommendations. For example, for an arid-zone red kangaroo population, the maximum sustained yield (MSY) is taken at a harvest rate of about 35% for the ratio-dependent model, 25% for a Ricker model and 13% for the interactive model (Pople 2003, 2008) yet each can yield a seductively good *post hoc* fit to observed population trajectories.

Caughley's (1987) model is mechanistic and so is supported by kangaroo ecology rather than model fit. Nevertheless, some assessment of the fit to time series data is needed. The model was therefore run using rainfall in the north-east pastoral SCB in northern South Australia (Fig. 18.1) over data from 1975–2003. Rainfall was first rescaled to Menindee's mean because that is where the model was parameterised. The model was fitted using least squares to the actual aerial survey, harvest and livestock (dry sheep equivalents) data by adjusting the parameter S, the proportion of the pasture consumed by livestock. The fit of this unstructured model to red kangaroo density in the north-east pastoral SCB was remarkably good (Fig. 18.2), given that only this single

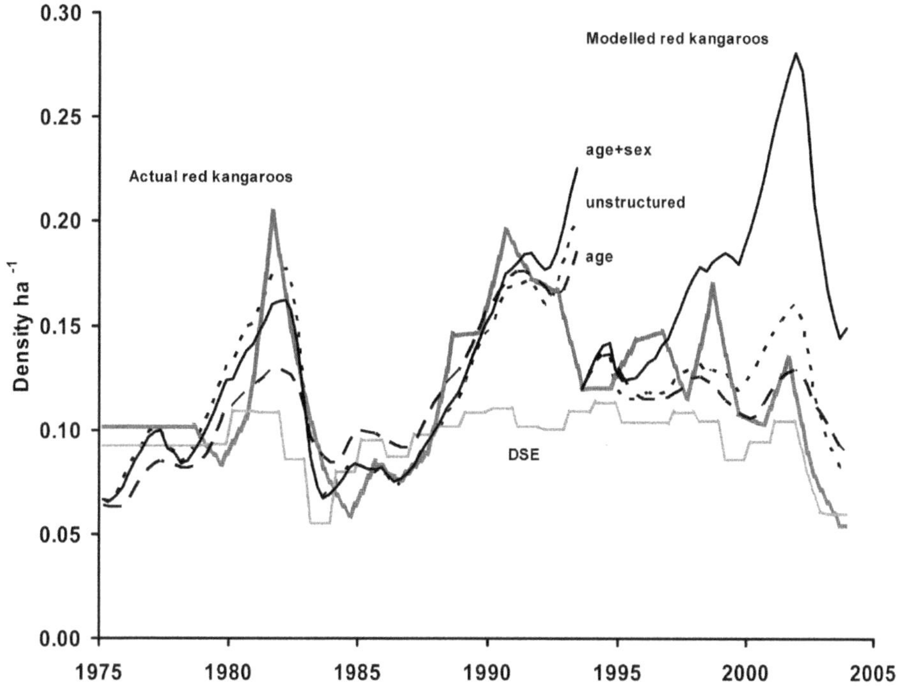

Figure 18.2: Caughley's (1987) interactive unstructured model (dotted line) fitted to red kangaroo (*M. rufus*) density estimated by aerial survey (thick solid line) in north-east pastoral SCB in the north-east of the study area in South Australia. Kangaroo rate of increase is modelled as a function of pasture biomass, a function of rainfall and grazing by kangaroos and livestock (converted to dry sheep equivalents, DSE; thin solid grey line). Models with age structure (dashed line) and age and sex structure (thin solid line) were also fitted to the data. Both modelled and actual populations were harvested. All modelled populations were reinitialised in 1993.

parameter was fitted (S = 2.2, sum of squares (SSQ) = 0.016) after initialising the model at the start of the series. Other model parameters were not fitted as they were based on field data collected in 1980–83 and the exercise was to test the model against another, longer, time series. However, the model needed to be reinitialised in 1993 when the population declined sharply whereas the modelled population increased markedly on the back of high pasture biomass. The fit was poorer following 1993 than for 1978–92, suggesting a change in pasture state. It is worth noting that livestock numbers fluctuated to a far lesser extent than red kangaroo numbers over the time period. Obvious future work is to fit this model to other time series.

Age- and sex-structured models

An alternative to including two periods of rainfall to predict rate of increase is to model juvenile and adult survival directly. This would allow for the relatively high rates of population increase following drought, the result of an unstable age distribution with few juveniles. Survival, in quarterly time-steps, was modelled as a logistic function of pasture biomass (McCallum 1995), replacing the numerical response in the interactive model:

$$S_A = c \times e^{(a+bV)} / [1 + e^{(a+bV)}]$$
$$S_J = g \times e^{(d+fV)} / [1 + e^{(d+fV)}]$$

where S_A is adult (≥2 years old) and subadult (1–2 years old) survival, S_J is juvenile (<1 year old) survival and V is pasture biomass as before. The three parameters in each survival function were estimated by fitting the resulting modelled density to the time series in red kangaroo density in the north-east pastoral SCB. Half the adults (even sex ratio assumed) gave birth to 1.56 juveniles each year, spread evenly among the quarterly time-steps. Only adults were harvested. The parameters were constrained by data on juvenile and adult survival (Pople 1996) so c was set to 0.99 and g to 0.80. The fitted model (SSQ = 0.017) had parameters a = –0.397, b = 0.020, d = –6.610 and f = 0.017. For ease of fitting, S was set at 2.2 as for the unstructured model.

The sexes were modelled separately as an additional step. This allowed the male bias in the harvest and the male-biased mortality in drought (Robertson 1986; Pople and Cairns 1995; Pople 1996) to be accommodated, creating further instability in the age structure. Adult and subadult survival was modelled separately for the two sexes:

$$S_M = c \times e^{(h+iV)} / [1 + e^{(h+iV)}]$$
$$S_F = c \times e^{(j+kV)} / [1 + e^{(f+kV)}]$$

where S_M is adult and subadult male, S_F is adult and subadult female survival and V is pasture biomass. c was again fixed to 0.99, g to 0.80 and S to 2.2. A single juvenile survival function was modelled for both sexes. The fitted model (SSQ = 0.070) had parameters $h = j = -0.121$, $i = k = 0.010$, $d = -2.929$ and $f = 0.009$ (common male and female survival functions).

Distinguishing the sexes enabled the model to closely match the unstructured model and actual data prior to 1994. The model with age structure alone could not capture the dramatic increase in 1981. The unstructured model appears a good approximation for the structured dynamics. The fit of both structured models was poor after 1993 even after reinitialisation, particularly for the model with sex, when harvest rates were relatively high. It is worth noting that the maximum exponential rate of increase for a red kangaroo population with a stable age distribution (r_m) is about 0.29, using the Euler-Lotka equation and assuming adult annual survival of 0.95, juvenile annual survival of 0.7 and fecundity of 0.78 (number of female young per female per year). However, an unstable age distribution with a female bias can generate a rate approaching 0.6 (Pople 1996).

DISTRIBUTION PATTERNS

Short-term: aggregations in droughts

The data, being collected in a spatially explicit manner, allowed an assessment of the regional dispersion of both species of kangaroos (Pople *et al.* 2007). At a scale of 1–16 ha, red and eastern grey kangaroos are reported to exhibit a clumped distribution (Caughley 1964) with an increased aggregation during dry periods around remnant pasture (Newsome 1965). This is in agreement with oft-repeated anecdotal reports from kangaroo harvesters that kangaroos become harder to shoot following rain because they have dispersed into wooded areas (Kirkpatrick and Amos 1985), but are easier to find during dry spells when they make greater use of open areas (Hill 1979). It also matches the anecdotal observations that in droughts kangaroo numbers are often reported to be in plague proportions, partly because they congregate in areas where there is green pick from road run-off and are more conspicuous. This aggregation in dry periods is, in turn, reflected in analyses of harvest returns (Prince 1984).

Kangaroos in the SAPZ are managed in regions based on SCBs that correspond roughly to the environmental provinces described by Laut *et al.* (1977). Six SCBs, covering most of the survey area (Fig. 18.1) were used in the analyses here. The degree of clumping in the red and western grey kangaroo populations at the scale of a 2 km² (= 200 ha) survey unit in each of the six SCBs was measured by the parameter b in Taylor's power law, $s^2 = am^b$ (Taylor 1961; Elliott 1977), where the variance (s^2) of a population is proportional to a fractional power (b) of the mean (m).

Parameters a and b were calculated from the intercept ($\log_{10}(a)$) and slope (b) of a regression of $\log_{10}(s^2)$ on $\log_{10}(m)$ of kangaroo densities in survey units within each SCB. Each year provided a single point in the analysis. For red kangaroos there were common slopes but separate intercepts for each SCB, indicating constant degrees of aggregation across SCBs (Fig. 18.3). Points for the higher rainfall SCBs in the south fall along lines elevated from the other SCBs. The residuals from this relationship were significantly correlated negatively to rainfall, most strongly to rain falling 12 months prior to survey ($r = -0.32$, 95%CI: −0.45, −0.17). Examining the scatter in the residuals, low rainfall frequently leads to greater aggregation, but the population is always more evenly dispersed following high rainfall. There were also common slopes for western grey kangaroos

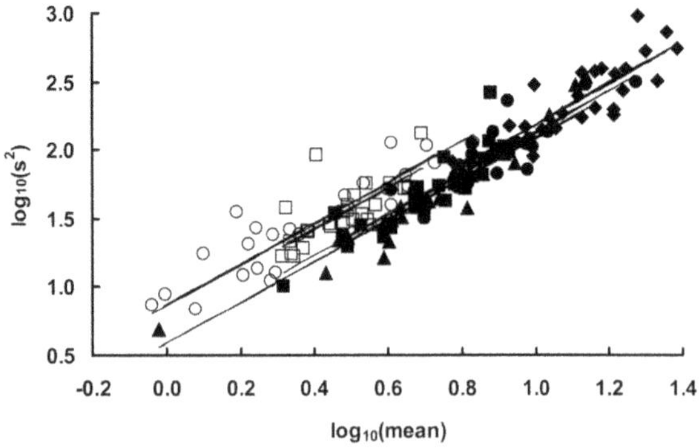

Figure 18.3: Relationship between $\log_{10}$(variance) and $\log_{10}$(mean) of red kangaroo (*M. rufus*) density in six soil conservation boards in the South Australian study for each year 1978–2003. Fitted regression lines with common slopes are shown. Eastern districts (○), Gawler Ranges (□), Kingoonya (■), Marree (●), North Flinders (▲), North-east pastoral (◆). After Pople *et al.* (2007).

indicating constant degrees of aggregation, but the residuals from these relationships were not correlated with recent rainfall.

Short-term: broad-scale movement

While the broad pattern of kangaroo distribution remained similar over the study period, there were year-to-year differences. One way of displaying this variation is to map the annual exponential rate of kangaroo increase, which was done for the SAPZ. In some areas and in some years, the increase in numbers is so great that it cannot be explained by recruitment and survival alone. This suggests an important effect of movement on population dynamics at a broad scale. Figure 18.4 shows an interpolated surface for the exponential rates of increase for red kangaroos in 1986–87. In the eastern SAPZ there is a large contiguous area where red kangaroo numbers have more than tripled. Measurement error presumably contributes to this fluctuation, but the fact that it occurs over such a large contiguous block and coincides with declines elsewhere suggests that it is real and is the result of movement by individual red kangaroos. Areas of extraordinary increases partially coincided with areas receiving above-average rainfall in the six months prior to the aerial survey in 1987, but the correlation was not significant at a scale of 25 km^2 (Pople *et al.* 2007).

Long-term: expansions in range in three species

Surveys by fixed-wing aircraft along the same transect lines (Fig. 18.1) over long periods also enabled longer-term shifts in the pattern of distribution to be assessed. To display this over the entire study area, annual exponential rates of increase (the differences between the logarithms of density + 0.001) were calculated between 1992–93 and 2001 in half-degree blocks (~2500 km^2), then interpolated by kriging (Fig. 18.5a, b, c). Rates of increase over longer but different periods were available for each state separately. It has already been reported that eastern grey kangaroos have expanded their range westwards into more arid areas (Caughley *et al.* 1984a). The assessment here shows that their pattern of distribution behind the expanding front has also shifted westwards. Red kangaroos have expanded eastwards into more mesic areas, in association with land

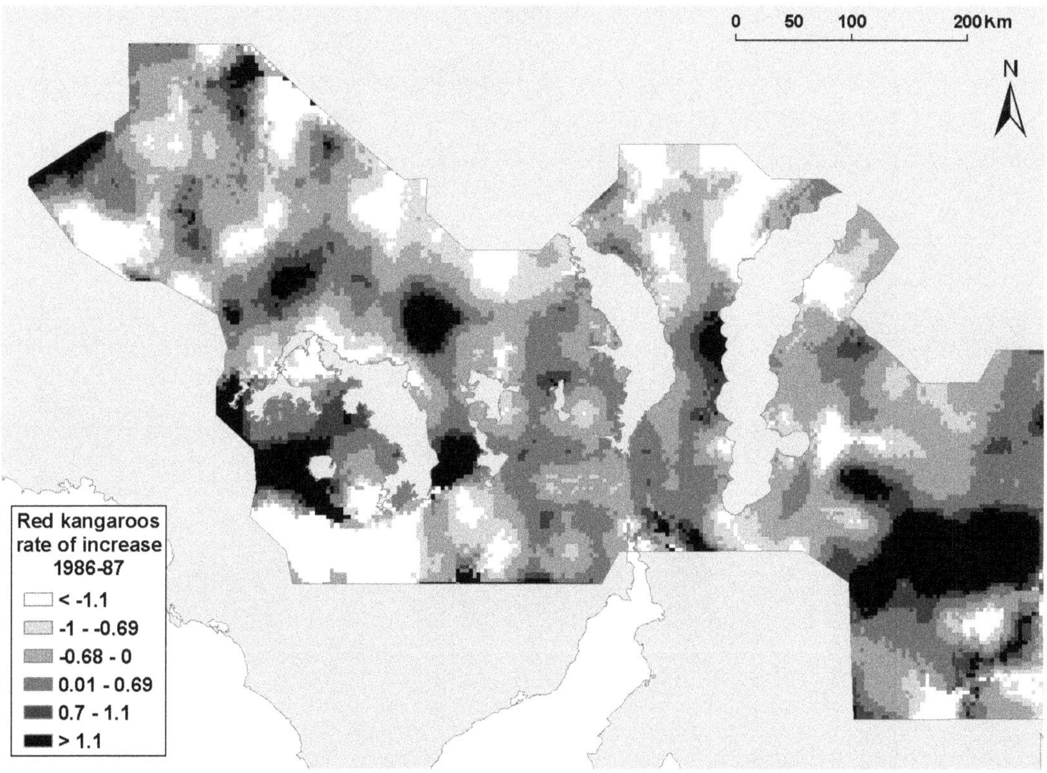

Figure 18.4: Annual exponential rate of increase for red kangaroos (*M. rufus*) in the South Australian pastoral zone, 1986–87. Calculated from annual densities in 1986 and 1987 in 5 x 5 km cells interpolated from densities in aerial survey segments using universal kriging. After Pople *et al.* (2007).

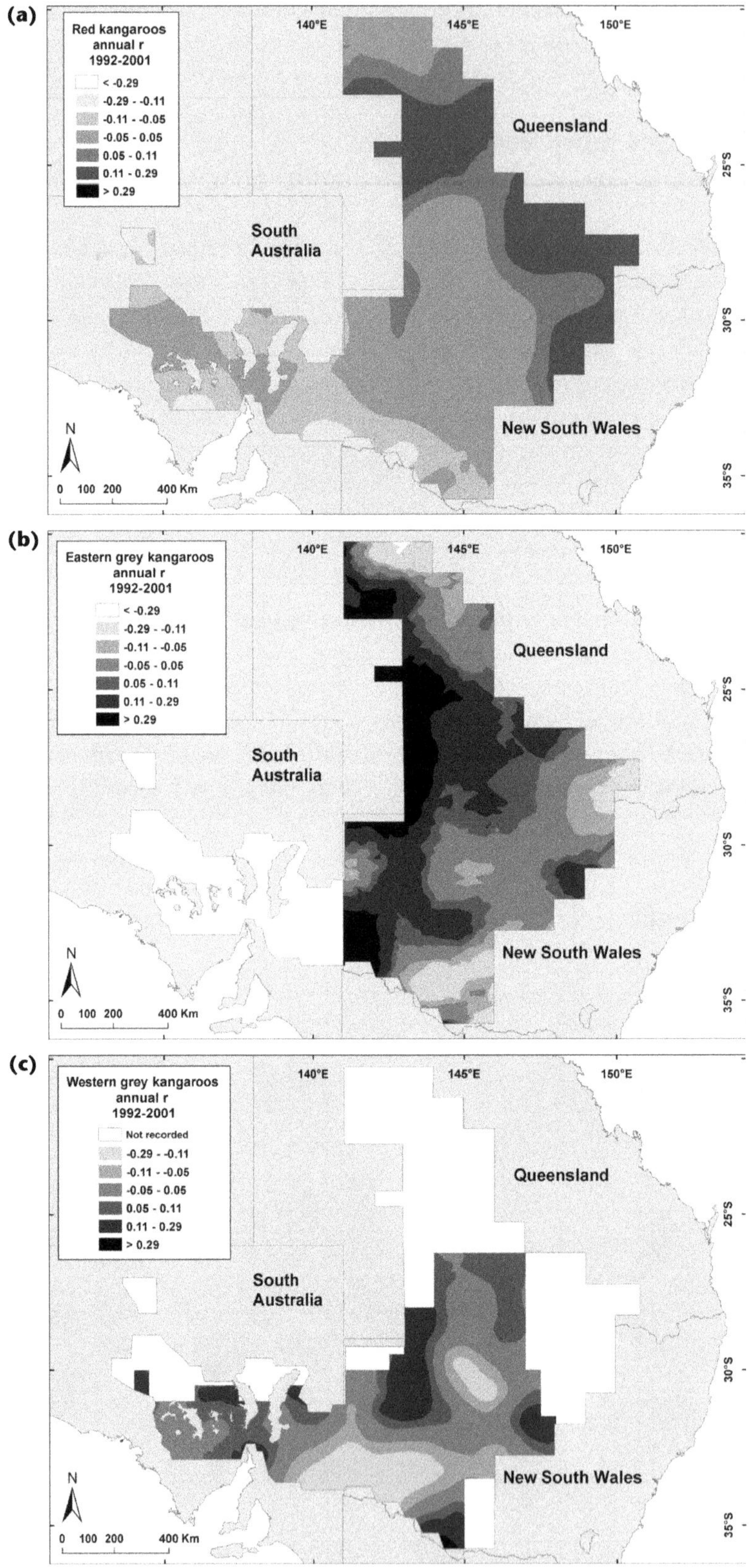

Figure 18.5: Annual exponential rate of increase of (a) red kangaroos (*M. rufus*), (b) eastern grey kangaroos (*M. giganteus*) and (c) western grey kangaroos (*M. fuliginosus*) over 1992–2001 within the sheep rangelands of eastern Australia. Rates of increase were first calculated in half-degree blocks (~50 km ´ 50 km) then interpolated using universal kriging. In South Australia, all grey kangaroos recorded during aerial survey were assumed to be western grey kangaroos.

clearing in Queensland based on the reduction in woody cover as shown in the state-wide land cover and trees study (DNRM 2003). In South Australia, western grey kangaroos have shifted their pattern of distribution northwards into more arid areas, perhaps being advantaged by the provision of artificial watering points.

SURVEY FREQUENCY AND PRECISION

Background

Models that accurately predict population size should allow reductions in the frequency and intensity of expensive surveys. However, there need to be guidelines for when to use a model prediction and when and how intensively to conduct a survey. Using decision theory to optimise survey frequency, Hauser *et al.* (2006) used a model to predict the future density of kangaroo population to be harvested. If the prediction was uncertain or the population was likely to be outside a desirable density range, a survey was worth undertaking. An alternative related approach, described here, is to determine survey frequency from the risk of over- or underharvesting (Pople 2008). Survey precision will also influence this risk and affect the cost of a survey, so is included in the assessment.

In wildlife management, the appropriate survey frequency and precision has generally been considered with the aim of detecting trends (Caughley 1979; Harris 1986; Gerrodette 1987). In kangaroo management, harvest regulation is primarily through setting harvest quotas that are calculated as proportions of absolute estimates of population size. Trends are considered of secondary importance. This harvesting strategy has been well studied and is considered relatively safe and efficient for a fluctuating population (Caughley 1987; Engen *et al.* 1997; McLeod and Pople 1998). However, a proportional harvesting strategy requires a regular estimate of abundance on which to set a quota. Imprecision in population estimates and infrequent estimates will risk applying a quota that is too high or too low (Pople 2003). By not harvesting at the desired rate (e.g. 15%), costs to the kangaroo industry are incurred through reduced and more variable yield and to farmers through increased competition with sheep and damage to crops. There is a social cost if kangaroos are reduced below some arbitrarily low density (i.e. quasiextinction) because of a perceived conservation threat and a desire to see kangaroos in abundance (Hauser *et al.* 2006). These costs must be balanced against the cost of more intensive and frequent surveys (i.e. there is a trade-off). Alternative harvest strategies can be considered, to reduce the risk of over- or underharvesting. These include harvesting at a different and even variable rate, regulating effort or incorporating spatial reserves.

Assessment of risk of lowered survey frequencies

Precision (0–200%) and frequency (one to five years) of density estimates and harvest rate (10% or 15%) were varied in sets of 1000 simulations of Caughley's (1987) interactive model over 100 years (Pople 2008). Two environments were modelled – Menindee (annual mean rainfall = 244 mm) and a more mesic environment at Yass in south-east New South Wales. For the latter, the population fluctuations in the interactive model required dampening by reducing the standard deviation in seasonal rainfall by 37% so that it was equivalent to that around Yass (annual mean rainfall = 652 mm). Altering the standard deviation in rainfall and not the mean avoided the 'paradox of enrichment', which would have increased the mean kangaroo density and amplitude of fluctuations in kangaroo density (Crawley 1983).

The appropriate survey frequency will depend on the costs of the surveys and the costs of compromising the harvest strategy through not always knowing population size. Imprecision can also incur costs, but this only became important when precision was >50%. To provide a first approximation of the trade-off, a simple estimate of survey cost was contrasted with quasiextinction risk for increasing survey frequency. Some fixed costs associated with aerial surveys do not change if surveys are conducted less frequently. These include labour costs and are assumed to be \$50 000 per annum. The variable costs primarily comprise aircraft running costs and are assumed to be \$50 000 per survey. The total five-year survey cost (fixed + variable × no. of surveys) is given in Figure 18.6, showing the ever-diminishing savings from reducing survey frequency. The sigmoidal increase in quasiextinction risk with increasing survey frequency is also shown, for a threshold density of two kangaroos per km^{-2}. It is markedly higher for the more arid population, but reducing harvest rate to 10% reduced quasiextinction risk dramatically. The actual probability should be interpreted

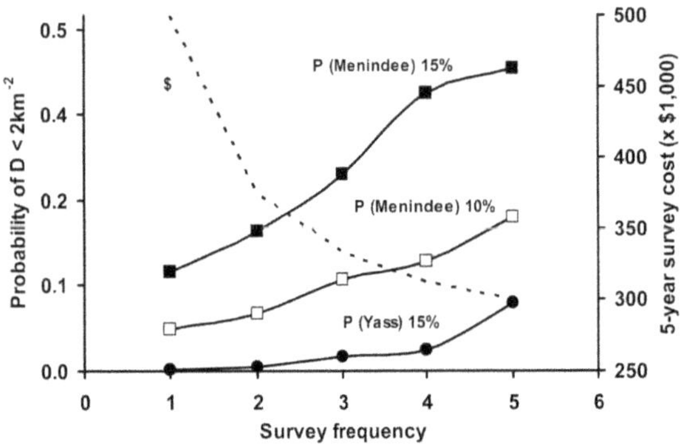

Figure 18.6: Probability of quasiextinction (solid lines) and survey cost over five years (dotted line) for survey frequencies of one to five years for Yass (closed circles) and Menindee (solid squares), where harvest rate was 15%, and Menindee (open squares), where harvest rate was 10%. Precision for surveys was 20%. D = kangaroo density. After Pople (2008).

cautiously as it is dependent on the model structure and parameters, which have not been assessed for this population. Furthermore, the acceptable level of risk is a value judgment.

Management implications

A number of management recommendations are reasonably clear from the risk assessment. First, in more mesic environments, reducing the survey frequency to once every three years with a precision of 20% would appear to increase the risk of overharvest only slightly, while substantially reducing survey costs. Whether the increase in risk is acceptable is a decision for management in consultation with stakeholders. Second, in arid environments, where precision may be as poor as 50%, harvesting close to the MSY (e.g. 13% for the unstructured interactive model) ideally requires estimates of population size to be determined annually. In some regions, such as north-west South Australia and north-west Queensland, where harvest rates are low (Pople 2006), there is unlikely to be a need for annual surveys. The results also suggest that reducing the quota to 10% would largely offset the increased risks resulting from surveying less frequently.

CONCLUSIONS AND FURTHER WORK

Rainfall can still be viewed as an indicator of changes in kangaroo numbers, particularly in drought, but it does not have universal application. Outside drought, the ability to predict fluctuations in density varies regionally and among species. Some of these predictive models should still be useful for harvest management. The risk assessment provides a framework in which to monitor populations indirectly between population surveys via harvest statistics (Ch. 32 this volume) or rainfall-driven models. In the risk assessment, quotas were not adjusted between surveys, but this was a conservative approach. In practice, quotas would be modified according to model predictions until the availability of the next survey-based estimate of population size, providing a reduced risk of over- or underharvesting.

Extensions of the work described above would be worthwhile, including fitting population models such as Caughley's (1987) to other time series. Spatial modelling is one area that also needs attention and had preliminary analyses in this project.

Spatial modelling produces continuous density surfaces, providing alternative estimates of population size to those derived from the averages of survey units. Integration under the density surface allows population size to be estimated for any area within the overall study area. These surfaces can be created by interpolating aerial survey data using techniques such as kriging, but habitat covariates can also be used.

In New South Wales and South Australia, annual quotas or tags or both are issued to properties within each region to put a ceiling on the number of kangaroos that can be harvested commercially. In all states, non-commercial culling (pest destruction) permits are also allocated to properties. All allocations are made largely on an *ad hoc* basis. A spatial model would therefore provide a more rigorous basis for allocating quotas and permits.

In Queensland, surveys of kangaroos by helicopter in blocks result in incomplete geographic coverage of the harvest area and sampling is non-random. Extrapolating the average kangaroo density in these survey blocks to the rest of the harvest area therefore requires some strong and perhaps unrealistic assumptions. A spatial model can overcome these limitations by providing regional estimates of density through integration under a density surface.

ACKNOWLEDGMENTS

This project was supported by an ARC Linkage grant to Hugh Possingham, Gordon Grigg and Stuart Phinn at the University of Queensland, with financial contributions from Queensland, New South Wales,

South Australian and Commonwealth conservation agencies and Packer Tanning. We are grateful to the South Australian Department of Environment and Heritage for its long-running support of the surveys in that state, and to all the industry partners for making data available. We also acknowledge the conscientious work by observers, on whose considerable skills many of these data depend.

REFERENCES

Bayliss P 1985a. The population dynamics of red and western grey kangaroos in arid New South Wales, Australia. I. Population trends and rainfall. *J. Anim. Ecol.* **54:** 111–125.

Bayliss P 1985b. The population dynamics of red and western grey kangaroos in arid New South Wales, Australia. II. The numerical response function. *J. Anim. Ecol.* **54:** 127–135.

Bayliss P 1987. Kangaroo dynamics. In *Kangaroos: Their Ecology and Management in the Sheep Rangelands of Australia.* (Eds G Caughley, N Shepherd and J Short) pp. 119–134. Cambridge University Press, Cambridge.

Bayliss P and Choquenot D 2002. The numerical response: rate of increase and food limitation in herbivores and predators. *Phil. Trans. Roy. Soc. Lond. B Biol. Sci.* **357:** 1233–1248.

Beard L 1999. Training observers. *Aust. Zool.* **31**: 287–291.

Burgman MA, Ferson S and Akcakaya HR 1993. *Risk Assessment in Conservation Biology.* Chapman and Hall, London.

Cairns SC 1989. Models of macropodid populations. In *Kangaroos, Wallabies and Rat-kangaroos.* (Eds G Grigg, P Jarman and I Hume) pp. 695–704. Surrey Beatty and Sons, Sydney.

Cairns SC and Grigg GC 1993. Population dynamics of red kangaroos (*Macropus rufus*) in relation to rainfall in the South Australian pastoral zone. *J. Appl. Ecol.* **30:** 444–458.

Cairns SC, Grigg GC, Beard LA, Pople AR and Alexander P 2000. Western grey kangaroos (*Macropus fuliginosus*) in the South Australian pastoral zone: the dynamics of populations at the edge of their range. *Wildl. Res.* **27:** 309–318.

Caughley G 1964. Density and dispersion of two species of kangaroo in relation to habitat. *Aust. J. Zool.* **12:** 238–249.

Caughley GJ 1979. Design for aerial censuses. In *Aerial Survey of Fauna Populations.* pp. 15–23. Australian Government Publishing Service, Canberra.

Caughley G 1987. Ecological relationships. In *Kangaroos: Their Ecology and Management in the Sheep Rangelands of Australia.* (Eds G Caughley, N Shepherd and J Short) pp. 159–187. Cambridge University Press, Cambridge.

Caughley GJ and Grigg GC 1981. Surveys of the distribution and density of kangaroos in the Pastoral Zone of South Australia, and their bearing on the feasibility of aerial survey in large and remote areas. *Aust. Wildl. Res.* **8:** 1–11.

Caughley G and Sinclair ARE 1994. *Wildlife Ecology and Management.* Blackwell Scientific, London.

Caughley GJ, Short J and Grigg GC 1983. How many kangaroos? *Search* **14:** 151–152.

Caughley G, Sinclair R and Scott-Kemmis D 1976. Experiments in aerial survey. *J. Wildl. Man.* **40:** 290–300.

Caughley G, Brown B, Dostine P and Grice D 1984a. The grey kangaroo overlap zone. *Aust. Wildl. Res.* **11:** 1–10.

Caughley J, Bayliss P and Giles J 1984b. Trends in kangaroo numbers in western New South Wales and their relation to rainfall. *Aust. Wildl. Res.* **11:** 415–422.

Caughley G, Grigg GC and Smith L 1985. The effect of drought on kangaroo populations. *J. Wildl. Man.* **49:** 679–685.

Caughley G, Shepherd N and Short J (Eds) 1987. *Kangaroos: Their Ecology and Management in the Sheep Rangelands of Australia.* Cambridge University Press, Cambridge.

Clancy TF, Pople AR and Gibson LA 1997. Comparison of helicopter line transects with walked line transects for estimating densities of kangaroos. *Wildl. Res.* **24:** 397–409.

Crawley MJ 1983. *Herbivory.* Blackwell Scientific, London.

Crawley MJ 2002. Statistical computing: an introduction to data analysis using S-Plus. Wiley, Chichester.

Cridland SW and Fitzgerald NJ 2001. Apparent stability in the rangelands from NDVI-derived indicators. In IEEE 2001 *International Geoscience and Remote Sensing Symposium.* Environmental Resources Information Network, Environment Australia, Sydney.

DNRM (Dept of Natural Resources and Mines) 2003. Land cover change in Queensland: a statewide landcover and trees report, January 2003. Dept of Natural Resources and Mines, Brisbane. http://www.nrm.qld.gov.au/slats/report.html.

Elliott JM 1977. Some methods for the statistical analysis of samples of benthic invertebrates. *Freshwater Biol. Assoc. Sci. Publ.* **25:** 1–142.

Engen S, Lande R and Saether B-E 1997. Harvesting strategies for fluctuating populations based on uncertain population estimates. *J. Theor. Biol.* **186:** 201–212.

Environment Australia 2000. Revision of the Interim Biogeographic Regionalisation of Australia (IBRA) and the Development of Version 5.1. Summary Report. Unpublished report to Dept of Environment and Heritage, Canberra.

Gaillard J, Festa-Bianchet M and Yoccoz NG 1998. Population dynamics of large herbivores: variable recruitment with constant adult survival. *Trends Ecol. Evol.* **13:** 58–63.

Gerrodette T 1987. A power analysis for detecting trends. *Ecology* **68:** 1364–1372.

Grigg GC, Pople AR and Beard LA 1997. Application of an ultra-light aircraft to aerial surveys of kangaroos on grazing properties. *Wildl. Res.* **24:** 359–372.

Grigg GC, Beard L, Alexander P, Pople AR and Cairns SC 1999. Aerial survey of kangaroos in South Australia 1978–1998: a brief report focusing on methodology. *Aust. Zool.* **31:** 292–300.

Harris RB 1986. Reliability of trend lines obtained from variable counts. *J. Wildl. Man.* **50:** 165–171.

Hauser CE, Pople AR and Possingham HP 2006. Should managed populations be monitored every year? *Ecol. Appl.* **16:** 807–819.

Hill GJE 1979. Application of LANDSAT to habitat analysis for the grey kangaroo. *Proceedings of New Zealand Geographical Conference* **10:** 293–296.

Hobbs TJ 1995. The use of NOAA-AVHRR NDVI data to assess herbage production in the arid rangelands of central Australia. *Int. J. Remote Sens.* **16:** 1289–1302.

Jonzen N, Pople AR, Grigg GC and Possingham HP 2005. Of sheep and rain: large-scale population dynamics of the red kangaroo. *J. Anim. Ecol.* **74:** 22–30.

Kirkpatrick TH and Amos PJ 1985. The kangaroo industry. In *The Kangaroo Keepers.* (Ed. HJ Lavery) pp. 75–102. University of Queensland Press, Brisbane.

Krebs CJ 1995. Two paradigms of population regulation. *Wildl. Res.* **22:** 1–10.

Krebs CJ 2002. Two complementary paradigms for analysing population dynamics. *Phil. Trans. Roy. Soc. Lond. B Biol. Sci.* **357:** 1211–1219.

Laut P, Heyligers PC, Keig G, Loffler E, Margules C and Scott RM 1977. *Environments of South Australia.* CSIRO Division of Land Use Research, Canberra.

McCallum HI 1995. Modelling translocation strategies for the bridled nailtail wallaby (*Onychogalea fraenata* Gould 1840): Effects of reintroduction size. In *Reintroduction Biology of Australian and New Zealand Fauna.* (Ed. M Serena) pp. 7–14. Surrey Beatty and Sons, Sydney.

McCarthy MA 1996. Red kangaroo (*Macropus rufus*) dynamics: effects of rainfall, density dependence, harvesting and environmental stochasticity. *J. Appl. Ecol.* **33:** 45–53.

McLeod S and Pople T 1998. Optimal exploitation strategies for kangaroo populations in variable environments. In *Proceedings of the 11th Australian Vertebrate Pest Conference*, 3–8 May 1998. Bunbury, Western Australia. pp. 375–380.

Menke N, Pople AR, Jonzén N, Cairns SC, McAlpine C, Grigg GC, Phinn SR and Possingham HP 2005. Using AVHRR satellite and herbivore time series data to model kangaroo dynamics at two scales in Australian rangelands. In *Remote Sensing and Geoinformation Processing in the Assessment and Monitoring of Land Degradation and Desertification.* (Eds A Röder and J Hill) pp. 340–347. Trier, Germany.

Newsome AE 1965. The distribution of red kangaroos, *Megaleia rufa* (Desmarest), about sources of persistent food and water in central Australia. *Aust. J. Zool.* **13:** 289–299.

Pople AR 1996. Effects of harvesting upon the demography of red kangaroos in western Queensland. PhD thesis. University of Queensland, Brisbane.

Pople A 2003. Harvest management of kangaroos during drought. Unpublished report to New South Wales National Parks and Wildlife Service, Dubbo, NSW.

Pople AR 2006. Modelling the spatial and temporal dynamics of kangaroo populations for harvest management. Final report to the Dept of Environment and Heritage, Canberra.

Pople AR 2008. Frequency and precision of aerial surveys for kangaroo management. *Wildl. Res.* **35:** 340–348.

Pople A and Cairns S 1995. Impact of harvesting on kangaroos. In *Conservation through Sustainable Use of Wildlife.* (Eds G Grigg, P Hale and D Lunney) pp. 224–229. Centre for Conservation Biology, University of Queensland, Brisbane.

Pople AR, Phinn SR, Menke N, Grigg GC, Possingham HP and McAlpine C 2007. Spatial patterns of kangaroo density across the South Australian pastoral zone over 26 years: aggregation during drought and suggestions of long distance movement. *J. Appl. Ecol.* **44:** 1068–1079.

Prince RIT 1984. Exploitation of kangaroos and wallabies in Western Australia. I. Exploitation and management of the red kangaroo: 1970–1979. Dept of Fisheries and Wildlife, Perth.

Robertson GG 1986. The mortality of kangaroos in drought. *Aust. Wildl. Res.* **13:** 349–354.

Robertson G 1987. Plant dynamics. In *Kangaroos: Their Ecology and Management in the Sheep Rangelands of Australia.* (Eds G Caughley, N Shepherd and J Short) pp. 50–68. Cambridge University Press, Cambridge.

Sibly RM and Hone J 2002. Population growth rate and its determinants: an overview. *Phil. Trans. Roy. Soc. Lond. B Biol. Sci.* **357:** 1153–1170.

Sinclair ARE and Krebs CJ 2002. Complex numerical responses to top-down and bottom-up processes in vertebrate populations. *Phil. Trans. Roy. Soc. Lond. B Biol. Sci.* **357:** 1221–1231.

Taylor LR 1961. Aggregation, variance and the mean. *Nature* **189:** 732–735.

Tucker CJ, Vanpraet CL, Sharman J and Van Ittersum G 1985. Satellite remote sensing of total herbaceous biomass production in the Senegalese Sahel: 1980–1984. *Rem. Sens. Environ.* **17:** 233–249.

19 Impact of human disturbance on the population dynamics and ecology of tammar wallabies on Garden Island, Western Australia

B.K. Chambers and R. Bencini

SUMMARY

Garden Island, Western Australia, supports an endemic population of tammar wallabies (*Macropus eugenii*) and hosts the HMAS Stirling Naval Base. The base is fenced and confined to the south of the island, leaving the north relatively free of disturbance. Irrigated lawns on the base provide the tammar wallabies with a year-round source of food. A causeway was built to facilitate vehicle access to the island and has resulted in up to 400 tammars per year being killed by vehicles. We hypothesised that the tammar wallaby populations in the disturbed and undisturbed parts of the island would differ in density, age structures, reproductive success and body condition and that the high number of road-kills would be biased towards young animals. Tammar wallabies were trapped at the naval base, our disturbed site, and in the northern undisturbed bushland, between May 2005 and May 2006 to estimate population density, age structure, reproductive success and body weight adjusted for *pes* length by ANCOVA as an indication of body condition. Road-killed tammar wallabies were also collected, aged and sexed. Tammars living on the base had higher reproductive success ($p < 0.05$) despite being at a density of 2.45 ± 0.30/ha, twice that of those living in the undisturbed areas (1.27 ± 0.12/ha). The age distribution of road-killed tammars was biased towards one-year-old animals, as expected. The population at the disturbed site was biased towards older males ($p < 0.05$). The condition of the animals at the two sites was not significantly different overall ($P = 0.06$), except for the winter months of 2005 when females at the disturbed site were approximately 0.5 kg heavier than those at the undisturbed site ($p < 0.05$). These results support the hypothesis that tammar wallabies living on the base would have higher reproductive success and survival, which was in part counteracted by the high number of road-kills.

INTRODUCTION

Human activities such as habitat modification and fragmentation usually affect populations of wild animals by decreasing their numbers, which can eventually result in local extinction. However, when the habitat modifications provide resources for the animals this can result in increases in numbers and populations becoming overabundant. Understanding the changes in population dynamics that drive these changes underpins our capacity to manage wildlife.

The tammar wallaby (*Macropus eugenii*) has become so abundant on Kangaroo Island that culling is now used to limit damage to agricultural crops (Wright and Stott 1999), while in Western Australia habitat fragmentation combined with predation by the European red fox (*Vulpes vulpes*) saw the species dramatically decline in range and abundance (Kinnear *et al.* 2002). In contrast, an endemic population of tammar wallabies persists in good numbers on Garden Island, 5 km off the Western Australian coast, 35 km south-west of Perth, possibly due to the absence of feral predators (Bradshaw 1988). In the 1970s the Commonwealth assumed control of Garden Island to build the largest naval base in Australia, HMAS Stirling. The base infrastructure is fenced and primarily confined to the southern end of the island, resulting in much of the island remaining relatively undisturbed. Construction involved the establishment of three large sporting ovals and the planting and irrigation of lawns around the buildings. These irrigated and fertilised lawns have provided tammar wallabies in this area with a continuous supply of food that is not available to the animals living in the undisturbed bushland and is possibly of greater nutritive value than the native vegetation (Wann and Bell 1997; Bennetts and Cocks 1999).

Tammar wallabies living in the undisturbed bushland on Garden Island were reported to be in the poorest condition in late autumn due to the small amount and poor quality of food available (Bradshaw 1988). The wallabies living on the base should not experience this seasonal food shortage and therefore we expected that they would live at higher densities and remain in better body condition throughout the year compared to those in the bushland areas of the island.

Tammar wallabies are seasonal breeders and on Garden Island give birth to their young in January by reactivating the dormant blastocyst in response to decreasing day length (Renfree and Tyndale-Biscoe 1973; Hinds and den Ottolander 1983). A second blastocyst can be reactivated if the young is lost, but only until the winter solstice in late June (Smith and Hinds 1995). Lactation continues until October–November, when the young are weaned. In red kangaroos (*Macropus rufus*), nutritional stress during lactation results in females being unable to support a pouch young (Sharman *et al.* 1964). We expected that access to a constant source of food would reduce the nutritional stress on lactating females and that female tammar wallabies living on the base would be more successful at raising pouch young to weaning.

The constant food supply on the base is also likely to affect the survival of wallabies in that area. The onset of winter is known to be a difficult period for tammar wallabies as they are in poor condition after the summer and autumn and many are unable to cope with the change to cold and wet weather, and consequently die (Inns 1982; Bradshaw 1988). This should not affect the tammars on the base, as they should be in better condition at the start of winter and mortality would thus be reduced. We expected that this reduction in mortality would result in the population having a greater proportion of older animals than the population in the bushland areas of the island.

As part of the base construction an 800 m causeway was built from Point Peron on the mainland to allow vehicle access to the island. This has resulted in road-kills of up to 400 tammars per year, of which 200–250 occur on the base at the southern end of the island (Walker 2002). In western New South Wales 45–46% of red, eastern (*Macropus giganteus)* and western (*Macropus fuliginosus*) grey kangaroos killed by vehicles were under three years of age (Lee *et al.* 2004). We expected that road-kills would show a similar pattern of bias towards younger animals on Garden Island, which would result in a change in the age structure of the population compared to the wallabies in undisturbed parts of the island.

METHODS

Study sites

Garden Island is located 5 km off the Western Australian coast, approximately 35 km south-west of the city of Perth (115°40'E, 32°16'S; Fig. 19.1). Two sites, disturbed and undisturbed, were used for the study. These were located on the naval base at the southern end of the island and at the far northern end of the island respectively (Fig. 19.1). The disturbed area had approximately 40% of the native vegetation cleared and replaced with buildings, roads and irrigated lawns. The northern end of the island, approximately 5.6 km from the base, was virtually undisturbed bushland dominated by *Acacia rostellifera*, *Callitris preissii* and *Melaleuca lanceolata* (Bell *et al.* 1987). There was no naval activity and the only public access was by boat during daylight, with the majority of visitors confining their activities to the beaches (T. Smith, *pers. comm.*).

Trapping

Tammar wallabies were trapped on six occasions over the study period, in May, July, September and

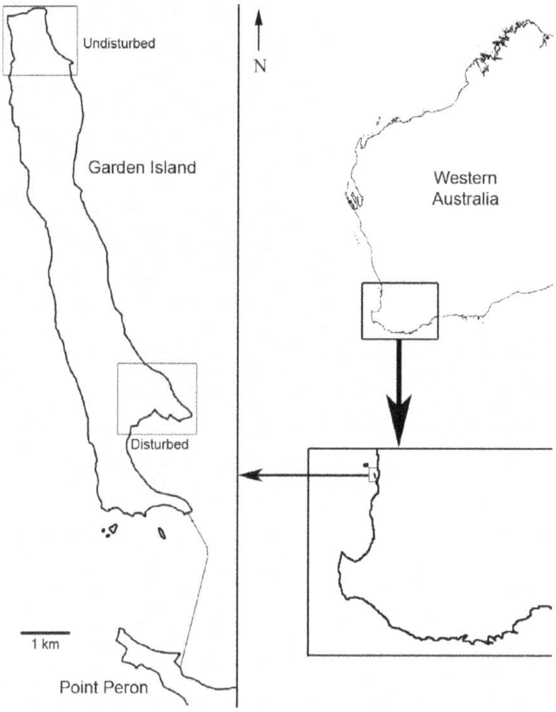

Figure 19.1: Location of Garden Island, Western Australia, and the disturbed and undisturbed study sites.

November 2005 and February and May 2006. Thirty Thomas soft-wall traps made with shade-cloth stretched over a steel frame (450 × 450 × 800 mm, Sheffield Wire Works, Welshpool, Western Australia), baited with a mixture of rolled oats, lupins, sunflowers seeds and molasses (Kangaroo Muesli, Thompson & Redwood Produce Supplies, Upper Swan, Western Australia), were set over the study areas in a grid pattern with 200 m between trap sites. Traps were set for three nights at each trapping occasion and were checked twice each night in the undisturbed site, at 2100 and 0600 hr, and three times each night at the disturbed site, at 2100, 0100 and 0430 hr. Traps were checked more frequently at the disturbed site, that was assumed to have a greater density of tammars, to increase the chances of catching a greater proportion of the population. This gave a total of 1560 trap clearances at the disturbed site and 1380 at the undisturbed site.

Captured tammar wallabies were weighed, the length of their pes (excluding the claw) was measured, their sex was recorded and, if previously tagged, their ear-tag numbers were recorded. For females, the presence of a pouch young or an elongated teat was recorded; large furred pouch young were removed from the pouch, weighed and sexed. They were then returned to the pouch and taped in to reduce the chance of females ejecting pouch young on release. Untagged animals were sedated using a mixture of 10 mg/kg of Ketamine (Ketamil, Ilium Troy Laboratories) and 1.25 mg/kg Xylazine (Xylazil, Ilium Troy Laboratories) given intramuscularly. They were then aged by molar eruption according to Inns (1982) and given a numbered ear-tag in each ear. These animals were held for two to three hours to allow them to recover from the sedation before being released at their site of capture.

Road-kills

The security and gardening staff on the island has been recording the date and location of all tammar wallaby road-kills over the whole island since 1993. Tammars killed by vehicles from February 2005 to February 2006 were also collected and stored so that their sex and age could be recorded. Age was determined by molar eruption using the same method as for the live animals (Inns 1982).

Data analysis

Population size in the two study areas was calculated from the capture histories of all animals caught using a full Jolly Seber model in POPAN 5 (Population Analysis Software Group, University of Manitoba and Simon Fraser University, Canada), which assumes open populations to include births and mortality. Due to the constraints of mark recapture analysis, the estimates of population size and density presented are those for February 2006. At the disturbed site, the grazing animals were highly visible so a spotlight count was done in February 2006 to verify the population estimate. This was done by driving on all the base roads and counting the tammars that were visible using a 100 W spotlight. Three ovals on the base were also surveyed. The spotlight count was started two hours after sunset and took one hour to complete. The probability of animals moving between areas and being counted more than once was therefore very low.

To determine population densities, the area of each study site was calculated by plotting the trap site locations in a GIS program (QGIS) then creating a polygon that enclosed all trap sites and extended 100 m past the edge of the trapping grid. The area of the base study site was 123 ha and that of the northern bushland was 121 ha. The density of the populations of tammar wallabies in each study site was calculated by dividing the population estimates by the area estimates.

Body weights of tammar wallabies at the two sites and at different trapping sessions were compared using one-way analysis of covariance (ANCOVA) with body weight as the dependent variable, location as the independent variable and pes length (mm, excluding the claw) as the covariate (SuperANOVA, Abacus Concepts Inc.). This yielded a least square means ± SE table of weights adjusted for pes length. Data from individual animals were used only once in the analysis, to ensure independence. The body weights of the male animals were log transformed before analysis, as the untransformed data did not conform to the assumption of homogeneous slopes as required in ANCOVA. The least square means were then back-transformed for ease of presentation.

Females were considered to be breeding if they were carrying a pouch young during the breeding season or if they had an elongated teat when caught in November, as this meant that they had recently weaned a pouch young. Comparisons between the numbers of females breeding in the two study sites and between different trapping events were made using chi square tests.

The age structure of the populations in the two study areas was determined from the captured animals, based on the assumption that animals of all ages were equally catchable. Males and females were considered separately, as previous studies of tammar wallabies showed that age-specific survival differs between males and females (Inns 1980). The age structures were compared between the two study sites using a chi square test. The same method was used to compare the age structure at the disturbed site to that of the animals killed by vehicles in that area.

Sex ratios of adult animals were calculated from separate male and female population estimates. They were then compared both between the two populations and to even sex ratios, to identify any sex bias using a G-test (JMP, SAS Institute Inc.).

All results presented are means, plus or minus their standard errors, unless otherwise indicated.

RESULTS

Population densities

We had a total of 1123 captures of 565 individual tammar wallabies, 618 on the base (327 individuals) and 505 at the undisturbed site (238 individuals). Of these, 372 were males and 193 were females. Capture rates were 42 ± 6.9% (SE, n = 6) on the base and 40 ± 8.4% in the undisturbed bushland.

The density of tammar wallabies based on mark recapture population size estimates at the disturbed site was 2.45 ± 0.30/ha, twice that at the undisturbed site (1.27 ± 0.12/ha). The density of tammars at the disturbed site based on the spotlight count taken at the same time as the February 2006 trapping was 4.65/ha.

Age structures and sex ratios

At the disturbed site the population was heavily biased towards males aged six years and older compared with the undisturbed site, with 45% of the males caught in this population in this age group ($p < 0.01$). The population at the undisturbed site had a greater proportion of two- and three-year-old males, but fewer one-year-olds than the population at the disturbed site (Fig. 19.2). The population at the disturbed site had just above 2% of four-year-old females compared with the 15% found in the population at the undisturbed site (Fig. 19.2; $p < 0.05$).

The population at the disturbed site also had a highly significant male bias, with an estimated male to female ratio of 2.37:1 ($p < 0.01$). This was significantly different from the adult sex ratio of 1.48:1 ($p < 0.01$) in the population at the undisturbed site. The sex ratio of pouch young at the disturbed site was

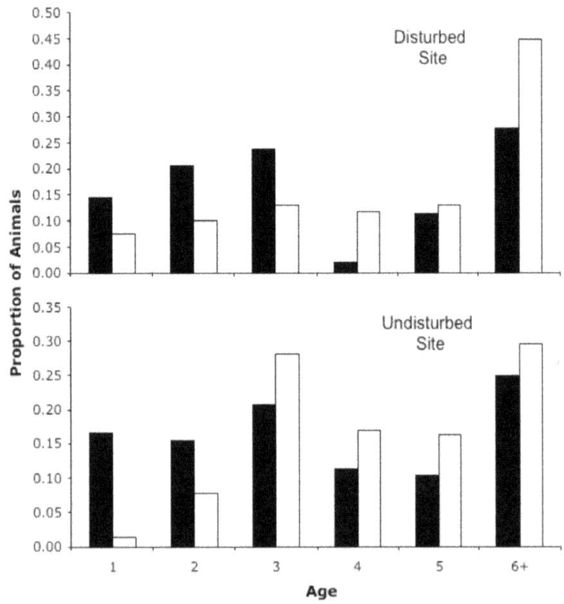

Figure 19.2: Age structure of live captured female (dark bars) and male (light bars) tammar wallabies (*M. eugenii*) at the disturbed site (top) and undisturbed site (bottom), Garden Island, Western Australia.

Table 19.1: Percentages of female tammar wallabies (*M. eugenii*) found to be breeding in the disturbed and undisturbed study sites on Garden Island, May 2005 to May 2006.

	Females with pouch young (%)			
Date	**Disturbed site**	**n**	**Undisturbed site**	**n**
May 2005	57	21	62	8
July 2005	80	20	46	33
Sept. 2005	45	9	13	37
Nov. 2005	94	17	29	28
Feb. 2006	67	30	83	23
May 2006	93	30	85	41

also skewed towards males at 1.33:1 although the bias was not statistically significant ($p > 0.05$; $n = 56$). The male to female sex ratio of pouch young from the undisturbed population was not significantly different from parity at 0.94:1 ($p > 0.1$; $n = 60$).

Reproductive success

The proportion of females that successfully weaned a pouch young in the 2005 breeding season was 94% at the disturbed site and 29% at the undisturbed site ($p < 0.05$; Table 19.1). There was no significant difference in the proportion of females with pouch young at the two sites in May 2005, but between May and July 2005 the proportion of females with pouch young at the undisturbed site fell significantly ($p < 0.05$; Table 19.1). In the early stages of the 2006 breeding season there was no significant difference in the proportion of females breeding in the two study areas.

Body condition

The adjusted weight of the males at the disturbed site was 5.1 ± 0.06 kg, no different from that of males at the undisturbed site (4.9 ± 0.08 kg; $p > 0.05$). The females at the disturbed site weighed 3.8 ± 0.06 kg, when weights were adjusted to a common pes length, while those at the undisturbed site weighed 3.5 ± 0.06 kg, but this difference was not significant ($p > 0.05$). However, females at the undisturbed site were approximately 0.5 kg lighter than those at the disturbed site in July and September 2005 respectively ($p < 0.05$; Fig. 19.3). By November 2005 the condition of the females at the undisturbed site had improved considerably and was better than those at the disturbed site (Fig. 19.3).

Road-kills

Over the study period, 300 tammar wallabies were killed by vehicles at the disturbed site. Their age distribution was significantly different from that of the live population that inhabited the same area ($p < 0.001$; Fig. 19.4). One-year-old females and males were present at rates 58% and 7% higher in the road-kills than in the live population.

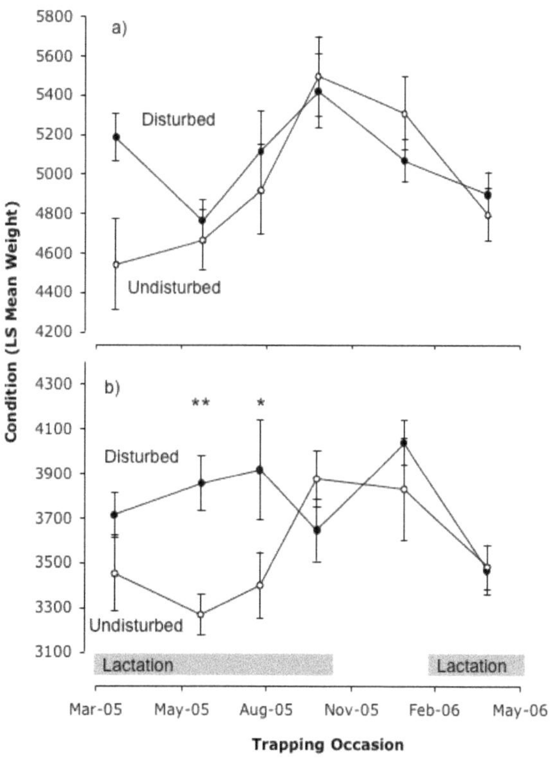

Figure 19.3: Adjusted body weights (least square means, kg ± SE) of (a) male and (b) female tammar wallabies (*M. eugenii*) from the disturbed (black circles) and undisturbed (white circles) populations between May 2005 and May 2006. Weights were compared using ANCOVA with pes length (mm, excluding claw) as the covariate (SuperANOVA, Abacus Concepts Inc.). Grey bars represent the lactation period and asterisks indicate trapping occasions when the weights were significantly different at $p < 0.001$ (**) or $p < 0.05$ (*). Note that y axes are different scales.

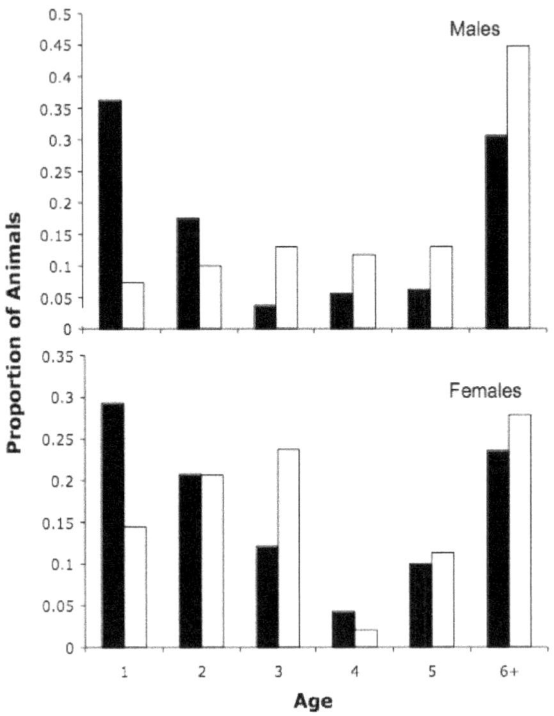

Figure 19.4: Proportion of live caught (light bars) and road-killed (dark bars) male (top) and female (bottom) tammar wallabies (*M. eugenii*) from the disturbed site on Garden Island, Western Australia.

DISCUSSION

Our results supported the hypothesis that the year-round supply of food provided by the fertilised and irrigated lawns in the disturbed site would result in animals in that area living at higher densities. The density of tammar wallabies observed at the disturbed site was similar to those of 1.6–3.4/ha observed on Kangaroo Island, where culling is used to reduce damage to crops (Wright and Stott 1999). The difference in the density estimates between the mark–recapture method and the spotlight count suggests that either there were individuals on the base that did not enter traps, or that some individuals were 'trap-happy'. Both instances would result in the mark–recapture analysis giving a lower estimate of population density than the spotlight count.

The condition of females at the disturbed site was remarkably constant over the 2005 breeding season compared with that of the undisturbed population, supporting our expectation that the increased food supply on the base would result in the tammars in that area being under less nutritional stress. Although we did not measure the nutritive value of the lawns accessible by the tammars on the base, previous studies have shown that tammars rely heavily on this food source (McMillan 2006). Western grey kangaroos are known to graze selectively on grasses in preference to native vegetation, presumably because of their higher nutritive value (Halford *et al.* 1984), so it is very likely that the lawns are a good source of nutrition for the tammars living on the base.

The significant difference in age structures between the two study sites also supported the hypothesis that survival rates would be greater at the disturbed site. The high proportion of animals, especially males, aged six years and older at the disturbed site indicates that animals in this area were surviving longer than those in the undisturbed site. This high proportion of older animals may also be a result of the very high road mortality of young animals, especially one-year-old males and one- and two-year-old females. It could be also due to these older larger males preventing access by younger males to this resource, as part of their mating strategy. By defending resource-rich areas they might gain better access to females (Rudd 1994). The observed male bias could also be due to the fact that male tammars tend to be trap-happy and enter traps more readily than females, as shown by the greater number of males captured throughout our study.

The small proportion of four-year-old females captured at the disturbed site could be a result of poor reproductive success in 2001, high road mortality of juveniles in 2002 or a combination of the two. The very small proportion of one-year-old males at the undisturbed site is difficult to explain, especially since the number of one-year-old females at the same site was much greater. Given that the sex ratio of pouch young in this area was close to parity, it seems that the mortality rate for juvenile males from the 2004 breeding season was very high. Alternatively, a large proportion of juvenile males may have dispersed out of the study area.

The hypothesis that reproductive success would be higher in the disturbed population was also supported. The high proportion of females losing pouch young in the undisturbed site may not be a regular pattern, as Inns (1980) found that pouch young mortality on Kangaroo Island was significant in only one of the eight years of study. The high reproductive success of females living on the base is most likely to result from their better condition at the start of winter. The apparent increase in the proportion of females carrying young at the disturbed site between

May and July 2005 reflects the capture of different females on different trapping trips. The start of the 2005 winter was particularly wet; 264 mm of rain fell in June, much more than the average 120 mm since weather observations began on the island in 2000 and the long-term average rainfall for June from the nearby town of Kwinana, of 162 mm (Bureau of Meteorology 2006). This could explain the high mortality of pouch young in the undisturbed areas. In contrast, Inns (1980) found that in tammars from Kangaroo Island pouch young mortality was very low, with more than 80% of females aged two years and older successfully rasing a pouch young each year.

The hypothesis that younger animals would be overrepresented in the road-kills was also supported, which suggests that young animals may be less aware of the danger posed by vehicles. This result is consistent with previous studies in macropodids where juveniles were found to be at greater risk of being killed by vehicles than adults (Lee *et al.* 2004). This bias may also be the result of younger animals crossing roads while dispersing, and therefore putting themselves at greater risk of being killed.

The very high road mortality of one- and two-year-old females in the disturbed population, combined with a bias towards males in the pouch young, is the most likely cause of the highly skewed sex ratios in the adult population. The bias towards male pouch young, although not significant, may also be related to the increased food supply on the base. Sunnucks and Taylor (1997) found that female tammar wallabies with a body mass of 3.2 kg or more were more likely to have a male pouch young, and those with a body mass over 4.2 kg had male pouch young almost 100% of the time. Having access to a better supply of food resulted in the females in the disturbed population being heavier than those in the undisturbed population and thus more likely to give birth to male pouch young, thereby further contributing to the male bias in the population.

The bias towards juvenile animals in the road-kills also seems to have an effect on the age structure of the disturbed population. While many authors have suggested that bias in road mortality may result in changes to sex ratios and age structures in affected populations, this has not been widely documented. One of the few published studies of sex bias demonstrated that in freshwater and terrestrial turtles in the US road-kills are biased towards females, which resulted in turtle populations near roads becoming increasingly male-biased over time (Steen and Gibbs 2004; Gibbs and Steen 2005; Steen *et al.* 2006). In one of the few studies of age bias in road-kills, Hauer *et al.* (2000) found that otter (*Lutra lutra*) road-kills were biased towards adult animals but that this did not have a significant effect on the age structure of the source populations. The results from our study therefore suggest that the juvenile bias observed in tammar road-kills may have a significant impact on the demographics of this population.

MANAGEMENT IMPLICATIONS

The results of this study demonstrate how disturbance by humans in the form of habitat modification and road mortality can dramatically change the population dynamics of a macropodid population.

Disturbance by humans has increased food available throughout the year to a population of tammars to the extent that the females are in better condition at crucial times and raise pouch young more successfully than those in undisturbed habitat. Natural mortality due to seasonal conditions is greatly reduced, but it seems that road mortality has taken over as the major cause of death in the population. It is quite plausible to expect that if the number of tammars killed by vehicles were to be significantly reduced without removal of the additional food sources, the population would grow rapidly and could become overabundant. Any management strategies to reduce road-kills in this population must therefore be combined with a reduction in access to the ovals and lawns on the base.

ACKNOWLEDGMENTS

We would like to thank the Australian Defence Force and the School of Animal Biology at the University of Western Australia for funding this research. Our thanks also go to Trevor Smith (Defence Public Area Ranger, Garden Island), Joanne Wann (Defence Regional Environmental Officer) and the security and grounds and gardens staff of HMAS Stirling.

Experimental protocols were approved by the University of Western Australia Animal Ethics Committee (approval no. 04/100/376) and all procedures followed the recommendations of the Australian National Health and Medical Research Council.

REFERENCES

Bell DT, Moredoundt JC and Loneragan A 1987. Grazing pressure by the tammar (*Macropus eugenii* Desm.) on the vegetation of Garden Island, Western Australia, and the potential impact on food reserves of a controlled burning regime. *J. Roy. Soc. West. Aust.* **69:** 89–94.

Bennetts SJ and Cocks PS (Eds) 1999. *Genetic Resources of Mediterranean Pasture and Forage Legumes.* Kluwer Academic, Netherlands.

Bradshaw SD 1988. Garden Island Tammar Research Program: an investigation into the physiological wellbeing of the tammar wallaby, *Macropus eugenii*, on Garden Island, in Western Australia. University of Western Australia, Perth.

Bureau of Meteorology 2006. Climate averages for Australian sites: averages for BP Kwinana Refinery. http://www.bom.gov.au/climate/averages/ tables/ cw_009064.shtml. Accessed 11 September 2006.

Gibbs JP and Steen DA 2005. Trends in sex ratios of turtles in the United States: implications of road mortality. *Conserv. Biol.* **19:** 552–556.

Halford DA, Bell DT and Loneragan WA 1984. Diet of the western grey kangaroo (*Macropus fuliginosus* Desm.) in a mixed pasture–woodland habitat of Western Australia. *J. Roy. Soc. West. Aust.* **66:** 119–128.

Hauer S, Ansorge H and Zinke O 2000. A long-term analysis of the age structure of otters (*Lutra lutra*) from eastern Germany. *Int. J. Mammal. Biol.* **65:** 360–368.

Hinds LA and den Ottolander RC 1983. Effect of changing photoperiod on peripheral plasma prolactin and progesterone concentrations in the tammar wallaby (*Macropus eugenii*). *J. Reprod. Fertil.* **69:** 631–639.

Inns RW 1980. Ecology of the Kangaroo Island wallaby (*Macropus eugenii*) in the Flinders Chase National Park. PhD thesis. University of Adelaide, Australia.

Inns RW 1982. Age determination in the Kangaroo Island wallaby, *Macropus eugenii* (Desmarest). *Aust. Wildl. Res.* **9:** 213–220.

Kinnear JE, Sumner NR and Onus ML 2002. The red fox in Australia: an exotic predator turned biocontrol agent. *Biol. Conserv.* **108:** 335–359.

Lee E, Klocker U, Croft DB and Ramp D 2004. Kangaroo–vehicle collisions in Australia's sheep rangelands, during and following drought periods. *Aust. Mammal.* **26:** 215–226.

McMillan A 2006. The utilisation of marine-derived nutrients by the tammar wallaby (*Macropus eugenii*) on Garden Island, Western Australia. BSc Hons thesis. University of Western Australia, Perth.

Renfree MB and Tyndale-Biscoe CH 1973. Intra-uterine development after diapause in the marsupial *Macropus eugenii*. *Dev. Biol.* **32:** 28–40.

Rudd CD 1994. Sexual behaviour of the male and female tammar wallabies (*Macropus eugenii*) at postpartum oestrus. *J. Zool. Lond.* **232:** 151–162.

Sharman GB, Frith HJ and Calaby JH 1964. Growth of pouch young, tooth eruption and age determination in the red kangaroo, *Megaleia rufa*. *CSIRO Wildl. Res.* **9:** 20–49.

Smith MJ and Hinds L 1995. Tammar wallaby, *Macropus eugenii*. In *The Mammals of Australia*. (Ed. R Strahan) pp. 329–331. Reed Books, Sydney.

Steen DA and Gibbs JP 2004. Effects of roads on the structure of freshwater turtle populations. *Conserv. Biol.* **18:** 1143–1148.

Steen DA, Aresco MJ, Beilke SG, Compton BW, Condon EP, Kenneth Dodd C Jr, Forrester H, Gibbons JW, Greene JL, Johnson G, Langen TA, Oldham MJ, Oxier DN, Saumure RA, Schueler FW, Sleeman JM, Smith LL, Tucker JK and Gibbs JP 2006. Relative vulnerability of female turtles to road mortality. *Anim. Conserv.* **9:** 269–273.

Sunnucks P and Taylor AC 1997. Sex of pouch young related to maternal weight in *Macropus eugenii* and *M. parma* (Marsupialia: Macropodidae). *Aust. J. Zool.* **45:** 573–578.

Walker L 2002. Factors affecting the road deaths of tammar wallabies (*Macropus eugenii*) on Garden Island. BSc Hons thesis. University of Western Australia, Perth.

Wann JM and Bell DT 1997. Dietary preferences of the black-gloved wallaby (*Macropus irma*) and the western grey kangaroo (*M. fuliginosus*) in Whiteman Park, Perth, Western Australia. *J. Roy. Soc. West. Aust.* **80:** 55–62.

Wright M and Stott P 1999. The Kangaroo Island tammar wallaby: assessing ecologically sustainable commercial harvesting. Rural Industries Research and Development Corporation. Publication No. 98/114.

20 Habitat partitioning among sympatric grey kangaroos and swamp wallabies in box-ironbark remnants

B. Schmidt, G. Coulson and J. Di Stefano

SUMMARY

Resource partitioning among sympatric populations of closely related or ecologically similar species reduces competition through niche diversion, which can allow multiple populations to coexist. Habitat partitioning is a major mode of reducing interspecific competition, by decreasing the spatial overlap of two populations. Populations of the eastern grey kangaroo (*Macropus giganteus*) and its sibling species the western grey kangaroo (*Macropus fuliginosus*) are sympatric over a broad zone in eastern Australia. We examined habitat use and possible segregation between these species and the swamp wallaby (*Wallabia bicolor*) in Wychitella Nature Conservation Reserve, north-west Victoria. We conducted line transect surveys on foot through a range of habitat types. At the specific location of each macropodid sighting, we measured food abundance and quality, recorded ecological vegetation class (EVC) and took digital photographs to calculate lateral cover. At the level of EVC, differences in habitat selection between the three species show that each species utilised habitats with a different structural composition. A significant degree of segregation also occurred at the microhabitat level along the gradient of lateral cover. Habitat utilised by western grey kangaroos contained twice the amount of lateral cover and a greater abundance of shrubs than habitat utilised by eastern grey kangaroos. These findings reinforce previous suggestions of habitat partitioning between grey kangaroos, and suggest fine-scale variation in habitat structure may be more important than broad habitat type.

INTRODUCTION

Grey kangaroos were separated into the sibling eastern grey (*Macropus giganteus*) and western grey (*Macropus fuliginosus*) kangaroo by Kirsch and Poole (1967, 1972) on the basis of morphological, reproductive and serological data, ending many years of taxonomic confusion. Their geographic distributions revealed a broad zone of sympatry throughout south-east South Australia, western Victoria and western New South Wales (Kirsch and Poole 1972). Isolated discoveries (Coulson 1980; Coulson and Hutchinson 1983; Shepherd 1982) and systematic surveys (Caughley *et al.* 1984; Poole 1977) following this primary division have expanded the known sympatric zone to include north-west Victoria, north-west New South Wales, and southern Queensland.

The broad distribution patterns of each species are related to climatic variables, which provide a rough index of forage production, water supply and habitat (Caughley *et al.* 1987). Grey kangaroos inhabit a variety of often dissimilar habitats. However, the presence of both species is strongly associated with forests and woodlands, which tend to offer a mosaic of vegetation containing two key elements: lateral cover for shelter from the elements and protection from predators, coupled with open grassy areas to feed in (Frith and Calaby 1969; Hill 1981). The ideal habitat of the eastern grey kangaroo is largely determined by the balance of lateral cover and grassy areas (Caughley 1964; Hill 1981; McAlpine *et al.* 1999). Studies of allopatric populations report a preference for open grassy forests with a sparse understorey (Taylor 1980; Southwell 1984, 1987; Moore *et al.* 2002). The western grey kangaroo occurs at highest densities where there is a high level of habitat heterogeneity (Short *et al.* 1983; Arnold and Steven 1988; Terpstra and Wilson 1989; Cairns *et al.* 1991; Coulson 1993), and has been shown to frequently use denser mallee habitats and habitats with a shrubby understorey (Arnold *et al.* 1992; Coulson 1993).

While there are reports of mixed mobs feeding at dusk (Emison *et al.* 1978; Coulson 1990, 1999), there is some evidence to suggest that differential habitat selection between the two species occurs during the day. In a heterogenous environment, differential habitat selection is one mechanism that allows ecologically similar, sympatric populations to coexist (MacArthur and Levins 1964; Rosenzweig 1981). The consequence is that, in sympatry, populations often occupy a narrower range of habitats than in allopatry.

The first observations of differences in habitat use between the two species were by Le Souef (1923), who noted that eastern grey kangaroos generally occupied open forests whereas western grey kangaroos tended to occur in long grass and scrub. Kirsch and Poole (1972) and Shepherd (1982) noted that eastern grey kangaroos were frequently seen in riverine habitat. However, they also noted that it was often difficult to see any difference in the habitat occupied by the two species. A faunal survey in the Grampians–Edenhope area by Emison *et al.* (1978) concluded that eastern grey kangaroos occurred in open forests and woodland with a ground layer of grass and herbs and sometimes bracken, but very few shrubs, while western grey kangaroos occurred in shrublands with a heath understorey. Menkhorst and Beardsell (1982) reported similar findings, but both studies reported difficulties in distinguishing the two species in the field.

Coulson's (1990) study in the Grampians, western Victoria, provided the first quantitative data on differential habitat use. He found that the densities of eastern grey kangaroos were highest in woodland containing an understorey of grass, bracken or shrubs; western grey kangaroos were frequently observed in heathland and low heath woodland, yet never in the habitats favoured by eastern grey kangaroos. Another study at Yathong Nature Reserve, western New South Wales, showed that eastern grey kangaroos preferred grassland and woodland whereas western grey kangaroos made most use of mallee (McCullough and McCullough 2000). Although these studies showed clear habitat partitioning in parts of the range of the two species, the applicability of these findings across the rest of the sympatric zone is unknown.

We studied eastern and western grey kangaroos in another part of the sympatric zone, the box-ironbark region in northern Victoria. This ecosystem is climatically intermediate between the Grampians National Park and Yathong Nature Reserve. The vegetation communities in box-ironbark differ from those in the Grampians and Yathong in terms of species composition, but have structural similarities (grassy woodland, mallee and heath) to the other two sites (Tzaros 2005). To test the generality of the patterns elsewhere in the sympatric zone, we assessed habitat use and possible resource segregation by sympatric populations of grey kangaroos in Wychitella Nature Conservation Reserve and surrounding land. Unlike previous studies, we adopted a multivariate approach and tested resource selection by each species at multiple scales. We collected additional data on the swamp wallaby (*Wallabia bicolor*) which is common throughout the study site, for comparison with the grey kangaroos.

METHODS

Study site

We conducted this study in the Wychitella Nature Conservation Reserve (36°16'S, 143°32'E to 36°26'S, 143°42'E) and surrounding private lands, in the box-ironbark region of northern Victoria between October 2005 and February 2006. The reserve is a complex of 10 discrete forest blocks, roughly arranged in a doughnut shape, fragmented by roads and farmland

(Fig. 20.1). It surrounds the town of Wedderburn and extends north to the township of Wychitella, covering the state's most significant area of box-ironbark mallee vegetation (ECC 1997; Tzaros 2005).

Much of the land around the reserve is low-lying alluvial plains and undulating sedimentary plains, crossed by many metamorphic ridges (LCC 1978). The sedimentary substrates form numerous gentle ridges, while granitic intrusions form more pronounced metamorphic ridges on the west and east margins. To the east, these ridges lead to a peak at Mt Egbert (259 m). To the west, the metamorphic rocks associated with these areas form the hilly landscape of Mt Kerang, the highest point at 398 m.

The study area has a temperate climate, with warm dry summers and cool wet winters. Charlton, which is representative of the Wedderburn area, has a mean daily maximum temperature of 30.4°C and minimum temperature of 13.8°C in January, and a mean daily maximum of 13.4°C and a minimum of 3.4°C in July (LCC 1978; Bureau of Meteorology). Rainfall throughout the area is winter-dominated, with a mean annual rainfall of 430 mm (LCC 1978; Bureau of Meteorology).

Although the reserve is broadly classified as box-ironbark it is a highly heterogenous landscape, which comprises a variety of different vegetation types. The Department of Natural Resources and Environment

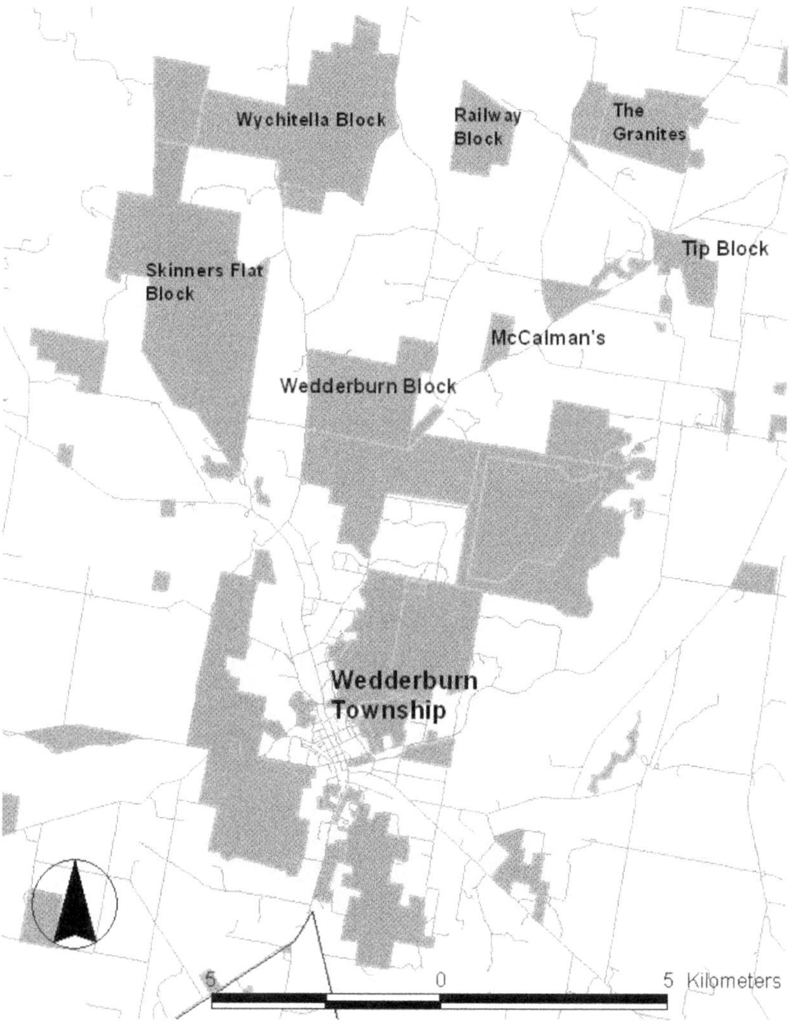

Figure 20.1: The Wychitella Nature Conservation Reserve, centred on the town of Wedderburn in the box-ironbark region of northern Victoria. The reserve is a complex of 10 discrete forest blocks (shaded) and surrounding private lands (unshaded).

Table 20.1: Summary of transect surveys conducted in Wychitella Nature Conservation Reserve October 2005 to February 2006.

Block	Area (ha)	Transects 500 m	Transects 150 m	Length (km)	Surveys
Railway Block	190	2	0	2.8	3
Tip Block	150	2	0	1.6	4
McCalman's	54	1	0	1	5
The Granites	332	5	0	7.6	1–2
Wedderburn Block	440	9	5	12.4	1–2
Wychitella Block	1120	3	0	3.5	2–3
Skinner's Flat Block	1236	21	6	27.0	1–4
Glenview	550	5	0	4.5	1–2

Includes the total area of the block, the number of transects 500 m and 150 m apart, their total length and the number of times the lines were surveyed.

has identified six ecological vegetation classes (EVCs) within the portion of the reserve surveyed (Table 20.1). The most widespread EVC is broombush mallee, dominated by bull mallee (*Eucalyptus behriana*), blue mallee (*Eucalyptus polybractea*) and green mallee (*Eucalyptus viridis*), with a predominantly sparse understorey. In some areas this EVC instead consists of a dense shrub layer of broombush (*Melaleuca uncinata*), totem-poles (*Melaleuca decussata*) and wallowa (*Acacia calamifolia*), with a heathy understorey. Areas of open forest are dominated by yellow gum (*Eucalyptus leucoxylon*) and grey box (*Eucalyptus microcarpa*), with golden wattle (*Acacia pycnantha*), gold-dust wattle (*Acacia acinacea*) and drooping cassinia (*Cassinia arcuata*) forming a sparse understorey. Grassy woodland communities intersect mallee and open forest. Mt Egbert supports areas of granitic outcrop, shrubland, herbland and open herb-rich woodlands (Tzaros 2005).

Macropodid surveys

Stake-out surveys

We conducted stake-out observations of macropodids in various blocks to determine the presence or absence of each species in different areas. Using binoculars and a spotting scope, we observed macropodids from a variety of vantage points at the forest–farmland ecotone. We observed macropodids as they emerged from cover at dusk and recorded the number of each species sighted just prior to dark, while it was still possible to make positive identifications.

Transects

We determined the presence or absence of the three species in a range of habitat types using sightings of macropodids on line transect surveys. Alternative methods such as faecal pellet counts were not feasible, due to the inability to distinguish eastern and western grey kangaroo pellets. Our line transects traversed seven blocks of public land and an area of private land (Table 20.1), which we selected during a pilot study in consultation with rangers from Parks Victoria and the Department of Sustainability and Environment. All the major EVCs within the reserve were represented in the areas surveyed (Table 20.2), and we specifically chose the area of private land as it contained an EVC not present in the reserve. We did not survey blocks in which extensive areas of dense vegetation restricted our movements and greatly reduced visibility.

We sampled each block using two sets of parallel transect lines. The first set ran across the block 500 m apart, an arbitrary distance that allowed us to sample the entire block on a broad scale. We surveyed 1–21 lines in each block, depending on the size and shape of the block (Table 20.1). It became apparent that most sightings of macropodids were within 300 m of forest–farmland interfaces. Therefore we positioned a second set of parallel lines, perpendicular to the forest edge and only 300 m long, inside two large blocks, Wedderburn and Skinners Flat (Table 20.1). To increase sampling intensity, we placed these lines 150 m apart. The transect lines were 60.4 km in total length, and we surveyed each line one to four times.

Once we had sighted a macropodid, we determined its specific identity with the aid of 8 × 42 Vixen binoculars. If an unknown macropodid moved out of sight before identification we discarded the sighting, but if a grey kangaroo moved before identification we recorded it as an unidentified kangaroo. For each

Table 20.2: Ecological vegetation classes within Wychitella Nature Conservation Reserve.

EVC	Structure	Main canopy species	Main shrub and ground layer species	% surveyed
Creekline grassy woodland	Open woodland	River red gum	Black wattle, silver wattle, common tussock-grass, tall sedge	n/a
Hills herb-rich woodland	Woodland	Hill red gum, red stringybark, yellow box, long-leaf box	Deane's wattle, black wattle, austral bugle, wallaby grasses	6.2
Hillcrest herb-rich woodland	Woodland	Yellow box, long-leaf box, grey box	Stinking pennywort, clustered everlasting, kangaroo grass	3.0
Box-ironbark forest	Open forest	Red ironbark, grey box, yellow gum	Golden wattle, gold-dust wattle, drooping cassinia	8.9
Metamorphic slopes shrubby woodland	Woodland	Grey box, drooping she-oak	Wallowa, sticky hop-bush, rough spear grass, tall raspwort	3.3
Heathy woodland	Low open woodland	Red ironbark, red box, grey box	Daphne heath, common fringe-myrtle, shiny everlasting	1.0
Plains grassy woodland	Open woodland or woodland	Grey box, buloke, yellow gum, river red gum, yellow box	Gold-dust wattle, hedge wattle, spear grasses, kangaroo grass, sundew, scaly buttons	n/a
Grassy woodland	Woodland	Grey box, white box	Golden wattle, twiggy bush-pea, saloop	n/a
Pine box woodland	Woodland	Yellow box, white cypress-pine, buloke	Golden wattle, mallee wattle, rough spear grass, sweet bursaria	n/a
Broombush mallee (gravelly sediment)	Open scrub to low open forest	Bull mallee, blue mallee, green mallee	Gold-dust wattle, broombush, totem-poles, mountain grevillea	72.9
Broombush mallee (Northern Goldfields)	Open scrub	Broombush, wallowa, green mallee	Totem-poles, shrubby dampiera, cranberry heath, grey everlasting	
Alluvial terraces herb-rich woodland mosaic	Woodland	Yellow box, grey box, river red gum, white box	Golden wattle, hedge wattle, slender goodenia, yellow rush-lily	n/a

Includes their structure and dominant species (adapted from ECC 1997) and the percentage of the total distance surveyed.

sighting, we recorded the number of each species and the exact location at which the macropodid was first sighted, using a handheld GPS unit (Garmin e-trex). In cases where we were unable to determine the macropodid's exact location, we simply noted its presence and made no habitat measurements. While observability may have varied between habitats, there was no reason to suspect that relative observability varied between species within the same area.

Habitat measurements

At each of the confirmed locations we made several habitat measurements. We estimated the percentage cover of five functional vegetation groups within a circular plot with a 2 m radius. These were forbs, grasses, sedges, trees and shrubs. Small eucalypt saplings and large acacias were classified as shrubs. We estimated the abundance of each functional group in increments of 5%, recording quantities <5% as 1%. We included only foliage up to 1.5 m high and excluded dead vegetation. We assessed the quality of each functional group of vegetation within each plot. We classified trees and shrubs as high-quality if they had new growth, evidenced by the different colour and shape of new foliage. Similarly, we recorded forbs and sedges as high-quality if they were completely green, and grasses as high-quality if they appeared healthy and were not trampled. If any of these signs were absent from the respective groups of vegetation within the plots, we classified that group as low-quality. On several occasions we observed macropodids under the shade of large trees, in a cleared hip-hole (Eldridge and Rath 2002). In such cases, no vegetation could be recorded.

To obtain a measure of lateral cover, we took digital photographs of a white screen (1 × 1 m) set upright in the centre of each plot. Using a digital camera (Sony Cybershot N50), we photographed the screen from a crouching position at bearings of 0°, 90°, 180° and 270° and at a height of 1 m, from 2 m and 5 m away, ensuring that the sheet filled as much of the camera screen as possible. Using Photoshop CS2, we converted each digital image to greyscale, so that the vegetation appeared black on a white background. To calculate lateral cover we determined the number of pixels of vegetation covering the screen, expressed as a percentage of the image of the screen. The average of each set of four photographs provided measures of lateral cover from 2 m and 5 m at each plot.

For each block surveyed, we mapped the GPS positions of each animal sighted using ArcView GIS v.3.3. On each of the maps we overlaid an EVC layer to analyse habitat use by EVC. We omitted any individuals situated on the border of two EVCs.

Statistical analyses

We analysed habitat selection at the level of EVC for each species using Manly's selection index. Individuals observed on the border of two EVCs were not included in the analysis. Statistical significance is determined at the 5% level. We made no correction for multiple comparisons to avoid a large type II error rate (Rothman 1990; Roback and Askins 2005).

To compare the amount of lateral cover in habitats occupied by eastern and western grey kangaroos, we performed a one-directional, independent samples t-test, with SPSS v.11 for Windows, after a Kolmogorov-Smirnov test confirmed normality of the proportional data. To compare the lateral cover of habitat occupied by all three species, we used a univariate analysis of variance (ANOVA) with a Tamhane's *post hoc* comparison for non-parametric data. Results are presented as means ± standard error and significance level for all tests was α.

We analysed measures of food abundance and food quality for each species together on PC-Ord v.4.25. We used the non-metric multidimensional scaling method of ordination and Bray-Curtis similarities to summarise the multivariate dataset. To reduce the influence of large variables, we square-root transformed the data and performed 30 runs with real data and 30 runs with randomised data, stepping down from six to one dimension. A preliminary analysis showed that shrub abundance and shrub quality were strongly correlated, so we removed shrub quality from the analysis. We assessed the dimensionality of the dataset using a scree plot, which indicated that a reduction in stress from three dimensions to two dimensions was not substantial. Therefore we used two dimensions in the final analysis. A Monte-Carlo test revealed that the difference in stress between the real data and randomised data was statistically significant ($p = 0.03$). Using the PC-Ord method (McCune and Grace 2002), we overlaid vectors on the ordination plot to illustrate the strength and direction of the correlations between the axes and measured variables, represented by r^2. Using a multi-response permutation procedure, we were able to make comparisons between species. This ordination diagram contained several plots from which no vegetation was recorded. The differences in ecological distances among individuals of each species between this analysis and one in which these plots were omitted, were not substantial.

RESULTS

In total, we sighted 255 eastern grey kangaroos, 78 western grey kangaroos and 103 swamp wallabies. A further 87 kangaroos were unidentified. Of these totals, we identified 151 eastern grey kangaroos, 28 western grey kangaroos and 90 swamp wallabies on the transect line surveys and were able to determine the precise location of 124, 24 and 69 respectively. We observed the remainder at the forest–farmland ecotone.

During the day, we observed macropodids only within forest blocks, not on open farmland. On line transect surveys we observed eastern grey kangaroos in groups of one to 11, while western grey kangaroos were either solitary or in pairs. Swamp wallabies were also solitary or in pairs. At dusk, both kangaroo species emerged to feed in open clearings and we recorded mixed mobs on several occasions in farmland surrounding Skinner's Flat Block and the Wedderburn Block. Swamp wallabies were less frequently seen feeding in farmland.

Grey kangaroos were present in all the forest blocks but we were unable to identify the species in the Railway Block and McCalman's Block. Both species were present in Skinner's Flat Block and Wedderburn Block. We also recorded one western grey kangaroo in The Granites. Western grey kangaroos were only found in blocks that also contained eastern grey kangaroos, yet eastern greys were present in blocks from which western grey kangaroos were absent. Swamp wallabies occurred in all the blocks.

Broad-scale habitat selection by EVC

Grey kangaroos occupied all the EVCs surveyed except heathy woodland (Table 20.2). Eastern grey kangaroos occurred in every other EVC, whereas western grey kangaroos were absent from three: metamorphic slopes shrubby woodland, grassy woodland and box-ironbark forest. Eastern grey kangaroos selected for hillcrest herb-rich woodland and hills herb-rich woodland, as well as box-ironbark forest, and selected against grassy woodland and broombush mallee (Table 20.3). Western grey kangaroos showed no selection for any EVC. Although we observed the majority of western grey kangaroos in broombush mallee, this habitat was used in proportion to its availability, as were hillcrest herb-rich woodland and hills herb-rich woodland (Table 20.3). Swamp wallabies occurred in every EVC, but selected for metamorphic slopes shrubby woodland and against hills and hillcrest herb-rich shrubby woodlands (Table 20.3).

Food abundance and food quality

The ordination plot (Fig. 20.2) of food abundance and quality at locations of each species had a stress value of 14.67, indicating that the plot reliably represents real differences among species. Axis 2, which was most strongly correlated with the abundance of shrubs, explained 86% of the variance in the original data; axis 1, correlated with the abundance of grasses, explained 54% of the variation in the original data.

There was a clear separation between the two kangaroo species (Fig. 20.2). Eastern grey kangaroos, placed higher on axis 2, were more strongly correlated with a higher abundance of grass; western grey kangaroos were associated with a greater abundance of shrubs. There were two main clusters of swamp wallabies: the larger occupied a similar ecological space as western grey kangaroos, and the second was situated where abundance of shrubs was greater. The

Table 20.3: Habitat selection for the three species of macropodids, showing the available proportion of each habitat, the proportion of the habitat used by each species and Manly's selection indices for each EVC.

Species	Habitat	Available proportion	Used proportion	Selection index (w)	w L95%	w U95%	Standardised index (B)*
Eastern grey kangaroo							
	Broombush mallee	0.74	0.13	0.18	0.06	0.30	0.02–
	Box-ironbark forest	0.10	0.21	2.13	1.51	2.76	0.19+
	Grassy woodlands	0.04	0.00	0.00	0.00	0.96	0.00–
	Metamorphic slopes shrubby woodland	0.02	0.03	2.08	0.46	3.70	0.18
	Hills and hillcrest herb-rich woodlands	0.09	0.62	7.0	6.33	7.65	0.61+
Western grey kangaroo							
	Broombush mallee	0.74	0.88	1.19	0.95	1.43	0.46
	Box-ironbark forest	0.10	0.00	0.00	0.00	1.21	0.00
	Grassy woodlands	0.04	0.00	0.00	0.00	1.87	0.00
	Metamorphic slopes shrubby woodland	0.02	0.00	0.00	0.00	3.14	0.00
	Hills and hillcrest herb-rich woodlands	0.09	0.13	1.40	0.12	2.69	0.54
Swamp wallaby							
	Broombush mallee	0.74	0.75	1.02	0.88	1.17	0.13
	Box-ironbark forest	0.10	0.12	1.24	0.51	1.98	0.16
	Grassy woodlands	0.04	0.03	0.70	0.00	1.83	0.09
	Metamorphic slopes shrubby woodland	0.02	0.08	4.81	2.90	6.71	0.61+
	Hills and hillcrest herb-rich woodlands	0.09	0.02	0.17	0.00	0.95	0.02–

+ = selection for a particular EVC. – = selection against an EVC. No symbol = no selection.

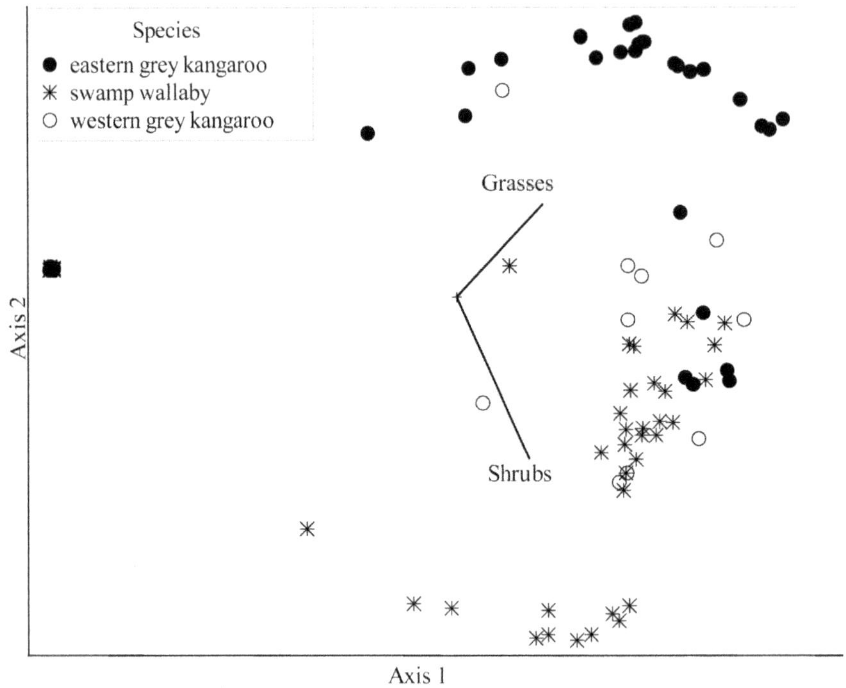

Figure 20.2: Multivariate differences in habitat variables between three macropodid species at Wychitella Nature Conservation Reserve. Vectors represent the direction and strength of the variables most strongly correlated with the axes.

outlying plot corresponded to locations where no vegetation was present. The multi-response permutation procedure (Table 20.4) indicated that the multivariate differences between eastern and western grey kangaroos, and those between eastern grey kangaroos and swamp wallabies, were greater than expected by chance. The effect size, measure (A), showed that these two differences were of approximately equal magnitude. Contrastingly, the differences in ecological space between western grey kangaroos and swamp wallabies were not significant.

Lateral cover

The mean (± SE) level of lateral cover at locations occupied by eastern grey kangaroos (11 ± 1.7%) was significantly less ($t_{17.3} = -2.106$, $p = 0.025$) than for western grey kangaroos (22 ± 4.9%) from 2 m away (Fig. 20.3). From a distance of 5 m, the mean level of lateral cover was 19 ± 1.9% for eastern grey kangaroo locations and 38 ± 6.8% for western grey kangaroo locations ($t_{16.2} = -2.693$, $p = 0.008$). Lateral cover from 5 m also differed significantly when all three species were included in the analysis ($F_{112,2} = 9.696$, $p < 0.001$; Fig. 20.3). Locations of western grey kangaroos had

Table 20.4: Statistical results for multivariate comparisons between species.

	A	P
Global	0.099	<0.001
Eastern grey/western grey	0.033	0.017
Eastern grey/swamp wallaby	0.117	<0.001
Western grey/swamp wallaby	0.002	0.282

A is a measure of effect that is largely independent of sample size.

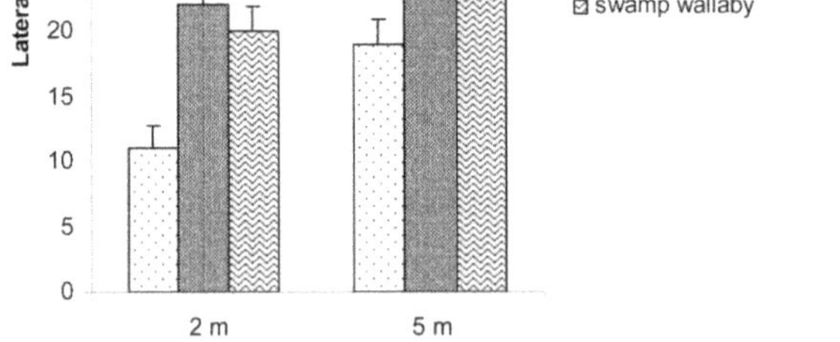

Figure 20.3: Mean lateral cover values for eastern grey kangaroos (*M. giganteus*), western grey kangaroos (*M. fuliginosus*) and swamp wallabies (*W. bicolor*) from a distance of 2 m and 5 m from the confirmed sighting locations in areas surveyed in Wychitella Nature Conservation Reserve, October 2005 to February 2006.

significantly greater cover than did locations of eastern greys (Tamhane's *post hoc*: $p = 0.046$), as did locations of swamp wallabies ($p < 0.001$), but there was no significant difference between locations of swamp wallabies and western grey kangaroos ($p = 0.738$). Lateral cover from only 2 m also differed among the three species ($F_{112,\,2} = 5.601$, $p = 0.005$). Although there was no significant difference between locations of the two kangaroo species (Tamhane's *post hoc*: $p = 0.142$) or between western grey kangaroos and swamp wallabies ($p = 0.979$), locations of eastern grey kangaroos had significantly less cover than did locations of swamp wallabies ($p = 0.004$).

DISCUSSION

This study confirms Le Souef's (1923) original observations of habitat partitioning between grey kangaroo species. We identified different patterns of broad-scale habitat use among the three macropodid species, according to EVC, and report habitat partitioning among grey kangaroos at a microhabitat level, in relation to habitat structure. Our findings concur with previous studies of grey kangaroos in other areas of the sympatric zone, which also identify structural differences in the habitats occupied by each species (Coulson 1990; McCullough and McCullough 2000), and with the typical habitat descriptions given by Emison *et al.* (1978).

Segregation by broad habitat type

At Wychitella Nature Conservation Reserve, eastern grey kangaroos selected strongly for hillcrest herb-rich woodland and hills herb-rich woodland, which are structurally open habitats with a sparse understorey and a high level of grass cover. They also selected box-ironbark forest, which has a similarly open structure. This preference for woodland habitats matched that reported elsewhere (Taylor 1985; Southwell 1987; Coulson 1990; McCullough and McCullough 2000). Although eastern grey kangaroos also occurred in habitats occupied by western greys, eastern grey kangaroos selected against broombush mallee. Western grey kangaroos occupied some areas selected for by eastern grey kangaroos, but were absent from box-ironbark forest. This was in agreement with Coulson (1990) and McCullough and McCullough (2000), who also found that eastern greys occurred where western grey kangaroos did at moderate densities, even though eastern grey kangaroos made most use of more open habitats. They also found that western grey kangaroos were more restricted in their use of habitat type and occurred at low densities or not at all in habitats preferred by eastern grey kangaroos.

Western grey kangaroos appeared to be mainly restricted to broombush mallee, the most prevalent EVC in Wychitella Nature Conservation Reserve. Western grey kangaroos did not select for this or any other EVC, instead using different habitats according to availability. However, caution should be applied to this conclusion, as the sample sizes for some EVCs were small and there is potential for some inaccuracy. Our findings bear some similarity to those of McCullough and McCullough (2000), who reported that western grey kangaroos at Yathong Nature Reserve made most use of mallee, which was only moderately used by eastern grey kangaroos. Western grey kangaroos have been reported to occupy heathland and heath woodland (Emison *et al.* 1978; Coulson 1990), but we did not observe any kangaroos within heathy woodland. This was not surprising given the proportionally low availability of this EVC (Table 20.3). As several other researchers have observed (Emison *et al.* 1978; Coulson 1990, 1999), there was no segregation at night when grey kangaroos emerged from cover and they often fed together in mixed mobs.

Swamp wallabies selected for metamorphic slopes shrubby woodland. These areas within the reserve were often rocky and contained many fallen trees that wallabies were often seen hiding behind or under. Swamp wallabies selected against hills and hillcrest herb-rich woodland. This concurred with previous studies, which suggest that swamp wallabies use dense and rocky areas for shelter (Floyd 1980; Merchant 1995; Tzaros 2005).

Segregation by microhabitat

Mapping habitat types on the scale of EVC provides only a broad indication of differences in vegetation. Therefore, only a coarse assessment of habitat segregation can be made by comparing the occurrence of each species in an EVC. There is often considerable variation within an EVC, as encountered by McCullough and McCullough (2000) at Yathong Nature Reserve. For example, hillcrest herb-rich woodland consisted of open grassland with scattered trees, but also contained denser patches and clumps of trees. Broombush mallee, in particular, contained the greatest structural variation. Although a large proportion of broombush mallee within the reserve was relatively open or of moderate density, a considerable amount was dense enough to restrict our movements on foot. Thus, each habitat type comprised a fine-scale mosaic within which kangaroos may differentially select favoured patches. Given this variation, it was essential to examine habitat use on a finer scale, as advocated by Southwell *et al.* (1999).

Lateral cover is one of the key habitat requirements for kangaroos (Frith and Calaby 1969; Hill 1981). When measured at the specific point of each macropodid sighting, it is likely to reflect patch selection at the microhabitat level and can thus be used to measure the relative preferences of each species. This study revealed a significant degree of habitat partitioning between the two kangaroo species, based on the amount of lateral cover. Western grey kangaroos used areas that afforded greater amounts of lateral cover than areas used by eastern grey kangaroos. This result was in agreement with the observations of Emison *et al.* (1978) and the quantitative findings of Coulson (1990) and McCullough and McCullough (2000). All those researchers reported that western grey kangaroos occurred in habitats that were of denser composition and with greater amounts of lateral cover, than were habitats used by eastern grey kangaroos.

Grey kangaroos are primarily grazers, relying heavily on grass in their diet (Barker 1987; Taylor 1980). Individual kangaroos will strongly select for areas containing good foraging habitat, both day and night (Moore *et al.* 2002), with the quality of the forage as important as its relative abundance (Hill 1982). Because herbage yields are inversely proportional to tree density (Beale 1973), a trade-off must occur between lateral cover for shelter and clearings for forage. The extremes of dense forests or completely cleared farmland rarely support eastern grey kangaroos (Short and Grigg 1982; Southwell 1987; McAlpine *et al.* 1999). Similarly, mallee vegetation, usually characterised by a shrubby understorey, has been described as unsuitable habitat for grey kangaroos because of the deficiency of forage (Short and Grigg 1982).

The two grey kangaroo species selected a different balance of cover and forage resources. Habitat patches selected by eastern greys had a greater abundance of grass than did those selected by western grey kangaroos. The distribution of western grey kangaroos was more strongly correlated with both the presence and quality of shrubs. Some evidence suggests that western grey kangaroos are not as reliant upon grass in their diet as eastern greys, and consume a greater proportion of shrubs (Barker 1987). This conforms with the observation of Emison *et al.* (1978) that herbs and short grasses for grazing, rather than shrubs for browsing, were prevalent in most eastern grey kangaroo habitats.

The pattern of habitat use by swamp wallabies was most similar to that of western grey kangaroos. Although at the broad scale swamp wallabies selected for different EVCs they occurred in areas with a similar level of lateral cover, in keeping with previous studies suggesting that dense cover is a critical determinant of swamp wallaby distribution (Floyd 1980; Merchant 1995). The patches selected by swamp wallabies were also in accordance with their generalist diet, consisting of a wide array of plant species (Hollis *et al.* 1986).

Eastern grey kangaroos occupied areas with a higher abundance of grass than did western grey kangaroos or swamp wallabies. Western grey kangaroos tended to occur in areas that contained fewer shrubs of better quality than swamp wallabies did, demonstrating a hierarchy of distribution among the three species along a gradient from a greater abundance of grass to a greater abundance of shrubs. This distributional hierarchy may be a product of partitioning according to levels of lateral cover, or may bear some relationship to the diet of each species. The swamp wallaby is more of a browser than larger macropodids (Hollis *et al.* 1986), and hence may select areas with a higher density of shrubs.

This study confirmed that habitat segregation occurs between the ecologically similar eastern and western grey kangaroo in another area of their sympatric zone, where the environmental conditions and vegetational structure and composition differed from areas previously studied. Several factors, such as foraging profitability, interspecific competition and predation risk, influence the habitat selection of animals at the broad spatial scale of habitat type and at the microhabitat scale of patch selection. The relative influence of these factors in habitat selection may vary at different spatial scales (Luck 2002). This study has demonstrated a separation of habitat use among sibling species of grey kangaroos in Wychitella Nature Conservation Reserve, which is evident at the broad habitat level and more clearly illustrated at the level of microhabitat.

ACKNOWLEDGMENTS

We are grateful to Nicholas Creed, Lupe Schmidt, Manfred Schmidt and Kate Hicks for their assistance with field work, much of which would not have been possible without their help. We thank Peter Morison from the Bendigo Department of Sustainability and Environment and Wedderburn Conservation Management Network, who provided much practical support and encouragement. We are indebted to Christine and Wig Chalmers for their personal support and for generously offering their hospitality. We also thank Roger Patterson for access to his property, and numerous other members of the Wedderburn community for making their property available for observations. Susan Campbell gave valuable input on data analysis and drafting this chapter. This was a collaborative project carried out in accordance with the Research Partners Panel Agreement between Parks Victoria and the University of Melbourne.

REFERENCES

Arnold CW and Steven DE 1988. Variations in distribution of western grey kangaroos, *Macropus fuliginosus ocydromus,* in Tutanning Nature Reserve and their impact on adjacent farmland. *Aust. Wildl. Res.* **15:** 119–128.

Arnold GW, Steven DE, Grassia A and Weeldenburg J 1992. Home-range size and fidelity of western grey kangaroos (*Macropus fuliginosus*) living in remnants of wandoo woodland and adjacent farmland. *Wildl. Res.* **19:** 137–143.

Barker RD 1987. The diet of herbivores in sheep rangelands. In *Kangaroos: Their Ecology and Management in the Sheep Rangelands of Australia*. (Eds G Caughley, N Shepherd and J Short) pp.74–83. Cambridge University Press, Cambridge.

Beale IF 1973. Tree density effects on yields of herbage and tree components in south-west Queensland mulga (*Acacia aneura* F.Muell) scrub. *Trop. Grassl.* **7:** 135–142.

Bureau of Meteorology website. http://www.bom.gov.au.

Cairns S, Pople AR and Grigg GC 1991. Density, distributions and habitat associations of red kangaroos, *Macropus rufus*, and western grey kangaroos, *M. fuliginosus*, in the South Australian pastoral zone. *Wildl. Res.* **18:** 377–402.

Caughley GJ 1964. Density and dispersion of two species of kangaroo in relation to habitat. *Aust. J. Zool.* **12:** 238–249.

Caughley G, Brown B, Dostine P and Grice D 1984. The grey kangaroo overlap zone. *Aust. Wildl. Res.* **11:** 1–10.

Caughley G, Short J, Grigg GC and Nix H 1987. Kangaroos and climate: an analysis of distribution. *J. Anim. Ecol.* **56:** 751–761.

Coulson G 1980. Extension of range of the western grey kangaroo, *Macropus fuliginosus*, in Victoria. *Vic. Nat.* **97:** 4–5.

Coulson G 1990. Habitat separation in the grey kangaroos, *Macropus giganteus* (Shaw) and *M. fuliginosus* (Desmarest) (Marsupialia: Macropodidae), in Grampians National Park, Western Victoria. *Aust. Mammal.* **13:** 33–40.

Coulson G 1993. Use of heterogeneous habitat by the western grey kangaroo, *Macropus fuliginosus. Wildl. Res.* **20:** 137–149.

Coulson G 1999. Monospecific and heterospecific grouping and feeding behaviour in grey kangaroos and red-necked wallabies. *J. Mammal.* **80:** 270–282.

Coulson G and Hutchinson C 1983. Extension of range of the eastern grey kangaroo, *Macropus giganteus*, in Victoria. *Vic. Nat.* **100:** 204–206.

ECC (Environment Conservation Council) 1997. Box-ironbark forests and woodlands investigation. Final report. Environment Conservation Council, Melbourne.

Eldridge DJ and Rath D 2002. Hip holes: kangaroo (*Macropus* spp.) resting sites modify the physical and chemical environment of woodland soils. *Austral Ecol.* **27:** 527–536.

Emison WB, Porter JW, Norris KC and Apps GJ 1978. Survey of the vertebrate fauna in the Grampians-Edenhope area of southwestern Victoria. *Mem. Nat. Mus. Vic.* **39:** 281–363.

Floyd RB 1980. Density of *Wallabia bicolor* (Desmarest) (Marsupialia: Macropodoidae) in eucalypt plantations of different ages. *Aust. Wildl. Res.* **7:** 333–337.

Frith HJ and Calaby JH 1969. *Kangaroos.* F.W. Cheshire, Melbourne.

Hill GJE 1981. A study of habitat preferences in the grey kangaroo. *Aust. Wildl. Res.* **8:** 245–254.

Hill GJE 1982. Seasonal movement patterns of the eastern grey kangaroo in southern Queensland. *Aust. Wildl. Res.* **9:** 373–387.

Hollis CJ, Robertshaw JD and Harden RH 1986. Ecology of the swamp wallaby (*Wallabia bicolor*) in north-eastern New South Wales. I. Diet. *Aust. Wildl. Res.* **13:** 355–365.

Kirsch JAW and Poole WE 1967. Serological evidence for speciation in the grey kangaroo, *Macropus giganteus* Shaw 1790 (Marsupialia: Macropodidae). *Nature* **215:** 1097–1098.

Kirsch JAW and Poole WE 1972. Taxonomy and distribution of the grey kangaroos, *Macropus giganteus* Shaw and *Macropus fuliginosus* (Desmarest), and their subspecies (Marsupialia: Macropodidae) *Aust. J. Zool.* **20:** 315–339.

LCC (Land Conservation Council) 1978. Report on the north central study area. Land Conservation Council, Victoria.

Le Souef AS 1923. The great grey kangaroo (*Macropus giganteus*) and its allies. *Aust. Zool.* **3:** 145–147.

Luck GW 2002. The habitat requirements of the rufous treecreeper (*Climacteris rufa*). I. Preferential habitat use demonstrated at multiple spatial scales. *Biol. Conserv.* **105:** 383–394.

MacArthur R and Levins R 1964. Competition, habitat selection, and character displacement in a patchy environment. *Proc. Natl Acad. Sci. USA* **51:** 1207–1210.

McAlpine CA, Grigg GC, Mott JJ and Sharma PP 1999. Influence of landscape structure on kangaroo abundance in a disturbed semi-arid woodland of Queensland. *Rangeland J.* **21:** 104–134.

McCullough DR and McCullough Y 2000. *Kangaroos in Outback Australia: Comparative Ecology and Behaviour of Three Coexisting Species.* Columbia University Press, New York.

McCune B and Grace BJ 2002. Analysis of ecological communities. MJM Software Design, Oregon.

Menkhorst PW and Beardsell CM 1982. Mammals of southwestern Victoria from the Little Desert to the coast. *Proc. Roy. Soc. Vic.* **94:** 221–247.

Merchant JC 1995. Swamp wallaby, *Wallabia bicolor.* In *The Mammals of Australia.* (Ed. R Strahan) pp. 261–262. Reed Books, Sydney.

Moore BD, Coulson G and Way S 2002. Habitat selection by adult female eastern grey kangaroos. *Wildl. Res.* **29:** 439–445.

Poole WE 1977. The eastern grey kangaroo, *Macropus giganteus*, in south-east South Australia: its limited distribution and need of conservation. Division of Wildlife Research Technical Paper No. 31. CSIRO Division of Wildlife Research, Canberra.

Roback PJ and Askins RA 2005. Judicious use of multiple hypothesis tests. *Conserv. Biol.* **19:** 261–267.

Rosenzweig ML 1981. A theory of habitat selection. *Ecology* **62:** 327–335.

Rothman KJ 1990. No adjustments are needed for multiple comparisons. *Epidemiology* **1:** 43–46.

Shepherd NC 1982. Extension of the known range of western grey kangaroos, *Macropus fuliginosus,* and eastern grey kangaroos, *M. giganteus*, in New South Wales. *Aust. Wildl. Res.* **9:** 389–391.

Short J, Caughley G, Grice D and Brown B 1983. The distribution and abundance of kangaroos in relation to environment in Western Australia. *Aust. Wildl. Res.* **10:** 435–451.

Short J and Grigg GC 1982. The abundance of kangaroos in suboptimal habitats: wheat, intensive pastoral, and mallee. *Aust. Wildl. Res.* **9:** 221–227.

Southwell CJ 1984. Variability in grouping in the eastern grey kangaroo, *Macropus giganteus.* I. Group density and group size. *Aust. Wildl. Res.* **11:** 423–435.

Southwell C 1987. Macropod studies at wallaby creek. II. Density and distribution of macropod species in relation to environmental variables. *Aust. Wildl. Res.* **14:** 15–33.

Southwell CJ, Cairns SC, Pople AR and Delaney R 1999. Gradient analysis of macropod distribution in open forest and woodland of eastern Australia. *Aust. J. Ecol.* **24:** 132–143.

Taylor RJ 1980. Distribution of feeding activity of the eastern grey kangaroo, *Macropus giganteus*, in coastal lowland of south-east Queensland. *Aust. Wildl. Res.* **7:** 317–325.

Taylor RJ 1985. Habitat use by the eastern grey kangaroo and wallaroo in an area of sympatry. *Mammalia* **49:** 173–186.

Terpstra JW and Wilson AD 1989. Grazing distribution of sheep and kangaroos in a semi-arid woodland. *Appl. Anim. Behav. Sci.* **24:** 343–352.

Tzaros C 2005. *Wildlife of the Box-ironbark Country.* CSIRO Publishing, Melbourne.

21 Diversity and distribution of helminth parasites in macropodoid marsupials

I. Beveridge, D.M. Spratt and P.M. Johnson

SUMMARY

Macropodoid marsupials harbour an extremely diverse array of helminth parasites. Nematodes dominate the helminth fauna, with smaller numbers of anoplocephalid, davaineid and hymenolepidid cestodes and two species of digenetic trematodes belonging to the family Paramphistomidae. Hypsiprymnodontids and potoroids are parasitised by relatively few parasites compared with the macropodids. Within the macropodids, the greatest diversity occurs among the nematodes belonging to the endemic subfamily Cloacininae found in the oesophagus and sacculated forestomach. Other significant nematode groups include the herpetostrongylids in the small intestine, the oxyuroids in the colon, the metastrongyloids in the lungs and the filarioids in the body cavities, intermuscular and subcutaneous connective tissues and vascular system. Comparisons of helminth communities in different macropodids indicate that each host species harbours its own relatively distinctive community of helminths, suggesting a high degree of host specificity. Exceptions to this pattern occur in the closely related species of rock-wallabies belonging to the *Petrogale penicillata* complex (*P. herberti*, *P. inornata*, *P. assimilis*, *P. godmani*), the pademelons *Thylogale stigmatica* and *T. thetis* and the closely related wallaroo species *Macropus robustus* and *M. antilopinus*. In addition, in instances where different macropodid species are sympatric, there is evidence that certain nematode and cestode parasites are capable of parasitising a number of different host species.

INTRODUCTION

Price (1980) estimated that in excess of 50% of the members of the animal kingdom could be considered parasitic and that each vertebrate had at least one species of parasite. The kangaroos, wallabies and rat-kangaroos (Macropodoidea) more than exemplify this phenomenon in harbouring an extraordinary range of helminth parasites (Spratt *et al.* 1991), with current estimates suggesting that they may in fact harbour more species of parasites than any other known group of mammals (Beveridge and Chilton 2001). Therefore, studies of the helminth parasites of macropodoids may not only contribute to the documentation of invertebrate biodiversity in Australia, but may also elucidate novel host–parasite relationships and provide insights into parasite evolution as well as the role of parasites as agents of disease in macropodoid populations.

Most published studies of the helminths of macropodoids have been based on opportunistic collections. The first systematic study of the parasites of any macropodoid was that of *Macropus rufus*

(Mykytowycz 1964), with a subsequent study of the same host species at a single locality in New South Wales (Arundel *et al.* 1979). Smales and Mawson (1978) conducted a systematic study of the nematode parasites of *Macropus eugenii* on Kangaroo Island, South Australia. Although knowledge of the helminths of macropodoids has increased significantly since the 1960s and 1970s, systematic studies are still lacking and most of the available information is based on opportunistic collections.

Most studies have concentrated on determining the species of parasite present in particular macropodoids, with relatively little attention paid to the life cycles of the parasites. Smales (1977a, b) studied the life cycle of the strongyloid nematode *Labiostrongylus eugenii* found in the stomach of *M. eugenii*, while Beveridge and Presidente (1978) described the various life-cycle stages of the strongyloid *Rugopharynx rosemariae* but did not support their observations with experimental infections as Smales (1977b) had done. In both species, larval development occurred in the environment and the third-stage larva was infective for the kangaroo host, as expected from studies with other strongyloid nematodes (Anderson 2000). The life-cycles of two species of filarioid nematode of macropodoids have been elucidated, the common subcutaneous filarioid *Pelecitus roemeri* (formerly *Dirofilaria roemeri*), which has tabanid flies as intermediate hosts (Spratt 1972), and *Breinlia macropi*, a parasite from the abdominal cavity of *Setonix brachyurus*, which is transmitted by mosquitoes (Yen 1983). The oxyurid nematode parasites occurring in the colon of macropodoids probably have direct life-cycles based on studies of other oxyurids (Anderson 2000). No life-cycle studies have been published of the anoplocephalid cestodes of macropodoids, but all known members of this group utilise oribatoid mites as intermediate hosts (Stunkard 1969). Likewise, known davaineid and hymenolepidid cestodes of mammals use arthropods as intermediate hosts although no studies have been undertaken on representatives in marsupials. Thus, knowledge of the life-cycles of the parasites of macropodoids is still extremely fragmentary.

Apart from a review of potential modes of evolution in one subfamily, the Cloacininae, parasitic in the stomachs of macropodoids (Beveridge and Chilton 2001), and an overview of the helminths of marsupials (Beveridge and Spratt 1996), there has been no previous attempt to review the diversity and host or geographic distribution of the helminth parasites of macropodoids.

METHODS

Data were taken from the published literature with appropriate modification of host and parasite names.

Hosts

Host nomenclature follows Strahan (1995) except for more recent changes in the nomenclature of rock-wallabies (Eldridge *et al.* 2001). The musky rat-kangaroo (*Hypsiprymnodon moschatus*) has recently been placed in a separate family, the Hypsiprymnodontidae (see Burk *et al.* 1998). Members of the Macropodidae are listed in tables by clade following Cardillo *et al.* (2004).

Parasites

Compilations of parasite species lists were based on Beveridge and Chilton (2001) and Spratt *et al.* (1991) and references contained therein, with the addition of more recent records by Appan *et al.* (2004), Beveridge (2002a, b), Beveridge and Durette-Desset (2004), Beveridge and Johnson (2004), Beveridge and Turni (2003), Chilton *et al.* (2002), Huby-Chilton *et al.* (2002) and Smales (2006). In the case of complexes of sibling species of parasites, identification of individual species and their hosts follows Beveridge and Chilton (2001) and Hu *et al.* (2005). Higher-level taxonomy for the nematodes follows Anderson *et al.* (1974) and, for the cestodes, Khalil *et al.* (1994).

Overall similarities between the helminth communities found in different host species were estimated using Sorenson's index (twice the number of shared species divided by the total number of helminth species present, expressed as a percentage) (Greig-Smith 1964). Sorenson's index values were converted to a dendrogram using the group average algorithm. Diversity within the helminth community of each macropodoid host was measured by the reciprocal of Simpson's index (the sum of the squares of prevalence of each species divided by the intensity squared) (Leong and Holmes 1981) as in Beveridge *et al.* (1989, 1992, 1998).

RESULTS AND DISCUSSION

Helminth diversity

Currently, 372 helminth species are known from macropodid hosts, 15 from potoroids and two from hypsiprymnodontids (Table 21.1). The current data undoubtedly underestimate parasite diversity, for several reasons. First, a number of macropodoid species have never been examined for parasites (three potoroids, six macropodids) (Beveridge and Chilton 2001), few individuals of rare species of macropodids have been examined, parasites may be tiny and occur in organs which are not easy to examine or are not frequently examined, thereby escaping detection, or taxonomic limitations may apply. These limitations arise from lack of taxonomic activity in describing species collected or from the morphological simplicity of parasites, such that biologically distinct species may be very similar and therefore may be overlooked or underestimated by taxonomists. Several recent studies on nematodes and cestodes of macropodoids

Table 21.1: Diversity of named, endemic helminth parasites of macropodoid marsupials and their distribution within macropodids, potoroids and hypsiprymnodontids.

	MACROPODIDAE		**POTOROIDAE**		**HYPSIPRYMNODONTIDAE**	
	No. of genera	**No. of species**	**No. of genera**	**No. of species**	**No. of genera**	**No. of species**
NEMATODA (nematodes)						
Strongylida						
Strongyloidea						
Chabertiidae						
Cloacininae	35	275	1	5	0	0
Phascolostrongylinae	4	8	0	0	1	1
Trichostrongyloidea						
Herpetostrongylidae	5	28	1	1	1	1
Metastrongyloidea						
Angiostrongylidae	2	4	0	0	0	0
Filaroididae	1	1	0	0	0	0
Rhabditida						
Strongyloididae[a]	1	1	1	1	0	0
Oxyurida						
Oxyuridae	1	2	1	1	0	0
Spirurida						
Spiruroidea	2	2	0	0	0	0
Filarioidea						
Onchocercidae	2	14	1	1	0	0
Muspiceoidea	1	1	0	0	0	0
CESTODA (cestodes or tapeworms)						
Cyclophyllidea						
Anoplocephalidae	2	31	1	1	0	0
Davaineidae	1	1	1	4	0	0
Hymenolepididae	0	0	1	2	0	0
Taeniidae	1	1[b]	0	0	0	0
DIGENEA (trematodes or flukes)						
Fasciolidae	1	1[b]	0	0	0	0
Paramphistomatidae	2	2	0	0	0	0
Total	61	372	7	16	2	2

a: Unnamed, but included to indicate presence of order.
b: Introduced parasites, normally parasitic in domestic livestock.

(Beveridge and Chilton 1999; Hu *et al.* 2005) suggest that cryptic species may be relatively common. Finally, the material used in most of the studies cited was collected opportunistically rather than in a systematic fashion, which may affect the results of the studies. All these factors suggest that the current census of parasite diversity is an underestimate of actual diversity.

Nevertheless, current estimates indicate an extraordinary level of diversity of parasites, with helminth communities dominated numerically by gastrointestinal nematodes. Smaller numbers of species occur in the lungs and body cavities or vessels. Cestodes are common in the small intestine and bile ducts with only three species (two endemic) of digeneans. All the helminths of macropodoids are endemic with the exception of the common liver fluke (*Fasciola hepatica*) and the hydatid, the metacestode or larval stage of the cestode *Echinococcus granulosus*, the only two species acquired from introduced livestock.

The distribution of helminth parasites between macropodoid families and organs presents several striking contrasts. The majority of helminth species are found in macropodids (Table 21.1), with relatively few species occurring in potoroids and hypsiprymnodontids. This difference is not ascribable to the fact that there are fewer species of potoroids or that fewer specimens of potoroids have been examined for parasites. Helminths occur in most organ systems of the macropodoid body but the centre of helminth diversity is clearly the sacculated forestomachs of macropodids, with some 291 species known to date (Table 21.2).

The large numbers of nematode species in the sacculated forestomachs of macropodids has been attributed to the size of the viscus, allowing large numbers of species of content-feeding nematodes to evolve in the absence of competition (Beveridge and Spratt 1996). The dramatic macropodid host radiation, further increasing the number of available niches for parasites, as well as the common occurrence of host colonisation or host switching by nematodes, appear to have been significant factors in the evolution of nematode diversity (Beveridge and Chilton 2001). The comparable lack of nematodes in the forestomachs of potoroids is surprising, but the structure and function of the potoroid forestomach differ dramatically from the macropodid stomach (Hume 1999) and the gastric environment in potoroids may be less suitable for strongyloid nematodes. Both morphological and molecular evidence support the hypothesis that the cloacinine nematodes of the stomach are derived from phascolostrongyline nematodes found in the caecum and colon of macropodids, potoroids and wombats (Chilton *et al.* 1997).

Helminth distribution

The distribution of helminth species within a community of macropodids has received limited attention. The distribution of species of a single nematode genus, *Cloacina*, has been studied by Beveridge *et al.* (2002) and limited data on host specificity for the Cloacininae were provided by Beveridge and Chilton (2001). From published studies, comparisons can be made between pairs of sympatric host species, namely *M. fuliginosus* and *M. eugenii* on Kangaroo Island, South Australia (Smales and Mawson 1978; Webley *et al.* 2004) and between *M. rufus* and *M. fuliginosus* in western New South Wales (Arundel *et al.* 1979). However, in both cases only two species of macropodids are involved.

The largest body of data on helminths of macropodoid marsupials is contained in a series of studies based on hosts collected in Queensland, including rock-wallabies (Begg *et al.* 1996; Beveridge *et al.* 1989; Bradley *et al.* 2000), pademelons (Griffith *et al.* 2000), various small macropodoids (Beveridge *et al.* 1992) and large wallabies and kangaroos (Beveridge *et al.* 1998). Although comparisons among hosts were made, they were limited to congeners (Beveridge *et al.* 1989, 1998) or, in one case (Beveridge *et al.* 1992), a series of only distantly related host species. The hosts examined from eastern and northern Queensland comprise a large community of macropodoids within which broader questions of parasite distributions can be examined. Data from the studies were therefore combined to address questions of parasite distribution and host specificity.

Data from a total of 25 species of macropodoids based on the examination of 436 individual animals from Queensland (Table 21.3) indicate that they harbour between two and 42 helminth species each. Sample sizes for some of the rare or threatened species (*Bettongia tropica*, *Onychogalea fraenata*, *Petrogale mareeba*, *P. sharmani*, *P. penicillata*) were small and in these instances the data may not present a reliable indication of numbers of helminth species present. In the case of *B. tropica*, examination of a larger series of *B. gaimardi* from Tasmania and

Table 21.2: Location and number of various endemic helminth species within the organs of their macropodoid hosts.

Host family	MACROPODIDAE	POTOROIDAE	HYPSIPRYMNODONTIDAE	Principal groups
No. of host species examined	50	3	1	N = nematode, C = cestode
Site in host				
Oesophagus	8	0	0	Strongyloidea (N)
Stomach	291	5	1	Strongyloidea (N) Trichostrongyloidea (N
Small intestine	46	9	1	Trichostrongyloidea (N) Anoplocephalidae (C) Davaineidae (C)
Lung[a]	5	0	0	Metastrongyloidea (N)
Large intestine	8	1	1	Strongyloidea (N) Oxyuroidea (N)
Liver/bile ducts[b]	3	0	0	Anoplocephalidae (C)
Body cavities/vessels/ intermuscular/subcutaneous	14	0	0	Filarioidea (N)

a: Excludes the metacestode of *Echinococcus granulosus*.
b: Excludes *Fasciola hepatica*.

B. penicillata from Western Australia (Spratt *et al.* 1991; unpubl. obs.) suggest that bettongs harbour only three to four helminth species, so that increasing the sample size of *B. tropica* may not substantially increase the number of helminth species recovered. Griffith *et al.* (2000) used a bootstrap method to calculate the likelihood of recovering all the helminth parasites from small samples of pademelons (*Thylogale* spp.). They found that a single pademelon was likely to yield 50% of the known helminths and a sample size of five animals would yield 87% of the known helminths. For this reason, estimates of diversity have not been presented for the sample sizes smaller than 10 (Table 21.3), which would have ensured a high probability of detecting most helminth species.

The available data suggest that potoroids harbour relatively few (two to seven) helminth species and that in the macropodids for which there were adequate sample sizes 13–42 helminth species were present in each host species. The exception within the Macropodidae was the tree-kangaroo, *Dendrolagus lumholtzi*, which harboured only four helminth species. Although only a limited number of helminth life-cycles have been elucidated in marsupials, those of the strongylid nematodes are essentially terrestrial, with larval development occurring in the faeces and on herbage (Anderson 2000; Smales 1977a, b). Arboreal macropodids may therefore avoid exposure to these parasites more than their terrestrial counterparts (Beveridge and Spratt 1996).

On average, parasite diversity, as expressed by the reciprocal of Simpson's index (Leong and Holmes 1968), was high in pademelons (*Thylogale* spp.) and in large wallabies and kangaroos (*Macropus*, *Wallabia*) but lower in small wallabies such as the rock-wallabies (*Petrogale* spp.), hare-wallabies (*Lagorchestes*) and nailtail wallabies (*Onychogalea*) (Table 21.3). The correlation between diversity and average body weight (Strahan 1995) for 18 of the host taxa revealed a correlation coefficient of 0.25 ($p = 0.29$), suggesting that the diversity of the helminth community was not closely related to body size. A similar result was reported by Beveridge *et al.* (1992), based on a smaller sample of host species.

Overall similarities between the helminth communities of 21 of the macropodoid species listed in Table 21.3 were calculated using Sorensen's index (Greig-Smith 1964; Fig. 21.1, Table 21.4). A total of 233 helminth species was included.

The hypsiprymnodontids and potoroids, *Aepyprymnus rufescens*, *B. tropica* and *H. moschatus*, together with *D. lumholtzi*, apart from having small helminth communities, exhibited virtually no similarity of helminth communities with those of other

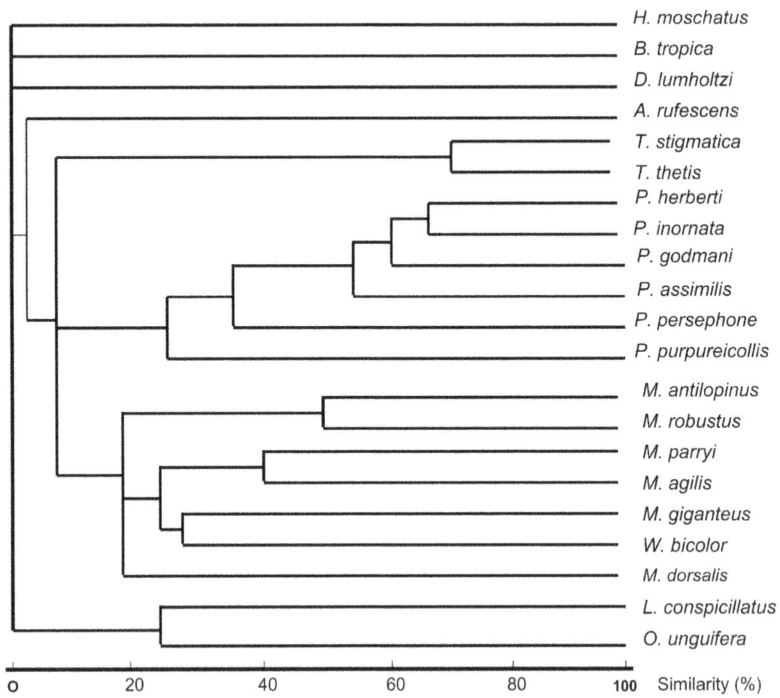

Figure 21.1: Overall similarity (%) of helminth communities of macropodoids from Queensland.

Table 21.3: Diversity of gastrointestinal helminth parasite communities in macropodoids from Queensland.

	No.examined	No. of helminth spp.	Diversity (1/S)[a]
HYPSIPRYNODONTIDAE			
Hypsiprymnodon moschatus	4	2	-
POTOROIDAE			
Aepyprymnus rufescens	18	7	4.2
Bettongia tropica	2	2	-
MACROPODIDAE			
Clade 2[b]			
Dendrolagus lumholtzi	10	4	2.8
Clade 3			
Thylogale stigmatica	45	33	21.8[d]
Thylogale thetis	10	32	22.7
Petrogale assimilis	36	29	10.9
Petrogale godmani	15	18	13.9
Petrogale herberti	11	13	10.6
Petrogale inornata	15	19	11.8
Petrogale mareeba	3	11	-
Petrogale penicillata	2	9	-
Petrogale persephone	20	19	14.3
Petrogale purpureicollis	12	13	9.9
Petrogale sharmani	4	14	-
Clade 4			
Lagorchestes conspicillatus	10	18	11.8
Clade 5			
Onychogalea unguifera	13	16	10.9
Onychogalea fraenata	2	4	-
Clade 7			
Macropus agilis	40	26	14.4
Macropus antilopinus	12	17	-[c]
Macropus dorsalis	39	42	26.6
Macropus giganteus	28	27	14.7
Macropus parryi	29	37	19.4
Macropus robustus	30	40	23.0
Wallabia bicolor	26	32	17.5

Based on the reciprocal of Simpson's index(1/S), including sampling effort and the total number of helminth species found in each host species.

a: Calculated only for sample sizes ≥10.

b: Clades from Cardillo *et al.* (2004).

c: Not calculated due to missing prevalence data for some parasites

d: Average of three populations.

macropodoids. The small wallabies *Lagorchestes conspicillatus* and *O. unguifera* shared a proportion of parasites, but their communities differed dramatically from those found in other macropodids. As noted by Beveridge *et al.* (1992), the two macropodid genera *Lagorchestes* and *Onychogalea* are broadly sympatric and occur in similar habitats, thus facilitating the switching of parasites between hosts.

Table 21.4: Sorenson's index (%) for the similarity of helminth communities in macropodoid hosts from Queensland.

	H.m	A.r	B.t	T.s	T.t	P.a	P.g	P.h	P.i	P.pe	P.pu	D.l	L.c	O.u	O.f	M.ag	M.ant	M.d	M.g	M.p	M.r	W.b
H. moschatus	-																					
A. rufescens	0	-																				
B. tropica	0	0	-																			
T. stigmatica	0	0	0	-																		
T. thetis	0	0	0	74	-																	
P. assimilis	0	7	0	0	3	-																
P. godmani	0	8	0	4	0	52	-															
P. herberti	0	0	0	-	4	50	65	-														
P. inornata	0	7	0	4	3	61	51	70	-													
P. persephone	0	10	0	21	14	20	23	21	26	-												
P. purpureicollis	0	0	0	0	0	20	23	21	21	7	-											
D. lumholtzi	0	0	0	0	0	0	0	0	0	0	0	-										
L. conspicillatus	0	0	0	0	0	0	0	0	0	0	0	0	-									
O. unguifera	0	0	0	0	0	0	0	0	0	0	0	0	24	-								
O. fraenata	0	0	0	0	0	0	0	0	0	0	0	0	0	24	-							
M. agilis	0	0	0	0	0	8	18	11	17	10	0	0	16	0	0	-						
M. antilopinus	0	0	0	0	0	10	0	0	0	0	13	0	11	7	0	0	-					
M. dorsalis	0	0	0	3	0	6	13	7	9	3	0	0	3	0	0	12	3	-				
M. giganteus	0	0	0	0	0	14	4	13	17	4	4	0	13	5	0	25	4	5	-			
M. parryi	0	0	0	0	0	10	15	12	17	4	0	0	16	4	0	34	16	20	18	-		
M. robustus	0	0	0	0	0	16	7	12	16	0	15	0	4	4	0	29	56	2	6	11	-	
W. bicolor	0	0	0	0	0	4	4	0	4	0	11	0	9	5	0	18	9	27	6	18	14	-

Helminth communities in *Thylogale* spp, *Petrogale* spp. and *Macropus* spp. (including *Wallabia*) were distinct, but exhibited some degree of overall similarity. The distinctiveness of the helminth communities in *T. stigmatica* and *T. thetis* is readily attributable to the large number of genera and species of both nematodes and cestodes unique to these hosts (Griffith *et al.* 2000). Both host species are primarily rainforest inhabitants (Strahan 1995) and their habitat overlaps with that of relatively few other macropodid species in Queensland. Communities of helminths in rock-wallabies (*Petrogale* spp.) were similar to one another, but those of *P. purpureicollis* and *P. persephone* were distinct from those found in members of the *P. penicillata* complex (*P. assimilis, P. godmani, P. herberti, P. inornata*) which occur along the eastern coast of Queensland in a parapatric distribution and which exhibited community similarities in excess of 54%.

The helminth community of *P. persephone* was investigated by Begg *et al.* (1995), who observed that differences from members of the *penicillata* complex were due to several parasites unique to *P. persephone* and to the acquisition of parasites (18% of the total) from the sympatric species *T. stigmatica. Petrogale persephone* is unusual in being the only rock-wallaby species to inhabit vineforest or rainforest (Maynes 1982), hence its habitat overlaps with *T. stigmatica*. It is also considered to be a basal member of the rock-wallaby clade (Cardillo *et al.* 2004), sharing similarities with its common ancestors, the pademelons (*Thylogale* spp.). The extent to which the composition of the helminth community of *P. persephone* can be explained by phylogenetic rather than ecological factors will depend upon the establishment of robust phylogenies for the various groups of parasites which are found in it and in *T. stigmatica*.

Petrogale purpureicollis, from western Queensland, also differs significantly in its helminth community from other rock-wallaby species found along the east coast of Queensland. It shares its habitat primarily with *M. robustus* and 46% of the parasites found in it also occur in sympatric *M. robustus* (Bradley *et al.* 2000). Hence, divergences within communities of parasites found in rock-wallabies can be explained principally by the acquisition of parasites from sympatric but distantly related macropodid hosts.

In the large wallabies and kangaroos (*Macropus, Wallabia*), Beveridge *et al.* (1998) concluded that most species harboured distinctive helminth communities, a result confirmed when communities from additional host taxa are added (Fig. 21.1). *M. robustus* and *M. antilopinus* have the most similar communities, but they are closely related phylogenetically and were collected in an area of sympatry (Beveridge *et al.* 1998). Similarities between the communities of helminths of *M. agilis* and *M. parryi* were also explained by host-switching of parasites in areas of host sympatry (Beveridge *et al.* 1998).

Specificity

Host specificity for the 233 parasite species occurring in macropodoids from Queensland is presented as a simple plot of frequency against the number of host species parasitised (Fig. 21.2). The pattern is similar to that described for species of *Cloacina* (Beveridge *et al.* 2002) and for the nematode subfamily Cloacininae (Beveridge and Chilton 2001). In both analyses, refinements such as considering only core species (Beveridge and Chilton 2001; Beveridge *et al.* 2002) or taking into account the occurrence of an individual parasite in two closely related host species (Beveridge and Chilton 2001) did not affect the overall pattern. As with the previous studies, the most frequent occurrence was that of a single host followed by a decline in the frequency of parasite species as the number of host species parasitised increased. Of the parasites occurring in two host species, 40 (17%)

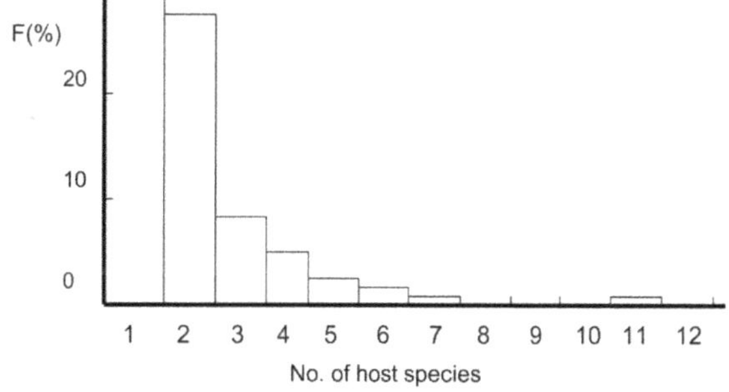

Figure 21.2: Host specificity of 233 species of helminths in macropodoids from Queensland presented as frequency of occurrence (F%) of parasites occurring in one to 12 host species.

occurred in closely related hosts such as the two species of *Thylogale* and *M. robustus* – *M. antilopinus*.

Among species of parasites occurring in three to six species of hosts are several nematodes parasitic in members of the *P. penicillata* complex. Unpublished data suggest that these may be species complexes. The species occurring in 10 or more host species are sometimes generalists or may be core species in one macropodid host but can occur in small numbers in virtually any sympatric macropodid species. All the qualifications discussed, if valid, have the effect of increasing the degree of host specificity beyond that shown in the distribution presented in Figure 21.2.

CONCLUSIONS

This review of the helminth parasites of macropodoid marsupials documents a high level of parasite diversity but suggests that available data greatly underestimate that extent. Although a wide range of parasite orders and families is present, diversity is dominated by the cloacinine nematodes present in the sacculated forestomachs of macropodid marsupials. In contrast, the hypsiprymnodontids and potoroids examined to date have a relatively depauperate helminth fauna.

Within the macropodids, greatest diversity occurs in the pademelons (*Thylogale* spp.) and in the large wallabies and kangaroos (*Macropus*, *Wallabia*), based on studies of the Queensland macropodid fauna. Smaller wallabies of the genera *Petrogale*, *Onychogalea* and *Lagorchestes* appear to have less diverse helminth communities. In general, the parasites appear to be relatively host-specific and most species of macropodids harbour distinct helminth communities. Obvious exceptions are the two species of *Thylogale* and the members of the *P. penicillata* complex, which have very similar helminth communities respectively. In instances where divergences from this general pattern occur, the most common explanation is the switching of parasites between sympatric species of macropodids.

This review is clearly provisional, with major limitations being the incomplete examination of a substantial number of macropodoids for parasites, taxonomic impediments in the description of parasites and an obvious concentration of study on hosts occurring on the east coast of Australia. All these deficiencies need to be remedied before general conclusions can be drawn concerning the helminth communities of parasites of macropodoids.

REFERENCES

Anderson RC 2000. *Nematode Parasites of Vertebrates: Their Development and Transmission*. 2nd edn. CAB International, Wallingford.

Anderson RC, Chabaud AG and Willmott S 1974. *CIH Keys to the Nematode Parasites of Vertebrates*. Commonwealth Agricultural Bureau, UK.

Appan A, Bergfeld J and Beveridge I 2004. New species of parasitic nematodes from macropodid marsupials in Western Australia. *Trans. Roy. Soc. S. Aust.* **128:** 77–84.

Arundel JH, Beveridge I and Presidente PJA 1979. Parasites and pathological findings in enclosed and free-ranging populations of *Macropus rufus* (Desmarest) (Marsupialia) at Menindee, New South Wales. *Aust. Wildl. Res.* **6:** 371–379.

Begg M, Beveridge I, Chilton NB, Johnson PM and O'Callaghan MG 1996. Parasites of the Proserpine rock wallaby, *Petrogale persephone* (Marsupialia: Macropodidae). *Aust. Mammal.* **18:** 45–53.

Beveridge I 2002a. New species and new records of *Cloacina* von Linstow, 1898 (Nematoda: Strongyloidea) parasitic in macropodid marsupials from Papua New Guinea. *Rec. S. Aust. Mus.* **35:** 1–32.

Beveridge I 2002b. Revision of *Coronostrongylus* (Nematoda: Strongyloidea) parasitic in the stomachs of macropodid marsupials. *Invert. Syst.* **16:** 893–917.

Beveridge I and Chilton NB 1999. Revision of the *Rugopharynx australis* (Moennig, 1926) complex (Nematoda: Strongyloidea) from macropodid marsupials. *Invert. Taxon.* **13:** 805–843.

Beveridge I and Chilton NB 2001. Co-evolutionary relationships between the nematode subfamily Cloacininae and its macropodid marsupial hosts. *Int. J. Parasitol.* **31:** 976–996.

Beveridge I and Durette-Desset M-C 2004. *Sutarostrongylus johnsoni* sp. n. (Nematoda: Trichostrongylina, Heligmosomoidea) from the pademelon, *Thylogale stigmatica* (Marsupialia) from Queensland, Australia. *Parasite* **11:** 141–148.

Beveridge I and Johnson PM 2004. Cestode parasites of tree kangaroos (*Dendrolagus* spp.: Marsupialia), with the description of two new species of *Progamotaenia* (Cestoda: Anoplocephalidae). *Trans. Roy. Soc. S. Aust.* **128:** 175–185.

Beveridge I and Presidente PJA 1978. *Rugopharynx rosemariae* new species (Nematoda: Pharyngostrongylidae) from grey kangaroos (*Macropus giganteus* and *M. fuliginosus*) with life cycle stages and associated pathology. *Int. J. Parasitol.* **8:** 379–387.

Beveridge I and Spratt DM 1996. The helminth fauna of Australasian marsupials: origins and evolutionary biology. *Adv. Parasitol.* **37:** 135–254.

Beveridge I and Turni C 2003. *Progamotaenia capricorniensis* sp. nov. (Cestoda: Anoplocephalidae) from wallabies (Marsupialia: Macropodidae) from Queensland, Australia. *Parasite* **10:** 309–315.

Beveridge I, Spratt DM, Close RL, Barker SC and Sharman GB 1989. Helminth parasites of rock wallabies, *Petrogale* spp. (Marsupialia) from Queensland. *Aust. Wildl. Res.* **39:** 691–702.

Beveridge I, Speare R, Johnson PM and Spratt DM 1992. Helminth parasite communities of macropodoid marsupials of the genera *Hypsiprymnodon, Aepyprymnus, Thylogale, Onychogale, Lagorchestes* and *Dendrolagus* from Queensland. *Wildl. Res.* **19:** 359–376.

Beveridge I, Chilton NB, Johnson PM, Smales LR, Speare R and Spratt DM 1998. Helminth parasite communities of kangaroos and wallabies (*Macropus* spp. and *Wallabia bicolor*) from north and central Queensland. *Aust. J. Zool.* **46:** 473–495.

Beveridge I, Chilton NB and Spratt DM 2002. The occurrence of species flocks in the nematode genus *Cloacina* (Strongyloidea: Cloacininae), parasitic in the stomachs of kangaroos and wallabies. *Aust. J. Zool.* **50:** 597–620.

Bradley C, Beveridge I, Chilton NB and Johnson PM 2000. Helminth parasites of the purple-necked rock wallaby, *Petrogale lateralis purpureicollis*, from Queensland. *Trans. Roy. Soc. S. Aust.* **124:** 37–40.

Burk A, Westerman M and Springer M 1998. The phylogenetic position of the musky rat-kangaroo and the evolution of bipedal hopping in kangaroos (Macropodidae: Diprotodontia). *Syst. Biol.* **47:** 457–474.

Cardillo M, Bininda-Edmonds ORP, Boakes E and Purvis A 2004. A species-level phylogenetic supertree of marsupials. *J. Zool. Lond.* **264:** 11–31.

Chilton NB, Gasser RB and Beveridge I 1997. Phylogenetic relationships of Australian strongyloid nematodes inferred from ribosomal DNA sequence data. *Int. J. Parasitol.* **27:** 1481–1494.

Chilton NB, Huby-Chilton F, Gasser RB and Beveridge I 2002. Review of *Papillostrongylus* Johnston & Mawson, 1939 (Nematoda: Strongyloidea) from wallabies and kangaroos (Marsupialia: Macropodidae) using morphological and molecular techniques, with the description of *P. barbatus* n. sp. *Syst. Parasitol.* **51:** 81–93.

Eldridge MDB, Wilson ACC, Metcalfe CJ, Dollin AE, Bell JN, Johnson PM, Johnston PG and Close RL 2001. Taxonomy of rock-wallabies, *Petrogale* (Marsupialia: Macropodidae). III. Molecular data confirms the species status of the purple-necked rock-wallaby *Petrogale purpureicollis* Le Souef, 1924. *Aust. J. Zool.* **49:** 323–343.

Grieg-Smith P 1964. *Quantitative Plant Ecology.* Butterworth, London.

Griffith JE, Beveridge I, Chilton NB and Johnson PM 2000. Helminth communities of pademelons, *Thylogale stigmatica* and *T. thetis* (Marsupialia: Macropodidae) from eastern Australia and Papua New Guinea. *J. Helminthol.* **74:** 307–314.

Hu M, Gasser RB, Chilton NB and Beveridge I 2005. Genetic variation in the mitochondrial cytochrome *c* oxidase subunit 1 within three species of *Progamotaenia* (Cestoda: Anoplocephalidae) from macropodid marsupials. *Parasitology* **130:** 117–129.

Huby-Chilton F, Beveridge I, Gasser RB and Chilton NB 2002. Redescription of *Zoniolaimus mawsonae* Beveridge, 1983 (Nematoda: Strongyloidea) and the description of *Z. latebrosus* n. sp. from the red kangaroo *Macropus rufus* (Marsupialia: Macropodidae) based on morphological and molecular data. *Syst. Parasitol.* **51:** 135–147.

Hume ID 1999. *Marsupial Nutrition.* Cambridge University Press, Cambridge.

Khalil LF, Jones A and Bray RA 1994. *Keys to the Cestode Parasites of Vertebrates.* CAB International, Wallingford.

Leong TS and Holmes JC 1981. Communities of metazoan parasites in open water fishes of Cold Lake, Alberta. *J. Fish Biol.* **18:** 693–713.

Maynes GM 1982. A new species of rock wallaby, *Petrogale persephone* (Marsupialia: Macropodidae), from Proserpine, central Queensland. *Aust. Mammal.* **5:** 47–58.

Mycytowycz R 1964. A survey of the endoparasites of the red kangaroo, *Megaleia rufa* (Desmarest). *Parasitology* **55:** 1–17.

Price PW 1980. *Evolutionary Biology of Parasites.* Princeton University Press, Princeton.

Smales LR 1977a. The life history of *Labiostrongylus eugenii*, a nematode parasite of the Kangaroo Island wallaby (*Macropus eugenii*): development and hatching of the egg and free-living stages. *Int. J. Parasitol.* **7:** 449–456.

Smales LR 1977b. The life history of *Labiostrongylus eugenii*, a nematode parasite of the Kangaroo Island wallaby (*Macropus eugenii*): the parasitic stages. *Int. J. Parasitol.* **7:** 457–461.

Smales LR 2006. New host records, redescriptions, and new species of Labiostrongylinea (Nematoda: Chabertiidae) from macropod marsupials, with a revised host–parasite checklist. *Trans. Roy. Soc. S. Aust.* **130:** 85–100.

Smales LR and Mawson PM 1978. Nematode parasites of the Kangaroo Island wallaby, *Macropus eugenii*. I. Seasonal and geographic distribution. *Trans. Roy. Soc. S. Aust.* **102:** 9–16.

Spratt DM 1972. Aspects of the life-history of *Dirofilaria roemeri* in naturally and experimentally infected kangaroos, wallaroos and wallabies. *Int. J. Parasitol.* **2:** 139–156.

Spratt DM, Beveridge I and Walter EL 1991. A catalogue of Australasian monotremes and marsupials and their recorded helminth parasites. *Rec. S. Aust. Mus. Monog. Ser.* **1:** 1–105.

Strahan R (Ed.) 1995. *The Mammals of Australia*. Reed Books, Sydney.
Stunkard HW 1969. *Panceriella* nom.n. for *Panceria* Fuhrmann, 1899, preoccupied by *Panceria* Chun, 1879, and systematic relations of the genus. *J. Parasitol.* **55:** 1162–1168.
Webley LS, Beveridge I and Coulson G 2004. Endoparasites of an insular subspecies of the western grey kangaroo, *Macropus fuliginosus*. *Aust. J. Zool.* **52:** 623–633.
Yen PKF 1983. Filariasis in the Australian marsupial, *Setonix brachyurus*. PhD thesis. University of Western Australia, Perth.

PART IV

MANAGEMENT

22 Returning the rat-kangaroos: translocation attempts in the family Potoroidae (superfamily Macropodoidea) and recommendations for conservation

G.R. Finlayson, S.T. Finlayson and C.R. Dickman

SUMMARY

The marsupial family Potoroidae (bettongs and potoroos) has been severely affected in Australia by European settlement. Of the 11 known contemporary species, three have become extinct and six are listed on the IUCN Red List 2008 with a range of status assignations from 'conservation dependent' to 'critically endangered'. Four of the species in this family are still in decline. Possible causes for the decline of potoroids include predation by the introduced European red fox (*Vulpes vulpes*) and cat (*Felis catus*), competition from introduced herbivores and loss of habitat. Translocation has become an extremely important and popular tool for the management of wildlife, particularly to increase the ranges of endangered species and return them to parts of their former distribution. In this chapter, we document and analyse the outcomes of all translocation attempts for species within the Potoroidae. We located a total of 50 translocation attempts in the literature for three potoroid species. Sixty percent of translocations occurred in Western Australia. Only 33.3% of all translocations have been successful, with limiting factors associated mainly with predation. Success rates have been higher on islands or when the source population was derived from captive individuals. This review highlights the need to publish the results of all translocation attempts to increase their likelihood of success, and recognises the benefits associated with translocations of members of this family in the future.

INTRODUCTION

The decline and extinction of mammals in Australia since European settlement has been well-documented (Burbidge and McKenzie 1989; Dickman 1994a, b; Short and Smith 1994; Short 1998; Burbidge and Manly 2002; McKenzie *et al.* 2007). Currently, 52 of the 189 extant species and subspecies of Australian marsupials are considered to be at some risk of extinction (Hilton-Taylor 2000). In the superfamily Macropodoidea, out of a total of 50+ described species 33 are considered stable, 11 have declined (to <50% of their former range) and six (12%) are classified as extinct (Johnson *et al.* 1989; Burbidge and Eisenberg 2006). The most devastating impact since European settlement has been on smaller macropodoids weighing 35 –5500 g. This size range was characterised as the critical weight range (CWR) by

Burbidge and McKenzie (1989), and critiqued and expanded by Cardillo and Bromham (2001). Within the Macropodoidea are the families Hypsiprymnodontidae and Potoroidae, which include 12 species of potoroos, bettongs and rat-kangaroos (Table 22.1), all of which fit within the CWR. Three of the species (25%) – *Bettongia pusilla, Caloprymnus campestris* (desert rat-kangaroo) and *Potorous platyops* (broad-faced potoroo) – have become extinct since European settlement. The remaining nine species have declined severely in both distribution and abundance, due to a range of threatening processes including hunting or harvesting, habitat clearing for agricultural purposes, altered fire regimes, competition from introduced herbivores (rabbits and stock) and introduced predators such as foxes and cats (Johnson *et al.* 1989).

Between 1883 and 1920 more than 2.5 million bounties were paid for potoroids in New South Wales (Short and Smith 1994). Jarman and Johnson (1977) reported that close to 46 000 'rat-kangaroos' were harvested in 1895 in western New South Wales alone. In Queensland, 27 million medium-sized terrestrial marsupials (including bettongs) were killed over a period of 53 years (1877–1930) under the dictates of the Marsupial Destruction Act by means of poisoning, trapping and shooting (Hrdina 1997; Dickman 2005).

The loss of suitable habitat as a result of land clearing for agriculture and/or industry is one of the most frequently cited causes for the declines of species within the Potoroidae (Calaby 1971; Johnson *et al.* 1989; Saunders 1994; Short and Smith 1994). Land clearing may certainly have played a role in the extinction of local populations of rat-kangaroos, with the disappearance of populations of both *P. platyops* and *B. penicillata* (brush-tailed bettong) from regions in the south-west of Australia that have been significantly cleared for agriculture (Johnson *et al.* 1989). However, it is possible that the destruction of habitat has worked in combination with other factors (e.g. the migration of introduced predators into areas following habitat clearing) to cause these declines. On St Francis Island off South Australia, for example, *B. penicillata* succumbed to the conversion of native vegetation into vegetable gardens and to the deliberate introduction of house cats (*Felis catus*) in the late 19th century (Wood Jones 1924).

Fire regimes throughout Australia have also changed dramatically since European settlement, with patch-burning giving way to broad-scale wildfires over large areas (Latz and Griffin 1978; Kimber 1983; Burbidge 1985). These land management changes reduced habitat patchiness in the broader landscape, and perhaps homogenised native vegetation so that the mosaics preferred by many species were no longer present. These changes are often thought to be detrimental, but only limited information is available on the direct effects of changed fire regimes on CWR mammals (Burbidge and McKenzie 1989). Certainly, many mammals declined when Aborigines left parts of the desert and the fire regimes changed (Johnson and Roff 1982; Burbidge *et al.* 1988), but cause and effect has yet to be established. Fire regimes have been shown to play a pivotal role in providing suitable habitat for *B. penicillata* (Christensen 1977) and their disruption, in combination with other threatening factors, may have contributed to the demise of other members within the Potoroidae.

Introduced herbivores such as sheep, goats, cattle, deer, pigs and rabbits have led to the reduction of food resources and availability of shelter across vast areas of Australia and have been implicated as the primary cause of declines of many CWR native mammals throughout Australia (Calaby and Grigg 1989; Johnson 1999; Morton 1990; Lunney 2001; Fisher *et al.* 2003). There is evidence that the destruction of critical habitat (for food resources and shelter) by rabbits and other introduced herbivores caused population and range reductions of many species within the Potoroidae and subsequently left them vulnerable to exposure from drought and introduced predators (Newsome 1975; Morton 1990; Letnic and Dickman 2006). *B. penicillata*, for example, disappeared from parts of New South Wales where sheep were overstocked but persisted for longer in areas where stocking was less intense (Marlow 1958; Dickman *et al.* 1993; Dickman 1994a).

Perhaps the most detrimental impact on Australian CWR mammals has been predation by introduced predators. The evidence for predation and impacts on populations of medium-sized mammals has been made clear (Dickman *et al.* 1993; Horsup and Evans 1993; Gibson *et al.* 1994; Salo *et al.* 2007). There is also strong evidence that the effective removal of red foxes (*Vulpes vulpes*) by baiting with 1080 poison results in an increase in the population sizes of many native medium-sized mammals (Kinnear *et al.* 2002; Mawson 2004). Fox and cat predation have been implicated in the declines of both

Table 22.1: Species within the Potoroidae and Hypsiprymnodontidae, their current status (IUCN 2008), trend in population since European settlement and recognised reason for decline.

Species	Common name	National status	Since European settlement	Reasons for decline
Family Potoroidae				
Bettongia pusilla	Nullarbor dwarf bettong	Extinct	Extinction – species discovered recently	Unknown
Caloprymnus campestris	Desert rat-kangaroo	Extinct	Extinction?	Unknown – probably foxes
Potorous platyops	Broad-faced potoroo	Extinct	Extinction	Unknown
Potorous gilbertii	Gilbert's potoroo	Critically endangered	Declined significantly (rediscovered 1994)	Foxes, cats, loss of habitat (*Phytophthora*)
Potorous longipes	Long-footed potoroo	Endangered	Declined	Habitat loss/ degradation
Bettongia tropica	Northern bettong	Endangered	Declined	Habitat loss (agriculture/fire)
Bettongia lesueur	Burrowing bettong	Near threatened	Extinct on mainland, survived on islands of Western Australia only	Exotic predators
Bettongia penicillata	Brush-tailed bettong	Critically endangered	Severely reduced range	Exotic predators, habitat alteration
Potorous tridactylus	Long-nosed potoroo	Least concern	Reduced range	Clearing of forests, predation
Bettongia gaimardi	Tasmanian bettong	Near threatened	Extinct on mainland	Foxes, habitat alteration
Aepyprymnus rufescens	Rufous bettong	Least concern	Reduced range	Habitat alteration, foxes
Family Hypsiprymnodontidae				
Hypsiprymnodon moschatus	Musky rat-kangaroo	Least concern	Reduced range	Rainforest clearing

Bettongia species and *Aepyprymnus rufescens* (rufous bettong) (Kaufmann 1974; Christensen 1980; Burbidge 1983; Rose 1986; Robertshaw and Harden 1989; Short *et al.* 2002), with predation likely to be a significant limiting factor for the remaining extant species in this family, particularly on the mainland.

Given the perilous state of many of Australia's marsupial species, particularly those in the CWR, effective conservation and management strategies are urgently required. One such strategy is translocation. The return of CWR species to parts of their former distribution using translocation or expansion of their existing ranges often relies on the effective control of the threatening processes, particularly the presence of introduced predators.

Translocation can be defined as the 'movement of native or introduced species to habitats within or outside their natural or previous distribution' (Griffith *et al.* 1989). Translocations have occurred for centuries throughout the world for various reasons other than conservation. For example, in the early 20th century in Australia several native species were translocated to islands as a source of food for shipwrecked mariners (Robinson 1989). In recent years, translocations have become a popular and increasingly important conservation and management tool for expanding the ranges of endangered species, as well as for establishing secure populations of these species for captive breeding and subsequent translocations (Thomas 1989; vanEmden 1989; Whalley 1989; Craig and Veitch 1990; Serena 1995). They also provide wildlife researchers with the opportunity to learn about the processes that led to the decline of a species in the first place and therefore can assist with subsequent conservation and management decisions (Johnson 1999). In Australia, translocation has been recommended as a conservation strategy in action plans for one-third of the nation's threatened vertebrate taxa (Serena and Williams 1995), including many species within the Potoroidae.

The aims of this chapter are to document translocations of species within the Family Potoroidae and identify what factors influence the success or failure of these translocations, the most appropriate release methods, which locations are most appropriate for translocations and what issues need to be considered for the application of translocation techniques to other species within this family. Apart from commenting briefly on the musky rat-kangaroo (*Hypsiprymnodon moschatus*), we follow recent convention and consider this species to be the sole member of its own family, Hypsiprymnodontidae (Eldridge 2006).

METHODS

Translocation

For the purpose of this chapter, we refer to three distinct forms of translocation:

- reintroduction – the translocation of a species back into part of its former distribution;
- introduction – the translocation of a species into an area not recognised as being within its historical distribution;
- supplementation – the translocation of individuals into an existing species population for the purpose of increasing the population size and/or genetic diversity (Griffith *et al.* 1989).

In this review we considered only translocations where animals were released into an area and were not supplementary fed. It was originally intended to compare release methods; however, there was insufficient published information for analysis. We classified multiple releases within a site, where animals from previous releases still survived, as a single translocation attempt. In the case of multiple releases at the same site where previous releases had failed, subsequent releases were considered a separate release. Where translocations are referred to broadly throughout the chapter we are referring specifically to intentional translocations of endangered native species only and not the deliberate introduction of either non-native or native species.

Data collection

We established a database on translocations of species within the Potoroidae using articles from journals, making web-based searches for government and NGO reports and contacting relevant agencies for identification and clarification of translocation attempts that had not been published. The following questions were asked for each translocation.

1 Which species was translocated?
2 What were the numbers and sex ratio of the released animals?
3 What type of translocation was carried out (introduction, reintroduction, supplementation)?
4 Where was the translocation carried out (state and region)?
5 What type of area was used (predator fence on mainland, island, baited area for predators on mainland, wild, none of the above)?
6 What was the release method (hard/soft release)?
7 What was the source population (wild/captive-reared and where from)?
8 Was the translocation a success and what factors, if any, contributed to the result?

Measuring success

We identified two levels of success for translocations. First, for a translocated population that had been released less than five years ago, success was assumed if the population was greater than the number of animals released and showed evidence of breeding and recruitment, or was the same size as the total number of animals released and showed evidence of recruitment. Second, for populations that had survived more than five years, the translocation was considered successful if the population was greater than the number of animals released and showed evidence of recruitment and breeding, was the same size as the released population with evidence of recruitment, was self-sustaining with breeding and recruitment or was meeting the requirements outlined in the original translocation proposal.

A translocation was considered to be a failure if the guidelines set in translocation proposals were not met or if all the animals in the translocated population did not survive and the population failed to establish.

Reporting translocations

For each translocation, we determined how the outcome was reported. This was either in a peer-reviewed journal, an internal unpublished report or no recorded report (*pers. comm.*). We also attempted to determine the degree to which each translocation was followed up or monitored after the initial publication of results.

Analysis and presentation of results

All results are presented in tables and histograms. Chi-squared analyses with Yates' correction in tests with one degree of freedom were used to determine differences between groups in association with the success and failure of each translocation attempt. Results that were pending were excluded from these analyses.

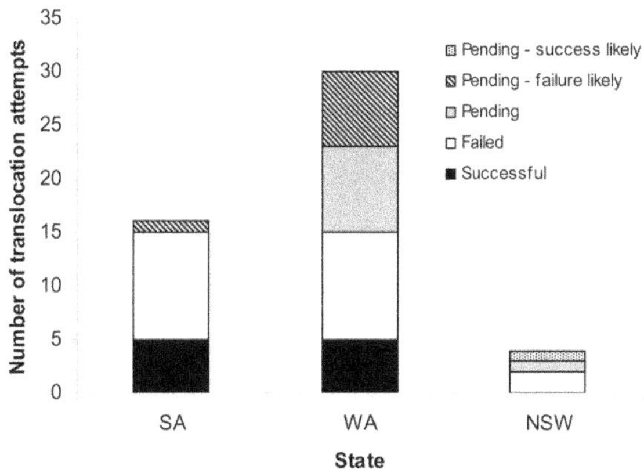

Figure 22.2: Outcome of translocation attempts for all species (*B. penicillata*, *B. lesueur* and *P. gilbertii*) according to the state in which the animals were released.

RESULTS

Throughout Australia we found a total of 55 translocations that had been attempted on three species within the Potoroidae. Of these, 45 translocations had been attempted for *B. penicillata* (Table 22.2, Fig. 22.1), nine for the burrowing bettong (*B. lesueur*) (Table 22.3, Fig. 22.1) and one for Gilbert's potoroo (*Potorous gilbertii*) (Table 22.4, Fig. 22.1). Most attempted translocations have been reintroductions (82%), with the remainder either introductions (16%) or supplementations (2%). Less than half of all translocations have been reported in peer-reviewed journal articles. Several translocations were reported in a single review in *Conservation Science Western Australia*, while review articles (Short *et al.* 1992; Morris 2000) have been used to communicate the findings of many translocation attempts. Numerous translocations have been reported in unpublished internal reports or not at all. Translocations have been attempted by both government and CSIRO, and non-government and community-based organisations (Tables 22.2-22.4). Research and monitoring to assess the outcome of translocations as well as other aspects of the life history and ecology of translocated populations have been conducted by scientists within these institutions or by those based at universities or in private consultancies.

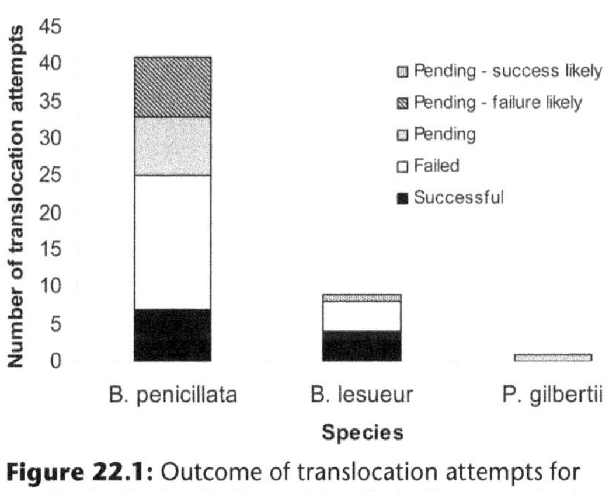

Figure 22.1: Outcome of translocation attempts for species in the family Potoroidae. Only three species have been translocated in Australia – *B. penicillata*, *B. lesueur* and *P. gilbertii.*

At the state level, we found that 35 translocations had occurred in Western Australia, 16 in South Australia and four in New South Wales (Fig. 22.2). Success rates were the same in both South Australia and Western Australia for translocations with a known result (33.3%); however, the results of 15 translocations in Western Australia are still pending with approximately half of these categorised as 'pending – likely to fail' (Mawson 2004). A number of translocations of *B. penicillata* presented in a review by Morris (2000) were then considered successful but, due to recent dramatic population declines in this species (Mitchell and Wayne 2008), the current state of populations at these sites is now unknown. These few cases were not included in the analysis. The causes associated with a failed translocation ranged from predation by foxes, cats or wedge-tailed eagles (*Aquila audax*), to possible unsuitable habitat, drought and included unidentified (or unknown) reasons (Table 22.5). The success rates of translocations of *B. penicillata* were influenced significantly by whether the source animals were transferred from the wild or from captivity (χ^2 = 9.73, d.f. = 1, p = 0.002), with translocations from captivity achieving higher success (Fig. 22.3).

The success rates of translocations, with all species combined, were significantly lower on the mainland compared to islands (χ^2 = 7.96, d.f. = 1, p = 0.005;

Table 22.2: Summary of translocations of brush-tailed bettongs (*B. penicillata*) in Australia.

Year	TP	Release location	Source	Org	Release site type – feral status	TNR	CP	Outcome	References
1977	RI	Yendicup Block, Perup, WA	Perup	CALM	Mainland – baited	52	0	F	Short *et al.* 1992; Morris 2000
1979	IN	Bird Club Island, SA	Captive-bred	DEH	Island	6	0	F	Short *et al.* 1992; Delroy *et al.* 1986; Nelson *et al.* 1992
1980	IN	Island A, Venus Bay, SA	Captive-bred	DEH	Island – feral-free	7	30	S	Short *et al.* 1992; Delroy *et al.* 1986; Nelson *et al.* 1992; J. Van Weenan, *pers. comm.*
1981	RI	St Francis Island, SA	Captive-bred	DEH	Island – feral-free	40	0	F	Short *et al.* 1992; Delroy *et al.* 1986; Nelson *et al.* 1992
1982	IN	Bairds Bay Island, SA	Island A	DEH	Island	10	0	F	Short *et al.* 1992; Delroy *et al.* 1986; Nelson *et al.* 1992
1982–83	RI	Collie, WA	Perup	CALM	Mainland – baited	52	0	F	Short *et al.* 1992; Morris 2000
1983	RI	St Johns Brook, Nannup, WA	Perup	CALM	Mainland	67	0	F	Short *et al.* 1992; Morris 2000
1983	IN	Wedge Island, SA	Captive-bred, Monarto Zoo	DEH/ CALM	Island – feral-free	11	~1000	S	Short *et al.* 1992; Delroy *et al.* 1986; Nelson *et al.* 1992; J. Van Weenan, *pers. comm.*
1984	RI	St Francis Island, SA	Captive-bred	DEH	Island – feral-free	42	0	F	Short *et al.* 1992; Delroy *et al.* 1986; Nelson *et al.* 1992
1987	RI	St Francis Island, SA	Captive-bred	DEH	Island – feral-free	48	0	F	Short *et al.* 1992; Delroy *et al.* 1986; Nelson *et al.* 1992
1989	IN	St Peter Island, SA	Captive-bred	DEH	Island	113	~2000	S	Short *et al.* 1992; Nelson *et al.* 1992; J. Van Weenan, *pers. comm.*
1991	RI	Yookamurra, SA	No information	ES (now AWC)	Mainland – fenced	80+	?	P	J. Bentley, *pers. comm.*

Year	TP	Release location	Source	Org	Release site type – feral status	TNR	CP	Outcome	References
1992	RI	Boyagin Nature Reserve	Dryandra	CALM	Mainland – baited	40	?	U	Morris 2000
1994	RI	Northern Jarrah Forest, WA, 33 sites	Dryandra	CALM	Mainland – WS	492	?	U	Morris 2000
1994	RI	Karakamia, WA	No information	AWC	Mainland – fenced	37	~400–500	S	J. Richards, *pers. comm.*
1994–95	RI	Venus Bay, CP	Dryandra Forest Reserve, WA	DEH/ CALM	Mainland peninsula – fenced	67	~1500	S	J. Van Weenan, *pers. comm.*
1995	RI	Julimar Forest	Perup	ECU/ CALM	Mainland – baited	39	?	U	Morris 2000
1996	RI	Hills Forest, WA	Dryandra Forest Reserve, WA	CALM	Mainland – WS	37	0	F	Mawson 2004
1997–98	RI	Lake Magenta NR, WA	Dryandra Forest Reserve, WA	CALM	Mainland – WS	35	0	F	Mawson 2004
1997–98	RI	Francois Peron NP, WA	Batalling Forest, Dryandra Forest, WA	CALM	Mainland – WS	147	0	F	Mawson 2004
1998	RI	St Johns Forest, WA	Perup Forest WA	CALM	Mainland – WS	40	0	F	Mawson 2004
1998	RI	Denmark Forest, WA	Perup Forest, WA	CALM	Mainland – WS	38	0	F	Mawson 2004
1998	S	Poorginup and Chitelup Forest, WA	Perup Forest, WA	CALM	Mainland – WS	40	0.70%	P	Mawson 2004
1998	RI	Centaur Forest, WA	Batalling Forest, WA	CALM	Mainland – WS	39	5.40%	P	Mawson 2004
1998	RI	Easter and Barlee SF, WA	Perup Forest, WA	CALM	Mainland – WS	40	?	S	Mawson 2004
1998	RI	Mt Lindesay	Perup	CALM	Mainland – WS	40	?	U	Morris 2000
1998–99	RI	Genaren Hill Sanctuary, NSW	Karakamia Sanctuary, Dryandra Forest, WA	GHLG	Mainland – fenced	24	0	F	Mawson 2004
1999	RI	Katarapko Island, SA	No information	KCAG	Island (inland)	21	0	F	S. Loffler, *pers. comm.*
1999	IN	Giants Forest, WA	Perup Forest, WA	CALM	Mainland – WS	40	1%	P	Mawson 2004
1999–2001	RI	Lincoln NP, SA	Venus Bay and St Peter Island	DEH	Mainland – baited	123	?	P	J. Van Weenan, *pers. comm.*; Martin *et al.* 2006

Table 22.2: Summary of translocations of brush-tailed bettongs (*B. penicillata*) in Australia (Continued)

Year	TP	Release location	Source	Org	Release site type – feral status	TNR	CP	Outcome	References
1999–2001	RI	Flinders Ranges, SA	Wedge Island, St Peter Island and Venus Bay CP, SA	NPWSA	Mainland – baited	97	0	F	Bellchambers 2001
2000	RI	Kalbarii NP, WA	Dryandra Forest Reserve, WA	CALM	Mainland – WS	32	?	P	Mawson 2004
2000	RI	Shannon NP, WA	Perup Forest, WA	CALM	Mainland – WS	40	?	P	Mawson 2004
2000	RI	Watts PP, WA	Batalling Forest, WA	CALM	Mainland – WS	19	51.60%	P	Mawson 2004
2000	RI	Tucker PP, WA	Batalling Forest, WA	CALM	Mainland – WS	19	?	P	Mawson 2004
2000	RI	Wellington NP, WA	Batalling Forest, WA	CALM	Mainland – WS	30	6.60%	P	Mawson 2004
2000	RI	Davis Forest, WA	Batalling Forest, WA	CALM	Mainland – WS	37	8.10%	P	Mawson 2004
2000	RI	Strickland Forest, WA	Perup Forest, WA	CALM	Mainland – WS	40	3%	P	Mawson 2004
2000	RI	Hadfield Forest, WA	Batalling Forest, WA	CALM	Mainland – WS	29	8.30%	P	Mawson 2004
2000–01	RI	Paruna Sanctuary, WA	Karakamia Sanctuary	AWC	Mainland – baited, fenced	200	~150–200	S	Mawson 2004
2001	RI	Yathong NR, NSW	Dryandra Forest Reserve, WA, Venus Bay CP and St Peter Island, SA	NPWS NSW	Mainland – baited	85	0	F	Priddel and Wheeler 2004
2002	RI	Thames Forest, WA	Perup Forest, WA	CALM	Mainland – WS	37	?	P	Mawson 2004
2002	RI	Flybrook, WA	Perup Forest, WA	CALM	Mainland – WS	21	?	P	Mawson 2004
2002	RI	Boyndaminup, WA	Perup Forest, WA	CALM	Mainland – WS	21	?	P	Mawson 2004
2004–05	RI	Scotia Sanctuary, NSW	Captive-bred, on-site	AWC	Mainland – fenced, feral-free	183	~80	P	Finlayson (unpubl. data)

Release sites classified as 'feral-free' indicate that the area is free of foxes, cats and rabbits. House mice may still be present. Fenced areas utilise predator exclusion fencing. Baiting in these areas usually means the use of 1080 baits.

Percentages quoted under current population indicate the most recent capture success rate.

TP = type of translocation; Org = organisation; TNR = total number of released animals (there may have been more than one release occasion at the same site); CP = current population (at last survey).

RI = reintroduction; IN = introduction; WS = part of Western Shield Program, which involves regular 1080 baiting; S = supplement; F = failure; S = successful; P = pending; U = unknown fate due to recent population declines.

DEH = Dept for Environment and Heritage South Australia; CALM = Conservation and Land Management Western Australia (now WA Dept of Environment and Conservation); AWC = Australian Wildlife Conservancy; ES = Earth Sanctuaries; KCAG = Katarapko Conservation Action Group; GHLG = Genaren Hills Landcare Group.

Table 22.3: Summary of translocations of burrowing bettongs (*B. lesueur*) in Australia.

Year	TP	Release location	Source	Org	Release site type – feral status	TNR	CP	Outcome	References
1924–27	IN	Flinders Chase NP, Kangaroo Island, SA	No information	-	Island	8	0	F	Harris 1974; SA DEP 1986; Copley 1995
1992	RI	Gibson Desert, WA	Barrow Island	CALM	Mainland – baited	40	0	F	Christensen and Burrows 1994
1992–93	RI	Heirisson Prong, WA	Dorre Island	CSIRO	Mainland peninsula – fenced, baited	30	45	S	Short *et al.* 2000, J. Short, *pers. comm.*
1993	RI	Boodie Island NR	Barrow Island	CALM	Island	36	?	S	Morris 2000
1999–2000	RI	Arid Recovery Reserve (Roxby Downs), SA	Heirisson Prong, Bernier Island, WA	ARR	Mainland – fenced, feral-free	30	~200	S	Mawson 2004; www.aridrecovery.org.au
2002	RI	Faure Island, WA	Heirisson Prong, WA	AWC	Island	17	~200	S	J. Richards, *pers. comm.*
2003	RI	Dryandra, WA	Captive-bred on-site (originally sourced from Dorre Island in 1998)	CALM	Mainland – WS	37	0	F	N. Thomas, *pers. comm.*; Thomas *et al.* 2005
2004–05	RI	Scotia Sanctuary, NSW	Captive bred on-site	AWC	Mainland – fenced, feral-free	120	~160	P	Finlayson (unpubl. data)
2005	RI	Yookamurra	Bernier Island then captive-bred on-site	ES, CALM and now AWC	Mainland – fenced (fox and cat incursions)	?	0	F	J. Bentley, *pers. comm.*

Release sites classified as 'feral-free' indicate that the area is free of foxes, cats and rabbits. House mice may still be present. Fenced areas utilise predator exclusion fencing. Baiting in these areas usually means the use of 1080 baits.

Percentages quoted under current population indicate the most recent capture success rate.

TP = type of translocation; Org = organisation; TNR = total number of released animals (there may have been more than one release occasion at the same site); CP = current population (at last survey).

RI = reintroduction; IN = introduction; WS = part of Western Shield Program, which involves regular 1080 baiting; F = failure; S = successful; P = pending.

CALM = Conservation and Land Management Western Australia (now WA Dept of Environment and Conservation); CSIRO = Commonwealth Scientific and Industrial Research Organisation; ARP = Arid Recovery Project; AWC = Australian Wildlife Conservancy; ES = Earth Sanctuaries.

Table 22.4: Summary of translocations of Gilbert's potoroo (*Potorous gilbertii*) in Australia.

Year	TP	Release location	Source	Org	Release site type – feral status	TNR	CP	Outcome	References
2005 (trial)	IN	Bald Island, WA	Two Peoples Bay, WA	CALM	Island – feral-Free	2	NA	S – animals survived for the duration of the 5-week trial	Friend 2005
2005	IN	Bald Island, WA	Two Peoples Bay, WA	CALM	Island – feral-Free	4	6	P – 1 PY born on island and 1 PY conceived on island	Friend 2006

Release sites classified as 'feral-free' indicate that the area is free of foxes, cats and rabbits. House mice may still be present.

TP = type of translocation; Org = organisation; TNR = total number of released animals (there may have been more than one release occasion at the same site); CP = current population (at last survey).

RI = reintroduction; IN = introduction; F = failure; S = successful; P = pending.

CALM = Conservation and Land Management Western Australia (now WA Dept of Environment and Conservation).

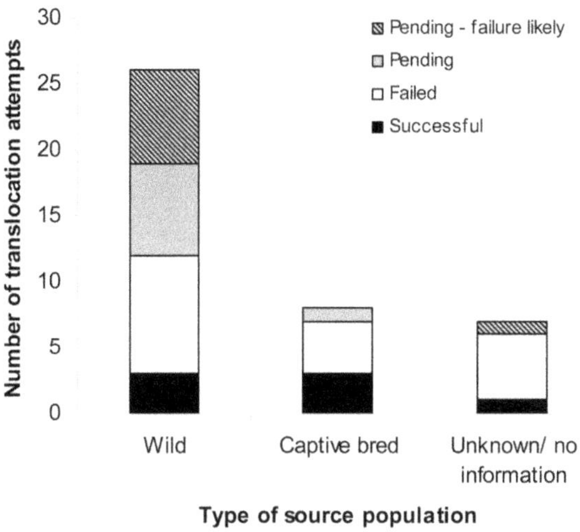

Figure 22.3: Outcome of brush-tailed bettong (*B. penicillata*) translocations according to type of source population.

Fig. 22.4). Translocations of the Potoroidae have occurred on the mainland and from the mainland to islands. On the mainland, translocation sites have included baited areas, fenced, fenced and baited areas, areas with no control of introduced pests, and areas with unknown conditions (not described or no available information). Island translocations could be classified as feral-free (no introduced predators or rabbits) or unknown. Translocation success was lower in predator-baited release sites than in feral-free areas, i.e. within fenced enclosures or on islands ($\chi^2 = 8.92$, d.f. = 1, $p = 0.003$; Fig. 22.5).

There was insufficient information for the analysis of release type (hard vs soft release), but there is evidence that there is no significant difference between the outcome of hard and soft releases for burrowing bettongs in the presence of introduced predators (N. Thomas, *pers. comm.*). There was also insufficient data to determine the influence of sex ratio on the success rates of translocations.

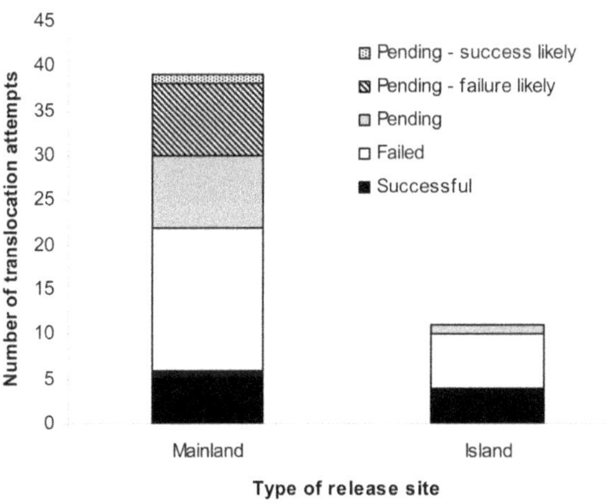

Figure 22.4: Outcome of translocation attempts for all species (*B. penicillata, B. lesueur* and *P. gilbertii*) according to type of release site – on the mainland or an island.

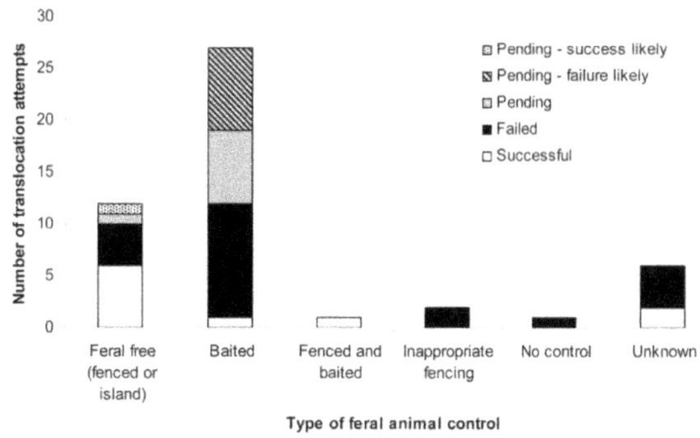

Figure 22.5: Outcome of translocation attempts for all species (*B. penicillata, B. lesueur* and *P. gilbertii*) according to type of feral animal control techniques. Sites that used inappropriate fencing were separated from the 'feral-free' group because incursions by foxes and/or cats occurred regularly.

DISCUSSION

The number of translocations of Potoroidae is high, especially considering that the reported figures are predominantly for only two species. There appear to be a number of benefits in choosing species such as *B. penicillata* for translocation. They are fast to establish, sexually mature at a young age (six months) and have high fecundity (de Tores and Start 2008). Moreover, it has been clearly demonstrated that *B. penicillata* plays a critical functional role in Australian environments by influencing the dispersal and regeneration of sandalwood (*Santalum spicatum*) (Murphy *et al.* 2005), as well as turning over large amounts of soil each year and dramatically altering soil water repellency and nutrient content (Garkaklis *et al.* 2000, 2003, 2004). Studies have shown that foraging pits created by *B. lesueur* are important for the functioning of healthy ecosystems and provide nutrient-rich patches which assist germination of seeds (James and Eldridge 2007). Some of these studies, and others that have led to an understanding of the role these species play in climatically unpredictable Australian ecosystems, have only been possible due to the presence of translocated populations. Such findings also accentuate the benefits of large collaborative projects between government and non-government agencies, and research institutes. These collaborations allow

Table 22.5: Summary of failed translocation attempts of potoroid species in Australia and possible adverse factors.

Year	Release location	Time to failure	Factors limiting success	References
B. penicillata				
1977	Yendicup Block, Perup, WA	13 yr	Foxes	Short *et al.* 1992; Morris 2000
1979	Bird Club Island, SA	1 animal >1 yr, others about 6 mth	Dogs	Short *et al.* 1992; Delroy *et al.* 1986; Nelson *et al.* 1992
1982	Bairds Bay Island, SA	12 yr	1994 fox came to island	Short *et al.* 1992; Delroy *et al.* 1986; Nelson *et al.* 1992
1982–83	Collie, WA	<2 yr	Foxes, insufficient ground cover, possibly quolls	Short *et al.* 1992
1983	St Johns Brook, Nannup, WA	<6 mth	Foxes	Short *et al.* 1992
1981	St Francis Island, SA	1 animal >3 yr, others <8 mth	Unknown – possibly vegetation unsuitable from heavy grazing, competition with bandicoots, predation from pythons and sea eagles	Short *et al.* 1992; Delroy *et al.* 1986; Nelson *et al.* 1992
1984	St Francis Island, SA	<1 yr	Unknown – possibly vegetation unsuitable from heavy grazing, competition with bandicoots, predation from pythons and sea eagles	Short *et al.* 1992; Delroy *et al.* 1986; Nelson *et al.* 1992
1987	St Francis Island, SA	<1 yr	Unknown – possibly vegetation unsuitable from heavy grazing, competition with bandicoots, predation from pythons and sea eagles	Short *et al.* 1992; Delroy *et al.* 1986; Nelson *et al.* 1992
1996	Hills Forest, WA	No information	Fox control	Mawson 2004
1983, 1998	St Johns Forest, WA	No information	No predator control	Morris 2000; Mawson 2004
1998	Denmark Forest, WA	No information		Mawson 2004
1999	Katarapko Island, SA	No information	Drought 2002	S. Loffler, *pers. comm.*
2001	Yathong NR, NSW	13 mth	Predation by cats	Priddel and Wheeler 2004
1999–2001	Flinders Ranges, SA	21 mth	Over-dispersion, predation by birds of prey, predation by introduced carnivorous mammals, malnutrition/exposure, snake bite and trap injury	Bellchambers 2001

Year	Release location	Time to failure	Factors limiting success	References
1997–98	Lake Magenta NR, WA	No information		Mawson 2004
1997–98	Francois Peron NP, WA	No information		Mawson 2004
1998–99	Genaren Hill Sanctuary, NSW	No information	Unknown	Mawson 2004
B. lesueur				
1924–27	Flinders Chase NP, Kangaroo Island, SA		Unknown	Harris 1974, SA DEP 1986, Copley 1995
1992	Gibson Desert, WA		Cats	Christensen and Burrows 1994
2003	Dryandra, WA	8 mth	Predation by wedge-tailed eagles and foxes. Possibly a behavioural problem because the bettongs failed to build their own burrows or stayed within the burrows, therefore more susceptible to predation	N. Thomas, *pers comm.*
2004	Yookamurra	~1 yr	Fence failed – incursions by cats and foxes. Animals removed in 2005 when a burrowing bettong was found dead due to predation	J. Bentley, *pers comm.*

wildlife managers to focus on the monitoring of populations and conservation scientists to look at aspects of the translocated populations that may assist future management decisions. Collaborations also lead to an increase in the available literature on translocated populations as well as increasing the likelihood of their success (Kleiman 1989).

An alarming outcome of the current review is the general lack of complete documentation associated with translocations. In a review of worldwide 'relocations', Fischer and Lindenmayer (2000) recognised the limited 'publication of results' as a major concern in the use of relocations as a tool for management and conservation. This concern is also prominent in translocations recorded here. The results that are immediately accessible to wildlife managers and conservation biologists are misleading because of the bias towards the publication of successful translocations. There are clear benefits associated with documenting failed translocation attempts, as they identify problems that were encountered and can therefore be avoided in future attempts. Such problems range from the use of inappropriate monitoring techniques, including radio-collars which may be damaging to the target species, or problems with baiting programs. A good example observed as part of this review is the study of Priddel and Wheeler (2004) which identified feral cats as the limiting factor in the re-establishment of *B. penicillata* into Yathong Nature Reserve in New South Wales, where fox baiting was prevalent and extremely effective in the control of local fox populations.

Although the success rate of translocations for potoroids was low, the importance of translocations as a conservation and management tool cannot be underestimated. In 1996 the IUCN status of *B. penicillata* changed from 'endangered' to 'lower risk (conservation dependent)' after its abundance increased markedly through successful fox control and translocations in Western Australia. However, more recently, both naturally occurring and translocated populations of this species in Western Australia have declined significantly (Wayne 2006), which has led to a review of their status. Preliminary studies are yet to comprehensively determine the causes of these declines (A. Wayne, *pers. comm.*). Such declines and the uncertain estimates for many of the translocated populations of *B. penicillata* reported here highlight the importance of long-term monitoring. Strict guidelines should be incorporated into translocation proposals for population monitoring and the collection of ecological and demographic data (e.g. changes in the population over a set time period, sex ratios, adult:juvenile ratios, recruitment) which may give an early indication of changes in success.

Our results suggest that translocations of species within the Potoroidae are more likely to succeed using captive-bred individuals, but this must be interpreted with caution as the number of captive-bred animals used as founders was far greater to islands and fenced populations whereas wild-caught founder stock was generally translocated to mainland sites. More comprehensive experimental reintroductions are required to determine the success rate of establishment of captive-bred or wild-caught founder animals. A captive-bred population may provide the opportunity to select specific sex ratios or individuals, which may optimise establishment once individuals have been translocated. Selection of appropriate animals from wild populations may be less effective because captures are more opportunistic, less selective and often more labour-intensive. Wild animals may also be more stressed by the capture, processing and relocation process than animals that have had closer contact with humans. The benefits of using captive populations also include predator avoidance training, which is becoming increasingly popular for translocations (Griffin *et al.* 2000) and for macropodoids in particular (Griffin *et al.* 2001; Blumstein *et al.* 2002). This technique has yet to be confirmed for species within the Potoroidae.

Our findings also suggest that a translocation is more likely to succeed when the target species is translocated into a feral-free area (fenced or island), compared to an area that is baited. As noted in Priddel and Wheeler (2004), baiting programs that are extremely effective for a specific target may not be effective for all introduced predators in a local area. In some circumstances, particularly in the arid zone, baiting programs established to target foxes and cats may eventually lead to an increase in the population of these species due to a reduction in beneficial non-target species such as dingoes (Johnson *et al.* 2007). It is apparent that, until a bait can be designed and developed to specifically target cats, future translocations of potoroids into parts of their former distributions, without the use of effective fencing, are likely to be doomed.

The results of this review highlight the importance of clear criteria for determining the success of a translocation. To determine the outcome of translocations, many criteria have been used. In a summary of translocations in Western Australia, Mawson (2004) described the range of criteria used to assess the outcomes. Of the 20 mammalian species described in his review, only 10 had clear guidelines for monitoring and defining translocation success. The criteria ranged considerably, from sighting rates of numbats (*Myrmecobius fasciatus*) of one to five per 100 km (Friend 1994), to average daily trap success of 7.5% for *B. penicillata* (Start *et al.* 1995), to breeding of F1 or F2 generations (Friend 2001; Orell and Dans 2002). The most widely recognised criterion for success is establishing a self-sustaining population (Morris *et al.* 2000). Given the poor success rate of all Australian fauna translocations and the results highlighted here, it seems imperative that national guidelines be established for all translocation programs. Failure to set such guidelines will merely continue Australia's poor success rate for translocation efforts. In this review we consider a successfully translocated population to have survived for over five years with numbers equal to or greater than those reintroduced. In the case of translocations that had run for less than five years, we suggest that if recruitment and breeding are apparent it could be considered a likely success, but long-term monitoring will be the only way to confirm ultimate success rates.

CONCLUSIONS

Translocation has been recommended for the protection and recovery efforts of other potoroids (Maxwell *et al.* 1996; Anon. 2002). It is imperative that results from previous recovery efforts be evaluated, to assist with these decisions. We propose a number of recommendations for the use of this effective management tool in subsequent translocation efforts of potoroids. First, translocation should be considered for all species of potoroids to increase the distribution and abundance of local populations and supplement existing populations. Supplementation is especially important when considering the limited genetic variation likely to be present in many of the current founder populations. Second, in some circumstances it may be appropriate to conduct trial translocations (Soderquist 1995; Soderquist and Serena 1995; Friend *et al.* 2005) to consider the impact of a larger-scale translocation, particularly where only a few animals are available for release. Third, we encourage the development of large-scale collaborative projects for all translocation efforts to enable fully resourced and effective management and research outcomes that should increase our understanding of the interaction between species and the environment. It is also imperative that relationships between agencies, both government and non-government, promote regular reporting of results and that there be continued monitoring of population dynamics to assist with management decisions.

ACKNOWLEDGMENTS

We thank Adrian Wayne, Jeff Short, Jason Van Weenen, Jacquie Richards, Tony Friend, Neil Thomas, Sandy Loffler and Joss Bentley for providing information on translocated populations of potoroids. We also thank two anonymous reviewers for their comments on an earlier draft of this chapter.

REFERENCES

Anon 2002. Approved recovery plan for the long-footed potoroo (*Potorous longipes*). NSW National Parks and Wildlife Service, Sydney.

Bellchambers K 2001. Brush-tailed bettong re-introduction and monitoring in the Flinders Ranges National Park and bushbird and emu monitoring in the Flinders Ranges and Gammon Ranges National Parks. Ardeotis Biological Consultants, Adelaide.

Blumstein DT, Daniel JC, Schnell MR, Ardron JG and Evans CS 2002. Antipredator behaviour of red-necked pademelons: a factor contributing to species survival? *Anim. Conserv.* **5:** 325–331.

Burbidge AA 1983. Burrowing bettong, *Bettongia lesueur.* In *The Complete Book of Australian Mammals.* (Ed. R Strahan) pp. 187–189. Angus and Robertson, Sydney.

Burbidge AA 1985. Fire and mammals in hummock grasslands of the arid zone. In *Fire Ecology and the Management of Ecosystems in Western Australia. Proceedings of the Fire Symposium.* (Ed. J Ford) pp. 91–94. WAIT Environmental Studies Group Report No. 14. Western Australian Institute of Technology, Perth.

Burbidge AA and Eisenberg JF 2006. Conservation and management. In *Marsupials.* (Eds PJ Armati, CR Dickman and ID Hume) pp. 299–330. Cambridge University Press, Cambridge.

Burbidge AA and Manly BFJ 2002. Mammal extinctions on Australian islands: causes and conservation implications. *J. Biogeogr.* **29:** 465–473.

Burbidge AA and McKenzie NL 1989. Patterns in the modern decline of Western Australia's vertebrate fauna: causes and conservation implications. *Biol. Conserv.* **50:** 143–198.

Burbidge AA, Johnson KA, Fuller PJ and Southgate RI 1988. Aboriginal knowledge of the mammals of the central deserts of Australia. *Aust. Wildl. Res.* **15:** 9–40.

Calaby JH 1971. The current status of Australian Macropodidae. *Aust. Zool.* **16:** 17–29.

Calaby JH and Grigg GC 1989. Changes in macropodoid communities and populations in the past 200 years, and the future. In *Kangaroos, Wallabies and Rat-kangaroos.* (Eds G Grigg, P Jarman and I Hume) pp. 813–820. Surrey Beatty and Sons, Sydney.

Cardillo M and Bromham L 2001. Body size and risk of extinction in Australian mammals. *Conserv. Biol.* **15:** 1435–1440.

Christensen PES 1977. The biology of *Bettongia pencillata* and *Macropus eugenii* in relation to fire. PhD thesis. University of Western Australia, Perth.

Christensen PES 1980. The biology of *Bettongia pencillata* (Gray, 1837) and *Macropus eugenii* (Desmarest, 1817), in relation to fire. *Forests Dept of WA Bull.* **91:** 1–90.

Christensen P and Burrows N 1995. Project Desert Dreaming: the reintroduction of mammals to the Gibson Desert, Western Australia. In *Reintroduction Biology of Australian and New Zealand Fauna.* (Ed. M Serena) pp. 199–207. Surrey Beatty and Sons, Sydney.

Copley PB 1995. Translocations of native vertebrates in South Australia: a review. In *Reintroduction Biology of Australian and New Zealand Fauna.* (Ed. M Serena) pp. 35–42. Surrey Beatty and Sons, Sydney.

Craig JL and Veitch CR 1990. Transfer of organisms to islands. In *Ecological Restoration of New Zealand Islands.* (Eds DR Towns, CH Daugherty and IAE Atkinson) pp. 255–260. NZ Dept of Conservation, Wellington.

Delroy LB, Earl J, Radbone I, Robinson AC and Hewett M 1986. The breeding and re-establishment of the brush-tailed bettong, *Bettongia penicillata*, in South Australia. *Aust. Wildl. Res.* **13:** 387–396.

de Tores PJ and Start AN 2008. Woylie *Bettongia penicillata* Gray, 1837. In *The Mammals of Australia.* 3rd edn. (Eds S Van Dyck and R Strahan) pp. 291–292. Reed New Holland, Sydney.

Dickman CR 1994a. Mammals of New South Wales: past, present and future. *Aust. Zool.* **29:** 158–165.

Dickman CR 1994b. Native mammals of western New South Wales: past neglect, future rehabilitation? In *Future of the Fauna of Western New South Wales.* (Eds D Lunney, S Hand, P Reed and D Butcher) pp. 81–91. Royal Zoological Society of NSW, Sydney.

Dickman CR 2005. Marsupials of the world: an introduction. In *Walker's Mammals of the World.* (Ed. RM Nowak) pp. 1–67. Johns Hopkins University Press, Baltimore.

Dickman CR, Pressey RL, Lim L and Parnaby HE 1993. Mammals of particular conservation concern in the Western Division of New South Wales. *Biol. Conserv.* **65:** 219–248.

Eldridge MDB 2006. Superfamily Macropodoidea. In *The Mammals of Australia.* 3rd edn. (Eds S Van Dyck and R Strahan) pp. 279–280. Reed New Holland, Sydney.

Fischer J and Lindenmayer DB 2000. An assessment of the published results of animal relocations. *Biol. Conserv.* **96:** 1–11.

Fisher DO, Blomberg SP and Owens IPF 2003. Extrinsic versus intrinsic factors in the decline and extinction of Australian marsupials. *Proc. Roy. Soc. Lond. B Biol. Sci.* **270:** 1801–1808.

Friend JA 1994. Recovery plan for the numbat (*Myrmecobius fasciatus*) 1995–2004. Western Australian Wildlife Management Program No. 18. Dept of Conservation and Land Management, Perth.

Friend JA 2001. Translocation proposal for dibbler (*Parantechinus apicalis*) from Perth Zoo to Peniup proposed Nature Reserve. Dept of Conservation and Land Management, Perth.

Friend JA 2006. Diets of native quokkas and introduced Gilbert's potoroos on Bald Island, Western Australia (abstract). *Newsl. Aust. Mammal Soc.* **October**: 64.

Friend JA, Hill SJE and Button TA 2005. Trial translocation of Gilbert's potoroo to Bald Island, Western Australia (abstract). *Newsl. Aust. Mammal Soc.* **October**: 20.

Garkaklis MJ, Bradley JS and Wooller RD 2000. Digging by vertebrates as an activity promoting the development of water-repellent patches in sub-surface soil. *J. Arid Environ.* **45:** 35–42.

Garkaklis MJ, Bradley JS and Wooller RD 2003. The relationship between animal foraging and nutrient rich patchiness in south-west Australian woodland soils. *Aust. J. Soil Res.* **41:** 665–673.

Garkaklis MJ, Bradley JS and Wooller RD 2004. Digging and soil turnover by a mycophagous marsupial. *J. Arid Environ.* **56:** 569–578.

Gibson DF, Lundie-Jenkins G, Langford DG, Cole JR, Clarke DE and Johnson KA 1994. Predation by feral cats, *Felis catus*, on the rufous hare-wallaby, *Lagorchestes hirsutus*, in the Tanami Desert. *Aust. Mammal.* **17:** 103–107.

Griffin AS, Blumstein DT and Evans CS 2000. Training captive-bred or translocated animals to avoid predators. *Conserv. Biol.* **14:** 1317–1326.

Griffin AS, Evans CS and Blumstein DT 2001. Learning specificity in acquired predator recognition. *Anim. Behav.* **62:** 577–589.

Griffith B, Scott IM, Carpenter JW and Reed C 1989. Translocation as a species conservation tool: status and strategy. *Science* **245:** 477–480.

Harris CR 1974. The national parks and reserves of South Australia. MA thesis. University of Adelaide, Australia.

Hilton-Taylor C 2000. 2000 IUCN red list of threatened species. IUCN, Gland, Switzerland.

Horsup A and Evans M 1993. Predation by feral cats *Felis catus* on an endangered marsupial, the bridled nailtail wallaby, *Onychogalea fraenata*. *Aust. Mammal.* **16:** 85–86.

Hrdina F 1997. Marsupial destruction in Queensland, 1877–1930. *Aust. Zool.* **30:** 272–286.

IUCN 2008. 2008 IUCN Red List of Threatened Species. www.iucnredlist.org.

James AI and Eldridge DJ 2007. Reintroduction of fossorial native mammals and potential impacts on ecosystem processes in an Australian desert landscape. *Biol. Conserv.* **138:** 351–359.

Jarman PJ and Johnson KA 1977. Exotic mammals, indigenous mammals, and land-use. *Proc. Ecol. Soc. Aust.* **10:** 146–166.

Johnson KA 1999. Recovery and discovery: where we have been and where we might go with species recovery. *Aust. Mammal.* **21:** 75–86.

Johnson KA and Roff AD 1982. The western quoll, *Dasyurus geoffroii* (Dasyuridae, Marsupialia) in the Northern Territory: historical records from venerable sources. In *Carnivorous Marsupials.* (Ed. M Archer) pp. 221–226. Royal Zoological Society of NSW, Sydney.

Johnson KA, Burbidge AA and McKenzie NL 1989. Australian Macropoidea: status, causes of decline and future management. In *Kangaroos, Wallabies and Rat-kangaroos.* (Eds G Grigg, P Jarman and I Hume) pp. 641–657. Surrey Beatty and Sons, Sydney.

Johnson CN, Isaac JL and Fisher DO 2007. Rarity of a top predator triggers continent-wide collapse of mammal prey: dingoes and marsupials in Australia. *Proc. Roy. Soc. Lond. B Biol. Sci.* **274**: 341–346.

Kaufmann JH 1974. The ecology and evolution of social organization in the kangaroo family (Macropodidae). *Amer. Zool.* **14:** 51–62.

Kimber R 1983. Black lightning: Aborigines and fire in central Australia and the western desert. *Archaeol. Ocean.* **18:** 35–48.

Kinnear JE, Sumner NR and Onus ML 2002. The red fox in Australia: an exotic predator turned biocontrol agent. *Biol. Conserv.* **108:** 335–359.

Kleiman DG 1989. Reintroduction of captive mammals for conservation. *Biosci.* **39:** 152–161.

Latz PK and Griffin GF 1978. Changes in Aboriginal land management in relation to fire and to food plants in central Australia. In *The Nutrition of Aborigines in Relation to the Ecosystem of Central Australia.* (Eds CBS Hetzel and HJ Frith) pp. 77–85. CSIRO Publishing, Melbourne.

Letnic M and Dickman CR 2006. Boom means bust: interactions between the El Nino/Southern Oscillation (ENSO), rainfall and the processes threatening mammal species in arid Australia. *Biodivers. Conserv.* **15:** 3847–3880.

Lunney D 2001. Causes of the extinction of native mammals of the Western Division of New South Wales: an ecological interpretation of the nineteenth century historical record. *Rangelands J.* **23:** 44–70.

Marlow BJ 1958. A survey of the marsupials of New South Wales. *CSIRO Wildl. Res.* **3:** 71–114.

Martin S, Ball S and Peeters P 2006. Reintroduction of the brush-tailed bettong (*Bettongia penicillata ogilbyi*) into Lincoln National Park: program review from September 1999 to July 2004. Unpublished internal report for Dept of Environment and Heritage, South Australia.

Mawson PR 2004. Translocations and fauna reconstruction sites: *Western Shield* review, February 2003. *Conserv. Sci. WA* **5:** 108–121.

Maxwell S, Burbidge AA and Morris K (eds) 1996. *1996 Action Plan for Australian Marsupials and Monotremes.* Wildlife Australia, Canberra.

McKenzie NL, Burbidge AA, Baynes A, Brereton RN, Dickman CR, Gordon G, Gibson LA, Menkhorst PW, Robinson AC, Williams MR and Woinarski JCZ 2007. Analysis of factors implicated in the recent decline of Australia's mammal fauna. *J. Biogeog.* **34:** 597–611.

Mitchell S and Wayne A 2008. Down but not out: solving the mystery of the woylie population crash. *Landscope* **25:** 10–15.

Morris K 2000. Fauna translocations in Western Australia 1971–1999: an overview. In *Biodiversity and the Re-introduction of Native Fauna at Uluru–Kata Tjuta National Park.* (Eds JS Gillen, R Hamilton, WA Low and C Creagh) pp. 64–74. Bureau of Rural Sciences, Canberra.

Morris K, Speldwinde P and Orell P 2000. Djoongari (Shark Bay Mouse) recovery plan. Wildlife Management Program No. 17. Dept of Conservation and Land Management, Perth.

Morton SR 1990. The impact of European settlement on the vertebrate animals of arid Australia: a conceptual model. *Proc. Ecol. Soc. Aust.* **16:** 201–213.

Murphy MT, Garkaklis MJ and Hardy GSJ 2005. Seed caching by woylies *Bettongia penicillata* can increase sandalwood *Santalum spicatum* regeneration in Western Australia. *Austral Ecol.* **30:** 747.

Nelson LS, Storr RF and Robinson AC 1992. Plan of management for the brush-tailed bettong, *Bettongia penicillata* (Marsupialia, Potoroidae) in South Australia. Dept of Environment and Planning, Adelaide.

Newsome AE 1975. An ecological comparison of the two arid-zone kangaroos of Australia, and their anomalous prosperity since the introduction of ruminant stock to their environment. *Q. Rev. Biol.* **50:** 389–424.

Orell P and Dans P 2002. Translocation proposal for black-flanked rock-wallaby (*Petrogale lateralis lateralis*) from Mount Caroline Nature Reserve and Querekin Rock to Walyunga National Park. Dept of Conservation and Land Management, Perth.

Priddel D and Wheeler R 2004. An experimental translocation of brush-tailed bettongs (*Bettongia penicillata*) to western New South Wales. *Wildl. Res.* **31:** 421–432.

Robertshaw JD and Harden RH 1989. Predation on Macropodoidea: a review. In *Kangaroos, Wallabies and Rat-kangaroos*. (Eds G Grigg, P Jarman and I Hume) pp. 735–753. Surrey Beatty and Sons, Sydney.

Robinson AC 1989. Island management in South Australia. In *Australian and New Zealand Islands: Nature Conservation Values and Management*. (Ed. AA Burbidge). Dept of Conservation and Land Management, Perth.

Rose RW 1986. The habitat, distribution and conservation status of the Tasmanian bettong, *Bettongia gaimardi* (Desmarest). *Aust. Wildl. Res.* **13:** 1–6.

Salo P, Korpimäki E, Banks PB, Nordström M and Dickman CR 2007. Alien predators are more dangerous than native predators to prey populations. *Proc. Roy. Soc. Lond. B Biol Sci.* **274:** 1237–1243.

Saunders DA 1994. The effects of habitat reduction and fragmentation on the mammals and birds of the Western Australian central wheatbelt: lessons for western New South Wales. In *Future of the Fauna of Western New South Wales*. (Eds D Lunney, S Hand, P Reed and D Butcher) pp. 99–105. Royal Zoological Society of NSW, Sydney.

Serena M (Ed.) 1995. *Reintroduction Biology of Australian and New Zealand Fauna*. Surrey Beatty and Sons, Sydney.

Serena M and Williams MR 1995. Wildlife conservation and reintroduction: an Australasian perspective. In *Reintroduction Biology of Australian and New Zealand Fauna*. (Ed. M Serena) pp. 247–252. Surrey Beatty and Sons, Sydney.

Short J 1998. The extinction of rat-kangaroos (Marsupialia: Potoroidae) in New South Wales, Australia. *Biol. Conserv.* **86:** 365–377.

Short J and Smith A 1994. Mammal decline and recovery in Australia. *J. Mammal.* **75:** 288–297.

Short J, Bradshaw SD, Giles J, Prince RIT and Wilson GR 1992. Reintroduction of macropods (Marsupialia: Macropodoidea) in Australia: a review. *Biol. Conserv.* **62:** 189–204.

Short J, Kinnear JE and Robley A 2002. Surplus killing by introduced predators in Australia: evidence for ineffective anti-predator adaptations in native prey species? *Biol. Conserv.* **103:** 283–301.

Soderquist TR 1995. The importance of hypothesis testing in reintroduction biology: examples from the reintroduction of the carnivorous marsupial *Phascogale tapoatafa*. In *Reintroduction Biology of Australian and New Zealand Fauna*. (Ed. M Serena) pp. 159–164. Surrey Beatty and Sons, Sydney.

Soderquist TR and Serena M 1995. An experimental reintroduction programme for brush-tailed phascogales (*Phascogale tapoatafa*): the interface between captivity and the wild. In *Creative Conservation: Interactive Management of Wild and Captive Animals*. (Eds PJS Olney, GM Mace and ATS Feister) pp. 431–438. Chapman and Hall, London.

Start T, Burbidge A and Armstrong D 1995. Woylie recovery plan. Dept of Conservation and Land Management, Perth.

Thomas JA 1989. Ecological lessons from the reintroduction of Lepidoptera. *The Entomol.* **108:** 56–68.

Thomas N, Marlow N and Friend T 2005. Reintroduction of the boodie (*Bettongia lesueur*) into Dryandra woodland (abstract). *Newsl. Aust. Mammal Soc.* **October**: 35.

vanEmden HF 1989. The ecology of insect introductions and encouragement of rare species: introductions. *The Entomol.* **108:** 3–4.

Wayne A 2006. The post-recovery collapse of the woylie: diagnosing their recent declines in south-western Australia (abstract). *Newsl. Aust. Mammal Soc.* **October**: 59.

Whalley P 1989. Principles and outcome of introductions. *The Entomol.* **108:** 69–76.

Wood Jones F 1924. *The Mammals of South Australia. I. The Monotremes and the Carnivorous Marsupials. II. The Bandicoots and the Herbivorous Marsupials*. Government Printer, Adelaide.

23 Cross-fostering and short-term pouch young isolation in macropodoid marsupials: implications for conservation and species management

D.A. Taggart, D.J. Schultz, T.P. Fletcher, J.A. Friend, I.G. Smith, W.G. Breed and P.D. Temple-Smith

SUMMARY

Cross-fostering is the rearing of young by a surrogate mother of a different taxon. In marsupials this technique has been used to study lactation as well as pouch young growth and development. Recently, cross-fostering and short-term pouch young isolation have been used by wildlife agencies and zoos to increase female reproductive rates and fecundity, and to manage the genetics of captive and wild animals as part of wider conservation and management strategies for macropodoid marsupials. Data on cross-fostering are now available for six potoroid and 13 macropodid species. Success of cross-fostering studies between donor and foster species varies widely from complete failure to an eight-fold increase in the production of pouch young annually. Factors implicated in successful cross-fostering include relative size of donor and surrogate females, size of pouch young at weaning, differences in length of pouch life between species and size differences between donor young and those of the surrogate species at transfer. Studies have shown that females regulate milk composition and production irrespective of pouch young age, and that transfer of donor young to species with more immature or advanced mammary glands will result in a slowing or an acceleration of pouch young growth and development, and affect the duration of pouch life. Small pouch young tolerate short-term isolation from the pouch at a range of temperatures, provided high humidity is maintained throughout the period of isolation. Maintenance of pouch young at temperatures lower than those that occur in the pouch (23°C compared with 37°C) during isolation reduces the pouch young's basal metabolic rate, oxygen consumption and evaporative water loss and thus improves survival rates of very small pouch young. The success of these techniques in managing population genetics and accelerating breeding in donor species within the Macropodoidea are enhanced by post-partum oestrus and mating after the removal of pouch young, and the reactivation and birth of the diapausing embryo. This chapter reviews the available data on cross-fostering and pouch young isolation in macropodoid marsupials, and examines use of these techniques in the conservation and management of species in this group.

INTRODUCTION

The development and application of assisted reproductive technology in marsupials is a comparatively new field and has only recently been applied to assist in conservation of threatened species (Taggart *et al.* 2005; Schultz *et al.* 2006; Taggart and Smith, unpubl. data; Taggart and Friend, unpubl. data). Although a variety of reproductive technologies are being examined, the techniques which offer the most immediate benefit to many marsupial species are those of pouch young isolation and cross-fostering. Cross-fostering is defined as the rearing of young by a surrogate mother of a different taxon. In marsupials, this technique has the distinct advantage of being used as early as day 1 of pouch life when the pouch young weighs <1 g. When applied to marsupial conservation, the aim is to circumvent the period of lactational anoestrus or lactational quiescence from the natural mother's reproductive cycle, thus permitting her to undergo a series of pregnancies with a minimum of intervening lactation and thereby enhance her reproductive rate and fecundity. Cross-fostering was initially proposed and developed approximately 40 years ago by Merchant and Sharman (1966). At that time they stressed the importance of cross-fostering for the conservation of threatened marsupial species, but despite their early successes in a variety of macropodids (Merchant and Sharman 1966), few studies (Johnson 1981) were published on this topic until the late 1990s (Table 23.1). There is still little information available on non-macropodoid species (Caton 2005; Finlayson *et al.* 2008).

Short-term pouch young isolation is a technique developed in marsupials to facilitate the collection of milk samples from mothers with pouch young, surgery on pouch young, or transport of young (Merchant and Sharman 1966; Shield 1966; Messer and Green 1979; Rose *et al.* 1998; Taggart *et al.* 2002). Recently this procedure has been used in association with cross-fostering for transporting small pouch young between donor and surrogate mothers. For the purposes of this discussion, it refers to the isolation of small pouch young from the teat for up to 12 hours.

A controversial aspect of cross-fostering in macropodoids is the selection of surrogate taxa that will ensure the most successful outcome. Currently, there are no simple and effective guidelines or criteria to direct the selection of the most appropriate surrogate taxon for cross-fostering, or to help predict the optimum time for pouch young transfer to maximise both survival of young and female reproductive rate. This chapter reviews the available data from the Macropodoidea, provides technical recommendations to enhance the likelihood of cross-fostering and pouch young isolation success and examines the potential application of these techniques as tools for conservation and management in this group.

METHODS

Cross-fostering

Selecting a donor species

In order to achieve accelerated breeding, and thus obtain the greatest conservation benefit from a cross-fostering program, the donor species (the species that is to be fostered) must display polyoestry and ideally cycle for a considerable portion of the year, if not year-round. Removal of the donor pouch young, and therefore the sucking stimulus, ends the mother's period of lactational anoestrus or lactational quiescence (Tyndale-Biscoe *et al.* 1974; Tyndale-Biscoe 1979). In most species with embryonic diapause, this acts to reinitiate development of the diapausing embryo and a young is usually born within 20–40 days (Tyndale-Biscoe *et al.* 1974; Tyndale-Biscoe 1979; Tyndale-Biscoe and Renfree 1987). Even in those species without diapause, pouch young removal provides an opportunity for the female to cycle, mate and give birth again within the same season (Finlayson *et al.* 2007, 2008).

Knowledge of the social organisation of the donor species (solitary or gregarious), breeding requirements and behaviour (space, den requirements, cover, diet etc.) and reproductive biology (time between mating and birth, length of lactation, age at sexual maturity) are relevant in maximising reproductive success and influencing the determination of the most appropriate surrogate species for the rearing of donor young.

Selecting a potential surrogate species

The following factors appear to be important in selecting a potential surrogate species for maximising the likelihood of a successful cross-fostering program and maximising breeding and fostering

Table 23.1: Summary of cross-fostering and pouch young isolation studies in the Macropodoidae.

Cross-fostering **Macropodids (M)/Potoroids (P)**	**Pouch young isolation** **Macropodids (M)/Potoroids (P)**
Published articles	**Published articles**
Merchant and Sharman 1966 – M	Bolliger and Pascoe 1953 – M
Johnson 1981 – M	Renfree and Tyndale-Biscoe 1978 – M
Bell and Close 1994 – M (abstract)	Green *et al.* 1980 – M
Smith 1998 – P	Crowley *et al.* 1988 – P
Taggart 2002 – M	Merchant *et al.* 1996 – M
Jones *et al.* 2004 – M	Rose *et al.* 1998 – P
Taggart *et al.* 2005 – M	Taggart *et al.* 2002 – M
Waite *et al.* 2005 – M	
Sterneberg 2005 – P	**Honours theses**
Schultz *et al.* 2006 – M	Collins 2002 – P
Menzies *et al.* 2007 – M	
Honours theses	
Collins 2002 – P	
Menzies 2003 – M	
O'Brien 2004 – M	

opportunities (Merchant and Sharman 1966; Johnson 1981; Jones *et al.* 2004; Taggart *et al.* 2005; Schultz *et al.* 2006; Menzies *et al.* 2007). Most criteria are self-explanatory:

1 quiet and tractable surrogate mothers of (roughly) similar size to the donors (Merchant and Sharman 1966; Johnson 1981; Jones *et al.* 2004; Taggart *et al.* 2005; Schultz *et al.* 2006; Menzies *et al.* 2007);
2 similar pouch life duration in donor and surrogate species (Merchant and Sharman 1966; Johnson 1981; Jones *et al.* 2004; Taggart *et al.* 2005; Schultz *et al.* 2006; Menzies *et al.* 2007);
3 no major mismatch in milk contents or the time each young receives critical milk components (Menzies *et al.* 2007);
4 a close taxonomic relationship between donor and foster species;
5 donor and foster species do not naturally occur within the same region/environment;
6 foster species are gregarious, easy to maintain and breed reliably in captivity (Merchant and Sharman 1966; Taggart *et al.* 2005).

Criterion 4 has been included as detailed studies examining the success of cross-fostering between distantly related macropodoids have yet to be undertaken. Preliminary trials (Taggart, unpubl. data, n = 2) involving transfers of long-nosed potoroo (*Potorous tridactylus*) pouch young to tammar wallaby (*Macropus eugenii*) surrogate mothers were unsuccessful. It is therefore recommended that, until more detailed studies have been undertaken and there is available data to suggest otherwise, a close taxonomic relationship between donor and surrogate species will result in the greatest likelihood of cross-fostering success.

When selecting a potential surrogate species it is not recommended that a species be chosen if it occurs in the same geographical region as the donor species (criterion 5), as this may lead to problems with species recognition upon reintroduction of foster-reared animals. Although there is no evidence to support or refute this idea, it would be wise to avoid any factor that may negatively affect the breeding success of foster-reared animals.

Pouch young transfer procedure

Pouch young removal

Hands should be cleaned in warm water without soap before handling pouch young and kept moist during fostering to prevent pouch young sticking to hands when being handled (Taggart *et al.* 2005; Schultz *et al.* 2006). Hand creams or any other substance that will leave hands perfumed or oily prior to handling pouch young should be avoided.

To remove the pouch young in preparation for transfer, the thumb and forefinger of one hand are used to grasp the teat while those of the other hand are positioned around the front of the young's mouth to prevent the head moving forward. The teat is then gently and slowly removed from the mouth using bouts of intermittent pressure and relaxation. This procedure places minimal pressure on the head and

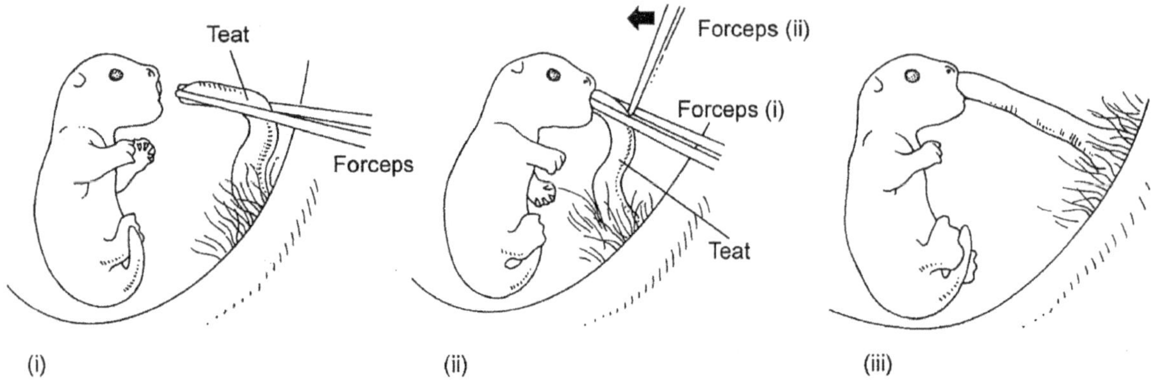

Figure 23.1: Attachment of a wallaby pouch young to a teat using fine blunt-nosed forceps. Note: Forceps grasp the end of the teat and the teat is inserted into the mouth of the young (i and ii). The teat is then held in place using a single blade of a second pair of forceps (ii) while the first pair of forceps are removed (ii and iii). The young is then returned to the pouch (iii). From Taggart *et al.* 2005, with permission.

mouth of the pouch young. Rapid removal of the teat from the young's mouth can result in damage to the buccal cavity and tearing of the pouch young's tongue. It is essential to confirm the presence and size of the pouch young on the surrogate mother prior to removal of the donor pouch young (Taggart *et al.* 2005; Schultz *et al.* 2006).

Pouch young reattachment

A physical examination of the surrogate mother is recommended before transferring any donor young. Particular attention should be paid to the oral cavity (to exclude the presence of disease) and pouch of the potential surrogate mother. A healthy tooth–gum margin and a clean moist pouch with no apparent odour are essential to maximise the chances of success (Schultz *et al.* 2006). Different methods of pouch young reattachment have been used, including the simple placing of the fostered young into the surrogate mother's pouch and allowing the young to attach to the teat unaided (Smith 1998; Collins 2002). The end of a match stick (Merchant and Sharman 1966), fine blunt-nosed forceps (Taggart 2002; Figs 23.1, 23.2) and the flexible plastic sheath of a hypodermic needle (Sterneberg 2005) have been used to assist attachment. The latter three aids help the operator to position the teat inside the foster young's mouth (Merchant and Sharman 1966; Taggart 2002; Taggart *et al.* 2005; Sterneberg 2005).

If young are placed within the surrogate mother's pouch and not physically attached to a teat, anaesthesia of the surrogate mother is not necessary (Smith 1998; Taggart Collins and Breed, unpubl. data; Taggart and Friend, unpubl. data). Although this method of teat attachment is not recommended for members of the Macropodidae, it has been successful in potoroos (long-nosed potoroo; Gilbert's potoroo, *Potorous gilbertii*) and bettongs (northern bettong, *Bettongia tropica*; brush-tailed bettong, *Bettongia penicillata*) using pouch young much less than 21 days old. Individuals of these species have been found to readily attach to the teat of their own mothers (Smith 1998; Taggart, Collins and Breed, unpubl. data; Taggart and Friend, unpubl. data) or surrogate mothers (Smith 1998; Taggart, Collins and Breed, unpubl. data). This may be due to the early development of the lateral margins of the lips in this group (Hill and Hill 1955; Sharman 1962; Hughes and Hall 1988).

For direct physical attachment of small pouch young to the teat in any species, chemical restraint (sedation) is required (Merchant and Sharman 1966; Taggart 2002; Sterneberg 2005; Taggart *et al.* 2005). This ensures the surrogate mother is immobile while physical attachment of the young is carried out.

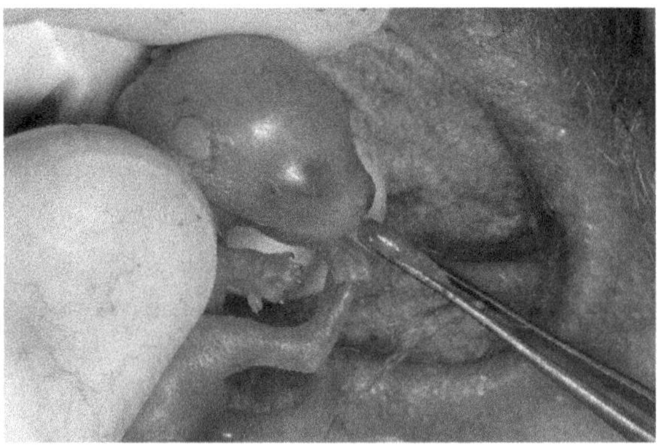

Figure 23.2: Cross fostering a brush-tailed rock-wallaby pouch young (~8 grams) to the teat of a tammar wallaby surrogate mother. Note the very narrow neck of the forceps and the position of the tip of the forceps in relation to the end of the teat.

Attachment is achieved by placing the tip of the teat at the opening of the mouth of the pouch young and gently pushing the teat into the young's mouth (Figs 23.1, 23.2; Merchant and Sharman 1966; Johnson 1981; Sterneberg 2005; Taggart 2002; Taggart *et al.* 2005). The use of fine blunt-nosed forceps in directing the teat into the mouth has proved useful (Taggart 2002; Taggart *et al.* 2005).

Successful attachment of the young to the surrogate mother's teat results in immediate suckling and the procedure takes very little time to perform (Merchant and Sharman 1966; Taggart *et al.* 2005). Moistening the hands prevents the young from sticking to the operator and facilitates the procedure (Schultz *et al.* 2006). The surrogate mother is kept anaesthetised for 15–20 minutes after the transfer and the foster young should be periodically checked to ensure that it remains attached to the teat. If it detaches, the procedure is repeated (Taggart *et al.* 2005; Schultz *et al.* 2006). Checking teat attachment should be rapid and with limited artificial light. Shining strong artificial light directly into the pouch and for prolonged periods is not recommended as it results in restlessness (squirming) of the pouch young and increased movement of the front legs, which may lead to disengagement from the teat. Likewise, the raising and separation of the outer pouch wall also causes squirming and may result in dislodgment (Taggart *et al.* 2005; Schultz *et al.* 2006). When pouch young have been separated from the teat for more than two hours, two or three drops of Hartmann's Solution (Baxter Healthcare Pty Ltd, Toongabbie, NSW) placed at the entrance to the young's mouth aids in reducing dehydration (Taggart and Schultz, unpubl. data). If pouch young have cooled prior to reattachment, their core temperature can be raised slowly by placing the young into the pouch and initially elevating the ventral pouch wall so that the young has contact with only one side of the pouch. After successful attachment of the young to the surrogate mother's teat has been confirmed, the surrogate mother can be returned to her enclosure and the fostered pouch young allowed to grow and develop in the pouch (Fig. 23.3) with as little interference as possible from researchers and keeping staff.

Figure 23.3: Cross-fostered brush-tailed rock-wallaby in the pouch of its yellow-footed rock-wallaby surrogate mother.

Short-term (<10 h) isolation of very small (1–40 g) pouch young

Preparing pouch young for short-term isolation

Olive oil (Johnson 1981) and humilac (an oil-free humectant; Taggart *et al.* 2005) have been used to help maintain skin moisture of unfurred pouch young following removal from their mother's pouch. In mild weather or over short periods, this is not necessary. Young are then generally wrapped in soft, smooth, moistened porous cloth such as linen, medical cotton swabs and/or surgical scarf material and transferred to an incubator for the period of isolation.

Incubator characteristics

The type of incubator used to maintain isolated, unfurred macropodoid pouch young depends on the purpose of the study, the desired length of isolation and the age of the young (Loh and Shield 1977; Rose *et al.* 1998; Taggart *et al.* 2002). The function of an incubator is to maintain young in an environment with a stable temperature and humidity. Laboratory studies that examined the onset of homeothermy typically described the unfurred pouch young, wrapped as outlined above, and placed within containers suspended within a water bath (Messer and

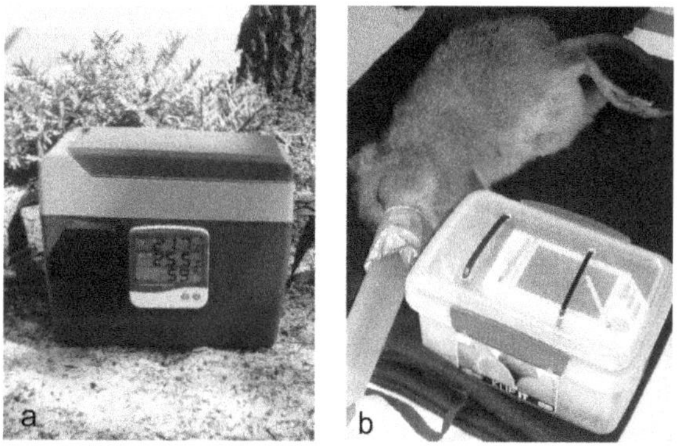

Figure 23.4: Two different types of incubator used for transporting pouch young. (a) Battery powered incubator for cooling and floor heating; (b) simple incubator design relying on the operator to regulate heating and cooling. Anesthetised long-nosed potoroo in background.

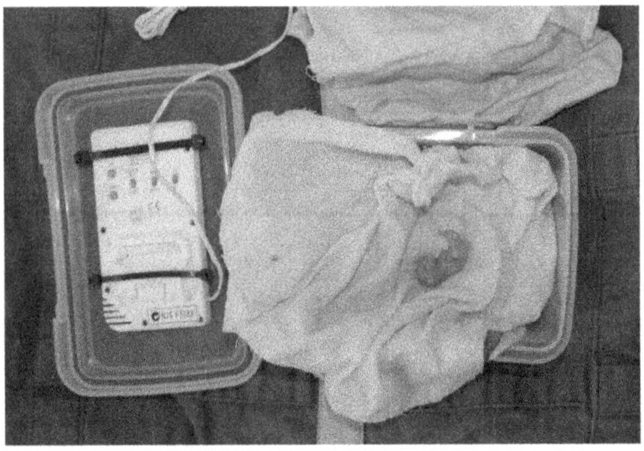

Figure 23.5: A 2 gram brush-tailed rock wallaby pouch young in a simple incubator following its transport from Canberra to Adelaide for cross-fostering to a yellow-footed rock wallaby foster mother.

Green 1979; Green *et al.* 1980, 1988; Gemmel and Johnson 1985).

Incubators used for the transport of pouch young from one site to another should be light-weight and portable (Figs 23.4, 23.5) and, if powered, run on a 12 V DC system (Taggart and Schultz, unpubl. data). In the authors' experience, power is generally only required in hot conditions for maintenance of a cooler environment for the pouch young. Careful planning of pouch young movements can avoid these conditions. If artificially heated, incubators should be floor-heated. There should not be any convection of warm air, as this dries the skin of the young and, in severe cases, causes death (Taggart and Schultz, unpubl. data). Small (17 × 12 × 9 cm) portable incubators (Fig. 23.4b) can be warmed very effectively using the body heat of the person involved in the isolation (held under clothing against the skin) and heat loss can be further minimised by placing the incubator in a soft insulated cooler.

Supplementary feeding/fluids

Supplementary feeding of pouch young is not required or advised if young are isolated from the pouch for ≤10 hours. Depending upon the size of the pouch young and the temperature and humidity of the environment inside the incubator or holding container, there may be a need to supplement the young with fluids (Taggart *et al.* 2002). Larger unfurred young, or those maintained in warmer conditions, are more likely to require a fluid supplement. A drop of Hartmann's Solution placed on the lips will generally be accepted without any side effects (Taggart and Schultz, unpubl. data).

OBSERVATIONS/DISCUSSION

Cross-fostering

Survival of cross-fostered young

Cross-fostering studies have been carried out with six potoroid and 13 macropodid species, with a total of 30 different crosses trialled (six potoroid and 21 macropodid; Tables 23.2 and 23.3). Survival of cross-fostered pouch young, to either the end of the experiment or permanent pouch exit, varied from 0–100% between species pairings. Most studies examined fostering within members of the families Macropodidae or Potoroidae; four of nine crosses in the Potoroidae and 14 of 21 crosses in the Macropodidae resulted in survival rates of cross-fostered young in excess of 75%. One study, however, performed crosses between a potoroid (long-nosed potoroo) and macropodid species (tammar wallaby); neither (n = 2) was successful (Taggart, unpubl. data). Failure of these pouch young to survive may have been due to milk incompatibilities or to problems with teat size or pouch size, or all of them. Further studies are required to explore the limits of this technique across families.

Differences in cross-fostering success rates between various interspecific crosses and different studies are hard to interpret. Differences between studies in the age and size of the donor young at fostering, the age and size of the donor and foster mother's own young at fostering, the length of time cross-fostered young were allowed to remain on the teat of their surrogate mother and sample sizes (Table 23.1) make comparisons difficult. Despite these challenges, the studies clearly show that young from one macropodoid species can be successfully reared to independence by the female of another and, if an appropriate foster species is available, the survival rates of fostered young can equal or better (Smith 1998; Schultz *et al.* 2006) those of young reared on their natural mother (Tables 23.2 and 23.3).

Timing/age of donor young at transfer

In marsupials, pouch young from day 1 of pouch life (<1 g) to 80% of the period of lactation, when the pouch young had already developed some body hair (Merchant and Sharman 1966; Taggart *et al.* 2005), have been successfully cross-fostered. Although most studies have not shown a period during lactation when cross-fostering is more likely to succeed (Merchant and Sharman 1966) there do appear to be some advantages, especially in endangered species, in transferring young early in pouch life. Unless the

Table 23.2: Cross-fostering studies of potoroid pouch young detailing survival rates of surrogate-reared young to the end of experimentation or permanent pouch exit.

Donor	Surrogate	n	% survival	Reference
Po. tridactylus	*B. penicillata*	7	43	Collins 2002
Po. tridactylus	*B. lesueur*	2	100	Taggart, unpubl. data
Po. tridactylus	*M. eugenii*	2	0	Taggart, unpubl. data
Po. tridactylus	*B. gaimardi*	7	86	Sterneberg 2005
Po. gilbertii	*Po. tridactylus*	5	60	Taggart, Hill and Friend, unpubl. data
B. gaimardi	*Po. tridactylus*	7	29	Sterneberg 2005
B. penicillata	*Po. tridactylus*	10	40	Collins 2002
B. lesueur	*Po. tridactylus*	2	100	Taggart, unpubl. data
B. tropica	*B. penicillata*	11	82	Smith 1998

researcher is very experienced, the use of pouch young <1.5 g for cross-fostering is not recommended because of physical constraints associated with handling and reattaching tiny young. Likewise, leaving young on their mothers until late in pouch life is not ideal as this lengthens the period of lactation of the donor female, increases the birth interval and reduces the donor female's reproductive rate and production of additional pouch young. Interestingly, a recent study of growth and survival of tammar wallaby

Table 23.3: Cross-fostering studies of macropodid pouch young detailing survival rates of surrogate reared young to the end of experimentation or permanent pouch exit.

Donor	Surrogate	n	% survival	Reference
M. eugenii (KI)	*M. rufus*	2	0	Merchant and Sharman 1966
M. eugenii (KI)	*T. billardierii*	6	20	O'Brien 2004
M. eugenii (KI)	*M. parma*	4	100	Menzies *et al.* 2007
M. eugenii (ML)	*M. eugenii* (KI)	87	78	Taggart and Smith, unpubl. data
M. rufus	*W. bicolor*	1	100	Merchant and Sharman 1966
M. rufus	*M. giganteus*	1	100	Merchant and Sharman 1966
M. parryi	*M. agilis*	9	77	Johnson 1981
M. parma	*M. eugenii* (KI)	8	25	Menzies *et al.* 2007
M. giganteus	*M. rufus*	4	100	Merchant and Sharman 1966
M. rufogriseus	*M. rufus*	1	0	Merchant and Sharman 1966
T. billardierii	*M. eugenii* (KI)	8	66	O'Brien 2004
L. conspicillatus	*Pe. inornata*	2	100	Johnson 1981
O. fraenata	*Pe. inornata*	1	100	Johnson 1981
W. bicolor	*M. rufus*	5	80	Merchant and Sharman 1966
W. bicolor	*M. agilis*	2	100	Johnson 1981
Pe. penicillata	*M. eugenii* (KI)	15+	83	Taggart *et al.* 2005; Schultz *et al.* 2006
Pe. penicillata	*Pe. xanthopus*	20+	82	Taggart *et al.* 2005; Schultz *et al.* 2006
Pe. xanthopus	*M. eugenii* (KI)	14	71	Schultz and Taggart, unpubl. data
Pe. xanthopus	*M. rufus*	2	0	Merchant and Sharman 1966
Pe. lateralis	*M. eugenii* (KI)	2	100	Jones *et al.* 2004
Pe. lateralis	*Pe. xanthopus*	15	87	Taggart and Smith, unpubl. data

ML = mainland; KI = Kangaroo Island.

young cross-fostered to parma wallabies (*Macropus parma*) reported an increased weaning success when young were transferred after day 30 of pouch life compared with those transferred before day 15 (Menzies *et al.* 2007).

When cross-fostering of brush-tailed rock-wallaby (*Petrogale penicillata*) pouch young was delayed until young were between day 40 and day 70 of age there was a significant mortality of donor pouch young 'prior to transfer'. No pouch young mortality was observed in donor females, however, when young were cross-fostered between eight and 20 days of age (Schultz *et al.* 2006). Schultz *et al.* (2006) showed that this pouch young mortality prior to transfer resulted from a behavioural abnormality in some mothers (ejection of the pouch young) that was not linked to cross-fostering. While the cause of pouch young losses could not be identified and it was not known if these losses were accidental or caused by hidden stressors influencing maternal behaviour, the timing of losses appeared to coincide with an increase in the activity of the young and/or in the mother's awareness of the presence of its young (Schultz *et al.* 2006). It is still unknown why the foster mothers were less accident-prone or less sensitive to the increased movement of the pouch young. A suggested starting-point for cross-fostering of pouch young from an untested species is for the donor young to be between days 10–20 post-partum, or 2–10 g in weight (Merchant and Sharman 1966; Taggart and Schultz, pers. obs.).

Age and weight of pouch young of donor and surrogate mothers: effects on survival of cross-fostered young

Most cross-fostering studies suggest that synchrony in the age of the pouch young transferred and that of the young removed is an important factor influencing the success of fostering. However, the data from fostering studies do not fully support this. Merchant and Sharman (1966) showed that six of 10 fostered young had retarded growth or died when the fostered young was older than the young removed from the surrogate mother. This contrasts with the findings of Johnson (1981), in whose study five pouch young transferred to surrogate mothers with younger pouch young survived and developed normally. Likewise, cross-fostering studies using long-nosed potoroos and Tasmanian bettongs (*Bettongia gaimardi*) (Sterneberg 2005) showed that age differences at transfer did not appear to influence mortality rates, although sample size in this study was small.

These studies highlight the age difference between the young being fostered but give no clear guidance on the effects of weight differences (Merchant and Sharman 1966; Johnson 1981; Smith 1998) between the donor and foster mother's own young. In studies of brush-tailed rock-wallabies, 13 of 16 pouch young that were 0–50% lighter at the time of transfer than the young they were replacing, survived and grew to independence, whereas two pouch young 5% and 68% heavier than the young of the surrogate species did not survive (Taggart *et al.* 2005). The differences in weight between the young of donor and surrogate mothers relate directly to the volume and perhaps composition of milk the young has been consuming and thus may be an indicator of the suitability of potential surrogate mothers as hosts for foster young. Cross-fostering is more likely to be successful if the pouch young being transferred is of similar or slightly smaller size/weight than the young of the foster species, thus increasing the likelihood that sufficient milk of appropriate composition is available for the fostered pouch young (Taggart *et al.* 2005). Studies on other macropodid species (Johnson 1981; Bell and Close 1994) suggest that this difference may become less critical when the transferred young are older/larger (>40 g). Ideally, preference should be given to fostering donor young that are similar in size or just slightly larger than young from a surrogate mother. Cross-fostering studies in black-footed rock-wallabies (*Petrogale lateralis*) used donor young weighing between 20% less and 10% more than the surrogate mother's own pouch young. Success rates to weaning in this study were 87% of young transferred (Taggart and Smith, unpubl. data; Table 23.3). Age and weight of pouch young of selected surrogate mothers are, however, not the only factors to consider when selecting an appropriate surrogate female.

Effects of fostering on growth and development

Growth rate, development and pouch life

Various effects of cross-fostering on growth and length of pouch life of transfer pouch young have been reported, from slowed growth and extended pouch life to accelerated growth and shortened pouch life (Table 23.4). In general, the data show that the growth rate and pouch life of foster-reared young approached that of their surrogate species after transfer, although there were exceptions. For example, although the pouch life of young agile wallabies (*Macropus agilis*) is shorter than that of whiptail wallabies (*Macropus*

Table 23.4: Growth and pouch life of macropodoid young on mothers and surrogate mothers.

Donor species	Surrogate species	Pouch life (donor)	Pouch life (surrogate)	No. of surrogate young (n)	Pouch life (surrogate young)	Effect of fostering on pouch life of donor	Effect of fostering on growth/ development of donor	Key reference
M. eugenii	*T. billardierii*	250–260	200	6	-	-	Accelerated	O'Brien 2004
T. billardierii	*M. eugenii*	200	250–260	8	-	-	Slow	O'Brien 2004
M. parryi	*M. agilis*	259–281	207–237	5	266–381	Normal	Normal	Johnson 1981
M. parryi	*M. agilis*	259–281	207–237	2	-	Increased	Slowed	Johnson 1981
M. eugenii	*M. pama*	250–260	212	4	-	-	Normal	Menzies *et al.* 2007
M. parma	*M. eugenii*	212	250–260	10	270	Increased	Slowed	Menzies *et al.* 2007
M. rufus	*M. rufogriseus*	235 (229–253)	270	1	-	-	Normal	Merchant and Sharman 1966
M. eugenii	*M. rufus*	250–260	235 (229–253)	1	~178	Decreased	Normal	Merchant and Sharman 1966
M. giganteus	*M. rufus*	319 (283–330)	235 (229–253)	1	-	-	Normal	Merchant and Sharman 1966
M. giganteus	*M. rufus*	319 (283–330)	235 (229–253)	1	374	Increased	Normal	Merchant and Sharman 1966
Pe. lateralis	*M. eugenii*	180–200	250–260	2	-	-	Slow	Jones *et al.* 2004
W. bicolor	*M. agilis*	256 (210–260)	207–237	2	~197 and ~204	Acc	?	Johnson 1981
W. bicolor	*M. rufus*	256 (210–260)	235 (229–253)	2	267	Normal	Accelerated	Merchant and Sharman 1966
Pe. xanthopus	*M. rufus*	190–200	235 (229–253)	1	~181	Normal	Normal	Merchant and Sharman 1966
Pe. penicillata	*Pe. xanthopus*	190–230	190–200		180–210	Normal	Accelerated	Taggart *et al.* 2005
Pe. penicillata	*M. eugenii*	190–230	250–260		255	Increased	Slow	Taggart *et al* 2005
L. conspicillatus	*Pe. inornata*	140–162	204	2	205 and 234	Increased	Normal	Johnson 1981
Po. tridactylus	*B. gaimardi*	130	109	7	-	-	Accelerated	Sterneberg 2005
B. gaimardi	*Po. tridactylus*	109	130	7	-	-	Slow	Sterneberg 2005
B. tropica	*B. penicillata*	100–109	90–100	9	104	Normal	Normal	Smith 1998

parryi), the duration of pouch life of whiptail wallaby young increased when they were cross-fostered to agile wallabies (Johnson 1981). Similarly, the growth rate of an eastern grey kangaroo (*Macropus giganteus*) young transferred to a red kangaroo (*Macropus rufus*) was normal, although pouch life in red kangaroos is some 70–80 days shorter than that of eastern grey kangaroos (Merchant and Sharman 1966).

Retarded growth has been noted in some species following cross-fostering. For example, two of seven young whiptail wallabies cross-fostered to agile wallabies had extended pouch lives, yet all survived to weaning (Johnson 1981). Spectacled hare-wallabies (*Lagorchestes conspicillatus*) transferred to unadorned rock-wallabies (*Petrogale inornata*) grew at normal rates but remained in the pouch for significantly longer periods (n = 2; Johnson 1981). In 2004, O'Brien reported that all eight Tasmanian pademelon (*Thylogale billardierii*) pouch young cross-fostered to tammar wallaby surrogates showed retarded growth, including head length, weight and pes length, compared to controls. Ten parma wallaby young transferred to tammar wallabies (Menzies *et al.* 2007) showed a similar pattern of slowed growth. Developmental milestones including eyes opening, eruption of vibrissae and pigmentation of nails and skin also occurred later in lactation in both cross-fostering studies (O'Brien 2004; Menzies *et al.* 2007). In contrast, swamp wallaby (*Wallabia bicolor*) pouch young transferred to red kangaroos (Merchant and Sharman 1966), and brush-tailed rock-wallaby young transferred to yellow-footed rock-wallabies (*Petrogale xanthopus*) grew at accelerated rates (head length and weight) (Taggart *et al.* 2005). In a study of cross-fostering between long-nosed potoroos and Tasmanian bettongs, undernourishment due to inappropriate milk quality or quantity, or both, led to delayed growth and development while transfers of pouch young to surrogate mothers with more advanced milk generally enhanced growth and development (Sterneberg 2005).

Time to weaning, milk composition/timing of milk components

In some species the time of weaning of foster young was similar to that of the surrogate species' natural young, particularly when young of species with short pouch lives were transferred into the pouches of a foster species that normally carried young for longer periods. Young of the brush-tailed rock-wallaby, for example, were weaned significantly later when reared on tammar wallaby surrogate mothers than when reared in the pouches of donor mothers or yellow-footed rock-wallaby surrogate mothers (Taggart *et al.* 2005). In other species, such as the parma wallaby (Menzies *et al.* 2007), weaning of fostered animals (~365 days) occurred significantly later than in both the donor (300 days) and the foster species (tammar wallaby [270–300]; Murphy and Smith 1970). Likewise, Merchant and Sharman (1966) reported that a young eastern grey kangaroo reared on a red kangaroo surrogate mother continued to be suckled for over 550 days even though weaning in red kangaroo young usually occurs around 360 days. They concluded that, by continuing to demand milk, the foster young imposed on the red kangaroo surrogate mother a condition like that normally found in eastern grey kangaroos, which usually wean young around 540 days (Kirkpatrick 1965).

Onset of sexual maturity

In cross-fostering studies of the long-nosed potoroo and Tasmanian bettong, reduced adult size and delayed onset of puberty in foster-reared animals compared to controls was rare (Sterneberg 2005). In contrast, Menzies *et al.* (2007) observed, from a small sample size of foster-reared female parma wallaby and tammar wallaby young, that the production of first pouch young occurred later than was typical for either species. However, Merchant and Sharman (1966) reported that a swamp wallaby young reared on a red kangaroo surrogate mother showed accelerated growth and early sexual maturity.

Gastrointestinal maturation

Waite *et al.* (2005) examined the effects on gastrointestinal maturation in tammar wallabies of providing younger pouch young with older-stage milk. They reported a significant increase in pouch young weight when donor young were supplied with older-stage milk, possibly due to higher concentrations of lipid in milk from the more advanced mammary glands. However, no difference was found in stomach or small intestine development between young reared on the appropriate milk for age and those reared on milk for more advanced pouch young (Waite *et al.* 2005).

Observations on the behaviour of surrogate mothers and fostered animals

Surrogate mothers readily accepted cross-fostered pouch young at least for short periods, during which growth occurred, regardless of species differences and differences in lengths of pouch life between species (Merchant and Sharman 1966; Johnson 1981; Smith 1998; Sterneberg 2005; Taggart *et al.* 2005; Schultz *et al.* 2006; Menzies *et al.* 2007). In addition, although the calls of donor and foster species are often quite different (Sterneberg 2005), surrogate mothers show normal maternal responses to the calls of their foster young if they became lost after pouch emergence or when separated in captivity (Merchant and Sharman 1966; Sterneberg 2005). Cross-fostered long-nosed potoroo young were able to recognise the calls of their surrogate mothers and both surrogate mother and fostered young showed increased vigilance when young left the pouch (Sterneberg 2005). Sterneberg (2005) reported that young cross-fostered potoroos and bettongs initially imitated the preferred locomotion of their surrogate mother (i.e. bipedal locomotion for bettongs and a quadrupedal locomotion for potoroos) before their species-specific locomotion began to dominate. Heightened aggression was observed when integrating bettongs reared in potoroo pouches with mother-reared bettongs: potoroo-reared male bettongs became the target of agonistic behaviour from adult females (Sterneberg 2005). This appeared to be due to the more solitary life-style of the mother-reared bettongs compared with those reared in potoroo pouches.

In contrast, from the time of pouch exit, young brush-tailed rock-wallabies foster-reared on tammar wallabies showed normal donor species behavioural traits, with young at foot using trees, rocks, hollows and other high vantage points when basking or moving between locations while their tammar wallaby surrogate mothers remained terrestrial (Taggart *et al.* 2005). Young foster-reared brush-tailed rock-wallabies also readily mixed with, and chose to den close to, members of their own species following separation from their surrogate mothers immediately after weaning. Den sites included caves, tree overhangs and rabbit burrows (Taggart 2002; Taggart *et al.* 2005; Taggart, unpubl. data).

Results from all studies indicate that, at sexual maturity, reproductive behaviour was normal for all cross-fostered animals for which data are available. These animals recognised and mated with their own species and, in the case of females, gave birth to and nursed their own young (swamp wallaby, Merchant and Sharman 1966; Johnson 1981; northern bettong, Smith 1998; brush-tailed rock-wallaby, Taggart 2002; Taggart *et al.* 2005; long-nosed potoroo and Tasmanian bettong, Sterneberg 2005; brush-tailed rock-wallaby and yellow-footed rock-wallaby, Schultz *et al.* 2006; tammar wallaby and parma wallaby, Menzies *et al.* 2007).

Effects of pouch young removal for fostering on donor female reproductive rate

The effects of pouch young removal on the reproductive rate of donor females can be dramatic. In some individuals, up to eight-fold increases in female reproductive rate have been observed (Table 23.5; Smith 1998; Taggart *et al.* 2005; Schultz *et al.* 2006). In the brush-tailed rock-wallaby, Schultz *et al.* (2006) reported a reduction in the birth interval in some females from ≥250 days to only 40 days where pouch young averaging just eight to 14 days old and weighing ~2 g were removed. The authors described one female that gave birth to 26 young in six seasons and another that produced 19 young in only four seasons. This is markedly higher than the life-time reproductive output of a normal female brush-tailed rock-wallaby, which is probably around four young and very unlikely to exceed seven young (Jarman and Bayne 1997; Hazlitt *et al.* 2006; Wynd *et al.* 2006,; Taggart, Reside and Martin, unpubl. data).

In the northern bettong, Smith (1998) found that transfer of young at 21 days of age to brush-tailed bettongs resulted in the birth of northern bettong offspring (which survived to weaning) at intervals of about 40 days (Smith 1998). This suggests, since all potoroid species are capable of breeding year-round, that cross-fostering could increase the production of young from the norm of two to three per year to as many as nine, an increment of about 300%. In addition to the benefit of a greater success rate of brush-tailed bettong surrogate mothers rearing northern bettong pouch young to weaning than their natural mothers (Smith 1998), this provides a very effective means of enhancing breeding and recruitment. Two critical elements in the success of this procedure and maintenance of the high pouch young production rates are (i) the continued successful breeding of both donor and foster species, and (ii) that the breeding program has suitable numbers of appropriate surrogate mothers.

Recommendations for maximising cross-fostering success and production of pouch young

Based on available data (Merchant and Sharman 1966; Johnson 1981; Smith 1998; Jones *et al.* 2004; Taggart *et al.* 2005; Schultz *et al.* 2006) the following

Table 23.5: Effects of cross-fostering on female reproductive rate in threatened macropodoid taxa.

Threatened donor taxon	Surrogate taxon	Donor female reproductive rate	PY survival to weaning (%)	Reference
Pe. penicillata	*Pe. xanthopus*	2–8-fold ↑	82	Taggart *et al.* 2005; Schultz *et al.* 2006
	M. eugenii	2–3-fold ↑	83	Taggart *et al.* 2005; Schultz *et al.* 2006
M. eugenii (ML)	*M. eugenii* (KI)	2–3-fold ↑	78	Taggart and Smith, unpubl. data
B. tropica	*B. penicillata*	2–9-fold ↑	82	Smith 1998
Pe. lateralis	*Pe. xanthopus*	>2-fold ↑	87	Taggart and Smith, unpubl. data
Pe. lateralis	*M. eugenii*	2–3-fold ↑	100	Jones *et al.* 2004
Po. gilbertii	*Po. tridactylus*	?	75	Taggart and Friend, unpubl. data

ML = mainland; KI = Kangaroo Island.

recommendations are made to ensure maximum cross-fostering success:

1 that the size (weight and head length) of the pouch young to be fostered is similar to, or slightly smaller than, that of the young removed from the surrogate mother's pouch (Jones *et al.* 2004; Taggart *et al.* 2005; Schultz *et al.* 2006);
2 that the selected surrogate mother has a pouch young of similar age to that of the intended foster young (Merchant and Sharman 1966; Johnson 1981; Smith 1998; Jones *et al.* 2004);
3 that teat size (particularly diameter) is similar between donor and surrogate mothers at the time of pouch young transfer (Merchant and Sharman 1966; Sterneberg 2005; Taggart, unpubl. data);
4 that cross-fostering is done early in pouch life when the young weighs 2–5 g;
5 that handling of surrogate mothers with cross-fostered pouch young is kept to a minimum, especially in the second half of pouch life.

Short-term isolation studies in small pouch young

Temperature regulation in small unfurred pouch young

At birth and during most of pouch life marsupial young are ectothermic, obtaining heat directly from the environment (Reynolds 1952; Shield 1966; Gemmel and Johnson 1985; Rose *et al.* 1998). In the tammar wallaby, as in other marsupials studied, pouch temperature equates roughly with the rectal temperature of the mother (~37°C) (Reynolds 1952; Bartholomew 1956; Gemmel and Johnson 1985). Early development of pouch young therefore generally occurs at a relatively constant temperature similar to the mother's body temperature. As development proceeds, macropodoid pouch young gradually increase their metabolic heat production, to adult levels by the time of pouch exit (Loh and Shield 1977; Rose *et al.* 1998).

Pouch young isolation studies in the Macropodoidea

Young of two potoroid and 11 macropodid species have been isolated from the pouch at a range of ages from one to 25 weeks and for various lengths of time from 30 minutes to 30 hours (Table 23.6). Most of the available data come from studies that examine changes in the composition of milk during development (Table 23.6; Shield 1966; Renfree and Tyndale-Biscoe 1978; Messer and Green 1979; Green *et al.* 1980, 1988; Gemmel and Johnson 1985; Merchant *et al.* 1996; Rose *et al.* 1998; Taggart *et al.* 2002). Successful short-term pouch young isolation has been reported at temperatures ranging from <4°C to 35°C with all studies indicating that humidity was maintained at high levels (>90% RH) during the period of isolation from the pouch (Renfree and Tyndale-Biscoe 1978; Gemmel and Johnson 1985; Rose *et al.* 1998; Taggart *et al.* 2002). Differences between the success rates of pouch young isolation at higher temperatures (≥30°C) may be explained by differences in study design, supplementation of young with fluids and/or pouch young age/size at the time of isolation (Taggart *et al.* 2002). Data indicate that, as their size increases, survival of pouch young after isolation from the teat at higher temperatures, and possibly for longer

Table 23.6: Short-term pouch young isolation studies using small macropodoid marsupial young.

Species	PY age (weeks)	Isolation temperature (°C)	Isolation humidity (%RH)	Hours isolated	Survival	References
Po. tridactylus	1–25	23–35	Humid (>95)	3–5; 10	Y	Crowley *et al.* 1988
B. gaimardi	1–25	22–35	>95	3–5	Y	Rose *et al.* 1998
Pe. penicillata	1–23	23–30	>95	6–13	Y (20–25°C)	Taggart *et al.* 2002
Pe. xanthopus	4–10	23–25	>95	6–8	Y (20–25°C)	Schultz and Taggart, unpubl. data
Pe. lateralis	2–8	20–25	>95	5.5–9	Y (20–25°C)	Taggart and Smith, unpubl. data
Pe. assimilis	10–22.5		Humid		Y	Merchant *et al.* 1996
M. eugenii	4–32	<4–35	>95	0.5; 3–8	Y	Renfree and Tyndale-Biscoe 1978; Green *et al.* 1980; Taggart *et al.* 2002
M. rufus	14+	?	Humid	3	Y	Lemon and Baker 1967
M. parryi	6–25	?	?	4–30	Y	Johnson 1981
L. conspicillatus	6–25	?	?	4–30	Y	Johnson 1981
O. fraenata	6–25	?	?	4–30	Y	Johnson 1981
W. bicolor	6–25	?	?	4–30	Y	Johnson 1981
S. brachyurus	1.2–25	20–40	Humid	9	Y	Shield 1966

periods, also increases (Green *et al.* 1988; Merchant *et al.* 1996; Taggart *et al.* 2002). This is probably related to differences in the surface area to body mass ratio between small and larger pouch young, and thus differences in rates of evaporative water loss.

Small, unfurred, ectothermic pouch young can become heat stressed and die very quickly if tempertures exceed that of pouch temperature (~37°C). However, the tolerance of these pouch young to temporary cooling during the first half of pouch life and their ability to recover and grow normally is well documented (Renfree and Tyndale-Biscoe 1978; Rose *et al.* 1998; Taggart *et al.* 2002). The success of pouch young isolation studies at lower temperatures is likely to be due to a combination of factors including lowered heart and metabolic rate, reduced evaporative water loss at lower temperatures, and slowing of the dehydration rate of the pouch young (Loh and Shield 1977; Rose *et al.* 1998; Taggart *et al.* 2002). A reduction by ~50% in the metabolic rate and heart rate of pouch young held at ~23°C has been reported (Bartholomew 1956; Loh and Shield 1977; Renfree and Tyndale Biscoe 1978; Rose *et al.* 1998; Shield 1966). All tammar wallaby and brush-tailed rock-wallaby pouch young, including one that weighed only 0.4 g, held at 23°C survived a six-hour isolation period from the pouch; when returned, they survived and developed normally (Taggart *et al.* 2002). The temporary lowering of metabolic rate for both small and large pouch young prevents excessive depletion of energy reserves while isolated from the teat (Loh and Shield 1977). In addition, as small unfurred pouch young are poikilothermic, lower ambient temperatures result in lower body temperatures and lower oxygen consumption rates (Bartholomew 1956; Loh and Shield 1977, Rose *et al.* 1998).

Recommendations for maximising pouch young survival after short-term isolation

Based on available data (Shield 1966; Renfree and Tyndale-Biscoe 1978; Messer and Green 1979; Green *et al.* 1980, 1988; Johnson 1981; Gemmel and Johnson 1985; Merchant *et al.* 1996; Rose *et al.* 1998; Taggart *et al.* 2002) and assuming small unfurred pouch young have been prepared and isolated as described, the following recommendations should maximise pouch young survival after short-term isolation:

1. all unfurred pouch young should be isolated at high humidity (>90% RH);
2. very small, unfurred pouch young should be isolated at temperatures around 20–25° C;
3. isolation of very small, unfurred pouch young should not exceed six to eight hours without supplementary fluids and, if data for that species is not available, without further research on that species.

IMPLICATIONS FOR CONSERVATION AND SPECIES MANAGEMENT

The use of procedures to maintain macropodoid young in isolation from the pouch for short periods, in combination with cross-fostering techniques (Merchant and Sharman 1966; Smith 1998; Taggart *et al.* 2002) gives conservationists a powerful tool for accelerating breeding in endangered species. Assuming suitable foster species are available, the techniques described here eliminate the need to bring wild-caught adults into captivity for genetic or breeding purposes. This has the advantage of not compromising the long-term reproductive capacity of threatened small populations in the wild and avoids the associated stressors that can potentially reduce the reproductive output in captivity of wild-born and raised adults. The techniques also provide an effective means of more safely moving animals, and their genes, between captive institutions to assist breeding and conservation objectives (Taggart and Schultz, unpubl. data) and eliminate welfare issues associated with bringing wild adults into captivity (Taggart *et al.* 2002).

Merchant and Sharman (1966) were the first to use this technique to transfer (transfer conditions unknown) a 78-day-old swamp wallaby, approximately 75 g (Bellamy 1992), from a mother shot in the field to a red kangaroo surrogate mother 177 km away. The pouch young was reared to independence. Based on their observations, Merchant and Sharman (1966) suggested that 'transfer of young of rare species to the pouches of common species of marsupials appeared to be a feasible method of increasing the numbers of rare species in captivity'. The combination of cross-fostering and pouch young isolation techniques has been used successfully to transfer pouch young as small as 1 g up to 1000 km to establish captive breeding colonies and/or supplement genetics in captive colonies of black-footed rock-wallaby, brush-tailed rock-wallaby, yellow-footed rock-wallaby and Gilbert's potoroo (Taggart, Smith, Schultz and Friend, unpubl. data).

Embryonic diapause in macropodoid species (Tyndale-Biscoe 1979; Tyndale-Biscoe and Renfree 1987) is an important but not essential factor in the success of cross-fostering for captive breeding and conservation. In species displaying embryonic diapause, pouch young removal causes cessation of lactation and reactivation and birth of the diapausing embryo. The capacity of most female macropodoids to rapidly enter a new cycle of post-partum oestrus and mating, to generate a new diapausing embryo, enhances the success of this procedure with captive and wild polyoestrous females. Removal of pouch young reduces birth intervals in macropodoid marsupials and significantly increases the number of pouch young available for cross-fostering. In this way pouch young removal, short-term isolation from the pouch and fostering not only benefit captive breeding programs but can be used to accelerate the production of young by females of endangered species in the wild.

Based on these studies, we suggest that the number of pouch young produced by females of many macropodoid species could be readily increased from one to as many as four to eight young per female annually depending upon the species and the frequency of cycling. Application of this technique could have a dramatic effect on animal numbers within a relatively short period, so maintaining a genetically diverse captive breeding group would therefore be a high priority.

CONCLUSIONS

The greatest current threat to many endangered macropodoid species is the small size and fragmentation of remaining populations. Small populations are highly vulnerable to local catastrophes, predation, inbreeding and the loss of genetic variation by genetic drift, and without an active program of intervention their continued decline is likely. As breeding in many of these species is slow, recovery of their populations through natural recruitment alone, even with significant protection and environmental manipulation, will likewise be slow. Cross-fostering and pouch young isolation techniques provide a means of rapidly and effectively alleviating the threat of small and declining population size without drastically affecting remaining wild populations. They also offer a subtle way of diversifying the genetic make-up of remnant isolated wild or captive populations. Thus, these techniques have major implications as tools for conservation and management of these species and endangered species from other marsupial groups (Caton 2005; Foster *et al.* 2006; Finlayson *et al.* 2008).

REFERENCES

Bartholomew GA 1956. Temperature regulation in the macropod marsupial, *Setonix brachyurus*. *Phys. Zool.* **29:** 26–40.

Bell JN and Close RL 1994. Fostering rock-wallaby pouch young: a method of accelerated production of young (abstract). *National Rock-wallaby Symposium*, p. 12. Oraparinna, South Australia.

Bellamy T 1992. *Urban Wildlife*. University of Sydney Press, Sydney.

Bolliger A and Pascoe JV 1953. Composition of kangaroo milk (wallaroo, *Macropus robustus*). *Aust. J. Sci.* **15:** 215–217.

Caton W 2005. The real deal, the good stuff. Mother's milk: you can't beat it! *Thylacinus* **29:** 2–4.

Collins M 2002. Cross-fostering and pouch isolation studies in the Potoroidae. BSc Hons thesis. University of Adelaide, Australia.

Crowley HM, Woodward DR and Rose RW 1988. Changes in milk composition during lactation in the potoroo, *Potorous tridactylus* (Marsupialia: Potoroinae). *Aust. J. Biol. Sci.* **41:** 289–296.

Finlayson GR, Taggart DA, Shimmin GA, White CR, Dibben R, Steele V, Paris M and Temple-Smith PD 2007. Pouch young removal and return to oestrus in wild southern hairy-nosed wombats (*Lasiorhinus latifrons*). *Anim. Reprod. Sci.* **100:** 216–222.

Finlayson G, Taggart DA, Steele VR, Shimmin GA and Temple-Smith PD 2008. The application of intra-species fostering techniques to the southern hairy-nosed wombat, *Lasiorhinus latifrons*. *Pac. Conserv. Biol.* **13:** 259–263.

Foster WK, Bradley AJ, Caton W and Taggart DA 2006. Comparison of growth and development of the red-tailed phascogale (*Phascogale calura*) in three captive colonies. *Aust. J. Zool.* **54:** 343–352.

Gemmel RT and Johnson G 1985. The development of thermoregulation and the emergence from the pouch of the marsupial bandicoot *Isoodon macrourus*. *Physiol. Zool.* **58:** 299–302.

Green B, Newgrain K and Merchant J 1980. Changes in milk composition during late lactation in the tammar wallaby (*Macropus eugenii*). *Aust. J. Biol. Sci.***33:** 35–42.

Green B, Merchant J and Newgrain K 1988. Milk consumption and energetics of growth in pouch young of the tammar wallaby, *Macropus eugenii*. *Aust. J. Zool.* **36:** 217–227.

Hazlitt SL, Sigg DP, Eldridge MDB and Goldizen AW 2006. Restricted mating dispersal and strong breeding group structure in a mid-sized marsupial mammal (*Petrogale penicillata*). *Mol. Ecol.* **15:** 2997–3007.

Hill JP and Hill WCO 1955. The growth stages of pouch young of the native cat (*Dasyurus viverrinus*) together with observations on the anatomy of the newborn young. *Trans. Zool. Soc. Lond.* **28:** 348–453.

Hughes RL and Hall LS 1988. Structural adaptations of the newborn marsupial. In *The Developing Marsupial: Models for Biomedical Research*. (Eds CH Tyndale-Biscoe and PA Janssens) pp. 8–27. Springer-Verlag, Berlin.

Jarman PJ and Bayne P 1997. Behavioural ecology of *Petrogale penicillata* in relation to conservation. *Aust. Mammal.* **19:** 219–228.

Johnson PM 1981. The rearing of marsupial pouch young by foster mothers of different species. *Int. Zoo Yearbook* **21:** 173–176.

Jones M, Taggart DA and Temple-Smith PD 2004. Age determination and growth in wild *Petrogale lateralis pearsoni* and captive *Petrogale lateralis* 'MacDonnell Ranges race'. *Aust. J. Zool.* **52:** 447–461.

Kirkpatrick TH 1965. Studies of Macropodidae in Queensland. III. Reproduction in the grey kangaroo (*Macropus major*) in southern Queensland. *Qld J. Agr. Anim. Sci.* **22:** 319–328.

Lemon M and Baker S 1967. Changes in milk composition of the red kangaroo, *Megaleia rufa* (Desmarest), during lactation. *Aust. J. Exp. Biol. Med. Sci.* **45:** 213–219.

Loh TT and Shield J 1977. Temperature regulation and oxygen consumption in the developing macropod marsupial, *Setonix brachyurus*. *J. Physiol.* **269:** 677–686.

Menzies BR, Shaw G, Fletcher TP and Renfree MB 2007. Perturbed growth and development in marsupial young after reciprocal cross-fostering between species. *Reprod. Fertil. Dev.* **19:** 976–983.

Merchant JC and Sharman GB 1966. Observations on the attachment of marsupial pouch young to the teats and on rearing of pouch young by foster-mothers of the same or different species. *Aust. J. Zool.* **14:** 593–609.

Merchant JC, Marsh H, Spencer P and Death G 1996. Milk composition and production in free-living allied rock-wallabies, *Petrogale assimilis*. *Aust. J. Zool.* **44:** 659–674.

Messer M and Green B 1979. Milk carbohydrates of marsupials. II. Quantitative and qualitative changes in milk carbohydrates during lactation in the tammar wallaby (*Macropus eugenii*). *Aust. J. Biol. Sci.* **32:** 519–531.

Murphy CR and Smith JR 1970. Age estimation of pouch young and juvenile Kangaroo Island wallabies. *Trans. Roy. Soc. South Aust.* **94:** 15–20.

O'Brien N 2004. Cross-fostering and its effects on the growth and development of pademelon (*Thylogale billardierii*) and tammar wallaby (*Macropus eugenii*) pouch young. BSc Hons thesis. University of Melbourne, Australia.

Renfree MB and Tyndale-Biscoe CH 1978. Manipulation of marsupial embryos and pouch young. In *Methods in Mammalian Reproduction*. (Ed. JC Daniel) pp. 307–331. Academic Press, London.

Reynolds HC 1952. Studies on reproduction in the opossum (*Didelphis marsupialis virginiana*). *Uni. Calif. Pub.* **52:** 223–283.

Rose RW, Kuswanti N and Colquhoun EQ 1998. Development of endothermy in a Tasmanian marsupial, *Bettongia gaimardi* and its response to cold and noradrenaline. *J. Comp. Physiol.* **168:** 359–363.

Schultz D, Whitehead P and Taggart DA 2006. Review of surrogacy program for endangered Victorian brush-tailed rock wallaby (*Petrogale penicillata*) with special reference to animal husbandry and veterinary considerations. *J. Zoo. Wildl. Med.* **37:** 33–39.

Sharman GB 1962. The initiation and maintenance of lactation in the marsupial *Trichosurus vulpecula*. *J. Endocrin.* **25:** 375–385.

Shield J 1966. Oxygen consumption during pouch development of the macropod marsupial *Setonix brachyurus*. *J. Physiol.* **187:** 257–270.

Smith MJ 1998. Establishment of a captive colony of *Bettongia tropicia* (Marsupialia: Potoroidae) by cross-fostering; and observations on reproduction. *J. Zool. Lond.* **244:** 43–50.

Sterneberg B 2005. The pouch advantage: cross-fostering for the conservation of rat kangaroos. *B.I.F. Futura* **20:** 245–253.

Taggart DA 2002. Use of pouch young removal and cross fostering techniques to accelerate breeding and recruitment in the threatened brush-tailed rock wallaby, *Petrogale penicillata*. *ANZCCART News* **15:** 7–9.

Taggart DA, Shimmin GA, Underwood G and Phillips K 2002. Survival of very small macropod pouch young following short-term isolation from the pouch at various environmental temperatures. *Animal Conserv.* **5:** 275–282.

Taggart DA, Schultz DJ, White C, Whitehead PJ, Underwood G and Phillips K 2005. Cross-fostering, growth and reproductive studies in the brush-tailed rock wallaby, *Petrogale penicillata* (Marsupialia: Macropodidae): efforts to accelerate breeding in a threatened marsupial species. *Aust. J. Zool.* **53:** 1–11.

Tyndale-Biscoe CH 1979. Hormonal control of embryonic diapause and reactivation in the tammar wallaby. In *Maternal Recognition of Pregnancy. Ciba Foundation Symposium* **64** (new series) pp. 173–190. Excerpta Medica, Amsterdam.

Tyndale-Biscoe CH and Renfree MB 1987. *Reproductive Physiology of Marsupials*. Cambridge University Press, Cambridge.

Tyndale-Biscoe CH, Hearn JP and Renfree MB 1974. Control of reproduction in macropodoid marsupials. *J. Endocrinol.* **63:** 589–614.

Waite R, Giraud A, Old J, Howlett M, Shaw G, Nicholas K and Familari M 2005. Cross-fostering in *Macropus eugenii* leads to increased weight but not accelerated gastrointestinal maturation. *J. Exp. Zoo.* **303(A):** 331–344.

Wynd NA, Sigg DP, Pople AR, Hazlitt SL and Goldizen AW 2006. Factors affecting female reproductive success and the survival of pouch young in the threatened brush-tailed rock-wallaby, *Petrogale penicillata*. *Aust. J. Zool.* **54:** 61–70.

24 Roads as drivers of change for macropodids

D. Ramp

SUMMARY

As an iconic and predominantly endemic Australian family of mammals, the Macropodidae are synonymous with Australia's faunal identity. Of the 53 species known prior to European settlement, six have since gone extinct and many more have suffered significant reductions in range and abundance. Although individual threats each take their toll (e.g. human modification of the landscape, hunting, the influx of invasive predators), it is the combined impact of multiple drivers of change that is of concern. This cumulative impact has the potential to expose even species typically considered common to extinction risk. Recent worldwide attention has been focused on the threat of roads to fauna, a threatening process now seen as a potential tipping-point for species struggling to survive. This chapter discusses what is known of the role that roads play in threatening species in the Macropodidae. It examines the direct and indirect effects of roads by collating published information of fatalities (road-kill), landscape modification and population persistence, contrasting impacts in urban and rural environments. It provides evidence that the impact of roads is more pervasive and potentially catastrophic for some species than previously thought, discusses attempts to mitigate the impacts of roads on fauna and explores the long-term conservation implications for the Macropodidae.

INTRODUCTION

The road environment, which includes the road, the road verge and surrounding habitat, is increasingly recognised as a major form of environmental disturbance, or 'driver of change' (*sensu* Millennium Ecosystem Assessment 2005). In the last 50 years, road networks around the world have been upgraded and expanded. Recognition of the potential for roads to severely affect species initially stemmed from collisions with large fauna, primarily ungulates, in Europe and North America. The effect of the road environment on biodiversity has spawned a new research field, termed 'road ecology' (Forman 1998; Forman and Alexander 1998). The two most important contributions this field has made are the documentation of how universal and sizeable the impacts on the environment are, and the documentation of how far impacts extend beyond the bitumen (Forman 2000; Ramp and Ben-Ami 2006). As research progresses, the considerable role roads play in causing animal fatalities, fragmenting and isolating populations and ultimately increasing extinction risk is becoming increasingly clear.

Recognition of the impacts of the road environment on Australia's biodiversity encompasses most taxa, but most conservation biologists and road safety researchers focus on collisions involving larger fauna. Much more research is needed, but it is becoming apparent that the effect of the road environment on medium and large-sized fauna in Australia, primarily populated by species in the family Macropodidae (with some notable exceptions such as the common wombat (*Vombatus ursinus*), Roger and Ramp 2009), has the potential to cause species decline. This chapter collates knowledge of how macropodids are affected by fatalities, landscape modification and other indirect effects on population persistence. It highlights the state of knowledge and identifies areas that require further research so that the threat to species and populations can be understood and managed at both local and regional scales.

DRIVERS OF CHANGE

The effect of roads must be considered within the context of the range of pressures facing biodiversity today. Unprecedented and multifaceted levels of anthropogenic drivers of change are a global challenge in efforts to conserve biodiversity (Pimm and Raven 2000). Landscapes of all types are changing to support an expanding human population (Imhoff *et al.* 2004; Foley *et al.* 2005) and extinction rates highlight the need to address threats to the effectiveness of conservation measures (Sekercioglu *et al.* 2004). The challenge for conservation biology is to identify drivers of biodiversity (Possingham and Wilson 2005; Wilson *et al.* 2005) and establish mechanisms to mitigate those drivers.

In Australia, the landscape has changed markedly since European settlement. Past changes in biodiversity have been driven primarily by extrinsic processes like long-term climate change and continental movement, but current driving processes of biodiversity loss are unquestionably linked to human activities. The most important anthropogenic drivers are loss of vegetation cover, invasive species, overexploitation, disease, pollution and rapid climate change (Mace *et al.* 2005). Drivers directly and indirectly affect ecosystems at many spatial and temporal scales, and with complex interactions. Understanding how these drivers and their interactions affect local populations, biodiversity and ecological functioning is essential to conservation and management efforts.

URBANISATION

One of the primary reasons for the increasing threat of roads is global urbanisation (Daily 1997). As urban areas expand, road density and traffic volume increase. Three billion people live in cities worldwide and this number is expected to double over the next 50 years (Crane and Kinzig 2005). The number of urban areas with over a million people is expected to grow by over 40% between 2000 and 2015 (Crane and Kinzig 2005). In the US, urbanisation has been identified as a primary cause, singly or in association with other factors such as roads, of declines in more than half the species listed as threatened or endangered under the US Endangered Species Act (Czech *et al.* 2000; Miller and Hobbs 2002).

Aside from dramatically altering vegetation cover, urbanisation has long been recognised as changing landscapes and isolating pockets of suitable habitat or refugia for many species (Fahrig 2003; Adams 2005). Urbanised regions have severely reduced wildlife populations, with many remaining species considered endangered (Flannery 2004). Environmental refugia in the urban landscape often provide the last refuge for species that were once widespread, and are important for local and regional biodiversity (Margules and Pressey 2000; Cornelis and Hermy 2004).

Australia is no exception – many protected areas close to urbanised areas are affected by increasing road densities and traffic. One example is the Royal National Park, just south of Sydney (New South Wales) – Australia's largest city, population 4.28 million in 2006. The oldest national park in the world (declared as 'The National Park' in 1879), it is surrounded by major highways and urban development and is criss-crossed by a network of roads. Calculations show that 58% of the park is within 1 km of sealed roads with speed limits of 60 kmh^{-1} or 80 kmh^{-1}. If unsealed roads and fire trails are included, 86% of the park is within 500 m of a road (Fig. 24.1). This important conservation reserve is a maze of roads, fragmenting wildlife and killing animals via collisions with vehicles. Macropodids that occur in such peri-urban reserves frequently struggle to maintain viable populations.

TRANSPORTATION NETWORKS AND RURAL LANDSCAPES

The role of transportation as a driver of change is widely accepted (Forman *et al.* 2003). There were 15.3 million registered vehicles in Australia in 2008 (Australian Bureau of Statistics), predominantly using urban roads but also moving between cities. Transportation corridors are a significant contributor to the impact of roads on wildlife. Road infrastructure is continually upgraded to allow for greater traffic volumes and speed, particularly on rural highways. The carcasses of large fauna, predominantly kangaroos, are commonly observed along these highways (Klöcker *et al.* 2006). Transportation of freight by large trucks and road trains is a

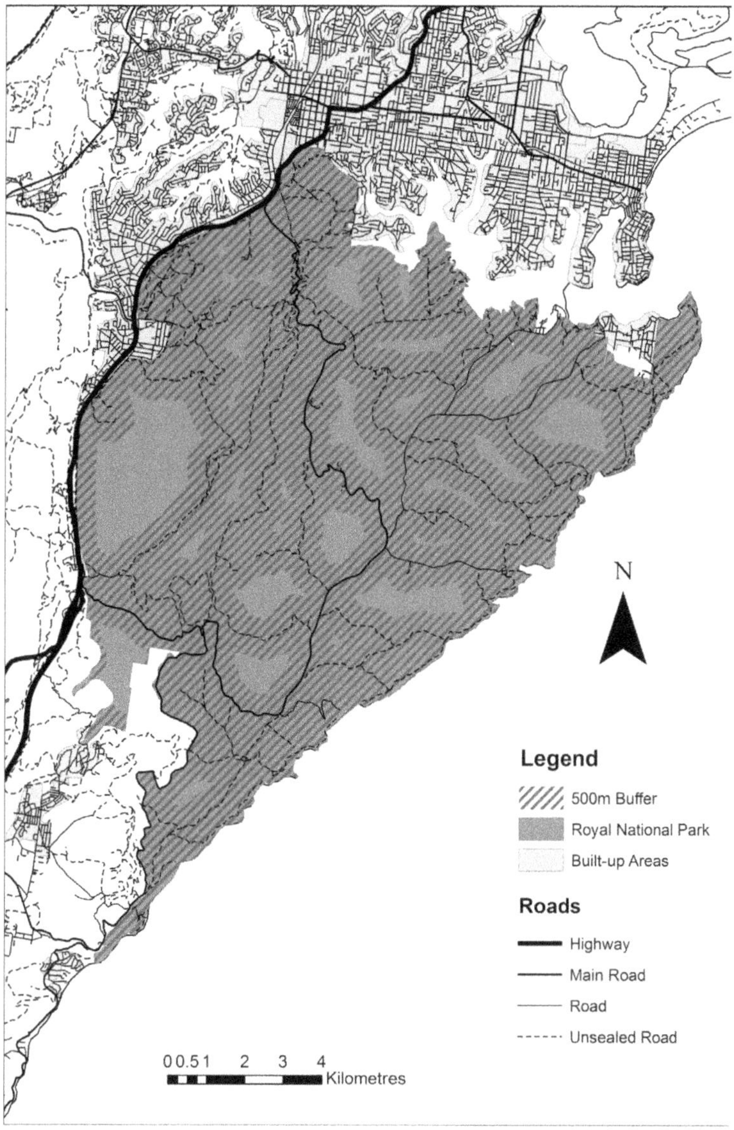

Figure 24.1: The road effect zone in the Royal National Park, New South Wales. The buffered 500 m area surrounding all roads, including unsealed roads, accounted for 86% of the entire reserve.

considerable contributor to this problem (although few data exist). Freight vehicles travel in excess of 157 000 million tonne-kilometres per year in Australia (Australian Bureau of Statistics), regularly on rural roads without slowing down or swerving to avoid wildlife.

In rural environments, the combination of vehicle frequency and speed is generally regarded as the primary cause of fauna fatalities on roads (Forman and Alexander 1998; Hubbard *et al.* 2000; Jones 2000; Trombulak and Frissell 2000; Dique *et al.* 2003; Seiler 2003). Research typically assumes a linear correlation between volume, speed and collision frequency, yet disturbance generated by busy roads can sometimes lead to a lower rate of fatalities than on less-travelled roads with similar speed limits (Clevenger *et al.* 2003). This pattern has been observed with kangaroos, where a rural highway with low traffic volume (60 vehicles per day) had a fatality rate (0.035 per km per day pre-drought) similar to roads with much higher traffic volumes (Lee *et al.* 2004). It follows that high fatality rates may be more likely to occur on roads with high speed limits and moderate to low traffic volumes adjacent to habitat containing abundant wildlife. It has long been recognised that road environments in Australia are unique, as often the only remaining habitat for wildlife exists along road corridors (Bennett 1991). This undoubtedly provides habitat for many species, but the proximity of habitat and fast-moving vehicles is of concern for macropodids and other fauna.

ROADS AS THREATENING PROCESSES

Most of the impacts of roads on the environment have been documented in literature unrelated to macropodids. Roads have been shown to affect microclimate (Ellenberg *et al.* 1981, cited in Forman *et al.* 2003), wind flow (Ahrens 1991), run-off and water flow (FHA 1996), to create noise pollution (Reijnen *et al.* 1997) and facilitate the dispersal of animals and plants (Tikka *et al.* 2001), including weeds (Ullmann 1998), feral animals (Seabrook and Dettmann 1996; Stiles and Jones 1998) and native species (Spooner *et al.* 2004). Roads, including the road surface and the road verge, result in a direct loss of habitat. Conversely, the road verge can provide resources and habitat for many species (Bellamy *et al.* 2000). However, negative impacts can be detected far from the road (Reijnen *et al.* 1995; Goosem 2000; Ortega and Capen 2002; Laurance 2004), leading to avoidance of these environments by many species (Forman *et al.* 2002) and changes in the behaviour of roadside populations (Foppen and Reijnen 1994). Roads often bisect habitat and, depending on the type of road, prevent or restrict movement. This 'barrier effect' has the potential to isolate populations and cause decline by preventing access to resources and mates (Richardson *et al.* 1997; Gerlach and Musolf 2000). Vehicles travelling on roads have perhaps the most significant impact on fauna, when they collide (Ramp *et al.* 2005a). If fatalities are biased towards particular sex–age classes (Coulson 1997a), there is the potential for the structure of local populations to be altered. Even when unbiased, fatalities have the potential to cause population decline (Mumme *et al.* 2000; Lopez 2004; Ramp and Ben-Ami 2006). Last, roads create welfare issues for the animals and humans injured and killed (Williams and Wells 2005; Ramp and Roger 2008; Rowden *et al.* 2008).

EVIDENCE OF IMPACT ON MACROPODIDS

Evidence for roads as drivers of change can be distinguished as direct or indirect. Indirect evidence comes in three forms. Large-scale information on crashes with vehicles is often gathered by government and insurance agencies. In the scientific literature, both small- and large-scale evidence is reported through the brief mention of road impacts on a species or population of interest. Last, community groups and concerned individuals often collect information at small scales that can prove invaluable for assisting the implementation of localised mitigation measures. Direct evidence comes through the targeted scientific study of the direct and indirect impacts of roads at both small and large scales.

General evidence

General information on the effects of roads on the macropodids varies among taxa. For the six species of kangaroos and wallaroos (*Macropus* spp.), the toll on individual animals, on local populations and across species' ranges can only be speculated upon. As their ranges are mostly extensive, except for the black wallaroo (*M. bernardus*), accurate population estimates are difficult to obtain. Despite considerable effort in conducting aerial surveys to estimate distributions and abundance (Pople *et al.* 2000), there is still uncertainty about actual population numbers. Likewise, comprehensive assessments of use of the road environment by kangaroos are yet to be conducted (some unpublished data exists). Piecing together how roads affect these species is a necessary but complicated task, particularly because the kangaroos must also contend with drought, habitat loss, shooting (including legal and illegal culling and harvesting) and impending transformations associated with climate change. There is a pressing need to assess how these threatening processes, including roads, combine to affect kangaroo populations at local and regional scales.

The kangaroos include the largest species encountered on roads in Australia. Adult male red kangaroos (*M. rufus*) can reach weights of 92 kg (Croft and Clancy 2008) and eastern grey kangaroos (*M. giganteus*) can reach 85 kg (Coulson 2008). Both can stand around 2 m in height. While not the same as colliding with moose (*Alces alces*) or elk (*Cervus elaphus*) (Seiler 2005), colliding with a large kangaroo is a dangerous experience for vehicle occupants although human fatalities are uncommon (Abu-Zidan *et al.* 2002). Insurance companies report the typical cost of repairing damage from a collision with a kangaroo as A$2000–7000. A vehicle-accessory industry has developed vehicle adornments in an effort to reduce vehicle damage (roo-bars) and the likelihood of a collision (Shu-Roo); none have much success (Grzebieta and Rechnitzer 2001; Bender 2003).

Indicative evidence of the impacts of roads on kangaroos is derived from reported crashes between animals and vehicles, such as those included in the Roads and Traffic Authority's traffic accident database system of New South Wales (TADS). TADS contains information only on crashes reported to New South Wales police, hence there are infrequent mentions of animal involvement. Crashes are reported to police only when human injuries occur or when vehicle damage requires the vehicle to be towed. However, the data are extremely useful because they cover a large area, are comparable among years and are often the only source of information. Ramp and Roger (2008) searched the TADS database between 1996 and 2005 and determined that animals were responsible for 22 human fatalities and 1462 human injuries out of 5119 reported crashes with animals. The majority (60%) of human fatalities were caused by kangaroos or wallabies (no distinction was made). Spatial analysis of these crashes identified hotspots for collisions with kangaroos or wallabies, which could be used to prioritise state-based mitigation strategies. The many limitations of these data, and suggestions for improvement, are discussed in detail by Ramp and Roger (2008).

Of the 11 wallabies in the genera *Macropus* and *Wallabia*, most have been documented as affected by roads. Given the extent of its range, the swamp wallaby (*W. bicolor*) is perhaps one of the more frequently observed; the red-necked wallaby (*M. rufogriseus*) is also prominent (Ramp *et al.* 2005a). Given the overlap in range and land utilised for human habitation, it is not surprising that fatalities on roads are a problem for many swamp wallaby populations (Ramp *et al.* 2005a). Populations are often isolated in peri-urban remnant habitat that is surrounded by major highways or criss-crossed by sealed roads (Ben-Ami *et al.* 2006; Ramp *et al.* 2006). Fatalities have been shown to be directly responsible for the decline of swamp wallaby populations, over and above impacts from predation (Ramp and Ben-Ami 2006). Ramp and Ben-Ami (2006) found that a predicted 90% decline in the population over the next 100 years could be reduced to only 10% if annual female swamp wallaby fatalities were reduced by 20%. To achieve a similar reversal of the predicted decline, the influence of fox predation on swamp wallabies would need to be reduced by 70%, an almost impossible target.

Of the smaller species, many of the rock-wallabies (*Petrogale* spp.) and the quokka (*Setonix brachyurus*) have been listed as threatened species for some time; pademelons (*Thylogale* spp.) are the exception, although they are listed as vulnerable in some areas. Habitat loss, fragmentation and predation by introduced fauna such as the red fox (*Vulpes vulpes*) are the primary reason for their precarious existence; however, fatalities on roads have been documented as having a significant impact on remaining populations. Hayward *et al.* (2005) identified that 40% of quokka deaths in a Western Australian study were attributable to collisions with vehicles. Likewise, Hazlitt *et al.* 2006 identified road-kill as a conservation concern for brush-tailed rock-wallabies (*Petrogale penicillata*). Their small and localised populations are susceptible to mortality arising from road-kill, increasing local extinction risk and reducing the abundance of rare alleles.

The highly endangered bridled nailtail wallaby (*Onychogalea fraenata*) is found only in Taunton National Park in central Queensland, which has unsealed tracks and is not open to the public. No cases of fatalities on roads have been documented within the park or at reintroduction sites, which also have only unsealed roads (D. Fisher, *pers. comm.*). In contrast, northern nailtail wallabies (*Onychogalea unguifera*) are reasonably common across much of northern Australia. Although their geographical distribution and abundance does not appear to have altered much since European settlement, the removal of habitat for agriculture is a major threat to their populations (Ingleby 1991). Despite their remoteness, they benefit from roads as these provide access to their favoured ground-layer herbs (S. Ingleby, *pers. comm.*). These plant species would typically have become available after fires or at run-on areas, circumstances mimicked by the road verge. As a consequence, road-killed northern nailtail wallabies are frequently observed on roads such as the Stuart Hwy. No population studies have been conducted to assess how these might contribute to local extinction risk.

Concern over the impact of roads has been noted for the tree-kangaroos (*Dendrolagus* spp.) in Australia's wet tropics. Newell (1999) confirmed that collisions with vehicles were a significant source of mortality, as were habitat loss and fragmentation.

Direct evidence

In Australia, scientific studies on the effect of roads on Macropodids target three primary areas of research, although they are not mutually exclusive or all-encompassing. Most research has focused on documenting collision rates for different lengths of road and for different species. More recently, research has begun to examine the fragmenting effects of roads as well as the impacts of roads on species populations.

Collisions with vehicles

Australia's macropodids present a novel group of species to study. Deer (*Odocoileus* spp.), moose and elk are responsible for most animal–vehicle collision research in the US, Canada and Europe. They are often killed on roads as they have large home ranges and often migrate (Groot Bruinderink and Hazebroek 1996; Seiler 2004). In contrast, none of the macropodids is migratory, being sedentary with fixed home ranges as adults (Jarman and Taylor 1983; Priddel 1988; Norbury *et al.* 1994; Dawson 1995). Exceptions exist, however, as subadult males often disperse from their natal range (Jarman 1991; Horsup 1994) and a small proportion of red kangaroos are known to disperse over many kilometres (Priddel *et al.* 1988). Most macropodids that encounter the road therefore do so for three primary reasons: they are moving through the landscape within their normal home range to find a mate or food or water, or to assess the state of neighbouring habitat, they are attracted to the road verge to gain access to resources on the verge or water from run-off, or they are dispersing. Positive selection for roadside habitat has been shown for kangaroos along a rural highway in far west New South Wales (Klöcker *et al.* 2006).

Being located on or next to the road makes an individual susceptible to colliding with a vehicle. The collection and reporting of statistics on the number of macropodids killed on roads is done by scientific researchers and concerned members of the community. While state road authorities and local councils regularly remove carcasses from roadsides, little or no data are retained. A number of small-scale road-kill studies employing systematic surveys have been reported, beginning with Coulson (1982) and followed by Osawa (1989). It was not until the late 1990s that interest in the issue gathered pace in an effort to broaden the information base on macropodid fatalities (Coulson 1997a; Cooper 1998; Lee *et al.* 2004; Taylor and Goldingay 2004; Ramp *et al.* 2005a, 2006; Klöcker *et al.* 2006;). Unpublished data also exist in reports and theses (Brown 2001; Magnus *et al.* 2004). Disappointingly, most cover less than 100 km of road and are not conducted over an appropriate length of time. There is a lack of comparability among studies as they are conducted within different environments, on different types of roads with different traffic volumes, and for different wildlife populations that have inherently different sizes. A number of papers have highlighted the disparity between surveys conducted under different climatic conditions (Coulson 1989; Lee *et al.* 2004). The frequency of surveys also confounds comparisons, as some studies are conducted daily while others are periodic. The calculation of daily rates from periodic surveys is problematic if the time between sampling allows fatalities to be missed (Coulson 1982; Hobday and Minstrell 2008).

These difficulties make the reporting of state-wide or even nation-wide statistics challenging. For the macropodids, published findings of daily fatality rates range between 0.0006 and 0.042/km/day (Table 24.1). The highest published rate comes from Kosciusko National Park where in 2003 eastern grey kangaroos, swamp wallabies and red-necked wallabies were reported to be killed at a rate of 0.044/km/day (Ramp *et al.* 2005a). It is tempting to extrapolate these rates. For example, New South Wales has 182 006 km of road, divided almost equally between bitumen and gravel roads. Fatality rates of 0.01 or 0.04 would result in 1820 or 7280 macropodids killed on New South Wales roads every day respectively (between 664 321 and 2 657 287 per year). These calculations do not account for different rates between bitumen and gravel roads, for variation in traffic volume and for differences between urban and rural roads. They assume that the ranges of species being killed are constant across the state and do not allow for changing climatic conditions. Given the large degree of spatial and temporal error associated with estimating state-wide or nation-wide totals from these small amounts of data, it is clear that more high-quality spatial and temporal data of fatality rates under all conditions is needed before we can be confident of our predictions.

Barrier effects

The barrier effect of roads refers to the segregation that can occur from the restriction of movement across the road. Hypothesised implications are the

Table 24.1: Fatality rates for macropodids in Australia.

Source	Time period	Sampling period	Frequency	Location	Climate	Environ-ment	Road type	Traffic volume (cars/day)	Speed (km/h)	Road surface	Species	Road-kill	Road length (km)	Average (km/day)
Vestjens 1973	1970–72	24 trips	Monthly	South-east NSW	Temperate	Rural	Highway	Various	100	Mostly sealed	EGK, RK	4	301	0.0006
Coulson 1982	1975–80	124 trips	Periodic	Heathcote, Vic.	Temperate	Rural	Highway	~1000	100	Sealed	EGK, SW	32	20	0.013
Osawa 1989	1981–82	365 days	Monthly	North Stradbroke Island, Qld	Tropical	Peri-urban	Road	Unknown	100	Sealed	SW	127	23	0.015
Taylor and Goldingay 2004	2000–01	140 days	Weekly	Byron Bay, NSW	Temperate	Rural	Highway, road	5000–20 000	60–100	Sealed	SW	13	100.3	0.0009
Ramp *et al.* 2005a	1998–2003	6 years	~twice daily	Kosciusko National Park, NSW	Temperate	Rural	Highway	~500	100	Sealed	EGK, SW, RNW	2750	40	0.031
Klöcker 2003	2002	168 days	Every two days	Fowlers Gap, NSW	Semi-arid	Rural	Highway	~50	100	Sealed	EGK, WGK, RK, E	125	21.2	0.035
Ramp *et al.* 2006	2003	143 days	Daily	Royal National Park, NSW	Temperate	Peri-urban	Road	~3000	60–80	Sealed	SW	14	22	0.0045
Hobday and Minstrell 2008	2001–04	154 trips	Periodic	Tasmania	Temperate	Peri-urban, rural	Highway, road	Various	Various	Mostly sealed	TP, BW	647	99.2[a]	0.042

a: Total distance travelled over 154 trips was 15 275 km, for which the average distance travelled per trip is 99.2.

EGK = eastern grey kangaroo (*Macropus giganteus*); WGK = western grey kangaroo (*M. fuliginosus*); RK = red kangaroo (*M. rufus*); E = euro (*M. robustus erubescens*); SW = swamp wallaby (*Wallabia bicolor*); RNW = red-necked wallaby (*M. rufogriseus banksianus*); TP = Tasmanian pademelon (*Thylogale billardierii*); BW = Bennett's wallaby (*M. rufogriseus rufogriseus*).

loss of genetic diversity; a few examples exist (Gerlach and Musolf 2000). The road environment can also alter species' behaviour, changing their intra- and interspecific interactions (Reijnen *et al.* 1997). There is great variation in the Macropodidae, not only in body size and morphology but also in behaviour and sociality (Coulson 2009). These differences have implications for how the road environment affects macropodids. In rural Australia, roads influence the movement and space use of kangaroos (Lee 2006). In these environments, dispersal and movement by individuals should protect kangaroo populations from genetic drift resulting from fragmentation effects, as suggested in the context of reserves by Tenhumberg *et al.* (2004).

For wallabies, the road environment may pose significant restrictions on movement and space use. Most research on the barrier effect has focused on small mammals, where road crossings are prevented when the road reaches a certain width (Goosem 2001). That wallabies are frequently killed on roads suggests that although crossing may be difficult, and often unsuccessful, crossing nevertheless does occur. There is little published research examining how wallaby species use roadside habitat (but see Ramp *et al.* 2006), although there is some evidence that wallabies living in proximity to major highways alter their use of space and behaviour to adapt to that environment (Ramp and Ben-Ami, unpubl. data).

Population effects

Surveys of the populations of the four large kangaroo species typically involve aerial survey techniques, as the species are large and often forage in open country (Caughley and Grigg 1981; Priddel 1988; Pople *et al.* 2000). Estimates suggest that kangaroos in Australia number in the millions; numbers are highly susceptible to annual variation, primarily dictated by climatic conditions. Kangaroos do not show an even distribution across the landscape; rather, they exhibit a patchy distribution in line with suitable habitat and resources at different scales (Ramp and Coulson 2002; Coulson 2009). At regional scales, their distributions comprise many local populations that are variously connected by dispersing individuals. It was first suggested by Coulson (1997a) that road-based fatalities might be influencing the demographic structure of populations. Reporting male bias in fatalities, he speculated whether males of some species come into contact with roads more often than females do, thereby altering population sex ratios. As there does not appear to be any intrinsic reason why males would be more likely to collide with a vehicle, bias presumably occurs through a higher propensity to be on the road. Spatial segregation, common in the macropodids (MacFarlane and Coulson 2005; Coulson *et al.* 2006), is also likely to affect the propensity for different sex or age classes to be killed on roads (Morrissey 2003; Lee *et al.* 2004).

Changes to population structure, isolation of subpopulations and loss of animals from collisions with vehicles have the potential to contribute to population decline (Ben-Ami and Ramp 2006). This is an under-explored area, not just for the macropodids but for most species (Forman *et al.* 2003). Hayward *et al.* (2005) reported that although fatalities of the threatened quokka in the northern jarrah forest of Western Australia are infrequent (two of eight recorded deaths over a 25 month period), the impact on the population was considerable. Recent evidence suggests that even for species considered common and relatively resilient to human disturbance, such as the swamp wallaby, roads can be a major contributor to population decline (Ben-Ami *et al.* 2006; Ramp and Ben-Ami 2006). While roads may not directly cause the extinction of macropodid species, there is good evidence to suggest that they may be causing localised extinctions of macropodid populations living adjacent to roads. Considered with other drivers of change, roads have the potential to significantly contribute to population decline.

MITIGATION

Much effort has been spent examining the effectiveness of various mitigation strategies for reducing animal fatalities on roads or improving the permeability of roads (Waring *et al.* 1991; Groot Bruinderink and Hazebroek 1996; Romin and Bissonette 1996). Given that the impact of the road environment on wildlife is considerable, research that identifies how to prevent collisions and minimise environmental impact is a high priority for conservation managers. Most research has been conducted overseas (Clevenger *et al.* 2001a, b; Dodd *et al.* 2004; Jaeger and Fahrig 2004), but Australian researchers have begun to make progress. This is important, as strategies implemented overseas have often proved unrewarding in Australia (Lintermans 1997).

One mitigation strategy that has been used in Australia for the last 20 years is the wildlife warning reflector (Strieter Corporation 2001). The devices are designed to reflect oncoming vehicle headlights into the eyes of wildlife on the roadside, to frighten them before the vehicle arrives. Despite their popularity, scientific evidence for their effectiveness in reducing collisions with deer suggests they have little or no effect (Waring *et al.* 1991; Ujvari *et al.* 1998). Recent evidence in Australia suggests that they may have limited efficacy with macropodids (Ramp and Croft 2006). Not only have trials in captive macropodids failed to detect a response that could lead to a reduction in collisions, it is also likely that the effectiveness of any mitigation strategy is going to vary considerably among species (Ramp *et al.* 2005b; Ch. 26 this volume).

As a mitigation strategy, scaring macropodids away from oncoming vehicles using sound as a deterrent has some merit. The use of sound as a warning cue is widespread in the animal kingdom and sounds are used by kangaroos and other macropodids in this fashion (Jarman 1991; Coulson 1997b; Bender 2005). There is a strong behavioural basis to encourage the development of an auditory repellent device in relation to roads. Devices that claim to do this, using ultrasonic frequencies, have been criticised (Bender 2003; Muirhead *et al.* 2006). A more fertile area of research would be to explore the response of kangaroos to natural sounds derived from themselves (e.g. foot thumps) or from predators (Bender 2005, 2006; Rose *et al.* 2006), as long as the repellent sound could be heard over normal vehicle sounds.

Physical structures along roads could assist, or prevent, animal movement across roads. There is a lack of data on the effectiveness and suitability of wildlife crossings for different species in different environments. The measures would not be cost-effective along the vast lengths of rural roads, but hotspots of high fatality rates on highways and major roads in peri-urban areas may provide suitable locations for physical structures. Some research in Australia has examined common strategies such as the installation of fences to prevent movement of wildlife and underpasses to facilitate the safe passage of wildlife (Mansergh and Scotts 1989; AMBS 1997; Jones 2000; Abson and Lawrence 2003; Taylor and Goldingay 2003). These studies report various levels of success, but often the definition of success is problematic (i.e. the recording of prints within underpasses without quantifying reductions in fatalities). To date, no study has linked the installation of a physical road structure with a reduction of macropodid fatalities or the facilitation of movement where none was previously possible, although evidence has been obtained for other species in Australia and elsewhere.

CONSERVATION IMPLICATIONS

Variation in response to the road environment may occur at both the species and individual level (Reijnen *et al.* 1996; Forman *et al.* 2003; Ramp *et al.* 2006; Ch. 26 this volume). This variation in suitability and susceptibility among species has implications for species survival and the functioning of ecosystems (Loreau 2000). Certain morphological, biological and ecological traits may confer suitability or susceptibility (most likely along a continuum) to road-affected environments (Forman and Alexander 1998). Presumably, highly susceptible species are lost soon after a road is built, while the suite of species with more subtle susceptibilities gradually reduces in richness and abundance (Findlay and Bourdages 2000). However, roads can also attract species because they facilitate movement (Getz *et al.* 1978; Bennett 1991) and provide food and water on the verge for species that can adapt to road environments. Identifying how different macropodids respond to roads will be crucial for defining the course of conservation efforts.

Although it is rare to find a population that is threatened solely because of roads, the combination of road impacts with other drivers of change may lead to decline and local extinction. It seems that where macropodid populations are isolated and relatively restricted in size, mortality associated with collisions with vehicles can jeopardise the long-term viability of local populations (Ramp and Ben-Ami 2006) and occasionally the entire species (Hayward *et al.* 2005). There is no doubt that the loss of individuals of common species, such as kangaroos and wallabies, is occurring at significant rates (Ramp *et al.* 2005a).

Until there are reliable statistics on how many macropodids are being killed, it is difficult to assess the implications of this loss for the long-term viability of this iconic group of species. It is imperative that systematic surveys of directly comparable fatality rates be collected under a range of conditions (environments, traffic volumes, speeds, road types, climate) so that reliable modelling of the numbers killed can be used to assess the overall impact on populations. With increasing pressure from landscape change, predation from feral animals, legal and illegal shooting and climate change, pressure from roads may be the final nail in the coffin for many macropodid populations. The opportunity exists for governments to legislate for changes to the way in which roads are designed, integrating environmental mitigation options such as fences and underpasses into the development phase of road-building.

REFERENCES

Abson RN and Lawrence RE 2003. Slaty Creek wildlife underpass study. Centre for Sustainable Regional Communities, La Trobe University, Bendigo, Australia.

Abu-Zidan FM, Parmar KA and Rao S 2002. Kangaroo-related motor vehicle collisions. *J. Trauma* **53:** 360–363.

Adams LW 2005. Urban wildlife ecology and conservation: a brief history of the discipline. *Urban Ecosys.* **8:** 139–156.

Ahrens CD 1991. *Meteorology Today: An Introduction to Weather, Climate, and the Environment.* West Publishing, Minnesota.

AMBS (Australian Museum Business Services) 1997. Fauna usage of three underpasses beneath the F3 Freeway between Sydney and Newcastle. Australian Museum Business Services Consulting/Roads and Traffic Authority, Sydney.

Bellamy PE, Shore RF, Ardeshir D, Treweek JR and Sparks TH 2000. Road verges as habitat for small mammals in Britain. *Mammal. Rev.* **30:** 131–139.

Ben-Ami D and Ramp D 2006. Modelling the effect of roads and other disturbances on wildlife populations in the peri-urban environment to facilitate long-term viability. In *On the Road to Stewardship. Proceedings of the 2005 International Conference on Ecology and Transportation.* 29 August–2 September. San Diego, California. (Eds CL Irwin, P Garrett and KP McDermott) pp. 317–322. ITRE, North Carolina State University, Raleigh.

Ben-Ami D, Ramp D and Croft DB 2006. Population viability assessment and sensitivity analysis as a management tool for the peri-urban environment. *Urban Ecosys.* **9:** 227–241.

Bender H 2003. Deterrence of kangaroos from agricultural areas using ultrasonic frequencies: efficacy of a commercial device. *Wildl. Soc. Bull.* **31:** 1037–1046.

Bender H 2005. Effectiveness of the eastern grey kangaroo foot thump for deterring conspecifics. *Wildl. Res.* **32:** 649–655.

Bender H 2006. Structure and function of the eastern grey kangaroo (*Macropus giganteus*) foot thump. *J. Zool.* **268:** 415–422.

Bennett AF 1991. Roads, roadsides and wildlife conservation: a review. In *Nature Conservation. II. The Role of Corridors.* (Eds DA Saunders and RJ Hobbs) pp. 99–117. Surrey Beatty and Sons, Sydney.

Brown BC 2001. Marsupial roadkill on the Kosciuszko Road in the Australian Alps: an investigation of factors influencing animal–vehicle collisions. Hons thesis. Australian National University, Australia.

Caughley G and Grigg GC 1981. Survey of the distribution and density of kangaroos in the pastoral zone of South Australia, and their bearing on the feasibility of aerial survey in large and remote areas. *Aust. Wildl. Res.* **8:** 1–11.

Clevenger AP, Chruszcz B and Gunson K 2001a. Drainage culverts as habitat linkages and factors affecting passage by mammals. *J. Appl. Ecol.* **38:** 1340–1349.

Clevenger AP, Chruszcz B and Gunson KE 2001b. Highway mitigation fencing reduces wildlife–vehicle collisions. *Wildl. Soc. Bull.* **29:** 646–653.

Clevenger AP, Chruszcz B and Gunson KE 2003. Spatial patterns and factors influencing small vertebrate fauna road-kill aggregations. *Biol. Conserv.* **109:** 15–26.

Cooper DW 1998. Road kills of animals on some New South Wales roads. Report on data collected by WIRES volunteers in 1997. Macquarie University, Sydney.

Cornelis J and Hermy M 2004. Biodiversity relationships in urban and suburban parks in Flanders. *Landscape Urban Plann.* **69:** 385–401.

Coulson G 1997a. Male bias in road-kills of macropods. *Wildl. Res.* **24:** 21–25.

Coulson G 1997b. Repertoires of social behaviour in captive and free-ranging grey kangaroos, *Macropus giganteus* and *Macropus fuliginosus* (Marsupialia: Macropodidae). *J. Zool.* **242:** 119–130.

Coulson G 2008. Eastern grey kangaroo, *Macropus giganteus.* In *The Mammals of Australia.* 3rd edn. (Eds S Van Dyck and R Strahan) pp. 335–338. Reed New Holland, Sydney.

Coulson G 2009. Behavioural ecology of red and grey kangaroos: Caughley's insights into individuals, associations and dispersion. *Wildl. Res.* **36:** 57–69.

Coulson G, MacFarlane AM, Parsons SE and Cutter J 2006. Evolution of sexual segregation in mammalian herbivores: kangaroos as marsupial models. *Aust. J. Zool.* **54:** 217–224.

Coulson GM 1982. Road-kills of macropods on a section of highway in central Victoria. *Aust. Wildl. Res.* **9:** 21–26.

Coulson GM 1989. The effect of drought on road mortality of macropods. *Aust. Wildl. Res.* **16:** 79–83.

Crane P and Kinzig A 2005. Nature in the metropolis. *Science* **308:** 1225.

Croft DB and Clancy TF 2008. Red kangaroo, *Macropus rufus.* In *The Mammals of Australia.* 3rd edn. (Eds S Van Dyck and R Strahan) pp. 352–354. Reed New Holland, Sydney.

Czech B, Krausman PR and Devers PK 2000. Economic associations among causes of species endangerment in the United States. *Bioscience* **50:** 593–601.

Daily GC 1997. *Nature's Services: Societal Dependence on Natural Ecosystems.* Island Press, Washington DC.

Dawson TJ 1995. *Kangaroos: Biology of the Largest Marsupials*. University of NSW Press, Sydney.

Dique DS, Thompson J, Preece HJ, Penfold GC, de Villiers DL and Leslie RS 2003. Koala mortality on roads in south-east Queensland: the koala speed-zone trial. *Wildl. Res.* **30:** 419–426.

Dodd CK, Barichivich WJ and Smith LL 2004. Effectiveness of a barrier wall and culverts in reducing wildlife mortality on a heavily traveled highway in Florida. *Biol. Conserv.* **118:** 619–631.

Ellenberg H, Muller K and Stottele T 1981. Strassen-Okologie: Auswirkun-gen von Autobahnen und Strasse auf Okosysteme deutscher Landschaften. In *Okologie und Strasse.* Broschurenreihe der Deutschen Strassenliga. pp. 19–122. Ausgabe 3, Bonn.

Fahrig L 2003. Effects of habitat fragmentation on biodiversity. *Ann. Rev. Ecol. Evol. Syst.* **34:** 487–515.

FHA (Federal Highway Administration) 1996. Evaluation and management of highway runoff water auality. US Dept of Transportation, Washington.. Publication FHWA-PD-96–032.

Findlay CS and Bourdages J 2000. Response time of wetland biodiversity to road construction on adjacent lands. *Conserv. Biol.* **14:** 86–94.

Flannery T 2004. *Country.* Text Publishing, Melbourne.

Foley JA, DeFries R, Asner GP, Barford C, Bonan G, Carpenter SR, Chapin FS, Coe MT, Daily GC, Gibbs HK, Helkowski JH, Holloway T, Howard EA, Kucharik CJ, Monfreda C, Patz JA, Prentice IC, Ramankutty N and Snyder PK 2005. Global consequences of land use. *Science* **309:** 570–574.

Foppen R and Reijnen R 1994. The effects of car traffic on breeding bird populations in woodland. II. Breeding dispersal of male willow warblers (*Phylloscopus trochilus*) in relation to the proximity of a highway. *J. Appl. Ecol.* **31:** 95–101.

Forman RTT 1998. Road ecology: a solution for the giant embracing us. *Landscape Ecol.* **13:** III–V.

Forman RTT 2000. Estimate of the area affected ecologically by the road system in the United States. *Conserv. Biol.* **14:** 31–35.

Forman RTT and Alexander LE 1998. Roads and their major ecological effects. *Ann. Rev. Ecol. Syst.* **29:** 207–231.

Forman RTT, Reineking B and Hersperger AM 2002. Road traffic and nearby grassland bird patterns in a suburbanizing landscape. *Environ. Man.* **29:** 782–800.

Forman RTT, Sperling D, Bissonette JA, Clevenger AP, Cutshall CD, Dale VH, Fahrig L, France R, Goldman CR, Heanue K, Jones JA, Swanson FJ, Turrentine T and Winter TC (Eds) 2003. *Road Ecology: Science and Solutions.* Island Press, Washington DC.

Gerlach G and Musolf K 2000. Fragmentation of landscape as a cause for genetic subdivision in bank voles. *Conserv. Biol.* **14:** 1066–1074.

Getz LL, Cole FR and Gates DL 1978. Interstate roadsides as dispersal routes for *Microtus pennsylvanicus*. *J. Mammal.* **59:** 208–212.

Goosem M 2000. Effects of tropical rainforest roads on small mammals: edge changes in community composition. *Wildl. Res.* **27:** 151–163.

Goosem M 2001. Effects of tropical rainforest roads on small mammals: inhibition of crossing movements. *Wildl. Res.* **28:** 351–364.

Groot Bruinderink GWTA and Hazebroek E 1996. Ungulate traffic collisions in Europe. *Conserv. Biol.* **10:** 1059–1067.

Grzebieta RH and Rechnitzer G 2001. Crashworthy systems: a paradigm shift in road safety design. II. *Transport Engineering in Australia.* Vol. 7, Nos 1 and 2, IE Aust., Australia.

Hayward MW, de Tores PJ, Augee ML and Banks PB 2005. Mortality and survivorship of the quokka (*Setonix brachyurus*) (Macropodidae: Marsupialia) in the northern jarrah forest of Western Australia. *Wildl. Res.* **32:** 715–722.

Hazlitt SL, Goldizen AW and Eldridge MDB 2006. Significant patterns of population genetic structure and limited gene flow in a threatened macropodid marsupial despite continuous habitat in southeast Queensland, Australia. *Conserv. Genet.* **7:** 675–689.

Hobday AJ and Minstrell ML 2008. Distribution and abundance of roadkill on Tasmanian highways: human management options. *Wildl. Res.* **35:** 712–726.

Horsup A 1994. Home-range of the allied rock-wallaby, *Petrogale assimilis. Wildl. Res.* **21:** 65–84.

Hubbard MW, Danielson BJ and Schmitz RA 2000. Factors influencing the location of deer–vehicle accidents in Iowa. *J. Wildl. Man.* **64:** 707–713.

Imhoff ML, Bounoua L, Ricketts T, Loucks C, Harriss R and Lawrence WT 2004. Global patterns in human consumption of net primary production. *Nature* **429:** 870–873.

Ingleby S 1991. Distribution and status of the northern nailtail wallaby, *Oncychogalea unguifera* (Gould, 1841). *Wildl. Res.* **18:** 655–675.

Jaeger JAG and Fahrig L 2004. Effects of road fencing on population persistence. *Conserv. Biol.* **18:** 1651–1657.

Jarman PJ 1991. Social behaviour and organisation in the Macropodoidea. *Adv. Stud. Behav.* **20:** 1–50.

Jarman PJ and Taylor RJ 1983. Ranging of eastern grey kangaroos and wallaroos on a New England pastoral property. *Aust. Wildl. Res.* **10:** 33–38.

Jones ME 2000. Road upgrade, road mortality and remedial measures: impacts on a population of eastern quolls and Tasmanian devils. *Wildl. Res.* **27:** 289–296.

Klöcker U 2003. An evaluation of factors leading to kangaroo–vehicle collisions in the arid zone of Australia. Graduate Diploma thesis. Bonn University.

Klöcker U, Croft DB and Ramp D 2006. Frequency and causes of kangaroo–vehicle collisions on an Australian outback highway. *Wildl. Res.* **33:** 5–15.

Laurance SGW 2004. Responses of understorey rainforest birds to road edges in Central Amazonia. *Ecol. Appl.* **14:** 1344–1357.

Lee E 2006. The ecological effects of sealed roads in arid ecosystems. PhD thesis. University of New South Wales, Sydney.

Lee E, Klöcker U, Croft DB and Ramp D 2004. Kangaroo–vehicle collisions in Australia's sheep rangelands, during and following drought periods. *Aust. Mammal.* **26:** 215–226.

Lintermans M 1997. A review of the use of Swareflex wildlife reflectors to reduce the incidence of road-kills in native fauna. In *Living with Eastern Grey Kangaroos in the ACT.* 3rd report, Appendix D. ACT Kangaroo Advisory Committee, Canberra.

Lopez RR 2004. Florida key deer (*Odocoileus virginiaus clavium*): effects of urban development and road mortality. In *Species Conservation and Management: Case Studies.* (Eds HR Akçakaya, MA Burgman, O Kindvall, CC Wood, P Sjögren-Gulve, JS Hatfield and MA McCarthy) pp. 450–458. Oxford University Press, New York.

Loreau M 2000. Biodiversity and ecosystem functioning: recent theoretical advances. *Oikos* **91:** 3–17.

Mace G, Masundire H and Baillie J 2005. Biodiversity. In *Ecosystems and Human Well-being: Current State and Trends.* (Eds R Hassan, R Scholes and N Ash) Vol. 1. Island Press, Washington DC.

MacFarlane AM and Coulson G 2005. Synchrony and timing of breeding influences sexual segregation in western grey and red kangaroos (*Macropus fuliginosus* and *Macropus rufus*). *J. Zool.* **267:** 419–429.

Magnus Z, Kriwoken LK, Mooney NJ and Jones ME 2004. Reducing the incidence of wildlife roadkill: improving the visitor experience in Tasmania. CRC for Sustainable Tourism, Gold Coast, Australia.

Mansergh IM and Scotts DJ 1989. Habitat continuity and social organization of the mountain pygmy-possum restored by tunnel. *J. Wildl. Man.* **53:** 701–707.

Margules CR and Pressey RL 2000. Systematic conservation planning. *Nature* **405:** 243–253.

Millennium Ecosystem Assessment 2005. *Ecosystems and Human Well-being: Biodiversity Synthesis.* World Resources Institute, Washington DC.

Miller JR and Hobbs RJ 2002. Conservation where people live and work. *Conserv. Biol.* **16:** 330–337.

Morrissey VK 2003. Wildlife–vehicle collisions in the Royal National Park, Sydney. Hons thesis. University of New South Wales, Sydney.

Muirhead S, Blache D, Wykes B and Bencini R 2006. Roo-Guard[R] sound emitters are not effective at deterring tammar wallabies (*Macropus eugenii*) from a source of food. *Wildl. Res.* **33:** 131–136.

Mumme RL, Schoech SJ, Woolfenden GW and Fitzpatrick JW 2000. Life and death in the fast lane: demographic consequences of road mortality in the Florida Scrub-Jay. *Conserv. Biol.* **14:** 501–512.

Newell GR 1999. Australia's tree-kangaroos: current issues in their conservation. *Biol. Conserv.* **87:** 1–12.

Norbury GL, Norbury DC and Oliver AJ 1994. Facultative behaviour in unpredictable environments: mobility of red kangaroos in arid Western Australia. *J. Anim. Ecol.* **63:** 410–418.

Ortega YK and Capen DE 2002. Roads as edges: effects on birds in forested landscapes. *Forest Sci.* **48:** 381–390.

Osawa R 1989. Road-kills of the swamp wallaby, *Wallabia bicolor*, on North Stradbroke Island, south-east Queensland. *Aust. Wildl. Res.* **16:** 95–104.

Pimm SL and Raven P 2000. Biodiversity: extinction by numbers. *Nature* **403:** 843–845.

Pople AR, Grigg GC, Cairns SC, Beard LA and Alexander P 2000. Trends in the numbers of red kangaroos and emus on either side of the South Australian dingo fence: evidence for predator regulation? *Wildl. Res.* **27:** 269–276.

Possingham HP and Wilson KA 2005. Biodiversity: turning up the heat on hotspots. *Nature* **436:** 919–920.

Priddel D 1988. Habitat utilisation by sympatric red kangaroos, *Macropus rufus*, and western grey kangaroos, *M. fuliginosus*, in western New South Wales. *Aust. Wildl. Res.* **15:** 413–421.

Priddel D, Wellard G and Shepherd N 1988. Movements of sympatric red kangaroos, *Macropus rufus*, and western grey kangaroos, *M. fuliginosus*, in western New South Wales. *Aust. Wildl. Res.* **15:** 339–346.

Ramp D and Ben-Ami D 2006. The effect of road-based fatalities on the viability of a peri-urban swamp wallaby population. *J. Wildl. Man.* **70:** 1615–1624.

Ramp D and Coulson G 2002. Density dependence in foraging habitat preference of eastern grey kangaroos. *Oikos* **98:** 393–402.

Ramp D and Croft DB 2006. Do wildlife warning reflectors elicit aversion in captive macropods? *Wildl. Res.* **33:** 583–590.

Ramp D and Roger E 2008. Frequency of animal–vehicle collisions in NSW. In *Too Close for Comfort: Contentious Issues in Human–wildlife Encounters.* (Eds D Lunney, A Munn and W Meikle) pp. 118–126. Royal Zoological Society of NSW, Sydney.

Ramp D, Caldwell J, Edwards KA, Warton D and Croft DB 2005a. Modelling of wildlife fatality hotspots along the Snowy Mountain Highway in New South Wales, Australia. *Biol. Conserv.* **126:** 474–490.

Ramp D, Russell BG and Croft DB 2005b. Predator scent induces differing responses in two sympatric macropodids. *Aust. J. Zool.* **53:** 73–78.

Ramp D, Wilson VK and Croft DB 2006. Assessing the impacts of roads in peri-urban reserves: road-based fatalities and road usage by wildlife in the Royal National Park, New South Wales, Australia. *Biol. Conserv.* **129:** 348–359.

Reijnen R, Foppen R and Meeuwsen H 1996. The effects of traffic on the density of breeding birds in Dutch agricultural grasslands. *Biol. Conserv.* **75:** 255–260.

Reijnen R, Foppen R, Terbraak C and Thissen J 1995. The effects of car traffic on breeding bird populations in woodland. III. Reduction of density in relation to the proximity of main roads. *J. Appl. Ecol.* **32:** 187–202.

Reijnen R, Foppen R and Veenbaas G 1997. Disturbance by traffic of breeding birds: evaluation of the effect and considerations in planning and managing road corridors. *Biodivers. Conserv.* **6:** 567–581.

Richardson JH, Shore RF, Treweek JR and Larkin SBC 1997. Are major roads a barrier to small mammals? *J. Zool.* **243:** 840–846.

Roger E and Ramp D 2009. Incorporating habitat use in models of fauna fatalities on roads. *Divers. Distrib.* **15:** 222–231.

Romin LA and Bissonette JA 1996. Deer–vehicle collisions: status of state monitoring activities and mitigation efforts. *Wildl. Soc. Bull.* **24:** 276–283.

Rose TA, Munn AJ, Ramp D and Banks PB 2006. Foot-thumping as an alarm signal in macropodoid marsupials: prevalence and hypotheses of function. *Mammal Rev.* **36:** 281–298.

Rowden P, Steinhardt D and Sheehan M 2008. Road crashes involving animals in Australia. *Accid. Anal. Prev.* **40:** 1865–1871.

Seabrook WA and Dettmann EB 1996. Roads as activity corridors for cane toads in Australia. *J. Wildl. Man.* **60:** 363–368.

Seiler A 2003. The toll of the automobile: wildlife and roads in Sweden. PhD thesis. University of Agricultural Sciences, Sweden.

Seiler A 2004. Trends and spatial patterns in ungulate–vehicle collisions in Sweden. *Wildl. Biol.* **10:** 301–313.

Seiler A 2005. Predicting locations of moose–vehicle collisions in Sweden. *J. Appl. Ecol.* **42:** 371–382.

Sekercioglu CH, Daily GC and Ehrlich PR 2004. Ecosystem consequences of bird declines. *Proc. Natl Acad. Sci. USA* **101:** 18042–18047.

Spooner PG, Lunt ID and Briggs SV 2004. Spatial analysis of anthropogenic disturbance regimes and roadside shrubs in a fragmented agricultural landscape. *Appl. Veg. Sci.* **7:** 61–70.

Stiles JH and Jones RH 1998. Distribution of the red imported fire ant, *Solenopsis invicta*, in road and powerline habitats. *Landscape Ecol.* **13:** 335–346.

Strieter Corp. 2001. Strieter-Lite: wild animal highway warning reflector system. Installation instructions. Strieter Corporation, Rock Island, Illinois.

Taylor BD and Goldingay RL 2003. Cutting the carnage: wildlife usage of road culverts in north-eastern New South Wales. *Wildl. Res.* **30:** 529–537.

Taylor BD and Goldingay RL 2004. Wildlife road-kills on three major roads in north-eastern New South Wales. *Wildl. Res.* **31:** 83–91.

Tenhumberg B, Tyre AJ, Pople AR and Possingham HP 2004. Do harvest refuges buffer kangaroos against evolutionary responses to selective harvesting? *Ecology* **85:** 2003–2017.

Tikka PM, Hogmander H and Koski PS 2001. Road and railway verges serve as dispersal corridors for grassland plants. *Landscape Ecol.* **16:** 659–666.

Trombulak SC and Frissell CA 2000. Review of ecological effects of roads on terrestrial and aquatic communities. *Conserv. Biol.* **14:** 18–30.

Ujvari M, Baagoe HJ and Madsen AB 1998. Effectiveness of wildlife warning reflectors in reducing deer–vehicle collisions: a behavioral study. *J. Wildl. Man.* **62:** 1094–1099.

Ullmann I 1998. Lateral differentiation and the role of exotic species in roadside vegetation in southern New Zealand. *Flora* **193:** 149–164.

Vestjens WJM 1973. Wildlife mortality on a road in New South Wales. *Emu* **73:** 107–112.

Waring GH, Griffs JL and Vaughn ME 1991. White-tailed deer roadside behaviour, wildlife warning reflectors, and highway mortality. *Appl. Anim. Behav. Sci.* **29:** 215–223.

Williams AF and Wells JK 2005. Characteristics of vehicle–animal crashes in which vehicle occupants are killed. *Traffic Injury Prev.* **6:** 56–59.

Wilson K, Pressey RL, Newton A, Burgman M, Possingham H and Weston C 2005. Measuring and incorporating vulnerability into conservation planning. *Environ. Man.* **35:** 527–543.

25 Speed limit, verge width and day length: major factors in road-kills of tammar wallabies on Garden Island, Western Australia

B.K. Chambers, R. Dawson, J. Wann and R. Bencini

SUMMARY

Up to 400 tammar wallabies (*Macropus eugenii*) are killed on the roads of Garden Island, Western Australia, every year, which corresponds to approximately 14–18% of the estimated population size. These road-kills are a result of the large volume of traffic associated with the HMAS Stirling Naval Base. This study investigated the relationship between the number of road-kills and verge width, fences, speed limit, roadside lighting and type of vegetation adjoining the road by subdividing the roads on the base into sections of approximately 500 m and fitting a generalised linear model to the data from each section. A GIS held by the Royal Australian Navy was used to obtain data on roadside fences and type of vegetation and to map numbers of road-kills, to identify the frequency of road-kills in each section. The impacts of day length, rainfall and number of personnel in port on the number of animals killed were determined by linear regression and analysis of covariance. Day length, verge width and speed limit were significantly correlated with the frequency of road-kills. Day length was negatively correlated with the number of road-kills ($P < 0.001$), while increased verge width was correlated with increased road-kills. Speed limits of 50 km/h resulted in significantly fewer road-kills per section than speed limits of 60 km/h or 80 km/h ($P < 0.001$). This work demonstrates the practical use of GIS technology to identify factors affecting road-kills and to determine possible mitigation strategies.

INTRODUCTION

Road mortality has the potential to be a serious threatening process for populations of wild animals, especially where the rate of deaths exceeds the population's recruitment rate (Jones 2000). Even when the number of animals killed by vehicles is not large enough to cause a decline in population size, road-kills can alter sex ratios and age structures within a population if animals of a particular sex or age are killed more frequently (Hauer *et al.* 2000; Steen and Gibbs 2004; Steen *et al.* 2006). On Garden Island, Western Australia, approximately 300–400 tammar wallabies (*Macropus eugenii*) are killed each year by vehicles associated with the HMAS Stirling Naval Base. As there are approximately 2200 tammars on the island (Bradshaw 1988), this number of deaths equates to 14–18% of the total population each year.

This chapter describes the use of a geographical information system (GIS) to gather data on the relationship between high numbers of road-kills and environmental variables. In the past GIS has been used to study habitat selection by wildlife (Stoms *et al.* 1993; Harvey and Hill 2003; Jedrzejewski *et al.* 2005) or to predict the distribution of wildlife in the landscape based on habitat structure variables (Austin *et al.* 1996; Lahm *et al.* 1998; Carrascal *et al.* 2002). More recently, GIS has been used to predict the distribution of wildlife in relation to environmental threats such as pollution (Chow *et al.* 2005) and habitat modification (Figala *et al.* 2001) including road construction (Treweek and Veitch 1996). Despite the fact that road mortality is now recognised as a threat for wildlife (Forman and Alexander 1998), few studies have used GIS to examine the factors associated with road-kills (Ramp *et al.* 2005, 2006).

This study aimed to find spatial and temporal variables that were correlated with the number of tammar wallabies killed on different roads and at different times of the year, to inform decisions on possible mitigation methods. We focused on day length, number of personnel in port, roadside fencing, width of road verges, roadside lighting, type of roadside vegetation and speed limit. We expected that day length would be negatively correlated with the number of road-kills, as shorter days are associated with personnel starting and finishing work closer to dawn and dusk when tammars come out to graze.

The number of personnel in port was expected to be positively correlated with the number of road-kills, as more personnel in port should result in greater traffic volumes on the base; increased traffic volumes have been shown to result in more animals being killed on roads (Ramp *et al.* 2006).

Roadside fencing has been shown to increase or decrease the number of animals killed by vehicles, depending on the extent to which the fences exclude animals from the road (Romin and Bissonette 1996; Putman 1997; Jaeger and Fahrig 2004). Fences that completely exclude animals from the road obviously result in much fewer animals being killed by vehicles but, when fences act as a barrier to startled animals exiting the road verge, they can increase the number of road-kills by causing the animals to cross the road in an attempt to escape. On Garden Island, roadside fencing does not completely exclude animals from the road verge but may act as a barrier to animals trying to escape from oncoming cars. We therefore expected that roadside fencing would be positively correlated with the numbers of road-kills.

Verge width was expected to be positively correlated with the number of animals killed on the roads as plants on the road verges of the base are a major food source for the tammars (McMillan 2006). Lee *et al.* (2004) and Ramp *et al.* (2006) found that more macropodids were killed by vehicles when there was more food available on the road verges. Wider verges provide more grass for the tammars to graze, and this should bring more animals to the roadside. We would expect that the greater number of animals at the roadside would result in more animals crossing the road, and therefore in more animals being killed by vehicles.

Tammars have been shown to associate increased levels of nocturnal lighting with safety by reducing the amount of time invested in vigilance activities (Beiebouw and Blumstein 2003). They have also been shown to recognise potential predators by sight rather than by sound or smell (Blumstein *et al.* 2000, 2002). Roads on the base were lit on one or both sides. We expected that the tammars grazing on road verges that were lit on both sides would see and respond to vehicles by moving away from the road sooner than animals grazing on road verges that were less well lit, thereby resulting in fewer road-kills.

The type of vegetation near the roadside was also expected to have an impact on the number of tammars killed by vehicles, as tammars have been shown to devote more time to vigilance behaviour the further they are from cover (Blumstein *et al.* 1999). We therefore expected that tammars grazing on verges that did not have nearby dense vegetation would be more flighty and therefore more likely to attempt to cross the road when a vehicle approached, therefore leading to more road-kills.

Vehicle speed has been shown on many occasions to be positively correlated with animal deaths on roads, as higher speeds reduce the amount of time a driver has to react to the animal's presence and avoid a collision (Jones 2000; Ramp *et al.* 2006). We therefore expected that road sections with higher posted speed limits would have higher numbers of road-kills than those with lower speed limits.

METHODS

We subdivided the roads of Garden Island into 500 m sections and used a GIS to identify the frequency of

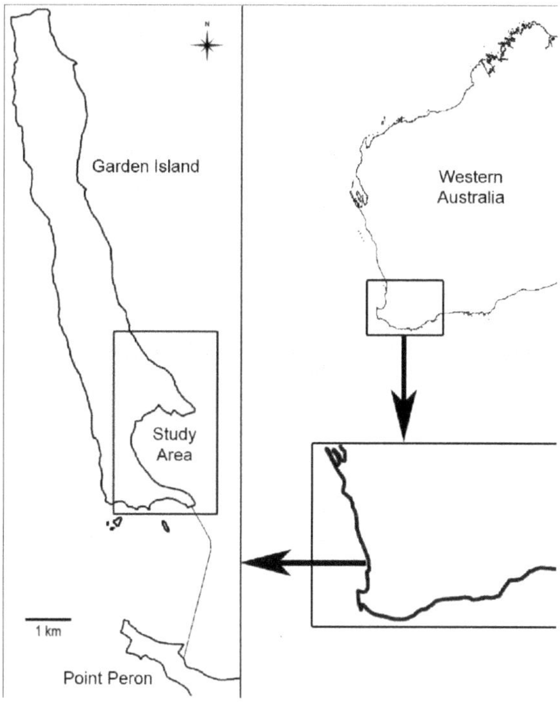

Figure 25.1: Location of Garden Island, Western Australia.

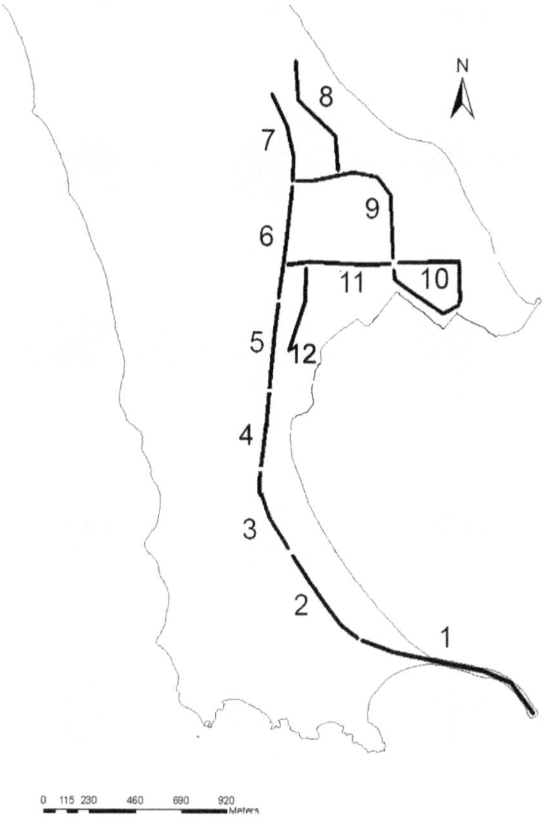

Figure 25.2: Road sections used in the analysis of tammar wallaby (*M. eugenii*) road-kill frequencies on Garden Island, Western Australia.

road-kills and the factors that might affect the frequency of road-kills for each section. We then used a generalised linear model to test the effects of individual factors on the frequency of road-kills.

Study site

Garden Island is located approximately 35 km south of the Western Australian city of Perth and approximately 5 km west of the mainland (Fig. 25.1). The island hosts Australia's largest naval base, HMAS Stirling, which was built during the 1970s. As part of the construction of the base a causeway was built from Point Peron on the mainland to the southern end of the island to allow access to vehicles on Defence Department related business. The infrastructure of the base is primarily in the southern part of the island and this is where the majority of tammars, approximately 200–250 per year, are killed by vehicles.

Road-kills

Between January 2000 and December 2004 the gardening and security staff and the ranger recorded the date and location of every tammar killed by vehicles on the island. The roads on the main area of the base were divided into 12 sections of approximately 500 m each (Fig. 25.2) and entered into ArcGIS (ESRI, New York) as a new shapefile. The road-kill data were added to each of these segments along with the explanatory variables. 500 m was chosen as the length of the road sections because data for the 500 m sections were present in the GIS database from a previous study (Walker 2002). The single road north of this area was not studied, as access is restricted between 6 p.m. and 6 a.m. Due to the restricted access, this road also has much less traffic than the roads on the main area of the base.

Spatial variables

Fences

Data on the presence or absence of fences was obtained from a GIS database held by the Australian Navy and managed by Sinclair Knight Merz (Perth, Western Australia). These fence data were used to obtain a table specifying the presence of fences and the length of each road segment that was fenced. A road was considered fenced if the open road verge ended in a fence within 10 m of the road. Based on this classification, the road sections were classed as being unfenced (0), fenced on one side (1) or fenced on both sides (2). For the purposes of the analysis, fencing was considered a categorical variable.

Roadside vegetation

Data on roadside vegetation were obtained from the GIS database and classified according to the presence or absence of *Melaleuca lanceolata–Acacia rostellifera* low forest or *Melaleuca huegelii* scrub (Brooker 1992). To do this, 25 m buffers were created either side of the road segments and clipped with the vegetation data (resolution 10 m) to determine which vegetation types occurred within the buffers. Where this vegetation was not found within 25 m of the road the area consisted of irrigated lawn, buildings or car parks – these areas were classified as disturbed. The three vegetation types were treated as separate categories in the analysis.

Verge width

Verge width (m) was measured for each road section at approximately 100 m, 250 m and 400 m along the section. The verge width was measured on both sides and the mean of all six measurements was used in the analysis.

Roadside lighting

The presence of roadside lighting was recorded from a survey of the roads, as these data were not present in the GIS database. All roads had lighting of some form and so were classified as being lit on one side (1) or both sides (2).

Speed

The location of changes in speed limits was recorded from a survey of the roads on the island. There were three speed zones – 50 km/h, 60 km/h and 80 km/h – and each road section had the same speed limit for its entire length. These three speed limits were treated as categorical variables in the analysis.

Temporal variables

Day length

Day length (hours) was calculated from sunrise and sunset times obtained from Geoscience Australia for the study period. The day length for the 15th day of each month was used as the measure of day length in the analysis. Day length was chosen, rather than hours to work closing or opening, as animals were run over in the evenings and early morning.

Personnel in port

The number of visiting personnel was calculated from records of ship movements obtained from the Fremantle Port Authority. The number of personnel on each ship was obtained from publicly available information on each naval vessel and the number of days for each month that the each ship was in port was multiplied by its number of crew to give the monthly number of personnel in port. The number of personnel in port each day was averaged for the entire month and this value was used in the analysis.

Data analysis

The effects of the spatial variables on the number of animals killed were analysed using a generalised linear model assuming a normal distribution and identity link function. The data on total road-kills were inspected visually prior to analysis to ensure that they were normally distributed. The model was initially constructed with the total number of road-kills per road section as the dependent variable and all the independent spatial variables included. Models with all combinations of the independent variables were then tested and the most parsimonious model was chosen based on Akaike Information Criterion (AIC) values.

The effects of the temporal variables of day length, rainfall and number of personnel in port were analysed by regression to determine if they were correlated with the number of animals killed. As rainfall and day length are also correlated with each other, their effect was determined using analysis of covariance (ANCOVA). The mean rainfall for each month of the year was calculated over the five years of the study, and the rainfall for each individual month was classified as above or below average. The number of road-kills over all road sections was analysed against rainfall (high or low) with day length as a covariate.

RESULTS

Dates and locations were recorded for a total of 1258 road-kills at a rate of 3.5 ± 0.31 deaths/km/month (mean ± SE). The rate of road-kills over the entire study area varied between 0 deaths/km/month in December 2002 to 10.17 deaths/km/month in June 2003. Over the five-year study period, the largest numbers of animals were killed between March and August (Fig. 25.3).

Spatial variables

The most parsimonious model included only verge width and speed, without any interaction, as fencing,

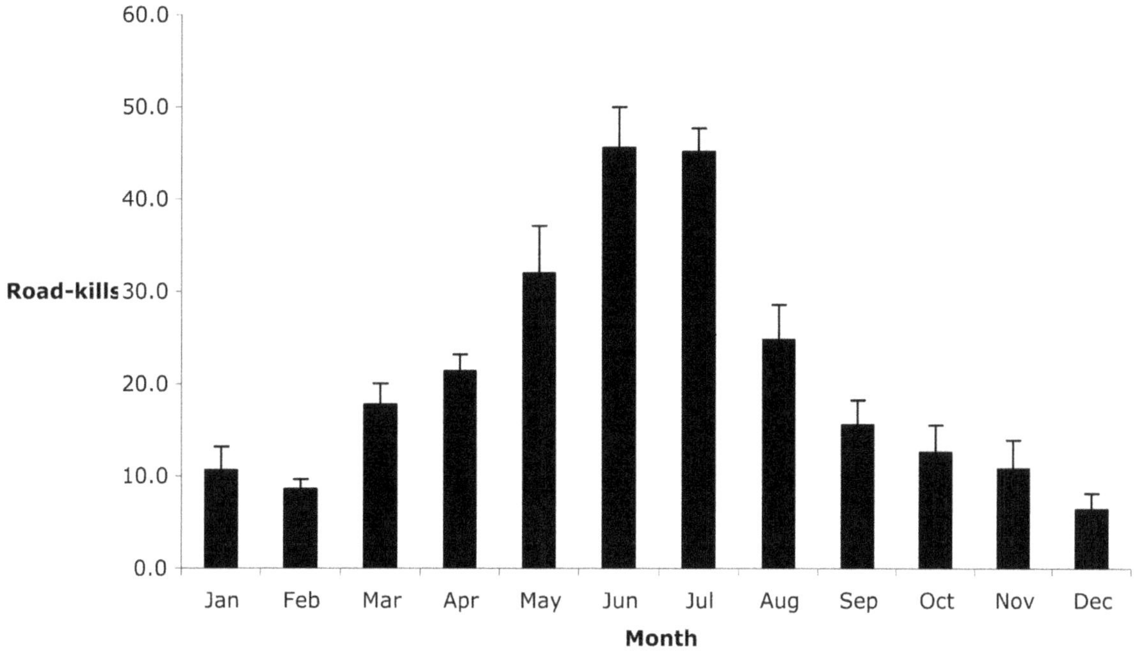

Figure 25.3: Mean number of tammar wallabies (*M. eugenii*) killed by vehicles on the HMAS Stirling Naval Base each month between January 2000 and December 2004.

roadside vegetation and lighting were not significant predictors of the number of road-kills. Verge width was positively correlated with the number of animals killed and more animals were killed in 60 km/h speed zones than in 80 km/h or 50 km/h zones (Table 25.1).

Temporal variables

Rainfall and day length were significantly correlated with the number of animals killed (rainfall r^2 = 0.436, P < 0.001; day length r^2 = 0.696, P < 0.001), while the number of personnel in port was not correlated significantly with the number of road-kills (r^2 = 0.004, P = 0.683). When rainfall was considered with day length as a covariate there was no significant effect of rainfall on road-kills (df = 1,56, F = 1.05, P = 0.311), but day length was highly significant with an additional 8.4 deaths per month for each one-hour reduction in day length (df = 1,56, F = 130.78, P < 0.001).

Table 25.1: The coefficients, standard errors and probability values of the explanatory variables that were significant in predicting the numbers of tammar wallaby (*M. eugenii*) killed on roads on Garden Island.

Variable	Coefficient	df	chi²	P
Verge width	6.98	1	3.03	0.081
Speed limit		2	8.85	0.012
50 km/h	–44.63			
60 km/h	37.45			
80 km/h	–7.18			

DISCUSSION

Day length was negatively correlated with the number of animals killed by vehicles, which supported our hypothesis. While day length was correlated with rainfall, the ANCOVA results demonstrated that rainfall did not have a significant effect on the number of animals killed. This is most likely the result of shorter days in winter resulting in more people arriving and leaving work before sunrise and after sunset.

Significantly fewer animals were killed on roads with a speed limit of 50 km/h compared with 60 km/h or 80 km/h. This supported our hypothesis that higher speed limits would be associated with more road-kills, and agrees with previous studies showing that vehicle speed is one of the most important causative factors in animal–vehicle collisions (Jones 2000; Ramp *et al.* 2006). However, the lack of difference between 60 km/h and 80 km/h zones was unexpected. When the road-kills hotspots were overlaid with the speed limit zones using the GIS maps, it became clear that the a large number of the tammars were killed on the road where the speed limit changed

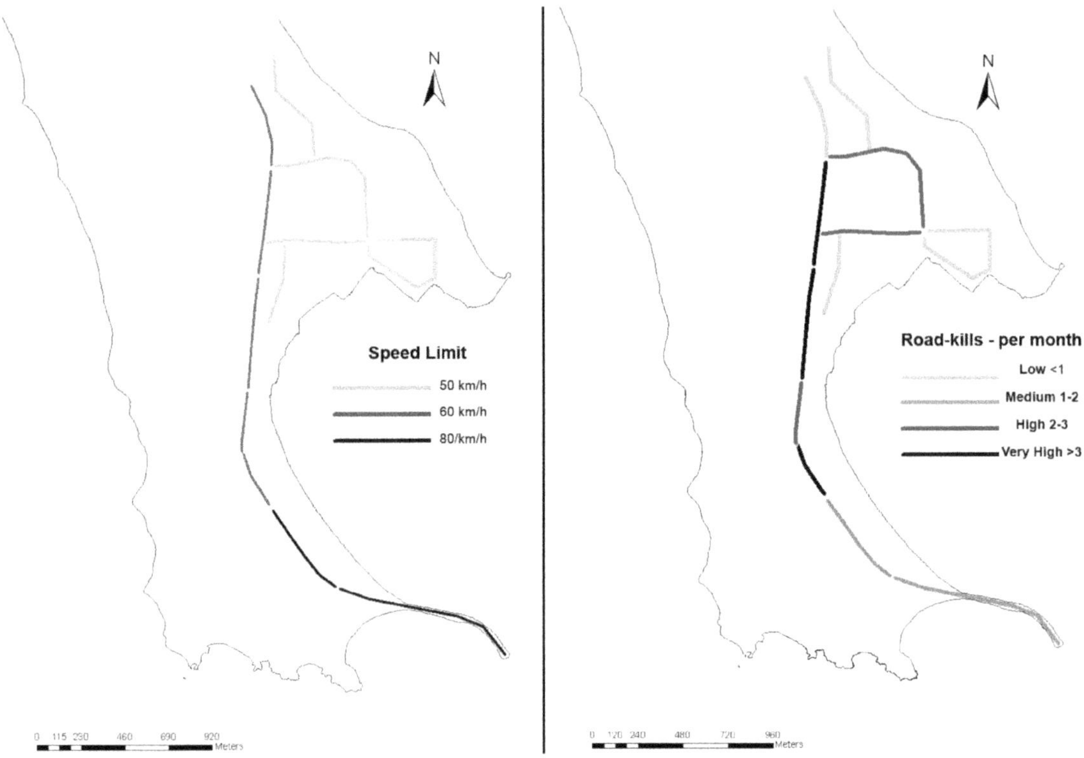

Figure 25.4: Comparison of posted speed limits and the number of tammar wallabies (*M. eugenii*) killed by vehicles on Garden Island, Western Australia.

from 80 km/h to 60 km/h (Fig. 25.4). Reports from the ranger on the island suggest that many people do not reduce their speed to comply with the 60 km/h speed limit and that this may cause the unexpected result (T. Smith, *pers. comm.*). Using actual mean vehicle speed rather than posted speed limits would probably have been a better predictor of the number of animals killed, but unfortunately it was not possible to measure the actual speed of vehicles.

The width of the road verge was also positively correlated with the number of animals killed by vehicles, which supported our hypothesis. We expected that wider verges would attract more animals to feed and that this would result in a greater number of animals killed by vehicles. Wider verges may also be related to different behaviours as tammars have been observed to dedicate more time to vigilance activities as they graze further from cover (Blumstein *et al.* 1999). The width of verges may also affect animal behaviour in relation to vehicles, so it would be worthwhile observing the behaviour of animals on road verges when vehicles approach.

The hypothesis that roadside fencing would increase the number of animals killed by vehicles was not supported, as fencing was not a significant predictor of the number of road-kills. This result suggests that the roadside fences on Garden Island do not act as a barrier to animals trying to leave the road verge in response to approaching vehicles. We observed many tammars exit the road verge by running parallel to the roadside fences until they come to a hole in the fence. This behaviour may be the reason why fences were not significantly correlated with the number of animals killed by vehicles.

The number of personnel in port was not related to increased numbers of tammars killed by vehicles, which did not support our hypothesis. We expected that an increase in the number of personnel in port would increase traffic on the island, but were unable to gather data on traffic volumes. It may be that the increased number of personnel in port did not increase traffic volumes or that traffic volumes were only increased during the day, when road-kills usually do not occur because the tammars are not active. It would have been advantageous to have access to traffic volume data; future work should attempt to address this.

Our hypotheses about the effect of roadside lighting and roadside vegetation were also not supported, as these variables were not significant predictors of road-kills. The lack of a significant relationship between lighting and road-kills may be due to the fact that all roads on the base were lit to some degree. Our model only distinguished between lighting on one side or both sides of the road and this may not actually produce the differences in vigilance behaviour reported in previous studies (Beiebouw and Blumstein 2003). The hypothesis that fewer road-kills would occur on roads with dense vegetation on the road verge was also not supported. This may have been a result of the highly disturbed nature of the environment on the base. We expected that the tammars would seek refuge in dense vegetation when vehicles approached, but it may be that tammars on the base perceive buildings or other man-made objects as potential shelter in the same way as dense vegetation. If so, this may explain the lack of correlation between type of vegetation and road-kills.

The road-kill frequencies observed in this study were exceptionally high, with a mean rate of 3.5 ± 0.31 deaths/km/month; rates of 1.2 deaths/km/month are normally regarded as high (Ramp *et al.* 2005). This extremely high frequency of road-kills has the potential to seriously affect the tammar population on the island, currently estimated at ~3500 individuals (B. Chambers, unpubl. data). The impact of the road-kills on the population dynamics is the subject of ongoing research.

We were successful in finding correlations between three factors and the number of animals killed by vehicles, but there are undoubtedly other factors that contribute to the chances of animals being killed by vehicles. The volume of traffic, actual vehicle speed and the number of animals grazing on road verges have been found to be significant factors in contributing to road mortality and would likely account for some of the variation in the number of road-kills (Lee *et al.* 2004; Ramp *et al.* 2005, 2006). Unfortunately we did not have access to these data for the study period, but future work should attempt to account for these variables.

MANAGEMENT IMPLICATIONS

The results of our analysis have shown that speed, verge width and day length were correlated with the number of road-kills and mitigation methods should therefore focus on these three factors. Significantly more tammars were killed on roads where the speed limit was greater than 50 km/h; reducing the speed limit to 50 km/h for all roads on the island, especially between dusk and dawn, is strongly recommended. The use of traffic calming devices such as speed bumps is impractical on Garden Island, as emergency vehicles require unhindered access to all areas of the base. The introduction and enforcement of a lower speed limit is the most practical way to attempt to reduce the number of deaths caused by vehicle speed. Enforcement of reduced speed limits would be particularly important – previous studies have found that, without enforcement, reduced speed limits do not result in lower actual vehicle speeds (Pojar *et al.* 1975; Dique *et al.* 2003).

Verge width was positively correlated with the number of animals killed by vehicles, but the exact reasons for this relationship are unclear. Further work should be done to determine how verge width might influence the aggregation of tammars and their behaviour, and the influence of these on the chances of animals being run over. If the presence of more animals on wider road verges is the main reason for this correlation then reducing the width of road verges or removing grasses from the verges would likely reduce the number of animals being killed by vehicles.

Obviously nothing can be done to change day length, but the results suggest that targeting management strategies to March–August would have the greatest impact on reducing the number of animals killed. At this time of year the majority of female tammars are carrying pouch young, so reducing the number of animal–vehicle collisions would also reduce the number of pouch young that are killed or need to be taken into care. Limiting the amount of non-essential traffic between dusk and dawn may also reduce the number of tammars killed. Educating drivers to be more aware of the problem of tammar road-kills may also be useful in reducing the number of animals killed.

ACKNOWLEDGMENTS

We would like to thank the Australian Defence Force for funding this research. Our thanks also go to Trevor Smith (Defence Public Area Ranger, Garden Island), and the security and grounds and gardens staff of HMAS Stirling.

REFERENCES

Austin GE, Thomas CJ, Houston DC and Thompson DBA 1996. Predicting the spatial distribution of buzzard *Buteo buteo* nesting areas using a geographical information system and remote sensing. *J. Appl. Ecol.* **33:** 1541–1550.

Beiebouw K and Blumstein DT 2003. Tammar wallabies (*Macropus eugenii*) associate safety with higher levels of nocturnal illumination. *Ethol. Ecol. Evol.* **15:** 159–172.

Blumstein DT, Evans CS and Daniel JC 1999. An experimental study of behavioural group size effects in tammar wallabies, *Macropus eugenii. Anim. Behav.* **58:** 351–360.

Blumstein DT, Daniel JC, Griffin AS and Evans CS 2000. Insular tammar wallabies (*Macropus eugenii*) respond to visual but not acoustic cues from predators. *Behav. Ecol.* **11:** 528–535.

Blumstein DT, Mari M, Daniel JC, Ardron JG, Griffin AS and Evans CS 2002. Olfactory predator recognition: wallabies may have to learn to be wary. *Anim. Conserv.* **5:** 87–93.

Bradshaw SD 1988. Garden Island Tammar Research Programme: an investigation into the physiological wellbeing of the tammar wallaby, *Macropus eugenii*, on Garden Island, in Western Australia. University of Western Australia.

Brooker M 1992. *Biological Survey of Garden Island.* CSIRO Division of Wildlife and Ecology, Canberra.

Carrascal LM, Palomino D and Lobo JM 2002. Patterns of habitat preference and distribution and abundance of wintering bird fauna in Central Spain. Analysis and prediction of the effect of ecological factors. *Anim. Biodivers. Conserv.* **25:** 7–40.

Chow TE, Gaines KF, Hodgson ME and Wilson MD 2005. Habitat and exposure modelling for ecological risk assessment: a case study for the raccoon on the Savannah River Site. *Ecol. Model.* **189:** 151–167.

Dique DS, Thompson J, Preece HJ, Penfold GC, Villiers DL and Leslie RS 2003. Koala mortality in south-east Queensland: the koala speed zone trial. *Wildl. Res.* **30:** 419–426.

Figala J, Prchalova J and Tester JR 2001. GIS assessment of the decline of gray partridge (*Perdix perdix*) nesting habitat in the Elbe River Lowlands, the Czech Republic, 1949–1996. *Ekologia (Bratislava)* **20:** 209–218.

Forman RT and Alexander LE 1998. Roads and their major ecological effects. *Ann. Rev. Ecol. Syst.* **29:** 207–231.

Harvey KR and Hill GJE 2003. Mapping the nesting habitats of saltwater crocodiles (*Crocodylus porosus*) in Melacca Swamp and the Adelaide River wetlands Northern Territory: an approach using remote sensing and GIS. *Wildl. Res.* **30:** 365–375.

Hauer S, Ansorge H and Zinke O 2000. A long-term analysis of the age structure of otters (*Lutra lutra*) from eastern Germany. *Int. J. Mammal. Biol.* **65:** 360–368.

Jaeger JG and Fahrig L 2004. Effects of road fencing on population persistence. *Conserv. Biol.* **18:** 1651–1657.

Jedrzejewski W, Niedzialkowska M, Mysiajek RW, Nowak S and Jedrzejewska B 2005. Habitat selection by wolves *Canis lupus* in the uplands and mountains of southern Poland. *Acta Theriol.* **50:** 417–428.

Jones ME 2000. Road upgrade, road mortality and remedial measures: impacts on a population of eastern quolls and Tasmanian devils. *Wildl. Res.* **27:** 289–296.

Lahm SA, Barnes RFW, Beardsley K and Cervinkas P 1998. A method for censusing the greater white-nosed monkey in northeastern Gabon using the population density gradient in relation to roads. *J. Trop. Ecol.* **14:** 629–643.

Lee E, Klocker U, Croft DB and Ramp D 2004. Kangaroo–vehicle collisions in Australia's sheep rangelands, during and following drought periods. *Aust. Mammal.* **26:** 215–226.

McMillan A 2006. The utilisation of marine-derived nutrients by the tammar wallaby (*Macropus eugenii*) on Garden Island, Western Australia. Hons thesis. University of Western Australia, Perth.

Pojar TM, Prosence RA, Reed DF and Woodard TN 1975. Effectiveness of a lighted animated deer crossing sign. *J. Wildl. Man.* **39:** 87–91.

Putman RJ 1997. Deer and road traffic accidents: options for management. *J. Environ. Man.* **51:** 43–57.

Ramp D, Caldwell J, Edwards KA, Warton D and Croft DB 2005. Modelling of wildlife fatality hotspots along the Snowy Mountain Highway in New South Wales Australia. *Biol. Conserv.* **126:** 474–490.

Ramp D, Wilson VK and Croft DB 2006. Assessing the impacts of roads in peri-urban reserves: road-based fatalities and road usage by wildlife in the Royal National Park, New South Wales, Australia. *Biol. Conserv.* **129:** 348–359.

Romin LA and Bissonette JA 1996. Deer–vehicle collisions: status of state monitoring activities and mitigation efforts. *Wildl. Soc. Bull.* **24:** 276–283.

Steen DA and Gibbs JP 2004. Effects of roads on the structure of freshwater turtle populations. *Conserv. Biol.* **18:** 1143–1148.

Steen DA, Aresco MJ, Beilke SG, Compton BW, Condon EP, Kenneth Dodd C Jr, Forrester H, Gibbons JW, Greene JL, Johnson G, Langen TA, Oldham MJ, Oxier DN, Saumure RA, Schueler FW, Sleeman JM, Smith LL, Tucker JK and Gibbs JP 2006. Relative vulnerability of female turtles to road mortality. *Anim. Conserv.* **9:** 269–273.

Stoms DM, Davis FW, Cogan CB, Painho MO, Duncan BW and Scepan J 1993. Geographical analysis of California condor sighting data. *Conserv. Biol.* **7:** 148–159.

Treweek J and Veitch N 1996. The potential application of GIS and remotely sensed data to the ecological assessment of proposed new road schemes. *Glob. Ecol. Biogeogr. Lett.* **5:** 249–257.

Walker LJ 2002. Factors affecting the road deaths of tammar wallabies *(Macropus eugenii)* on Garden Island. Hons thesis. University of Western Australia, Perth.

26 Flight response as a causative factor in kangaroo–vehicle collisions

E. Lee, D.B. Croft and D. Ramp

SUMMARY

The reactions of large animals to vehicles approaching along roads and the role these reactions play in the likelihood of collisions with vehicles are not well understood. This study examined the flight responses of kangaroos (*Macropus* spp.) to an approaching vehicle to determine what factors influence flight responses, and whether flight responses influence collision likelihood. We recorded the behaviour of four kangaroo species (*M. rufus*, *M. giganteus*, *M. fuliginosus* and *M. robustus erubescens*) towards an approaching four-wheel-drive single-cab utility vehicle across a range of vehicle speeds, and compared kangaroo responses to the incidence of kangaroo–vehicle collisions on the same road over a two-year period. The study was conducted on the Fowlers Gap Research Station in north-west New South Wales. We found that the tendency to take flight was higher at night than during the day, and higher for *M. rufus* than for *M. robustus erubescens* during the night. Flight tendency was also higher in the absence of cover and in small groups for grey kangaroos. We found positive relationships between kangaroo flightiness and the likelihood of collisions and between species flightiness and collision frequency, suggesting that the behaviour of kangaroos in response to approaching vehicles has a direct bearing on their susceptibility to collisions with those vehicles. *M. rufus* were more susceptible to being hit than were *M. robustus erubescens*, while *M. giganteus* and *M. fuliginosus* had intermediate fatality rates. Kangaroos have evolved flight in an unpredictable direction along an erratic path to deal with high-level threats, as opposed to stationary vigilance behaviour for lower threat levels. We suggest that the degree to which oncoming vehicles invoke flight behaviour is a causative factor in collisions between kangaroos and vehicles.

INTRODUCTION

The effects of roads and traffic on fauna are many and varied (Forman *et al.* 2003). Mortality of fauna via collisions with motor vehicles (often referred to as wildlife road-kill) is probably the most obvious effect, influencing species from a range of faunal groups worldwide. Large numbers of wildlife road-kill have been documented (Forman *et al.* 2003; Ramp *et al.* 2005). Coupled with small or rare populations that may be additionally fragmented by roads (Jaeger *et al.* 2005), road-kill can negatively affect faunal populations (Jones 2000; Mumme *et al.* 2000; Dique *et al.* 2003; Aresco 2005) and consequently affect changes at the community level (Lee *et al.* 2004).

Many investigations have been conducted on the patterns and causes of road-kill for the development of strategic management frameworks mitigating collisions (Bender 2001; Forman *et al.* 2003; Aresco 2005). Various spatial and temporal factors have been identified as influencing the likelihood of collisions, including roadside and landscape features (Aresco 2005; Ramp *et al.* 2005; Klöcker *et al.* 2006), traffic volumes and vehicle speeds, faunal densities and fauna migration patterns, and weather conditions (Lee *et al.* 2004; Seiler 2005). However, the reactions of fauna to approaching vehicles have not been comprehensively investigated as a causal factor for collisions (Forman *et al.* 2003).

For collisions to occur, the presence of both fauna and vehicles is necessary and so, given the presence of fauna at roads, their reactions towards vehicles may affect their susceptibility to being hit (Horejsi 1981). Anecdotal accounts of differences in faunal behaviour towards approaching vehicles abound, suggesting that there may be some consistency in faunal responses to approaching vehicles. Research has indirectly observed differences in faunal behaviour towards approaching vehicles that may relate to the likelihood of collisions (Jones 2000; Brown 2001). Formal investigations addressing the consistency of faunal responses to approaching vehicles, how reactions vary between species or between age and sex classes within species, and how these reactions subsequently affect the likelihood of a collision, are long overdue.

This study used four sympatric large kangaroos as model species to examine the behaviour of fauna to approaching vehicles and the factors that influence these behaviours. We tested the hypothesis that an animal's behaviour towards an approaching vehicle influences the likelihood of collisions. The species were red kangaroo (*Macropus rufus*), eastern grey kangaroo (*Macropus giganteus*), western grey kangaroo (*Macropus fuliginosus*) and the euro (*Macropus robustus erubescens*). We combined *M. giganteus* and *M. fuliginosus* as grey kangaroos as there were few *M. giganteus* in the area and they appeared to behave in similar ways to *M. fuliginosus*. These kangaroos are good model species as they are morphologically similar, yet exhibit very different responses to perceived threats (Dawson 1995). The study focused on variations in the tendency of these kangaroos to take flight from an approaching vehicle and the relationship between kangaroo flightiness and road-kill frequency. Flightiness was regarded as the overall responsiveness of kangaroos to a stimulus or threat (e.g. variable climatic conditions, an approaching vehicle), with an increase in responsiveness signalling a relative increase in flightiness.

METHODS

Study area

The study was conducted at the University of New South Wales Arid Zone Research Station, Fowlers Gap (31°05'S, 141°43'E), in north-west New South Wales, Australia. Fowlers Gap is located on the Silver City Hwy approximately 110 km north of Broken Hill and covers an area of 38 888 ha. A 21.2 km bitumised section of the Silver City Hwy is contained within Fowlers Gap. This section is a dual carriageway c. 6.5 m wide which runs from the south-west boundary of the station to its north-east boundary. The topography of Fowlers Gap is diverse: the western section of the station includes part of the Barrier Ranges (180–240 m) while the eastern section consists of alluvial floodplains (140–170 m) (Mabbutt 1973). Thus the highway traverses areas of both high and low relief. The dominant vegetation in the areas of higher relief is low woody perennial shrubs (<1 m), chiefly of the Chenopodiaceae family. The floodplains are dominated by various tussock grasses (c. 50 cm; Burrell 1973).

Kangaroo behavioural data

Data were collected opportunistically over two years between March 2003 and February 2005 along the 21.2 km stretch of the Silver City Hwy. Only one observer collected data, to prevent introduction of variation into the dataset by differences between observers. To control for habituation by kangaroos towards an approaching vehicle on the highway, data were also collected along a second road. That road was a parallel single-lane dirt track approximately 3 m wide that was used infrequently by station staff and is located approximately 500 m from the highway (hinterland). Drives along roads (n = 299) were conducted at three times of day (dawn, dusk and night – two hours after sunset) and at different speeds (20–80 km/h) in a standard four-wheel-drive diesel vehicle (Nissan Navara single-cab tray-top pick-up). No visual aids were used to observe behavioural responses at night. This was an unavoidable limitation of recording behavioural responses at night, but night-time records were only used in select analyses investigating the tendency to take flight. The highway is rated at 100 km/h but a lower 80 km/h speed was the fastest speed that could be safely driven while collecting data. All kangaroos within sight of the vehicle were candidates for observation, but only kangaroos that were initially undisturbed by the vehicle's approach were considered. This may have biased sampling towards kangaroos that have a higher tolerance to vehicle presence, but was necessary in order to collect reliable response-distance data.

If flight occurred, the vehicle was stopped to record the distance between the kangaroo and vehicle (the initial distance to flight), the initial angle of the kangaroo from the vehicle and the location of the vehicle on the road. Final distances and angles of kangaroos from the vehicle after flight were also recorded. These measurements were made so that the total distance travelled during flight could be later calculated using a custom Microsoft Visual Basic application and simple trigonometry. Here we use only the initial distance to flight and the total distance travelled during flight for analyses. A Bushnell 2000 rangefinder was used to record distances, a protractor was used to record angles and a Garmin GPS II receiver was used to record the vehicle's location on the road.

If flight did not occur, the vehicle was stopped when the kangaroo was alongside the vehicle (i.e. perpendicular to the direction of the road) to record the kangaroo's closest distance to the vehicle. The behaviour of kangaroos during approach was recorded following Croft's (1980) behavioural definitions for *M. rufus*.

Other variables measured, for both kangaroos that took flight and kangaroos that did not take flight, were the species and age/sex class of the kangaroo, the size of the group of kangaroos and whether the kangaroo was initially in cover, defined as within 2 m of vegetation or rocks that obscured more than 30% of a kangaroo. As subjects were not individually identified they may have been included more than once in this study but, given that behavioural observations were conducted over relatively large spatial scales, pseudo-replication was likely to effect minimal bias (Oksanen 2004).

Road-kill data

Data for road-killed kangaroos along the highway were collected over 21 months between February 2003 and November 2004. The highway was surveyed every second day from a vehicle travelling at 40 km/h. For each road-kill, the date, location (Garmin GPS II), species and sex were recorded. When present and relatively undamaged, the heads of kangaroos were collected for an estimation of age by molar progression (after Kirkpatrick 1965).

Statistical analyses

Kangaroo flight response

Two types of analyses were used to determine if kangaroos differed in their flight responses (tendencies to take flight): chi-square analyses and logistic regression analysis. For the chi-square tests, exact tests with 10 000 Monte Carlo simulations were used to compare the frequencies of behaviours exhibited by kangaroos towards an approaching vehicle between the day and night (pooled across species), and between species during the day and night (only two behaviours: flight and no flight examined). The likelihood ratio statistic was used to account for the small sample size, and analyses were conducted using SPSS for Windows V13.0.

For the logistic regression, night-time observations were excluded as there were too many categories which did not have an adequate number of observations. Predictors entered into the model were time of day (dawn and dusk), road type (highway and hinterland), season (summer, autumn, winter, spring), speed of the approaching vehicle (continuous variable), distance from the approaching vehicle when flight occurred/closest distance to the approaching vehicle if no flight occurred (continuous variable), kangaroo species, kangaroo age/sex (adult/subadult male, adult/subadult female, young at foot), group size (one, two, greater than two) and presence of cover (binary). Data were excluded where the gender of the subject was unknown. A stepwise forward likelihood method in SPSS for Windows V13.0 was used, and predictors included in the final model were those which were individually significant at $p < 0.05$ and included terms for which elimination would have reduced the explanatory power of the model (R_L^2 used, Quinn and Keough 2002).

Significance statistics for Wald were used to determine if the levels within categorical predictors differed from each other in predicting the probability of flight (for predictors that had more than two categories). Contrasts between all levels within predictors were achieved by changing reference categories from first to last for predictors with three categories, and by recoding categories to contrast remaining categories against each other for predictors with four categories.

To determine the effect of a continuous predictor on flight, odds ratios (Exp β) were used, with values greater than one indicating that the predictor increased the probability of kangaroo flight and values less than one indicating the opposite. Interactions were difficult to interpret from the regression output; thus chi-square tests were used to determine differences within predictors influencing the probability of flight, with the signs of standardised residuals used to determine significant increases or decreases in flight probability. Exact tests with the likelihood ratio statistic were used, and checks of model accuracy and assumptions for logistic regressions were made after running the regression.

Kangaroo flightiness and the risk of road collision

Due to the opportunistic way that kangaroo behavioural data were recorded, it was difficult to directly relate the probability of a road-kill occurring with the tendency of a kangaroo to take flight in a manner that was also compatible with relating the probability of road-kill occurring with temporal factors known to influence road-kill. However, relationships between kangaroo flightiness and the probability of a road-kill occurring were inferred with two assumptions:

1 flight responses, including the initial distance to flight, the total distance travelled during flight and the tendency of a kangaroo to take flight are consistent between species;
2 strong winds increase the tendency of a kangaroo to take flight.

Given these assumptions, relationships between kangaroo flightiness and road-kill could be inferred because two of the three measures of kangaroo flightiness (the initial distance to flight and the total distance travelled during flight) could be directly related with road-kill frequency. Further, temporal factors, including windiness, could be directly related with the probability of a road-kill occurring. The assumption that strong winds increase flightiness in kangaroos has a good basis as kangaroos are known to exhibit a higher level of flightiness on windy days (Croft 1980; Southwell 1987).

Univariate GLM (SPSS for Windows V13.0) was used to determine if the initial distance to flight and the distance travelled by a kangaroo were influenced by species. These analyses tested the assumption that flight responses were consistent between species and thus justify inferences about the relationship between species flightiness and road-kill frequency. Other predictors such as time of day, road type, age/sex, group size, the presence of cover, season and speed of an approaching vehicle were also included in the models to account for their effects. For the model of the distance travelled by a kangaroo, the initial distance at which flight was elicited was included as a dependent variable. Night-time observations and kangaroos of indeterminate age/sex were excluded. Model terms were added into GLM using a stepwise technique, with all terms and interactions initially entered into models and dropped if not significant at $p < 0.05$. The final models were those which included terms for which elimination would have reduced the explanatory power of the models (adjusted R^2 used, Quinn and Keough 2002). Bonferroni corrected post-hoc comparisons were used to determine differences among species and all assumptions for GLM were checked.

Binary logistic regression was used to determine whether road-kill probability (kill/no kill) was influenced by days with strong winds. As road-kill probability is also influenced by various temporal factors (Lee *et al.* 2004; Seiler 2005; Klöcker *et al.* 2006), these were included in the regression model. Thus, the predictors used in the model were the night-time densities of kangaroos along the highway estimated per week, the average number of kangaroo highway crossings per week, the average day-time and night-time speeds of vehicles per week, the average night-time numbers of vehicles per week, total current weekly rainfall, total rainfall over the last two weeks and weekly weather components (see below).

Data for predictors, other than the average number of kangaroo highway crossings per week, were collected as outlined in Lee *et al.* (2004) (for post-drought data). For the average number of kangaroo highway crossings per week, data were collected using Trailmasters (Goodson and Associates Inc., Kansas) mounted at known kangaroo crossing points along a fenceline that ran parallel to the highway and approximately 50 m away from it. Trailmasters are infrared devices which detect the passing of animals and record the date and time of the events. They were configured with a five-second delay to prevent repeated counts of individual kangaroos. Only night-time crossings were included in analyses. Weekly weather variables (maximum and minimum temperatures, humidity, dewpoint, barometric pressure, maximum heat index, minimum windchill and maximum wind speed) were reduced to components using principal component analysis prior to analysis as they were highly correlated (PCA, SPSS for Windows V13.0). Three weather components were extracted: weather component 1 was positively associated with temperature, maximum heat index and minimum windchill, and negatively associated with humidity; weather component 2 was positively associated with maximum wind speed and negatively associated with barometric pressure; and weather component 3 was positively associated with dewpoint. Due to the association of maximum wind speed with weather component 2, this was the weather component of interest in the analysis.

Note that by including the densities of kangaroos along the highway in the logistic regression, the different susceptibilities of the species to being killed were considered. Also, note that the results for the effects of the other temporal factors on the probability of a kill are not presented in this chapter. The effects of spatial factors on the probability of a kill are not the focus of this chapter. Readers should refer to Lee *et al.* (2004) and Lee (2006) for the effects of other temporal and spatial factors on road-kill probability. Klöcker *et al.* (2006) detail the spatio-temporal factors affecting road-kill along the Silver City Highway.

RESULTS

Influence of time of day and species on flight response

Kangaroos took flight more often and exhibited relatively less vigilance or other behaviours during the night than the day ($\chi^2 = 193.72$, df = 2, $p < 0.001$, n = 5482; Fig. 26.1). During the day *M. rufus* (34%) took flight more often than grey kangaroos (23%; $\chi^2 = 57.45$, df = 2, $p < 0.001$, n = 5132); euros did not differ from other species (35%), while during the night *M. rufus* (81%) took flight more often than *M. robustus erubescens* (19%; $\chi^2 = 75.88$, df = 2, $p < 0.001$, n = 360). Grey kangaroos did not differ from other species (68%).

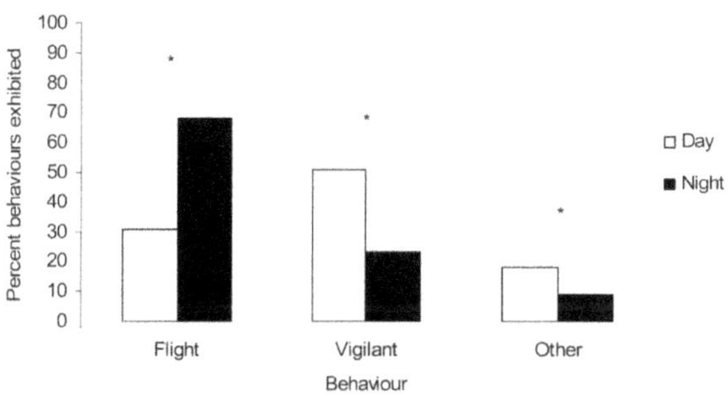

Figure 26.1: Proportion of behaviours exhibited by kangaroos (*Macropus* spp.) towards an approaching vehicle during the day and night (pooled across species, n = 5482). * denotes significant differences within a behaviour type between day and night.

Influences of other factors on flight response

Four predictors (road type, season, vehicle speed and vehicle distance) and two interaction terms (species*cover and species*group size) had significant effects on the probability of a kangaroo taking flight (Table 26.1). The differences in the odds of a kangaroo taking flight were more marked for road type and the two interaction terms, and less marked for the other significant predictors.

The odds that a kangaroo took flight in the hinterland was almost three times higher than a similar area along the highway (exp $\beta = 2.93$). The probability of flight was highest in spring (spring vs summer: Wald = 14.71, df = 1, $p < 0.001$; spring vs autumn: Wald = 15.66, df = 1, $p < 0.001$; spring vs winter: Wald = 20.85,

Table 26.1: Final logistic regression modelling the probability of a kangaroo (*Macropus* spp.) taking flight.

Model terms	G^2	Score	df	P values
Time of day		0.37	1	p = 0.543
Road type	128.48		1	p < 0.001
Season	24.23		3	p < 0.001
Speed of approaching vehicle	36.17		1	p < 0.001
Distance from approaching vehicle	47.34		1	p < 0.001
Species		0.21	2	p = 0.899
Age/sex		0.45	2	p = 0.798
Group size		4.45	2	p = 0.108
Cover		0.70	1	p = 0.405
Species*cover	21.93		2	p < 0.001
Species*group size	22.29		4	p < 0.001

Non-significant interactions not shown; non-significant predictors show scores not G^2 values.

The final model accounted for 6% of variation ($R_L^2 = 0.063$, $\chi^2 = 256.13$, df = 12, $p < 0.001$); n = 3276.

df = 1, $p < 0.001$; no differences between autumn, summer and winter: autumn vs summer: Wald = 0.02, df = 1, p = 0.893; autumn vs winter: Wald = 2.23, df = 1, p = 0.135; summer vs winter: Wald = 1.72, df = 1, p = 0.190), although the odds of flight were not much higher in spring than in the other seasons (exp β only slightly higher than 1 for all comparisons). The probability of flight was higher when the approaching vehicle was travelling at lower speeds, and when kangaroos were closer to the approaching vehicle (exp β = 1.01 and 0.99, respectively).

A lack of cover had a significant effect on the probability of grey kangaroos taking flight ($\chi^2 = 14.28$, df = 1, $p < 0.001$; Fig. 26.2), although cover had no effect on the probability of *M. rufus* ($\chi^2 = 0.17$, df = 1, p = 0.688) or *M. robustus erubescens* ($\chi^2 = 1.80$, df = 1, p = 0.209) taking flight. Similarly, group sizes of one and two significantly increased the probability of grey kangaroos taking flight ($\chi^2 = 18.28$, df = 2, $p < 0.001$; Fig. 26.3), while group size had no effect on the probability of *M. rufus* ($\chi^2 = 3.61$, df = 2, p = 0.166) or *M. robustus erubescens* ($\chi^2 = 4.37$, df = 2, p = 0.118) taking flight.

Numbers, species, sex and age of road-kills

A total of 92 road-killed kangaroos were found during the study period. *M. rufus* made up the majority of kills, followed by *M. fuliginosus*, *M. robustus erubescens* and *M. giganteus* (Table 26.2). Kills of *M. rufus* and *M. fuliginosus* were divided equally between sexes, with a slight female bias for *M. fuliginosus*. In contrast, more males than females were killed among *M. giganteus* and *M. robustus erubescens*.

Table 26.2: Percentage of kangaroo (*Macropus* spp.) species killed and males and females killed per species (%).

Species	Total (n = 92)	Females (n = 35)[a]	Males (n = 42)[a]
M. rufus	48	50	50
M. fuliginosus	39	53	47
M. giganteus	4	25	75
M. robustus erubescens	9	0	100

a: Excludes individuals where sex was indeterminate due to condition of carcass (n = 15).

Road-killed kangaroos were aged between one and 10 years (median = 2, mode = 1). When examined per species, road-killed *M. rufus* and *M. fuliginosus* mostly comprised young individuals, with 83% and 71% of all *M. rufus* and *M. fuliginosus* (n = 34 and 27 for road-killed individuals of known age, respectively) aged between one and three. Road-killed *M. robustus erubescens* were aged three and four years old (25% and 75%, respectively; n = 4 for individuals of known age). Ages of *M. giganteus* were unknown.

Kangaroo flightiness and the risk of road collision

Both the initial distance to flight and the total distance travelled during flight varied among species

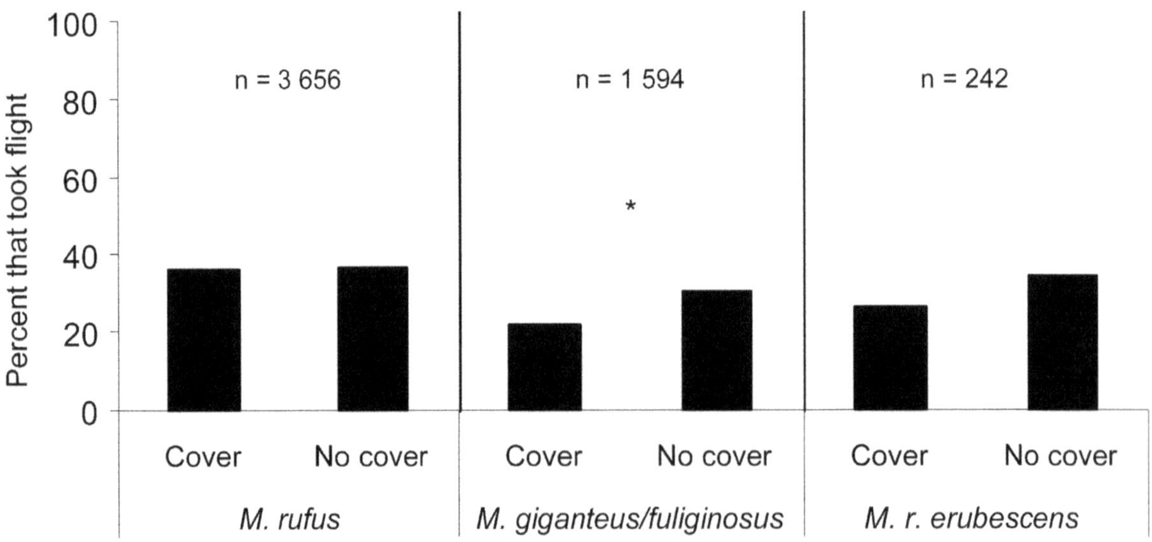

Figure 26.2: Percentage of kangaroos (*Macropus* spp.) which took flight in two categories of cover for each species. * denotes a significant difference between cover categories within a species. Results are from separate χ^2 tests per species testing the effect of cover on the probability of flight that followed significant results for species*cover interactions ($G^2 = 21.93$, df = 2, $p < 0.001$).

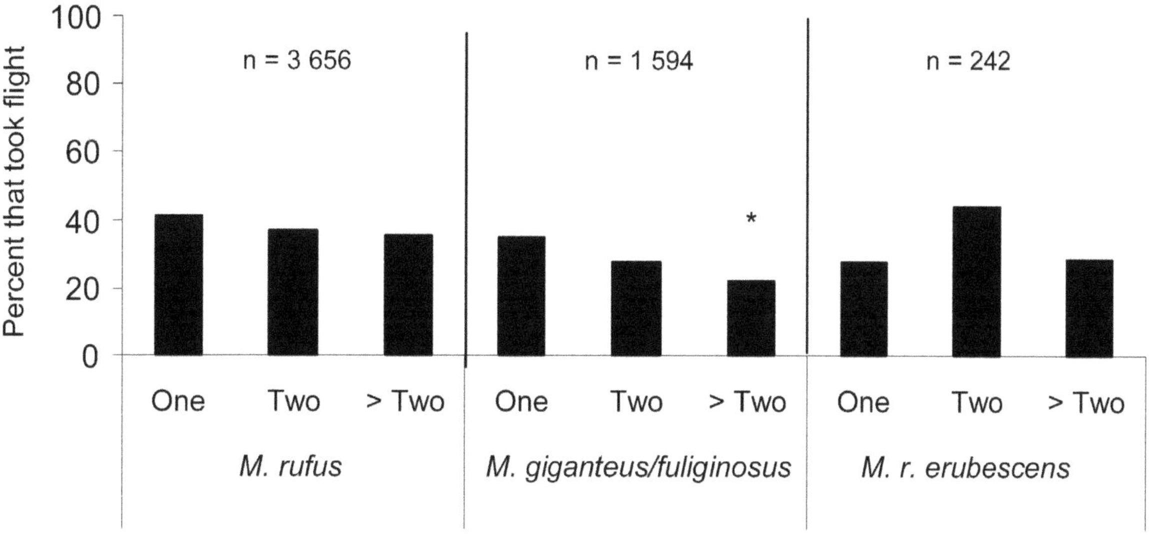

Figure 26.3: Percentage of kangaroos (*Macropus* spp.) which took flight in three categories of group size for each species. * denotes a significant difference between the labelled and other group size categories within a species. Results are from separate χ^2 tests per species testing the effect of group size on the probability of flight that followed significant results for species*group size interactions ($G^2 = 22.29$, df = 4, $p < 0.001$).

($F_{(2,3263)} = 36.61$, $p < 0.001$ and $F_{(2,900)} = 3.69$, $p = 0.025$, respectively). Bonferroni tests revealed that *M. rufus* had greater initial distances to flight and travelled further during flight than *M. robustus erubescens*, with grey kangaroos intermediate between the other two species (Table 26.3). These results are consistent with *M. rufus* showing a greater tendency to take flight than *M. robustus erubescens* and grey kangaroos showing intermediate flight tendencies during the night (Table 26.3).

Kangaroo species with the lowest average initial distances to flight had the lowest kill frequency, while species with higher average initial distances to flight had higher kill frequencies (Fig. 26.4). Likewise, kangaroo species with the lowest average distances travelled during flight had the lowest kill frequency, while species with higher average distances travelled during flight had higher kill frequencies. Thus, the tendency for kangaroos to take flight compared to exhibiting other behaviours probably related to kill frequencies in this pattern.

Higher wind speed (strong winds) increased the probability of a kill occurring ($G^2 = 6.23$, df = 1, $p = 0.013$; $\mathrm{Exp}\beta = 2.23$; $n = 92$). The full model equation is:

$$y = 1/(1 + e^{-5.76 + 0.41 * \text{night-time density} + 0.37 * \text{night–time traffic volume} + 0.80 * \text{Component2}})$$

This accounted for 6% out of the total of 38% of the variation explained by the final model. Thus, the probability of a road-kill likely grew with increases in kangaroo flightiness.

DISCUSSION

Kangaroo flight response

This study revealed that kangaroo flight responses towards an approaching vehicle were quite consistent and that among the factors influencing the tendency to take flight, time of day, road type and species were strong predictors for flight. A likely reason for the increased tendency to take flight during the night is

Table 26.3: Percentage of kangaroos (*Macropus* spp.) that took flight from an approaching vehicle, average initial distance to flight from an approaching vehicle and average distance travelled during flight per species.

Model terms	% took flight	Average initial flight distance (m)	Average distance travelled (m)
M. rufus	81 (208)	152 ± 2 (2 481)	107 ± 4 (692)
M. fuliginosus/M. giganteus	68 (95)	146 ± 4 (614)	101 ± 7 (156)
M. robustus erubescens	19 (57)	77 ± 4 (181)	61 ± 8 (57)

Sample sizes are shown in brackets; errors represent standard errors.

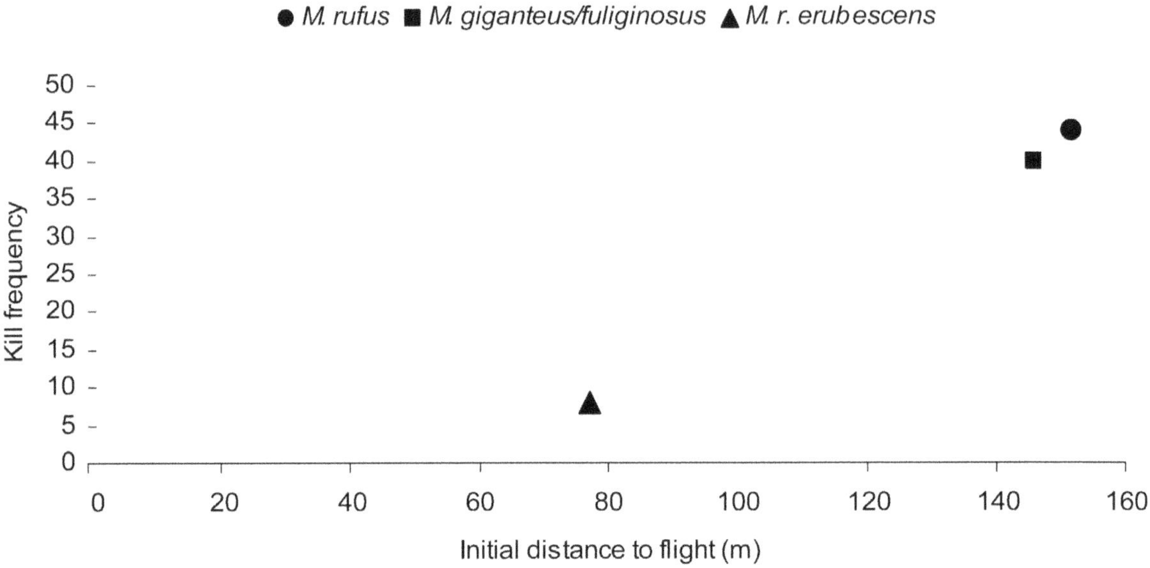

Figure 26.4: Relationship between kill frequency and a kangaroo species' average initial distance to flight.

that threats can be seen and detected by kangaroos at greater distances during the day; at night, only vehicle headlights are seen. Vehicles that could be detected more clearly during the day may have been perceived as less threatening. Further, average night-time traffic volume (mean ± SE = 15 ± 2 vehicles) was lower than average daytime traffic (45 ± 2 vehicles) along the road and animals often invest higher levels of anti-predator effort in situations where threats are brief and infrequent (Lima and Bednekoff 1999). Similarly, the greater tendency for kangaroos to take flight along the hinterland road than along the highway may relate to temporal variations in danger. In this study, traffic volumes were low in the hinterland, consisting mainly of motorbikes that traversed the hinterland road infrequently, while traffic volumes were higher along the highway.

The greater tendency of *M. rufus* to take flight compared with *M. robustus erubescens*, and the intermediate tendency of grey kangaroos, were most likely related to the habitat preferences of the species and their various responses to predators or perceived risks. *M. rufus* generally prefers open habitats that have occasional patches of shrubs and shade trees (Croft 1991a; Dawson 1995), grey kangaroos prefer habitats with shrub or tree cover (Southwell 1987; Coulson 1993) and *M. robustus erubescens* prefers rough hilly habitats (Croft 1991b). As the open habitat preferred by *M. rufus* provides little cover, *M. rufus* is easily detected by predators and so may rely on early flight and long flight distances, plus superior speed in open country, to escape from predators or situations perceived as risky (Dawson 1995).

In contrast, the preferred habitats of grey kangaroos provide more cover, decreasing the likelihood of being detected by predators. Grey kangaroos often remain vigilant in cover rather than taking early flight, since flight may do little to increase their safety (as found for other animals; Frid and Dill 2002). This response was confirmed in this study as they took flight less often when in cover. When not initially in cover, the flight behaviour of grey kangaroos was similar to that of *M. rufus*, although grey kangaroo movements have often been observed to be towards cover rather than across open country (Dawson 1995). Flight behaviour in grey kangaroos was also influenced by group size, with individual or paired kangaroos taking flight more often than kangaroos in larger groups. This has previously been documented for grey kangaroos, as group size is known to influence anti-predator behaviour (Jarman and Wright 1993) and the risk of predation is higher for smaller groups (Banks 2001; Lett *et al.* 2004).

Macropus robustus erubescens frequently occupies rocky habitats and employs crypsis and avoids detection by predators by remaining still (Croft 1981). Flight typically occurs once animals perceive they have been detected. The short distances travelled during flight are likely to be related to the rough terrain that may deter predators from pursuit but poses no movement problems for *M. robustus erubescens* (Dawson 1995).

Kangaroo flightiness and the risk of road collision

The positive relationships between a kangaroo's overall flightiness and the likelihood of collisions, and between a species' flightiness and its road-kill frequency, suggested that increasing kangaroo flightiness towards an approaching vehicle increased collision likelihood. These findings imply that flight by kangaroos may not be an effective defence against vehicles. Rather, vigilance behaviour may be a more suitable anti-predator response to the threat of collisions with vehicles as the kangaroo's stationary behaviour is more predictable to the driver. Vigilance and flight are fundamental to a kangaroo's survival and are both performed primarily to minimise a kangaroo's close encounter with predators (Southwell 1987; Coulson 1996). For large kangaroos of open habitat, flight is generally a more effective anti-predator behaviour against threats than stationary vigilance and crypsis.

When a kangaroo takes flight it may increase its susceptibility to becoming road-kill by moving, often unpredictably, into the path of an oncoming vehicle; a stationary kangaroo does not increase its susceptibility provided, of course, that it is not on the road in the path of the vehicle. Risk from taking flight increases if initiated at closer distances to approaching vehicles, as this prevents drivers from taking evasive action, and if the path is across rather than away from the direction of an approaching vehicle. We confirmed the former risk factor as the probability of flight increased as a function of the nearness of the vehicle. Lee (2006) found that during the night, when kangaroos displayed more flight behaviour than during the day, higher proportions of kangaroos fled across the path of an oncoming vehicle. This exacerbates risk, as the driver's vision is relatively impaired at night. Likely reasons for greater kangaroo flight behaviour during the night have been discussed above. A possible reason for why kangaroos take flight when vehicles approach to closer distances is that approaching vehicles appear larger and are louder at close distances, and predators or threats that are large or that reach particular threshold distances from animals induce high levels of anti-predator behaviour (Frid and Dill 2002).

Since kangaroo road-kill frequency was positively related to kangaroo flightiness, any factors increasing kangaroo flightiness may by inference be causal in road-kill frequency. Two factors – the speed of an approaching vehicle and the road type on which the approaching vehicle was travelling – influenced the probability of flight for kangaroos, albeit very slightly. Even so, an increase in road-kill frequency may not have occurred in situations where these factors increased the probability of flight (at low speeds and in the hinterland). This is because, when travelling under these conditions, drivers would be able to influence the likelihood of collisions by reacting to and avoiding kangaroos before collisions occurred. This clearly illustrates that driver behaviour can influence road-kill frequency, an observation that has been made in other studies (Jones 2000; Dique *et al.* 2003). This study did not explore the influence of driver behaviour but clearly it can be targeted as a form of road-kill mitigation and deserves further investigation. Our findings that flight responses of kangaroos are tempered by environmental conditions could inform how driver behaviour could be targeted. As wind speed increased, the likelihood of flight increased. Drivers should be educated to be particularly aware of the heightened risk of animal–vehicle collisions in inclement weather. Such weather is likely to compromise driving ability and thus increase the risk of road-kill.

Contrary to expectations, the age or sex of kangaroos did not influence the majority of the measures of kangaroo flightiness. Flight response is therefore unlikely to account for the greater proportion of young kangaroos killed, particularly for *M. rufus* and *M. fuliginosus*, or the slight male bias in road-killed *M. giganteus* and *M. robustus erubescens*. Young at foot were expected to be more flighty than adult kangaroos as they are the class most sensitive to predation (Banks 2001; Dawson 1995). Differences in flightiness between age/sex classes may not have been detected in this study due to lack of differentiation of reproductive condition. For example, reproductive females, females with young at foot and females without young behave differently (Croft 1980; Colagross and Cockburn 1993; Banks 2001), with reproductive females with heavy pouch young investing more of their time in vigilance (Colagross and Cockburn 1993).

IMPLICATIONS FOR THE MITIGATION OF ROAD-KILL

This study highlights that kangaroo behaviour towards an approaching vehicle is consistent in form, varying with a small number of factors such as species, and that the behaviour of kangaroo species is a potential causative factor for collisions with vehicles. As such, the behaviour of the different kangaroo species should be incorporated into predictive models of road-kill incidence along with spatio-temporal factors known to be important in the risk of road collision. This would improve the accuracy of such models for future mitigation.

Results also shed new light on the effectiveness of abatement technologies available for mitigating kangaroo–vehicle collisions and the rationale behind them, where flightiness in kangaroos towards approaching vehicles is encouraged. Given the positive relationship between kangaroo flightiness and road-kill frequency, where flightiness may place kangaroos at a disadvantage relative to vigilance, abatement technologies that encourage flight on vehicle approach may be counterproductive. There are some ultrasonic devices which attempt to encourage animals to move away from approaching vehicles (e.g. Shu Roo). There is evidence that these are not effective in reducing road-kill due to the unresponsiveness of kangaroos (Bender 2001), and the logic of encouraging flight responses from kangaroos in advance of the vehicle may not be the best approach.

ACKNOWLEDGMENTS

This study was supported by an Australian Research Council Linkage grant (LP0346925) in collaboration with NSW National Parks and Wildlife Service, the International Fund for Animal Welfare, the NSW Wildlife Information and Rescue Service and Roe Koh and Associates. Enhua Lee was supported by an Australian Postgraduate Award. Thanks to Dr Andrew Russell for statistical advice. Further assistance was provided by the NSW Road and Traffic Authority. The study was conducted under approval AC03/57 of the UNSW Animal Care and Ethics Committee.

REFERENCES

Aresco MJ 2005. The effect of sex-specific terrestrial movements and roads on the sex ratio of freshwater turtles. *Biol. Conserv.* **123:** 37–44.

Banks PB 2001. Predation-sensitive grouping and habitat use by eastern grey kangaroos: a field experiment. *Anim. Behav.* **61:** 1013–1021.

Bender H 2001. Deterrence of kangaroos from roadways using ultrasonic frequencies: efficacy of the Shu Roo. Report to NRMA Insurance Ltd, RACV, RTA and Transport South Australia. University of Melbourne, Australia.

Brown BC 2001. Marsupial road-kill on the Kosciuszko Road in the Australian Alps: an investigation of factors influencing animal–vehicle collisions. Hons thesis. Dept of Resource Management and Environmental Science, Australian National University, Canberra.

Burrell JP 1973. Vegetation of Fowlers Gap Station. In *Lands of Fowlers Gap Station, New South Wales.* (Ed. JA Mabbutt) pp. 175–195. University of New South Wales, Sydney.

Colagross AML and Cockburn A 1993. Vigilance and grouping in the eastern grey kangaroo, *Macropus giganteus. Aust. J. Zool.* **41:** 325–334.

Coulson G 1993. Use of heterogeneous habitat by the western grey kangaroo, *Macropus fuliginosus. Wildl. Res.* **20:** 137–149.

Coulson G 1996. Anti-predator behaviour in marsupials. In *Comparison of Marsupial and Placental Behaviour.* (Eds DB Croft and U Ganslosser) pp. 158–186. Filander Verlag Fürth.

Croft DB 1980. Behaviour of red kangaroos, *Macropus rufus* (Desmarest, 1822) in northwestern New South Wales, Australia. *Aust. Mammal.* **4:** 5–58.

Croft DB 1981. Social behaviour of the euro, *Macropus robustus* (Gould), in the Australian arid zone. *Aust Wildl. Res.* **8:** 13–49.

Croft DB 1991a. Home range of the red kangaroo *Macropus rufus. J. Arid Environ.* **20:** 83–98.

Croft DB 1991b. Home range of the euro, *Macropus robustus erubescens. J. Arid Environ.* **20:** 99–111.

Dawson TJ 1995. *Kangaroos: Biology of the Largest Marsupials.* University of New South Wales Press, Sydney.

Dique D, Thompson J, Preece HJ, Penfold GD, de Villiers DL and Leslie RS 2003. Koala mortality on roads in south-east Queensland: the koala speed-zone trial. *Wildl. Res.* **30:** 419–426.

Forman RTT, Sperling D, Bissonette JA, Clevenger AP, Cutshall CD, Dale VH, Fahrig L, France R, Goldman CR, Heanue K, Jones JA, Swanson FJ, Turrentine T and Winter TC 2003. *Road Ecology: Science and Solutions.* Island Press, Washington DC.

Frid A and Dill LM 2002. Human-caused disturbance stimuli as a form of predation risk. *Conserv. Ecol.* **6:** 11–16.

Horejsi BL 1981. Behavioural response of barren ground caribou to a moving vehicle. *Arctic* **34:** 180–185.

Jaeger JAG, Bowman J, Brennan J, Fahrig L, Bert D, Bouchard J, Charbonneau N, Frank K, Gruber B and Tluk von Toschanowitz K 2005. Predicting when animal populations are at risk from roads: an interactive model of road avoidance behaviour. *Ecol. Model.* **185:** 329–348.

Jarman PJ and Wright SM 1993. Macropod studies at Wallaby Creek. IX. Exposure and responses of eastern grey kangaroos to dingoes. *Wildl. Res.* **20:** 833–843.

Jones ME 2000. Road upgrade, road mortality and remedial measures: impacts on a population of eastern quolls and Tasmanian devils. *Wildl. Res.* **27:** 289–296.

Kirkpatrick TH 1965. Studies of Macropodidae in Queensland. II. Age estimation in the grey kangaroo, the red kangaroo, the eastern wallaroo and the red-necked wallaby, with notes on dental abnormalities. *Qld J. Agr. Anim. Sci.* **22:** 301–317.

Klöcker U, Croft DB and Ramp D 2006. Frequency and causes of kangaroo–vehicle collisions on an Australian outback highway. *Wildl. Res.* **33:** 5–15.

Lee E 2006. The ecological effects of sealed roads in arid ecosystems. PhD thesis. School of Biological Earth and Environmental Sciences, University of New South Wales, Sydney.

Lee E, Klöcker U, Croft DB and Ramp D 2004. Kangaroo–vehicle collisions in Australia's sheep rangelands, during and following drought periods. *Aust. Mammal.* **26:** 215–226.

Lett C, Auger P and Gaillard JM 2004. Continuous cycling of grouped versus solitary strategy frequencies in a predator–prey model. *Theor. Popul. Biol.* **65:** 263–270.

Lima S and Bednekoff PA 1999. Temporal variation in danger drives antipredator behaviour: the predation risk allocation hypothesis. *Am. Nat.* **153:** 649–659.

Mabbutt JA 1973. Geomorphology of Fowlers Gap Station. In *Lands of Fowlers Gap Station, New South Wales.* (Ed. JA Mabbutt) pp. 85–122. University of New South Wales Press, Sydney.

Mumme RL, Schoech SJ, Woolfenden GE and Fitzpatrick JW 2000. Life and death in the fast lane: demographic consequences of road mortality in the Florida scrub-jay. *Conserv. Biol.* **14:** 501–512.

Oksanen L 2004. The devil lies in details: reply to Stuart Hurlbert. *Oikos* **104:** 598–605.

Quinn GP and Keough MJ 2002. *Experimental Design and Data Analysis for Biologists.* Cambridge University Press, Cambridge.

Ramp D, Caldwell J, Edwards KA, Warton D and Croft DB 2005. Modelling of wildlife fatality hotspots along the Snowy Mountain Highway in New South Wales Australia. *Biol. Conserv.* **126:** 474–490.

Seiler A 2005. Predicting locations of moose–vehicle collisions in Sweden. *J. Appl. Ecol.* **42:** 371–382.

Southwell C 1987. Activity pattern of the eastern grey kangaroo, *Macropus giganteus. Mammalia* **51:** 211–223.

27 Advances in fertility control technologies for managing overabundant macropodid populations

C.A. Herbert, M.B. Renfree, G. Coulson, G. Shaw, T.E. Trigg and D.W. Cooper

SUMMARY

Kangaroo populations frequently reach high densities under favourable conditions, particularly near urban areas, where sociopolitical constraints limit management options. This study investigates new approaches to kangaroo management that focus on controlling the fertility of females. In particular, it investigates long-acting contraceptive implants containing the gonadotrophin-releasing hormone agonist deslorelin (Suprelorin®) or the synthetic progestin, levonorgestrel. It shows that both contraceptive implants are effective in a model marsupial species, the tammar wallaby (*Macropus eugenii*), and gives a detailed understanding of the effects of these two agents on the physiology of this species. Captive and field investigations on the eastern grey kangaroo (*Macropus giganteus*) have shown that long-term fertility control can be achieved in that species using either of these contraceptive implants, with no detectible side effects. However, these two contraceptive implants have different modes of action. Levonorgestrel treatment has no effect on follicular development but the luteinising hormone (LH) surge, and hence ovulation, is inhibited. Suprelorin, on the other hand, inhibits the synthesis and release of LH and follicle-stimulating hormone from the pituitary, thereby inhibiting follicular development and ovulation. The methods of administration and contraceptive duration also differ. These differences can be exploited to tailor kangaroo management programs to specific populations to ensure long-term contraception and population management.

INTRODUCTION

Since European settlement, humans have altered Australian landscapes dramatically through the processes of urbanisation, habitat clearance, introduction of exotic species (competitors, predators and pathogens) and alterations to fire regimes. Not surprisingly, this modification of the landscape has substantially altered faunal assemblages across the continent. In the case of macropodid marsupials, it is generally accepted that the smaller species have declined in both range and abundance, while the larger species (kangaroos) have increased in abundance, and possibly range, as a result of protection from predation by the 'dingo fence' and the provision of watering points and improved pasture (Maxwell *et al.* 1996; Pople and Grigg 1999). Management of both threatened and overabundant populations is critical for the conservation of these species and the habitats they occupy. In this chapter we discuss the macropodid species that are actively managed to control population density, evaluate the potential application of fertility control technology to these situations and describe recent investigations on the use of long-acting contraceptive technology in macropodid species. We also discuss the likely efficacy of applying fertility control to the management of wild macropodid populations.

OVERABUNDANT MACROPODID SPECIES

Several macropodid species are actively managed to reduce population densities and/or to mitigate damage in some locations (Table 27.1). The rationale for management has been reviewed elsewhere (Coulson 1998; Herbert 2007). Briefly, the reasons for macropodid management usually centres around four main themes: threats to human life or livelihood, depression of the density of a favoured species (e.g. one of conservation significance), decline in body condition and reproduction, and loss of equilibrium between plants and animals (after Coulson 1998). The kangaroos and larger wallabies are the species that most frequently require active management (Table 27.1) and are listed in the IUCN category for widespread and abundant taxa, i.e. 'lower risk (least concern)' (IUCN 2004). For these species, population management is generally conducted by culling at a local or regional scale under the regulation of the relevant state and territory government wildlife

Table 27.1: Macropodid species currently requiring management to reduce population densities in some locations, and the potential applicability of fertility control methods.

Species	Female av. weight (range) (kg)[a]	IUCN status[b]	General change since European settlement[c]	Current management (location)[c,d,e]	Potential for fertility control
Petrogale lateralis lateralis	3.5 (3.1–3.8)	VU	Range reduction, local increases (WA)	Translocation (WA wheat belt)	Yes
Thylogale billardierii	3.9 (2.4–10)	LR(lc)	Extinct on mainland, probable increase in Tas.	Culling (Tas.)*	No
Macropus eugenii	5.5 (4–6)	LR(nt)	Range reduction, increase on some islands	Culling (Kangaroo Is.), translocation (WA)	No (Kangaroo Is.) Yes (WA)
Macropus parryi	11 (7–15)	LR(lc)	Local eruptions	Culling (Qld)	?
Macropus dorsalis	6.5 (6–7.5)	LR(lc)	Range reduction, local eruptions (Qld)	Culling (Qld)	?
Macropus rufogriseus rufogriseus	14 (11–15.5)	LR(lc)	Little change, local eruptions	Culling (Tas.)*	No
Macropus agilis	11 (9–15)	LR(lc)	Little change or increase	Culling (Qld)	No
Macropus giganteus	3.2–32	LR(lc)	Increase in numbers and possibly range	Culling (Vic., NSW,* Qld*)	Yes (urban) No (rural)
Macropus fuliginosus	4.5–27.5	LR(lc)	Increase in numbers and possibly range	Culling (Vic., NSW,* SA,* WA*)	Yes (urban) No (rural)
Macropus robustus	6.25–25	LR(lc)	Little change or increase	Culling (NSW, Qld, SA)*	No
Macropus rufus	26.5 (17–35)	LR(lc)	Increase in numbers and possibly range	Culling (NSW, Qld, SA, WA)*	No
Setonix brachyurus	2.9 (2.7–3.5)	VU	Reduced range, probable increase on Rottnest Is.	Vegetation exclosures (Rottnest Is.)	Yes

* Commercial harvesting for export.
VU = vulnerable; LR(lc) = lower risk (least concern); LR(nt) = lower risk (near-threatened).
(a) Strahan 1995; (b) IUCN 2004; (c) Pople and Grigg 1999; (d) Herbert 2007; (e) DEH 2006.

management agencies. The larger macropodid species listed in Table 27.1 are considered to be common and widespread throughout their range, but some smaller wallaby species exhibit geographic and/or temporal variations in their status. These include the tammar wallaby (*Macropus eugenii*), black-footed rock-wallaby (*Petrogale lateralis lateralis*) and quokka (*Setonix brachyurus*). On a national scale these taxa are considered 'vulnerable' or 'near-threatened' (Table 27.1). Despite their overall threatened status, populations of the tammar wallaby and black-footed rock wallaby in the wheat belt of Western Australia (e.g. at Tutanning Reserve) have increased markedly over the last decade in response to widespread poisoning of introduced European red foxes (*Vulpes vulpes*), resulting in high population densities at the local scale (Mawson 2004). In addition, some species have flourished on offshore islands in the absence of predators and/or in response to increased nutritional resources associated with European settlement, for example tammar wallabies on Kangaroo, Garden and North Islands, and quokkas on Rottnest Island (Herbert 2007). These species are predominantly managed by non-lethal techniques (with the exception of tammar wallabies on Kangaroo Island), such as translocation to areas within their historic range (tammars and black-footed rock wallabies; Mawson 2004), or by excluding them from sensitive areas of revegetation (quokkas; RIA 2003). For these vulnerable or near-threatened species, lethal control techniques are generally not deemed appropriate.

FERTILITY CONTROL AS A POPULATION MANAGEMENT TOOL

The management of overabundant native animals has become a contentious issue in recent years, with lethal control techniques becoming sociopolitically unacceptable in many situations (Garrott *et al.* 1993). This is especially the case with highly regarded mammalian species, such as kangaroos and wallabies (Coulson 1998; Adderton Herbert 2004). Overabundant macropodid populations are frequently managed by culling (Table 27.1) but the use of lethal control techniques is facing mounting criticism, particularly in the context of managing urban kangaroo populations or local eruptions of otherwise vulnerable or near-threatened species. In rural areas, hundreds of thousands of kangaroos are culled each year to reduce conflict with primary producers and to harvest meat and skins. However, the use of culling to control kangaroo populations in urban areas is controversial. In addition, the use of high-velocity projectiles to humanely kill kangaroos (CONCOM 1998) could pose a serious human safety issue near urban areas. Therefore, there has been increasing interest in the use of fertility control to manage urban kangaroo populations (Adderton Herbert 2004).

The management of local eruptions of populations of vulnerable macropod species poses an interesting management challenge. Because the long-term viability of these populations cannot be guaranteed, the management approach should be cautious. The use of a reversible fertility control agent would be more desirable than culling as it would give managers greater flexibility over time and could be coupled with other non-lethal methods, such as translocation and exclosure fencing (Herbert 2007).

MACROPODID REPRODUCTION AND CONTRACEPTION

Kangaroos and wallabies are monovular and polyoestrous. The length of gestation is almost the same as the length of the oestrous cycle and post-partum oestrus and ovulation occurs in most species (Tyndale-Biscoe and Renfree 1987). If fertilisation is successful, subsequent development of the corpus luteum (CL) is inhibited by the suckling stimulus; the embryo remains as a unilaminar blastocyst of about 100 cells in embryonic diapause (lactational quiescence). If the first young is lost or weaned, the removal of the suckling stimulus stimulates the reactivation of the blastocyst in diapause (Tyndale-Biscoe and Renfree 1987). This reproductive pattern is common to most kangaroos and wallabies, with the exception of the eastern grey kangaroo (*Macropus giganteus*), parma wallaby (*Macropus parma*), whiptail wallaby (*Macropus parryi*) and western grey kangaroo (*Macropus fuliginosus*). In the first three, post-partum oestrus does not occur but females may come into oestrus and mate during the later stages of lactation and the embryo will enter diapause. In the western grey kangaroo, embryonic diapause does not occur (Tyndale-Biscoe and Renfree 1987). In two taxa, the tammar wallaby and Bennett's wallaby (*Macropus rufogriseus rufogriseus*), a photoperiodic control of diapause is superimposed on the lactational control after the autumn equinox (Tyndale-Biscoe and Renfree 1987; Renfree 1994).

Two different types of long-acting contraceptive implants have been tested on macropodids: silastic implants containing the synthetic gestagen levonorgestrel and lipid-based implants containing the gonadotrophin-releasing hormone (GnRH) agonist deslorelin (Suprelorin®, Peptech Animal Health Pty Ltd, Macquarie Park, NSW). The effects of both implants have been studied extensively in the tammar wallaby, a model macropodid species, to gain an understanding of the action mechanisms in this family of marsupials. Additional trials have been conducted in the eastern grey kangaroo, which is one of the potential target species for this form of fertility control.

LEVONORGESTREL IMPLANTS

Levonorgestrel implants consist of a thin-walled silastic tube (44 × 2.4 mm outside diameter) filled with 70 mg of levonorgestrel in a poly (dimethylsiloxane) elastomer (at a 1:1 ratio), sealed with silastic adhesive. In human clinical trials, the implants initially release 30 μg levonorgestrel per day, stabilising at 15 μg per day after 12 months (Sivin 1988; Sivin *et al.* 1997). Levonorgestrel exerts its contraceptive effect in women by reducing the concentration of follicle-stimulating hormone (FSH) and preventing the pre-ovulatory luteinising hormone (LH) surge, thereby inhibiting follicular development and ovulation in many instances (Segal *et al.* 1991). If ovulation does occur, changes to the cervical mucus prevent the migration of spermatozoa through the cervical canal and interference with the development of the endometrium inhibits implantation (Segal *et al.* 1991; Sivin 1994).

Levonorgestrel implants are inserted subcutaneously between the shoulder-blades through a 3–10 mm incision under local anaesthesia in sedated kangaroos and wallabies. After insertion of the implant(s), the incision is sealed with a single absorbable suture. A prophylactic dose of long-acting penicillin is usually administered to minimise the chance of infection, and the incision is dusted with antibiotic powder (Nave *et al.* 2002a).

Effects of levonorgestrel in the tammar wallaby

Treatment of tammar wallabies with one 70 mg levonorgestrel implant results in successful contraception for at least 48 months (Nave *et al.* 2002b). Follicular development continues during levonorgestrel treatment but ovulation is inhibited, suggesting that levonorgestrel has no effect on FSH secretion but suppresses the LH surge required for ovulation (Hynes *et al.* 2007). It is not surprising that levonorgestrel, a synthetic progestin, does not inhibit follicular development, as folliculogenesis in the tammar occurs in a high-progesterone environment (Tyndale-Biscoe *et al.* 1983). In addition to inhibiting ovulation, levonorgestrel also inhibits oestrus and mating in most animals. Nave *et al.* (2000) found no evidence that levonorgestrel-treated females (n = 14) mated, but in a subsequent study Hynes *et al.* (2007) found that mating occurred in 30% of treated females (n = 10). In those animals that did mate, there was a more sustained period of interest by males and mating occurred significantly later post-partum than in control animals (Hynes *et al.* 2007). Although follicular development occurs in all treated animals, the failure of most levonorgestrel-treated females to mate suggests significant endocrine disturbance that affects behaviour has occurred in the peri-ovulatory period.

Insertion of levonorgestrel implants at the time of removal of a pouch young (RPY) does not appear to affect reactivation from diapause of an existing dormant blastocyst (Nave *et al.* 2000). However, a higher proportion of pregnancies failed in treated than in control animals, possibly as a result of gestagen-induced changes in the uterine environment and asynchrony between the stage of development of the endometrium and blastocyst about the time of reactivation (Hynes *et al.* 2007). In addition, the gestation period in levonorgestrel-treated tammars is generally shorter than in control animals (Nave *et al.* 2000; Hynes *et al.* 2007). This may reflect an early reactivation of the dormant blastocyst as a result of levonorgestrel treatment at the time of RPY (Nave *et al.* 2000; Hynes *et al.* 2007), as observed after exogenous progesterone treatment for either 10 days after RPY or for a few days before RPY in the tammar (Renfree and Tyndale-Biscoe 1973; Fletcher and Renfree 1988; Tyndale-Biscoe *et al.* 1988). Despite the shorter gestation period, the neonates are normal and levonorgestrel treatment has no effect on the establishment and maintenance of early lactation (Nave *et al.* 2000).

The contraceptive duration for levonorgestrel implants in tammar wallabies is at least four years (Nave *et al.* 2002b) and it is effective in 100% of animals (Nave *et al.* 2000). The contraceptive effects of levonorgestrel implants are reversible, as demonstrated by the resumption of breeding in female tammar wallabies after implant removal (Nave *et al.* 2000).

Effects of levonorgestrel in the eastern grey kangaroo

Levonorgestrel implants were tested for contraceptive effectiveness in free-ranging female eastern grey kangaroos in two populations in Victoria – at Portland Aluminium Smelter (38°23'S, 141°37'E) and Woodlands Historic Park (37°44'S, 144°54'E). At Portland, 18 female kangaroos received two 70 mg levonorgestrel implants and eight females received control implants between February and April 1999 (Nave *et al.* 2002a). At Woodlands, 25 females received two 70 mg levonorgestrel implants between June 1999 and April 2000 (Nave 2002). Any pouch young present at the time of treatment were removed. Animals at each site were individually identified by a unique combination of reflective coloured eartags, and at Portland each also had a colour-coded collar. Visual monitoring was conducted each spring at both sites to record the reproductive status of each female, i.e. whether she had a pouch young (>50 days old, visible as a bulge in the pouch) or a young at foot.

The longevity of contraception varied greatly between individuals at both sites (Fig. 27.1). At Portland, three of 16 treated females gave birth within the first year, presumably as a result of the reactivation of a dormant blastocyst present at the time the implant was administered. All these females had large pouch young (>280 days old) removed at the time of levonorgestrel treatment (Nave *et al.* 2002a); female eastern grey kangaroos are capable of coming into oestrus and mating when the young occupying the pouch is greater than 160 days old if conditions are favourable (Poole and Catling 1974). Reactivation of a dormant blastocyst may also account for four of the young observed at Woodlands within the first year. These results suggest that levonorgestrel does not prevent reactivation of a blastocyst in diapause, embryo development, parturition or the initiation of lactation in the eastern grey kangaroo (Nave *et al.* 2002a). However, the studies do not show if pregnancy failures occurred in some levonorgestrel-treated eastern grey kangaroos, as occurs in some tammars (Nave *et al.* 2000; Hynes *et al.* 2007). Levonorgestrel did not prevent successful lactation, since all females that gave birth during their first year of treatment at Portland raised their young to weaning some 500 days later (Nave *et al.* 2002a).

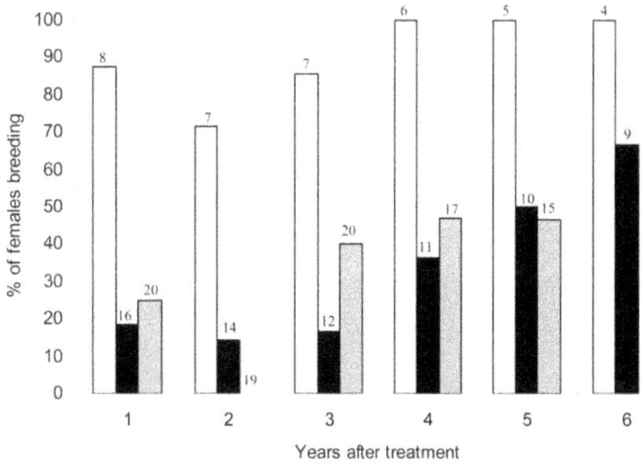

Figure 27.1: Percentage of female eastern grey kangaroos (*M. giganteus*) observed with pouch young at yearly intervals after treatment with two 70 mg levonorgestrel or two control implants. Animals at Portland were treated with control (open bars) or levonorgestrel (black bars) between February and April 1999. Animals at Woodlands were treated with levonorgestrel implants (grey bars) between June and November 1999 or March and April 2000. Observations of the reproductive status were made each spring (Nave *et al.* 2002a; Coulson *et al.* 2008). The numbers above each bar refer to the number of animals observed from each group (not all animals could be observed in all years and the fate of some is unknown).

The response of the eastern grey kangaroos to levonorgestrel treatment (Nave *et al.* 2002a; this study) was much more variable than that of tammar wallabies (Nave *et al.* 2002b). Tammar wallabies were infertile for at least four years after treatment with one 70 mg implant, with a 100% success rate (Nave *et al.* 2002b). At Portland, two animals may have been resistant to the contraceptive effects of levonorgestrel, as both animals bred within the second year of treatment. With the study design it is impossible to be certain if these animals were 'non-responders' or if they represent animals with a very short duration of infertility following treatment. Assuming the former, this gives a conservative contraceptive success rate of 87% (n = 15) for levonorgestrel implants in eastern grey kangaroos. In subsequent years, progressively more females returned to fertility at both sites, though by year 5 there was still 50% infertility in the levonorgestrel-treated females (Fig. 27.1). It would be interesting to determine if larger doses of levonorgestrel produce more prolonged periods of contraception in this species.

Effect of levonorgestrel on macropodid behaviour

The effects of levonorgestrel on macropodid behaviour have been studied in most detail in the eastern grey kangaroo (Poiani *et al.* 2002), with some observations on the effects at the time of post-partum oestrus in the tammar wallaby (Hynes *et al.* 2007). Levonorgestrel contraception had no effect on the amount of time free-ranging female eastern grey kangaroos spent feeding, scanning, grooming, moving or interacting with males, but control females spent more time feeding in open paddocks over a 24-hour period than levonorgestrel-treated animals did, presumably because of the energetic burden of lactation (Poiani *et al.* 2002). Also, males preferred to spend more time with groups containing a higher proportion of control females, suggesting that males may discriminate between control and levonorgestrel-treated females (Poiani *et al.* 2002). In tammar wallabies, the majority of females did not mate at the end of their first cycle post-treatment. In females that did mate, there was a more sustained period of interest by males than there was for control females (Hynes *et al.* 2007).

SUPRELORIN IMPLANTS

Each Suprelorin implant contains 4.7 mg of the GnRH agonist deslorelin (D-Trp^6-Pro^9-des-gly^{10}-GnRH ethylamide). The implants are manufactured by a proprietary method that involves extrusion of deslorelin with a matrix consisting of low-melting point lipids and biological surfactant (Trigg *et al.* 2001). The implant formulation comprises 2–10% (w/w) deslorelin (on an active basis), 0.5–2.5% (w/w) lecithin and 85–97.5% (w/w) stearin. Lecithin is a phosphatidylcholine and stearin is partially dehydrogenated palm oil (Walsh and Trigg 2001). A 4.7 mg implant is 2.3 mm in width and 12.5 mm in length and is supplied pre-sterilised in a large-bore (12 gauge) hypodermic needle-implanter along with a reusable applicator. Deslorelin exerts its contraceptive effects in eutherian mammals by decreasing the synthesis and release of LH and FSH from the pituitary. Folliculogenesis is inhibited, as FSH concentrations are reduced below the threshold required for follicular development, and steroid hormone concentrations are reduced, which inhibits the oestrogen-induced LH surge required for ovulation (Fraser 1993; Herbert and Trigg 2005).

Suprelorin implants are administered subcutaneously between the shoulder-blades in kangaroos and wallabies. In some small wallaby species (e.g. tammar wallabies) this procedure is easy to perform in conscious animals and does not require sedation or local anaesthesia, but in the larger species (e.g. eastern grey kangaroos) sedation and/or anaesthesia may be required for capture and/or ease of handling and to reduce the occurrence of capture-induced myopathy (Vogelnest 1999). No prophylactic antibiotic treatment has been necessary.

Effects of Suprelorin in the tammar wallaby

Treatment of female tammar wallabies with one 5 mg Suprelorin implant (earlier implant batches contained 5 mg of deslorelin, instead of the standard 4.7 mg dose) inhibited reproduction for 515 ± 87 days (mean ± SE), but there was a high degree of variation in the duration of infertility (range 344 to >781 days, n = 7) between animals (Herbert *et al.* 2005). This may indicate that the 5 mg dose is close to the threshold required to induce infertility in this species. A higher dose of Suprelorin may achieve a more consistent duration of contraception between animals, as has been suggested for the eastern grey kangaroo (Herbert *et al.* 2006).

Suprelorin exerts its contraceptive effect in macropodids by inhibiting follicular development, presumably resulting from a reduction in the synthesis and release of FSH from the pituitary (Herbert *et al.* 2004) and a reduction in basal LH concentrations (Herbert *et al.* 2005). Oestrogen concentrations most likely remain at basal levels, as the pre-ovulatory follicle is the source of the elevated oestradiol concentrations associated with behavioural oestrus in this species (Shaw and Renfree 1984; Harder *et al.* 1985). The failure of the majority of treated females to mate at the time of expected oestrus supports this suggestion (Herbert *et al.* 2004). The inhibition of follicular development and failure to ovulate means that there is no luteinisation of follicular cells to form the CL, so that progesterone concentrations remain at basal levels during the breeding season (Herbert *et al.* 2005).

When Suprelorin implants were administered at the time of RPY in animals presumed to be carrying dormant blastocysts, there was a significant reduction in the number of treated animals that gave birth compared with control animals (Herbert *et al.* 2004, 2005). This result was surprising because the tammar CL is thought to be autonomous once formed (Tyndale-Biscoe *et al.* 1974; Tyndale-Biscoe and Renfree 1987). Total hypophysectomy has no effect on the growth and secretory functions of an already formed CL (Hearn 1974). Fetal development continues, but parturition is inhibited (Hearn 1974). Similarly, passive immunisation against GnRH has no effect on blastocyst and CL reactivation after RPY, or on parturition (Short *et al.* 1985). The reduced proportion of neonates detected in the pouch of Suprelorin-treated animals suggests that chronic GnRH agonist treatment has an effect on some aspect of blastocyst survival, luteal reactivation, pregnancy, parturition or the initiation of lactation (Herbert *et al.* 2005); this is the subject of ongoing investigations. In terms of population management, this failure to give birth post deslorelin treatment is advantageous, as it means that there is an immediate reduction in the capacity of the females to produce an additional young as a result of an already conceived blastocyst. If Suprelorin is administered to females suckling a young, there is no effect on the maintenance of lactation or on the growth and development of the young (C.A. Herbert, unpubl. data).

The contraceptive effects of Suprelorin treatment in the tammar wallaby are reversible. LH and FSH concentrations return to pre-treatment concentrations within four weeks of surgical excision of the implant (Herbert 2002). If the implant is left *in situ*, contraception lasts for one or two breeding seasons and the strict seasonality of the tammar is maintained (i.e. treated females will not resume breeding out of season) (Herbert *et al.* 2005). Longer contraceptive durations may be attainable with higher dosages of Suprelorin (Herbert *et al.* 2004), as has been reported in dogs (Trigg *et al.* 2001) and heifers (D'Occhio *et al.* 2000).

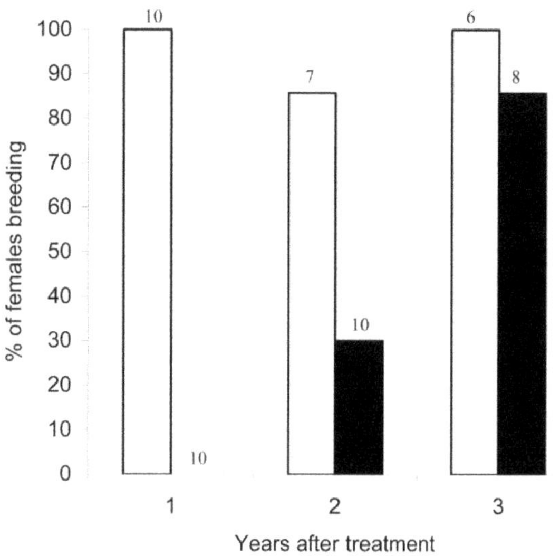

Figure 27.2: Percentage of female eastern grey kangaroos (*M. giganteus*) with pouch young at yearly intervals after treatment with two 4.7 mg Suprelorin (black bars) or two placebo implants (open bars). The animals were caught and their pouches inspected for young (Woodward *et al.* 2006, C.A. Herbert, unpubl. data). The numbers above each bar indicate the number of animals captured from each group.

Effects of Suprelorin in the eastern grey kangaroo

Suprelorin implants are an effective contraceptive in captive female eastern grey kangaroos (Herbert *et al.* 2006; Woodward *et al.* 2006). A 9.4 mg dose provided contraception for a minimum of 510 days (n = 10), with a mean (± SE) contraceptive duration of 734 ± 65 days (n = 7; Fig. 27.2; updated data from the study reported by Woodward *et al.* 2006). Of the remaining three animals from that trial, one had still not bred by 942 days post-treatment and the others died approximately 650 and 900 days post-treatment, without giving birth. Suprelorin treatment effectively inhibits reproduction in 96% of female eastern grey kangaroos (n = 23) (Herbert *et al.* 2006; Woodward *et al.* 2006). It is not known why one female has failed to respond to Suprelorin treatment even at a variety of different dosages (Herbert *et al.* 2006). There is also likely to be a small proportion of non-responding female tammars. It is important to determine if the continued reproduction in these individuals is the result of insufficient dosage or whether it reflects a more fundamental resistance to GnRH agonist-induced pituitary desensitisation. If the latter, the heritability of the non-response must be measured, as this may influence the long-term contraceptive success rates within a population. Likewise, the mechanism underlying unresponsiveness of the small number of individuals to the levonorgestrel implant needs to be determined. Despite the small number of potential non-responders to both contraceptive implants, the success rate is very high, probably higher than immunological forms of contraception because of the inherently variable nature of the mammalian immune response (Cooper and Larsen 2006).

Behavioural effects of Suprelorin treatment in macropodids

In a study on the effects of Suprelorin treatment on the crepuscular time budget in captive eastern grey kangaroos, Woodward *et al.* (2006) could not detect any changes in the amount of time treated animals spent foraging, scanning, grooming, moving or interacting with other animals. The most significant behavioural effect was that simultaneous Suprelorin treatment and RPY in eastern grey kangaroos induced a behavioural oestrus within approximately three days, possibly as a result of the initial agonist-induced hyperstimulation of LH and FSH within the first 24–48 hours post-treatment (Herbert and Trigg 2005). Females would mate with males at this time, but mating was apparently unsuccessful (Woodward *et al.* 2006).

CONTRACEPTIVE EFFECTS AND MANAGEMENT IMPLICATIONS

A summary of the contraceptive effects and relevant characteristics of each of the two long-acting contraceptive implants is given in Table 27.2. There are some fundamental differences between the two contraceptives. Suprelorin, a protein hormone analogue, inhibits reproduction by acting at the level of the pituitary, effectively shutting down gonadotrophin release. As a result, follicular development is halted, steroid hormone concentrations are reduced and oestrus is inhibited. Levonorgestrel, a synthetic steroid hormone analogue, causes only limited suppression of gonadotrophin secretion by the pituitary. The females effectively continue to cycle, as follicular development continues, but the majority of animals do not mate, so it appears to inhibit only LH and not FSH (Hynes *et al.* 2007). The effects of levonorgestrel implants on circulating steroid hormone concentrations have been investigated in the tammar wallaby, and the plasma profiles were identical to those of a normal cycle (Nave 2002).

There are differences in the duration of contraception at the dosages investigated. Levonorgestrel implants render tammar wallabies infertile for a minimum of four years, whereas Suprelorin affords only one to two years of contraception. In eastern grey kangaroos the difference is not nearly as great (Table 27.2). Although some levonorgestrel-treated eastern grey kangaroos remain infertile for up to five years, up to 40% may be fertile within their third breeding season post-treatment (Fig. 27.1).

Grigg (1997) presented a model of the proportion of female red kangaroos (*Macropus rufus*) rendered infertile, and the attendant risk of extinction. He concluded that infertility levels as low as 10% could cause a decline in population density, and more than 20–30% could result in local extinction. There are no other published models of fertility control targets in macropodid populations, but the target for the proportion of females to be treated in most medium to large mammals is much higher, in the order of 60–80% (e.g. koalas, *Phascolarctos cinereus*, Duka and Masters 2005; common brushtail possums, *Trichosurus vulpecula*, Barlow 1994; white-tailed deer, *Odocoileus virginianus*, Porter *et al.* 2004). Wildlife managers would need to reapply treatment before contraceptive rates fell below this level, so the contraceptive duration for levonorgestrel in eastern grey kangaroos has been specified as three years (Table 27.2).

Contraceptive duration alone cannot be used as a measure of the potential efficacy of a fertility control agent. The relationship between the duration of contraception and the reproductive life-cycle of individuals within the population must be taken into account. This relationship can be used to establish the number of times that an individual might need to be treated within its lifetime, assuming that ongoing population control is necessary. These calculations would allow direct comparisons of the relative cost-efficiency of different contraceptive/fertility control methods. Other variables, such as the method of delivery, reproductive biology of the species and management goals for the population, must also be considered.

There may also be a trade-off between the ease of administration and the longevity of contraception. If the animals need to be captured to allow implant delivery a longer-lasting method of fertility control is likely to be the most cost-effective, because animal capture is expensive and time-consuming and poses the risk of injury to animals or handlers (particularly with large and often flighty species such as eastern grey kangaroos). If efficient remote delivery methods can be developed, the frequency of treatment becomes less important in cost-efficiency analyses. The small size of the Suprelorin implant makes it more amenable to the development of a remote delivery dart system, which would negate the need to physically capture animals. Such a system would ideally involve a method that delivers the implant and simultaneously marks the animal in some way, to avoid repeat dosing of individuals.

The management of overabundant native species requires a balance between overcoming the negative impacts of the species yet maintaining an adequate population base with genetic integrity to ensure species survival within the area. This may be especially important for some of the smaller macropodid

Table 27.2: Levonorgestrel and Suprelorin contraceptive implants and their effects on the tammar wallaby (*M. eugenii*) and eastern grey kangaroo (*M. giganteus*).

	Levonorgestrel	Suprelorin®
Implant details		
Active ingredient	Synthetic progestin	GnRH agonist (deslorelin)
Implant dimensions	44 × 2.4 mm	12.5 × 2.3 mm
Implant dose	70 mg	4. 7 mg
Trial dose	70 mg (wallaby)[a] 140 mg (kangaroo)[b]	4.7 mg (wallaby)[c] 9.4 mg (kangaroo)[d]
Administration	Small incision and SC injection under LA,[b] potential for SC injection	SC injection[d,e,f]
Mechanisms of action/physiological effects[#]		
Follicular development	No effect[g]	Inhibited[e]
FSH secretion	Minimal or no effect[*,g]	Reduced[*,e]
LH secretion	Surge inhibited[*,g]	Reduced[c]
Ovulation	Inhibited[g]	Inhibited[e]
Blastocyst survival	Slightly reduced[a,g]	Reduced[c,e]
Oestrus and mating	Suppressed by: 100% (n = 14)[a] 70% (n = 10)[g]	Suppressed by 88% (n = 8)[e]
Ovarian progesterone	Suppressed[*,g]	Suppressed[c]
Lactation	Not affected[h]	Not affected[f]
Contraceptive duration		
Tammar wallaby	>4 years[h]	1–2 years[c]
Eastern grey kangaroo	3 years[i]	2 years[d]
Contraceptive success		
Tammar wallaby	100% (n = 18)[a]	93% (n = 15)[c,e]
Eastern grey kangaroo	91% (n = 33)	96% (n = 23)[d,f]
Other		
Behavioural effects	Minimal[j]	Minimal[d]
Negative side effects	None detected	None detected
Potential for remote delivery	Not in present form	Yes

[#] Determined from studies in the tammar wallaby.

* Presumed effect based on other observations.

LA =local anaesthetic; SC =subcutaneous.

(a) Nave *et al.* 2000; (b) Nave *et al.* 2002a; (c) Herbert *et al.* 2005; (d) Woodward *et al.* 2006; (e) Herbert *et al.* 2004; (f) Herbert *et al.* 2006; (g) Hynes *et al.* 2007; (h) Nave *et al.* 2002b; (i) Coulson *et al.* 2008; (j) Poiani *et al.* 2002.

species that are threatened or vulnerable on a national scale, but have temporal or localised population eruptions. In these situations, a reversible contraceptive may be more suitable than a more permanent fertility control method (e.g. surgical sterilisation), as it allows flexibility for future population management and enables more individuals within the population to breed at some stage, limiting the loss of genetic diversity from the population. The extent to which contraception might lead to a loss of genetic variation has not been studied until recently. In an initial study, Tanaka *et al.* (submitted) have concluded that regular contraception, in which all females are allowed to reproduce at some stage, will not affect effective population size (*Ne* in the genetic sense). They suggest that a suitable approach (to reduce population size yet retain genetic diversity within a population) could be to prevent reproduction of all females older than a particular age. The importance of retaining genetic variability will depend upon population history – the extent to which it has been through bottlenecks – and the proposed management outcomes.

To date there have been no negative side effects detected following the use of levonorgestrel or Suprelorin contraceptive implants in macropodids, that is, none of the observed physiological or behavioural responses have resulted in a measurable decline in animal health or are likely to have a deleterious effect on the well-being of individual animals. Treated animals maintain good body condition, there is no effect on the growth and development of an existing young in the pouch, and behavioural changes are minimal. The research on the use of these two contraceptive methods in macropodids over the last 10 years has generated a large body of data on the physiological and endocrine effects of these agents, the contraceptive success rate and the longevity of contraception in different species. More emphasis needs to be placed on evaluating the efficacy of fertility control in field situations and on fine-tuning the technology.

Fertility control may be more applicable as a management tool in certain environments, such as closed or semi-closed populations, or areas where animals are habituated to human presence and/or are easy to capture. The ecological consequences of fertility control are not yet understood for free-ranging, wild macropodid populations. Does the demography change in response to a decreased birth rate by increased compensatory survival of adults or juveniles? Does it alter the rate of immigration/emigration in open systems? What are the effects on male behaviour? These and many more questions need to be addressed, as they may affect the success of fertility control as a method of population management. It must also be borne in mind that the nature of fertility control is to effect a population decline initially by halting population growth, with subsequent population reduction because the mortality rate exceeds the recruitment rate. As such, fertility control may be most effective if combined with an initial population reduction, such as translocation or culling, in areas where this is feasible and socially acceptable.

The research reported here demonstrates the potential for long-acting contraceptive technology to play a role in the management of macropodid populations in certain environments. Further research should be devoted to refining this technology and understanding the interactions between fertility control and the ecology of the target species. Final assessment of efficacy awaits the results of such research.

ACKNOWLEDGMENTS

At Portland: John Hill, Sarah Holland, Ron Jeffries, Kirstin Long, Ken O'Connor, Aldo Poiani, Dylan Saloman, Kevin Saunders. At Woodlands: Mike Cusack, Bron Davies, Will MacAulay, Fiona Smith. Captive kangaroo and tammar trials at Macquarie University and the University of New South Wales: Ron Claassens, Anne Mouland, James Cook, Emily Miller and Rachel Woodward. Much of this research was funded by the Koala and Kangaroo Contraception Program, an Australian Research Council Linkage grant (LP LP0560344) awarded to the authors and others.

REFERENCES

Adderton Herbert C 2004. Long-acting contraceptives: a new tool to manage overabundant kangaroo populations in nature reserves and urban areas. *Aust. Mammal.* **26:** 67–74.

Barlow ND 1994. Predicting the effect of a novel biocontrol agent: a model for viral-vectored immunocontraception of New Zealand possums. *J. Appl. Ecol.* **31:** 454–462.

CONCOM 1998. Code of Practice for the Humane Shooting of Kangaroos. Environment Australia, Canberra.

Cooper DW and Larsen E 2006. Immunocontraception of mammalian wildlife: ecological and immunogenetic issues. *Reproduction* **132:** 821–828.

Coulson G 1998. Management of overabundant macropods: are there conservation benefits? In *Managing Marsupial Abundance for Conservation Benefits*. (Ed. A Austin) pp. 37–48. CRC for the Conservation and Management of Marsupials, Sydney.

Coulson G, Nave CD, Shaw G and Renfree MB 2008. Long-term efficacy of levonorgestrel implants for fertility control of eastern grey kangaroos (*Macropus giganteus*). *Wildl. Res.* **35:** 520–524.

DEH (Department of Environment and Heritage) 2006. Background information: commercial kangaroo and wallaby harvest quotas. Dept of Environment and Heritage, Canberra.

D'Occhio MJ, Fordyce G, Whyte TR, Aspden WJ and Trigg TE 2000. Reproductive responses of cattle to GnRH agonists. *Anim. Reprod. Sci.* **60–61:** 433–442.

Duka T and Masters P 2005. Confronting a tough issue: fertility control and translocation for over-abundant koalas on Kangaroo Island, South Australia. *Ecol. Man. Restor.* **6**:172–181.

Eymann J, Herbert CA, Thomson BP, Trigg TE, Cooper DW and Eckery DC 2007. Effects of deslorelin implants on reproduction in the common brushtail possum (*Trichosurus vulpecula*). *Reprod. Fertil. Dev.* **19:** 899–909.

Fletcher TP and Renfree MB 1988. Effects of corpus luteum removal on progesterone, estradiol-17 beta and LH in early pregnancy of the tammar wallaby, *Macropus eugenii. J. Reprod. Fertil.* **83:**185–191.

Fraser HM 1993. GnRH analogues for contraception. *Brit. Med. Bull.* **49:** 62–72.

Garrott RA, White PJ and Vanderbilt White CA 1993. Overabundance: an issue for conservation biologists. *Conserv. Biol.* **7:** 946–949.

Grigg GC 1997. A crossroads in kangaroo politics. *Aust. Biol.* **10:** 12–22.

Harder JD, Hinds LA, Horn CA and Tyndale-Biscoe CH 1985. Effects of removal in late pregnancy of the corpus luteum, graafian follicle or ovaries on plasma progesterone, oestradiol LH, parturition and post-partum oestrus in the tammar wallaby, *Macropus eugenii. J. Reprod. Fertil.* **75:** 449–459.

Hearn JP 1974. The pituitary gland and implantation in the tammar wallaby, *Macropus eugenii. J. Reprod. Fertil.* **39:** 235–241.

Herbert CA 2002. Contraception of marsupials using slow release implants containing the GnRH agonist deslorelin. PhD thesis. Macquarie University, Sydney.

Herbert CA 2007. From the urban fringe to the Abrolhos Islands: management challenges of burgeoning marsupial populations. In *Pest or Guest: The Zoology of Overabundance.* (Eds D Lunney, P Eby, P Hutchings and S Burgin) pp. 129–141. Royal Zoological Society of NSW, Sydney.

Herbert CA and Trigg TE 2005. Applications of GnRH in the control and management of fertility in female animals. *Anim. Reprod. Sci.* **88:** 141–153.

Herbert CA, Trigg TE and Cooper DW 2004. Effect of deslorelin implants on follicular development, parturition and post-partum oestrus in the tammar wallaby (*Macropus eugenii*). *Reproduction* **127:** 265–273.

Herbert CA, Trigg TE, Renfree MB, Shaw G, Eckery DC and Cooper DW 2005. Long-term effects of deslorelin implants on reproductive success and plasma progesterone and LH concentrations in the female tammar wallaby. *Reproduction* **129:** 361–369.

Herbert CA, Trigg TE and Cooper DW 2006. Fertility control in female eastern grey kangaroos using the GnRH agonist deslorelin. I. Effects on reproduction. *Wildl. Res.* **33:** 41–46.

Hynes EF, Nave CD, Shaw G and Renfree MB 2007. The effects of levonorgestrel on ovulation and oestrous behaviour in the female tammar wallaby. *Reprod. Fertil. Dev.* **19:** 335–340.

IUCN 2004. IUCN Red List of Threatened Species. http://www.iucnredlist.org/.

Mawson PR 2004. Translocations and fauna reconstruction sites: Western Shield review, February 2003. *Conserv. Sci. W. Aust.* **5:** 108–121.

Maxwell S, Burbidge AA and Morris K 1996. Action plan for Australian marsupials and monotremes. IUCN/SSC Australasian Marsupial and Monotreme Specialist Group. Endangered Species Program, Environment Australia, Canberra.

Nave CD 2002. Fertility control in the eastern grey kangaroo, *Macropus giganteus.* PhD thesis. University of Melbourne, Australia.

Nave CD, Shaw G, Short RV and Renfree MB 2000. Contraceptive effects of levonorgestrel implants in a marsupial. *Reprod. Fertil. Dev.* **12:** 81–86.

Nave CD, Coulson G, Poiani A, Shaw G and Renfree MB 2002a. Fertility control in the eastern grey kangaroo using levonorgestrel implants. *J. Wildl. Man.* **66:** 470–477.

Nave CD, Coulson G, Short RV, Poiani A, Shaw G and Renfree MB 2002b. Long-term fertility control in the kangaroo and wallaby using levonorgestrel implants. *Reprod. Supp.* **60:** 71–80.

Poiani A, Coulson G, Salamon D, Holland S and Nave CD 2002. Fertility control of eastern grey kangaroos: do levonorgestrel implants affect behavior? *J. Wildl. Man.* **66:** 59–66.

Poole WE and Catling PC 1974. Reproduction in the two species of grey kangaroos, *Macropus giganteus* Shaw and *M. fuliginosus* (Desmarest). I. Sexual maturity and oestrus. *Aust. J. Zool.* **22:** 277–302.

Pople T and Grigg G 1999. Commercial harvesting of kangaroos in Australia. Environment Australia, Canberra.

Porter WF, Underwood HB and Woodard JL 2004. Movement behavior, dispersal, and the potential for localised management of deer in a suburban environment. *J. Wildl. Man.* **68:** 247–256.

Renfree MB 1979. Initiation of development of diapausing embryo by mammary denervation during lactation in a marsupial. *Nature* **278:** 549–551.

Renfree MB 1994. Endocrinology of pregnancy, parturition and lactation of marsupials. In *Marshall's Physiology of Reproduction.* 4th edn. Vol. 3, *Pregnancy and Lactation.* (Ed. GE Lamming) pp. 677–766. Chapman and Hall, London.

Renfree MB and Tyndale-Biscoe CH 1973. Intrauterine development after diapause in the marsupial, *Macropus eugenii. Dev. Biol.* **32:** 28–40.

RIA (Rottnest Island Authority) 2003. Rottnest Island management plan 2003–2008. Rottnest Island Authority, Government of Western Australia.

Segal SJ, Alvarez-Sanchez F, Brache V, Faundes A, Vilja P and Tuohimaa P 1991. Norplant implants: the mechanism of contraceptive action. *Fertil. Steril.* **56:** 273–277.

Shaw G and Renfree MB 1984. Concentrations of oestradiol-17 b in plasma and corpora lutea throughout pregnancy in the tammar, *Macropus eugenii*. *J. Reprod. Fertil.* **72:** 29–37.

Short RV, Flint AP and Renfree MB 1985. Influence of passive immunisation against GnRH on pregnancy and parturition in the tammar wallaby, *Macropus eugenii*. *J. Reprod. Fertil.* **75:** 567–575.

Sivin I 1988. International experience with NORPLANT and NORPLANT-2 contraceptives. *Stud. Fam. Plann.* **19:** 81–94.

Sivin I 1994. Contraception with NORPLANT implants. *Hum. Reprod.* **9:** 1818–1826.

Sivin I, Lahteenmaki P, Ranta S, Darney P, Klaisle C, Wan L, Mishell DR Jr, Lacarra M, Viegas OA, Bilhareus P, Koetsawang S, Piya-Anant M, Diaz S, Pavez M, Alvarez F, Brache V, LaGuardia K, Nash H and Stern J 1997. Levonorgestrel concentrations during use of levonorgestrel rod (LNG ROD) implants. *Contraception* **55:** 81–85.

Strahan R (Ed.) (1995). *The Mammals of Australia*. Reed New Holland, Sydney.

Tanaka MM, Cristescu R and Cooper DW 2009. Effective population size of koala populations under different population management regimes. *Wildl. Res.* In press.

Trigg TE, Wright PJ, Armour AF, Williamson PE, Junaidi A, Martin GB, Doyle AG and Walsh J 2001. Use of a GnRH analogue implant to produce reversible long-term suppression of reproductive function in male and female domestic dogs. *J. Reprod. Fertil. Suppl.* **57:** 255–261.

Twigg LE, Lowe TJ, Martin GR, Wheeler AG, Gray GS, Griffin SL, O'Reilly CM, Robinson DJ and Hubach PH 2000. Effects of surgically imposed sterility on free-ranging rabbit populations. *J. Appl. Ecol.* **37:** 16–39.

Tyndale-Biscoe CH and Renfree MB 1987. *Reproductive Physiology of Marsupials*. Cambridge University Press, Cambridge.

Tyndale-Biscoe CH, Hearn JP and Renfree MB 1974. Control of reproduction in macropodid marsupials. *J. Endocrinol.* **63:** 589–614.

Tyndale-Biscoe CH, Hinds LA, Horn CA and Jenkin G 1983. Hormonal changes at oestrus, parturition and post-partum oestrus in the tammar wallaby (*Macropus eugenii*). *J. Endocrinol.* **96:** 155–161.

Tyndale-Biscoe CH, Hinds LA and Horn CA 1988. Fetal role in the control of parturition in the tammar, *Macropus eugenii*. *J. Reprod. Fertil.* **82:** 419–428.

Vogelnest L 1999. Chemical restraint of Australian native fauna. In *Wildlife in Australia: Healthcare and Management*. (Ed. DI Bryden) pp. 149–188. Post Graduate Foundation in Veterinary Science, University of Sydney, Dubbo, Australia.

Walsh JD and Trigg TE 2001. Formulations for peptide release. US Patent and Trademark Office, Washington DC.

Woodward R, Herberstein M and Herbert CA 2006. Fertility control in female eastern grey kangaroos using the GnRH agonist deslorelin. II. Effects on behaviour. *Wildl. Res.* **33:** 47–55.

28 Darting eastern grey kangaroos: a protocol for free-ranging populations

M.W. Roberts, L.E. Neaves, R. Claassens and C.A. Herbert

SUMMARY

We describe a simple step-by-step approach for the capture of free-ranging kangaroos by chemical immobilisation delivered using projectile darts ('darting'). Before attempting to capture an animal, consideration must be given to the environmental conditions suitable for darting, the health of animals in the target population, the type of immobilisation agent to be used and the appropriate tranquilliser firearm (in terms of its range, dart impact and accuracy). This field protocol is divided into six primary stages: dart preparation, approaching the animal, taking the shot, stages of sedation, animal processing, and recovery. The optimum darting distance of the tranquilliser firearm (Pneu-Dart® Inc., Williamsport) was 23–28 m, based on measures of precision and dart impact. The anaesthetic of choice, a combination of tiletamine hydrochloride and zolazepam hydrochloride at a 1:1 ratio (Zoletil 100®), has a wide safety margin. For simplicity, all females and subadults received a total of 125 mg of Zoletil 100® (equating to an average dosage of 4.7 mg/kg) and all large males received 250 mg of Zoletil 100® (average dosage 4.9 mg/kg). No relationship between body weight and either the time to recumbency or recovery was observed. We evaluate this protocol by assessing the survivorship of captured individuals and behavioural changes of kangaroos following darting activities. The use of this protocol to capture 123 eastern grey kangaroos (*Macropus giganteus*) in Sydney's water supply catchment indicated that it provided an efficient, safe and applicable methodology for immobilising free-ranging medium to large macropodids to facilitate routine scientific sample collection.

INTRODUCTION

There is an ongoing requirement to capture free-ranging macropodids for their conservation (Horsup 1994; Telfer and Griffiths 2006) and management (Nave *et al.* 2002; Viggers and Hearn 2005), or for research purposes (Moore *et al.* 2002). A wide variety of capturing techniques have been reported in the literature (Table 28.1). The choice of technique is primarily determined by the need to maximise human safety and animal welfare, as well as environmental and population attributes that influence the relative efficiency of different capture methods.

Chemical immobilisation of animals can be achieved using projectile darts for intramuscular administration of an anaesthetic agent. This remotely delivered form of chemical immobilisation (commonly referred to as darting) can meet all the necessary human safety and animal welfare requirements, and is often considered superior to capture techniques that rely on physical restraint (Higgingbottom 1989).

Table 28.1: Capture techniques used on macropodids.

Capture technique	Species	Stress to animal	Human risk	Effectiveness
Hand net (Kitchener 1973)	Quokka (*Setonix brachyurus*)	Moderate	Low	Selective Not suitable for larger macropodids in open settings
Trap yards (Jarman and Taylor 1983)	Eastern grey kangaroo (*M. giganteus*)	Moderate to high	Moderate	High capture rates Construction of fencing required Attractant needed to lure animals
Cannon nets (Edwards *et al.* 1994)	Red kangaroo (*M. rufus*)	Moderate	Moderate	Highly selective captures Requires specific sites where food and water availability are low
Oral baiting (Arnold *et al.* 1986)	Western grey kangaroo (*M. fuliginosus*)	Minimal	Low	Non-target animal concerns Potential overdoses, bait aversion and lengthy care required
Stunning (Moss and Croft 1999)	Red kangaroo (*M. rufus*)	Usually moderate	High	Highly selective, limited distance <50 m High degree of accuracy required Restricted to open habitats
Drawstring traps (Coulson 1996)	Eastern grey kangaroo (*M. giganteus*)	Minimal	Low	Selective and yields high capture rates Requires fencing
Darting (Higgingbottom 1989)	Red-necked wallaby (*M. rufogriseus*)	Minimal	Low	Highly selective, limited distance <50 m Dart injuries from high-velocity dart impact Slow induction times

Stress to the animal is gauged by the potential for post-capture myopathy as minimal, moderate and high.
Human risk is gauged by the method's risk of injury to the handlers during the capturing process as low, moderate and high.
Effectiveness outlines the pros and cons of each method.
Adapted from Coulson (1996) and personal experience.

There is a mounting body of evidence to suggest that capture techniques involving physical restraint lead to higher mortality rates, primarily from capture myopathy, a disease induced by excessive muscular activity, often associated with capture-related exertion and fear, which results in necrosis of heart and skeletal muscle (Vogelnest 1999).

The remote delivery of immobilising drugs has a unique place in numerous situations where other methods are unsuitable. The use of drawstring traps (in fences), for instance, is not applicable in circumstances where there are no fences, or where fence construction would be impractical. Similarly, oral baiting is difficult to use in situations where food and water are readily available, where there are non-target animals and where targeted capture of individuals is required. Capture via darting is a highly selective method applicable in a variety of environments where a clear path for the projectile can be obtained. It causes limited disturbance to the target animals and therefore does not compromise future observations (Higgingbottom 1989).

Some disadvantages have been associated with the use of darting to capture macropodids (Coulson 1996). These include injuries from high-velocity dart impact, loss of the darted animal because of slow induction times, and the short maximum effective range of the projectiles (<50 m). These disadvantages initially precluded the wider use of remote delivery systems in favour of other methods to capture free-ranging macropodids (Arnold *et al.* 1986; Coulson 1996). However, improvements in projectile technology and anaesthetic agents have to some extent reduced the first two concerns.

Chemical immobilising agents can be remotely administered to animals via a blowpipe or some form of tranquilliser firearm (pistol or rifle). Blowpipe systems have the advantage of being cheap, but their limited range (10–15 m) means that the method is typically unsuitable for free-range situations. It also relies heavily on the user's skill to overcome inconsistencies in the amount of air expelled to propel a dart the required distance and to mitigate wind impacts on the dart's trajectory.

Tranquilliser firearms (specifically rifles) can overcome the shortfalls of blowpipes as they have a larger effective range (20–40 m) and higher accuracy and reliability, making them suitable for the capture of free-ranging animals that are less habituated. Several workers have used tranquilliser firearms to capture larger species of kangaroos with considerable success, both alone (Stirrat 1997; Viggers and Hearn 2005) and in conjunction with other methods (Moore *et al.* 2002). This success has been assisted by pharmacological developments, with immobilising agents such as Zoletil 100® (Virbac Australia Pty Ltd) that produce rapid and smooth recumbency (Stirrat 1997), thus reducing the chance of losing contact with the darted animal.

This chapter describes a simple step-by-step approach for the capture of free-ranging kangaroos by chemical immobilisation delivered using projectile darts (hereafter referred to as darting). The development of a darting protocol eventuated from the capture of eastern grey kangaroos (*Macropus giganteus*) for the analysis of their zoonotic disease risk (e.g. *Cryptosporidium* and *Giardia*) to Sydney's water supply (Hawkins *et al.* 2000; Power *et al.* 2004).

STUDY SITE AND POPULATION

Eastern grey kangaroos were captured at three sites within the Wollondilly River Catchment – Murphy's Crossing, Jooriland and Douglas Scarp (34°11'1"S, 150°18'22"E) from June to September 2005 and June to September 2006. These sites lie adjacent to Lake Burragorang, within Blue Mountains and Nattai National Parks in the central tablelands of New South Wales, Australia, approximately 95 km southwest of Sydney (Fig. 28.1). This remote area is a major part of Sydney's hydrological catchment, supplying approximately 80% of the city's potable water. Unauthorised public access to the area is prohibited, to protect water quality.

Eastern grey kangaroos reside in this region at high densities (4–6 kangaroos per hectare; Roberts, unpub. data). These densities are fuelled by relatively low levels of competition and predation and a high availability of food. The latter is a result of previous grazing activities that ceased in most areas in the late 1950s after the construction of Warragamba Dam.

The study area consists of a heterogeneous matrix of grassland (comprising *Themeda australis*, *Microlaena stipoides* and *Eragrostis curvula*) and woodland

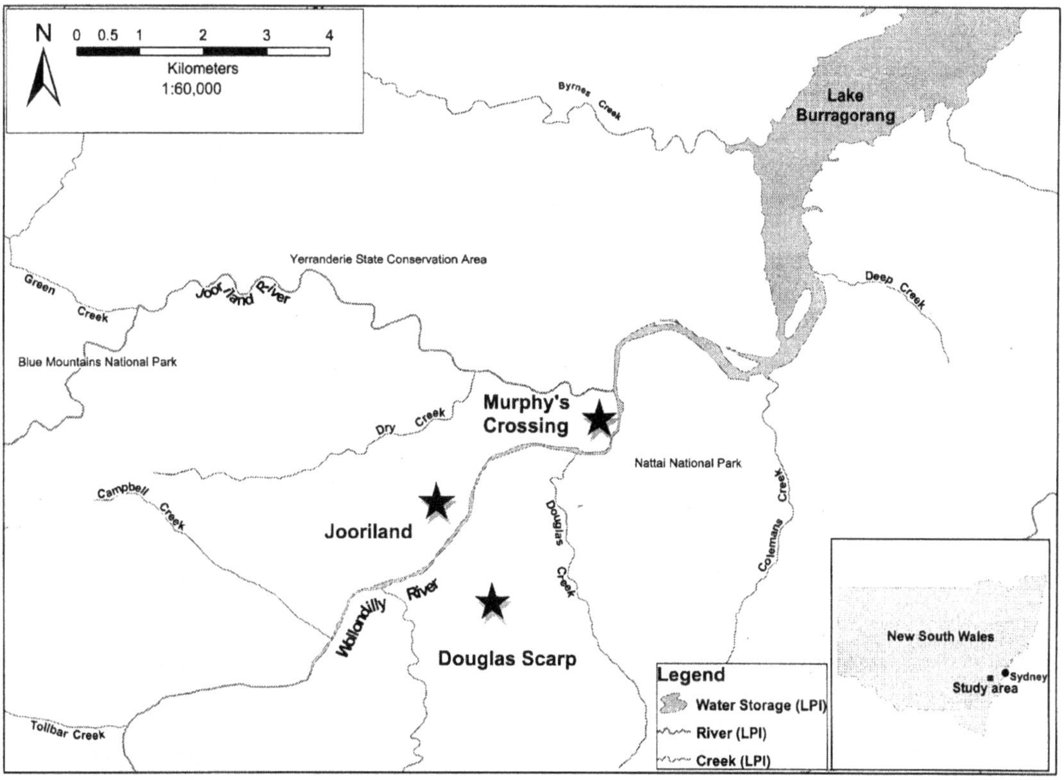

Figure 28.1: The three sites – Murphy's Crossing, Jooriland and Douglas Scarp – in the Wollondilly River catchment where eastern grey kangaroos (*M. giganteus*) were darted.

associations (*Eucalyptus tereticornis* and *Eucalyptus moluccana*), on open flood plains and undulating hills, with relatively intact riparian vegetation (*Casuarina cunninghamiana* subsp. *cunninghamiana* and *Tristaniopsis laurina*) adjacent to water courses. The only other macropodid species that occupy the area are the red-necked wallaby (*Macropus rufogriseus*), the swamp wallaby (*Wallabia bicolor*) and common wallaroo (*Macropus robustus*), but these occur at low densities and with minimal overlap with populations of eastern grey kangaroos. Red foxes (*Vulpes vulpes*), wild dogs (*Canis lupus familiaris/dingo*) and wedge-tailed eagles (*Aquila audax*) also occur in the catchment area. These sites have an annual average rainfall of 571 mm; during 2005 and 2006 the region was experiencing drought conditions.

A large number of kangaroos were captured to research home range and habitat utilisation, population demography and zoonotic disease prevalence. The capture method needed to minimise the risks to animals and handlers and minimise stress to the animals, to avoid impeding subsequent observations and to facilitate the capture of animals within certain age and sex classes. For these reasons, oral baiting was not suitable. The kangaroos showed a reluctance to cross through old farm fences in the area, which meant that drawstring traps were not feasible (Coulson 1996). Darting with a tranquilliser firearm was the most feasible option to meet the site attributes and capture requirements.

This protocol documents the successful use of free-ranging darting to capture a large number (n = 123) of eastern grey kangaroos. It outlines the practice methods, field stages of darting and methods of evaluating the darting process. The outcome is a repeatable, safe and effective technique that encourages its wider use.

CAPTURED KANGAROOS

Of the 123 animals captured, 89 were females and 34 were males. The average body mass for adult females (n = 83) was 28.3 ± 2.8 kg (mean ± SD) and for adult males (n = 24) it was 53.3 ± 5.2 kg. These were within the upper ranges of adult body masses reported for this species (Poole 1995), and corroborated observations that suggested that animals within the population were in good condition. The average body mass for subadult females (n = 6) was 18.8 ± 4.0 kg and for subadult males (n = 10) it was 29.0 ± 10.0 kg. Young at foot were not darted. Forty-nine percent of females were carrying pouch young; the mass of young was subtracted from the female body mass using calculations derived from the head, leg and foot measurements (Poole *et al.* 1982).

METHODS

Conditions suitable for darting

Darting was conducted during daylight hours in temperatures less than 25°C, in the cooler months (June–September). Temperature was monitored throughout the day, and if it rose above 25°C darting was restricted to early mornings and late afternoons. Darting was abandoned if high temperatures persisted, which at times was evidenced by kangaroos licking their forearms. This precaution was taken to reduce the chances of hyperthermia and post-capture myopathy (Shepherd 1981; Williams and Thorne 1996). Darting was not attempted at night due to the high risk associated with decreased visibility, the increased chance of losing contact with the darted animal, increased activity of predators and the potential risk of injury to personnel.

Darting was conducted only in calm conditions. This reduced the likelihood of the wind altering the dart's trajectory and the subsequent potential for injuries to animals or for missed shots. In addition, it was more difficult to approach animals during moderate to strong winds as they showed an increased flight response. Darting was not conducted during heavy rain, but light rain actually made the process easier because animals were often found grazing in open areas. Darting could still be conducted in misty and foggy conditions, but estimating the distance to the animal was more difficult.

Tranquilliser firearm and darts

There are a number of tranquilliser firearm systems. They can be divided into two broad categories based on the method of dart propulsion – systems powered by a gas (CO_2) canister or by a blank cartridge (powder charge, modified .22 calibre rifle). The gas-powered rifles are usually purpose-built (e.g. models from Telinject, Daninject or Pneudart). In the case of the modified .22 calibre rifle, a .50 calibre (13 mm) barrel sits on top of the conventional barrel and a blank .22 charge provides air pressure that is transferred into the top barrel, projecting the dart towards the target.

There are two broad categories of commercially available darts – reusable pressurised darts and disposable darts with a detonating charge. The former have two chambers; the front chamber is loaded with the drug and the rear chamber is pressurised with air using a 10 cc syringe. A flight stabiliser is added to the rear of the dart. The force of the dart hitting the animal causes the pressure in the rear of the dart to expel the drug into the animal. These darts can be washed and reused. The darts with a detonating charge do not require pressurisation by the operator. The drug is loaded into the dart through the needle. The rear of the dart contains a detonating charge which is activated by the deceleration that occurs as the dart hits the animal, propelling the drug into the animal. These darts cannot be reused. The type of system used (tranquilliser firearm and dart) depends on personal preference and budget.

We used a Pneu-Dart® Model 193 (Pneu-Dart® Inc., Williamsport, US) tranquilliser firearm with a 4 × 32 mm optical sight (Silver Antler, Tasco® US). This is essentially a modified .22 rifle (as described above) with a five-shot box magazine for the .22 calibre charges. The distance is controlled by a five-stage velocity setting and the charges are available in three different powers, denoted by colours (green, yellow and red, from lowest to highest power). The second-lowest velocity setting on the rifle and green-powered charges were used, as these suited our distance requirements and minimised the risk of injury from the dart's impact. Disposable Type C Pneu-Dart® darts (1CC, ¾", PA 17703) were used, which are supplied either barbed or unbarbed. We used unbarbed darts to ensure that the dart dislodged if the animal failed to become immobilised.

Immobilising agent

The immobilising agent of choice for this study was Zoletil 100® (Virbac Australia Pty Ltd), which contains zolazepam HCl and tiletamine HCl at a 1:1 ratio. Zolazepam (a benzodiazepine) is a muscle relaxant with anticonvulsant and anxiolytic properties. Tiletamine produces dissociative anaesthesia with immobilisation and analgesia (Booth 1988). Zoletil 100® produces a rapid and smooth induction with very few side effects and it has been shown to have a wide safety margin in a variety of species, including macropodids (Stirrat 1997; Vogelnest 1999; Mauthe von Degerfield 2005). Zoletil 100® is ideal for remote delivery as it is supplied as a freeze-dried powder, which can be reconstituted with sterile water to make high concentrations (and thus lower volumes) that are ideal for darting (Vogelnest 1999). It also has a wide safety margin in comparison with other available sedatives, which is important when darting animals of unknown weight in the field (Higgins *et al.* 2002; Mauthe von Degerfield 2005).

During processing, animals received diazepam (Pamlin®, ~0.5 mg/kg body weight; IM, Parnell Laboratories Pty Ltd, Alexandria, NSW) if they became excessively restless, were panting heavily or had an increased body temperature. Diazepam is an excellent muscle relaxant and reduces stress by calming the animal during handling (English 2004).

Licensing requirements

A tranquilliser firearm permit was obtained from the New South Wales Firearms Registry for the possession and use of a tranquilliser gun under the *Firearms Act 1996 (NSW)*. Firearms safety awareness training and testing was required. Additional approval was acquired from joint land managers, the Sydney Catchment Authority and the Department of Environment and Climate Change. Licence requirements may vary over time or between states, so advice on licensing matters should be sought from the state firearms registry before use of a tranquilliser firearm.

Pamlin® and Zoletil 100® are listed as Schedule 4 (Prescription Animal Remedies – PARs) drugs under the *Poisons and Therapeutic Goods Act 1966*. The drugs were prescribed by a veterinarian. The use of these drugs is governed by the *Veterinary Practice Act 2003* in New South Wales (and similar Acts in other states and territories), which stipulates the conditions under which a non-veterinarian may use them.

Euthanasia

When capturing wild animals, there is always the potential for injuries and post-capture myopathy in addition to the risks associated with anaesthesia, even in healthy animals (Arnemo *et al.* 2006). It is important to be prepared for these by having training in the appropriate methods and access to equipment and/or drugs to humanely euthanise an animal if necessary. Fortunately, these events did not occur during our study.

PRACTICE

Practising is essential to gain familiarity with the operation of the tranquilliser firearm, the dart trajectory and the dart impact. Darting practice was conducted before every darting session to take account of prevalent weather conditions and unexpected changes in the accuracy of the firearm and to increase the confidence of the darter. The barrel of the tranquilliser firearm was regularly cleaned using gun oil and a cleaning rod, to ensure that there was no carbon residue to impede dart accuracy.

Measures of accuracy and the target zone

Practice consisted of firing darts into high-density foam to assess the group formation and the distance over which darting could accurately and safely take place. 'Group' is a ballistics term used to describe the pattern made by the fall of darts fired with the same point of aim (English 2004). This was essentially a practical measure of accuracy and animals were not darted until three shots out of three formed a group approximately 12–15 cm (the diameter of a circle drawn around the fall of shots), which is considered the target zone of an adult kangaroo. The tranquilliser firearm was zeroed using this process at 26 m. This distance was chosen as it was at the midpoint of the optimum range for darting kangaroos (15–34 m), as described by English (2004).

The target zone of a kangaroo is the large muscle mass on the upper hind leg, caudal thigh or rump (Fig. 28.2). Aiming at this region ensures the delivery of a deep intramuscular injection, avoids subcutaneous fatty deposits and evades bones such as the spine, femur and tibia. If the dart drops at this point of aim, it avoids sensitive anatomical structures and misses the animal as it falls to the ground between the tail and leg. If the dart were to move left or right at this point of aim, the chance of animal injury is reduced given the large muscle mass in this zone.

Figure 28.2: The target zone of an eastern grey kangaroo (*M. giganteus*).

Assessment of the dart path

It was important to assess the parabolic movement of the dart from the darter to the target zone, given the high degree of accuracy required. The rise and fall of the dart path was measured as the difference between the geometric centre of a group of darts fired from varying distances, and a marked line on the foam. During this process, the same velocity settings of the tranquilliser firearm and the same powered .22 calibre charges were used. The distances (15–34 m) were based on the optimum range of darting kangaroos (English 2004) and the closest distance that it was possible to approach an animal in the field (M. Roberts, pers. obs.).

After numerous iterations, the trial results suggested that the dart would consistently hit the target at points below the marked line as the distance from the target increased (Fig. 28.3). The knowledge of the dart path obtained during this assessment meant that regular adjustments to the firearm between shots (e.g. rezeroing or adjusting velocity setting) were not required. For example, if a shot could not be taken at the distance that it was zeroed (26 m), adjustments could simply be made to the point of aim. Horizontal movement of the dart was considered negligible because darting was to take place in minimal wind conditions, inside the optimum range and after the firearm was sighted in correctly.

Measure of dart impact

It was important to consider the assessment of dart impact across varying distances before the darting process began, to reduce animal injury and to ensure that darts would successfully penetrate the skin and activate the drug discharge mechanism. As a surrogate measure of dart impact, the relative depth of dart penetration into the high-density foam was measured from repeated shots (n = 5) taken at varying distances between 15 m and 34 m (Fletcher, *pers. comm.*). Velocity settings of the firearm and the power of .22 calibre blanks remained constant.

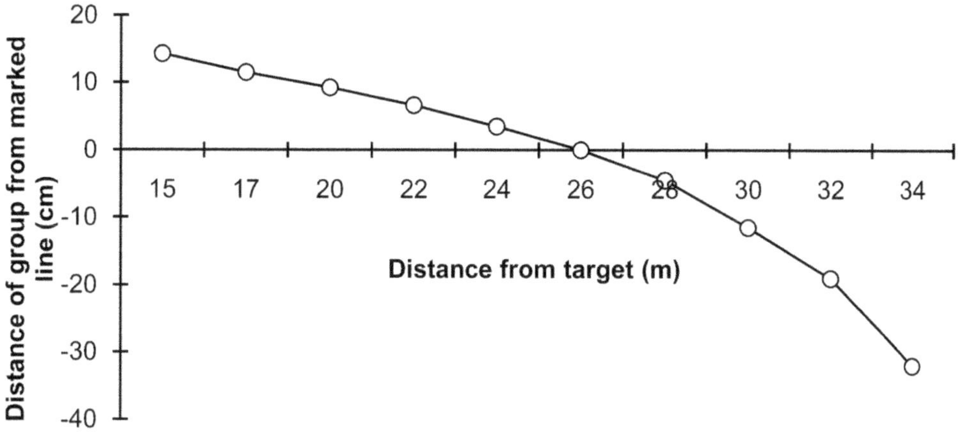

Figure 28.3: Assessment of the dart trajectory of practice shots taken at varying distances from 15–34 m, showing how the dart hits the target lower with increasing distance from the target. 0 on the y-axis is the marked line.

Dart penetration was minimal between the distances of 20 m and 34 m (Fig. 28.4). At these distances, there was a negligible difference in the depth of dart penetration as the distance from the target increased. An increase in dart penetration at distances <17 m was noted, which resulted in larger deeper holes in the foam compared to the shallower and narrower holes from shots taken further away.

Optimal darting distance

Although the tranquilliser firearm was zeroed at 26 m, information obtained from the assessment of the dart path and the dart impact measurements during practice were used to determine the optimal distance threshold for darting in the field. The optimal distance was 23–28 m, determined by the high degree of accuracy and relatively low levels of dart impact achieved within this range.

FIELD STAGES OF DARTING

From the initial captures of kangaroos, the darting process was broken into six primary stages. This was a key facet of the protocol as it permitted the process to be easily repeated, making subsequent captures more efficient. We conducted darting operations in crews of three people, one darter (MR) and two experienced handlers.

Stage 1: Dart preparation

This involved loading the darts with an immobilising agent. Darts were prepared in two concentrations

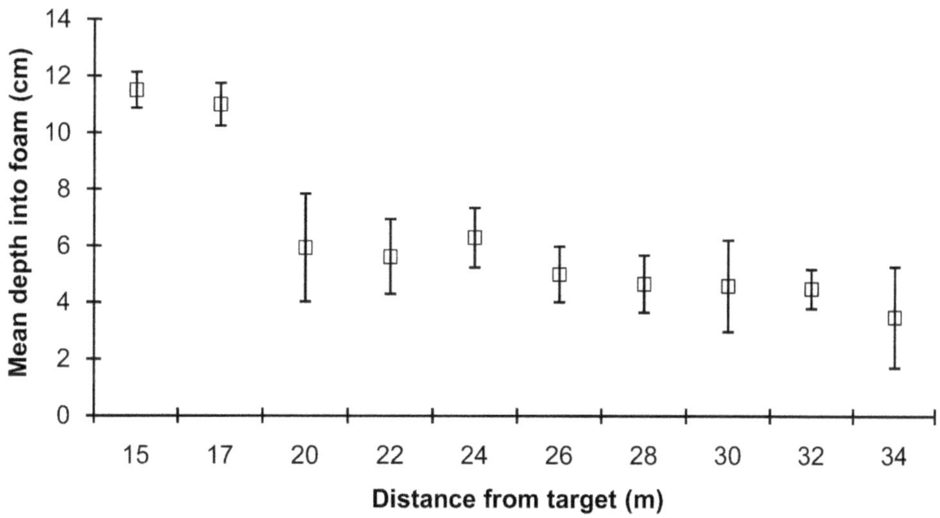

Figure 28.4: Mean (± SE) depth of dart penetration into high-density foam from shots taken at varying distances from 15–34 m (n = 5 for each distance). This was essentially a surrogate measure of dart impact on animals.

based on the field-estimated weights for females and subadult males (approximately 25 kg) and adult males (approximately 50 kg). All females, subadults and small males received a total of 125 mg of Zoletil 100®, while all adult males received a total dose of 250 mg of Zoletil 100®. These doses were prepared by reconstituting one bottle of Zoletil 100® (containing 500 mg of tiletamine:zolazepam) with either 4 mL or 2 mL of sterile water for injection to give a concentration of 125 mg/mL and 250 mg/mL for females and males, respectively. These concentrations of Zoletil 100® were chosen with the aim of achieving a final dose rate of approximately 5 mg/kg without having to calculate dosages for individual animals in the field. In terms of field practicality, we felt that using two broad quantities of Zoletil increases efficiency by eliminating the need for weight estimates and calculations of dosage rates during the minimal time available between animal selection and dart firing. The simplicity of this approach is also likely to reduce unintended errors in dosage calculation.

Once reconstituted, Zoletil 100® was injected into each dart using 1.5" 18 gauge needles, which reduced the occurrence of air bubbles and drug overflow from the dart needle. Darts were then sealed with Vaseline® to avoid drug leakage and labelled with the date and category (125 or 250) and stored in an Esky®. Once reconstituted with sterile water, Zoletil 100® can be kept for up to four days at room temperature, or two weeks at 4°C (Vogelnest 1999).

Stage 2: Approach

The target animal or mob was approached on foot or in a vehicle. It was desirable to get as close to the subject as possible, especially within the optimum darting distance defined during practice (23–28 m). This minimised the time taken to make the necessary adjustments to the firearm, and the inefficiency of choosing a different powered .22 calibre charge for a different distance. The challenge was to develop a method of approach that overcame the flight response of the subject animals. An increased level of flight response was typically illustrated by kangaroos in larger mobs (up to 80 individuals), whereas minimal levels were shown by individuals in smaller mobs or when animals were solitary. This was an important consideration when undertaking darting; extra precautions were taken when approaching a target animal within a larger mob to overcome increased vigilance behaviour.

Darting on foot

The terrain of the study area dictated that the majority of darting was conducted on foot. We found that the minimum flight distance (distance between the darter and the target animal at which the target animal first takes flight) was achieved when animals were approached in a meandering motion at a slow and gentle pace (Fig. 28.5). Distance was frequently measured using a rangefinder (Bushnell Yardage Pro®, range 12–300 m, accuracy ±1 m). Eye contact with the subject was kept to a minimum when moving towards it, and if the target animal or another in the same mob showed evidence of a flight response (e.g. stopped grazing or foot-stomped) the darter's direction of travel was changed or halted so it appeared that the pursuit was completed. If kangaroos did not move away upon approach, the pursuit would resume in the same meandering motion until the distance between the darter and the target kangaroo was satisfactory for a shot to be taken. If kangaroos appeared to be too flighty, darting was abandoned in that particular area.

Darting from a vehicle

In flatter areas of the study area, darting was undertaken on individual kangaroos from a vehicle using available service or fire trails. The flight distance was generally much shorter when approaching in a vehicle and a much more direct line of approach could be taken before any flight responses were observed. The focus was on achieving appropriate orientation of the vehicle for a shot to be taken.

Stage 3: Taking the shot

Once the distance between the darter and the target animal was in the optimum range (23–28 m), there was only a very short amount of time for a shot to be taken (up to three seconds for the most flighty animals). Considerations of the firing position and dart path, and assessing potential obstructions (e.g. other animals, natural features), were paramount at this stage.

The firing position that was adopted depended on whether darting was conducted on foot or from a vehicle. When darting on foot, a higher level of accuracy was obtained by taking the shot from a crouching position, supporting both arms on both knees. However, this was not possible when shooting in long grass, down slight declines in the landscape or in situations when animals were too flighty to permit

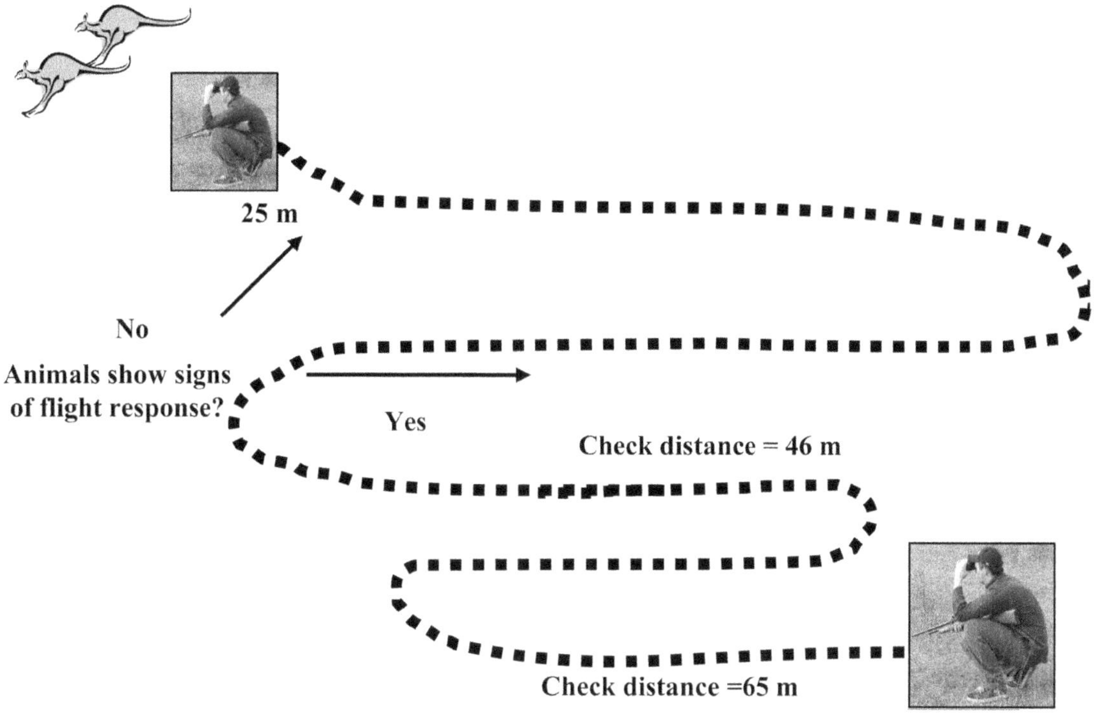

Figure 28.5: The meandering approach taken to the target kangaroo when darting was conducted on foot.

time for the darter to crouch. In these circumstances, shooting was conducted from a standing position and, where possible, the firearm was supported using fence posts or trees if available. When darting was conducted from a vehicle, a higher degree of accuracy was achieved as support could be obtained from the bonnet or the cab.

Knowledge of the dart path was increasingly important before taking the shot. If it was not possible to get within the optimum range (23–28 m from the target animal), the point of aim was slightly elevated to achieve an accurate rump shot. This overcame lengthy adjustments to the firearm and nullified the need for further advances to the target animal to reduce distance, especially important when animals were flighty and there was limited time to take the shot. Other tranquilliser firearms, such as the CO_2 projectile systems, DanInject, Telinject and Pneu-Dart, permit continuous adjustment to range to overcome this problem.

Caution was taken before the shot to ensure that the dart path to the animal's target zone was unobstructed. Obstructions such as tree branches and long grass have the potential to deflect the dart and cause injury to the animal. Other kangaroos in the same mob may also be a form of obstruction (Fig. 28.6). Therefore, the target animal and its likely movement paths must be clear of other kangaroos as the shot is taken.

Stage 4: Stages of immobilisation

The time of dart impact was noted and any movement of the darted animal was traced with binoculars. The darter crouched down at the position from which the dart was fired, to avoid causing unnecessary disturbance to the animal before the anaesthetic

Figure 28.6: Potential obstructions by other kangaroos in the same mob as the target animal. The kangaroos in the foreground could move in the direction of the dart path or the dart may miss and accidentally strike animals in the background.

had taken effect. Animals generally moved less than 60 m before the first signs of ataxia were noted.

Three major stages of immobilisation were noted:

1 the kangaroo showed signs of instability, uncoordination and 'star-gazing';
2 the kangaroo fell to the ground but could still struggle to its feet;
3 the kangaroo could not rise to its feet and its head was down.

The time to immobilisation (induction time) was considered to be the time from when the animal was darted until stage 3 was reached. The capture team maintained visual contact with the darted kangaroo, but did not approach until approximately 5–10 minutes after stage 3 was reached.

The field-estimated body mass for captured kangaroos was very similar to their actual body masses. The average body mass (± SD) of females and subadult males that received 125 mg of Zoletil 100® was 27.3 ± 4.0 kg (range 13–36 kg) and for adult males that received 250 mg it was 51.7 ± 5.9 kg (range 37–68 kg). The resultant dosage rate for adult males was 4.9 ± 0.6 mg/kg (range 3.7–6.8 mg/kg) and for subadult animals and adult females it was 4.7 ± 0.9 mg/kg (range 3.5–9.6 mg/kg).

The mean (± SD) induction time for kangaroos receiving the 125 mg of Zoletil 100® was 7.76 (± 4.42) minutes, whereas for animals receiving 250 mg induction time it was 9.57 (± 3.09) minutes (Fig. 28.7). There was no significant relationship between induction time and weight for either group (125 mg group $y = -0.05x + 9.1$, $r^2 = 0.004$, $F = 0.37$, $P = 0.54$; 250 mg group $y = -0.08x + 13.86$, $r^2 = 0.01$, $F = 0.31$, $P = 0.59$). This highlights the wide safety margin of Zoletil 100® and the effectiveness of the two broad categories of Zoletil 100® employed in this study. It also emphasises that there are large differences in effects between individual animals, which may be related to the degree of excitation before induction of anaesthesia, the placement of the intramuscular injection (Vogelnest 1999) and inherent differences in the susceptibility of different individuals to this type of anaesthesia.

Stage 5: Processing the animal

A darted kangaroo was approached quietly and cautiously by two people, it was blindfolded and its tail and feet were held firmly. After the animal was restrained, it was placed immediately into a large hessian bag (Animal Rescue Kits Pty Ltd, Wimbledon, NSW). Processing was conducted as quickly and quietly as possible to minimise undue stress to the animal. The kangaroo's head was placed upright at all times to permit saliva to drain away from the airway, as Zoletil anaesthesia has the potential to cause excess salivation (Pitt *et al.* 2006). The animal's eyes remained covered at all times, as Zoletil anaesthesia causes the animal's eyes to remain open.

Heart rhythm and the respiratory patterns of each kangaroo were monitored frequently during processing. Ten animals showed increased levels of stress when first handled, evidenced by increased respiratory and heart rates. These kangaroos were administered diazepam (Pamlin®, 0.5 mg/kg body weight IM; Parnell Laboratories Australia Pty Ltd)

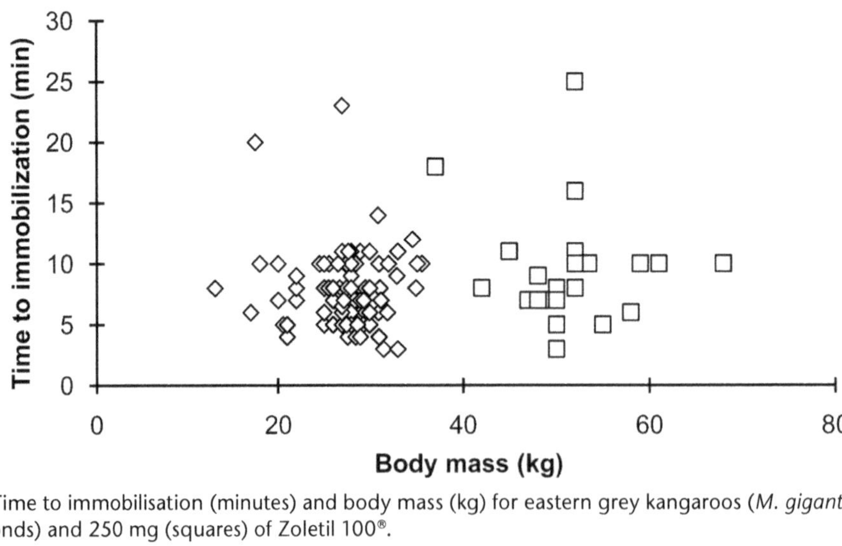

Figure 28.7: Time to immobilisation (minutes) and body mass (kg) for eastern grey kangaroos (*M. giganteus*) receiving 125 mg (diamonds) and 250 mg (squares) of Zoletil 100®.

soon after initial handling, which restored a more regular heart and respiratory pattern (Williams and Thorne 1996). Two individuals still did not settle (i.e. increased heart rate and excessive struggling) after diazepam was administered. These individuals posed a serious safety risk to handlers and to themselves, so an additional half-dose of Zoletil 100® was given (after Vogelnest 1999). The decision to top up (based on cardiac and respiratory signs) depended on the stage of processing: if processing was almost complete and the signs were related to a lighter plane of anaesthesia, top-ups were avoided in favour of completing processing rapidly and releasing the animal as soon as possible. A top-up with Zoletil 100® was employed in only 2% of captures and these animals were not included in the analysis of recovery times. All captured kangaroos were administered vitamin E with selenium (Vitamin-E-Selen, 0.025 mL/kg IM, Hoechst Roussel Vet, Melbourne), which may reduce the risk of post-capture myopathy (Vogelnest 1999).

The dart wound was cleaned and a topical broad-spectrum antibiotic aerosol was applied (Terramycin Pink Eye, active agent oxytetracycline hydrochloride 2.0 mg/g, Pfizer, Australia) and if necessary a long-acting antibiotic was administered as a prophylactic measure (Benecillin; Troy Laboratories Pty Ltd, Allhank, Australia; 0.1 mL/kg body weight IM).

Kangaroos were weighed in the hessian sack using hanging scales (Wedderburn hanging scale, maximum 100 kg, increments (d) = 0.5 kg) and morphometric measurements of pes (foot), leg and head were made (Poole *et al.* 1982). Approximately 6 mL of blood was collected from the lateral tail vein of each kangaroo using a 21 gauge needle and 10 mL syringe. A tissue sample (3 mm diameter) was taken from the right ear of each kangaroo; the ear was cleaned in the same way as the dart wound. The pouch of females was inspected for pouch young and the condition of the pouch was noted (clean, dirty, teats elongated/everted/not everted). Pes, leg and head measurements were taken on all pouch young using vernier callipers. A unique colour combination of ear-tags (Leader, Australia) was applied to all captured kangaroos and radio-collars and GPS collars (Sirtrack, NZ) were fitted to a subsample (n = 32). Females with pouch young had the opening of their pouch taped (Fixomul®, Smith & Nephew) after the young were returned to the pouch, to prevent loss. Defecation, a common side effect of Zoletil anaesthesia, was observed in 51% of kangaroos captured (Taylor *et al.* 1989).

Stage 6: Recovery

One person continued to observe the animal from a distance to monitor recovery and to prevent potential predator attacks. The recovering kangaroo was placed in the shade with its legs and tail out of the hessian bag, while its head remained covered and elevated to permit saliva drainage. Maintaining cover over the kangaroo's head served two purposes: it avoided corneal drying and avoided an early response to light stimuli, as this generally leads to premature movements and the potential for injuries. Full recovery was judged to be when the animal could maintain enough coordination to string together 10 or more consecutive hops. Kangaroos would often resume feeding before a full recovery.

The mean (± SD) recovery time for kangaroos that received 125 mg of Zoletil was 128.49 (± 42.23) minutes and for animals that received 250 mg recovery was 116.59 (± 37) minutes (Fig. 28.8). There was no significant difference in the recovery times for animals that received 125 mg or 250 mg of Zoletil (Wilcoxon ranks test, $Z = -1.02$, $P > 0.05$). There was no significant relationship between body weight and recovery time for both groups (125 mg group, $y = -1.34x + 165.02$, $r^2 = 0.02$, $F = 1.55$, $P > 0.05$; 250 mg group, $y = 2.05x + 10.85$, $r^2 = 0.11$, $F = 3.06$, $P > 0.05$). This again supports the effectiveness and safety of the two broad categories of Zoletil employed in this study. It should be noted that higher dosages of Zoletil are safe but will generally result in extended recovery times (L. Vogelnest, *pers. comm.*). This should be taken into account before darting animals of unknown weight/body condition.

The long recovery periods (approx. two hours) may create logistical issues if it is necessary to capture a large number of kangaroos per day (e.g. >15/ day). The recovery period could be reduced by combining Zoletil with medetomidine. This means that the effective dose of Zoletil could be reduced by as much as 50–75%, and the medetomidine could be reversed with atipamezole. This method has been used successfully when capturing koalas (Lynch and Martin 2003), but there are no published reports of its use in macropodids.

The recovery times may have been overestimated in this study. Lewandowski *et al.* (2002) defined full recovery as time from induction of anaesthetic to the

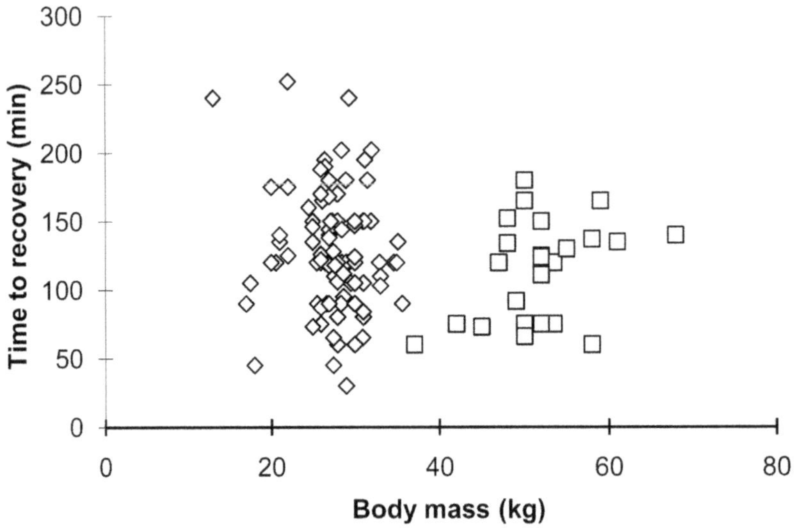

Figure 28.8: Time for recovery (minutes) and body mass (kg) for eastern grey kangaroos (*M. giganteus*) receiving 125 mg (diamonds) and 250 mg (squares) of Zoletil 100®.

return of standing and ambulating but with some minor residual anaesthetic-related ataxia. Therefore, caution needs to be exercised when recovery times are compared between studies.

ADVERSE REACTIONS

Incomplete immobilisation

If there were no signs of immobilisation after a 15-minute period, the kangaroo was redarted with the full dose of Zoletil (English 2004). This occurred on three occasions where the dart hit the animal and no obvious signs of ataxia were noted. If the darted animal demonstrated the first signs of drug action (stage 1– instability, uncoordination and star-gazing) but was not recumbent after 30 minutes or rose to its feet when approached, it was darted again with a half-dose of Zoletil (English 2004). This occurred on one occasion (M055). On that occasion, a male (57 kg) was darted in the caudal thigh and appeared to have reached stage 3 of immobilisation after 16 minutes. When he was approached he stood up and showed no apparent signs of ataxia. He was redarted with a 125 mg dart (250 mg + 125 mg) and was immobilised after 15 minutes. He was reapproached another 15 minutes later and was very difficult to restrain. Due to the risk of injury to the animal or the handlers, he was given an additional 125 mg of Zoletil (250 mg + 125 mg + 125 mg = 500 mg, 8.8 mg/kg total dose) intramuscularly using a hand syringe (intravenous injection was preferred but was unsafe). After a further 12 minutes, it was possible for him to be held and for procedures to be carried out. This kangaroo recovered in 55 minutes. This example demonstrates the high degree of variability in animal responses to sedation.

There are a couple of plausible reasons why incomplete immobilisation occurred in this instance. First, the initial full dose may not have fully discharged from the dart and the induction time may vary depending on the placement of the IM injection (Vogelnest 1999). However, these factors are unlikely to have been in play on the two instances that this animal was darted. Second, the higher metabolic rate of males may influence the time to recumbency (Pitt *et al.* 2006), but the use of top-up dosages at the time of processing was necessary to permit safe handling equally among males and females. An additional factor affecting the induction time and depth of anaesthesia is the state of the animal at the time of darting. If the animal is excited, increased heart rate and adrenaline can counter the effects of the immobilisation agent. This is unlikely to be the explanation for this kangaroo as, in general, animals that were in a state of obvious excitation were not darted because they were more flighty and could not be approached. It has been widely reported that individual responses to anaesthetics, including Zoletil, are highly variable even when all other variables are equal (Fournier-Chambrillon *et al.* 2000), and this is most likely reflected in this instance.

Prolonged recovery

An extended recovery was observed for one kangaroo, an adult female (F014) that was in good condition and received a Zoletil dosage of 4.2 mg/kg. There were several instances during recovery where she had to be roused after she had fallen into a deep sleep. Four hours after being darted and with nightfall encroaching, she was only able to stand unaided for relatively short periods and could not hop with coordination. Her safety was in question: there were signs of wild dog activity (e.g. howling) and the temperature was beginning to fall (minimum 5.5°C). Consequently, the kangaroo was placed in the hessian bag and was taken back to camp. She was kept warm overnight and was given an additional 1 mL vitamin E with selenium IM. The kangaroo was successfully released the next morning and was resighted two months after her capture. The reasons for the prolonged recovery in this animal remain uncertain.

EVALUATION OF THE DARTING PROTOCOL

Survivorship

No dart injuries, drug or handling mortality occurred during the capture of kangaroos, but it was still essential to assess post-release survivorship of captured kangaroos. This was measured by resighting ear-tagged individuals and by radio-tracking collared kangaroos. Ear-tagged kangaroos were resighted along transects and opportunistically during other field work activities. Attempts were made to locate animals over a one-week period per month, and animals captured in the same month of the resighting survey were not included in the analysis. Radio- (n = 22) and GPS- (n = 6) collared kangaroos were tracked for a week after capture across four periods of the day (6 a.m., 12 p.m., 6 p.m. and 12 a.m.) and mortality sensors were fitted to each collar to permit rapid detection of animal death if there was no movement within 24 hours. This timeframe was considered adequate to monitor mortality from severe cases of post-capture myopathy, because post-capture myopathy-related death is reported to typically occur within hours and up to a month after restraint (Berringer *et al.* 1996; Haulton *et al.* 2001).

On average, over 80% of ear-tagged kangaroos were resighted one month after capture (Fig. 28.9). These high survivorship rates were corroborated by radio- and GPS-collared kangaroos, who all survived more than six months after capture (n = 28). A major limitation of resighting marked animals to assess survivorship was the undetermined fate of kangaroos that were not sighted. However, the home ranges of

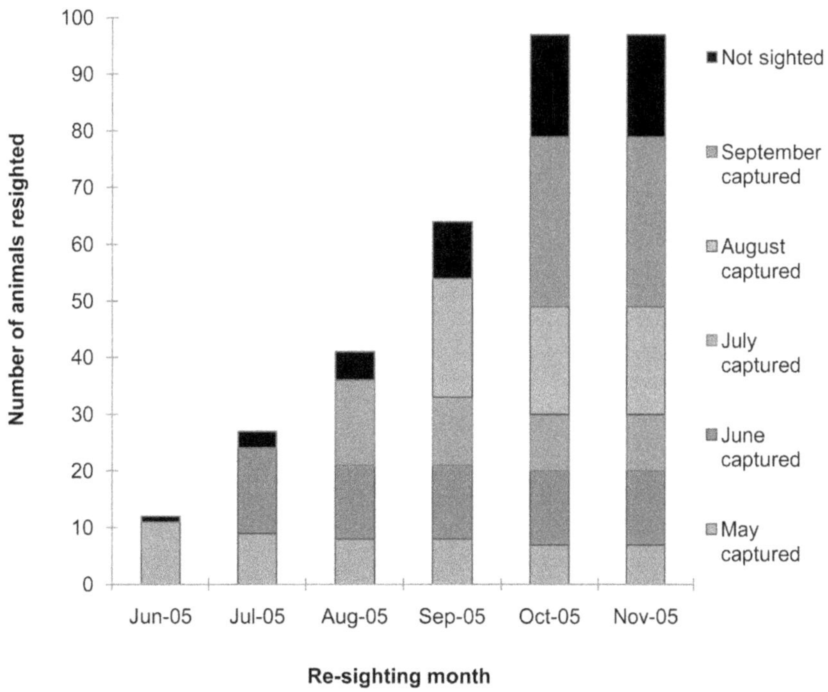

Figure 28.9: Number of captured eastern grey kangaroos (*M. giganteus*) (out of a subsample of 100 animals) resighted in surveys June–November 2005.

collared animals suggested that kangaroos captured in the study area ranged to areas where they could not be observed from transects or opportunistically during routine field work. This was probably the most plausible explanation for not sighting some ear-tagged animals. There was no indication that animals were not sighted as a result of capture-related mortality.

Of the 123 animals captured, only three mortalities (F013, F046 and M056) were detected after several months of capture. The length of time between these mortalities and the individual captures suggests that they were unrelated to the capturing process. The pressure exerted by the drought conditions is a more likely explanation, as kangaroo mortality in the population was elevated during this period.

Post-capture survivorship of pouch young was an important consideration in evaluating the darting protocol. All females that were carrying pouch young (49% of adult females caught) retained them during the immobilisation and recovery period. Medical tape was applied over the pouch to prevent the loss of pouch young during recovery. Longer-term survivorship of pouch young was assessed as part of a larger study on the home ranges of collared kangaroos. All the collared females carrying pouch young retained them one week after capture and 80% retained them for up to six months. Again there was no direct evidence that the loss of the pouch young was related to animal capture, but the harsh environmental conditions probably caused an increase in pouch young mortality above the longer-term average.

Behaviour of animals

Anecdotal observations suggested that there were no changes in the behaviour of kangaroos in the study area, either through habituation or increased flight, following the onset of darting activities. All the kangaroos demonstrated the same level of awareness when approached as they did at the commencement of darting. However, the number of missed shots decreased from over 50% of shots taken, to 3% in the later stages. This was probably due to an increased level of confidence in the darter, following lengthy hours of practice and increased familiarity with kangaroo behaviour, which permitted an easier approach before taking the shot.

The refinement of techniques was further demonstrated in the recapture of the GPS-collared kangaroos. These kangaroos showed a higher level of awareness of the darter's movements, because they had been initially captured approximately 60 days beforehand. Extra care in the timing of darter movements during the approach to these animals was required to prevent an animal fleeing before it was within darting range. Using this strategy, all these animals were successfully captured after a minimal number of attempts.

CONCLUSION

This study has shown that it is possible to dart and immobilise a large number of free-ranging eastern grey kangaroos safely and efficiently by employing a logical step-by-step protocol. The key elements of this protocol centred on practice, familiarisation with the equipment and knowledge of the target animal, and are vital when darting kangaroos in any environment. The high survival rates, minimal adverse reactions and limited impact on the target animals and associated population are evidence of the usefulness of the protocol in the darting of free-ranging macropodids.

Although this protocol was developed in response to a specific set of conditions, many of the principles are applicable to the capture of free-ranging macropodids in other situations.

ACKNOWLEDGMENTS

This work would not have been possible without the assistance of numerous volunteers including D. Brennan (Macquarie University) and M. Gilmour, J. Ray, M. Krogh and I. Wright (Sydney Catchment Authority, SCA). We thank C. Banffy (Dept of Environment and Climate Change) and D. Ashton and A. Simson (SCA) for assisting with the approval process for tranquilliser use in the study area. We are very appreciative of L. Gallen, G. Capararo, B. Waldron, U. Manna and T. Kondek (SCA) for permitting and facilitating access to the study area. We acknowledge L. Vogelnest (Taronga Zoo) and I. Wright (University of Western Sydney) for comments on the manuscript. This work was conducted under a NPWS scientific licence (11002) and Macquarie University Ethics approval (2003/016). This project was supported by SCA funding to Macquarie University (D.W. Cooper).

REFERENCES

Arnemo JM, Ahlqvist P, Andersen R, Berntsen F, Ericsson G, Odden J, Brunberg S, Segerström P and Swenson JE

2006. Risk of capture-related mortality in large free-ranging mammals: experiences from Scandinavia. *Wildl. Biol.* **12:** 109–113.

Arnold GW, Steven D, Weeldenburg J and Brown OE 1986. The use of alpha-chloralose for the repeated capture of western grey kangaroos, *Macropus fuliginosus. Aust. Wildl. Res.* **13:** 527–533.

Berringer J, Hansen LP, Wilding W, Fischer J and Sheriff SL 1996. Factors affecting capture myopathy in white-tailed deer. *J. Wildl. Man.* **60:** 373–380.

Booth NH 1988. Intravenous and other parenteral anaesthetics. In *Veterinary Pharmacology and Therapeutics.* (Eds NH Booth and LE McDonald) pp. 212–274. Iowa State University Press, Iowa.

Coulson G 1996. A safe and selective draw-string trap to capture kangaroos moving under fences. *Wildl. Res.* **23:** 621–627.

Edwards GP, Croft DB and Dawson TJ 1994. Observations of differential sex/age class mobility in red kangaroos (*Macropus rufus*). *J. Arid Environ.* **27:** 169–177.

English A 2004. The capture and handling of wild animals in Australia. University of Sydney, Sydney.

Fournier-Chambrillon C, Vogel I, Fournier P, Thoisey B and Vie J 2000. Immobilisation of free-ranging nine banded and great long-nosed armadillos with three anaesthetic combinations. *J. Wildl. Dis.* **36:** 131–140.

Haulton SM, William WF and Rudolph BA 2001. Evaluating 4 methods to capture white-tailed deer. *Wildl. Soc. Bull.* **29:** 255–264.

Hawkins PR, Swanson P, Warnecke M, Shanker SR and Nicholson C 2000. Understanding the fate of *Cryptosporidium* and *Giardia* in storage reservoirs: a legacy of Sydney's water contamination incident. *J. Water Supply Res. Technol.* **496:** 289–306.

Higgingbottom K 1989. Macropod studies at Wallaby Creek. VII. Capture of wild red-necked wallabies by 'blow-darting'. *Aust. Wildl. Res.* **16:** 173–178.

Higgins DP, Rogers TL, Irvine AD and Hall-Aspland SA 2002. Use of midazolam/pethidine and tiletamine/zolazepam combinations for the chemical restraint of leopard seals (*Hydrurga leptonyx*). *Mar. Mamm. Sci.* **18:** 483–500.

Horsup A 1994. Home range of the allied rock wallaby *Petrogale assimilis. Wildl. Res.* **21:** 65–84.

Jarman PJ and Taylor RJ 1983. Ranging of eastern grey kangaroos and wallaroos on a New England pastoral property. *Aust. Wildl. Res.* **10:** 33–38.

Kitchener DJ 1973. Notes on the home range and movement in two small macropods, the potoroo (*Potorous apicalis*) and the quokka (*Setonix brachyurus*). *Mammalia* **37:** 231–240.

Lewandowski AH, Bonar CJ and Evans SE 2002. Tiletamine-zolazepam ketamine, and xylazine anaesthesia of captive cheetah (*Acininyx jubatus*) *J. Zoo Wildl. Med.* **33:** 332–336.

Lynch M and Martin R 2003. Capture of koalas (*Phascolarctos cinereus*) by remote injection of tiletamine–zolazepam (Zoletil®) and medetomidine. *Wildl. Res.* **30:** 255–258.

Mauthe von Degerfield M 2005. Personal experiences in the use of Zoletil® for anaesthesia of the red-necked wallaby (*Macropus rufogriseus*). *Vet. Res. Commun.* **29:** 297–300.

Moore BD, Coulson G and Way S 2002. Habitat selection by adult female eastern grey kangaroos. *Wildl. Res.* **29:** 439–445.

Moss GL and Croft DB 1999. Body condition of the red kangaroo (*Macropus rufus*) in arid Australia: the effect of environmental condition, sex and reproduction. *Aust. J. Ecol.* **24:** 97–109.

Nave C, Coulson G, Poiani A, Shaw G and Renfree M 2002. Fertility control in the eastern grey kangaroo using levonorgestrel implants. *J. Wildl. Man.* **66:** 470–477.

Pitt J, Larivière S and François M 2006. Efficacy of Zoletil® for field immobilisation of raccoons. *Wildl. Soc. Bull.* **34:** 1045–1048.

Poole WE 1995. Eastern grey kangaroo *Macropus giganteus.* In *The Mammals of Australia.* (Ed. R Strahan) pp. 335–338. Reed New Holland, Sydney.

Poole WE, Carpenter SM and Wood JT 1982. Growth of grey kangaroos and the reliability of age determination from body measurements. I. The eastern grey kangaroo, *Macropus giganteus. Aust. Wildl. Res.* **9:** 9–20.

Power M, Slade M, Sangster N and Veal D 2004. Genetic characterisation of *Cryptosporidium* from a wild population of eastern grey kangaroos *Macropus giganteus* inhabiting a water catchment. *Infect. Genet. Evol.* **4:** 59–67.

Shepherd NC 1981. Capture myopathy in the red kangaroo (*Macropus rufus*). In *Wildlife Diseases and Other Diseases in the Pacific Basin and Other Countries.* (Ed. ME Flower) pp. 239–247. Wildlife Diseases Association, Sydney.

Stirrat S 1997. Behavioural responses of agile wallabies (*Macropus agilis*) to darting and immobilisation with tiletamine hydrochloride and zolazepam hydrochloride. *Wildl. Res.* **24:** 89–95.

Taylor WP, Reynolds HV and Ballard WB 1989. Immobilisation of grizzly bears with tiletamine hydrochloride and zolazepam hydrochloride. *J. Wildl. Man.* **53:** 978–981.

Telfer WR and Griffiths AD 2006. Use of space, habitats and shelters by the short-eared rock wallaby, *Petrogale brachyotis*, in the monsoon tropics. *Wildl. Res.* **33:** 207–214.

Viggers KL and Hearn JP 2005. The kangaroo conundrum: home range studies and implications for land management. *J. Appl. Ecol.* **42:** 99–107.

Vogelnest L 1999. Chemical restraint of Australian native fauna. In *Wildlife in Australia: Healthcare and Management.* (Ed. DI Bryden) pp. 149–188. Post Graduate Foundation in Veterinary Science, University of Sydney, Dubbo.

Williams ES and Thorne ET 1996. Exertional myopathy. In *Non-infectious Diseases of Wildlife.* (Eds A Fairbrother, LN Locke and GL Hoff) pp. 181–193, Iowa State Press, Iowa.

29 Monitoring the fate of translocated eastern grey kangaroos at the Gold Coast

K. Higginbottom and S. Page

SUMMARY

Translocation of common macropodids has recently emerged as a management option for addressing impacts of destruction of their habitat in urban areas. However, little is known of the likely success of such translocations. Eastern grey kangaroos (*Macropus giganteus*) were translocated from a residential development site at the Gold Coast (Queensland) to a nearby 288 ha council conservation area. Objective criteria were developed as a basis for site selection. Ten adults were radio-tracked for up to a year after release or until death (if earlier). The majority of the animals soon moved outside the partially fenced conservation area into adjacent vegetated areas, to a distance of up to 6 km. Animals moved in a somewhat random fashion through suitable habitat then, after a variable period of time, generally settled into fairly stable home ranges. Out of the translocated animals, an estimated 80% were still alive four months after release and an estimated 60% were still alive after one year. All surviving animals appeared to be in good condition and health. While the translocation was reasonably successful in the short term, the site is considered far from ideal due to threats associated with urban development in adjacent areas. However, it was very difficult to find a site meeting all the necessary criteria. Capture, release and monitoring were expensive in time and money, especially considering the small number of animals involved. It is concluded that translocation is not likely to be a widely applicable option for dealing effectively with destruction of eastern grey kangaroo habitat, associated with development.

INTRODUCTION

In urban areas around Australia, it is increasingly common to have a situation where isolated habitat fragments are converted to forms of development that are incompatible with wildlife persistence. One such case involved a decision to clear a patch of wildlife habitat surrounded by urban development at Harbour Town North on the Gold Coast, Queensland, to make room for a residential estate. There were only two wildlife management options: to destroy or translocate the resident animals. The former was not considered politically acceptable. Given a paucity of data on the likely success of carefully managed translocation in such circumstances, Queensland Parks and Wildlife Service (responsible for approving the management of wildlife in such cases) saw the present study as an opportunity to gather data to inform future decisions about whether translocation should be used and, if so, how it should be carried out.

The scientific literature on reintroductions and translocations, which focuses mainly on threatened species, shows that many attempts at translocation or reintroduction have failed (Griffith *et al.* 1989; Kleiman 1989; Short *et al.* 1992; Wolf *et al.* 1996; Fischer and Lindenmayer 2000). On the other hand, it indicates that they are often successful in cases where wild-caught animals are translocated into suitable habitat within the species' geographic range, using suitable methodology, and where key threatening processes are adequately addressed (Delroy *et al.* 1986; Griffith *et al.* 1989). This is perhaps best illustrated in South Africa, where 85% of translocations of ungulates into national parks that met these criteria were considered successful (Novellie and Knight 1994).

Many studies have monitored the outcomes of reintroduction of threatened species of macropodoids (Delroy *et al.* 1986; Short *et al.* 1992; Copley 1994; Fischer and Lindenmayer 2000; Pople *et al.* 2001). Copley's (1994) review of such reintroductions concluded that only 11% of reintroductions to mainland sites and 60% of reintroductions to island sites had been successful.

In comparison, translocations of common species of large macropodids in Australia have been rare, and thus there is limited experience on which to draw. Further, policies on animal translocations generally fail to cover common species (IUCN 1987; NSW NPWS 2001) or do not provide detailed advice about translocation of such species (DEH 2003). The available information indicates that translocations can be highly successful in terms of population persistence. The largest recorded translocation of kangaroos involved the Tasmanian subspecies of the eastern grey kangaroo (*Macropus giganteus tasmaniensis*) (Tanner and Hocking 2001), between 1971 and 1978. More than 1900 animals were translocated to 18 mainland and island sites around Tasmania. At all but two of these sites resident breeding populations became established, although not all populations have persisted until the present day. Two translocations of unknown numbers of western grey kangaroos (*Macropus fuliginosus*) to islands were successful (Short *et al.* 1992; Copley 1994). In several of the cases that involved translocations to islands, populations increased to the point of causing vegetation damage (Short *et al.* 1992; Copley 1994; Tanner and Hocking 2001). None of these translocations involved monitoring the fate of individuals.

There have been numerous anecdotal reports of foster-reared macropodids being successfully released into the wild. The one published study on such animals (Campbell and Croft 2001) showed that 100% of foster-reared eastern grey kangaroos (*M. giganteus giganteus*) survived the first year after release, and that this applied whether the animals were hard- or soft-released (involving acclimatisation in a large fenced yard) and whether or not they were familiar with other individuals in the release group. Hard-release techniques have been shown to be successful for other macropodoid species (released into suitable habitat) when the source animals are from wild populations (Delroy *et al.* 1986; Griffith *et al.* 1989; Pople *et al.* 2001).

Based on a literature review of factors expected to affect translocation success, we developed a set of criteria for release site selection to facilitate the best chance of survival for the translocated animals. These criteria included the site being geographically close to the source site, being large, containing similar habitat to the source site, meeting habitat requirements of the translocated species, having appropriate land tenure, being well connected to other areas of suitable habitat, including minimal major roads and other barriers to natural movement, imposing minimal threats from anthropogenic sources, and containing resident populations below carrying capacity (see Higginbottom 2004).

Using these criteria, Pimpama River Conservation Area (henceforth 'Pimpama' or the 'conservation area') was chosen as the only suitable release site for eastern grey kangaroos and various other species occurring at Harbour Town North. It is a 288 ha area managed primarily for conservation by the Gold Coast City Council in the Pimpama region of the northern Gold Coast, 14 km from Harbour Town North. The area consists largely of mature and regrowth open forest (mostly eucalypt-dominated), a large disturbed grassland area (dominated by exotic grasses) and an intertidal area. The area was bordered along its north-west and south-east boundaries by a high fence which was macropodid-proof for most of its length, and by a river on most of the remainder of the boundary. It was hoped that this fence, while not likely to prevent all movement of macropodids, would inhibit their movements to a large extent in the months immediately following translocation.

Policies of governments and conservation organisations and the literature all support the idea that translocated animals should be properly monitored (Short *et al.* 1992; Fischer and Lindenmayer 2000; DEH 2003), particularly where current knowledge of likelihood of success, and factors affecting it, is poor. This allows lessons to be learned about if and how subsequent translocations should be conducted. It is also considered good practice in modern conservation management to monitor the outcomes of wildlife management actions as an integral part of the management process, to inform adaptive management.

METHODS

Over a two-month period from mid-May 2004, attempts were made to capture all eastern grey kangaroos present at Harbour Town North. The capture process did not involve the present researchers; details are given in Dickinson (2004). First, animals were attracted (by baiting) or gently driven along funnel fencing into capture yards. All 13 kangaroos (excluding pouch young) present at the site were confined in this way. Two adults and one young at foot died due to physical injuries as a result of the capture process. The remaining 10 adults (six males, two females with pouch young, two females with no pouch young) were successfully captured by darting with Zoletil. After being clinically assessed by a veterinarian as fit for release, they were tagged and radio-collared, transported in a padded air-conditioned van to Pimpama and released under shady trees in open woodland within a few hours of capture.

Each released animal was placed under or on a hessian bag in a shady spot near the edge of the forest where trees were relatively sparse, and observed quietly from a distance of 7–15 m until it moved out of sight. Releases occurred in the morning wherever possible, through to mid-afternoon, providing time for the animal to recover from the influence of the drugs before dusk (when there was expected to be enhanced risk of predation and of injury through hopping into obstructions). At this observation distance, and given the continuing influence of the drugs, the animal generally appeared unaware of the observer's presence. The observer did not pursue the animal once it moved out of sight, to avoid further disturbance.

All translocated kangaroos were monitored using radio-tracking. Unfortunately, it was not practically feasible to capture and radio-track any resident animals at the release site. It was planned that each translocated animal would be radio-tracked until either it died, it lost its collar or one year after release, whichever came first. The supplier (Titley Electronics) indicated that the transmitters should continue to emit detectable signals for at least one year. The transmitters had mortality switches so that if an animal were stationary for a sustained period (indicating death) the transmitter would transmit a distinct signal. The collars were fitted with weak links so that they would break after about a year or if the collar became stuck in vegetation, thus avoiding mortality due to strangulation. No mortality or injuries occurred due to the collars.

Radio-tracking was initially carried out using only a mobile hand-held antenna. To increase range, a mobile 7 m mast and a vehicle-mounted antenna were used from July 2004, and in August 2004 three fixed 10 m high antennae were installed in the conservation area. All these methods were valuable in different circumstances and were subsequently used in combination.

Fixes were obtained with a hand-held GPS using a combination of triangulation (from at least three points where possible) and direct sightings of animals. Attempts to track each animal were made daily for the first week, then every effort was made to locate each animal at least fortnightly for the remainder of the period (although this was often not possible). Obtaining signals from some animals proved difficult due to their moving substantial distances, or due to obstruction by dense vegetation in the forest areas. Thus, most tracking attempts were restricted to obtaining fixes through triangulation. A fix was considered sufficiently accurate for use in data analysis if the intercept of the three triangulation lines gave an estimated error of no more than 50 m in any direction. In cases where an animal's radio signal had not been detected for two weeks, extended searching was undertaken in all locations considered feasible based on previous records of that animal.

All reasonable efforts were made to obtain a sighting of each individual once every six weeks. Priority was given to finding animals that had not been sighted recently. The aim of sightings was to observe the health and condition of the animal, the state of its collar, any interactions with other eastern grey kangaroos and, where applicable, the presence and size class of young. Condition was scored according to a modification of an unpublished system developed for eastern grey kangaroos by Menna Jones and Peter

Jarman for this species at Wallaby Creek, New South Wales (M. Jones, *pers. comm.*).

Tracking for the purpose of obtaining fixes was generally carried out between two hours after first light and two hours before last light. Thus this information generally describes where animals were when resting. Attempts to sight animals were mostly carried out in the first two or last two hours of daylight, when most animals were foraging, since they were easiest to observe during this period. If animals moved away from the observer unseen, they were not pursued unless there was a good opportunity to sight them without causing flight. This was to avoid the animals being 'chased' long distances and thus potentially causing them to unnaturally shift their home range, and to avoid their becoming increasingly wary of observers over time.

A median of 14 fixes was obtained per individual. Signals were no longer detected after a varying period of four to 10 months. Various lines of evidence indicate that this was probably due in all cases to premature failure of the transmitter batteries (Higginbottom *et al.* 2006). One individual continued to be sighted regularly for seven months after its transmitter stopped working, five months after release. Thus the pool of animals on which data were based decreased progressively over time (Fig. 29.1).

This work was conducted through a combination of university research and private consultancy, and was fully funded by the developer of the Harbour Town North site. Accurate and full details of costings are not available due to commercial confidentiality and since monitoring of eastern grey kangaroos was only part of a larger project. However, consultancy fees to the principal investigator and casual salaries to research assistants comprised approximately 79% of the budget, with the remainder spent mainly on travel and equipment. A total of approximately 1400 person hours was spent on aspects of the project involving monitoring of these eastern grey kangaroos and 10 other radio-tracked macropodids, with approximately 73% of this time spent on field work.

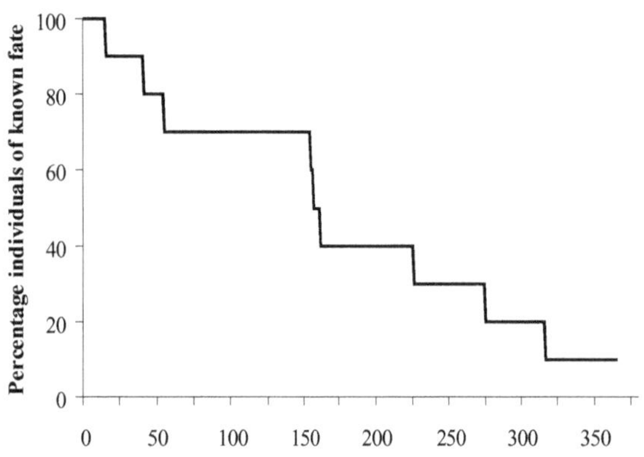

Figure 29.1: Eastern grey kangaroos (*M. giganteus*) of known fate over time since first release.

RESULTS

Survival

During recovery from anaesthesia, behavioural signs of stress appeared to be minimal and no injuries were observed to occur. None of the released eastern grey kangaroos died of post-capture stress or myopathy. Figure 29.2 presents the best available estimates of survivorship of translocated animals, with 80% of animals surviving at least four months after release (possible range 70–80%), and 60% surviving at least one year after release (possible range 40–70%). Two of the animals found dead had been predated or severely scavenged, probably by feral dogs (precluding determination of whether predation was the primary cause of death, as seemed likely). One animal was killed by collision with a vehicle and one was found dead after having become stuck in mud in the mangroves, with nearby tracks indicating it may have been pursued by dogs into that habitat.

Locations and movements

Only two individuals (a large male and a small male) remained inside the conservation area for most of the study period. The others moved outside the area after varying amounts of time; an example is illustrated in Figure 29.3. Data were insufficient to draw conclusions about differences in distance or pattern of movement between males and females, and there were no consistent differences in movement patterns between the three small males and the three large males.

Of the individuals that survived and continued to be tracked for most of the period, the majority appeared to settle into a fairly stable home range after a variable period, although this cannot be verified given the relatively small number of such individuals with sufficiently frequent sightings. The directions of movement varied, and seemed consistent with the idea that animals moved in a somewhat random fashion through suitable habitat.

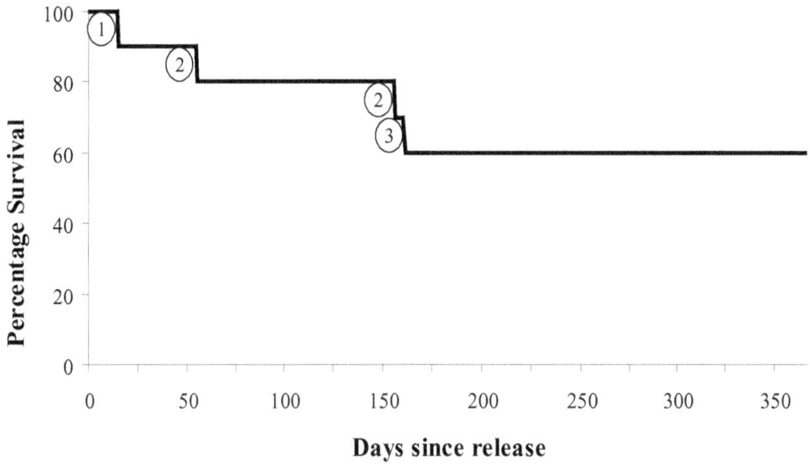

Figure 29.2: Percentage survival of eastern grey kangaroos (*M. giganteus*) tracked since time of release. The data are based on all animals of known fate at each point in time and therefore are the best available estimate of survival. The level of confidence in the estimates decreases as the number of tracked animals decreases. The fate of two animals (one considered dead and one alive) is based on probable rather than confirmed information. Circled numbers represent the proximate cause of death: 1 = stuck in mangrove mud, perhaps pursued there by predators; 2 = predation or scavenging, probably by feral dogs; 3 = road-kill.

The greatest linear distance moved from the release point by any animal was approximately 6 km, with the animal successfully crossing the Pacific Hwy (M1) (before being killed by a vehicle on a service road). There appeared to be a reduction in distance moved over time, but sample sizes were insufficient to justify statistical analysis.

Given the directional movements of many animals soon after release, the area occupied by individuals throughout the study could not legitimately

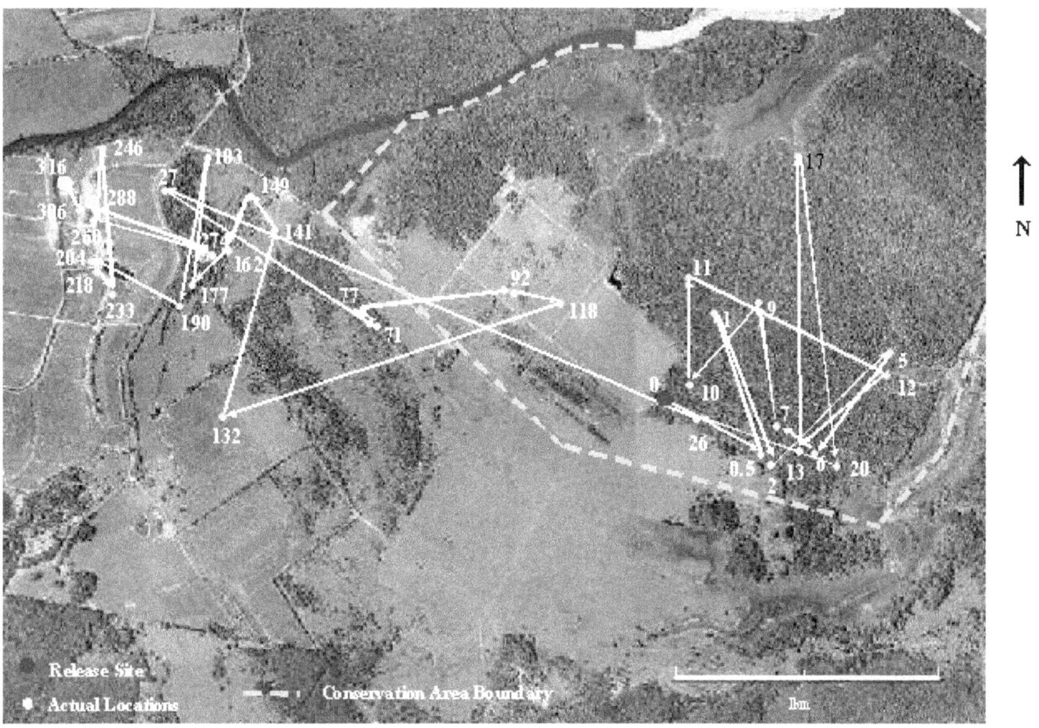

Figure 29.3: Locations of an adult female eastern grey kangaroo (*M. giganteus*) over the tracking period. Numbers represent days since release, arrows represent direction of movement. 'Actual location' is a sighting or an estimate of location using triangulation of signals.

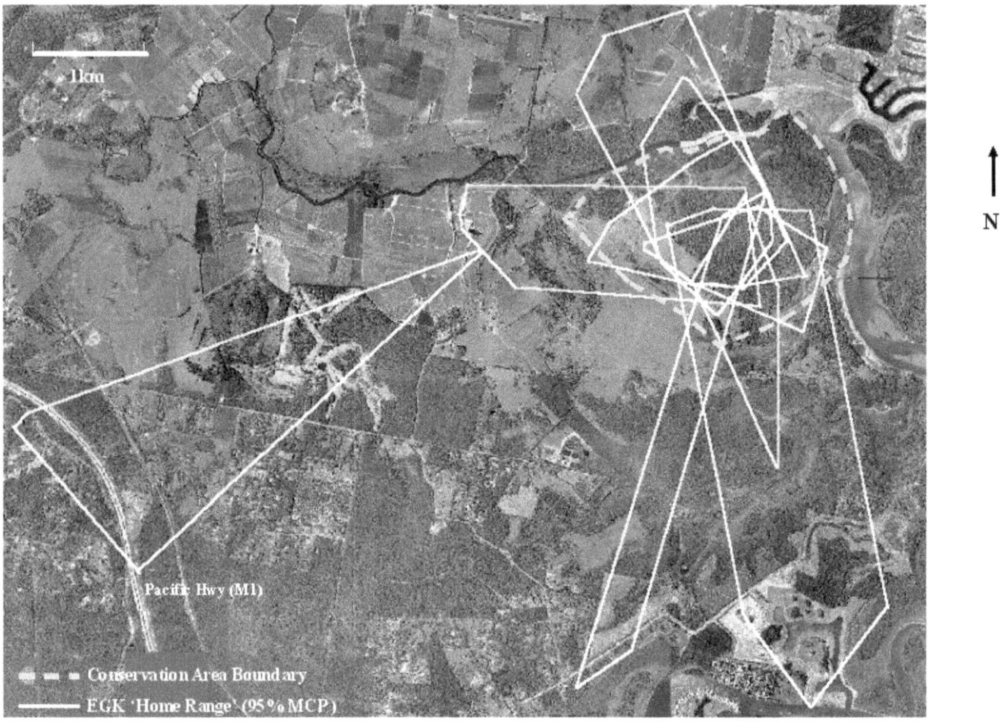

Figure 29.4: Minimum convex polygons (95%) for all translocated eastern grey kangaroos (*M. giganteus*).

be considered home ranges. However, 95% minimum convex polygons were constructed to provide an image of the areas covered (Fig. 29.4). The map probably overstates the degree of overlap between individuals' use of space, as many of the overlapping areas were used at different times. The median size of these minimum convex polygons was 164 ha, and the mean was 245 ha.

Condition and reproductive status

All sighted animals were moving normally, were in fair to good condition and had no observable injuries. Of the four females released, only one is known to have survived for at least 11 months after release. In this case, her young, an unfurred pouch young with eyes closed at translocation, survived to permanent emergence from the pouch and by the end of the study had spent several months as a young at foot.

Interaction with resident conspecifics

Right from the day of release, translocated animals were often found grazing with resident conspecifics and there was no evidence of any unusual agonistic or avoidance behaviour. One translocated male apparently became the alpha male in the conservation area (based on observations of male interactions around females approaching oestrus) within a few months of release, and was often still found there a year after his release.

DISCUSSION

Finding a suitable release site that would meet all the criteria proposed in this study was difficult. Even though Pimpama appeared to be a good option, it became apparent that it was less than ideal because, contrary to expectations, the fencing did not significantly inhibit animals from dispersing shortly after release, including to areas with busy roads and new residential development, and because feral predators appeared to be more abundant than expected. With hindsight, it may have been better to find an inland (rather than coastal) site with fewer threats from development, as the habitat differences are unlikely to be a problem for eastern grey kangaroos, and resident animals would still come from the same gene pool.

The translocation of eastern grey kangaroos in this study was partially successful, in that a significant proportion of the animals survived for one year and there was no evidence of adverse behavioural or physical impacts from translocation. The extent to which translocation had any negative effect on the animals concerned cannot be definitively assessed in the absence of a control group of resident animals.

Nevertheless, the study seems to show that eastern grey kangaroos are capable of adapting and surviving when translocated to a new area meeting the site selection criteria. It seems likely that the degree of success will depend largely on the extent to which the criteria are met.

On the other hand, there were a number of mortalities. In three cases, there is no reason to believe that these were related to the translocation, since deaths due to collision with vehicles and predation by feral dogs also occurred among the resident population, and the deaths of translocated animals from those causes occurred at least 54 days after release. It does, however, demonstrate that the fate of any animals translocated to this area is threatened by expanding residential development. Release sites for future translocations should be surrounded by large areas of suitable habitat with little prospect of encroaching development. The remaining mortality, in which the animal became stuck in mangrove mud only 14 days after release, may be directly related to lack of familiarity with the habitat, since the research team never encountered any resident animals in this predicament.

The fact that the majority of kangaroos moved out of the conservation area could indicate that it did not provide suitable habitat for additional eastern grey kangaroos or that they were seeking to avoid resident conspecifics. However, it seems more likely that the movements represented responses to an unfamiliar environment. The areas occupied by the study animals are much larger than have been reported in other substantial studies of eastern grey home range size (Jaremovic and Croft 1987; Viggers and Hearn 2005), probably reflecting these exploratory movements. The apparent overlap in areas used by different individuals is not surprising, given that eastern grey kangaroos are not territorial and overlapping home ranges are the norm (Jarman and Taylor 1983).

There was no evidence that translocated animals displaced or otherwise disturbed resident animals. Although there was a slight increase in population density at the conservation area over the study period, this was apparently not primarily due to the translocation (Higginbottom *et al.* 2006). This does not, however, preclude the possibility that there may have been some subtle and unobserved effects of translocated animals on resident animals. Radio-tracking of resident animals would have been required to confirm that conclusion, but it was not practicable in this study.

CONCLUSION

The experience gained in this study demonstrates that, given adherence to the necessary site criteria, translocation is a feasible option for saving some wild animals, such as eastern grey kangaroos, that would otherwise die as a result of urban development. However, translocation should not be considered a widespread solution to the impact of development on animals, since there are unlikely to be many sites that meet the necessary criteria and it is resource-intensive in terms of time and money (see also Fischer and Lindenmayer 2000). Further, since translocation inevitably stresses the animals and capture can often cause their death, a preferable option is to avoid development or to design it so that there is minimal habitat destruction.

ACKNOWLEDGMENTS

This research was funded by Lewiac Pty Ltd and carried out in collaboration with Gold Coast City Council and Queensland Parks and Wildlife Service.

REFERENCES

Campbell L and Croft DB 2001. Comparison of hard and soft release of hand-reared eastern grey kangaroos. In *Veterinary Conservation, Biology, Wildlife Health and Management in Australasia. Proceedings of International Joint Conference.* Taronga Zoo, Sydney. (Eds A Martin and L Vogelnest) pp. 173–180.

Copley PB 1994. Translocations of native vertebrates in South Australia: a review. In *Reintroduction Biology of Australian and New Zealand Fauna.* (Ed. M Serena) pp. 35–42. Surrey Beatty and Sons, Sydney.

DEH (Dept of Environment and Heritage Queensland) 2003. Relocation or re-establishment of native animals. *QNPWS Procedure Notice* 26/11/2003. Notice No. 019.

Delroy LB, Earl J, Radbone I, Robinson AC and Hewett M 1986. The breeding and re-establishment of the brush-tailed bettong, *Bettongia penicillata*, in South Australia. *Aust. Wildl. Res.* **13:** 387–396.

Dickinson M 2004a. Harbour Town North fauna report. Unpublished consultancy report to Lewis Land Group. Gold Coast, Queensland.

Dickinson M 2004b. Harbour Town North fauna report 2004. Unpublished report to Queensland Parks and Wildlife Service. 16 September 2004.

Fischer J and Lindenmayer DB 2000. An assessment of the published results of animal relocations. *Biol. Conserv.* **96:** 1–11.

Griffith B, Scott JM, Carpenter JW and Reed C 1989. Translocation as a species conservation tool: status and strategy. *Science* **245:** 477–480.

Higginbottom K 2004. Relocation of mammals from Harbour Town North. Unpublished consultancy report to Lewis Land Group. Gold Coast, Queensland.

Higginbottom K, Page S and Streatfield C 2006. Final Report to QPWS on monitoring of translocation of macropods, possums and gliders to Pimpama Wetlands Conservation Area. Unpublished report to Queensland Parks and Wildlife Service.

IUCN 1987. IUCN Position Statement on translocation of living organisms: introductions, re-introductions and re-stocking. IUCN, Gland, Switzerland.

Jaremovic RV and Croft DB 1987. Comparison of techniques to determine eastern grey kangaroo home range. *J. Wildl. Man.* **51**: 921–930.

Jarman PJ and Taylor RJ 1983. Ranging of eastern grey kangaroos and wallaroos on a New England pastoral property. *Aust. Wildl. Res.* **10**: 33–38.

Kleiman DG 1989. Reintroduction of captive mammals for conservation. *Bioscience* **39**: 152–161.

Novellie PA and Knight M 1994. Repatriation and translocation of ungulates into South African national parks: an assessment of past attempts. *Koedoe* **37**: 115–119.

NSW NPWS 2001. Policy for the translocation of threatened fauna in NSW. Policy and Procedure Statement No. 9. http://www.environment.nsw.gov.au/resources/nature/policyFaunaTranslocation.pdf. Accessed on 17 October 2008. NSW National Parks and Wildlife Service, Sydney.

Pople AR, Lowry J, Lundie-Jenkins G, Clancy TF, McCallum HI, Sigg D, Hoolihan D and Hamilton S 2001. Demography of the bridled nailtail wallabies translocated to the edge of their former range from captive and wild stock. *Biol. Conserv.* **102:** 285–299.

Short J, Bradshaw SD, Giles J, Prince RIT and Wilson GR 1992. Reintroduction of macropods (Marsupialia: Macropodoidea) in Australia: a review. *Biol. Conserv.* **62:**189–204.

Tanner Z and Hocking G 2001. Status and management of the forester kangaroo in Tasmania 2000. Nature Conservation Report 2001/2. Dept of Primary Industries, Water and Environment, Hobart.

Viggers KL and Hearn JP 2005. The kangaroo conundrum: home range studies and implications for land management. *J. Appl. Ecol.* **42:** 99–107.

Wolf CM, Griffith B, Reed C and Temple SA 1996. Avian and mammalian translocations: update and reanalysis of 1987 survey data. *Conserv. Biol.* **10:** 1142–1154.

30 Managing a kangaroo population by culling to simulate predation: the Wyperfeld trial

D.G. Morgan and P. Pegler

SUMMARY

After predators and stock were removed from Wyperfeld National Park, its western grey kangaroo (*Macropus fuliginosus*) population showed marked fluctuation, with high mortality and vegetation degradation during and after population peaks. The population was managed over an eight-year period in a way intended to simulate conceivable natural predation patterns, using a population growth model that made it possible to predict densities a year or more ahead. The previous extremes in resource-related mortality have not occurred since commencing the management program, and some improvements to vegetation condition have been measured. Although there were assumptions about an appropriate cull level, and mismatches between the observed and predicted kangaroo populations during two years, the cull strategy did allow population densities to vary in a way that seemed to correspond more closely to a pre-settlement situation.

INTRODUCTION

Kangaroo populations have been managed in the so-called sheep rangeland areas of inland Australia for over 30 years for a variety of reasons (Caughley 1981), largely through commercial harvesting programs (Caughley *et al.* 1987). In more recent years, macropodid populations have also been controlled in several Australian national parks and reserves, usually to restrict population size when kangaroo overpopulation and consequent vegetation degradation have been identified as issues (Coulson 1998, 2001, 2007). The management of kangaroo numbers in Hattah-Kulkyne National Park by Parks Victoria is an example (Sluiter *et al.* 1997), as intensive kangaroo browsing there was preventing the regeneration of woody plants.

Park managers often reduce population density by periodically culling towards a predetermined target density that is either fixed at a level thought to be typical for the ecosystems concerned (as at Hattah-Kulkyne) or that represents an appropriate carrying capacity (Shepherd and Caughley 1987). Cull numbers may also be varied from time to time according to some measure of vegetation response. This chapter sets out an alternative approach that draws on analysis of time series data to manage a population of the western grey kangaroo (*Macropus fuliginosus*) within Wyperfeld National Park, in the Victorian Mallee, where high kangaroo densities were limiting vegetation regeneration. It was based on the observation that perceived kangaroo overpopulation occurred in reserves, largely with a pastoral history, where dingo (*Canis lupus dingo*) predation and kangaroo hunting no longer took place. The approach was an adaptive experimental one, involving varying cull levels to match likely changes in a kangaroo predator's population density and monitored vegetation condition.

STUDY AREA

Wyperfeld National Park is approximately 350 km north-west of Melbourne, near the eastern boundary of the Big Desert, an aeolian dune system of Quaternary age, and centred on the normally dry watercourse and floodplains of Outlet Creek, an overflow of the Wimmera River (Fig. 30.1).

The Wimmera River arises in the northern Grampians Range and Stawell district about 200 km to the south and continues northward across the Wimmera district plains to Lake Hindmarsh. Following periods of above-average rainfall, the Wimmera may overflow Lake Hindmarsh and run north along a normally dry watercourse as Outlet Creek, eventually petering out within Wyperfeld National Park in a large, currently dry lakebed known as the Wirrengren Plain.

Climate

The Wyperfeld climate is semi-arid with highly variable rainfall. Mean annual precipitation is approximately 340 mm, distributed throughout the year but with a slight bias towards winter rainfall. Winters are relatively cool and summers hot and dry, with high evaporation rates. Surface water is ephemeral. There are no permanent streams or other water bodies.

Vegetation

Before European settlement the sandy dunes, swales and sandplains that cover much of the area were vegetated by a mosaic of heathland communities and tall shrubland dominated by mallee eucalypts, with a species composition related to local site and soil characteristics. That vegetation remains largely intact. On sandy soils closer to Outlet Creek were woodlands of slender cypress-pine (*Callitris gracilis*) with occasional buloke trees (*Allocasuarina luehmannii*).

The vegetation bordering Outlet Creek and its floodplains is either low woodland dominated by black box (*Eucalyptus largiflorens*) or woodland dominated by river red gum (*Eucalyptus camaldulensis*), with a sporadic understorey of moonah (*Melaleuca lanceolata*). In former years it carried a varied ground cover. On the very heavy, often gypseous soils of the watercourse and lakebeds into which water flowed during floods, then evaporated, was low shrubland.

The distribution of vegetation types is complex. Because Outlet Creek followed a meandering course as it flowed northward through the sandhills, a compass line in almost any direction near the watercourse crosses a variety of vegetation in a relatively short distance.

RECENT HISTORY

Before the arrival of Europeans, the Wyperfeld area had for thousands of years been the land of the Wotjobaluk (Wergaia) people, who had relied on food-gathering and hunting, and on trading with other peoples nearby. The Wirrengren Plain, at the end of Outlet Creek, had been a meeting and trading centre (Allen 1975).

European settlement

The first European known to arrive at Wyperfeld was the explorer Edward John Eyre, who searched the area north of Lake Hindmarsh in 1838 for possible pasture, initially on horseback. Finding little forage and no surface water that year, he eventually turned the horses loose and returned on foot (Eyre 1838). In 1847 a squatter, James Maxwell Clow, drove sheep along Outlet Creek as far north as the Wirrengren Plain, where he set up the Pine Plains run. Years later one of his companions provided a glimpse of the area in a poem:

To look for runs back in the bush,
I went with Mr Clow ...
Wild dogs in thousands they were there,
To hear them howl at night;
And sheep, though watched,
when morning came it was a dreadful sight ...
Emus in scores out on each plain,
Wild turkeys feeding too;
While in the forest skirts the plain,
Swarming with kangaroo ...
And natives when they saw us come,
They soon did disappear.

D. Hughes, from *The Old Pioneer's Last Resource* (Allen 1975)

The picture we gain, allowing for poetic licence, is nevertheless informative: there were abundant dingoes, kangaroos and 'wild turkeys' – Australian bustards (*Ardeotis australis*), a species now very rare in

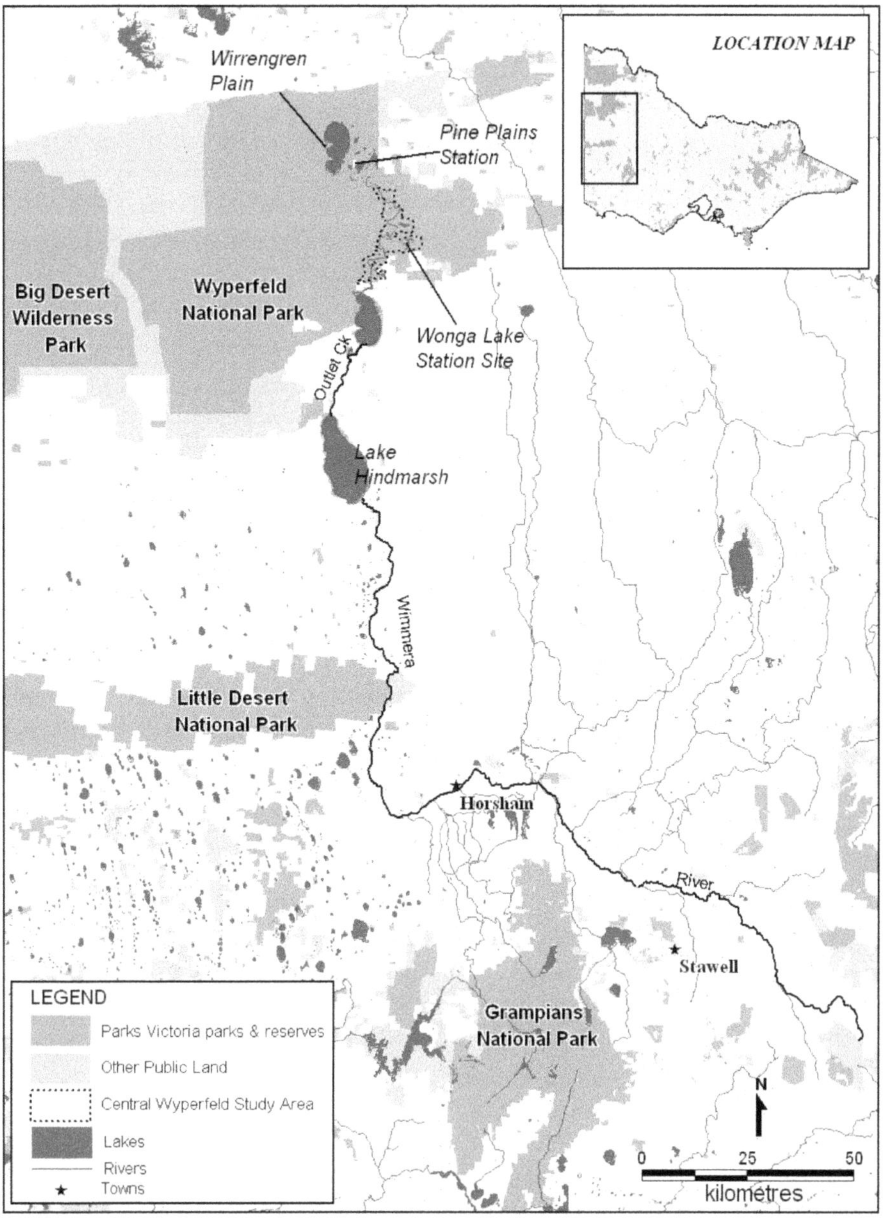

Figure 30.1: The study area (dotted outline) surrounds part of the watercourse of Outlet Creek within Wyperfeld National Park.

the district but associated with habitats where there is at least some abundant ground cover – and the presence of local people who depended partly on hunting for their livelihoods. Also, the writer's use of the word 'forest' to describe the vegetation around the lakes suggests denser vegetation than exists today.

1848–80

Following Clow's taking up the Pine Plains run, others established the Wonga Lake run along Outlet Creek to the south. The variety of landholders who followed experienced good seasons, floods and droughts. Initially sheep stocking was moderate, with about 4000 sheep in the Outlet Creek floodplains, but this grew to 20 000 by the 1870s (Allen 1975), by which time rabbits had also arrived. The heavy grazing of the floodplain and surroundings clearly had an impact: faced with poor hunting and destruction of their wells by stock, the Wotjobaluk people left the area in the 1860s. The Wonga Lake run itself was eventually in such poor condition that it was abandoned in 1880. Pine Plains Station, which regularly cropped some of its land as well as grazing sheep and, later, cattle, remains today a relatively small area of freehold land surrounded by national park.

From the beginning of settlement there was some clearing of the pine/buloke areas, creating open dunefield. Occasional severe wildfires contributed to this process. There have been continuing efforts to remove dingoes by trapping and shooting, resulting in very low dingo numbers in the district.

Floods

There is evidence of major floods flowing down Outlet Creek from Lake Hindmarsh approximately every 20 years from 1830 onward, the last major flood reaching the Wirrengren Plain in 1917–18. From the late 1880s, some of the headwaters of the Wimmera River were dammed and much of its water diverted into thousands of kilometres of open channels to farms across the Wimmera and southern Mallee, channels which are being replaced by pipelines. After 1918, Outlet Creek rarely flooded and never did so in a major way; only two floods reached Wyperfeld itself, one in 1956 and one in the summer of 1975–76. However, relatively little of the floodplain was affected. There was seedling regeneration of black box and river red gum along margins of the floodwaters but, on the rest of the floodplain, virtually no regeneration occurred and the progressive death of older trees continued.

Park reservation

An area of 9600 acres (3885 ha) in the former Wonga Lake Station area was reserved as a sanctuary in 1909 (*Victorian Government Gazette*, 20 October 1909). It was extended in 1921–22 to become Wyperfeld National Park, then further extended over many years to its 2006 size of 356 800 ha. The more northerly parts, including Pine Plains, were the last to be incorporated into the park. Wyperfeld National Park now includes much of the final course of Outlet Creek, all floodplain areas associated with it and an extensive area of dunes, swales and sandplains. The section of the park of interest in this study is a 98 km^2 sector in the southern portion, centred on Outlet Creek around the old Wonga Lake Station site near the present campground.

Park management

Sheep and cattle grazing of the floodplains along much of Outlet Creek did not end with the abandonment of Wonga Lake Station. Grazing leases were issued and the Outlet Creek watercourse was used as a stock route. The Wyperfeld Committee of Management, a voluntary body that usually met annually, issued the grazing leases from 1925 to 1958, when the National Parks Authority took over park administration. Until 1948, the lessees had been the Pine Plains landholders, who also served as honorary park rangers. The last lessee was a local farmer who, in 1958, became Wyperfeld's first paid ranger and continued to run stock in the southern portion of the park until the late 1960s. In the Pine Plains area, grazing continued until late 1997 when it was finally phased out after nearly 160 years of intermittent sheep pasturing on the floodplain. Wyperfeld National Park is now administered and staffed by Parks Victoria.

With the changes to park administration, the removal of stock and recognition of the degraded condition of much of the vegetation along the watercourse, the 1960s saw the introduction of a management program aimed at halting further degradation and restoring the floodplain and adjacent areas as far as possible towards their pre-European condition. Rehabilitation efforts included revegetation of some open dunefields, an extensive rabbit control program and some weed control. Following stock removal, however, kangaroo numbers seemed to become particularly high, noticeably on the dry lakebeds where there was also little vegetation regeneration. A kangaroo population monitoring program was begun in the early 1970s.

KANGAROO POPULATION MONITORING

The western grey kangaroo is the only macropodid species at Wyperfeld and, with domestic stock removed and rabbit numbers kept low, the primary vertebrate herbivore in the area. Because western grey kangaroos are known to browse seedlings of various woody plants as well as their favoured grasses (Hayes 1993; Sluiter *et al.* 1997), the possibility existed that high densities of kangaroos were having a negative impact on vegetation regeneration at Wyperfeld.

A kangaroo monitoring program using line transect distance methods began in central Wyperfeld in 1972 and has continued annually in all but three years since then, until recently in mid-winter (Morgan 2001). A typical census involved pairs of

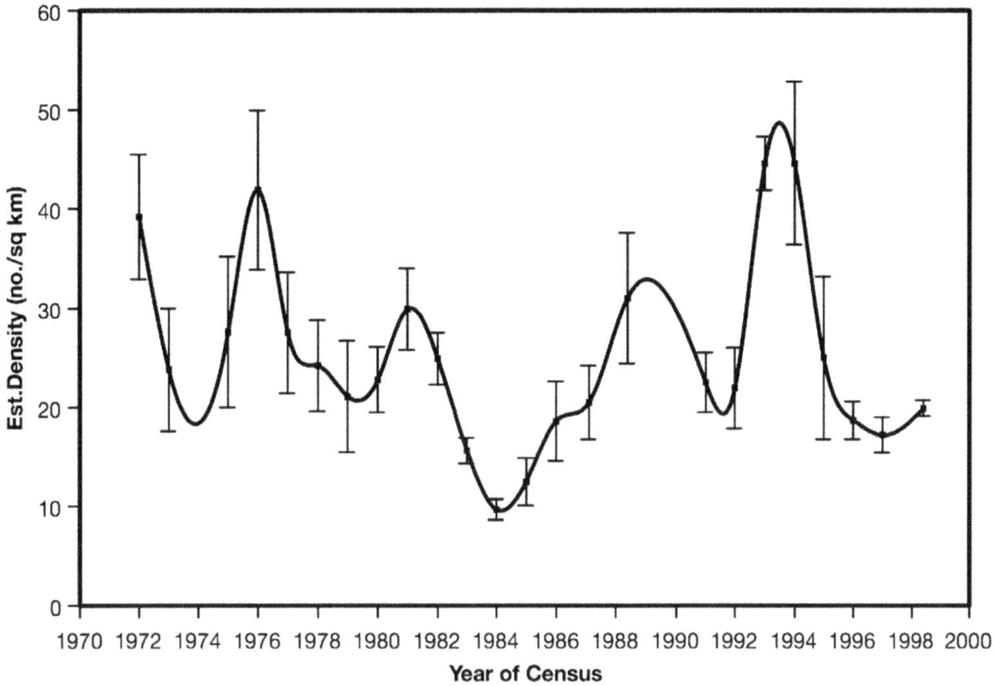

Figure 30.2: Changes in the western grey kangaroo (*M. fuliginosus*) population at Wyperfeld National Park from 1972 to 1998. All measurements are estimates, and the error bars represent plus or minus one standard error. The curve is a cubic spline fit to the estimates.

observers walking about 50 km of transects through an area centred on the floodplain but including all major vegetation types. Kangaroos at Wyperfeld are found in all vegetation types, in concentrations related to time of day, weather conditions and resource availability, so transects were sited accordingly. Transects were subdivided by habitat and density estimates made for each habitat type, combined using stratification methods to achieve an overall estimate, then multiplied by study site area to derive a total population estimate.

Most feeding was observed in the floodplain woodlands, lakebeds and nearby open dunefields. In seasons when kangaroo population densities were low and preferred foods (especially introduced grasses) abundant, kangaroos tended to be scattered throughout the area singly and in very small groups. When densities were high and food was scarce, they commonly aggregated into larger groups on sites that offered the remaining forage, such as open lakebeds. Lakebed densities at such times sometimes exceeded four per hectare. Kangaroo population density estimates for 1972–98 are shown in Figure 30.2.

These results show that, between 1972 and 1997, there were several periods of population fluctuation where kangaroo numbers at Wyperfeld rose to a peak, then declined; each involved two or more years of good rainfall followed by an unusually dry period. The years 1977, 1982, 1990 and 1994 were all dry years that marked the beginning of the population crash that ended each period. Typically, the population density grew exponentially following higher rainfall until density was high relative to the amount of available feed; as the dry hot part of the late summer/early autumn period was reached, preferred foods ran out and animals began to browse relatively unpalatable woody shrubs, tree trunks and exposed parts of surface roots. Vegetation damage was often severe, with little or no even slightly palatable ground cover remaining in many parts of the floodplains, and considerable damage to shrubs and trees. There were no water sources – surface water, palatable grasses with high water content or early morning dewfall – and kangaroos were noticeably emaciated and listless. Many kangaroos died during hot summer afternoons or the first severe frosts of autumn, with hyperthermia and hypothermia arising from dehydration and starvation the likely causes of death. The period of population decline that followed each population peak in 1972–97 varied from two to four years in length.

VEGETATION CONDITION

Changes to vegetation associated with European settlement – especially heavy stock grazing, partial clearing of the pine/buloke woodlands, the introduction of rabbits and many non-indigenous plant species and the near-cessation of the flooding regime – seem to have been mainly on the floodplains associated with Outlet Creek and in the adjacent pine/buloke woodlands. Mallee and heathland communities well away from Outlet Creek appear to have been much less affected.

In 1997 Parks Victoria commissioned the Centre for Environmental Management at the University of Ballarat to measure and report on the vegetation condition of the red gum/black box woodlands, dry lakebeds and the nearby pine/buloke woodlands along Outlet Creek (Ch. 31, this volume). Comparison with benchmarks derived from undisturbed reference sites showed that vegetation condition, particularly the ground cover of native species and especially shrubs, was in relatively poor condition in all three vegetation types studied.

MANAGING THE WYPERFELD KANGAROO POPULATION

Despite achieving very low rabbit densities through its management efforts and the spread of rabbit haemorrhagic disease post-1995, Parks Victoria was increasingly concerned about the continuing decline in floodplain vegetation condition and the periodic recurrence of starving and dying kangaroos. In 1998 Parks Victoria decided to manage the kangaroo population to allow the progressive recovery of key woody perennials (DNRE 1996). This involved a culling program to substitute for the effects of the natural predation and human hunting of former years.

It is widespread practice, when overpopulation by one species is identified as a management problem, to cull that population (usually by controlled shooting) to some predetermined constant level thought to represent an average natural density. However, there is long-standing overseas work on predator–prey relationships that suggested predator numbers may be relevant (Korpimaki and Krebs 1996); for example, in North America, predation by grey wolves (*Canis lupus*) has been shown to limit moose (*Alces alces*) populations (Messier 1994; Peterson 1999). It was expected that dingo numbers might rise and fall with kangaroo numbers but lag behind them; cull numbers should change in the same way. Parks Victoria therefore decided to change the cull rate from year to year in a way that tried to simulate the way in which predation rates might have changed in former years as predator densities rose and fell.

Dingoes as predators of kangaroos

If dingo predation is to be simulated in part by culling, what is known of any relationship between dingo and kangaroo numbers? There have been a number of studies of the food of dingoes in various parts of Australia (Shepherd 1981; Newsome *et al.* 1983; Robertshaw and Harden 1985; Corbett 1989, 1995a; Marsack and Campbell 1990; Thomson 1992), reviewed and summarised by Corbett (1995b). They have involved both relatively pristine ecosystems and highly modified agricultural systems, and have highlighted common features and regional differences in the dingo's diet.

Dingoes are both predators and scavengers, with a highly varied diet ranging from kangaroos to arthropods, with an apparent preference for small to medium-sized animals and a capacity to switch food items as availability alters. In districts where dingoes find them readily available, red kangaroos *(Macropus rufus)*, euros *(Macropus robustus)* and swamp wallabies *(Wallabia bicolor)* make up a major part of the diet and their densities are affected by dingo predation, particularly in certain combinations of rainfall, population density and availability of alternative foods (Corbett and Newsome 1987). Dingo numbers in turn vary with fluctuations in prey numbers (Caughley *et al.* 1980) when alternative foods such as rabbits are unavailable or scarce, though the precise relationship is unknown. Unfortunately, none of these studies involved ecosystems in which western grey kangaroos are the most abundant macropodid species.

The earliest accounts of Wyperfeld suggest that dingoes and western grey kangaroos were abundant, in habitats with a vegetation cover denser than at present. In addition, various medium-sized marsupials, such as bandicoots, rat-kangaroos and at least two of the smaller wallabies, were known to be present in north-west Victoria at that time, presumably including Wyperfeld; it is reasonable to assume that some would have been important elements in the dingo diet. Based on the dingo's known capacity to hunt and kill red kangaroos, especially older animals and juveniles, it is reasonable to assume that western grey kangaroos would have been eaten too, especially at times such as droughts when smaller prey would have been less widely available. Dingoes are likely to have dropped in numbers when food ran short. In addition to dingo predation, human hunting would have reduced the kangaroo population, but the likely scale of predation is again unknown.

POPULATION GROWTH MODEL

Population change results from recruitment, mortality and migration. Western grey kangaroo recruitment is annual and varies greatly from zero (in a poor season) upwards, depending on the body condition of the females (Shepherd 1987). Mortality is observable throughout the year and is most frequent in the mid to late summer dry weather and during winter frosts, particularly when animals are in poor physical condition, as is noticeable after persistent low rainfall when population densities are high. Predation is no longer a factor: following near-complete removal of the dingo population and the end of human hunting, the kangaroo population was effectively without predators. Because the Outlet Creek system at Wyperfeld (Fig. 30.1) is relatively isolated from other suitable western grey kangaroo habitats and the species is relatively sedentary (Priddel 1987), there is likely to be little migration.

To see whether the pattern of kangaroo population change could be explained and perhaps predicted, the rate of population growth was modelled. After initial exploration of possibilities (e.g. Cairns 1989), a relatively simple model based on Lotka's classic (1907) intrinsic growth rate model proved adequate:

$$N_t = N_0 e^{rt} \quad \text{(Equation 1)}$$

where N_0 = number of individuals at time 0, N_t = number of individuals at time t, r = growth rate and t = time duration.

The growth rate *r* of a population over a selected time period can be estimated from population size at the beginning and end of the period, with *r* calculated as follows:

$$r = \left(\ln(N_t) - \ln(N_0)\right)/t$$

What factors determine growth rate? Population density is well known as having an influence on population growth rate, with individuals competing more for resources when densities increase. In semi-arid regions with variable rainfall, rainfall may be a major determinant of ecosystem change. Studies of red and western grey kangaroo populations in western New South Wales (Bayliss 1987; Caughley *et al.* 1987) have shown that rainfall, acting through consequent vegetation growth, can be a major influence on kangaroo population growth, and variable rainfall has been used as a surrogate for pasture biomass in modelling studies (McCarthy 1996).

Critical rainfall period

If rainfall and density potentially affect growth rate, over what periods of time are they likely to exert effects? Although rainfall records are commonly aggregated on an annual basis, no data specifically link 12 months of cumulative rainfall to the population growth rate. The critical rainfall time period may be longer or shorter than 12 months and, in addition, begin and end at particular times of year. Sets of cumulative precipitation data were therefore compared with the annually calculated population growth rates for the various census years; the time period with the highest correlation to growth rate was assumed to approximate the influential period.

Although there was considerable variation in the correlation coefficient (Fig. 30.3) the result was fairly clear: a period of about 24 months, beginning about 30 months before the mid-year census and ending about six months before it, was most highly correlated with annually calculated population growth rate. Presumably rainfall at Wyperfeld acts by determining the total pasture and water available to kangaroos over that period, their body condition and their chances of reproduction and death (Shepherd 1987).

Growth rate model

With the intention of producing a model that was essentially simple, the relationship between population growth rate and the suggested explanatory variables – rainfall and density – was examined using a multiple linear regression (MLR) analysis, with growth rate as the dependent variable. Because the precise length of the critical period was uncertain, as was the number of years of previous population density data that could be influential, several versions of the predictor variables were supplied to the analysis and a backward stepwise procedure used to select the best solution. Average densities over the previous one and two years were used, together with a series of different sets of cumulative rainfall for periods from 24–26 months. After checking that the main assumptions of MLR were met, 64% of the variance in kangaroo population growth rate was accounted for by two variables: cumulative rainfall over a 25-month

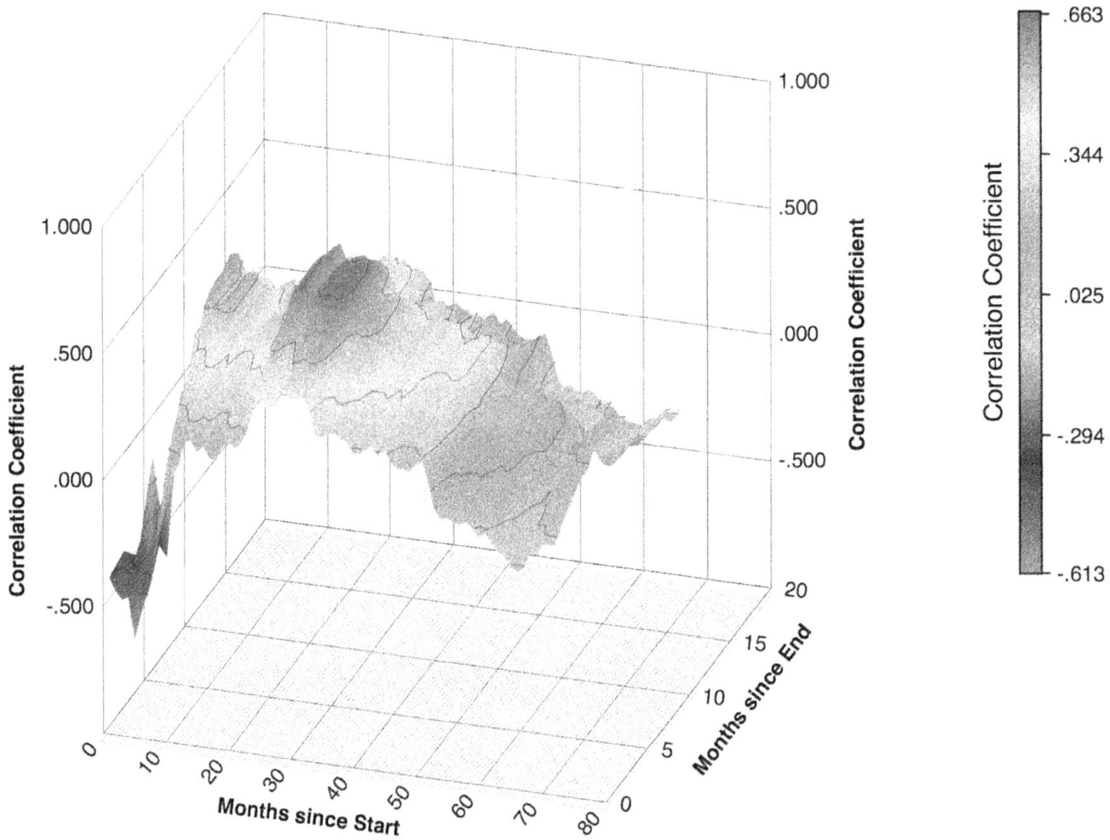

Figure 30.3: Correlations between cumulative rainfall over various time intervals prior to a census and the growth rate of the western grey kangaroo (*M. fuliginosus*) population. Highest correlation between time-cumulative rainfall and growth rate (distribution peak) was for a period that began about 30 months before a midwinter census and ended about six months before it.

period (beginning in October three years before a census and finishing at the end of October the year before the census) and mean population density over the two years prior to the census.

The regression equation yielded was as follows:

$$r = 0.001567p - 0.023692d - 0.563335 \quad \text{(Equation 2)}$$

where r = growth rate, p = cumulative precipitation over 25 months from October(–3 yr) to October(–1 yr), in mm, and d = mean kangaroo density in no./km^2 in the two years prior to the end of the growth period.

Using *r* values calculated from Equation 2 in Equation 1b produced the fit between the model and the data shown in Figure 30.4.

Although uncertainty in the density estimates is known to influence such correlations, the model's fit to the data was considered acceptable for the purposes of this study. Given appropriate rainfall and population density data, it should be possible to predict approximate kangaroo densities for the following year by estimating *r* from Equation 2 and applying it to Equation 1, using an initial density estimate D_0 derived directly from current census data or from a 'virtual' density estimate made a year earlier. For 1998–2005, kangaroo numbers were predicted from virtual estimates alone and cull targets were based on these estimates. At the same time, the annual census program was maintained.

With the likely scale of predation unknown, it was decided to reduce the population to an arbitrary target number calculated as half the 'virtual' modelled density on the floodplain areas. Linking the population target to the modelled density in this way allowed it to fluctuate with seasonal conditions as measured by cumulative rainfall. Actual cull numbers were calculated as the difference between the population estimated by survey and the calculated target, plus half the number estimated to occur in adjacent mallee buffer zones, to account for animals that were assumed to use floodplain habitats to feed but may have been temporarily sheltering in mallee habitats during the surveys.

Figure 30.4: Predicted western grey kangaroo (*M. fuliginosus*) population growth rate for the study period 1972–97 (dashes) compared to the growth rate estimated from field data (solid line). Agreement is acceptably close for every year for which a comparison is possible, accounting for an overall 64% of the variance in *r*.

EVALUATING THE STRATEGY

The trial was reviewed in 2006, after the management program had run for eight years and it was possible to evaluate both the management strategy and the changes in vegetation condition (Ch. 31, this volume). Figure 30.5 shows the predicted densities based on virtual estimates (dashed line) and those estimated from field data (continuous line). If the strategy were working well the observed densities should follow the trend of predicted densities, with population density behaving similarly in the modelled and real populations. This was the case for a number of years but, in 2004 and 2005, the observed density dropped to a level well below the estimated target density.

One possible explanation of the discrepancy lies in setting too high a cull target as a result of overestimating the number of kangaroos using adjacent mallee as shelter; field data from these habitats show few observations and high variances in the population estimates. Another possibility arises from using 'virtual' populations as the baseline for calculations of predicted numbers. These are subject to any effects of error in the model that could carry forward from year to year and produce, for example, the sequence of relatively high predicted numbers of more recent years. A third possibility lies in a change in survey technique from 2004, from observers travelling the park on foot to helicopter-based aerial surveys that produce far fewer observations. It has yet to be determined that the relationship between the estimates produced by the two methods is consistent and corresponding (Cairns 2006). Although there was clearly a problem with the aerial survey in 2005, which produced very little usable data, and the 2004 discrepancy is anomalous, in Wyperfeld census data collected since then the observed and predicted numbers correspond acceptably well.

Setting cull targets

The current cull target is based on a proportion of the predicted population occurring in floodplain habitats. In setting this proportion at 50%, an arbitrary estimate was made of the (unknown) scale of predation that may have applied to the kangaroo population. This simulated level of predation has allowed population densities to vary with interannual changes in seasonal conditions and there is evidence of progress towards meeting the principal management goal of vegetation rehabilitation on the Wyperfeld floodplain (Ch. 31, this volume). There has been no recurrence of mass kangaroo starvation in the park. The scale of management intervention has been reduced, with less culling required, particularly following wetter years, than would have been necessary with a simple fixed target density. However, it is not clear how vegetation recovery would

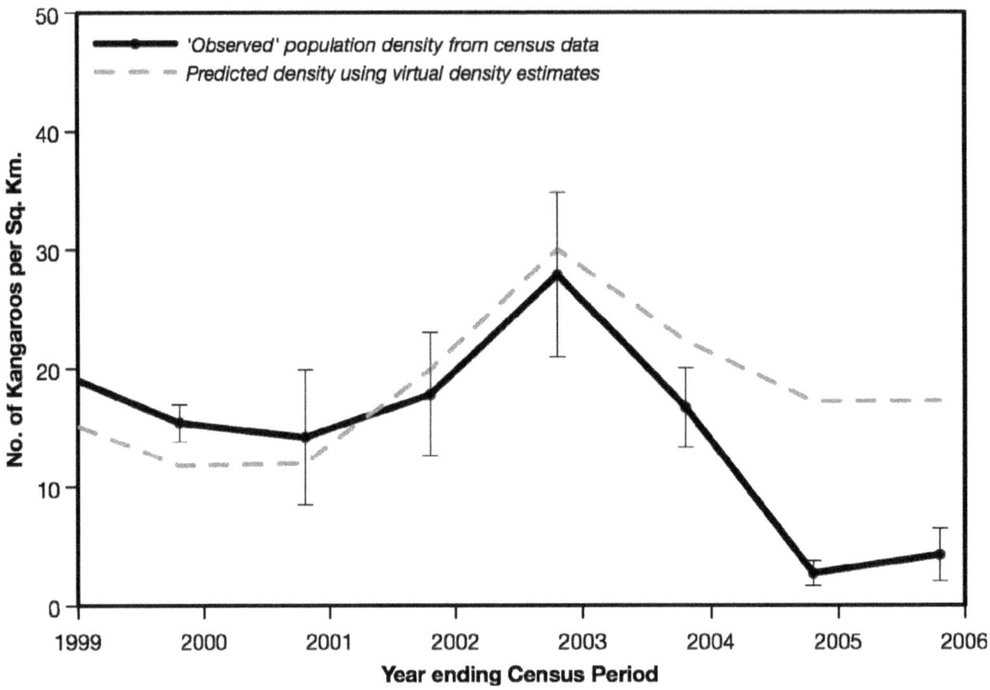

Figure 30.5: Western grey kangaroo (*M. fuliginosus*) densities estimated from census data (solid line) compared to densities predicted using the previous 'virtual' density estimate (dashes). Error bars are plus or minus one standard error.

have changed if different scales of predation had been applied. Determining the proportional level that optimises vegetation recovery while minimising culling would require periodic review over the long term as vegetation condition is reassessed.

A related issue in using a model that simulates predation is that the recovery of vegetation, particularly woody perennials, may be adversely affected by a lag between an interannual decline in vegetation productivity and a decline in predicted and target population densities. As the predicted (and target) population density follows the cumulative precipitation of previous years, the onset of interannually drier conditions is not reflected in reduced target densities until a year or more later. This increases the risk of high browsing pressure on woody perennial seedlings as the availability of favoured grasses diminishes (Sluiter *et al.* 1997). This issue could be addressed by increasing the proportion culled in years of declining rainfall.

CONCLUSION

This study provides support for cull strategies that allow population densities to vary in a way that corresponds more closely to the natural situation. While there are some problems with the 'predator simulation' approach, we believe they could be overcome by refining both its implementation and the population monitoring methods used, and continuing to monitor its effectiveness in achieving overall management goals.

REFERENCES

Allen TG 1975. *The History of Station and Settlement and the Flora and Fauna of Wyperfeld National Park*. Allen, Hopetoun, Vic.

Allen LR and Sparkes EC 2001. The effect of dingo control on sheep and beef cattle in Queensland. *J. Appl. Ecol.* **38:** 76–87.

Bayliss P 1987. Population dynamics. In *Kangaroos: Their Ecology and Management in the Sheep Rangelands of Australia*. (Eds G Caughley, N Shepherd and J Short) pp. 119–134. Cambridge University Press, Cambridge.

Cairns SC 1989. Models of macropodid populations. In *Kangaroos, Wallabies and Rat-kangaroos*. (Eds G Grigg, P Jarman and I Hume) pp. 695–704. Surrey Beatty and Sons, Sydney.

Cairns SC 2006. Monitoring and managing kangaroos in Victoria's Mallee National Parks. Unpublished report to Parks Victoria.

Caughley G 1981. Overpopulation. In *Problems in the Management of Locally Abundant Wild Animals*. (Eds PA Jewell, S Holt and D Hart) pp. 7–19. Academic Press, New York.

Caughley G, Grigg GC, Caughley J and Hill GJ 1980. Does dingo predation control the densities of kangaroos and emus? *Aust. Wildl. Res.* **7:** 1–12.

Caughley G, Shepherd N and Short J (Eds) 1987. *Kangaroos: Their Ecology and Management in the Sheep Rangelands of Australia.* Cambridge University Press, Cambridge.

Centre for Environmental Management 1998. Condition of vegetation in the riverine woodlands at Wyperfeld National Park. Report to Parks Victoria. University of Ballarat, Victoria.

Corbett LK 1989. Assessing the diet of dingoes from feces: a comparison of 3 methods. *J. Wildl. Man.* **53:** 343–346.

Corbett LK 1995a. Does dingo predation or buffalo competition regulate feral pig populations in the Australian wet-dry tropics? An experimental study. *Wildl. Res.* **22:** 65–74.

Corbett LK 1995b. *The Dingo in Australia and Asia.* University of NSW Press, Sydney.

Corbett LK and Newsome AE 1987. The feeding ecology of the dingo. *Oecologia* **74:** 215–227.

Coulson G 1998. Management of overabundant macropods: are there conservation benefits? In *Managing Marsupial Abundance for Conservation Benefits: Issues in Marsupial Conservation and Management.* (Ed. A Austin) pp. 37–48. Occasional Paper No. 1, Marsupial CRC, Sydney.

Coulson G 2001. Overabundant kangaroo populations in southeastern Australia. In *Wildlife Land and People: Priorities for the 21st Century. Proceedings of the 2nd International Wildlife Management Congress.* (Eds R Field, RJ Warren, H Okarma and PR Sievert) pp. 238–242. Wildlife Society, Bethesda, Maryland.

Coulson G 2007. Exploding kangaroos: assessing problems and setting targets. In *Pest or Guest: The Zoology of Overabundance.* (Eds D Lunney, P Eby, P Hutchings and S Burgin) pp. 174–181. Royal Zoological Society of NSW, Sydney.

Dawson TJ 1995. *Kangaroos.* University of NSW Press, Sydney

DNRE (Dept of Natural Resources and Environment) 1996. Mallee Parks management plan. Dept of Natural Resources and Environment, Melbourne.

Eyre EJ 1838. Letter to the editor. *South Australian Gazette*, 14 July 1838.

Hayes B 1993. Factors affecting the regeneration of woody perennial species within the Hattah-Kulkyne National Park. BSc Hons thesis. Dept of Zoology, University of Melbourne, Australia.

Korpimaki E and Krebs CJ 1996. Predation and population cycles of small mammals. *BioScience* **46:** 754–764.

Marsack P and Campbell G 1990. Feeding behaviour and diet of dingoes in the Nullarbor Region, Western Australia. *Aust. Wildl. Res.* **17:** 349–357.

McCarthy MA 1996. Red kangaroo *(Macropus rufus)* dynamics: effects of rainfall, density dependence, harvesting and environmental stochasticity. *J. Appl. Ecol.* **33:** 529–542.

Messier F 1994. Ungulate population models with predation: a case study with the North American moose. *Ecology* **75:** 478–488.

Morgan DG 2001. Wyperfeld National Park kangaroo populations November 2000. Report for Parks Victoria. Dept of Zoology, University of Melbourne.

Newsome AE, Catling PC and Corbett LK 1983. The feeding ecology of the dingo. II. Dietary and numerical relationships with fluctuating prey populations in south-eastern Australia. *Aust. J. Ecol.* **3:** 345–366.

Newsome AE, Catling PC, Cooke BD and Smyth R 2001. Two ecological universes separated by the dingo barrier fence in semi-arid Australia: interactions between landscapes, herbivory and carnivory, with and without dingoes. *Rangel. J.* **23:** 71–98.

Peterson RO 1999. Wolf–moose interaction on Isle Royale: the end of natural regulation? *Ecol. Appl.* **9:** 10–16.

Pople AR, Grigg GC, Cairns SC, Beard LA and Alexander P 2000. Trends in the numbers of kangaroos and emus on either side of the South Australian dingo fence: evidence for predator regulation? *Wildl. Res.* **27:** 269–276.

Priddel D 1987. The mobility and habitat utilisation of kangaroos. In *Kangaroos: Their Ecology and Management in the Sheep Rangelands of Australia.* (Eds G Caughley, N Shepherd and J Short) pp. 100–118. Cambridge University Press, Cambridge.

Robertshaw JD and Harden RH 1985. The ecology of the dingo in north-eastern New South Wales. *Aust. Wildl. Res.* **12:** 39–50.

Shepherd NC 1981. Predation of red kangaroos, *Macropus rufus*, by the dingo, *Canis familiaris dingo* (Blumenbach), in north-western New South Wales. *Aust. Wildl. Res.* **8:** 255–262.

Shepherd N 1987. Condition and recruitment of kangaroos. In *Kangaroos: Their Ecology and Management in the Sheep Rangelands of Australia.* (Eds G Caughley, N Shepherd and J Short) pp. 135–158. Cambridge University Press, Cambridge.

Shepherd N and Caughley G 1987. Options for management of kangaroos. In *Kangaroos: Their Ecology and Management in the Sheep Rangelands of Australia.* (Eds G Caughley, N Shepherd and J Short) pp. 188–219. Cambridge University Press, Cambridge.

Sluiter IRK, Allen GG, Morgan DG and Walker IS 1997. Vegetation responses to stratified grazing pressure at Hattah-Kulkyne National Park, 1992–96. Flora and Fauna Technical Report No.149. Dept of Natural Resources and Environment, Melbourne.

Thomson PC 1992. The behavioural ecology of dingoes in north-western Australia. III. Hunting and feeding behaviour, and diet. *Wildl. Res.* **19:** 531–541.

31 Changes in vegetation condition following kangaroo population management in Wyperfeld National Park

S.A. Gowans, M.S. Gibson, M.E. Westbrooke and P. Pegler

SUMMARY

Annual western grey kangaroo (*Macropus fuliginosus*) management programs have been implemented in Wyperfeld National Park, Victoria, as part of a strategy for the conservation and rehabilitation of vegetation communities. Due to public sensitivity to the culling of native animals it is essential to demonstrate that such actions result in environmental improvement. An assessment of the condition of the river red gum/black box woodland, pine/buloke woodland and lakebed herbfield vegetation communities was undertaken in 1998. Floristic and structural data were recorded from 162 quadrats across the park. In 2004, these quadrats were revisited to evaluate progress towards achieving targets for vegetation condition. An overall improvement in vegetation condition was found for all three vegetation communities assessed. Over half the quadrats sampled (52.2%) showed an increase in their condition indices since 1998. Results indicate that a general improvement, in relation to the management targets (increase in grazing-sensitive species richness and cover, native species ground cover and shrub layer richness) has occurred with this period of kangaroo population management. While a range of interacting factors may have influenced the changes, it is likely that management of grazing pressure is playing a significant role in the detected changes as no major rainfall event occurred between 1998 and 2004 to trigger a significant recruitment event.

INTRODUCTION

Wyperfeld National Park (357 017 ha) is located approximately 450 km north-west of Melbourne in the Mallee region of Victoria (Fig. 31.1). The park protects semi-arid vegetation growing on dunefields, plains and floodplains (DNRE 1996). Mallee vegetation covers most of the eastern part of the park on the dunes, while heath vegetation is common on the dunefields in the west. River red gum and black box woodlands cover the floodplains of Outlet Creek, while pine/buloke woodlands grow on nearby dunes (LCC 1987). The lakebeds associated with Outlet Creek, currently dry, typically support low shrubs, grasses and small herbs (LCC 1987). The vegetation has been categorised into three broad vegetation types (BVT): mallee (42%); mallee heath (50%) and riverine grassy woodland (8%) (DNRE 1997).

The state-wide cover of the riverine grassy woodland BVT has been reduced to 21% of that prior to

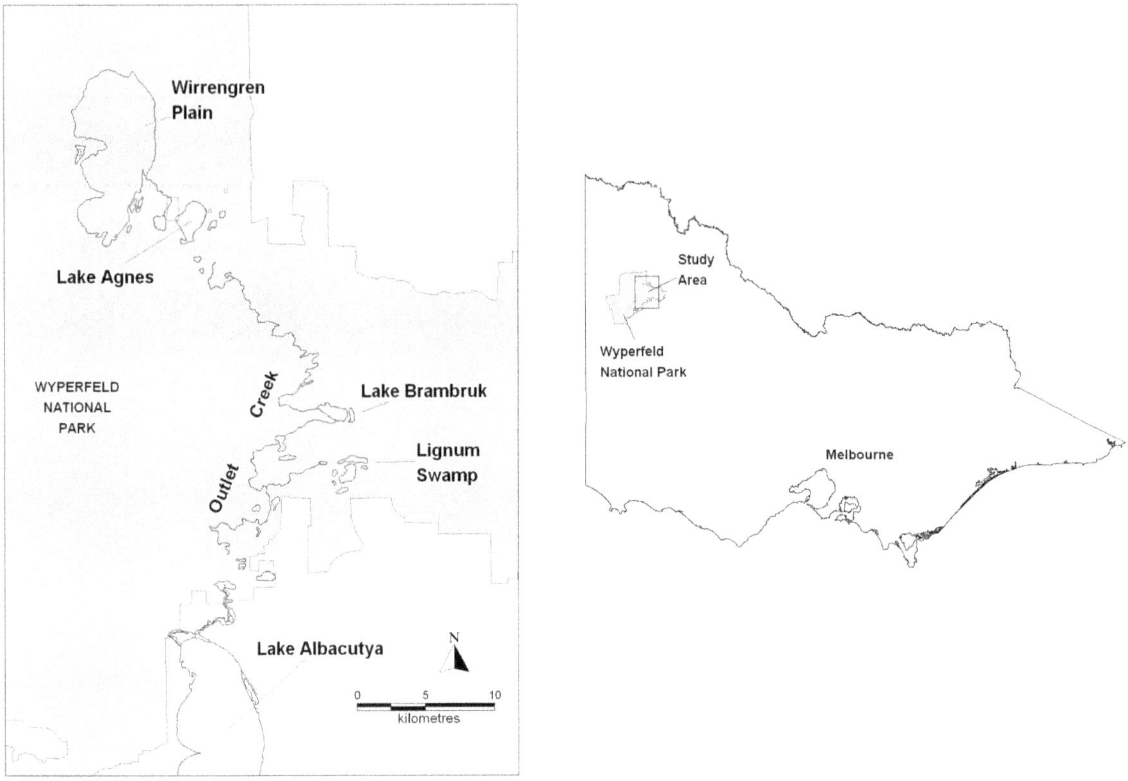

Figure 31.1: Locality map of Wyperfeld National Park.

European settlement. Outlet Creek, which supports the riverine grassy woodland BVT, last flowed between 1975 and 1977 and the whole system, including associated lakes, has not flooded since 1920 (LCC 1987; DNRE 1996). The conservation objectives for Wyperfeld National Park as set out in the Mallee Parks Management Plan (DNRE 1996) include a range of integrated strategies for the conservation and rehabilitation of vegetation communities and the management of the processes affecting them. Parks Victoria has undertaken management strategies in the riverine grassy woodland BVT to assist in reducing total grazing pressure, including closure of ground tanks and culling of western grey kangaroos (*Macropus fuliginosus*) to reduce population density to approximately five kangaroos per square kilometre (Ch. 30, this volume). Prior to the initiation of this kangaroo management program, an assessment of the condition of the riverine grassy woodland BVT in the park was required, to provide the basis for environmental performance measures and valuation of natural assets. To allow assessment of the success of strategies aimed at restoring the BVT in the park, Parks Victoria set short-, medium- and long-term targets for the condition of the vegetation (Wouters *et al.* 2000).

In 1998, the Centre for Environmental Management, University of Ballarat, in collaboration with Parks Victoria developed a method for the quantitative determination of vegetation condition for parks and reserves managed by Parks Victoria (Parks Victoria 1998). The method needed to be repeatable to allow for assessment of current condition, as well as detection of changes in condition across time.

Assessments of the riverine grassy woodland BVT within Wyperfeld National Park were undertaken in 1998 (Miller *et al.* 1998) and 2004 (Gowans *et al.* 2005a). This chapter focuses on detected changes between the two survey periods and evaluates the progress towards achieving targets for vegetation condition. Other studies relating to grazing animal abundance and their impacts have been summarised by Morgan and Pegler (Ch. 30, this volume).

METHODS

Study area

The study area was Wyperfeld National Park in north-west Victoria (141°49'S, 35°36'E). The median annual rainfall in the study area ranges from 342 mm at Patchewollock in the north-east to 367 mm at

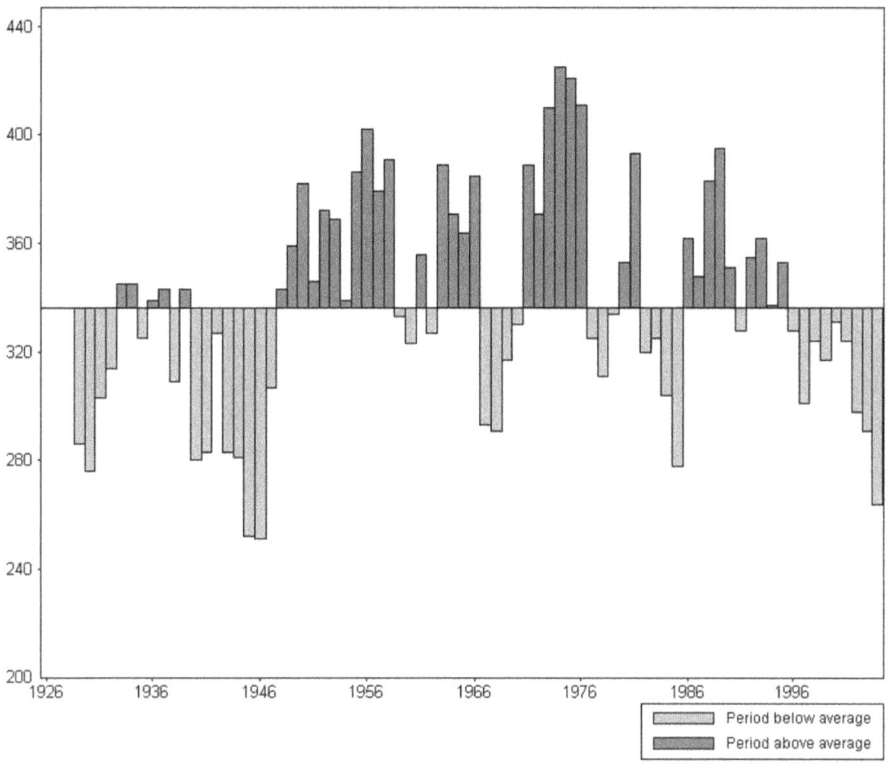

Figure 31.2: Three-year moving average annual rainfall at Patchewollock.

Rainbow in the south-east (Clewett *et al.* 2003). The long-term average annual rainfall for Patchewollock is 336 mm (horizontal line, Fig. 31.2). A three-year moving average which gives a more realistic indication of the availability of moisture for regeneration is given for Patchewollock (Clewett *et al.* 2003; Fig. 31.2). The data show high rainfall in the early 1970s, which led to flows in Outlet Creek. The data also show the extended drought from 1995, the period to which data presented in this chapter refers.

Sampling strategy

Riverine grassy woodland BVT covers 28 660 ha of Wyperfeld National Park. Three communities within this BVT – river red gum/black box woodland, pine/buloke woodland and lakebed herbfield – were assessed in January and March 1998 and March 2004. A quadrat-based sampling strategy was employed, quadrats for the 2004 field survey being the same as those in the 1998 assessment (Miller *et al.* 1998). All quadrats located in Wyperfeld in the initial 1998 survey were randomly located and subjectively placed within the vegetation community to avoid ecotones. Reference quadrats located outside the park (in neighbouring reserves) in the initial survey were subjectively placed within vegetation communities based on their perceived high condition status. All quadrat locations in the 1998 survey were recorded using a GPS receiver in the Australian map grid projection (Zone 54) and the 1966 Australian geodetic datum. Quadrats were relocated during the 2004 survey using a GPS receiver set in the same projection and datum.

Number of samples

The number of samples for the 1998 assessment was determined by analysis using the variation in preliminary samples to predict the power of subsequent comparisons based on a range of sample sizes. Each vegetation community was considered separately. The sample size for each vegetation community delivered a compromise between cost, effective size, probability of incorrectly identifying a change and probability of failing to identify a change. The analyses were performed by Parks Victoria and can be found in Miller *et al.* (1998). The sampling effort in each vegetation community is shown in Table 31.1.

One hundred and sixty-two quadrats were assessed. For each vegetation community three quadrats (to allow for natural variation between

Table 31.1: Number of quadrats sampled from each vegetation community within Wyperfeld National Park.

Vegetation community	Area (ha)	Area (%)	Quadrats (no.)	Quadrats (%)
river red gum/black box woodland	8075	41	40	25
pine/buloke woodland	4974	25	55	34
lakebed herbfield	6603	34	67	41
Total	**19 652**	**100**	**162**	**100**

sites) were chosen as reference sites. These sites were selected from areas considered representative of the vegetation community (i.e. having minimal historical disturbances) and were located both within the park and in neighbouring reserves. These quadrats were selected as representing the best condition of vegetation that could be expected, given the overriding impacts since settlement.

Size and orientation of samples

Different quadrat sizes were used for recording various parameters. Samples in both surveys consisted of a 100 × 100 m quadrat (1 ha) with a nested quadrat of 20 × 50 m (1000 m^2). The nested 20 × 50 m quadrats were oriented either north–south or east–west and the south-west corner marked with a numbered aluminium tag. All quadrats were permanently marked to ensure that they could be reliably relocated for future assessment. For the woodland vegetation community quadrats (river red gum/black box woodland and pine/buloke woodland), a tree marked the south-west corner; for the lakebed herbfield vegetation community, a metal post marked the south-west corner.

Data recorded

The date, recorders' names, a unique quadrat identification number, Australian map grid co-ordinates (determined using a GPS receiver with an accuracy of approximately 10 m), description of the site location and the vegetation community were recorded. All vascular flora present in the 20 × 50 m nested quadrat were recorded, identified to species level where possible and genus or family level in all other cases, and their conservation status determined (endemic or introduced). Nomenclature and conservation status followed Walsh and Entwisle (1994, 1996, 1999). All vascular flora were categorised according to life-forms as identified in the Victorian Flora Information System (DSE 2004), and given a cover/abundance value, modified from the Braun-Blanquet scale as cited in Kershaw and Looney (1985). Many annual species had completed their growth cycles at the time of the field surveys. Where dead material was identifiable it was recorded to family, genus or species level where possible.

A visual estimate of the projected foliage cover of native species in the ground layer was recorded in the 20 × 50 m nested quadrat. The presence of seedlings and/or juveniles was noted for all shrub species. Regeneration of indigenous perennial shrub species was recorded in the 20 × 50 m nested quadrat on a three-point scale:

0 no regeneration;
1 one or two species regenerating with each species represented by fewer than 10 young individuals;
2 regeneration of more than two species, or one or two species with each species represented by 10 or more young individuals;

In quadrats supporting an overstorey, dieback was recorded in the 100 × 100 m quadrat on a four-point scale:

0 more than two dead trees and more than 10 dead branches projecting from the canopy;
1 one to two dead trees and/or more than 10 dead branches projecting from the canopy;
2 up to 10 dead branches projecting from the canopy;
3 healthy, with no dead trees or dead branches projecting from the canopy.

Reproductive age classes were determined as senescent (large old trees with severely damaged crowns and usually large trunk hollows), mature (species flowers and/or fruit), juvenile (young species not yet capable of reproduction, >1 cm trunk diameter) and seedling (recently germinated, <1 cm trunk diameter). Age classes were considered present if one live individual tree from the class was present in the 100 × 100 m quadrat. The number of age classes present was then recorded for each 100 × 100 m quadrat.

Table 31.2: Comparison of 1998 and 2004 parameter values for pine/buloke woodland.

	1998		2004		
Condition parameter	**Mean**	**SE**	**Mean**	**SE**	**Wilcoxon P**
1. Native vascular plant species richness	8.1	0.43	11.9	0.54	<0.001
2. Native ground cover	6.8	0.85	8.3	1.33	n.s.
3. Regeneration of native shrub species	0.8	0.08	1.3	0.11	<0.001
4. Overstorey dieback	1.6	0.12	1.2	0.09	<0.05
5. Overstorey age classes	1.8	0.11	2.1	0.11	n.s.
6. Strata intactness	0.5	0.08	1.5	0.09	<0.001

n.s. = not significant.

The intactness of each stratum (Gowans *et al.* 2005b) expected to be present in each vegetation community was assessed to determine if the expected components (species) were present and in a density similar to that of the reference site. The number of intact strata was recorded for each 100 × 100 m quadrat, based on expert opinion.

All quadrats were photographed from 10 m outside the 20 × 50 m nested quadrat on the long axis from the marked end to provide a visual record. Photographs were taken to provide a record of the site appearance and were not intended to provide permanent monitoring points. Field recordings for both 1998 and 2004 assessments were preceded by field training and calibration of parameter estimates and co-ordinated by the same principal botanist.

Analysis of changes in vegetation condition

Vegetation condition

Six parameters were used to determine the condition of the river red gum/black box woodland and pine/buloke woodland vegetation communities and four condition parameters were used to determine the condition of the lakebed herbfield vegetation community. The condition of the three vegetation communities across their range within the park was assessed by making comparisons of individual parameters for quadrats sampled within the park against reference quadrats or expert opinion.

From the data recorded in the 1998 and 2004 surveys, vegetation parameters were used to assess the changes in vegetation condition. Table 31.2 outlines the vegetation parameters included in the calculation of the 2004 and 1998 condition for each vegetation community.

All data collected from the 100 × 100 m quadrats or the 20 × 50 m nested quadrats for the 2004 and 1998 surveys were analysed separately on a vegetation community basis across the study area. The value of the parameter at each quadrat was divided by the benchmark value, resulting in a score from zero to one (Gowans *et al.* 2005b). If a quadrat parameter value was greater than the benchmark value, the score was truncated to a maximum score of one.

Data analysis

A condition index for each quadrat for the 2004 and 1998 surveys was calculated as the average score of all parameters, where all parameters were weighted equally. An overall condition index for each vegetation community for the 2004 and 1998 survey was then calculated from the average value of the quadrat condition indices.

Change in parameter values and condition indices between 1998 and 2004 were analysed using pairwise statistical tests. Wilcoxon matched pairs tests were used to examine differences in raw parameter values, as some parameters were measured on an ordinal scale and not all parameters were expected to follow a normal distribution. Paired T-tests were used to test for differences in quadrat condition scores between the two sampling periods.

Vegetation condition targets

Wouters *et al.* (2000) lists the short- (five years), medium- and long-term vegetation condition targets. As the 2004 survey was conducted six years after the original survey, only the short-term targets could be examined in detail. The short-term targets were categorised as targets following an average to below average rainfall period, targets following a single year of

Table 31.3: Mean and standard error (1998 and 2004) for variables relating to short-term vegetation condition targets for pine/buloke woodland.

		Park quadrats		Reference quadrats	
Target	**Year**	**Mean**	**SE**	**Mean**	**SE**
Target 1 – increase in number of grazing-sensitive species	1998	5.67	0.28	11.67	1.76
	2004	7.48	0.35	14.33	0.88
Target 2 – increase in cover of grazing-sensitive species	1998	14.01	0.71	29.17	4.41
	2004	27.50	1.45	48.33	5.07
Target 3 – increase in ground cover of native species	1998	6.82	0.84	35.00	7.64
	2004	8.28	1.32	9.00	3.79
Target 4 – increase in shrub species richness	1998	1.22	0.22	6.67	0.88
	2004	2.46	0.24	7.33	0.67

significantly above-average rainfall, and targets following two consecutive years of significantly above-average rainfall with late summer or autumn storms.

An analysis of rainfall in the six years was conducted using data from Walpeup (north of the park) and Rainbow (south of the park) rainfall stations (Clewett *et al.* 2003). As the data indicate that the survey followed several years of below-average rainfall, only the first set of targets was assessed (Tables 31.3, 31.5, 31.7).

Two of the targets relate to the richness and cover of grazing-sensitive species. Wouters *et al.* (2000) described grazing sensitive species as 'those species that have been identified as palatable to kangaroos that exhibit changes in abundance that can be directly attributed to kangaroo grazing, and include native perennial grasses and palatable shrubs'. Grazing-sensitive species were determined based on data from Cunningham *et al.* (1981), Landsberg *et al.* (1999), Sluiter *et al.* (1997), Sandell *et al.* (2002) and Westbrooke (1998). Landsberg *et al.* (1999), Sluiter *et al.* (1997) and Sandell *et al.* (2002) define species in the pasture layer which are tolerant of intense grazing pressure (unpalatable) as increasers; species which are intolerant of intense grazing pressure (palatable) are defined as decreasers. Westbrooke (1998) identified species palatable to kangaroos based on many years of observation in the field. Cunningham *et al.* (1981) included notes for most species on their acceptability to stock (palatability and forage values). Palatability is only expressed in relation to the acceptability of other forage available to the stock at the same time (Cunningham *et al.* 1981). If a species were referred to as being intolerant of intense grazing pressure or acceptable to stock or native herbivores for foraging in any reference, it was regarded as a grazing-sensitive species. The cover of grazing-sensitive species was calculated for each quadrat by substituting the Braun-Blanquet cover abundance values with the midpoints of each cover class and summing these percent cover values for each grazing-sensitive species recorded within the quadrat. While it is recognised that this substitution is not ideal, it supplies a close approximation and any disadvantages are outweighed by the advantage of the availability of historical data recorded using the Braun-Blanquet scale, which was the standardised methodology for Victorian vegetation surveys based on Gullan (1978).

Table 31.4: Comparison of 1998 and 2004 parameter values for river red gum/black box woodland.

	1998		2004		
Condition parameter	**Mean**	**SE**	**Mean**	**SE**	**Wilcoxon P**
1. Native vascular plant species richness	7.4	0.46	11.4	0.54	<0.001
2. Native ground cover	4.2	0.59	10.4	1.17	<0.001
3. Regeneration of native shrub species	0.2	0.05	0.6	0.09	<0.001
4. Overstorey dieback	1.8	0.08	1.0	0.07	<0.001
5. Overstorey age classes	1.5	0.09	2.3	0.11	<0.001
6. Strata intactness	1.0	0.07	1.8	0.11	<0.001

Table 31.5: Mean and standard error (1998 and 2004) for variables relating to short-term vegetation condition targets for river red gum/black box woodland.

		Park quadrats		Reference quadrats	
Target	**Year**	**Mean**	**SE**	**Mean**	**SE**
Target 1 – increase in number of grazing-sensitive species	1998	4.58	0.26	10.67	0.33
	2004	6.47	0.42	13.67	0.67
Target 2 – increase in cover of grazing-sensitive species	1998	11.92	0.90	26.67	0.83
	2004	30.38	2.03	60.10	8.78
Target 3 – increase in ground cover of native species	1998	3.92	0.67	16.67	7.26
	2004	9.79	1.28	20.00	7.64
Target 4 – increase in shrub species richness	1998	0.79	0.20	6.00	1.00
	2004	1.42	0.21	7.33	1.76

RESULTS

Changes in pine/buloke woodland condition

Vascular plant species

More vascular plant species were recorded in the 2004 survey than in the 1998 survey. During the 1998 survey, a total of 63 native and 35 introduced flora species were recorded for pine/buloke woodland. During the 2004 survey, a total of 86 native and 45 introduced flora species were recorded for the pine/buloke woodland vegetation community. A total of 42 native and 25 introduced flora species were recorded during both the 1998 and 2004 surveys. Forty-five native and 20 introduced flora species were recorded during 2004 only, and 21 native and 10 introduced flora species were recorded during the 1998 survey but not during the 2004 survey.

Comparison of 1998 and 2004 parameters

Table 31.2 lists the descriptive statistics of the six condition parameters values for the 1998 and 2004 sets of data. It also indicates the Spearman rank order correlation between the two sets of parameters. The mean 2004 condition score (0.49 ± 0.02) was higher than the 1998 condition score (0.35 ± 0.01).

Short-term targets

Table 31.3 presents results for vegetation condition parameters in 1998 and 2004 to allow assessment of progress towards the four short-term vegetation condition targets.

Changes in river red gum/black box woodland condition

Vascular plant species

More vascular plant species were recorded in the 2004 survey than in the 1998 survey. In 1998 50 native and 30 exotic flora species were recorded, increasing to 69 and 37 species respectively in 2004. Thirty-five native and 22 exotic flora species were recorded during both the 1998 and 2004 surveys. Fifteen native and eight exotic flora species were recorded during 2004 only, and 34 native and 15 exotic flora species were recorded in 1998 only.

Comparison of 1998 and 2004 parameters

Table 31.4 lists the descriptive statistics of the six condition parameters values for the 1998 and 2004

Table 31.6: Comparison of 1998 and 2004 parameter values for lakebed herbfield.

	1998		2004		
Condition parameter	**Mean**	**SE**	**Mean**	**SE**	**Wilcoxon P**
1. Native vascular plant species richness	4.0	0.36	6.9	0.39	<0.001
2. Native ground cover	10.9	1.56	20.1	1.96	<0.001
3. Regeneration of native shrub species	0.7	0.07	0.6	0.09	n.s.
6. Strata intactness	0.3	0.09	1.3	0.09	<0.001

n.s. = not significant.

Table 31.7: Mean and standard error (1998 and 2004) for variables relating to short-term vegetation condition targets for lakebed herbfield.

Target	Year	Park quadrats		Reference quadrats	
		Mean	SE	Mean	SE
Target 1 – increase in number of grazing-sensitive species	1998	2.25	0.20	3.67	0.67
	2004	4.08	0.23	7.00	0.58
Target 2 – increase in cover of grazing-sensitive species	1998	5.55	0.51	9.17	1.67
	2004	17.50	1.45	21.67	5.46
Target 3 – increase in ground cover of native species	1998	10.14	1.41	21.67	6.01
	2004	19.15	1.80	34.00	6.66
Target 4 – increase in shrub species richness	1998	1.03	0.19	3.33	0.88
	2004	1.98	0.19	4.00	1.15

sets of data. It also indicates the Spearman rank order correlation between the two sets of parameters. The mean 2004 condition score (0.44 ± 0.02) was higher than the 1998 condition score (0.33 ± 0.02).

Short-term targets

Table 31.5 presents results for vegetation condition parameters in 1998 and 2004 to allow assessment of progress towards short-term vegetation condition targets.

Changes in lakebed herbfield condition

Vascular plant species

More vascular plant species were recorded in 2004 than in 1998. During the 1998 survey, 35 native and 22 introduced flora species were recorded for this vegetation community compared to 66 native and 38 introduced in 2004. Twenty-three native and 16 introduced species were recorded during both the 2004 and 1998 surveys. Forty-three native and 21 introduced flora species were recorded during the 2004 survey but not during the 1998 survey, and 12 native and six introduced species were recorded in the 1998 survey but not in 2004.

Comparison of 1998 and 2004 parameters

Table 31.6 lists the descriptive statistics of the six condition parameters values for the 1998 and 2004 sets of data. It also indicates the Spearman rank order correlation between the two sets of parameters. The mean 2004 condition score (0.47 ± 0.02) was higher than the 1998 condition score (0.36 ± 0.03).

Short-term targets

Table 31.7 presents results for vegetation condition parameters in 1998 and 2004 to allow assessment of progress towards the four short-term vegetation condition targets.

DISCUSSION

Changes in vegetation condition

Just over half the riverine grassy woodland BVT quadrats sampled (52.5%) showed an increase in condition indices since 1998. Notable increases were evident in the southern part of the park around Lake Brambruk, around Lignum Swamp and in the north at Lake Agnes. Almost half the quadrats (40.1%) showed no change or little change. These quadrats were scattered throughout the study area, although the southern region of the Wirrengren Plain and the area north-east of the plain included a concentration of quadrats showing little change. Quadrats showing a decrease in condition indices since 1998 were mostly sampled from the lakebed herbfield vegetation community of the Wirrengren Plain.

Factors influencing change

It is encouraging that almost half the quadrats sampled remained relatively stable over the six years and just over half showed an increase in their condition indices. A range of interacting factors could have influenced the changes in individual parameters across the park, such as management of grazing pressure, management of weeds or rainfall events, or a combination of these. It is likely that management of grazing pressure is playing a large role in the changes to native vascular plant species richness. No major rainfall event occurred between 1998 and 2004 that would trigger a significant recruitment event (Fig. 31.2), but low levels of recruitment occurred for some species. For each of the three communities, more vascular plant species were recorded in the 2004 survey than the 1998 survey. Successful establishment of seedlings requires reduced grazing pressure, weed control and adequate rainfall.

It is more difficult to explain the changes in the parameters relating to the overstorey. As the overstorey of river red gum/black box woodland relies on flooding for regeneration and there was no such event during the relevant six years, the differences observed would have been expected to be quite low. However, a significant difference was observed, with an increase in overstorey age classes since 1998. Despite the increase in overstorey age classes, the health of overstorey species significantly decreased during the study period for river red gum/black box woodland and for pine/buloke woodland. The interacting factors influencing these changes were not explored in this study but are likely to be independent of grazing pressure.

Vegetation condition targets

Short-term (five-year) targets

As the 2004 survey followed several years of below-average rainfall, there was no expectation of widespread regeneration of overstorey species such as *Callitris gracilis* subsp. *murrayensis* and *Allocasuarina luehmannii* within the pine/buloke vegetation community. Relevant short-term targets for the three communities assessed related to increases in richness of grazing-sensitive species, cover of grazing-sensitive species, percentage cover of native species in the ground layer and richness of perennial native species in the shrub layer.

For all vegetation communities, the mean values of these parameters were higher in 2004 than in 1998. The mean number of grazing-sensitive species increased by over 30% within river red gum/black box woodland and pine/buloke woodland, and by over 80% in lakebed herbfield quadrats. The mean percentage cover of grazing-sensitive species increased by approximately 100% within river red gum/black box woodland and over 150% in pine/buloke woodland and lakebed herbfield quadrats. The mean percentage cover of native species in the ground layer increased by 150% in river red gum/black box woodland, by 90% in lakebed herbfield and by only 20% in pine/buloke woodland. For all three vegetation communities, the mean richness of perennial native woody species in the shrub layer increased by over 80%. These results indicate a significant improvement, in relation to these four targets, since the 1998 study.

Medium- and longer-term targets

The medium-term targets relate to the successful establishment and survival of regenerating plants 10 years after above-average rainfall. Regenerating *Callitris gracilis* subsp. *murrayensis* plants, with stem diameters of less than or equal to 1 cm, were recorded in 12 of the 56 (21%) quadrats sampled within the pine/buloke woodland vegetation community. Regenerating *Allocasuarina luehmannii* plants were only located in one pine/buloke woodland quadrat. While the presence of regenerating overstorey trees may not lead to successful establishment and survival to maturity of these species, it is anticipated that adequate rainfall combined with ongoing maintenance of low total grazing pressure will lead to recovery in the condition of these woodlands, including their floristic and structural diversity (Sandell *et al.* 2002).

ACKNOWLEDGMENTS

This study was conducted under the Parks Victoria Research Partnership agreement with the Centre for Environmental Management, University of Ballarat. The project team would like to thank Marie Keatley, Peter Sandell and John Wright of Parks Victoria who provided overall management and support for this project, and Kim Schramm and Rhianna Bell, also of Parks Victoria, for providing support during the field survey. The project team would like to acknowledge John Miller, Peter Wilcock and Glenda Adler of the 1998 project team, and the 2004 project team of Kate Callister, Janet Leversha, Marion O'Keefe and Lia Gray.

REFERENCES

Clewett JF, Clarkson NM, George DA, Ooi SH, Owens DT, Partridge IJ and Simpson GB 2003. Rainman streamflow (version 4.3): a comprehensive climate and streamflow analysis package to assess seasonal forecasts and manage climatic risk. CD QI03040. Dept of Primary Industries, Queensland.

Cunningham GM, Mulham WE, Milthorpe PL and Leigh JH 1981. *Plants of Western New South Wales.* Soil Conservation Service of New South Wales.

DNRE (Dept of Natural Resources and Environment) 1996. Mallee Parks management plan. Dept of Natural Resources and Environment, Melbourne.

DNRE (Dept of Natural Resources and Environment) 1997. Victoria's biodiversity directions in management. Dept of Natural Resources and Environment, Melbourne.

DSE (Dept of Sustainability and Environment) 2004. Flora Information System, January 2004. Dept of Sustainability and Environment, Melbourne.

Gowans S, Gibson M and Westbrooke M 2005a. Vegetation condition assessment: Wyperfeld National Park. Report to Parks Victoria. Centre for Environmental Management, University of Ballarat, Victoria.

Gowans S, Callister K, Westbrooke M and Gibson S 2005b. Vegetation condition assessment of the semi-arid woodlands of Murray-Sunset National Park, Victoria. *Vic. Nat.* **122:** 85–93.

Gullan P 1978. Vegetation of the Royal Botanic Gardens Annex at Cranbourne, Victoria. *Proc. Roy. Soc. Vic.* **90:** 225–240.

Kershaw K and Looney J 1985. *Quantitative and Dynamic Plant Ecology*. Edward Arnold, London.

Landsberg J, James CD, Morton SR, Hobbs TJ, Stol J, Drew A and Tongway H 1999. The effects of artificial sources of water on rangeland biodiversity. Biodiversity Technical Paper No 3. Final Report to the Biodiversity Convention and Strategy Section, Biodiversity Group, Environment Australia, Canberra.

LCC (Land Conservation Council) 1987. Mallee Area review. Land Conservation Council, Melbourne.

Miller J, Gibson M, Westbrooke M, Wilcock P and Brown G 1998. Condition of vegetation in the riverine woodlands of Wyperfeld National Park. Report to Parks Victoria. Centre for Environmental Management, University of Ballarat, Victoria.

Parks Victoria 1998. Vegetation condition: draft protocol for the assessment of vegetation condition in Victoria's parks and reserves. Parks Victoria, Melbourne.

Sandell P, Ballentine M and Horner G 2002. Vegetation recovery in the Victorian Mallee Parks 1991–1998. Parks Victoria, Melbourne.

Sluiter IRK, Allen GG, Morgan DG and Walker IS 1997. Vegetation responses to stratified kangaroo grazing pressure at Hattah-Kulkyne National Park (1992–96). Flora and Fauna Technical Report No 149. Dept of Natural Resources and Environment, Melbourne.

Walsh N and Entwistle T (Eds) 1994. *Flora of Victoria*. Vol. 2. *Ferns and Allied Plants, Conifers and Monocotyledons*. Inkata Press, Melbourne.

Walsh N and Entwistle T (Eds) 1996. *Flora of Victoria*. Vol. 3. *Dicotyledons Winteraceae to Myrtaceae*. Inkata Press, Melbourne.

Walsh N and Entwistle T (Eds) 1999. *Flora of Victoria*. Vol. 4. *Dicotyledons Cornaceae to Asteraceae*. Inkata Press, Melbourne.

Westbrooke ME 1998. The ecology and regeneration status of Belah Woodlands in south-eastern Australia. PhD thesis. Dept of Botany, LaTrobe University, Melbourne.

Wouters C, Wright J and Pegler P 2000. Ecological rationale for kangaroo management in the Pine Plains area of Wyperfeld National Park. Parks Victoria, Melbourne.

32 Using harvest statistics to monitor temporal variation in kangaroo density and harvest rate

A.R. Pople, M. Evans, L. Farroway, J. Gilroy, G.C. Grigg, G. Lundie-Jenkins and N. Payne

SUMMARY

Management of the commercial harvest of kangaroos relies on quotas set annually as a proportion of regular estimates of population size. Surveys to generate these estimates are expensive and, in the larger states, logistically difficult; a cheaper alternative is desirable. Rainfall is a disappointingly poor predictor of kangaroo rate of increase in many areas, but harvest statistics (sex ratio, carcass weight, skin size and animals shot per unit time) potentially offer cost-effective indirect monitoring of population abundance (and therefore trend) and status (i.e. under- or over-harvest). Furthermore, because harvest data are collected continuously and throughout the harvested areas, they offer the promise of more intensive and more representative coverage of harvest areas than aerial surveys do.

To be useful, harvest statistics would need to have a close and known relationship with either population size or harvest rate. We assessed this using long-term (11–22 years) data for three kangaroo species (*Macropus rufus*, *M. giganteus* and *M. fuliginosus*) and common wallaroos (*M. robustus*) across South Australia, New South Wales and Queensland. Regional variation in kangaroo body size, population composition, shooter efficiency and selectivity required separate analyses in different regions. Two approaches were taken. First, monthly harvest statistics were modelled as a function of a number of explanatory variables, including kangaroo density, harvest rate and rainfall. Second, density and harvest rate were modelled as a function of harvest statistics. Both approaches incorporated a correlated error structure. Many but not all regions had relationships with sufficient precision to be useful for indirect monitoring. However, there was no single relationship that could be applied across an entire state or across species. Combined with rainfall-driven population models and applied at a regional level, these relationships could be used to reduce the frequency of aerial surveys without compromising decisions about harvest management.

INTRODUCTION

Kangaroo management in Australia primarily aims to maintain viable populations of the harvested species across their current range, while allowing for commercial exploitation and reductions in populations for damage mitigation (Pople and Grigg 1998). Regulation of the kangaroo harvest is principally through quotas set as proportions of estimates of population size determined mainly by aerial survey (Pople 1999, 2004). These surveys are costly and logistically difficult in remote areas, so it is worth considering alternatives such as monitoring populations indirectly by rainfall and other environmental data, or through harvest statistics. Despite showing promise in studies in the 1980s, rainfall is an inconsistent predictor of kangaroo rate of increase across Australia (Pople 2006; Ch. 18, this volume).

Throughout Australia, government agencies responsible for kangaroo harvest management have routinely collected a number of harvest statistics. These data may provide information about population abundance (and therefore trend) and status. Status includes the desired harvest rate as well as deleterious effects of harvesting, such as reduced body size. From a shooter's point of view, harvest data indicate the quality of the product (e.g. average size or range of sizes). These data provide a more extensive coverage of harvest areas than direct monitoring methods such as aerial survey, and the data are collected continuously as opposed to the infrequent, 'snapshot' estimates of density. Harvest data include sex ratio, carcass weight, skin size and animals shot per unit time (catch-per-unit-effort, CPUE) and are generally recorded for each property where kangaroos are harvested. In the larger states of Western Australia and Queensland, because aerial surveys are expensive and logistically difficult, greater emphasis has been placed on harvest data to monitor populations (Anon. 1984; Southwell 1989). Changes in harvest statistics such as harvest sex ratio and carcass weight can trigger management action such as quota reduction or additional aerial surveys (CALM 2002; QPWS 2003). However, as pointed out by Southwell (1989), there has been no empirical investigation to validate the use of harvest parameters as a monitoring tool.

The use of harvest data to monitor populations is common in wildlife management, particularly in fisheries and small fledgling operations where direct monitoring is either not feasible or not affordable. In fisheries management, CPUE is a notoriously poor index of population size for a variety of reasons, including improved catching efficiency and targeting of fish stock (Hilborn and Walters 1992). Kangaroo harvesting is less prone to some of these problems, so the potential of CPUE and other harvest statistics to monitor kangaroo populations indirectly is worth exploring.

This study sought relationships between harvest statistics and a range of variables, particularly kangaroo density and harvest rate. These relationships are needed by kangaroo managers and others to interpret changes in harvest statistics and use them to predict kangaroo density or harvest rate. The theoretical basis for these relationships and past data analyses are given first, to support the subsequent analyses.

THEORETICAL CONSIDERATIONS

Catch-per-unit-effort

An index I is useful only if the proportionality constant β connecting it to actual population size X ($I = \beta X$) remains constant between locations or over time (Lancia *et al.* 1996; McCallum 2000). For CPUE to monitor population abundance or harvest rate effectively, a number of assumptions must therefore be satisfied (Caughley 1977). Southwell (1989) identified several potential violations of these assumptions when monitoring kangaroo populations.

1 *Harvesting equipment must be standardised.* Unlike fisheries, kangaroo harvesting equipment (vehicles, rifles, spotlight) has changed little over the past 30 years.
2 *Harvesting efficiency must be standardised.* There is variation between shooters (e.g. part-time vs full-time shooters) in the speed in which they can shoot and process animals. This factor will be less pronounced as data are pooled over larger areas.
3 *Harvesting conditions (e.g. weather, access to animals, market prices) must be standardised.* As with any index of population size, CPUE is likely to be more useful in comparing estimates of population abundance over time rather than between areas (McCallum 2000). Regional differences in vegetation, road network and distance between towns invariably leads to regional differences in CPUE. Shooter success also varies with weather conditions. Prices paid to shooters for carcasses and skins also vary over time, as do fuel and other overhead costs for shooters.
4 *Harvesting of each animal is independent.* The relationship between CPUE and population abundance is unlikely to be linear because of time saturation (Caughley 1977), whereby CPUE even-

tually reaches a plateau at some level of population density because the capacity for harvesting reaches a limit. Changes in abundance in this zone of saturation will not be reflected in the CPUE.

5 *Kangaroos do not learn to avoid shooters.* Kangaroo behavioural response to shooters is unlikely to change from year-to-year and so is unlikely to be a problem for monitoring year-to-year changes in abundance.

Harvest composition

These assumptions must hold if attributes of harvest composition (e.g. sex ratio, carcass weight) are to be useful indices of population size or status. Harvest composition will change in quite different ways from CPUE and so it warrants separate discussion. Changes in the composition of the harvest will reflect one or more of three factors:

1 changes in shooter selectivity due to the environment, prices or population density;
2 changes in population composition due to harvesting;
3 changes in population composition due to the environment.

Kangaroos show considerable sexual size dimorphism, with males growing to two to three times the maximum size of females (Jarman 1989). Growth, particularly for males, extends well into adulthood and beyond the age of sexual maturity. Shooters are generally paid by carcass weight (although some shooting for human consumption will be selective for intermediate sizes) or skin size so, not surprisingly, there is selection for larger, older, predominantly male kangaroos (Pople 1996). As population density declines, the rationale is that shooters will not be able to maintain this selectivity and will therefore take fewer males, smaller animals and, overall, fewer animals per unit time. Similarly, as harvest rate increases, the age structure and sex ratio of the population will change. The sex ratio of the harvest should therefore shift towards females and average carcass weight should decline.

However, the composition of the population (size structure and therefore age structure and sex ratio) will also vary in response to the environment regardless of the rate at which it is being harvested. It is well documented that kangaroo populations in arid areas have unstable age distributions, a consequence of pulses of recruitment during good seasons and no recruitment during drought (Frith and Sharman 1964; Kirkpatrick and McEvoy 1966; Newsome 1966, 1977). This is compounded by male-biased mortality during drought (Pople and Cairns 1995; Pople 1996), a mortality that may be also biased towards older animals (Robertson 1986).

Regional variation in population composition is only one factor that leads to differences in harvest composition between areas. Adult sex ratios have been determined in a number of harvested red kangaroo (*Macropus rufus*) populations throughout eastern Australia (Pople and Cairns 1995; A.R. Pople and S.C. Cairns, unpubl. data) and can be compared with the sex ratio of the harvest from the same population (Fig. 32.1). The population sex ratio is restricted to animals older than two years, the age when red kangaroos are large enough to be harvested, which has been recorded in samples of harvest age structure (Wilson 1975; Pople 1996) and reflected in the minimum carcass weights that are often determined by the kangaroo industry. The comparison shows considerable variation in the strength of selection for males, indicated by departures from the line of no selection shown in Figure 32.1.

Shooter selectivity is likely to be influenced by environmental factors affecting shooter access, population density of all kangaroo species, harvesting

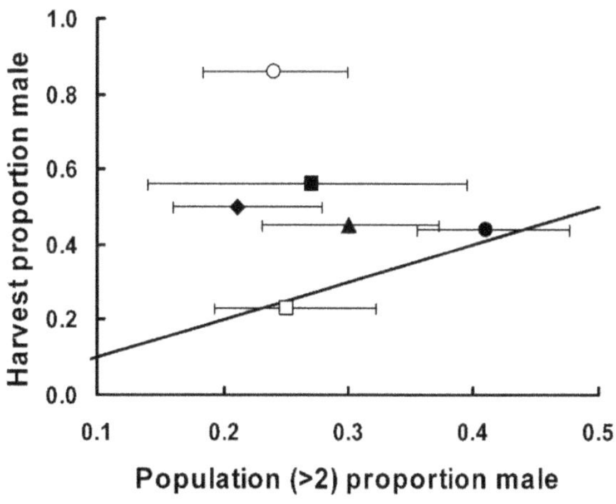

Figure 32.1: Relationship between adult (>2 years old) sex ratio (± 95% bootstrapped confidence intervals) and harvest sex ratio in six populations of red kangaroos (*M. rufus*) (Pople and Cairns 1995; A.R. Pople and S.C. Cairns, University of New England, Armidale, unpubl. data). Populations were located in far northern Queensland (closed square), northern Queensland (closed triangle), central Queensland (closed diamond), southern Queensland (open square), eastern South Australia (open circle) and western South Australia (closed circle).

costs and market prices. There will also be individual variation among shooters. Finally, there are regional differences in kangaroo morphology, particularly body weight and overall size (Yom-Tov and Nix 1986; Pople 1996), which will be reflected in the composition of the harvest.

One assumption common to the ability of all harvest statistics to monitor population size or harvest rate is that the harvest in a region is a representative sample of all harvestable animals in the region. This is likely to be violated if shooters only access one part of a region. Data could be collected from a heavily harvested population and so not reflect the dynamics of the broader population that is largely unharvested.

Empirical support

Until recently, there has been limited analysis of kangaroo harvest data. Analyses of historical data, prior to the imposition or influence of harvest quotas, focused on the factors determining harvest offtake. Kirkpatrick and Amos (1985) argued that both market demand and prevailing weather conditions determined the size of the kangaroo harvest in Queensland. Prince (1984b) found that a large component of the variation in the historical offtake of kangaroos in Western Australia could be explained by previous rainfall, and that market fluctuations accounted for only a minor component. Prince (1984a, b) suggested that offtake was impaired by rainfall in the short term because of its influence on access to animals, and increased with rainfall in the long-term because of resultant increases in population size.

Regional differences and temporal variation in harvest sex ratio and carcass weight in Western Australia were reported by Prince (1984a), who speculated on the factors responsible. Population modelling by Nance (1985) suggested that harvest sex ratio could effectively monitor the status of an eastern grey kangaroo (*Macropus giganteus*) population. This modelling, supplemented by samples of population composition from harvested and unharvested populations, supported the management program for kangaroos in Queensland prior to the 1990s (Kirkpatrick and Amos 1985). Under that program, harvest statistics were used to monitor the status of kangaroo populations and determine the size of harvest quotas (Anon. 1984; QNPWS 1989). Nance's (1985) model made a number of unsupported assumptions about regulation in kangaroo populations and shooter behaviour. Nevertheless, it hypothesised that sex ratio can monitor harvest rate, which can be assessed in studies such as the present one. Notably, all these previous studies lacked data on population abundance to which to relate harvest statistics.

With the benefit of estimates of population size, Pople (1996) examined the relationship between harvest statistics and harvest rate across several regions in Queensland for two years in the early 1990s. For red kangaroos, harvesting appeared to lead to a decrease in the size of male skins with an increase in harvest rate. However, the effect of drought on the population sex ratio was thought to have overshadowed any relationship between harvest sex ratio and harvest rate. For eastern grey kangaroos, carcass weight, skin size and harvest sex ratio all appeared to be potential indicators of harvest rate. The results supported Nance's (1985) modelling and refuted the argument of de la Mare (1988) that shooter selectivity when taking skins would be too weak to adequately monitor the population's status. However, the suggestion that Nance's (1985) model could be extended to other species (Kirkpatrick and Nance 1985; QNPWS 1989), particularly red kangaroos, was not supported (Pople 1996).

METHODS

Study area

The study area encompassed the sheep rangelands of eastern Australia in which kangaroo density has been estimated by annual aerial survey since the late 1970s (Figs 32.2, 32.3). This coincides with the main harvest areas in each of the three states covered by this area. The separate management regions within New South Wales and South Australia were the units of analyses. In Queensland, the 10 helicopter survey blocks (Fig. 32.2) were the units of analyses. Separate quotas are set within the core survey area in Queensland and areas outside it (QPWS 2003). Variation between regions in the value of harvest statistics and their relationship with density, harvest rate and environmental variables was expected as explained above, thus separate analyses were undertaken for each region and species. This is also consistent with the scale of management. The primary interest was temporal variation within each region.

Analysis

Two approaches were taken in the analysis. First, harvest statistics were used as response variables and

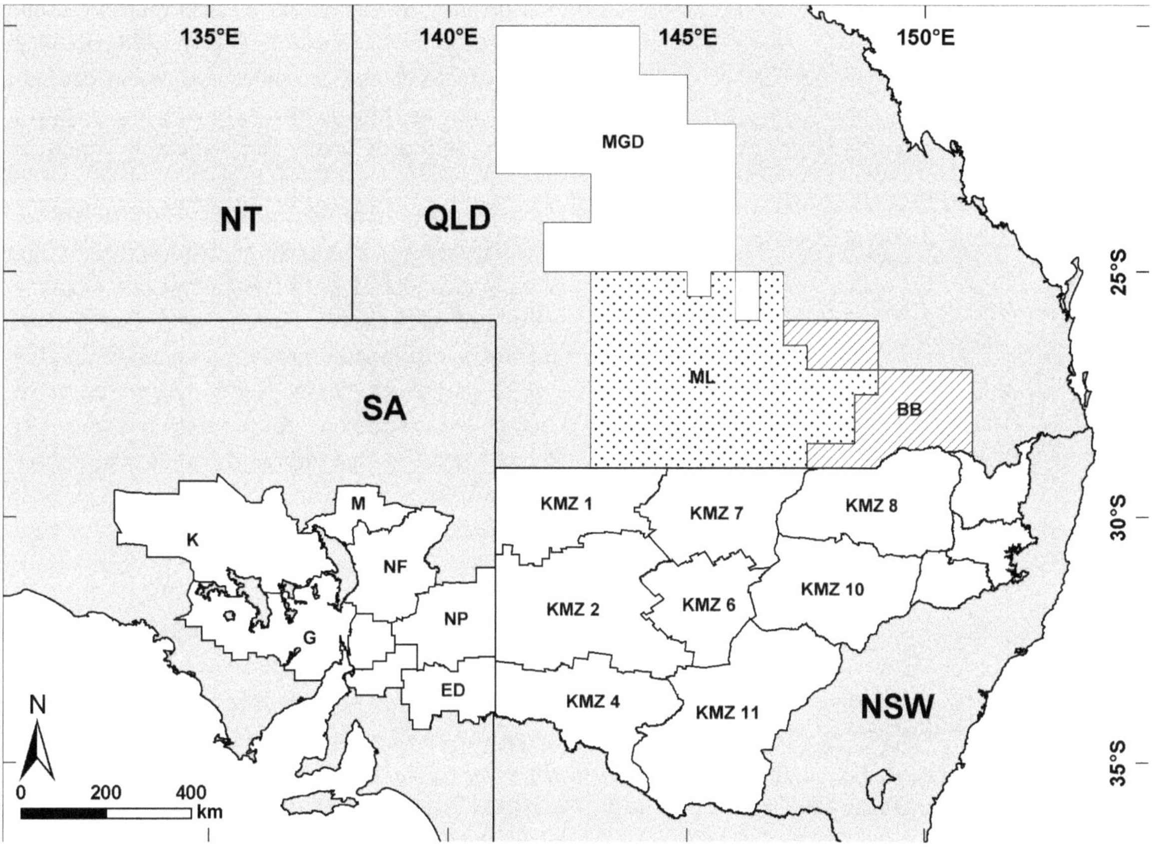

Figure 32.2: The survey area within the sheep rangelands of eastern Australia. This area has been broken up into regions in which kangaroo harvests are managed in New South Wales and South Australia. Biogeographic boundaries (Environment Australia 2000) adjusted to coincide with ½° latitude and longitude lines are shown in the Queensland core survey area. Unlabelled regions in southern South Australia had estimates of density that were considered too imprecise for analysis. Similarly, unlabelled regions in eastern New South Wales were not examined because they lack long-term aerial survey data. Salt lakes dissect the study area in South Australia. K = Kingoonya; M = Marree; NF = North Flinders Ranges; NP = North-east Pastoral; ED = Eastern Districts; G = Gawler; BB = Brigalow Belt; ML = Mulga Lands; MGD = Mitchell Grass Downs.

modelled as a function of a number of explanatory variables including kangaroo density, harvest rate and rainfall. This follows current management practice involving monitoring harvest statistics to infer changes in kangaroo density or harvest rate (e.g. QPWS 2003). This approach should also be useful to the kangaroo industry as it provides an understanding of product quality and quantity, such as carcass weight and catch rates. The second approach was to use density and harvest rate as response variables and use harvest statistics as the explanatory variables. This would allow managers to predict density or harvest rate to set harvest quotas in years when direct monitoring such as aerial survey is not conducted. The first approach allows greatest use of the monthly harvest data, while the second approach is restricted to the smaller sample size of annual estimates of population size and harvest rate and density.

Monthly harvest data were available for South Australia for 1980–2001, for New South Wales for 1993–2003 and for Queensland for 1991–2003. Three harvest statistics were used in the analyses: the percentage of the harvest comprising males (*%male*), CPUE and average male carcass weight (*male wt*). Skin size was not examined, because skin-only shooting is currently restricted to Queensland within the study area and is declining there as a proportion of the overall harvest. In Queensland, harvest sex ratio was restricted to carcasses as proportionally more females are taken as skins (Pople 1996, 2006). CPUE was calculated as the harvest in a region divided by the number of shooter nights. The structure of the harvest database for South Australia meant that effort was calculated as the number of shooter nights when males were harvested. This was determined separately for each species. In New South

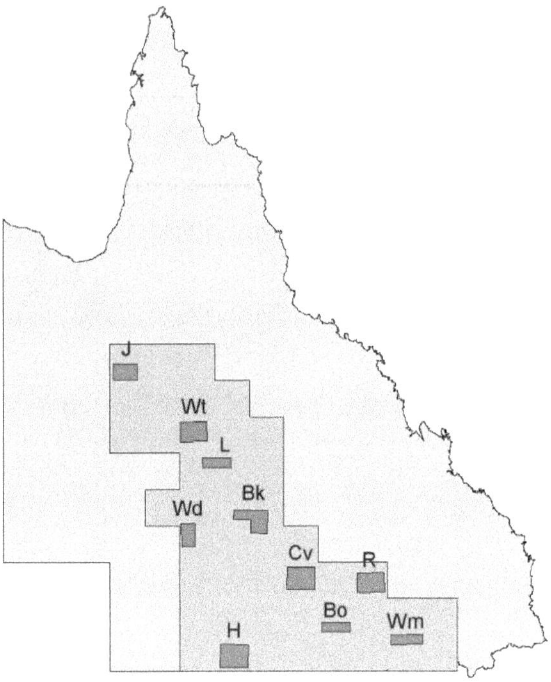

Figure 32.3: Helicopter survey blocks (black) used to monitor kangaroo populations in the core survey area (dark grey) in Queensland since 1992. J = Julia Creek; Wt = Winton; L = Longreach; Wd = Windorah; Bk = Blackall; Cv = Charleville; H = Hungerford; Bo = Bollon; R = Roma; Wm = Westmar.

Wales, CPUE was only available quarterly and only from 1997, and so was not incorporated into analyses. In Queensland, effort was calculated as the number of shooter nights regardless of species harvested. Male rather than female weight was considered a more sensitive indicator of changes in density and harvest rate as it spans a greater range, asymptotes at an older age and males are generally targeted by shooters. Season closures because quotas had been reached were uncommon, but occurred for up to six months for red kangaroos and five months for common wallaroos (*Macropus robustus*) in Queensland in the early 1990s, but only briefly for one month for eastern grey kangaroos. There were brief season closures of one to three months for each species in New South Wales and no closures in South Australia over the study period.

Explanatory variables used to predict harvest statistics included kangaroo density and density squared to allow for a saturation effect, instantaneous harvest rate, effort (but not with CPUE), seasonality (comprising $sin(2 \times \pi \times time)$ and $cos(2 \times \pi \times time)$ terms), dummy variables to account for harvest seasons and regulatory changes, and a number of rainfall terms. Seasonality was introduced into the models because harvesting effort and offtake (e.g. shooting occurs at night which is longer in winter, seasonal rainfall can limit access to animals which can be less aggregated), carcass and skin value and kangaroo condition, among other influences on harvest statistics, can all be expected to vary within years. Harvest rate was calculated as the annual harvest divided by the winter density estimate. Density and harvest rate were only estimated once a year and so were interpolated, assuming exponential growth, to provide monthly values. Rain that fell in the month of harvest data collection was expected to reduce access and selectivity, thereby resulting in a drop in the value of all harvest statistics. Rain falling in the three months prior to collection was also expected to have a negative effect on CPUE and *%male* by increasing food supply, resulting in a more even dispersion of kangaroos that make less use of open accessible areas (Newsome 1965; Pople *et al.* 2007). Rain falling in the 12 months lagged one month prior to collection was expected to improve kangaroo body condition, thereby increasing *male wt.* Shooters appear to be less size-selective when taking only skins (Pople 1996) and can potentially harvest more animals per night. In Queensland, the proportion of the harvest comprising skins was included as a variable as this may have a knock-on effect on the composition of the carcass harvest and CPUE. In New South Wales, a dummy variable was included to account for skin-only shooting, which occurred up to mid 1996. The harvest of each species is likely to be influenced by availability of other species to harvest (Pople 1996). This is manifest as spatial variation in the harvest. However, temporal variation in the harvest is likely to be influenced by season closures for other species, so the closure of the red kangaroo season in Queensland over the study period was included as a dummy variable in the analyses for eastern grey kangaroos and wallaroos. Other closures were not considered to be influential. In South Australia, carcasses were dressed to a smaller weight prior to 1997. Furthermore, an initial examination of the data revealed no differences between male and female carcass weights up to 1987. Dummy variables were therefore included to account for these changes. A likely influence that was not incorporated in these models was economics, such as prices paid for carcasses, which were not available.

Serial correlation in the data was expected and a correlated regression structure (first-order autoregressive, AR1) was included in the regression models using generalised least squares (gls) in R 1.9.1 (R Development Core Team 2003). Models were simplified by stepwise removal of non-significant (P > 0.05) terms from the full model (Crawley 2002). Finally, the most parsimonious model was compared with the equivalent model without the correlated error structure, using a likelihood ratio test (Maindonald and Braun 2003). *%male* could have been modelled using logistic regression, but was instead arcsine transformed and analysed in a linear regression to take advantage of the correlated error structure in gls. Only monthly harvests of >50 animals of both sexes combined were used in analyses.

Explanatory variables used to predict population statistics (density and harvest rate) were restricted to CPUE, *%male* and *male wt*, because sample size was limited to annual estimates of density and harvest rate, unlike the monthly harvest statistics. Harvest statistics were smoothed using a three-point moving average centred on the winter aerial survey estimate of density. This was done to avoid using an anomalous value for each harvest statistic that may have coincided with each winter aerial survey. Data were also analysed using multiple regression with a correlated error structure if needed. Kangaroo density was log transformed to improve linearity. The purpose of the analyses with harvest statistics as response variables was to identify their likely determinants. For the analyses with population statistics as response variables, the purpose was prediction. 95% prediction intervals (Maindonald and Braun 2003) were therefore determined for fitted values. Approximate precision (SE/mean, Andrew and Mapstone 1987) of predictions of density and harvest rate using these relationships were calculated as a quarter of the size of the prediction interval (*95%PI*) divided by the intercept.

All minimal adequate models (Crawley 2002) were subject to routine checks (Crawley 2002; Maindonald and Braun 2003) of the adequacy of model fit to the data. This included assessment of outliers and multicollinearity.

RESULTS

South Australia

Standardised parameter estimates for models predicting harvest and population statistics in South Australia are shown in Table 32.1. The ability to model fluctuations in harvest statistics varied greatly among soil conservation boards (SCBs). The apparent good fit for *male wt* is largely influenced by the change in average weight due to the change in carcass dressing being explained by the dummy variable. Among SCBs and species, it was often the model with the lower R^2 that provided the more impressive and useful fit to the data, particularly if it captured declines in harvest statistics during drought and large increases. Adjusted R^2 cannot be compared directly between regions as the range of explanatory variables differs. Examples of model fit are shown in Figures 32.4a and 32.5a. In the North-east Pastoral SCB, trends in all three harvest statistics were tracked reasonably well by the fitted models. In particular, the decline in red kangaroo numbers during the 1982–83 drought was accompanied by a sharp decline in the proportion of males in the harvest. For western grey kangaroos (*Macropus fuliginosus*) in Gawler SCB, the relationships were poorer with much greater scatter and lack of trend in all three statistics.

Kangaroo density and harvest rate were generally important influences on harvest statistics for both species. As expected, higher densities were usually associated with a higher proportion of males, higher catch rates and greater weights. Where the coefficient for density was negative, there was usually a curvilinear term indicating an inverted parabola. A negative coefficient alone is counterintuitive, but was recorded for western grey kangaroos in North-east Pastoral SCB where their densities are relatively low and shooters are targeting the more abundant red kangaroos. It is important to be cautious about interpreting regression coefficients. A coefficient with an unexpected sign may reflect the overriding influence of another explanatory variable (Maindonald and Braun 2003). Density had a saturating effect in a number of cases.

Higher harvest rates were generally associated with lower values of *%male* and *male wt* and higher catch rates, again in line with expectations. Rainfall had a smaller influence on harvest statistics as gauged by the size of the standardised coefficients and the few models where terms were included. Three months rainfall was not incorporated in any of the models. The effect of recent rainfall was negative, while 12 months rainfall had a positive effect on male carcass weight. Effort also had a minor positive influence on red kangaroo *%male* in North-east Pastoral SCB and male red kangaroo carcass weight in

Table 32.1: Standardised parameter estimates with standard errors (in brackets) and adjusted R^2 for the most parsimonious (according to backwards elimination) regression models predicting (a) red kangaroo (*M. rufus*) harvest statistics, (b) western grey kangaroo (*M. fuliginosus*) harvest statistics, (c) red kangaroo population statistics and (d) western grey kangaroo population statistics.

(a) Red kangaroos (South Australia)

			Explanatory variables								
SCB	**Response**	**Intercept**	**D**	***D*²**	***HR***	***rn*.1.0**	***rn*.12.1**	**Effort**	**Carcass**	**Cycle**	**Adj. *R*²**
NP	*%male*	91.64 (1.60)	5.19 (1.57)		–4.86 (1.52)			1.07 (0.52)			0.49
	CPUE	28.01 (081)	2.94 (0.75)		2.80 (0.81)	–1.00 (0.22)			4.72 (1.87)	0.07 (0.46) 0.94 (0.46)	0.52
	male wt	17.71 (0.40)	–2.93 (1.28)	3.12 (1.22)	1.02 (0.23)		0.54 (0.14)		1.05 (0.58) 7.19 (0.66)	–0.06 (0.12) 0.32 (0.12)	0.77
NF	*%male*	83.60 (0.53)	2.07 (0.55)		–2.36 (0.55)						0.32
	CPUE	23.97 (0.42)	8.20 (1.99)	–6.11 (1.97)	3.23 (0.39)				3.40 (0.96)		0.51
	male wt	18.04 (0.34)			–0.82 (0.21)				0.73 (0.45) 7.77 (0.54)		0.83
M	*%male*	78.29 (1.46)			–4.99 (1.34)	–1.24 (0.45)					0.14
	CPUE	35.05 (1.59)				–1.17 (0.41)					0.00
	male wt	17.73 (0.50)	0.74 (0.29)						1.38 (0.64) 13.28 (0.80)		0.82
K	*%male*	80.61 (0.92)									–
	CPUE	30.86 (0.82)	18.20 (5.59)	–13.59 (5.51)							0.41
	male wt	18.07 (0.32)					0.33 (0.13)	0.13 (0.07)	0.42 (0.42) 8.19 (0.53)	–0.28 (0.11) 0.21 (0.11)	0.87
G	*%male*	86.86 (0.97)	17.89 (7.45)	–17.86 (7.45)						–2.50 (0.76) 0.55 (0.75)	0.10
	CPUE	25.01 (0.88)	21.51 (6.18)	–17.43 (6.14)	6.85 (0.87)				–9.49 (2.07)		0.51
	male wt	17.84 (0.22)							0.41 (0.29) 8.11 (0.35)	–0.35 (0.11) 0.21 (0.11)	0.85
ED	*%male*	91.50 (1.28)							–17.87 (2.79)		0.24
	CPUE	15.83 (0.34)	5.40 (1.73)	–4.20 (1.73)						0.97 (0.33) –0.06 (0.33)	0.13
	male wt	19.05 (0.54)	3.99 (1.09)	–3.33 (0.94)	–1.00 (0.23)	–0.24 (0.12)			–0.97 (0.85) 5.03 (0.73)	–0.30 (0.15) 0.02 (0.15)	0.54

(b) Western grey kangaroos (South Australia)

SCB	Response	Intercept	Explanatory variables								Adj. R^2
			D	D^2	*HR*	*rn*.1.0	*rn*.12.1	Effort	Carcass	Cycle	
NP	*%male*	107.1 (1.50)	–5.29 (1.48)		–6.50 (1.38)				14.03 (3.53)		0.34
	CPUE	8.09 (0.25)	1.30 (0.23)						2.31 (0.57)		0.37
	male wt	19.33 (0.42)			–1.39 (0.26)				0.43 (0.58) 3.42 (0.66)	–0.38 (0.19) 0.08 (0.18)	0.46
NF	*%male*	101.9 (1.70)								–4.12 (1.63) –0.24 (1.63)	0.03
	CPUE	7.14 (0.21)	3.27 (0.92)	–2.12 (0.89)	0.71 (0.25)					0.46 (0.20) 0.58 (0.20)	0.22
	male wt	18.72 (0.34)					0.47 (0.18)		0.81 (0.44) 4.85 (0.48)	–0.78 (0.18) 0.25 (0.17)	0.52
K	*%male*	101.2 (3.65)									–
	CPUE	9.12 (1.21)	2.39 (1.13)						–6.02 (2.35)	–1.61 (0.69) 1.53 (0.71)	0.13
	male wt	17.26 (0.81)	–1.59 (1.36)	2.75 (1.24)			0.78 (0.34)		2.05 (0.95) 7.02 (1.29)		0.59
G	*%male*	89.91 (0.77)							–5.38 (1.69)	–3.65 (0.56) 1.26 (0.56)	0.30
	CPUE	19.83 (0.63)			–1.73 (0.60)					–1.17 (0.52) –0.86 (0.52)	0.12
	male wt	17.77 (0.31)			–0.55 (0.19)				0.47 (0.44) 6.40 (0.49)	–0.52 (0.13) 0.33 (0.13)	0.78
ED	*%male*	95.69 (1.81)			–5.99 (1.80)						0.12
	CPUE	13.42 (0.36)	0.99 (0.36)		3.18 (0.36)					–0.41 (0.35) 0.71 (0.34)	0.37
	male wt	18.33 (0.48)	2.93 (1.06)	–2.71 (1.04)	–0.85 (0.32)				1.18 (0.57) 5.18 (0.92)		0.44

Table 32.1: (Continued)

(c) Red kangaroos (South Australia)

			Explanatory variables			
SCB	**Response**	**Intercept**	**CPUE**	***pmale***	**Adj. R^2**	***95%PI***
NP	log*D*	2.5 (0.05)	0.15 (0.05)	0.21 (0.05)	0.55	2.54
	HR%	12.82 ((0.70)	2.41 (0.70)	–2.21 (0.70)	0.49	14.8
NF	log*D*	2.13 (0.05)	0.13 (0.06)	0.21 (0.06)	0.36	2.26
	HR%	12.36 (0.87)	2.96 (0.89)		0.33	17.6
M	log*D*	1.58 (0.10)		0.32 (0.11)	0.28	2.57
	HR%	19.46 (2.15)		–4.76 (2.20)	0.15	40.9
K	log*D*	1.65 (0.07)			–	–
	HR%	11.23 (0.71)	1.71 (0.73)		0.17	14.6
G	log*D*	1.03 (0.06)	0.03 (0.01)		–	–
	HR%	11.41 (0.89)	2.04 (0.91)		0.16	18.1
ED	log*D*	0.76 (0.13)			–	–
	HR%	13.32 (0.19)			–	–

(d) Western grey kangaroos (South Australia)

			Explanatory variables			
SCB	**Response**	**Intercept**	**CPUE**	***pmale***	**Adj. R^2**	***95%PI***
NP	log*D*	0.23 (0.22)			–	–
	HR%	16.09 (3.92)	–14.05 (5.53)	–13.73 (5.53)	0.21	57.1
NF	log*D*	–1.14 (0.33)			–	–
	HR%	51.8 (30.1)			–	–
K	log*D*	–2.65 (0.37)			–	–
	HR%	8.89 (3.41)			–	–
G	log*D*	1.83 (0.08)			–	–
	HR%	4.45 (0.38)	–0.93 (0.39)		0.18	7.6
ED	log*D*	–1.91 (0.39)	0.03 (0.01)	3.08 (0.34)	0.81	0.79
	HR%	8.05 (1.05)	3.91 (1.08)		0.40	17.5

K = Kingoonya; M = Marree; NF = North Flinders Ranges; NP = North-east Pastoral; ED = Eastern Districts; G = Gawler.

The size of the 95% prediction interval (*95%PI*) averaged across all data points is also shown for models predicting population statistics. Models are shown for soil conservation boards (SCBs) in the South Australian pastoral zone (Fig. 32.1). Population statistics are log transformed kangaroo density (log*D*) and harvest rate expressed as a percentage (*HR%*). Harvest statistics are catch-per-unit-effort (CPUE), the percentage of the harvest comprising males (*%male*, arcsine transformed) and average male carcass weight (*male wt*). Explanatory variables include untransformed density (*D*), harvest rate expressed as a proportion (*HR*), CPUE, *male wt*, the proportion of the harvest comprising males (*pmale*, arcsine transformed), rain falling in the same month as collection of the harvest statistic (*rn*.1.0), three months rain lagged one month prior to collection (*rn*.3.1), 12 months rain lagged one month prior to collection (*rn*.12.1), the number of shooter-nights (*effort*), seasonal *cycle* comprising *sin*(2 × π × *time*) and *cos*(2 × π × *time*) terms, and two or three dummy variables (*carcass*) representing periods when carcasses were dressed differently (*male wt* and CPUE) and when carcass weight was not distinguished between the sexes (*male wt*). The parameter for *sin*(2 × π × *time*) is given above the parameter for *cos*(2 × π × *time*).

Kingoonya SCB. The dummy variable for the change in carcass dressing was incorporated in many models for %*male* and CPUE, but the sign was not always consistent. Seasonal cycles were incorporated into many models, but these were not strong. There was no consistent cyclical pattern within years.

Predictions of population statistics were best in the SCBs with higher kangaroo density, which are in the north-east for red kangaroos and the south for western grey kangaroos. For red kangaroos these estimates of precision of predictions ranged 0.25–0.53 and for western grey kangaroos they ranged 0.43–0.89.

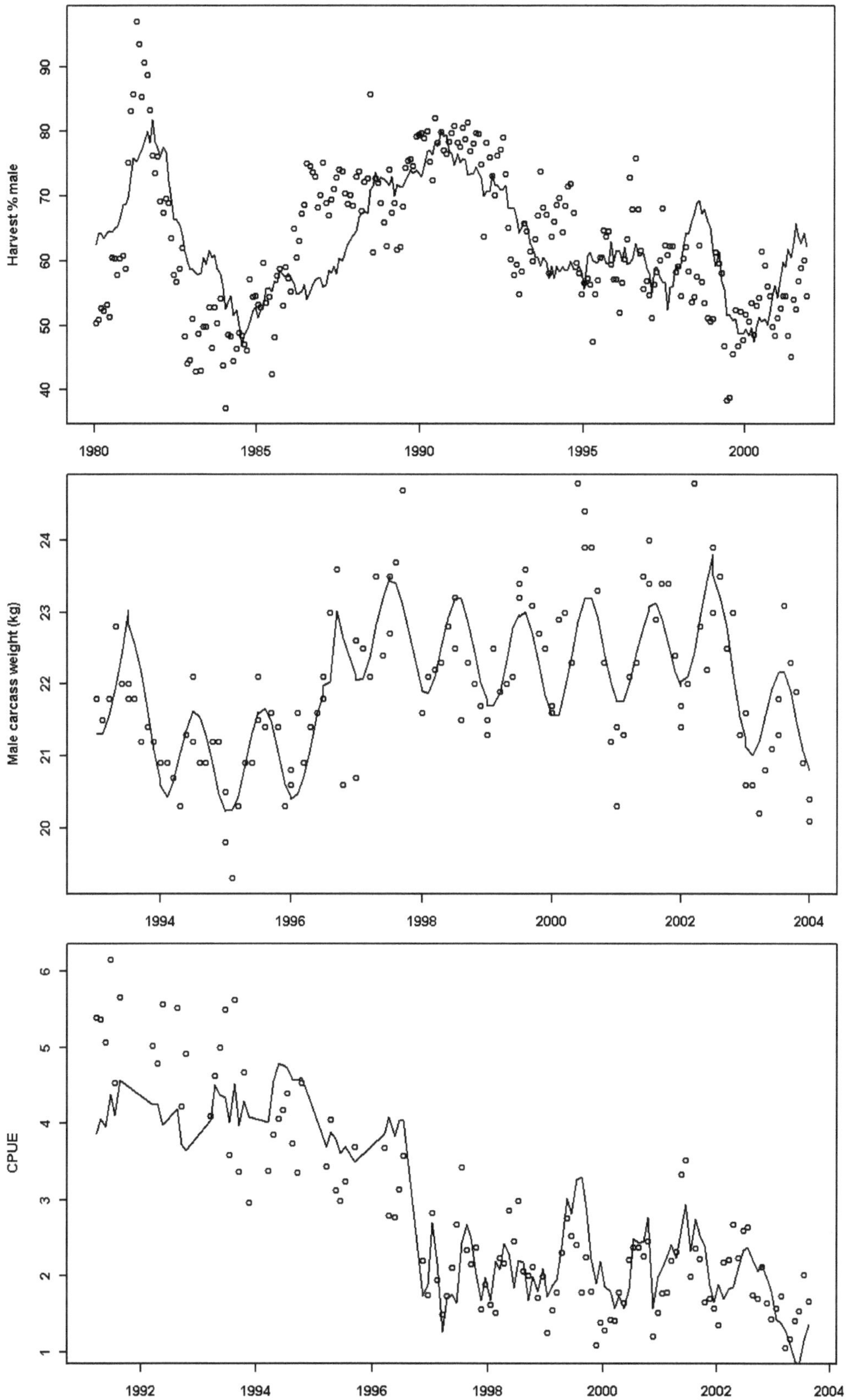

Figure 32.4: Monthly (a) *%male* harvested red kangaroos (*M. rufus*) in North-east Pastoral SCB in South Australia, (b) average male carcass weight for eastern grey kangaroos (*M. giganteus*) in KMZ8 in New South Wales and (c) CPUE for red kangaroos at Blackall in Queensland. Fitted models (Tables 32.1, 32.2 and 32.3) are shown as solid lines.

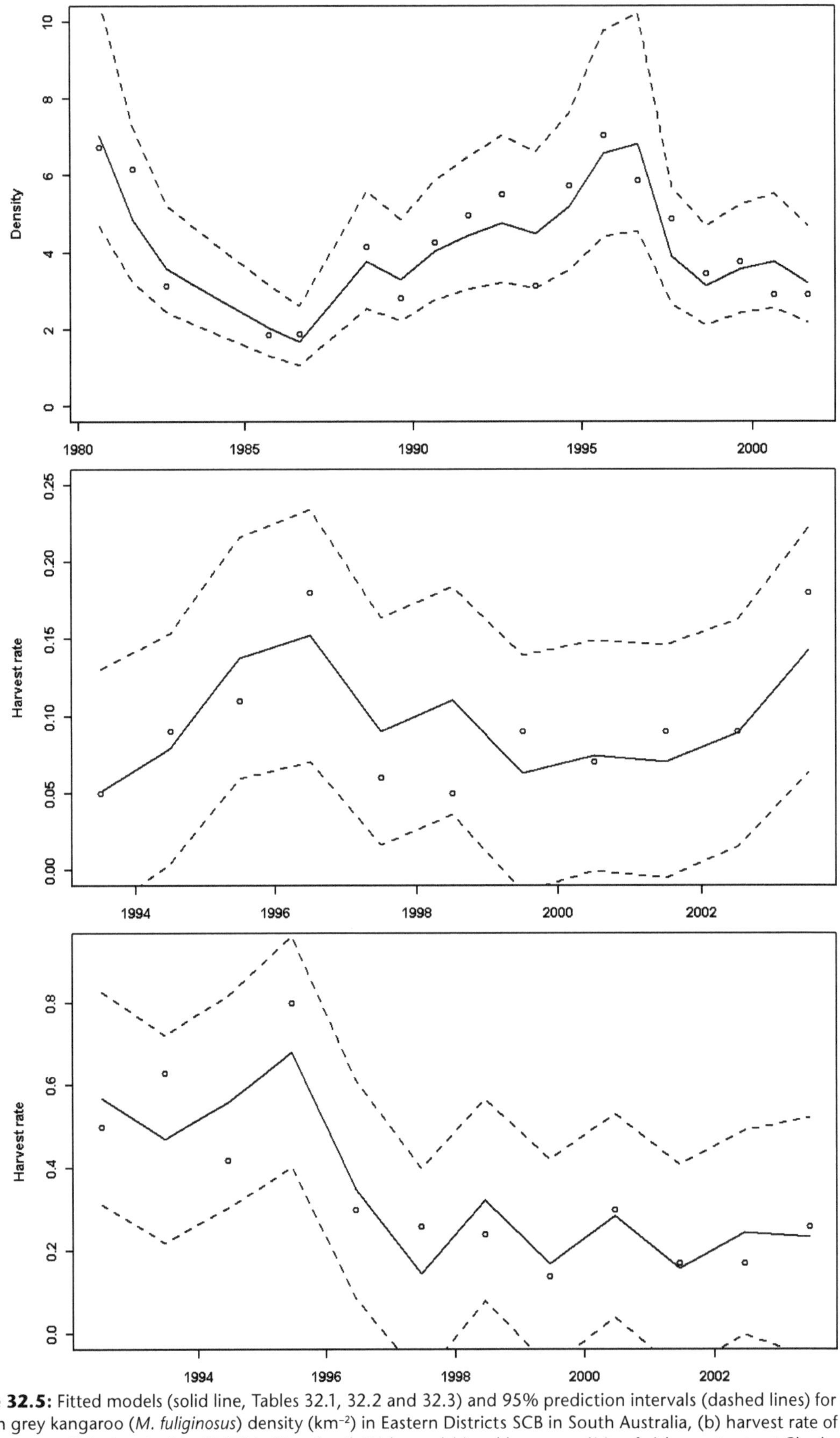

Figure 32.5: Fitted models (solid line, Tables 32.1, 32.2 and 32.3) and 95% prediction intervals (dashed lines) for (a) western grey kangaroo (*M. fuliginosus*) density (km^{-2}) in Eastern Districts SCB in South Australia, (b) harvest rate of eastern grey kangaroos (*M. giganteus*) in KMZ8 in New South Wales and (c) red kangaroo (*M. rufus*) harvest rate at Charleville in Queensland.

New South Wales

Standardised parameter estimates for models predicting harvest and population statistics in New South Wales are shown in Table 32.2. Striking features of the data are the decline in male weight during the 2002–03 drought and drops in *%male* at different times during the time series depending on the zone and species. Examples of model fit are shown in Figures 32.4b and 32.5b

As with South Australia, kangaroo density and harvest rate were generally important influences on harvest statistics for all three species. Recent rainfall entered into the models in some zones, but only as a minor influence. Inclusion of 12 months rainfall as an explanatory variable allowed the decline in carcass weight during the recent drought to be modelled. However, it was included in only seven of 23 cases. The dummy variable for skin shooting was incorporated in many models for *%male*, but not always as an expected negative effect. Seasonal cycles were incorporated into many models and, while more marked than in South Australia, their overall effect was small. For red kangaroos in kangaroo management zone 1, more males tended to be taken in winter. For eastern grey kangaroos in the higher-density zones, average male carcass weight tended to be higher in winter.

Predictions of red kangaroo population statistics were only possible for density and only in three zones. Models could be fitted for both statistics for eastern grey kangaroos, but again only in three zones. Western grey kangaroos fared better, with models fitted in five of seven zones. Approximate precision of predictions of density and harvest rate ranged 0.27–0.28 for red kangaroos, 0.18–0.43 for eastern grey kangaroos and 0.24–0.86 for western grey kangaroos.

Queensland

Standardised parameter estimates for models predicting harvest and population statistics in Queensland are shown in Table 32.3. There were considerable fluctuations in all three harvest statistics over the 13-year study period in Queensland, coinciding with a period of drought in the early 1990s and good rainfall in the late 1990s followed by dry seasons. Examples of model fit are shown in Figures 32.4c and 32.5c.

Kangaroo density and harvest rate were important influences on harvest statistics for all three species. Recent rainfall entered into the models for eastern grey kangaroos and wallaroos as minor negative influences in some blocks, but not for red kangaroos. Twelve months rainfall was a predictor for red kangaroo carcass weight in half the blocks, but only in three cases for the other species. The number of shooter nights (i.e. effort) influenced harvest composition in some cases, but the sign of the coefficient varied. This is likely to reflect variation in selectivity among shooters. Skin-only harvesting generally had a positive influence on *%male* and *male wt*, indicating that smaller and often female animals were being taken for skins. There are likely to be economic reasons for this, that are not explored here. *pskin* also had a positive effect on CPUE indicating that skin-only shooting allows higher catch rates. The dummy variable for the harvest season for red kangaroos (*red.hs*) had a strong negative coefficient for CPUE for eastern grey kangaroos and wallaroos in most blocks. The dummy variable generally had a positive coefficient for *%male* and *male wt*, suggesting that shooters were more selective when three rather than two species could be shot. The blocks where *red.hs* was not important had low densities of eastern grey kangaroos or wallaroos. Seasonal cycles were incorporated into many models, being more marked than in the southern states. For example, average male carcass weights of eastern grey kangaroos at Blackall were consistently higher in winter. CPUE was also higher in summer for eastern grey kangaroos in a number of blocks, particularly in the early 1990s, which coincided with the closure of the red kangaroo season. In the Charleville block, CPUE for red kangaroos tended to be higher in the winter months.

Predictive models for population statistics were possible in the higher-density blocks for all species. Approximate precision of predictions of density and harvest rate ranged 0.26–0.74 for red kangaroos, 0.27–0.79 for eastern grey kangaroos and 0.33–0.59 for wallaroos.

For all models in Tables 32.1–32.3, diagnostic plots suggested model fit was adequate and there was no evidence of multicollinearity. Likelihood ratio tests supported a correlated error structure for all models predicting harvest statistics. However, correlated error structures did not improve any models predicting population statistics.

DISCUSSION

Until now, the long-held view that kangaroo harvest statistics tracked changes in kangaroo density and harvest rate had not been challenged with data. The results now need to be translated into management

Table 32.2: Standardised parameter estimates with standard errors (in brackets) and adjusted R^2 for the most parsimonious (according to backwards elimination) regression models predicting (a) red kangaroo (*M. rufus*) harvest statistics, (b) eastern grey kangaroo (*M. giganteus*) harvest statistics, (c) western grey kangaroo (*M. fuliginosus*) harvest statistics, (d) red kangaroo population statistics, (e) eastern grey kangaroo population statistics and (f) western grey kangaroo population statistics.

(a) Red kangaroos (New South Wales)

			Explanatory variables								
KMZ	**Response**	**Intercept**	**D**	***D*²**	***HR***	***rn*.1.0**	***rn*.3.1**	***rn*.12.1**	**Skin**	**Cycle**	**Adj. *R*²**
1	*%male*	92.40 (0.73)	10.61 (3.16)	–10.49 (3.16)					–8.46 (1.38)	–0.87 (0.55) –1.38 (0.56)	0.33
	male wt	22.65 (0.27)	1.19 (0.27)								0.32
2	*%male*	98.45 (0.75)	2.63 (0.61)						–7.36 (1.30)		0.45
	male wt	21.50 (0.17)	0.54 (0.17)								0.25
4	*%male*	86.18 (0.90)							3.24 (1.48)		0.13
	male wt	21.24 (0.17)	–0.32 (0.20)		–0.49 (0.19)			0.55 (0.16)			0.45
6	*%male*	110.01 (2.2)									–
	male wt	21.44 (0.11)	2.21 (0.82)	–1.83 (0.82)		–0.18 (0.06)					0.29
7	*%male*	115.88 (1.0)	11.64 (3.82)	–9.17 (3.79)		–1.87 (0.50)	–1.71 (0.67)		–18.75 (1.75)		0.75
	male wt	22.17 (0.25)						0.66 (0.18)	–1.07 (0.42)	–0.15 (0.14) –0.45 (0.15)	0.45
8	*%male*	108.75 (1.4)	17.12 (4.87)	–14.30 (4.66)					6.47 (2.59)	–0.80 (0.72) –1.83 (0.80)	0.35
	male wt	23.47 (0.49)								–0.32 (0.15) –0.36 (0.17)	0.02
10	*%male*	117.85 (0.9)	14.53 (4.34)	–9.19 (4.29)	5.98 (1.08)						0.45
	male wt	22.49 (0.40)									–
11	*%male*	84.20 (0.73)							5.31 (1.23)		0.35
	male wt	24.00 (0.10)						0.60 (0.10)			0.41

(b) Eastern grey kangaroos (New South Wales)

			Explanatory variables								
KMZ	Response	Intercept	D	D^2	*HR*	*rn*.1.0	*rn*.3.1	*rn*.12.1	Skin	Cycle	Adj. R^2
1	*%male*	100.09 (1.9)									–
	male wt	22.47 (0.20)	0.96 (0.20)						3.15 (0.73)		0.31
2	*%male*	97.75 (1.60)			–4.45 (1.56)				–9.17 (3.40)		0.08
	male wt	21.48 (0.17)	0.39 (0.17)							–0.42 (0.17) 0.19 (0.17)	0.08
4	*%male*	83.09 (0.46)	–15.96 (3.48)	17.51 (3.43)	2.88 (0.93)		–0.88 (0.41)				0.37
	male wt	23.20 (0.14)	–4.35 (1.05)	4.61 (1.07)				0.63 (0.15)		–0.22 (0.12) –0.31 (0.12)	0.52
6	*%male*	109.64 (1.9)	3.42 (1.68)						8.58 (3.52)		0.12
	male wt	20.56 (0.14)	0.46 (0.14)			–0.13 (0.06)				–0.08 (0.10) –0.22 (0.11)	0.29
7	*%male*	122.36 (1.0)			–2.81 (0.77)				–11.46 (1.64)		0.43
	male wt	21.56 (0.21)	2.50 (1.10)	–2.44 (1.07)	–0.58 (0.20)	–0.18 (0.08)			–1.09 (0.37)	–0.16 (0.13) –0.34 (0.14)	0.50
8	*%male*	98.15 (0.73)	2.16 (0.81)		–3.01 (0.78)					–0.39 (0.53) –1.17 (0.57)	0.50
	male wt	22.40 (0.10)	–1.67 (0.49)	1.73 (0.48)	–0.33 (0.10)				–1.21 (0.18)	–0.15 (0.08) –0.52 (0.08)	0.60
10	*%male*	112.24 (0.8)	1.59 (0.75)								0.11
	male wt	21.28 (0.29)			–0.44 (0.22)					–0.12 (0.09) 0.38 (0.10)	0.19
11	*%male*	88.08 (0.64)	–13.87 (5.92)	14.07 (5.80)	2.51 (0.68)	–0.55 (0.21)			2.91 (1.12)		0.47
	male wt	24.19 (0.18)	3.75 (1.74)	–3.42 (1.73)				0.32 (0.14)		–0.16 (0.09) –0.25 (0.10)	0.47

Table 32.2: (Continued)

(c) Western grey kangaroos (New South Wales)

			Explanatory variables								
KMZ	**Response**	**Intercept**	**D**	***D*²**	***HR***	***rn*.1.0**	***rn*.3.1**	***rn*.12.1**	**Skin**	**Cycle**	**Adj. R^2**
1	*%male*	102.5 (1.57)								–0.78 (1.50) –3.44 (1.50)	0.04
	male wt	20.96 (0.97)			–0.78 (0.26)						0.15
2	*%male*	95.01 (0.72)	1.69 (0.66)						–6.85 (1.39)		0.19
	male wt	21.53 (0.16)	0.98 (0.18)		0.50 (0.15)			0.34 (0.11)		–0.22 (0.10) 0.17 (0.10)	0.41
4	*%male*	86.80 (0.63)	2.91 (0.62)								0.35
	male wt	20.63 (0.18)	–2.90 (1.17)	3.71 (1.17)							0.51
6	*%male*	109.10 (1.3)	–33.71 (13.2)	33.53 (13.19)		1.35 (0.68)					0.10
	male wt	20.65 (0.11)	0.55 (0.11)			–0.21 (0.06)				–0.23 (0.09) –0.18 (0.10)	0.40
7	*%male*	107.73 (1.7)	–25.80 (6.41)	27.91 (6.60)					–7.64 (3.20)		0.47
	male wt	21.26 (0.20)	0.63 (0.20)								0.17
10	*%male*	130.66 (1.6)									–
	male wt	21.89 (1.73)							–7.34 (1.21)		0.05
11	*%male*	92.70 (0.79)	24.08 (10.21)	–24.03 (10.07)	2.32 (1.09)						0.20
	male wt	22.05 (0.06)	–3.39 (0.92)	3.71 (0.91)	0.59 (0.09)			0.71 (0.07)			0.71

Table 32.2: (Continued)

(d) Red kangaroos (New South Wales)

			Explanatory variables			
KMZ	**Response**	**Intercept**	***male wt***	***pmale***	**Adj. R^2**	***95%PI***
1	log*D*	2.94 (0.08)	0.30 (0.08)		0.56	3.27
	HR%	9.55 (0.76)			–	–
2	log*D*	2.57 (0.07)	0.21 (0.07)		0.42	2.76
	HR%	13.09 (1.17)			–	–
4	log*D*	1.80 (0.10)			–	–
	HR%	13.91 (1.18)			–	–
6	log*D*	1.55 (0.07)	0.20 (0.07)		0.41	1.72
	HR%	11.09 (1.46)			–	–
7	log*D*	1.70 (0.14)			–	–
	HR%	9.36 (0.94)			–	–
8	log*D*	1.81 (0.20)			–	–
	HR%	7.36 (1.35)			–	–
10	log*D*	1.50 (0.14)			–	–
	HR%	8.91 (1.63)			–	–
11	log*D*	1.40 (0.08)			–	–
	HR%	10.18 (1.14)			–	–

(e) Eastern grey kangaroos (New South Wales)

			Explanatory variables			
KMZ	**Response**	**Intercept**	***male wt***	***pmale***	**Adj. R^2**	***95%PI***
1	log*D*	0.94 (0.10)	0.32 (0.12)	0.32 (0.12)	0.46	1.61
	HR%	3.40 (0.47)	–2.97 (1.19)	3.45 (1.19)	0.42	4.6
2	log*D*	1.15 (0.12)			–	–
	HR%	6.36 (0.10)			–	–
4	log*D*	1.09 (0.09)		–0.25 (0.09)	0.37	1.59
	HR%	9.18 (0.41)		1.31 (0.43)	0.46	6.7
6	log*D*	1.99 (0.07)			–	–
	HR%	5.90 (1.22)			–	–
7	log*D*	2.21 (0.13)			–	–
	HR%	5.36 (0.64)			–	–
8	log*D*	2.78 (0.12)		0.34 (0.12)	0.41	3.60
	HR%	9.64 (0.94)		–3.45 (0.99)	0.53	14.9
10	log*D*	3.27 (0.07)			–	–
	HR%	6.27 (0.86)			–	–
11	log*D*	2.29 (0.06)			–	–
	HR%	8.78 (0.72)			–	–

Table 32.2: (Continued)

(f) Western grey kangaroos (New South Wales)

			Explanatory variables			
KMZ	**Response**	**Intercept**	***male wt***	***pmale***	**Adj. R^2**	***95%PI***
1	log*D*	−0.14 (0.12)	0.33 (0.13)		0.42	0.48
	HR%	9.33 (1.73)			–	–
2	log*D*	1.59 (0.07)	0.34 (0.08)		0.65	1.85
	HR%	12.36 (1.97)			–	–
4	log*D*	2.03 (0.12)			–	–
	HR%	10.27 (0.78)			–	–
6	log*D*	2.37 (0.05)	0.17 (0.06)		0.45	2.30
	HR%	4.30 (0.56)			–	–
7	log*D*	1.24 (0.09)	0.35 (0.10)		0.54	1.79
	HR%	5.36 (0.87)			–	–
10	log*D*	1.05 (0.11)			–	–
	HR%	1.55 (0.17)		0.44 (0.18)	0.34	2.6
11	log*D*	0.70 (0.08)			–	–
	HR%	11.10 (1.44)			–	–

The size of the 95% prediction interval (*95%PI*) averaged across all data points is also shown for models predicting population statistics. Models are shown for kangaroo management zones (KMZs) in New South Wales (Fig. 32.1). Population statistics are log transformed kangaroo density (log*D*) and harvest rate expressed as a percentage (*HR%*). Harvest statistics are the percentage of the harvest comprising males (*%male*, arcsine transformed) and average male carcass weight (*male wt*). Explanatory variables include untransformed density (*D*), harvest rate expressed as a proportion (*HR*), *male wt*, the proportion of the harvest comprising males (*pmale*, arcsine transformed), rain falling in the same month of collection of the harvest statistic (*rn*.1.0), three months rain lagged one month prior to collection (*rn*.3.1), 12 months rain lagged one month prior to collection (*rn*.12.1), seasonal *cycle* comprising $sin(2 \times \pi \times time)$ and $cos(2 \times \pi \times time)$ terms, and a dummy variable (*skin*) representing the period when there was skin-only shooting. The parameter for $sin(2 \times \pi \times time)$ is given above the parameter for $cos(2 \times \pi \times time)$.

action such as quota adjustment. The models presented here show that kangaroo harvest statistics change over time in response to changes in population density and harvest rate, most noticeably at a regional scale. Usefully for management, these relationships are sufficiently precise (Pople 2008) to provide a tool by which populations could be monitored indirectly. They could be used in two ways. First, a consistent shift in a statistic over a number of months may indicate a decline in density or an increase in harvest rate. If the likely density or harvest rate is of concern, then management action such as quota revision may be triggered. Other factors such as rainfall and season would need to be taken into account. Whether harvest rate or density or both were involved would need to be teased apart. The second use is for quota setting when direct estimates of population size (e.g. from aerial survey) are unavailable. The estimates of precision from this second set of models are comparable to direct estimates of population size using aerial survey.

Important caveats to the preceding conclusions are that not all regions and blocks had useful relationships between harvest and population statistics and there was considerable variation among regions in the form of the relationship. There is certainly no one relationship that could be applied across an entire state or across species. The single state-wide trigger-point (i.e. a value that triggers management action) for harvest sex ratio for all species that is used in Queensland (Kirkpatrick and Nance 1985; QPWS 2003) is therefore inappropriate. Regional and species-specific trigger-points, as discussed above, could be used in many cases.

Unfortunately, economic factors such as prices paid for skins and carcasses could not be incorporated into the analyses as they were not readily available. They are not collected by management agencies and should be sought from the kangaroo industry for future analyses. Prices fluctuate with demand for product and availability of kangaroos to shooters, and should influence CPUE and the selectivity of

Table 32.3: Standardised parameter estimates with standard errors (in brackets) and adjusted R^2 for the most parsimonious (according to backwards elimination) regression models predicting (a) red kangaroo (*M. rufus*) harvest statistics, (b) eastern grey kangaroo (*M. giganteus*) harvest statistics, (c) common wallaroo (*M. robustus*) harvest statistics, (d) red kangaroo population statistics, (e) eastern grey kangaroo population statistics and (f) common wallaroo population statistics.

(a) Red kangaroos (Queensland)

			Explanatory variables									Adj. R^2
Block	**Response**	**Intercept**	**D**	***D***^{2}	***HR***	***rn*.1.0**	***rn*.3.1**	***rn*.12.1**	**Effort**	***pskin***	**Cycle**	
J	*%male*	90.08 (1.60)	5.09 (1.61)									0.25
	CPUE	11.64 (0.40)								2.37 (0.40)		0.25
	male wt	26.14 (0.20)	1.74 (0.35)		1.63 (0.45)						−0.20 (0.22) −0.75 (0.21)	0.24
Wt	*%male*	93.18 (1.11)										–
	CPUE	4.72 (0.15)	0.94 (0.25)		1.13 (0.25)							0.29
	male wt	25.83 (0.25)	3.86 (1.63)	−3.43 (1.59)						0.80 (0.26)		0.20
L	*%male*	104.52 (2.5)			−6.86 (2.11)					2.32 (1.08)		0.35
	CPUE	2.94 (0.12)								0.40 (0.10)	0.07 (0.09) −0.23 (0.11)	0.37
	male wt	24.77 (0.32)								0.70 (0.26)	−0.50 (0.25) 0.08 (0.28)	0.18
Wd	*%male*	111.93 (2.7)			−13.70 (2.65)							0.53
	CPUE	6.65 (0.29)	0.69 (0.31)							1.36 (0.30)		0.20
	male wt	24.83 (0.26)	−13.89 (2.2)	13.79 (2.18)				2.24 (0.28)		1.22 (0.29)	−0.30 (0.27) 0.66 (0.26)	0.72
Bk	*%male*	98.51 (1.22)	−16.07 (8.7)	19.23 (8.58)	−8.03 (1.37)							0.62
	CPUE	2.76 (0.12)	2.57 (0.81)	−2.18 (0.80)	0.42 (0.15)					0.77 (0.10)		0.74
	male wt	21.59 (0.16)	−3.91 (1.35)	3.01 (1.32)	−1.08 (0.28)			0.37 (0.18)		1.37 (0.24)	−0.05 (0.15) −0.03 (0.17)	0.40
Cv	*%male*	115.41 (2.4)	14.68 (2.31)						−1.92 (0.91)			0.67
	CPUE	2.45 (0.04)	0.14 (0.07)		0.45 (0.07)						−0.00 (0.04) −0.10 (0.04)	0.73
	male wt	22.04 (0.23)	0.59 (0.27)					0.57 (0.23)		0.71 (0.22)		0.32

Table 32.3: (Continued)

(a) Red kangaroos (Queensland) (Continued)

			Explanatory variables									Adj. R^2
Block	**Response**	**Intercept**	**D**	D^2	*HR*	*rn*.1.0	*rn*.3.1	*rn*.12.1	**Effort**	*pskin*	**Cycle**	
H	%male	106.94 (2.1)	6.15 (2.04)									0.18
	CPUE	7.16 (0.43)	–4.82 (2.23)	5.88 (2.17)	1.43 (0.49)					0.69 (0.30)		0.30
	male wt	23.86 (0.37)				0.57 (0.24)				1.23 (0.30)		0.18
Bo	*%male*	119.08 (2.7)	60.74 (13.8)	–41.97 (14.0)						8.15 (2.61)		0.51
	CPUE	2.35 (0.09)			0.27 (0.09)						0.17 (0.07) –0.12 (0.08)	0.20
	male wt	21.96 (0.24)	0.57 (0.24)					0.97 (0.23)				0.28
R	*%male*	110.4 (3.7)	–5.45 (2.37)									0.16
	CPUE	0.56 (0.02)	0.42 (0.11)	–0.30 (0.10)	0.19 (0.03)					0.17 (0.02)		0.63
	male wt	22.12 (0.22)	–0.64 (0.24)							–1.01 (0.23)		0.18
Wm	*%male*	107.47 (3.2)										–
	CPUE	0.46 (0.03)								0.14 (0.03)	0.10 (0.03) 0.04 (0.03	0.22
	male wt	24.18 (0.33)						0.90 (0.33)		0.70 (0.32)		0.13

Table 32.3: (Continued)

(b) Eastern grey kangaroos (Queensland)

Block	Response	Intercept	Explanatory variables										Adj. R^2
			D	D^2	*HR*	*rn*.1.0	*rn*.3.1	*rn*.12.1	Effort	*pskin*	Cycle	*red.hs*	
J	*%male*	115.59 (4.91)								−4.44 (1.89)			−0.03
	CPUE	50.76 (7.05)	−81.17 (25)	47.50 (23.6)	−29.25 (7.19)						−11.48 (4.02) −5.25 (4.55)		0.40
	male wt	25.62 (0.17)									−0.40 (0.18) −1.04 (0.18)		0.22
Wt	*%male*	90.62 (4.31)	21.33 (7.36)	−20.26 (7.3)			−3.47 (1.23)		2.48 (1.04)			10.57 (4.30)	0.30
	CPUE	18.93 (2.75)										−10.01 (2.66)	0.08
	male wt	23.59 (0.59)	−2.34 (0.82)	3.21 (0.83)	0.64 (0.20)						−0.22 (0.16) −0.58 (0.17)	1.73 (0.62)	0.45
L	*%male*	102.49 (2.82)	−34.03 (9.8)	29.77 (9.64)	−5.28 (2.00)							−9.69 (2.52)	0.40
	CPUE	30.75 (2.23)	−32.84 (5.5)	26.05 (5.38)						−2.19 (0.91)		−20.74 (2.35)	0.67
	male wt	23.05 (0.42)	−3.73 (0.90)	3.13 (0.89)	−1.03 (0.22)					0.59 (0.18)	−0.63 (0.14) −0.55 (0.15)	1.39 (0.45)	0.54
Wd	*%male*	126.76 (0.94)	10.02 (1.16)			−4.32 (1.52)	−4.66 (1.76)				0.39 (1.82) 4.57 (1.33)		0.71
	CPUE	80.78 (7.69)	−154.9 (46)	136.7 (46.1)	−20.87 (9.25)								0.27
	male wt	26.22 (0.44)			−2.46 (0.44)								0.38
Bk	*%male*	101.90 (4.54)			−46.37 (11.98)					16.88 (4.69)		6.42 (2.21)	0.50
	CPUE	4.84 (0.32)			0.42 (0.17)	−0.29 (0.09)					−0.55 (0.13) 0.16 (0.14)	−2.73 (0.33)	0.60
	male wt	19.80 (0.25)									−0.13 (0.09) −0.32 (0.10)	0.76 (0.27)	0.26

Table 32.3: (Continued)

(b) Eastern grey kangaroos (Queensland) (Continued)

			Explanatory variables										
Block	**Response**	**Intercept**	**D**	***D***^{2}	***HR***	***rn*.1.0**	***rn*.3.1**	***rn*.12.1**	**Effort**	***pskin***	**Cycle**	***red.hs***	**Adj. R^2**
Cv	*%male*	112.51 (1.40)	8.17 (2.08)		–14.31 (2.06)					2.23 (0.82)			0.87
	CPUE	7.47 (0.39)			1.27 (0.19)							–4.70 (0.40)	0.77
	male wt	19.88 (0.31)	0.73 (0.23)		–1.17 (0.23)					0.79 (0.12)		0.72 (0.31)	0.78
H	*%male*	107.61 (4.25)	13.59 (3.20)	–9.71 (3.09)								15.13 (4.52)	0.38
	CPUE	59.10 (7.66)	–29.10 (6.5)	23.22 (6.38)						–4.59 (1.82)		–34.45 (7.65)	0.40
	male wt	21.46 (0.58)	4.84 (1.32)	–4.67 (1.34)				0.81 (0.26)		0.55 (0.19)		2.66 (0.60)	0.45
Bo	*%male*	110.82 (3.04)	79.27 (8.27)	–61.12 (7.55)	4.41 (1.98)							8.84 (3.19)	0.80
	CPUE	34.40 (2.06)	–15.61 (4.7)	12.90 (4.31)	2.38 (1.13)						0.27 (0.68) 3.31 (0.74)	–20.97 (2.2)	0.76
	male wt	20.75 (0.16)	0.69 (0.23)		–0.76 (0.26)			0.46 (0.19)	0.34 (0.15)				0.56
R	*%male*	93.67 (2.07)	–4.82 (1.93)		–10.10 (1.87)	–0.66 (0.30)	–1.47 (0.53)					2.71 (1.21)	0.73
	CPUE	8.37 (0.57)			1.02 (0.35)	–0.41 (0.14)					0.01 (0.24) 0.69 (0.27)	–4.07 (0.58)	0.62
	male wt	20.48 (0.30)			–0.34 (0.15)			0.45 (0.12)		–0.49 (0.14)	–0.32 (0.11) –0.32 (0.12)	1.26 (0.32)	0.62
Wm	*%male*	103.09 (1.82)	5.30 (1.97)		–10.21 (1.85)	–1.54 (0.77)	–3.48 (1.13)			3.69 (1.49)			0.48
	CPUE	19.80 (2.11)	–2.87 (0.98)							2.33 (1.04)	2.33 (0.83) 3.84 (0.95)	–759 (2.29)	0.48
	male wt	21.66 (0.16)	6.63 (1.05)	–5.87 (1.06)	–1.33 (0.18)				0.43 (0.18)	0.48 (0.18)			0.57

Table 32.3: (Continued)

(c) Common wallaroos (Queensland)

			Explanatory variables									
Block	**Response**	**Intercept**	**D**	***D*²**	***HR***	***rn*.1.0**	***rn*.3.1**	***rn*.12.1**	**Effort**	***pskin***	**Cycle**	***red.hs***	**Adj. *R*²**
Wt	*%male*	151.65 (1.15)											–
	CPUE	1.66 (0.33)										–0.82 (0.33)	0.03
	male wt	23.40 (0.20)	–1.37 (0.57)	1.21 (0.58)					0.62 (0.19)				0.26
L	*%male*	149.83 (1.35)							–2.63 (0.45)			4.51 (1.45)	0.17
	CPUE	3.61 (0.29)	–2.14 (0.91)	2.73 (0.90)								–1.36 (0.29)	0.38
	male wt	23.10 (0.27)				–0.28 (0.13)							0.04
Bk	*%male*	146.13 (1.74)			–4.84 (1.00)				–3.02 (0.69)	2.05 (0.77)		6.30 (1.87)	0.64
	CPUE	7.26 (0.22)	1.08 (0.15)		0.34 (0.15)		–0.22 (0.08)					–3.47 (0.22)	0.74
	male wt	19.94 (0.12)			–0.74 (0.16)					0.81 (0.15)			0.27
Cv	*%male*	146.82 (2.54)	3.90 (0.99)							3.31 (0.98)	–0.79 (0.95) 2.11 (0.93)	7.00 (2.73)	0.29
	CPUE	1.31 (0.07)	0.64 (0.12)	–0.40 (0.12)			–0.06 (0.03)				0.05 (0.03) 0.01 (0.02)	–0.44 (0.08)	0.59
	male wt	20.88 (0.16)	3.80 (0.74)	–2.84 (0.74)		–0.40 (0.11)							0.47
R	*%male*	150.60 (0.95)											–
	CPUE	0.12 (0.02)								–0.03 (0.01)		0.06 (0.02)	0.37
	male wt	21.83 (0.12)	0.69 (0.46)	–0.90 (0.45)					–0.36 (0.16)	–1.09 (0.21)	–0.23 (0.13) –0.56 (0.14)		0.30

Table 32.3: (Continued)

(d) Red kangaroos (Queensland)

			Explanatory variables				
Block	**Response**	**Intercept**	**CPUE**	***pmale***	***male wt***	**Adj. R^2**	***95%PI***
J	log*D*	1.88 (0.23)				–	–
	HR%	14.00 (2.14)				–	–
Wt	log*D*	1.67 (0.16)				–	–
	HR%	16.33 (1.74)	6.68 (1.85)			0.60	26.3
L	log*D*	2.55 (0.10)				–	–
	HR%	10.30 (0.88)	2.99 (1.07)	−4.01 (1.07)		0.60	14.5
Wd	log*D*	2.14 (0.12)				–	–
	HR%	8.43 (0.90)	16.01 (5.64)	−21.35 (5.6)		0.87	8.8
Bk	log*D*	2.24 (0.09)		0.25 (0.09)		0.39	2.67
	HR%	33.36 (4.88)		−19.39 (5.1)		0.57	68.6
Cv	log*D*	1.67 (0.09)		0.41 (0.10)		0.61	2.26
	HR%	34.92 (2.94)	15.74 (3.08)		−8.46 (3.08)	0.75	48.6
H	log*D*	1.56 (0.10)		0.35 (0.10)		0.50	2.18
	HR%	13.10 (3.13)				–	–
Bo	log*D*	1.46 (0.16)		0.50 (0.17)		0.44	2.61
	HR%	23.36 (5.79)	15.17 (6.07)			0.34	69.6
R	log*D*	0.08 (0.18)				–	–
	HR%	25.70 (7.61)				–	–
Wm	log*D*	−1.07 (0.16)				–	–
	HR%	57.11 (12.2)				–	–

(e) Eastern grey kangaroos (Queensland)

			Explanatory variables				
Block	**Response**	**Intercept**	**CPUE**	***pmale***	***male wt***	**Adj. R^2**	***95%PI***
J	log*D*	0.58 (0.27)				–	–
	HR%	7.33 (0.95)		−4.97 (1.04)		0.81	12.9
Wt	log*D*	1.40 (0.13)				–	–
	HR%	11.38 (3.17)				–	–
L	log*D*	2.32 (0.10)	−0.32 90.11)			0.45	3.06
	HR%	10.18 (0.89)	4.61 (1.00)	−3.69 (1.00)		0.71	18.0
Wd	log*D*	1.55 (0.11)				–	–
	HR%	3.43 (0.96)		−2.68 (1.04)		0.48	10.8
Bk	log*D*	2.21 (0.09)	0.34 (0.10)	0.37 (0.10)		0.64	2.81
	HR%	31.58 (4.63)	−18.80 (4.8)			0.56	65.7
Cv	log*D*	2.67 (0.10)		0.51 (0.11)		0.68	3.41
	HR%	23.36 (1.60)		−13.53 (1.7)		0.86	25.2
H	log*D*	1.57 (0.19)				–	–
	HR%	8.09 (2.07)				–	–

(e) Eastern grey kangaroos (Queensland) (Continued)

			Explanatory variables				
Block	**Response**	**Intercept**	**CPUE**	***pmale***	***male wt***	**Adj. R^2**	***95%PI***
Bo	log*D*	3.18 (0.08)		0.45 (0.09)		0.70	3.68
	HR%	12.75 (2.08)			–6.25 (2.17)	0.40	26.9
R	log*D*	3.03 (0.14)				–	–
	HR%	17.33 (2.47)		–9.69 (2.58)		0.54	35.4
Wm	log*D*	3.26 (0.10)				–	–
	HR%	16.09 (1.45)		–5.51 (1.52)		0.55	21.7

(f) Common wallaroos (Queensland)

			Explanatory variables				
Block	**Response**	**Intercept**	**CPUE**	***pmale***	***male wt***	**Adj. R^2**	***95%PI***
Wt	log*D*	–0.20 (0.22)				–	–
	HR%	12.44 (2.99)				–	–
L	log*D*	2.09 (0.22)				–	–
	HR%	16.91 (3.13)		–9.11 (3.29)		0.40	39.4
Bk	log*D*	2.93 (0.11)	0.53 (0.11)	0.48 (0.11)		0.76	3.92
	HR%	33.00 (5.30)		–21.94 (5.5)		0.57	73.4
Cv	log*D*	1.11 (0.17)	0.73 (0.18)			0.61	2.60
	HR%	22.00 (3.19)	–7.80 (3.35)			0.31	46.1
R	log*D*	–0.41 (0.57)				–	–
	HR%	9.67 (1.89)				–	–

J = Julia Creek; Wt = Winton; L = Longreach, Wd = Windorah; Bk = Blackall; Cv = Charleville; H = Hungerford; Bo = Bollon; R = Roma; Wm = Westmar.

The size of the 95% prediction interval (*95%PI*) averaged across all data points is also shown for models predicting population statistics. Models are shown for survey blocks in the Queensland core harvest area (Fig. 32.1). Population statistics are log transformed kangaroo density (log*D*) and harvest rate expressed as a percentage (*HR%*). Harvest statistics are catch-per-unit-effort (CPUE), the percentage of the carcass harvest comprising males (*%male*, arcsine transformed) and average male carcass weight (*male wt*). Explanatory variables include untransformed density (*D*), harvest rate expressed as a proportion (*HR*), CPUE, *male wt*, the proportion of the carcass harvest comprising males (*pmale*, arcsine transformed), rain falling the same month as collection of the harvest statistic (*rn*.1.0), three months rain lagged one month prior to collection (*rn*.3.1), 12 months rain lagged one month prior to collection (*rn*.12.1), the number of shooter-nights (*effort*), seasonal *cycle* comprising $sin(2 \times \pi \times time)$ and $cos(2 \times \pi \times time)$ terms, the proportion of the harvest that was skin-only (*pskin*) and a dummy variable (*red.hs*) for months when the harvest season for red kangaroos was open. The parameter for $sin(2 \times \pi \times time)$ is given above the parameter for $cos(2 \times \pi \times time)$.

shooters. Inclusion of economic data would be expected to tighten the relationships described here. The fact that useful relationships were found without economic data highlights the strength of the connection between harvest and population statistics.

These models need further assessment through comparison with independent datasets. Consideration was given to splitting the data into two sets, one for model development and the other for model testing (Maindonald and Braun 2003). However, while the time series span 11–22 years, they contained a limited number of major droughts or runs of good seasons leading to sustained population increase. These have far-reaching consequences for kangaroo abundance and population dynamics (Caughley 1987) and shooter behaviour (Kirkpatrick and Amos 1985). As a result, it was decided to fit models to the full regional datasets. The resulting models should therefore be considered exploratory rather than confirmatory (Burnham and Anderson 1998). Validation by comparing a model fitted to one region's data to another region's data was not considered appropriate *a priori* and was confirmed by the regional differences recorded here.

A further limitation is that the modelling has not accounted for all the uncertainty in the data, in particular the error associated with the density estimates. This could be incorporated in future analyses by taking a hierarchical Bayesian approach (Clark 2005).

If harvest statistics are to be used in kangaroo management as a surrogate for population abundance or harvest rate they must be considered as supplementing direct monitoring, which may therefore be reduced in frequency to save effort and money, and compared with the alternative of using rainfall to predict rate of increase. Reductions in survey frequency are most appropriate in the lower-harvest regions and more mesic regions within each state (Pople 2008). Rainfall models were poor predictors of rate increase in central and northern Queensland, in lower-density regions in South Australia and to some extent New South Wales (Pople 2006; Ch. 18, this volume). Harvest statistics can fill some but not all these gaps and can supplement rainfall models in regions where both forms of indirect monitoring have useful relationships. Harvest statistics are most likely to be useful in Queensland. The relationships between harvest and population statistics were not explored outside the core survey area in Queensland. In these areas survey frequency is already less than annual. However, the relationships described here cannot be readily extrapolated to new areas, as demonstrated by the extent of regional variation in the relationships.

ACKNOWLEDGMENTS

This project was supported by an ARC SPIRT (now ARC Linkage) grant to Hugh Possingham, Gordon Grigg and Stuart Phinn at the University of Queensland, with financial contributions from Queensland, New South Wales, South Australian and Commonwealth conservation agencies and Packer Tanning. Norbert Menke kindly prepared the maps.

REFERENCES

Andrew NL and Mapstone BD 1987. Sampling and the description of spatial pattern in marine ecology. *Oceanogr. Mar. Biol.* **25:** 39–90.

Anon 1984. Kangaroo management programs of the Australian states. Green Apple Graphics, Canberra.

Burnham KP and Anderson DA 1998. *Model Selection and Inference: A Practical Information-Theoretic Approach.* Springer, Berlin.

CALM (Dept of Conservation and Land Management) 2002. Red kangaroo (*Macropus rufus*) management plan for Western Australia 2003–2007. Dept of Conservation and Land Management, Perth.

Caughley G 1977. *Analysis of Vertebrate Populations.* Wiley and Sons, London.

Caughley G 1987. Ecological relationships. In *Kangaroos: Their Ecology and Management in the Sheep Rangelands of Australia.* (Eds G Caughley, N Shepherd and J Short) pp. 159–187. Cambridge University Press, Cambridge.

Clark JS 2005. Why environmental scientists are becoming Bayesians. *Ecol. Lett.* **8:** 2–14.

Crawley MJ 2002. *Statistical Computing: An Introduction to Data Analysis using S-Plus.* Wiley, Chichester.

de la Mare WK 1988. Simulation studies on the use of the sex ratio for monitoring the effects of exploitation on kangaroo populations. Unpublished paper. National Advisory Committee on Kangaroos, Canberra.

Environment Australia 2000. Revision of the Interim Biogeographic Regionalisation of Australia (IBRA) and the development of version 5.1: summary report. Unpublished report. Dept of Environment and Heritage, Canberra.

Frith HJ and Sharman G 1964. Breeding in wild populations of the red kangaroo, *Megaleia rufa. CSIRO Wildl. Res.* **9:** 86–114.

Hilborn R and Walters CJ 1992. Quantitative fisheries stock assessment: choice, dynamics and uncertainty. Chapman and Hall, New York.

Jarman PJ 1989. Sexual dimorphism in Macropodoidea. In *Kangaroos, Wallabies and Rat-kangaroos.* (Eds G Grigg, P Jarman and I Hume) pp. 433–447. Surrey Beatty and Sons, Sydney.

Kirkpatrick TH and Amos PJ 1985. The kangaroo industry. In *The Kangaroo Keepers.* (Ed. HJ Lavery) pp. 75–102. University of Queensland Press, Brisbane.

Kirkpatrick TH and McEvoy JS 1966. Studies of Macropodidae in Queensland. V. Effects of drought on reproduction in the grey kangaroo (*Macropus giganteus*). *Qld J. Agri. Anim. Sci.* **23:** 439–442.

Kirkpatrick TH and Nance CA 1985. Conservation. In *The Kangaroo Keepers.* (Ed. HJ Lavery) pp. 161–187. University of Queensland Press, Brisbane.

Lancia RA, Nichols JD and Pollock KH 1996. Estimating the number of animals in wildlife populations. In *Research and Management Techniques for Wildlife and Habitats.* (Ed. TA Bookhout) pp. 215–253. Wildlife Society, Bethesda, Maryland.

Maindonald JH and Braun J 2003. *Data Analysis and Graphics using R: An Example-based Approach.* Cambridge University Press, Cambridge.

McCallum H 2000. *Population Parameters.* Blackwell Science, London.

Nance CA 1985. Macropodid management of cereal and pasture regions of Queensland. Unpublished report. Reserve Bank of Australia, Canberra.

Newsome AE 1965. The distribution of red kangaroos, *Megaleia rufa* (Desmarest), about sources of persistent food and water in central Australia. *Aust. J. Zool.* **13:** 289–299.

Newsome AE 1966. The influence of food on breeding in the red kangaroo in central Australia. *CSIRO Wildl. Res.* **11:** 187–196.

Newsome AE 1977. Imbalance in the sex ratio and age structure of the red kangaroo, *Macropus rufus*, in central Australia. In *The Biology of Marsupials*. (Ed. D Gilmour) pp. 221–233. Macmillan, London.

Pople AR 1996. Effects of harvesting upon the demography of red kangaroos in western Queensland. PhD thesis. University of Queensland, Brisbane.

Pople AR 1999. Aerial surveys for kangaroo management. *Aust. Zool.* **31:** 266–320.

Pople A 2004. Population monitoring for kangaroo management. *Aust. Mammal.* **26:** 37–44.

Pople AR 2006. Modelling the spatial and temporal dynamics of kangaroo populations for harvest management. Final report. Dept of Environment and Heritage, Canberra.

Pople AR 2008. Frequency and precision of aerial surveys for kangaroo management. *Wildl. Res.* **35:** 340–348.

Pople A and Cairns S 1995. Impact of harvesting on kangaroos. In *Conservation through Sustainable Use of Wildlife*. (Eds G Grigg, P Hale and D Lunney) pp. 224–229. Centre for Conservation Biology, University of Queensland, Brisbane.

Pople AR and Grigg GC 1998. Commercial harvesting of kangaroos in Australia. Environment Australia, Canberra. http://www.ea.gov.au/biodiversity/trade-use/wild-harvest/kangaroo/harvesting/index.html.

Pople AR, Phinn SR, Menke N, Grigg GC, Possingham HP and McAlpine C 2007. Spatial patterns of kangaroo density across the South Australian pastoral zone over 26 years: aggregation during drought and suggestions of long-distance movement. *J. Appl. Ecol.* **44:** 1068–1079.

Prince RIT 1984a. Exploitation of kangaroos and wallabies in Western Australia. I. Exploitation and management of the red kangaroo, 1970–1979. Dept of Fisheries and Wildlife, Perth.

Prince RIT 1984b. Exploitation of kangaroos and wallabies in Western Australia. I. Review to 1970, with special emphasis on the red and western grey kangaroos. Dept of Fisheries and Wildlife, Perth.

QNPWS (Queensland National Parks and Wildlife Service) 1989. Background information for the 1989 Queensland kangaroo management program. Unpublished report. Australian Nature Conservation Agency, Canberra.

QPWS (Queensland Parks and Wildlife Service) 2003. Queensland wildlife trade management plan for export: commercially harvested macropods, 2003–2007. Unpublished report. Dept of Environment and Heritage, Canberra.

R Development Core Team 2003. R: A language and environment for statistical computing. R Foundation for Statistical Computing, Vienna. http://www.R-project.org.

Robertson GG 1986. The mortality of kangaroos in drought. *Aust. Wildl. Res.* **13:** 349–354.

Southwell C 1989. Techniques for monitoring the abundance of kangaroo and wallaby populations. In *Kangaroos, Wallabies and Rat-kangaroos*. (Eds G Grigg, P Jarman and I Hume) pp. 659–693. Surrey Beatty and Sons, Sydney.

Wilson GR 1975. Age structures of populations of kangaroos (Macropodidae) taken by professional shooters in New South Wales. *Aust. Wildl. Res.* **2:** 1–9.

Yom-Tov Y and Nix H 1986. Climatological correlates for body size of five species of Australian mammals. *Biol. J. Linn. Soc.* **29:** 245–262.

Appendix: Macropod species list

Mark D.B. Eldridge

Recent species of kangaroo, wallaby and rat-kangaroo

Superfamily Macropodoidea[a]		Distribution	Status
Family Hypsiprymnodontidae[b]			
Musky rat-kangaroo	*Hypsiprymnodon moschatus* Ramsay 1876	Australia	LC
Family Potoroidae			
Rufous bettong	*Aepyprymnus rufescens* (Gray 1837)	Australia	LC
Tasmanian bettong	*Bettongia gaimardi* (Desmarest 1822)	Australia	NT
Burrowing bettong	*Bettongia lesueur* (Quoy & Gaimard 1824)	Australia	NT
Brush-tailed bettong	*Bettongia penicillata* Gray 1937	Australia	CR
Northern bettong	*Bettongia tropica* Wakefield 1967	Australia	EN
Desert rat-kangaroo	*Caloprymnus campestris* (Gould 1843)	Australia	EX
Gilbert's potoroo	*Potorous gilbertii* (Gould 1841)	Australia	CR
Long-footed potoroo	*Potorous longipes* Seebeck & Johnston 1980	Australia	EN
Broad-faced potoroo	*Potorous platyops* (Gould 1844)	Australia	EX
Long-nosed potoroo	*Potorous tridactylus* (Kerr 1792)	Australia	LC
Family Macropodidae			
Subfamily Macropodinae			
Black dorcopsis	*Dorcopsis atrata* Van Deusen 1955	New Guinea	CR
White-striped dorcopsis	*Dorcopsis hageni* Heller 1897	New Guinea	LC
Grey dorcopsis	*Dorcopsis luctuosa* (d'Albertis 1874)	New Guinea	VU
Brown dorcopsis	*Dorcopsis muelleri* (Schlegel 1866)	New Guinea	LC
Macleay's dorcopsis	*Dorcopsulus macleayi* (Mikluoho-Maclay 1885)	New Guinea	LC

Superfamily Macropodoidea[a]		**Distribution**	**Status**
Small dorcopsis	*Dorcopsulus vanheurni* (Thomas 1922)	New Guinea	NT
Bennett's tree-kangaroo	*Dendrolagus bennettianus* De Vis 1887	Australia	NT
Doria's tree-kangaroo	*Dendrolagus dorianus*[c] Ramsay 1883	New Guinea	VU
Goodfellow's tree-kangaroo	*Dendrolagus goodfellowi*[d] Thomas 1908	New Guinea	EN
Grizzled tree-kangaroo	*Dendrolagus inustus* Müller 1840	New Guinea	VU
Lumholtz's tree-kangaroo	*Dendrolagus lumholtzi* Collett 1884	Australia	LC
Matschie's tree-kangaroo	*Dendrolagus matschiei* Förster & Rothschild 1907	New Guinea	EN
Dingiso	*Dendrolagu mbasio* Flannery, Boeadi & Szalay 1995	New Guinea	CR
Tenkile	*Dendrolagus scottae* Flannery & Seri 1990	New Guinea	CR
Lowland tree-kangaroo	*Dendrolagus spadix* Troughton & Le Soeuf 1936	New Guinea	LC
Vogelkop tree-kangaroo	*Dendrolagus ursinus* Müller 1840	New Guinea	VU
Central hare-wallaby	*Lagorchestes asomatus* Finlayson 1943	Australia	EX
Spectacled hare-wallaby	*Lagorchestes conspicillatus* Gould 1842	Australia, New Guinea	LC
Rufous hare-wallaby	*Lagorchestes hirsutus* Gould 1844	Australia	VU
Eastern hare-wallaby	*Lagorchestes leporides* (Gould 1841)	Australia	EX
Western grey kangaroo	*Macropus* (*Macropus*) *fuliginosus* (Desmarest 1817)	Australia	LC
Eastern grey kangaroo	*Macropus* (*Macropus*) *giganteus* Shaw 1790	Australia	LC
Agile wallaby	*Macropus* (*Notamacropus*) *agilis* (Gould 1842)	Australia, New Guinea	LC
Black-striped wallaby	*Macropus* (*Notamacropus*) *dorsalis* (Gray 1837)	Australia	LC
Tammar wallaby	*Macropus* (*Notamacropus*) *eugenii* (Desmarest 1817)	Australia	LC
Toolache wallaby	*Macropus* (*Notamacropus*) *greyi* Waterhouse 1845	Australia	EX
Western brush wallaby	*Macropus* (*Notamacropus*) *irma* (Jourdan 1837)	Australia	LC
Parma wallaby	*Macropus* (*Notamacropus*) *parma* Waterhouse 1845	Australia	NT
Whiptail wallaby	*Macropus* (*Notamacropus*) *parryi* Bennett 1835	Australia	LC
Red-necked wallaby	*Macropus* (*Notamacropus*) *rufogriseus* (Desmarest 1817)	Australia	LC

Superfamily Macropodoidea[a]		Distribution	Status
Antilopine wallaroo	*Macropus* (*Osphranter*) *antilopinus* (Gould 1842)	Australia	LC
Black wallaroo	*Macropus* (*Osphranter*) *bernardus* Rothschild 1904	Australia	NT
Common wallaroo	*Macropus* (*Osphranter*) *robustus* Gould 1841	Australia	LC
Red kangaroo	*Macropus* (*Osphranter*) *rufus* (Desmarest 1822)	Australia	LC
Bridled nailtail wallaby	*Onychogalea fraenata* (Gould 1841)	Australia	EN
Crescent nailtail wallaby	*Onychogalea lunata* (Gould 1841)	Australia	EX
Northern nailtail wallaby	*Onychogalea unguifera* (Gould 1841)	Australia	LC
Allied rock-wallaby	*Petrogale assimilis* Ramsay 1877	Australia	LC
Short-eared rock-wallaby	*Petrogale brachyotis* (Gould 1841)	Australia	LC
Monjon	*Petrogale burbidgei* Kitchener & Sanson 1978	Australia	NT
Cape York rock-wallaby	*Petrogale coenensis* Eldridge & Close 1992	Australia	NT
Nabarlek	*Petrogale concinna* Gould 1842	Australia	DD
Godman's rock-wallaby	*Petrogale godmani* Thomas 1923	Australia	LC
Herbert's rock-wallaby	*Petrogale herberti* Thomas 1926	Australia	LC
Unadorned rock-wallaby	*Petrogale inornata* Gould 1842	Australia	LC
Black-footed rock-wallaby	*Petrogale lateralis* Gould 1842	Australia	NT
Mareeba rock-wallaby	*Petrogale mareeba* Eldridge & Close 1992	Australia	LC
Brush-tailed rock-wallaby	*Petrogale penicillata* (Gray 1825)	Australia	NT
Proserpine rock-wallaby	*Petrogale persephone* Maynes 1982	Australia	EN
Purple-necked rock-wallaby	*Petrogale purpureicollis* Le Souef 1924	Australia	LC
Rothschild's rock-wallaby	*Petrogale rothschildi* Thomas 1904	Australia	LC
Sharman's rock-wallaby	*Petrogale sharmani* Eldridge & Close 1992	Australia	NT
Yellow-footed rock-wallaby	*Petrogale xanthopus* Gray 1855	Australia	NT
Quokka	*Setonix brachyurus* (Quoy & Gaimard 1830)	Australia	VU
Tasmanian pademelon	*Thylogale billardierii* (Desmarest 1822)	Australia	LC

Superfamily Macropodoidea[a]		**Distribution**	**Status**
New Guinea pademelon	*Thylogale browni*[e] (Ramsay 1877)	New Guinea	VU
Dusky pademelon	*Thylogale brunii* (Schreber 1778)	New Guinea	VU
Calaby's pademelon	*Thylogale calabyi* Flannery 1992	New Guinea	EN
Red-legged pademelon	*Thylogale stigmatica* (Gould 1860)	Australia, New Guinea	LC
Red-necked pademelon	*Thylogale thetis* (Lesson 1827)	Australia	LC
Swamp wallaby	*Wallabia bicolor*[f] (Desmarest 1804)	Australia	LC
Subfamily Sthenurinae?[g]			
Banded hare-wallaby	*Lagostrophus fasciatus* (Péron & Lesueur 1807)	Australia	EN

Taxonomy follows Flannery (1995a, b) and Van Dyck and Strahan (2008). For more detail on species distribution see Flannery (1995a, b) and Van Dyck and Strahan (2008).

Conservation status (IUCN 2009): EX = extinct, CR = critically endangered, EN = endangered; VU = vulnerable, NT = near threatened, LC = least concern, DD = data deficient.

a: Several different higher-order taxonomic arrangements have been proposed. Aplin and Archer (1987) recognised two families (Macropodidae, Potoroidae) within the superfamily Macropodoidea. Kirsch *et al.* (1997) proposed the suborder Macropodiformes, containing the superfamily Macropodoidea which they divided into two families (Hypsiprymnodontidae, Macropodidae), with the Macropodidae further separated into the subfamilies Macropodinae and Potoroinae. Groves (2005) recognised three families (Hypsiprymnodontidae, Potoroidae, Macropodidae) within the suborder Macropodiformes.

b: Although previously regarded as a potoroid (Calaby and Richardson 1988), *Hypsiprymnodon* is now placed in a separate family (Hypsiprymnodontidae) as a consequence of its phylogenetic distinctiveness (Burk *et al.* 1998; Burk and Springer 2000) and many other unique features (see Johnson *et al.* 2005 and references in Burk *et al.* 1998).

c: Groves (2005) and IUCN (2009) recognise populations from the central mountains of Papua New Guinea as a distinct species, *Dendrolagus stellarum* (VU). In addition, the IUCN (2009) recognises populations from the Central Cordillera of West Papua and western Papua New Guinea as *Dendrolagus notatus* (EN) and from the Wondiwoi Peninsula of West Papua as *Dendrolagus mayri* (CR).

d: Groves (2005) and IUCN (2009) recognise northern coastal ranges population as a separate species, *Dendrolagus pulcherimus* (CR).

e: Groves (2005) and IUCN (2009) recognise the high-altitude (>3000 m) Huon Peninsula population as separate species, *Thylogale lanatus* (EN).

f: Molecular analyses have supported the monophyly of *Wallabia* and *Macropus* (Richardson and McDermid 1978; Burk and Springer 2000; Westerman *et al.* 2002), with many placing *Wallabia* within *Macropus* as currently defined (Kirsch *et al.* 1997; Cardillo *et al.* 2004; Bulazel *et al.* 2007; Meredith *et al.* 2008, 2009). This implies that *Macropus* is currently paraphyletic and that some taxonomic revision is required, either placing *Wallabia* as a fourth subgenus within *Macropus*, as suggested by Meredith *et al.* (2008), or splitting *Macropus* by elevating each of the existing subgenera (*Macropus, Notamacropus, Osphranter*) to full generic status.

g: The taxonomic position of *Lagostrophus fasciatus* within the Macropodidae remains unclear. Based on morphological data, Flannery (1983) has argued that *Lagostrophus* is the sole surviving member of the macropodid subfamily Sthenurinae, but this has been challenged (Prideaux 2004). Molecular data show that *Lagostrophus* is not closely related to extant macropodines or potoroids, although a sister taxon relationship with macropodines appears likely (Westerman *et al.* 2002; Nilsson 2006; Meredith *et al.* 2009). The comparison of *Lagostrophus* DNA with ancient DNA extracted from extinct fossil sthenurines is eagerly awaited.

REFERENCES

Aplin KP and Archer M 1987. Recent advances in marsupial systematics with a new syncretic classification. In *Possums and Opossums: Studies in Evolution.* (Ed. M Archer) pp. xv–lxxii. Surrey Beatty and Sons, Sydney.

Bulazel K, Ferreri GC, Eldridge MDB and O'Neill RJ 2007. Species-specific shifts in centromere sequence composition are coincident with breakpoint reuse in karyotypically divergent lineages. *Genome Biol.* **8:** R170.

Burk A and Springer MS 2000. Intergeneric relationships among Macropodoidea (Metatheria: Diprotodontia) and the chronicle of kangaroo evolution. *J. Mamm. Evol.* **7:** 213–237.

Burk A, Westerman M and Springer MS 1998. The phylogenetic position of the musky rat-kangaroo and the evolution of bipedal hopping in kangaroos (Macropodidae: Diprotodontia). *Syst. Biol.* **47:** 457–474.

Calaby JH and Richardson BJ 1988. Potoroidae. In *Zoological Catalogue of Australia. 5. Marsupialia.* (Ed. DW Walton) pp. 53–59. Australian Government Publishing Service, Canberra.

Cardillo M, Bininda-Emonds ORP, Boakes E and Purvis A 2004. A species-level phylogenetic supertree of marsupials. *J. Zool., Lond.* **264:** 11–31.

Flannery TF 1983. Revision in the macropodid subfamily Sthenurinae (Marsupialia: Macropodidae) and the relationships of the species *Troposodon* and *Lagostrophus*. *Aust. Mammal.* **6:** 15–28.

Flannery TF 1995a. *Mammals of New Guinea.* Reed Books, Sydney.

Flannery TF 1995b. *Mammals of the South-west Pacific and Moluccan Islands.* Reed Books, Sydney.

Groves CP 2005. Order Diprotodontia. In *Mammal Species of the World: A Taxonomic and Geographic Reference.* 3rd edn. (Eds DE Wilson and DM Reeder) pp. 43–70. Johns Hopkins University Press, Baltimore.

IUCN 2009. Red List of Threatened Species. Version 2009.1. www.iucnredlist.org. Accessed 3 June 2009.

Johnson PM, Lloyd S, Vallance T and Eldridge MDB 2005. First record of quadruplets in the musky rat-kangaroo *Hypsiprymnodon moschatus. Aust. Mammal.* **27:** 95–97.

Kirsch JAW, Lapointe FJ and Springer MS 1997. DNA-hybridisation studies of marsupials and their implications for metatherian classification. *Aust. J. Zool.* **45:** 211–280.

Meredith RW, Westerman M and Springer MS 2008. A phylogeny and timescale for the living genera of kangaroos and kin (Macropodiformes: Marsupialia) based on nuclear DNA sequences. *Aust. J. Zool.* **56:** 395–410.

Meredith RW, Westerman M and Springer MS 2009. A phylogeny of Diprotodontia (Marsupialia) based on sequences for five nuclear genes. *Mol. Phylogenet. Evol.* **51:** 554–571.

Nilsson MA 2006. Phylogenetic relationships of the banded hare-wallaby (*Lagostrophus fasciatus*) and a map of the kangaroo mitochondrial control region. *Zool. Scr.* **35:** 387–393.

Prideaux GJ 2004. Systematics and evolution of the Sthenurine kangaroos. *Uni. Calif. Publ. Geol. Sci.* **146:** 1–623.

Richardson BJ and McDermid EM 1978. A comparison of genetic relationships within the Macropodidae as determined from allozyme, cytological and immunological data. *Aust. Mammal.* **2:** 43–51.

Van Dyck S and Strahan R (Eds) 2008. *The Mammals of Australia.* 3rd edn. Reed New Holland, Sydney.

Westerman M, Burk A, Amrine-Madsen HM, Prideaux GJ, Case JA and Springer MS 2002. Molecular evidence for the last survivor of an ancient kangaroo lineage. *J. Mamm. Evol.* **9:** 209–223.

Species index

Subject index

www.ingramcontent.com/pod-product-compliance
Lightning Source LLC
LaVergne TN
LVHW070847210926
848482LV00016B/881
* 9 7 8 0 6 4 3 0 9 6 6 2 2 *